BIOCHEMISTRY

THIRD EDITION

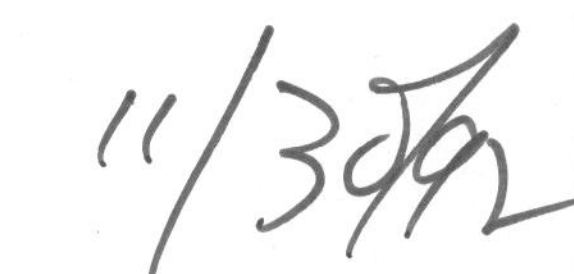

BIOCHEMISTRY

THIRD EDITION

GEOFFREY ZUBAY
COLUMBIA UNIVERSITY

Book Team

Editor *Kevin Kane*
Developmental Editor *Megan Johnson*
Production Editor *Sherry Padden*
Designer *Mark Elliot Christianson*
Art Editor *Miriam J. Hoffman/Janice M. Roerig*
Photo Editor *Lori Gockel*
Permissions Editor *Vicki Krug*
Visuals/Design Developmental Consultant *Marilyn A. Phelps*

Wm. C. Brown Publishers
A Division of Wm. C. Brown Communications, Inc.

Vice President and General Manager *Beverly Kolz*
National Sales Manager *Vincent R. Di Blasi*
Assistant Vice President, Editor-in-Chief *Edward G. Jaffe*
Director of Marketing *John W. Calhoun*
Marketing Manager *Carol J. Mills*
Advertising Manager *Amy Schmitz*
Director of Production *Colleen A. Yonda*
Manager of Visuals and Design *Faye M. Schilling*
Design Manager *Jac Tilton*
Art Manager *Janice Roerig*
Publishing Services Manager *Karen J. Slaght*
Permissions/Records Manager *Connie Allendorf*

Wm. C. Brown Communications, Inc.

Chairman Emeritus *Wm. C. Brown*
Chairman and Chief Executive Officer *Mark C. Falb*
President and Chief Operating Officer *G. Franklin Lewis*
Corporate Vice President, President of WCB Manufacturing *Roger Meyer*

Biochemistry: © Orion/West Light
Artistic rendition of the double helix
Volume One: Energy, Cells, and Catalysis: Photograph by A. M. Lesk
Model for hemoglobin
Volume Two: Catabolism and Biosynthesis: Photograph by A. M. Lesk
Model for phosphofructokinase
Volume Three: Genetics and Physiology: Photograph by A. M. Lesk
Model for λ cro protein binding to DNA

Copyedited by *Emily Arulpragasam*

Freelance permissions editor *Karen Dorman*

The credits section for this book begins on page C-1 and is considered an extension of the copyright page.

Library of Congress Catalog Card Number: 91–76759

ISBN 0–697–14267–1 **Biochemistry** (Casebound)
0–697–14878–5 **Volume One: Energy, Cells, and Catalysis**
0–697–14879–3 **Volume Two: Catabolism and Biosynthesis**
0–697–14880–7 **Volume Three: Genetics and Physiology**
0–697–14877–7 **Biochemistry** (Boxed Set)

Printed in the United States of America by Wm. C. Brown Communications, Inc., 2460 Kerper Boulevard, Dubuque, IA 52001

10 9 8 7 6 5 4 3 2 1

Publisher's Note

Biochemistry, third edition, is available as a full-length casebound text, boxed set, or as three paperbound separates

Binding Option	Description	ISBN
Biochemistry (Casebound)	The full-length text, Chapters 1–36	0-697-14267-1
Biochemistry, Volume One (Paperbound)	Chapters 1–11	0-697-14878-5
Biochemistry, Volume Two (Paperbound)	Chapters 12–24	0-697-14879-3
Biochemistry, Volume Three (Paperbound)	Chapters 25–36	0-697-14880-7
Biochemistry, Boxed Set (Paperbound)	The full-length text in an attractive boxed set of all three paperback "splits".	0-697-14877-7

This book is dedicated to Ed Jaffe, the late Editor-in-Chief of the William C. Brown Publishing Company.

Ed passed away suddenly just before this text went to press. I am deeply saddened by his passing and that he will not see the text he backed so strongly. He was a man of wisdom and integrity who maintained an excellent rapport with the WCB staff and with its authors.

Brief Contents

Volume One

PART 1

An Overview of Biochemistry and Energy Considerations 1

PART 2

Structure and Function of Major Components of the Cell 45

PART 3

Catalysis 197

Volume Two

PART 4

Catabolism and the Generation of Chemical Energy 305

PART 5

Biosynthesis of the Building Blocks 473

Volume Three

PART 6

Storage and Utilization of Genetic Information 689

PART 7

Physiological Biochemistry 929

Contents

PART

Structure and Function of Major Components of the Cell 45

Chapter 7

Lipids and Membranes 163

Dennis E. Vance, Gary R. Jacobson, and Milton H. Saier, Jr.

PART

Catalysis 197

Chapter 8

Enzyme Kinetics 199

William W. Parson

Chapter 9

Mechanisms of Enzyme Catalysis 219

William W. Parson

Chapter 10

Regulation of Enzyme Activities 254

William W. Parson

Chapter 14

The Tricarboxylic Acid Cycle 354

Daniel E. Atkinson and Geoffrey Zubay*

Chapter 15

Electron Transport and Oxidative Phosphorylation 379

William W. Parson

*Revised by Geoffrey Zubay from material in the preceding edition by Daniel E. Atkinson.

Chapter 16

Photosynthesis and Other Processes Involving Light 414

William W. Parson

Chapter 17

Metabolism of Fatty Acids 444

Dennis E. Vance

PART

Biosynthesis of the Building Blocks 473

Chapter 18

Biosynthesis of Amino Acids 475

H. Edwin Umbarger and Geoffrey Zubay

Chapter 21

Chapter 22

Chapter 23

Chapter 27

DNA Manipulation and Its Applications 757

Geoffrey Zubay

Chapter 28

RNA Synthesis and Processing 791

Richard R. Burgess, Ann Baker Burgess, and Geoffrey Zubay

Chapter 34

Carcinogenesis and Oncogenes 979

Geoffrey Zubay

Chapter 35

Neurotransmission 993

Gary R. Jacobson and Milton H. Saier, Jr.

Chapter 36

Vision 1012

William W. Parson

List of Supplementary Material

List of Contributors

Raymond L. Blakley
Chapter 20
Division of Biochemical and Clinical Pharmacology
St. Jude Children's Research Hospital
Memphis, TN 38101

James W. Bodley
Chapter 29
Biochemistry Department
University of Minnesota
Minneapolis, MN 55455

Ann Baker Burgess
Chapter 28
McArdle Laboratory for Cancer Research
University of Wisconsin
Madison, WI 53706

Richard R. Burgess
Chapter 28
McArdle Laboratory for Cancer Research
University of Wisconsin
Madison, WI 53706

Perry A. Frey
Chapter 11
Institute for Enzyme Research
University of Wisconsin
Madison, WI 53706

Irving Geis
Chapter 5
4700 Broadway
Apt. 4B
New York, NY 10040

Lloyd L. Ingram (Retired)
Chapter 2
Department of Biochemistry and Biophysics
University of California at Davis
Davis, CA 95616

Gary R. Jacobson
Chapters 7, 32, 35
Department of Biology
Boston University
Boston, MA 02215

Julius Marmur
Chapters 25, 26, 31
Department of Biochemistry
Albert Einstein College of Medicine
Bronx, NY 10461

Richard Palmiter
Chapter 24
Department of Biochemistry
University of Washington
Seattle, WA 98195

William W. Parson
Chapters 2, 8, 9, 10, 15, 16, 36
Department of Biochemistry
University of Washington
Seattle, WA 98195

Milton H. Saier, Jr.
Chapters 7, 32, 35
Department of Biology
University of California at San Diego
Lajolla, CA 92093

Pamela Stanley
Chapters 6, 21
Department of Cell Biology
Albert Einstein College of Medicine
Bronx, NY 10461

H. Edwin Umbarger
Chapters 18, 19
Department of Biological Sciences
Purdue University
West Lafayette, IN 47907

Dennis E. Vance
Chapters 7, 17, 22, 23
Lipid and Lipoprotein Group
Department of Biochemistry
University of Alberta
Edmonton, Alberta
Canada, T6G 2C2

Geoffrey Zubay
Chapters 1, 2, 3, 4, 5, 6, 12, 13, 14, 18, 19, 21, 24, 25, 26, 27, 28, 30, 31, 33, 34
Department of Biological Sciences
Columbia University
New York, NY 10027

Guided Tour through the Biochemistry Learning System

(A.) CHAPTER OUTLINES

Each chapter begins with an outline. These will allow students to tell at a glance how the chapter is organized and what major topics have been included in the chapter.

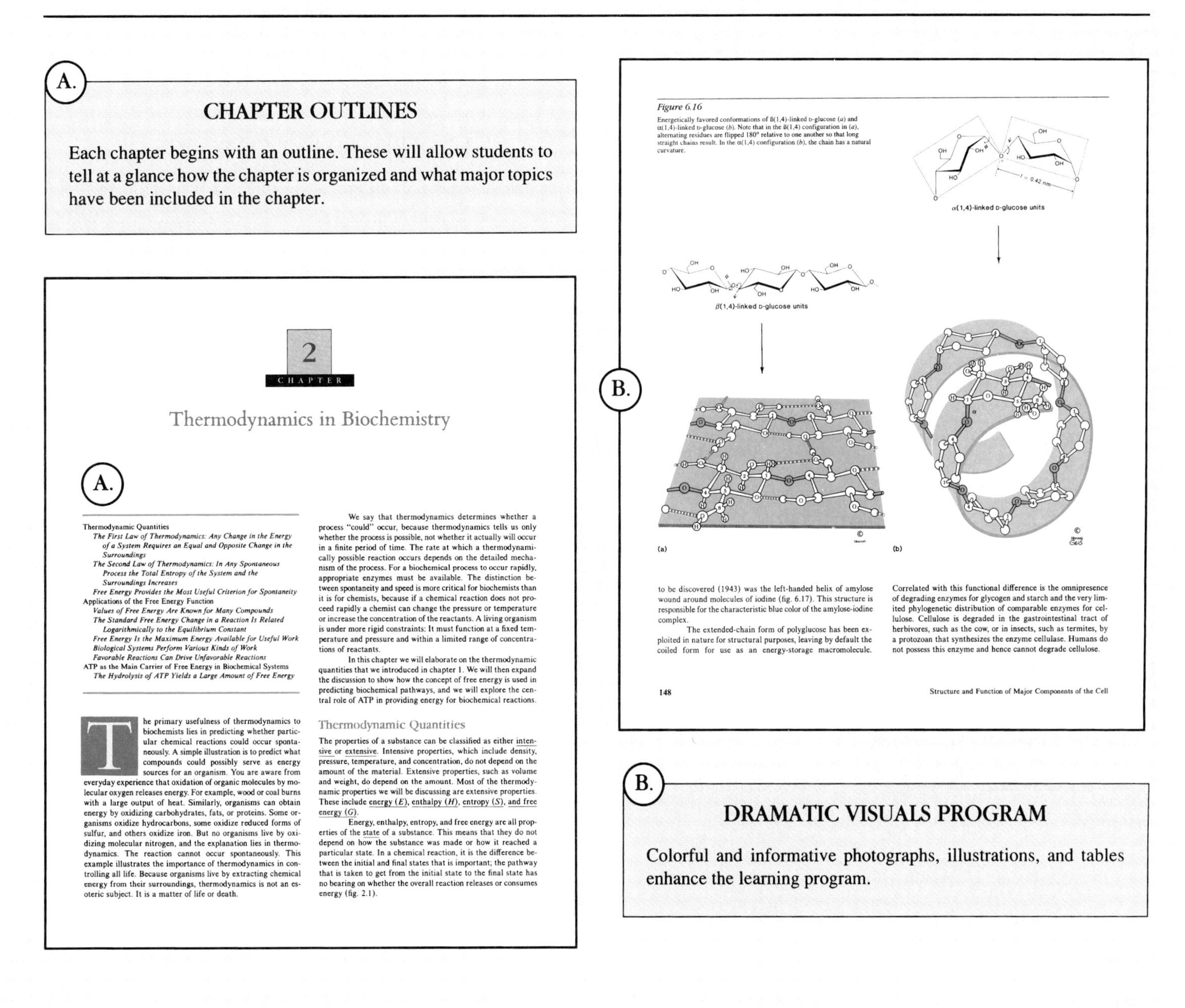

2

CHAPTER

Thermodynamics in Biochemistry

(A.)

We say that thermodynamics determines whether a process "could" occur, because thermodynamics tells us only whether the process is possible, not whether it actually will occur in a finite period of time. The rate at which a thermodynamically possible reaction occurs depends on the detailed mechanism of the process. For a biochemical process to occur rapidly, appropriate enzymes must be available. The distinction between spontaneity and speed is more critical for biochemists than it is for chemists, because if a chemical reaction does not proceed rapidly a chemist can change the pressure or temperature or increase the concentration of the reactants. A living organism is under more rigid constraints: It must function at a fixed temperature and pressure and within a limited range of concentrations of reactants.

In this chapter we will elaborate on the thermodynamic quantities that we introduced in chapter 1. We will then expand the discussion to show how the concept of free energy is used in predicting biochemical pathways, and we will explore the central role of ATP in providing energy for biochemical reactions.

The primary usefulness of thermodynamics to biochemists lies in predicting whether particular chemical reactions could occur spontaneously. A simple illustration is to predict what compounds could possibly serve as energy sources for an organism. You are aware from everyday experience that oxidation of organic molecules by molecular oxygen releases energy. For example, wood or coal burns with a large output of heat. Similarly, organisms can obtain energy by oxidizing carbohydrates, fats, or proteins. Some organisms oxidize hydrocarbons, some oxidize reduced forms of sulfur, and others oxidize iron. But no organisms live by oxidizing molecular nitrogen, and the explanation lies in thermodynamics. The reaction cannot occur spontaneously. This example illustrates the importance of thermodynamics in controlling all life. Because organisms live by extracting chemical energy from their surroundings, thermodynamics is not an esoteric subject. It is a matter of life or death.

Thermodynamic Quantities

The properties of a substance can be classified as either intensive or extensive. Intensive properties, which include density, pressure, temperature, and concentration, do not depend on the amount of the material. Extensive properties, such as volume and weight, do depend on the amount. Most of the thermodynamic properties we will be discussing are extensive properties. These include energy (E), enthalpy (H), entropy (S), and free energy (G).

Energy, enthalpy, entropy, and free energy are all properties of the state of a substance. This means that they do not depend on how the substance was made or how it reached a particular state. In a chemical reaction, it is the difference between the initial and final states that is important; the pathway that is taken to get from the initial state to the final state has no bearing on whether the overall reaction releases or consumes energy (fig. 2.1).

Figure 6.16

Energetically favored conformations of β(1,4)-linked D-glucose (*a*) and α(1,4)-linked D-glucose (*b*). Note that in the β(1,4) configuration in (*a*), alternating residues are flipped 180° relative to one another so that long straight chains result. In the α(1,4) configuration (*b*), the chain has a natural curvature.

to be discovered (1943) was the left-handed helix of amylose wound around molecules of iodine (fig. 6.17). This structure is responsible for the characteristic blue color of the amylose-iodine complex.

The extended-chain form of polyglucose has been exploited in nature for structural purposes, leaving by default the coiled form for use as an energy-storage macromolecule. Correlated with this functional difference is the omnipresence of degrading enzymes for glycogen and starch and the very limited phylogenetic distribution of comparable enzymes for cellulose. Cellulose is degraded in the gastrointestinal tract of herbivores, such as the cow, or in insects, such as termites, by a protozoan that synthesizes the enzyme cellulase. Humans do not possess this enzyme and hence cannot degrade cellulose.

148 Structure and Function of Major Components of the Cell

(B.) DRAMATIC VISUALS PROGRAM

Colorful and informative photographs, illustrations, and tables enhance the learning program.

C.

Measurement of Ultraviolet Absorption in Solution

The general quantitative relationship that governs all absorption processes is called the Beer-Lambert law:

$$I = I_0\, 10^{-\epsilon c d}$$

where I_0 is the intensity of the incident radiation, I is the intensity of the radiation transmitted through a cell of thickness d (in centimeters) that contains a solution of concentration c (expressed either in moles per liter or in grams per 100 ml), and ϵ is the extinction coefficient, a characteristic of the substance being investigated (see fig. 3.7).

Light absorption is measured by a spectrophotometer as shown in the illustration. The spectrophotometer usually is capable of directly recording the absorbance A, which is related to I and I_0 by the equation

$$A = \log_{10} (I_0/I)$$

Hence $A = \epsilon c d$, and A is a direct measure of concentration. We can see from figure 3.7 that the ϵ values are largest for tryptophan and smallest for phenylalanine.

Since protein absorption maxima in the near-ultraviolet (240–300 nm) are determined by the content of the aromatic amino acids and their respective values, most proteins have absorption maxima in the 280-nm region. By contrast, absorption in the far-ultraviolet (around 190 nm) is shown by all polypeptides regardless of their aromatic amino acid content. The reason is that absorption in this region is due primarily to the peptide linkage.

Figure 1

Schematic diagram of a spectrophotometer for measuring light absorption. Laboratory instruments for making measurements are much more complex than this, but they all contain the same basic components: a light source, a monochromator, a sample, and a detector. λ is the wavelength of the light, I_0 and I are the incident light intensity and the transmitted light intensity, respectively, and d is the thickness of the absorbing solution.

Source — Monochromator — Sample — Detector

Another convention for referring to configurations is called the *R, S* convention. As the *R, S* convention is not as popular for amino acids or sugars as it is for other types of biomolecules, such as lipids, we will not discuss this notation until chapter 11 (see box 11A).

D.

Peptides and Polypeptides

Amino acids can link together by a covalent peptide bond between the α-carboxyl end of one amino acid and the α-amino end of another. Formally, this bond is formed by the loss of a water molecule, as shown in figure 3.9. The peptide bond has partial double-bond character owing to resonance effects; as a result, the C—N peptide linkage and all of the atoms directly connected to C and N lie in a planar configuration called the amide plane. In the following chapter we will see that this amide plane, by limiting the number of orientations available to the polypeptide chain, plays a major role in determining the three-dimensional structures of proteins.

Any number of amino acids can be joined by successive peptide linkages, forming a polypeptide chain. The polypeptide chain, like the dipeptide, has a directional sense. One end, called the N-terminal or amino-terminal end, has a free α-amino group, whereas the other end, the C-terminal or carboxyl-terminal end, has a free α-carboxyl group. The sequence of main-chain atoms from the N-terminal end to the C-terminal end is C_α—C—N—C_α, etc., and in the opposite direction it is C_α—N—C—C_α, etc. Short polypeptide chains, up to a length of about 20 amino acids, are called peptides or oligopeptides if they are fragments of whole polypeptide chains. A small protein molecule may contain a polypeptide chain of only 50 amino acids; a large protein may contain chains of 3,000 amino acids or more. One of the larger single polypeptide chains is that of the muscle protein myosin, which consists of approximately 1,750 amino acid residues. Figure 3.10 shows a section of a polypeptide chain as a linear array with α carbons and planar amides alternating as repeating units of the main chain. Different side chains are attached to each α carbon.

In addition to the covalent peptide bonds formed between adjacent amino acids within a polypeptide chain, covalent disulfide bonds can be formed within the same polypeptide chain or between different polypeptide chains (fig. 3.11). Such disulfide linkages have an important stabilizing influence on the structures formed by many proteins (see chapter 4).

54 Structure and Function of Major Components of the Cell

C.

SUPPLEMENTARY BOX MATERIAL

Throughout the text, supplemental information on experimental procedures, mechanisms, and methods of evaluation are presented in boxes

D.

UNDERLINED KEY CONCEPTS

Important terms and concepts are highlighted by underlining.

E.

END-OF-CHAPTER SUMMARIES

Each summary offers a concise review of the material covered in the chapter. Students can use the summary to review the chapter or to preview the important topics.

F.

SELECTED READINGS

Each chapter concludes with carefully selected references that contain further information on the topics covered in that chapter.

G.

END-OF-CHAPTER PROBLEMS

These problems will challenge students' mastery of the chapter's basic concepts. Odd-numbered problems are answered briefly in the back of the text.

E.

Summary

In this chapter we have dealt with some of the fundamental properties of amino acids and polypeptide chains. The following points are especially important.

1. Nineteen of the twenty amino acids commonly found in proteins have a carboxyl group and an amino group attached to an α-carbon atom; they differ in the side chain attached to the same α carbon.
2. All amino acids have acidic and basic properties. The ratio of base to acid form at any given pH can be calculated from the p*K* with the help of the Henderson-Hasselbach equation.
3. All amino acids except glycine are asymmetric and therefore can exist in at least two different stereoisomeric forms.
4. Peptides are formed from amino acids by the reaction of the α-amino group from one amino acid with the α-carboxyl group of another amino acid.
5. Polypeptide formation involves a repetition of the process involved in peptide synthesis.
6. The amino acid composition of proteins can be discovered by first breaking down the protein into its component amino acids and then separating the amino acids in the mixture for quantitative estimation.
7. The amino acid sequences of proteins can be discovered by breaking down the protein into polypeptide chains and then partially degrading the polypeptide chains. For each polypeptide chain fragment, the sequence is determined by stepwise removal of amino acids from the amino-terminal end of the polypeptide chain. Two different methods of forming polypeptide chain fragments are used so as to produce a map of overlapping fragments, from which the sequence of undegraded polypeptide chains in the proteins can be deduced.
8. Polypeptide chains with a predetermined amino acid sequence can be synthesized by chemical methods involving carboxyl-group activation.

F.

Selected Readings

Barrett, G. C. (ed.). *Chemistry and Biochemistry of Amino Acids.* New York: Chapman and Hall, 1985. A recent and authoritative volume on this classical subject.

Gray, W. R., End group analysis using dansyl chloride. *Methods in Enzymology* 25:121–138, 1972. This volume of *Methods in Enzymology* contains several chapters on end-group analysis.

Hunkapiller, M. W., J. E. Strickler, and K. J. Wilson, Contemporary methodology for protein structure determination. *Science* 226:304–311, 1984.

Kent, S. B. H., Chemical synthesis of peptides and proteins. *Ann. Rev. Biochem.* 57:957–989, 1988. Comprehensive and up-to-date.

Merrifield, B., Solid phase synthesis. *Science* 232: 341–347, 1986.

Sanger, R., Sequences, sequences and sequences. *Ann. Rev. Biochem.* 57:1–28, 1988.

G.

Problems

1. (a) A 10-mM solution of a weak monocarboxylic acid has a pH of 3.00. Calculate K_a and pK_a for this carboxylic acid.
   (b) You add 0.06 g NaOH (M_r = 40) to 1,000 ml of the acid solution in part (a). Calculate the final pH, assuming no volume change.
2. Given the pK_a values in the text, predict how the titration curves for glutamic acid and glutamine would differ.
3. You have 50 ml of 10-mM fully protonated histidine. How many millimoles of base must be added to bring the histidine solution to a pH that is equivalent to the pI?
4. Calculate the isoelectric point for histidine, aspartic acid, and arginine. Calculate the fractional charge for each ionizable group on aspartate at pH equal to pI. Do the results verify the isoelectric point of aspartic acid?
5. Which of the naturally occurring amino acid side chains are charged at pH 2? pH 7? pH 12? (Consider only those amino acids whose side chains have >10% charge at the pH indicated.)
6. Amino acids are sometimes used as buffers. Indicate the appropriate pH value(s) of a buffer containing aspartic acid, histidine, and serine.
7. Ten ml of a 10-mM solution of lysine was adjusted to pH 11.20. Draw the structures of the principal ionized forms present in solution. Use the pK_a values shown in table 3.3 and calculate the concentration of each principal form.
8. For the tripeptide shown below, the numbers in parentheses are the pK_a values of the ionizable groups.

   $\overset{\oplus}{NH_3}$ (10.0) — CH(CH$_2$COO$^{\ominus}$ (4.0)) — C(=O) — NH — CH((CH$_2$)$_4$NH$_3^{\oplus}$ (10.0)) — C(=O) — NH — CH(CH$_2$SH (9.0)) — COO$^{\ominus}$ (3.0)

   (a) Estimate the net charge at pH 1 and pH 14.
   (b) Estimate the isoelectric pH.
9. Polyhistidine is insoluble in water at pH 7.8 but is soluble at pH 5.5. Explain the observation. Would you expect the polymer to be soluble at pH 10?

The Building Blocks of Proteins: Amino Acids, Peptides, and Polypeptides 67

Preface

hat makes biochemistry exciting? What makes biochemistry unique? What makes biochemistry so important? The answer to all three of these questions is virtually the same. Each and every reaction in biochemistry serves a function; that function is the maintenance and propagation of the living system. To understand biochemistry is to appreciate the place that each reaction occupies in the system.

Students of the previous generation discovered the principles that govern biochemistry, the basic reactions, and an immense body of facts consistent with these principles. Future biochemists to emerge from this generation of students will uncover the mysteries of developmental biology and physiology. They will find cures for cancers, cystic fibrosis, sickle cell anemia and many other diseases. My goal in putting this text together has been to convey my enthusiasm for this subject to this generation of students, to acquaint them with what is known and to prepare them for the discoveries of the future.

While the main principles of biochemistry are understood, the number of new facts that keep adding to the subject is immense. The field is becoming more and more subdivided as the volume of knowledge increases. No single person could directly cope with all the new information that is continuously appearing in the scientific literature. For the purpose of writing an authoritative up-to-date textbook on biochemistry, it would be ideal to involve a team of experts with hands-on experience. The contributing authors of this text are all research specialists who have made their mark in a specific area of biochemistry; they are teachers as well. Each contributing author could write an entire book on the chapter he/she has contributed to in this text. Recognizing that the primary goal of this text is to meet the needs of students facing the subject for the first time, each contributing author has had to pare down what he/she knows emphasizing principles, but not ignoring important facts. Each chapter has been carefully crafted with that goal in mind. My responsibility as coordinating author of this text has been to mold the contributions into a smooth reading, well integrated product.

This is the third edition of *Biochemistry*. Whereas I have had the same goal throughout this project, it is only with this edition, that the long-hoped-for product has finally emerged from our team approach. Reviewers teil us that this edition is not only consistently authoritative, it now reads like a cohesive, consistent single author textbook.

Each chapter emphasizes principles. Since the overriding principle is that every reaction serves a function, there is a strong emphasis on analysis of biochemical pathways and how different pathways are interrelated. In this pursuit balanced emphasis is given to chemistry and mechanisms, bioenergetics, metabolic regulation, and methods of biochemical analysis.

We have organized this text along conventional lines for teaching purposes. In Part 1 an overview is followed by an explanation of the importance of thermodynamics in biochemistry. In Part 2 we consider the structure and function of the major components of the cell. In Part 3 we discuss enzymes and coenzymes and the mechanisms of catalysis. Part 4 deals with energy producing catabolic processes, and Part 5 deals with energy consuming synthetic processes. In Part 6 we focus on nucleic acids and the way in which the biochemical information encoded in the chromosome is translated into the amino acid sequences found in proteins. Finally, in Part 7 we approach some of the applications of biochemistry to physiology.

Organizational Changes

The overall organization of the previous edition has been changed significantly:

- The chapter on methods for characterization and purification of proteins (formerly chapter 3) has been eliminated. Much of the material from this chapter has been disseminated into other chapters, appearing at the points where it is most relevant. The same goes for the discussion of methods in general, and there is a special appendix indicating where specific methods are discussed.

- The chapter on nucleic acid structure (formerly chapter 7) has been moved to a later part of the text (chapter 25). This has been done because nucleic acids are not discussed in any depth until Part 6. The general significance of nucleic acids is adequately discussed in chapter 1 so this should create no problem for the rare student who has not encountered this subject already.
- The chapter on origin of life has been eliminated. A few points about this subject are now made in chapters 1 and 28. This is a delightful subject and very close to my major research interests. However, it simply is not mainstream biochemistry and space would not allow us to keep this chapter.
- After many years of agonizing appraisal I have decided to place thermodynamics as the second chapter. Bioenergetics is so central to everything in biochemistry that I thought it should be discussed as soon as possible. At the same time an esoteric chapter on this subject could easily turn off many students. To avoid this we have considerably simplified this chapter to make it as palatable as possible without losing any of the significance of a more esoteric chapter.
- A new chapter on functional diversity of proteins (chapter 5) has been added. This gives us an opportunity to dwell on the very important subject of the relationship between protein structure and function. In addition we have used this chapter as a location to discuss the subject of protein purification.
- The chapters on lipids and membrane structures have been combined into one chapter. Students are more attentive to a description of lipids if they can see the important role they play in membrane structure.
- The chapters on lipid catabolism and synthesis have been combined in this edition (chapter 17). This makes it easier to discuss regulation of lipid metabolism.
- At the end of the treatment of intermediary metabolism (chapter 24) we have added a chapter on the integration of metabolism. In this chapter we have included most of the coverage of hormone action that was presented much later in the second edition. It is more relevant here as hormones are the key to the integration of metabolism in multicellular organisms.
- Two short chapters have been added to Part 7; chapter 33, on immunobiology, and chapter 34 on carcinogenesis. These are two hot topics that are of great interest to students. The chapters are kept short and simple for those who would choose to use them in a two term course. It would be best to consider them after the core chapters of Part 6 (i.e., chapters 25, 26, 28 and 29).

Content Features of This Edition

The coverage of biochemistry in this text is determined by the consensus feedback that we have received as well as the limitations of the present state of our knowledge. Coverage of the biochemistry of bacteria and mammals is emphasized over plant biochemistry.

Part 1 is entitled "An Overview of Biochemistry and Energy Considerations." The first chapter "Overview of Biochemistry," is very similar to the introductory chapter in the second edition. It presents the basic principles in chemistry and biology that relate to biochemistry. The second chapter, "Thermodynamics in Biochemistry," has been considerably rewritten and simplified over the corresponding chapter in the second edition, which had appeared as chapter 12. Chapter 2 elaborates on the thermodynamic quantities introduced in chapter 1, to show how the concept of free energy is used to predict the possibility that given reactions can occur. It also describes the central role of ATP in providing energy for biochemical reactions.

Part 2, "Structure and Function of Major Components of the Cell," contains five chapters describing the structure and function of proteins, carbohydrates, and lipids. The contents of Part 2 are similar to the contents of Part 1 in the previous edition, except that the discussion of nucleic acids has been moved to Part 6, and a new chapter 5, "Functional Diversity of Proteins," has been introduced.

Chapters 3, 4, and 5 present the basic structural and chemical properties of proteins. Chapter 3, "The Building Blocks of Proteins: Amino Acids, Peptides, and Polypeptides," is very similar to the comparable chapter in the second edition. Chapter 4, "The Three-Dimensional Structure of Proteins," describes the secondary, tertiary, and quaternary structures of proteins, the rules for which govern the folding of polypeptide chains and the ways in which these structures are determined. In this edition there is more coverage of the relationship between symmetry and quaternary structures than in the second edition. Chapter 5, "Functional Diversity of Proteins," focuses on the question of how protein structure relates to function and describes techniques used to isolate proteins.

Chapter 6, "Carbohydrates, Glycoproteins, and Cell Walls," describes the structural properties of sugars, oligosaccharides, and polysaccharides, along with the various roles that these compounds play as energy-storage molecules, components of glycoproteins, and in cell walls. This chapter's segment on glycoproteins has been considerably updated. Finally, in chapter 7, "Lipids and Membranes," two chapters from the second edition have been fused into one cohesive chapter. In it, the structural properties of lipids are discussed with reference to the roles lipids play as energy-storage molecules and in membrane structures.

Part 3, "Catalysis," is divided into four chapters, the first three of which have been completely rewritten, while the last has been simplified considerably, with discussions of lipid-soluble vitamins and the role of metals as cofactors added.

Chapter 8, "Enzyme Kinetics," features a description of kinetic methods for analyzing enzyme-catalyzed reactions. Chapter 9, "Mechanisms of Enzyme Catalysis," focuses on the way in which the structure of the enzyme active site is related to enzyme activity under different conditions. In this chapter a general discussion of factors determining enzyme specificity is followed by several detailed examples. Chapter 10, "Regulation of Enzyme Activities," describes enzymes regulated by structural modifications or the binding of specific factors at allosteric sites, and chapter 11, "Vitamins, Coenzymes, and Metal Cofactors," covers the function of small molecules acting in conjunction with enzymes as cocatalysts.

Part 4 "Catabolism and the Generation of Chemical Energy," is divided into six chapters. Since both Parts 4 and 5 are concerned with intermediary metabolism, it's not easy, or even appropriate, to separate the subject into artificially tidy discussions of catabolism and anabolism. Suffice it to say that the emphasis in Part 4 is on catabolism and the generation of chemical energy, while in Part 5 the emphasis is on biosynthesis. Chapter 12, "Metabolic Strategies," which initiates Part 4, relates strongly to both Parts 4 and 5, as does chapter 24, "Integration of Metabolism and Hormone Action," which concludes Part 5.

Chapter 12, "Metabolic Strategies," deals with the general ways in which metabolism is organized and the ways in which it is regulated. Chapters 13, 14, and 15 are devoted to carbohydrate metabolism in relationship to the generation of ATP. Chapter 13, "Glycolysis, Gluconeogenesis, and the Pentose Phosphate Pathway," describes anabolic as well as catabolic processes—namely, the synthesis of simple sugars and simple sugar polymers, as well as their breakdown. Both synthesis and breakdown are covered in the same chapter in order to facilitate meaningful discussion of the regulation of these two processes, which are intimately related.

Chapter 14, "The Tricarboxylic Acid Cycle," describes the further catabolism of sugars possible under aerobic conditions. Chapter 15, "Electron Transport and Oxidative Phosphorylation," complements chapter 14 because it shows how electrons released by the aerobic oxidation of sugars are channeled down the electron-transport system, ultimately to synthesize ATP. Chapter 16, "Photosynthesis and Other Processes Involving Light," appears immediately after the chapter on electron transport and the production of ATP because both chapters describe energy-generating metabolism. The complex light-powered processes are explained so that a minimal background in physics and chemistry is needed.

Part 4 concludes with Chapter 17, "Metabolism of Fatty Acids," which represents a fusion of two chapters from the second edition. The reasons for discussing both synthesis and breakdown of lipids in the same chapter are exactly the same as the reasons for doing this with sugars (as in chapter 13).

In many schools a one-semester treatment of biochemistry would end with chapter 17. Even though the first half of the text is tightly structured, instructors may still wish to seek additional material or skip around. Two recommendations involve chapter 25 and chapter 32. Chapter 25, "Structures of Nucleic Acids and Nucleoproteins," could be treated as a chapter in Part 2. If this is done, instructors should make sure students read the opening pages of chapter 20, which deal with the structure of nucleotides. Likewise, some instructors may wish to assign readings from chapter 32, which deals with mechanisms of membrane transport, after they have assigned chapter 7 on lipids and membranes.

Part 5, "Biosynthesis of the Building Blocks," contains seven chapters, mainly describing the biosynthesis and utilization of amino acids, nucleotides, complex carbohydrates, and lipids. Although there is a great deal of useful information in these chapters, most of Part 5, except for chapter 24, can be skipped if there is a strong desire to get to Part 6 as quickly as possible.

Chapter 18, "Biosynthesis of Amino Acids," describes the biosynthesis of amino acids. Because twenty amino acids found in proteins are made in bacteria and plants (whereas only a few amino acids are synthesized *de novo* in mammals), and because these pathways are better understood in bacteria, most of this chapter focuses on biosynthesis in bacteria. Nonprotein amino acids are discussed briefly. Finally, the problem of nitrogen fixation in biological systems is examined.

Chapter 19 is entitled "The Metabolic Fate of Amino Acids." It discusses the way in which amino acids, when in excess or when present as the only available carbon or energy source, are reduced to carbohydrates, which can serve as a source of energy or carbon skeletons for the synthesis of other molecules. A few products derived from amino acid metabolism are also discussed in detail, including porphyrins, biological active amines, glutathione, and peptide antibiotics.

Chapter 20, "Nucleotides," deals with the structures, synthesis, and breakdown of nucleotides, with considerable attention devoted to the role of nucleotide derivatives as inhibitors of nucleic acid synthesis and the important role they play in the medical field. The biosynthesis of nucleotide-containing coenzymes is also discussed in this chapter.

Chapter 21, "Biosynthesis of Complex Carbohydrates," describes the synthesis of complex carbohydrates that serve structural roles and the synthesis of oligosaccharides that exist in combination with proteins. A great deal of new information on this subject appears in this chapter, as does a detailed discussion of bacterial cell-wall synthesis.

Chapter 22, "Biosynthesis of Membrane Lipids and Related Substances," describes the metabolism of complex lipids. Phospholipids and sphingolipids are viewed in terms of their important structural roles in membranes, while eicosanoids are described in terms of functioning local hormones.

Chapter 23, "Metabolism of Cholesterol," describes the biosynthesis of cholesterol and some of its derivatives—bile acids and steroids. The chapter devotes considerable attention to the transport of these compounds, along with the problems associated with defects in cholesterol metabolism.

Chapter 24, "Integration of Metabolism and Hormone Action," concludes the two parts on intermediary metabolism. This chapter emphasizes the way in which intermediary metabolism is regulated in multicellular organisms.

Part 6, "Storage and Utilization of Genetic Information," features seven chapters describing the metabolism of nucleic acids and proteins. The entire section has been updated to keep pace with this rapidly changing area. Where useful, key genetic phenomena have also been explained.

This part begins with chapter 25, "Structures of Nucleic Acids and Nucleoproteins," which describes structures and functions of DNA and deoxyribonucleoproteins. It has intentionally been placed after structures of nucleotides have been discussed (in chapter 20) and before the structures of RNA and ribonucleoproteins are discussed (in chapter 28).

In chapter 26, "DNA Replication, Repair, and Recombination," the emphasis is on the metabolism of replication. The treatment is equally balanced between replication in prokaryotes and eukaryotes including animal viruses.

Chapter 27, "DNA Manipulation and Its Applications," begins with a discussion of the procedures for determining DNA sequence, proceeds with a discussion of the techniques for manipulating DNA sequences, and concludes by describing two important applications of DNA manipulation—the mapping of the globin gene family and the mapping of a gene responsible for the dreaded disease cystic fibrosis.

Chapter 28, "RNA Synthesis and Processing," which describes RNA synthesis and processing equally, presents a considerable amount of new information on transcription factors in animals cells and a variety of ways in which RNA is processed.

Chapter 29, "Protein Synthesis, Targeting, and Turnover," has been significantly expanded over the corresponding chapter in the second edition, particularly in the area of targeting, where many new discoveries have been made recently.

Chapter 30, "Regulation of Gene Expression in Prokaryotes," includes a discussion of several well-characterized gene systems found in *E coli.* It also features a treatment of the sequence of regulatory events involving bacteriophage λ infection and lysogeny.

Chapter 31, "Regulation of Gene Expression in Eukaryotes," begins by discussing several different gene systems in the intensively studied unicellular eukaryote *Saccharomyces cerevisiae.* Then it shifts to multicellular eukaryotes, focusing on the complex assemblages of protein factors involved in regulating various genes. Both this chapter and the previous one cover in considerable detail the structures of proteins that regulate gene expression and the ways in which they interact with DNA. The chapter concludes with a special section on the ways in which gene expression is regulated during early development.

Part 7, "Physiological Biochemistry," contains five chapters: "Mechanisms of Membrane Transport" (chapter 32), "Immunobiology" (chapter 33), "Carcinogenesis and Oncogenes" (chapter 34), "Neurotransmission" (chapter 35), and "Vision" (chapter 36). Except for the first one, all of these chapters are intentionally short, so that it will be convenient to assign them at different points in the course, depending on the taste of the instructor. The titles of the chapters are self-explanatory.

Learning Aids

We have included a number of in-text learning aids to help both the professor and the student navigate their way through the text.

The goal of this edition is to stress, not merely the facts, but especially the principles of biochemistry. In as many cases as is practical, explanations are given for how certain facts have been revealed by scientific investigations. Biochemists are sleuths always devising new methods to discover Nature's secrets. We felt it best to integrate discussion of the methods of biochemistry directly in the text at points where the methods are actually employed in biochemical investigations. To aid in location of a particular method description, we have placed a section entitled *A Student's Guide to Methods of Biochemical Analysis* in the appendices.

Great care has been taken to make explanations clear and to inject the maximum significance into topical headings. *Key terms* are always explained the first time they are introduced and *Key concepts* are underlined . Nevertheless, it is easy to forget the meaning of a particular term on occasion. To overcome this frustration we have included a *glossary* just before the index and a list of *common biochemical abbreviations,* we have placed a list just inside the back book cover.

As an additional learning aid, a *brief introduction* to each of the book's seven parts is included. Furthermore, each chapter begins with an *outline of contents* and an *introduction.*

All chapters have *summaries* at the end, and the longer chapters have periodic summaries within them. For those who would like to know more about any specific topic, *extensive references* are included at the end of each chapter.

The *exercise problems* at the end of the chapters are intended to reinforce what has been learned by the reading. *Brief answers* to selected problems are given at the end of the book, and instructors can obtain a *Solutions Manual* that provides complete solutions to all of the problems in the text (see the following section on ancillary materials).

Ancillary Material

In choosing the order of our presentation, we have been strongly influenced by extensive feedback from instructors of biochemistry. As always, however, some instructors would have preferred a different ordering of the subject matter. For their benefit we have suggested alternative possibilities for using the text in an accompanying instructor's manual. This manual also includes additional problems and solutions not offered in the text.

As an aid to course presentations, a set of 100 overhead color transparencies and 200 black-and-white transparency masters is available upon adoption of the text. Also available upon request to professors is a solutions manual, including worked-out solutions to all of the end-of-chapter problems. This manual is available for sale to students, but only at the instructor's request.

Acknowledgments

This preface is full of "I"s. That's because I, the coordinating author, wrote it. By contrast, the text itself is the brainchild of some twenty biochemists, of which I was only one. Nevertheless, because the manuscripts submitted by the contributing authors were modified by me, I must take the blame for any errors. If you should find errors, please notify me so that I can make corrections in the second printing. I would be most grateful for your input.

I would like to make a special note that Chapters 12, 13, and 14 were written by me based on Dr. Daniel Atkinson's material from the second edition of *Biochemistry.* So again, any errors should be considered my responsibility.

My fellow authors and I consulted many experts during the writing of each specific chapter, largely on an informal, "over the phone" basis. Although the list of their names is too long to include here, we are indebted to all of them. In addition, we had the benefit of formal reviews of chapters and whole sections of the text. I am deeply grateful to each of the reviewers, whose aid we will certainly wish to solicit for future editions. The reviewers included:

Teh-hui Kao *Penn State University*
Todd P. Silverstein *Willamette University*
Paul Melius *Auburn University*
Galen Mell *University of Montana*
Edward J. Miller *University of Alabama at Birmingham*
Rodney Cate *Midwestern State University*
Ricki Lewis *SUNY Albany*
Ronald A. MacQuarrie *University of Missouri-Kansas City*
Gerald W. Hart *Johns Hopkins University School of Medicine*
Larry Kirk *California State University at Chico*
Ralph C. Jacobson *California State University-San Luis Obispo*
Peter D. Jones *University of Tennessee*
C. Reynold Verret *Tulane University*
Arthur L. Haas, Ph.D *Medical College of Wisconsin*
Hans Gunderson *Northern Arizona University*
Ezio A. Moscatelli *University of Missouri at Columbia*
Gail Dinter-Gotleib *Drexel University*
David Gross *University of Massachusetts*
Edye E. Groseclose *Southeastern University of Health Sciences*
Maria O. Longas *Purdue University Calumet*
Bruce Howard Weber *California State University, Fullerton*
Donald R. Halenz *Pacific Union College*
Anthony T. Tu *Colorado State University*
Gary D. Small *University of South Dakota School of Medicine*
Celia Marshak *San Diego State University*
D. J. Davis *University of Arkansas*
Karl A. Wilson *SUNY at Binghamton*
Harry R. Matthews *University of California, Davis School of Medicine*
Richard E. Ebel *Virginia Polytechnic Institute and State University*
Malcolm Potts *Virginia Polytechnic Institute and State University*
Robert Lindquist *San Francisco State University*
Jeff Velten *New Mexico State University*

I am also very grateful to many workers at the Wm. C. Brown Publishing Company. First of all to Kevin Kane, who shared my vision and made a tremendous investment in our text. I am overwhelmed by the tremendous help received from Meg Johnson, who gave me advice on many aspects of the text, who obtained a multitude of top-notch reviews that were most useful in preparing the final manuscript, and who stuck by me during the hectic period of production. At times production problems became so great that it wasn't easy to keep going. I was very lucky to have a production coordinator, Sherry Padden, from whom I could get valuable advice on a day-by-day basis. She functioned well beyond the normal responsibilities of her job to help me and the other production workers to pull this project together and bring it to a successful conclusion.

Last but not least I once again had the opportunity to benefit from the superb copy editing of Emily Arulpragasam, who smoothed out bumps in the prose and helped me spot errors before they could be set in type. Her understanding of the subject was a major asset in making changes in the text that were always scientifically accurate.

An Overview of Biochemistry and Energy Considerations

1

PART

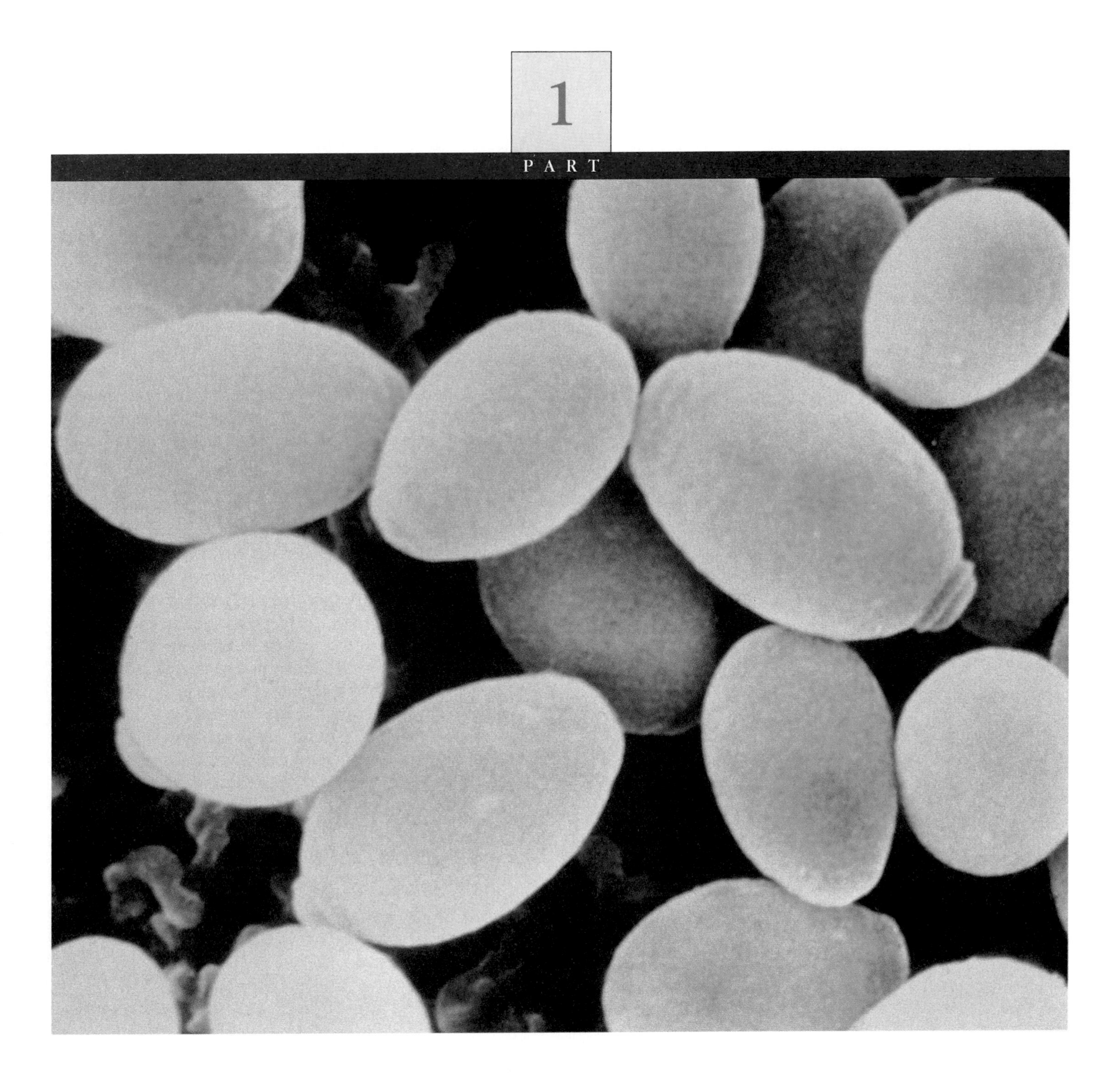

Scanning electron micrograph of growing yeast cells. On some of the cells, buds of small daughter cells can be seen to be forming. As these buds approach the size of the mature cells, they pinch off and the process of growth and budding off repeats itself. These seemingly simple spherical structures depict the miracle of the living cell. In each microscopic microcosm, thousands of reactions are simultaneously taking place, all promoting growth and duplicaltion. (Science VU-UFC/Visuals Unlimted.)

Biochemistry is such a diverse field of study that workers in one area are often unaware of significant discoveries being made by workers in another. Nevertheless, it is a single discipline governed by common principles. By integrating the many facts and concepts of biochemistry in a single text, we aim to help you to appreciate both the unity and the diversity that exist in this most rapidly developing area of the biological sciences.

The first chapter in this part of the text begins by describing current thinking on the origin and evolution of life, and it continues by presenting, in condensed form, some of the basic principles that apply throughout the study of biochemistry. As you become immersed in the details of specific topics in later chapters, you will find it helpful to keep in mind the broad perspective presented in chapter 1.

What topic merits being the first one presented in detail? The choice is not difficult. No subject is more basic to all aspects of biochemistry than thermodynamics (chapter 2). Thermodynamics tells us which reactions require energy and which reactions produce energy. For that reason, thermodynamics affects all chemical reactions in the universe and can be discussed from many vantage points. In chapter 2 we present the topic in a way that is appropriate for students of biochemistry. Our emphasis is on how to estimate equilibrium constants from thermodynamic quantities and on what strategies are used by living cells to assure that sufficient energy will be available for essential reactions.

1

CHAPTER

Overview of Biochemistry

Sometime between three and four billion years ago, a chemical change occurred on the young earth that was to have a profound effect on the planet's future. In the ancient seas, one of the several types of macromolecules that spontaneously formed from free-floating precursors acquired a striking new talent—the ability to replicate itself, functioning as a template to assemble component parts into a faithful copy of the original. This self-replicating molecule had another important characteristic: It was composed of a sequence of building blocks, so that its replication was at the same time an act of duplication and one of transmission of information. A likely candidate for this first molecule leading to life was ribonucleic acid (RNA), which is the only known molecule today that can indeed replicate itself without assistance from another molecule.

Yet a single type of molecule, no matter how complex its properties, was not in itself sufficient to qualify as the first example of the thing we call life. Over the millenia the RNA, or whatever the primordial molecule was, associated with other informational polymers—most likely its chemical cousin deoxyribonucleic acid (DNA) and various proteins. An aggregation between nucleic acid and protein may represent a bridge of sorts between the nonliving and the living—a structure that persists today in viruses, which consist of just a nucleic acid wrapped in a protein coat.

At first, the molecules that led to life obtained their building blocks—nutrients—from their surroundings. As more proteins, particularly enzymes, joined the nucleic acid–protein assemblages, these early protocells became capable of capturing energy, perhaps from the covalent bonds of carbohydrates, and of using that energy to convert one nutrient molecule into another. Gradually, metabolic pathways arose. When a third type of macromolecule, the lipids, joined the activity, the raw material for building membranes was introduced to the recipe for life. And finally—perhaps through random, trial-and-error combinations—from proteins, lipids, carbohydrates, and the founding nucleic acids there arose the first cells, the first units of life.

So successful was the simplest cell—a membrane housing nucleic acids, proteins, carbohydrates, and lipids, with some arranged as additional membranes—that it persists today in the form of bacteria and other prokaryotes. However, the development of life did not stop there. The original theme was elaborated on, first to generate the more complex eukaryotic cells, distinguished by their "sacs-within-sacs" division of labor, and then, much later, to produce multicelled organisms.

From chemistry arose life, and life today remains based on chemistry, studied in the subdiscipline of biochemistry. Our cells today must obtain nutrients, must extract energy from their

Figure 1.1

Cells are the fundamental units in all living systems, and they vary tremendously in size and shape. All cells are functionally separated from their environment by the plasma membrane that encloses the cytoplasm. Generalized representations of the internal structures of animal and plant cells (eukaryotic cells). Plant cells have two structures not found in animal cells: a cellulose cell wall, exterior to the plasma membrane, and chloroplasts. The many different types of bacteria (prokaryotes) are all smaller than most plant and animal cells. Bacteria, like plant cells, have an exterior cell wall, but it differs greatly in chemical composition and structure from the cell wall in plants. Like all other cells, bacteria have a plasma membrane that functionally separates them from their environment. Some bacteria also have a second membrane, the outer membrane, exterior to the cell wall.

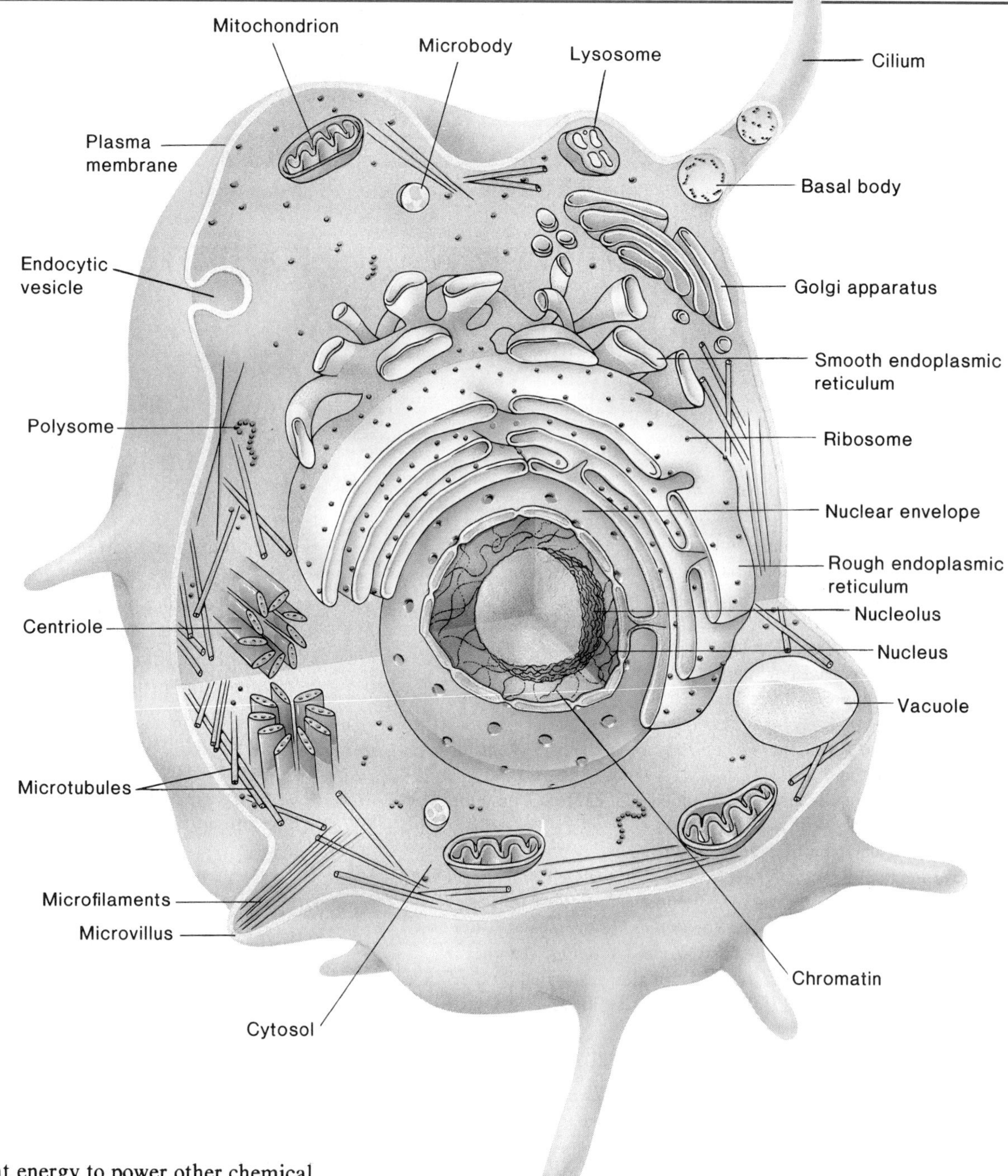

chemical bonds, must use that energy to power other chemical reactions, and, like that very first protocell, must replicate themselves, passing along, in the language of biochemistry, the information that distinguishes the living from the nonliving.

In the introduction to this part of the text, we said that certain common principles underlie all of biochemistry. Let's look briefly now at some of these principles.

The Cell Is the Fundamental Unit of Life

Microscopic examination of any organism will reveal that it is composed of membrane-enclosed structures called cells. The enclosing membrane is called the cell membrane or the plasma membrane. Cells vary enormously in size and shape. Figure 1.1 shows prototypical animal and plant cells, along with some common shapes and sizes of bacteria. Bacteria are single-celled organisms, but sometimes they are connected into long chains. In multicellular organisms the cells associate to form specialized tissues.

The plasma membrane is a delicate, semipermeable, sheetlike covering for the entire cell. By forming an enclosure it prevents gross loss of the intracellular contents; its semipermeable character permits the selective absorption of nutrients and the selective removal of metabolic waste products. In many plant and bacterial (but not animal) cells, a cell wall encompasses the plasma membrane. The cell wall is a more porous

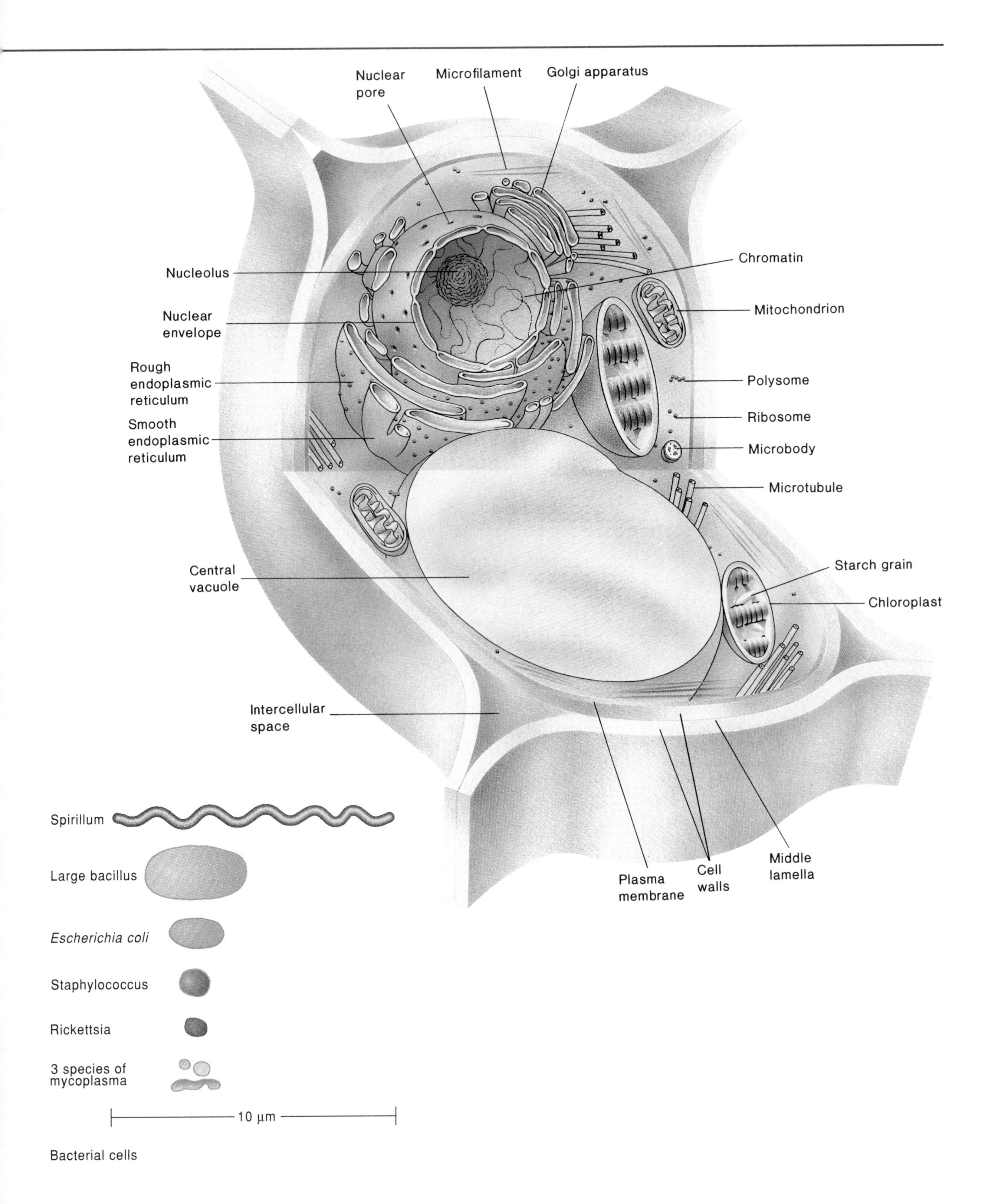
Nuclear pore
Microfilament
Golgi apparatus
Chromatin
Mitochondrion
Nucleolus
Nuclear envelope
Rough endoplasmic reticulum
Smooth endoplasmic reticulum
Polysome
Ribosome
Microbody
Microtubule
Starch grain
Chloroplast
Central vacuole
Intercellular space
Plasma membrane
Cell walls
Middle lamella
Spirillum
Large bacillus
Escherichia coli
Staphylococcus
Rickettsia
3 species of mycoplasma
10 μm
Bacterial cells

Figure 1.2

Specialized cell types found in the human. Although all cells in a multicellular organism have common constituents and functions, specialized cell types have unique chemical compositions, structures, and biochemical reactions that establish and maintain their specialized functions. Such cells arise during embryonic development by the complex processes of cell proliferation and cell differentiation. Except for the sex cells, all cell types contain the same genetic information, which is faithfully replicated and partitioned to daughter cells. Cell differentiation is the process whereby some of this genetic information is activated in some cells, resulting in the synthesis of certain proteins and not other proteins. Thus specialized cells come to have different complements of enzymes and metabolic capacities.

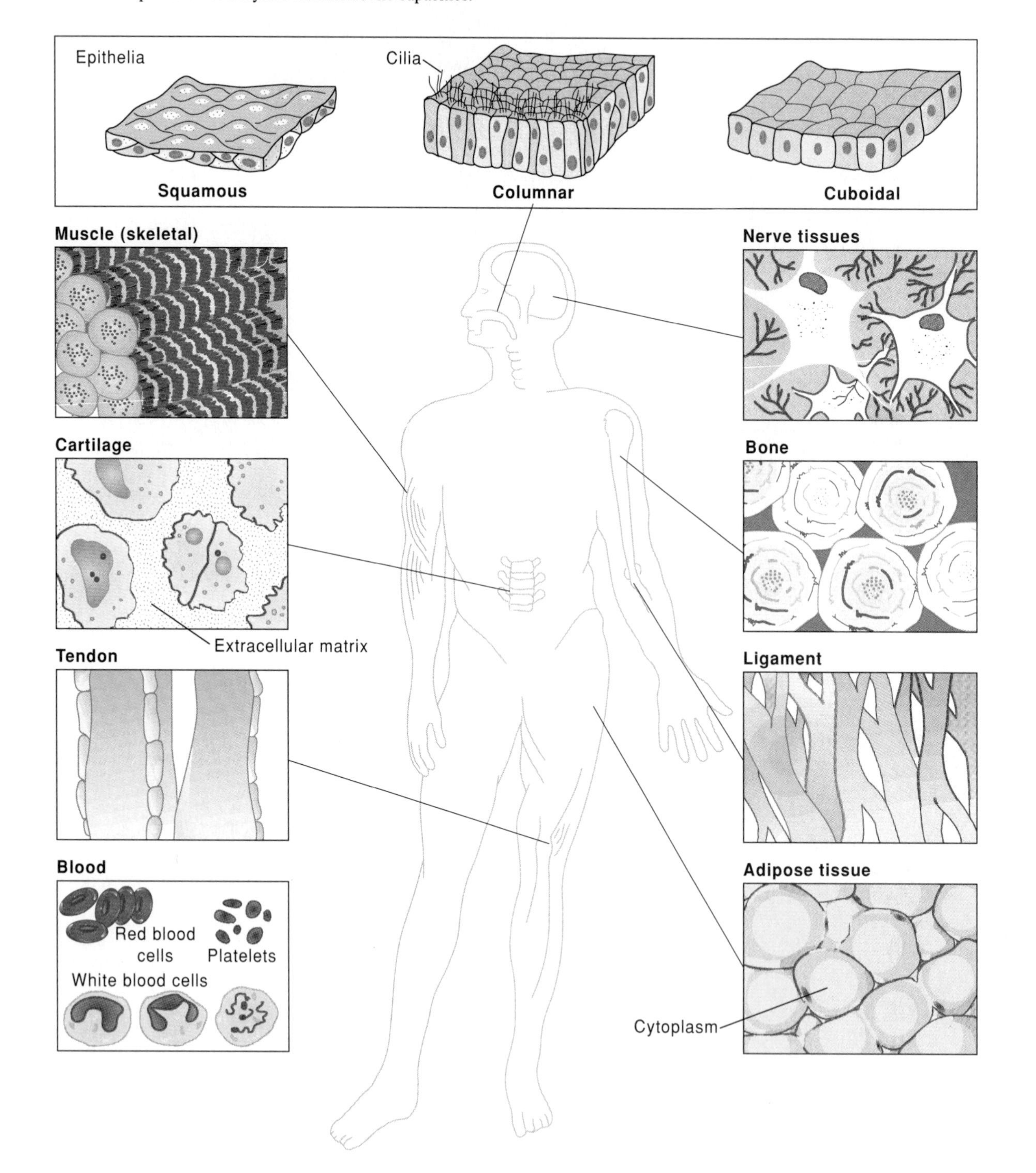

structure than the plasma membrane, but it is mechanically stronger because it is constructed of a covalently cross-linked, three-dimensional network. The cell wall maintains a cell's three-dimensional form when it is under stress.

The contents enclosed by the plasma membrane constitute the cytoplasm. The purely liquid portion of the cytoplasm is called the cytosol. Within the cytoplasm are a number of macromolecules and larger structures, many of which can be seen by high-power light microscopy or by electron microscopy. Some of the structures are membranous and are called organelles. Organelles commonly found in plant and animal cells include the nucleus, the mitochondria, the endoplasmic reticulum, the Golgi apparatus, the lysosomes, and the peroxisomes (see fig. 1.1). Chloroplasts are an important class of organelles found in many plant cells but never in animal cells. Each type of organelle is a specialized biochemical factory in which certain biochemical products are synthesized. In addition to organelles, animal and plant cells contain a collection of filamentous structures termed the cytoskeleton, which is important in maintaining the three-dimensional integrity of the cell.

As we will see, the evolutionary tree is bisected into a lower, or prokaryotic, domain and an upper, or eukaryotic, domain. The terms prokaryote and eukaryote refer to the most basic division between cell types. The fundamental difference is that in eukaryotes the cell contains a nucleus, whereas in prokaryotes it does not. The cells of prokaryotes usually lack most of the other membrane-bounded organelles as well. Plants, fungi, and animals are eukaryotes, and bacteria are prokaryotes. The biochemical functions associated with organelles are frequently present in bacteria, but they are carried out in a different way.

Cells are organized in a variety of ways in different living forms. Prokaryotes of a given type produce cells that are very similar in appearance. A bacterial cell replicates by binary fission, a process in which two identical daughter cells arise from an identical parent cell. Simple eukaryotes can also exist as single nonassociating cells. Eukaryotes of increasing complexity can contain many cells with specialized structures and functions. For example, humans contain about 10^{14} cells of more than a hundred different types. Specialized cells make up the skin, connective tissue, nervous tissue, muscle, blood, sensory functions, and reproductive organs (fig. 1.2). In such a complex organism, the capacity of different cells for replication is limited. When a skin cell or a muscle cell precursor replicates, it makes more cells of the same type. The only cells capable of reproducing an entire organism are the germ cells, that is, the sperm and the egg.

Cells Are Composed of Small Molecules, Macromolecules, and Organelles

Of the many different types of molecules in the various organelles and the cytosol that constitute the living cell, water is by far the most abundant, constituting about 70% by weight of most living matter (table 1.1). As a result, most other components are essentially in an aqueous environment.

Table 1.1
The Approximate Chemical Composition of a Bacterial Cell

	Percent of Total Cell Weight	Number of Types of Each Molecule
Water	70	1
Inorganic ions	1	20
Sugars and precursors	3	200
Amino acids and precursors	0.4	100
Nucleotides and precursors	0.4	200
Lipids and precursors	2	50
Other small molecules	0.2	~200
Macromolecules (proteins, nucleic acids, and polysaccharides)	22	~5,000

Except for water, most of the molecules found in the cell are macromolecules, which can be classified into four different categories: lipids, carbohydrates, proteins, and nucleic acids. Each type of macromolecule possesses distinct chemical properties that suit it for the functions it serves in the cell.

Lipids are primarily hydrocarbon structures (fig. 1.3). They tend to be poorly soluble in water, and are therefore particularly well suited to serve as a major component of the various membrane structures found in cells. Lipids also serve as a compact means of storing chemical energy to drive the metabolism of the cell.

Carbohydrates, like lipids, contain a carbon backbone, but they also contain many polar hydroxyl (—OH) groups and are therefore very soluble in water. Large carbohydrate molecules called polysaccharides consist of many small, ringlike sugar molecules, the sugar monomers, attached to one another by glycosidic bonds in a linear or branched array to form the sugar polymer (fig. 1.4). In the cell, such polysaccharides often form storage granules that may be readily broken down into their component sugars. With further chemical breakdown these sugars release chemical energy and may also provide the carbon skeletons for the synthesis of a variety of other molecules. Important structural functions are also served by polysaccharides. Linear polysaccharides form a major component of plant cell walls, and bacterial cell walls are composed of linear polysaccharides that are cross-linked by polypeptide chains.

Proteins are the most complex macromolecules found in the cell. They are composed of linear polymers called polypeptides, which contain amino acids connected by peptide bonds

Figure 1.3

The structures of common lipids. (*a*) The structures of saturated and unsaturated fatty acids, represented here by stearic acid and oleic acid. (*b*) Three fatty acids covalently linked to glycerol by ester bonds form a triacylglycerol. (*c*) The general structure for a phospholipid consists of two fatty acids esterified to glycerol, which is linked through phosphate to a polar head group. The polar head group may be any one of several different compounds—for example, choline, serine, or ethanolamine.

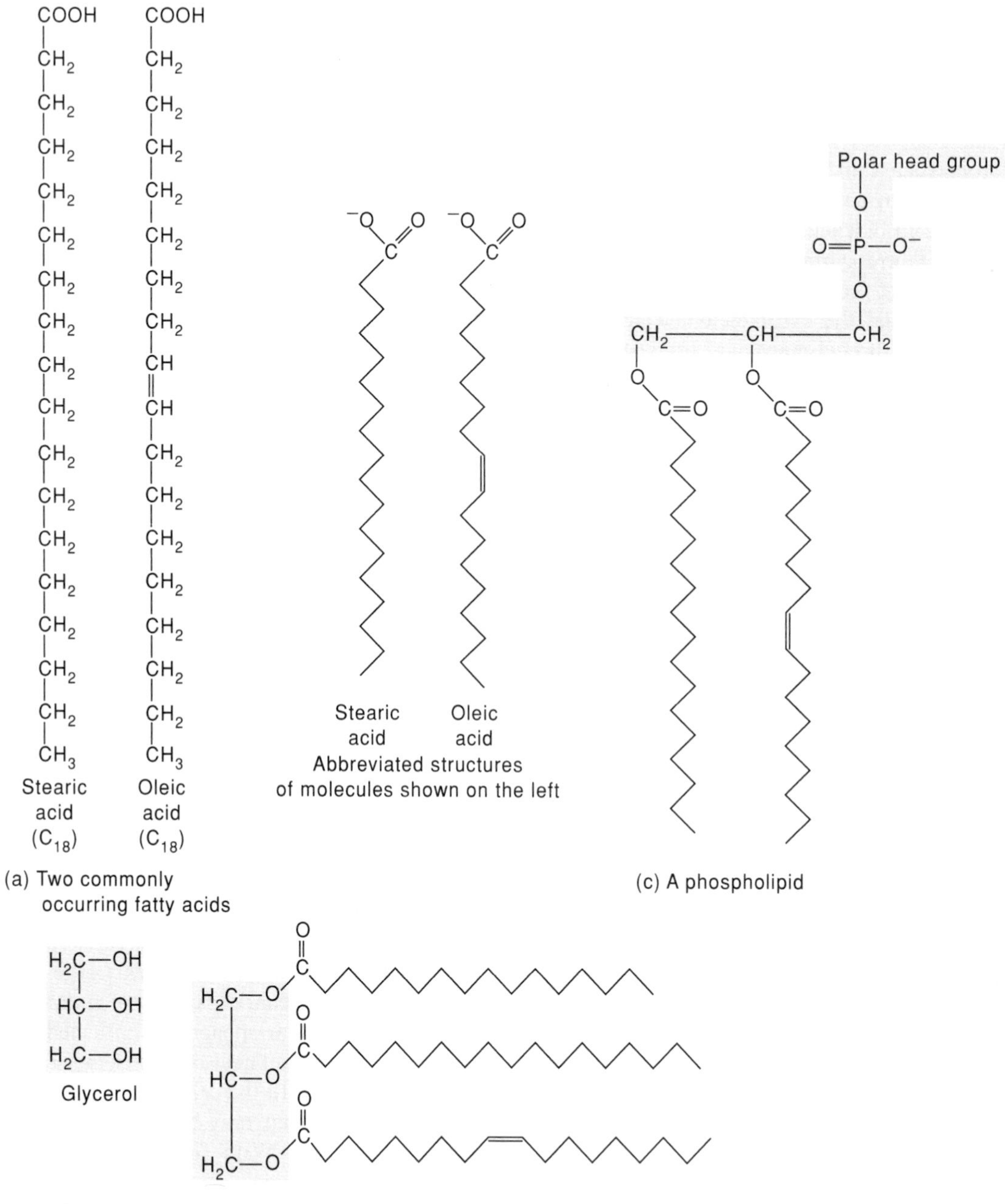

(fig. 1.5). Each amino acid contains a central carbon atom attached to four substituents: (1) a carboxyl group, (2) an amino group, (3) a hydrogen atom, and (4) an R group. The R group gives each amino acid its unique characteristics. There are twenty different amino acids in proteins. Some R groups are charged, some are neutral but still polar, and some are apolar.

The linear polypeptide chains of a protein fold in a highly specific way that is determined by the sequence of amino acids in the chains. Many proteins are composed of two or more polypeptides. Certain proteins function in structural roles. Some structural proteins interact with lipids in membrane structures. Others aggregate to form part of the cytoskeleton that gives the

Figure 1.4

Monomers and polymers of carbohydrates. (*a*) The most common carbohydrates are the simple six-carbon (hexose) and five-carbon (pentose) sugars. In aqueous solution, these sugar monomers circularize to form ring structures. (*b*) Polysaccharides are usually composed of hexose monosaccharides covalently linked together by glycosidic bonds to form long straight-chain or branched-chain structures.

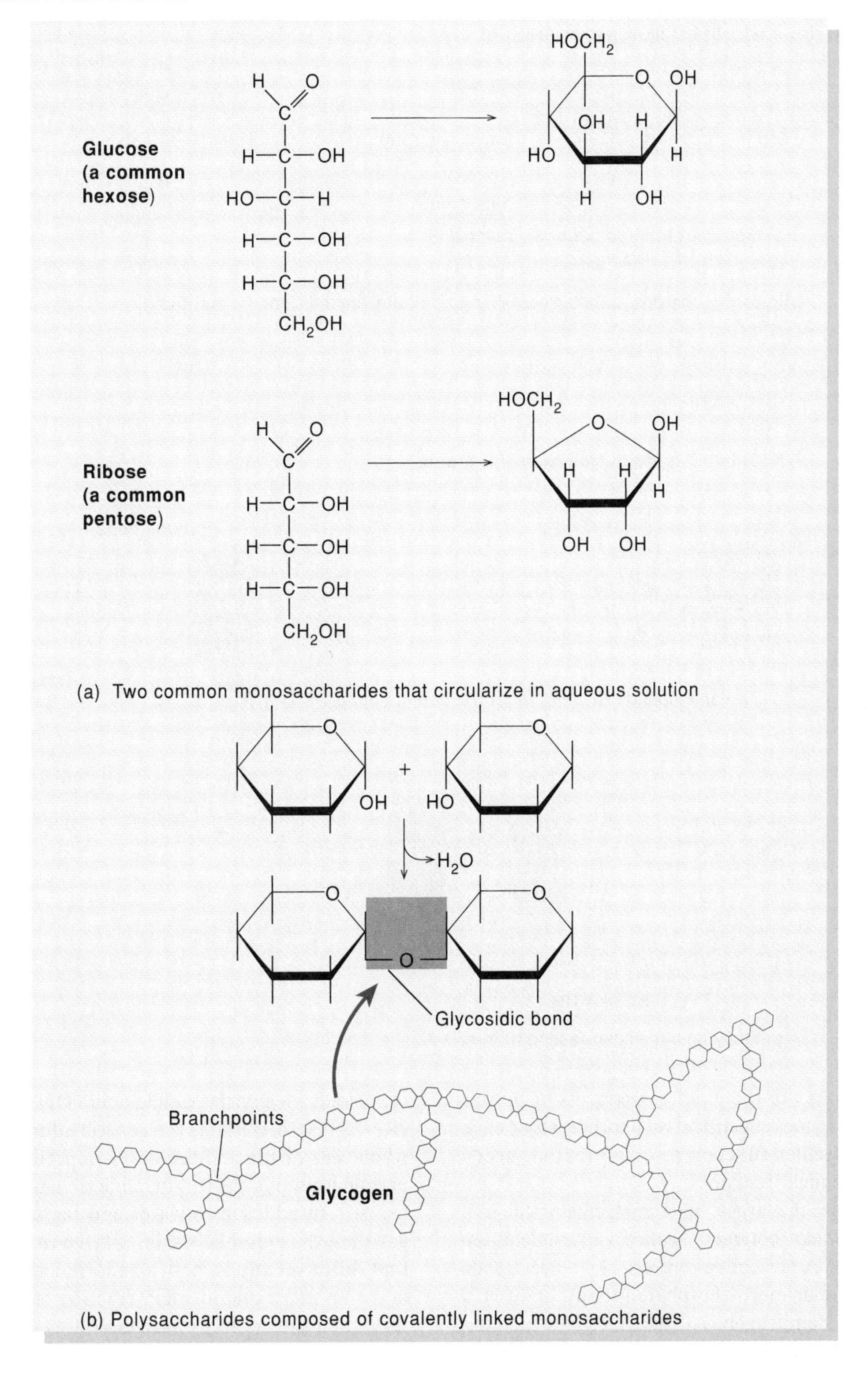

Figure 1.5

Amino acids and the structure of the polypeptide chain. Polypeptides are composed of L-amino acids covalently linked together in a sequential manner to form linear chains. (*a*) The generalized structure of the amino acid. The zwitterion form, in which the amino group and the carboxyl group are ionized, is strongly favored. (*b*) Structures of some of the R groups found for different amino acids. (*c*) Two amino acids become covalently linked by a peptide bond, and water is lost. (*d*) Repeated peptide bond formation generates a polypeptide chain, which is the major component of all proteins.

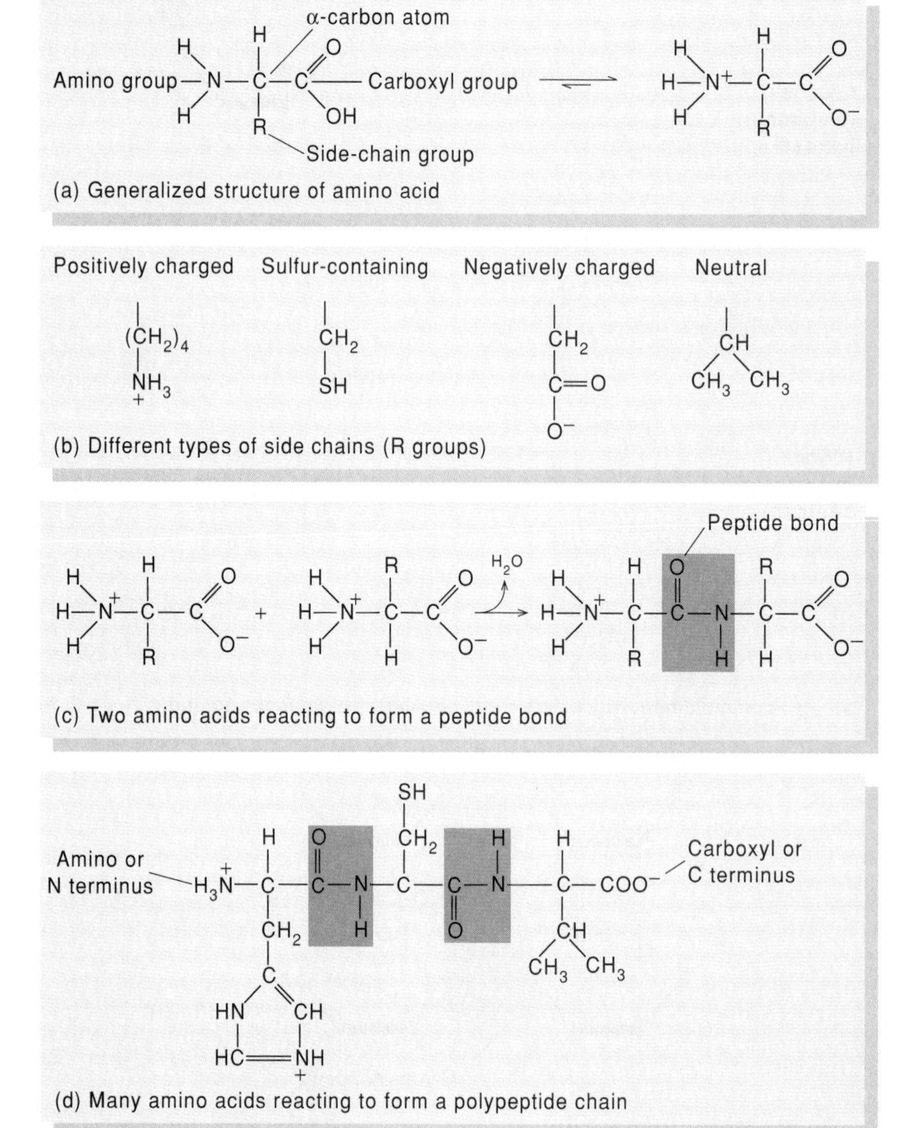

cell its shape. Still others are the chief components of muscle or connective tissue. Enzymes constitute yet another major class of proteins. These molecules function as catalysts that accelerate and direct biochemical reactions.

Nucleic acids are the largest macromolecules in the cell. They are very long linear polymers, called polynucleotides, composed of nucleotides. A nucleotide contains (1) a five-carbon sugar molecule, (2) one or more phosphate groups, and (3) a nitrogenous base. It is the nitrogenous base that gives each nucleotide a distinct character (fig. 1.6). There are five different types of nitrogenous bases found in the two main types of nucleic acids, deoxyribonucleic acid (DNA) and ribonucleic acid (RNA). DNA contains the genetic information that is inherited when cells divide and organisms reproduce. This genetic information is used in the cell to make ribonucleic acids and proteins.

In addition to water and the macromolecules and organelles, the cytosol contains a large variety of small molecules that differ greatly in both structure and function. These never make up more than a small fraction of the total cell mass despite their great variety (see table 1.1). One class of small molecules consists of the monomer precursors of the different types of macromolecules. These monomers are derived by chemical

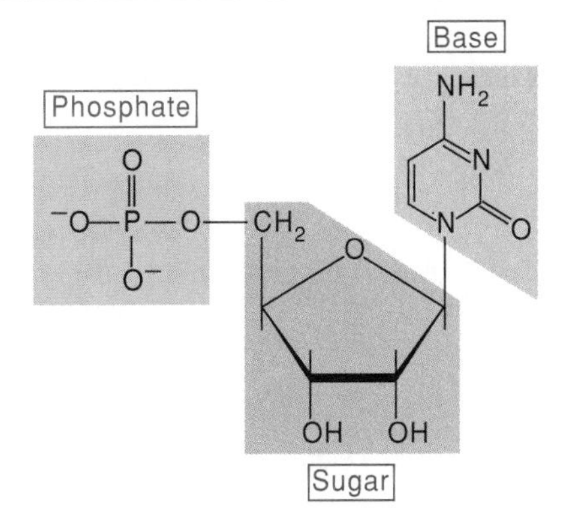

(a) Generalized structure of a nucleotide

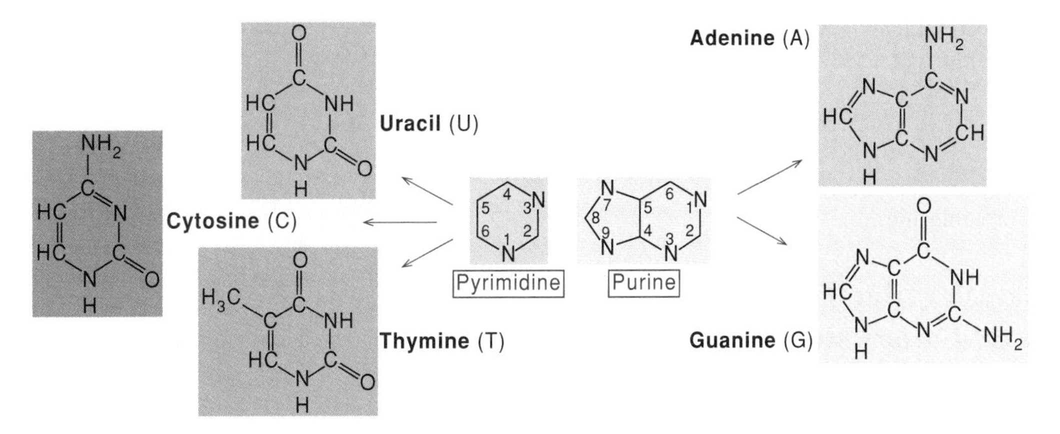

(b) Different bases found in nucleotides

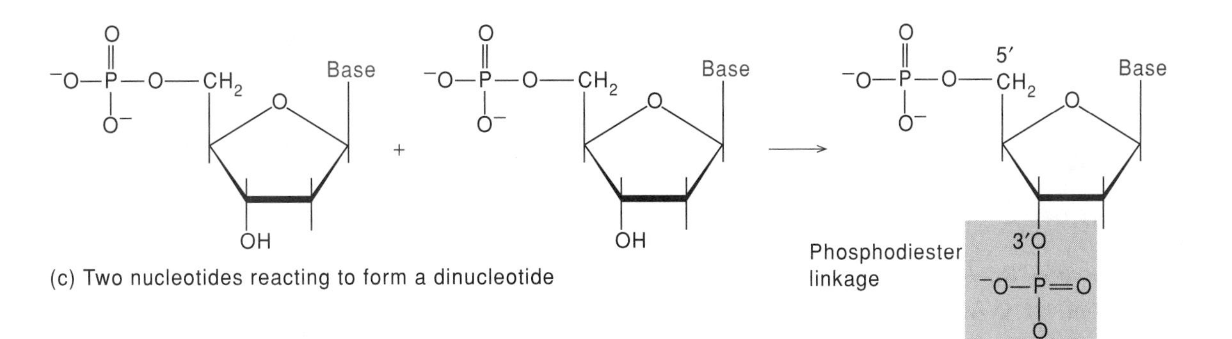

(c) Two nucleotides reacting to form a dinucleotide

Figure 1.6

The structural components of nucleic acids. Nucleic acids are long linear polymers of nucleotides, called polynucleotides. (*a*) The nucleotide consists of a five-carbon sugar (ribose in RNA or deoxyribose in DNA) covalently linked at the 5′ carbon to a phosphate, and at the 1′ carbon to a nitrogenous base. (*b*) Nucleotides are distinguished by the types of bases they contain. These are either of the two-ring purine type or of the one-ring pyrimidine type. (*c*) When two nucleotides become linked they form a dinucleotide, which contains one phosphodiester bond. Repetition of phosphodiester bond formation leads to a polynucleotide.

modification from the nutrients absorbed through the cell membrane. Rarely are the nutrients themselves the actual monomers used by the cell. As a rule each nutrient must undergo a series of enzymatically catalyzed alterations before it is suitable for incorporation into one of the biopolymers. The intermediate molecules between nutrients and monomers are also present in small concentrations in the cytosol. Another varied class of molecules found in the cytosol includes molecules formed as side products in important synthetic reactions and as breakdown products of the macromolecules. Finally, the cytosol contains small bioorganic molecules known as coenzymes, which act in concert with the enzymes in a highly specific manner to catalyze a wide variety of reactions.

All Biochemical Processes Obey the Laws of Thermodynamics

Reactions can occur only if there is an overall decrease in free energy (ΔG). You may recall from earlier courses that ΔG is a composite of two very different terms that dictate the spontaneity with which a transition between two states can occur:

$$\Delta G = \Delta H - T\,\Delta S$$

In this equation T is the temperature, ΔH is the enthalpy, which is approximately equal to the difference in bond energies, and ΔS is the entropy, which is a measure of the disorder in the system. For a given transition, a negative ΔH (favorable) indicates that the system is going from a weaker set to a stronger set of bonded interactions. A positive ΔS (favorable) indicates that the system is going from a more ordered to a less ordered state.

Thermodynamics plays a role in determining which reactions will proceed spontaneously, how biochemical reaction systems are designed, and which complex folded structures will be adopted by biological macromolecules.

Noncovalent Intermolecular Forces Determine the Three-Dimensional Folding Pattern of Macromolecules

The complex folding of biomolecules rarely entails making or breaking covalent linkages, but noncovalent linkages (to be described shortly) are important. The thermodynamic parameters ΔH and ΔS for the transition from unfolded to folded states have rarely been measured for macromolecules. However, in the few cases studied, it appears that the overall entropy of folding is slightly negative (unfavorable) and the overall enthalpy is also slightly negative (favorable). Thus, thermodynamically, folding is opposed by entropy but favored by the enthalpy change and occurs because the latter factor outweighs the former.

Interaction with Water Is a Primary Factor in Determining the Type of Structures that Form

Water, as we have seen, is the major component of living systems, and it interacts with many biomolecules. Some molecules are water-loving or hydrophilic, others are water-abhorring or hydrophobic, and still others are amphipathic or in between. What properties of a molecule make it hydrophilic or hydrophobic? First, consider the molecular properties of water and how water interacts with itself.

An individual water molecule has a significant dipole that is due to the greater electronegativity of the oxygen atom over the hydrogen atoms (fig. 1.7). This dipole leads to strong interactions between water molecules, in the form of hydrogen bonds. A hydrogen bond is a noncovalent interaction between polar molecules, one of which is an unshielded proton. In solid water or ice the polar forces hold the individual molecules together in a regular three-dimensional lattice (fig. 1.8). Most of the hydrogen bonds present in ice are also present in liquid water. Hence water is a highly hydrogen-bonded structure, not too different from ice, but with a somewhat less regular structure in which the individual molecules have greater mobility.

The dipolar properties of water molecules affect the interaction between water and other molecules that dissolve in water. For example, a favorable interaction accounts for the high solubility of sodium chloride in water (fig. 1.9). The kinds of

Figure 1.7

The structure of water and the interaction of water with other water molecules.

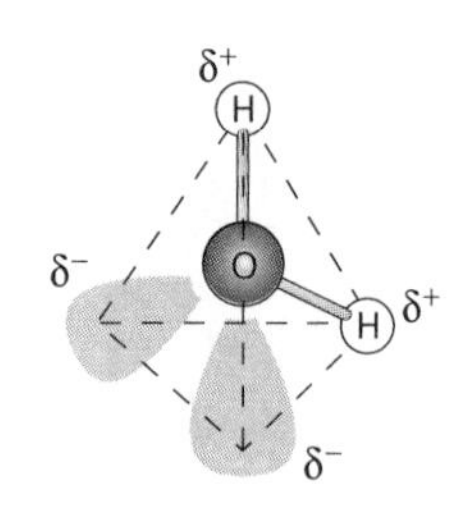

(a) Single water molecule

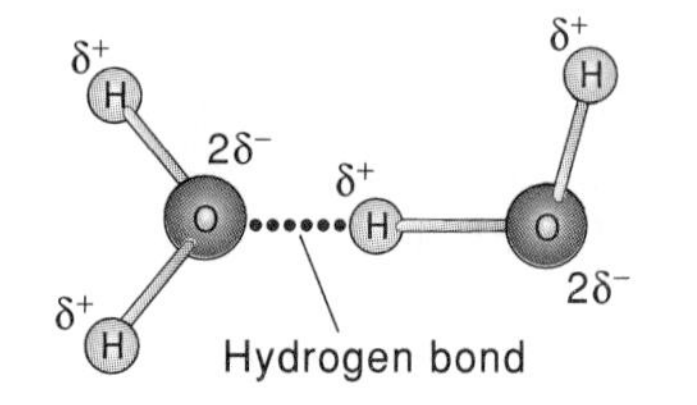

(b) Two interacting water molecules

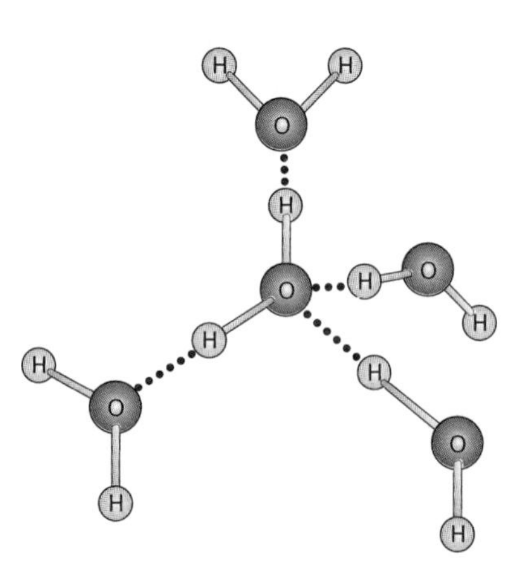

(c) Cluster of interacting water molecules

Figure 1.8

The arrangement of molecules in an ice crystal. Water molecules are orientated so that one proton along each oxygen–oxygen axis is closer to one or the other of the two oxygen atoms.

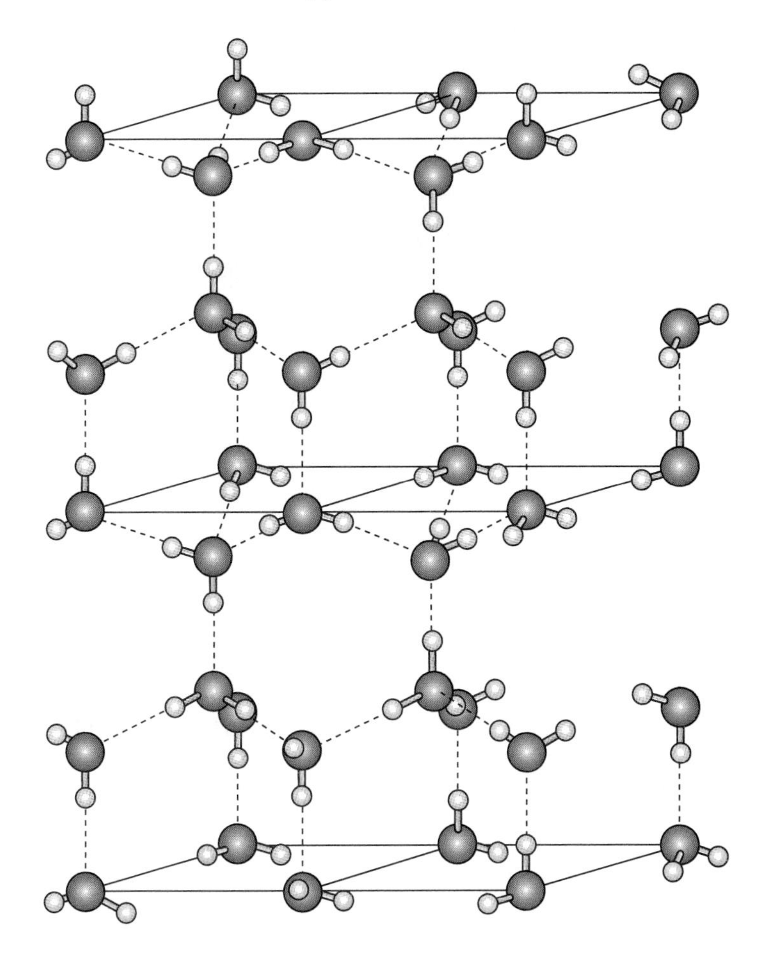

ion–dipole interactions that take place between water and simple ions such as Na^+ and Cl^- are also important in the interactions between the charged or polar groups on biomolecules and water. Thus biomolecules that contain charged residues, hydrogen-bond-forming substituents, or other kinds of polar groups are hydrophilic. In the form of small molecules such groups tend to be very soluble in water. When attached to biopolymers they determine which parts of the molecule will be oriented on the exposed surface, where they will make contact with water.

Apolar groups such as neutral hydrocarbon side chains do not contain significant dipoles or the capacity for forming hydrogen bonds. Consequently, they have nothing to gain by interacting with water, as evidenced by their poor solubility in water. When such hydrophobic molecules are present in water, the water forms a rigid clathrate (cagelike) structure around them (fig. 1.10). Apolar groups in biopolymers tend to bury themselves within the structure of the biopolymer, where they will be in the proximity of other apolar groups and will avoid contacts with water.

Figure 1.9

The water molecule is composed of two hydrogen atoms covalently bonded to an oxygen atom with tetrahedral (sp^3) electron orbital hybridization. As a result, two lobes of the oxygen sp^3 orbital contain pairs of unshared electrons, giving rise to a dipole in the molecule as a whole. The presence of an electric dipole in the water molecule allows it to solvate charged ions because the water dipoles can orient to form energetically **favorable** electrostatic interactions with charged ions.

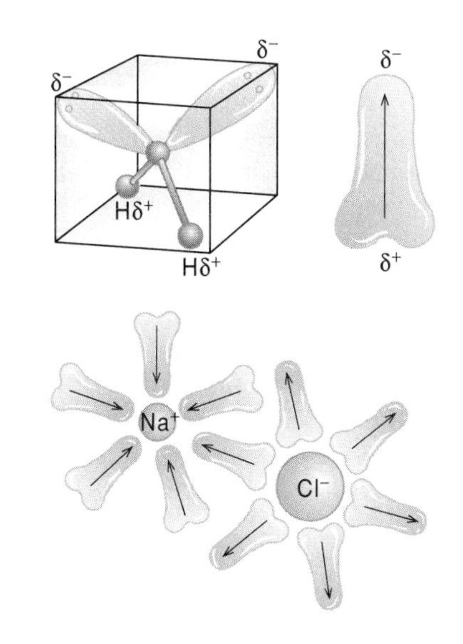

Figure 1.10

Clathrate structures are ordered cages of water molecules around hydrocarbon chains. A portion of the cage structure of $(nC_4H_9)_3$ S^+F^-, 23 H_2O is shown. The trialkyl sulfur ion nests within the hydrogen-bonded framework of water molecules. In the intact framework, each oxygen is tetrahedrally coordinated to four others. One such oxygen atom and its associated hydrogens are shown by the arrow.

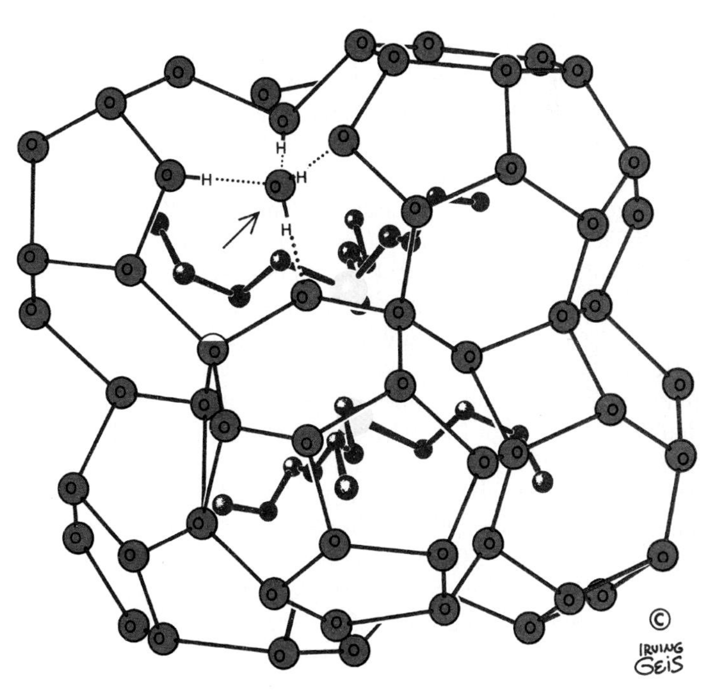

Some structures illustrating these principles for macromolecular interaction are shown in figures 1.11 through 1.13. Phospholipids (see fig. 1.3*c*), which have a hydrophilic polar group on one end and long hydrophobic side chains attached to it, form multimolecular aggregates in an aqueous environment (fig. 1.11). These phospholipid aggregates form monomolecular layers at the air–water interface, or micelles or bilayer vesicles within the water. In all of these structures, the polar head groups of the lipid are in contact with water, whereas the apolar side chains are excluded from the solvent structure.

As another example of polarity effects on macromolecular structure, consider polypeptide chains, which usually contain a mixture of amino acids with hydrophilic and hydrophobic side chains. Enzymes fold into complex three-dimensional globular structures with hydrophobic residues located on the inside of the structure and hydrophilic residues located on the surface, where they can interact with water (fig. 1.12).

DNA forms a complementary structure of two helically oriented polynucleotide chains (fig. 1.13). The polar sugar and phosphate groups are on the surface, where they can interact with water; the nitrogenous bases from the two chains form intermolecular hydrogen bonds in the core of the structure.

Figure 1.11

Structures formed by phospholipids in aqueous solution. Phospholipids may form a monomolecular layer at the air–water interface, or they may form spherical aggregations surrounded by water. A vesicle consists of a double molecular layer of phospholipids surrounding an internal compartment of water.

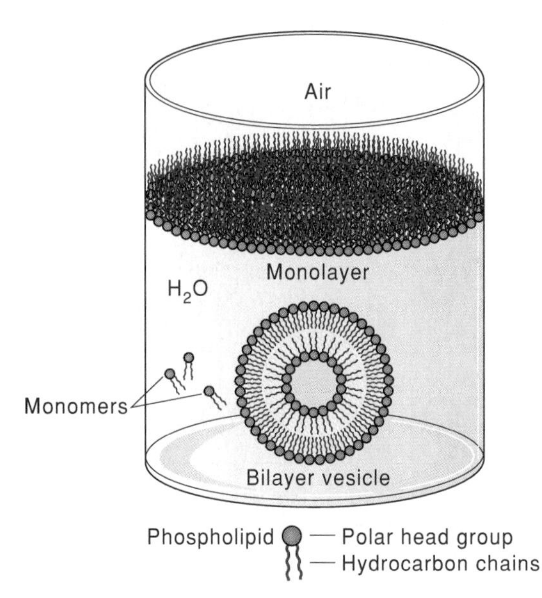

Biochemical Reactions Are a Subset of Ordinary Chemical Reactions

Even though the total number of biochemical reactions is very large, it is still much smaller than the potential number of reactions that occur in ordinary chemical systems. This simplification results partly from the fact that only a limited number of elements account for the vast majority of substances found in living cells. The elements of major importance, in order of decreasing numerical abundance, are hydrogen (H), carbon (C), oxygen (O), nitrogen (N), phosphorus (P), and sulfur (S). Certain metal ions are also important; these include Na^+, K^+, Mg^{2+}, Ca^{2+}, Zn^{2+}, and Fe^{2+} or Fe^{3+}. Other metals and elements that are needed in very small amounts are iodine, cobalt, molybdenum, selenium, vanadium, nickel, chromium, tin, fluorine, silicon, and arsenic. In some cases we don't know the biological roles of these "trace elements" but only that they are needed by some organisms for normal growth or development.

The types of covalent linkages most commonly found in biomolecules are also quite limited (table 1.2). Only sixteen different types of linkages account for more than 95% of the linkages found in biomolecules. All the elements form single or double bonds, except for hydrogen, which can form only single bonds; all the elements exist primarily in a single valence state except for carbon and sulfur, which are found in more than two valence states (table 1.3). Despite this overall simplicity, many other valence states can be found in unusual cases, and some of these are very important. For example, the biochemistry of nitrogen involves consideration of all the valence states of nitrogen from +5 to 0 to −3. A major source of nitrogen available to biosystems is gaseous nitrogen found in the atmosphere (valence state 0). Biochemical reactions convert gaseous nitrogen into other forms of nitrogen. These reactions are very complex and occur only in a select group of microorganisms that possess the necessary enzyme systems.

Biochemical reactions involving the different classes of substances use a limited number of functional groups, which are illustrated in figure 1.14. Most of the reactive groups in biomolecules contain one or more of these functional groups or closely related ones. Many cellular reactions involving these functional groups are closely related to reactions that take place outside the cell under different conditions and are studied in organic chemistry.

For example, alcohols, which contain a hydroxyl functional group, can undergo dehydration reactions with either carboxylic or phosphoric acid to form esters.

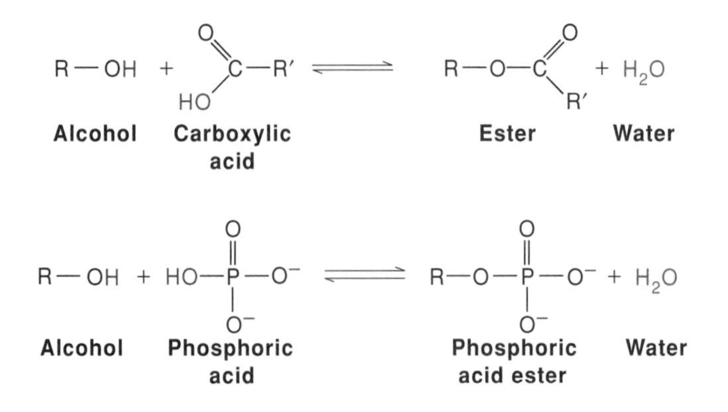

Figure 1.12

A graphic representation of a three-dimensional model of the protein cytochrome *c*. Amino acids with nonpolar, hydrophobic side chains (color) are found in the interior of the molecule, where they interact with one another. Polar, hydrophilic amino acid side chains (gray) are on the exterior of the molecule, where they interact with the polar aqueous solvent.

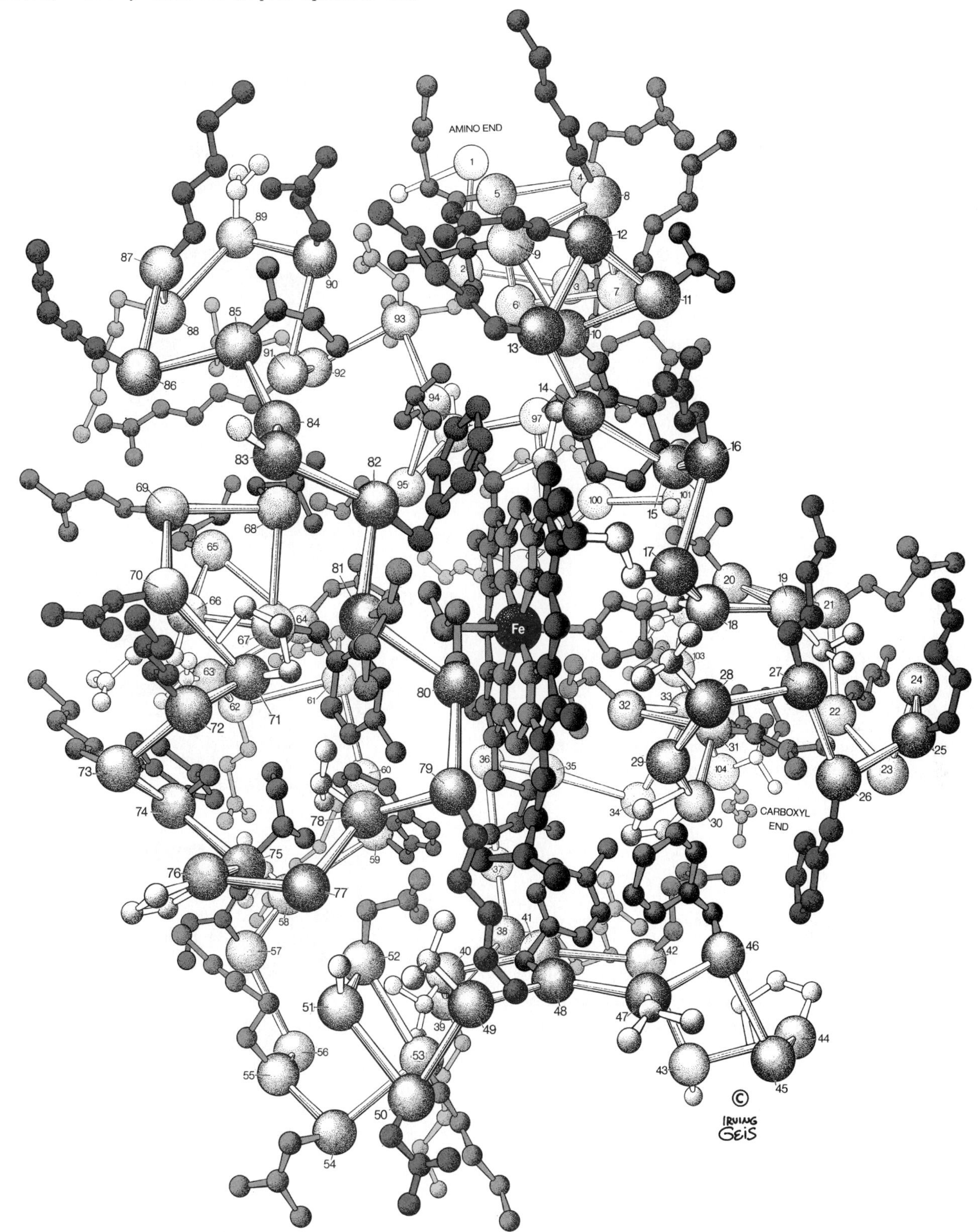

Figure 1.13

The right-handed helical structure of DNA. DNA normally exists as a two-chain structure held together by hydrogen bonds (···) formed between the bases in the two chains. Along the chain the planar surfaces of these bases interact and, together with the hydrogen bonds, contribute to the stability of the two-chain structure. The negatively charged phosphate groups are on the outside of the structure, where they interact with water, ions, or charged molecules.

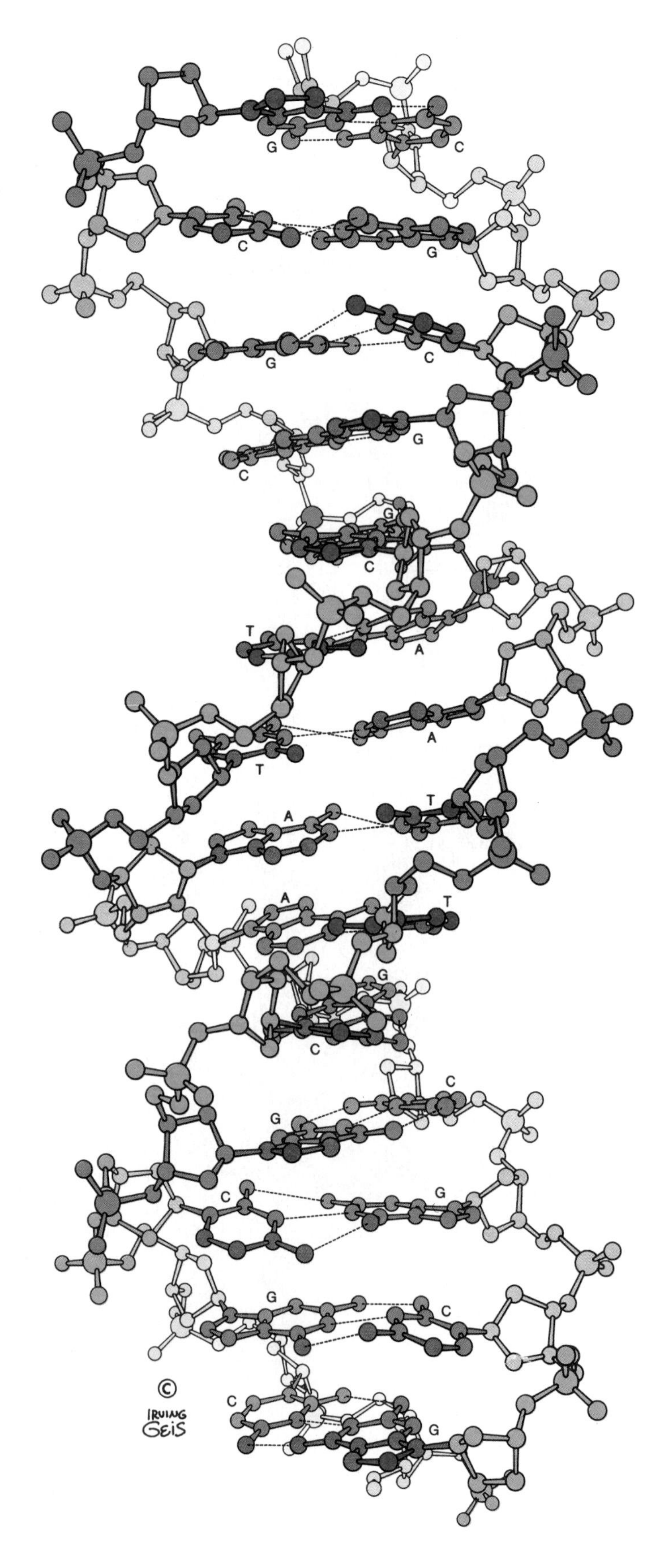

Thiols, containing sulfhydryl groups (—SH), can substitute for alcohols in some reactions, leading to the formation of thiol esters.

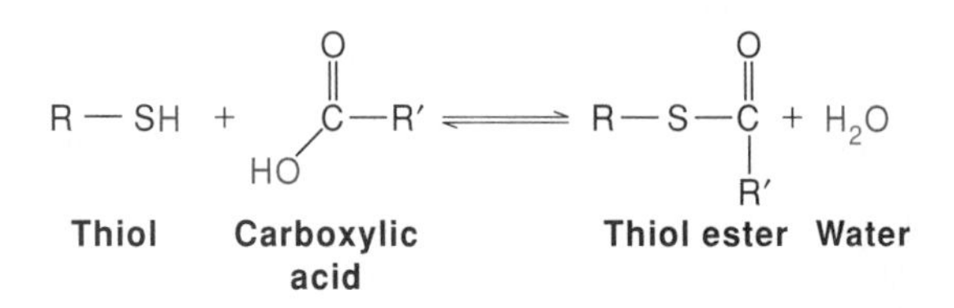

Two alcohols can react with one another to form an ether.

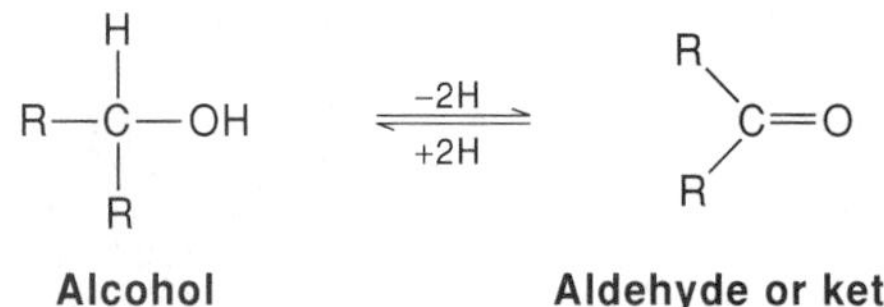

Alcohols can also undergo a dehydrogenation reaction to form a carbonyl derivative (aldehyde or ketone).

$$\text{R—CH(R)—OH} \underset{+2\text{H}}{\overset{-2\text{H}}{\rightleftharpoons}} \text{R}_2\text{C=O}$$

Alcohol **Aldehyde or ketone**

Amines undergo reactions with carboxylic acids comparable to the formation of esters from alcohols. The product is known as an amide.

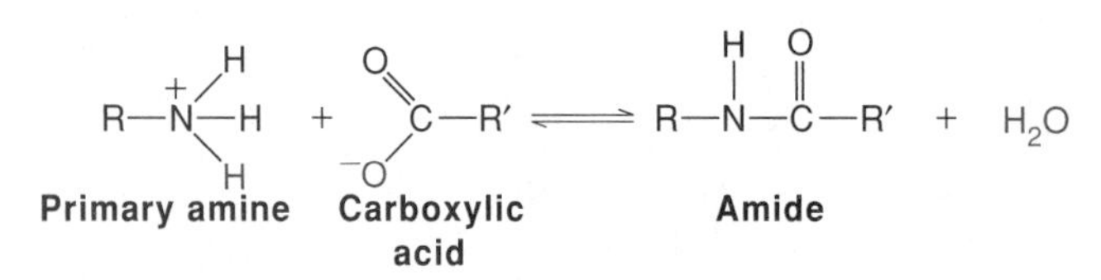

Amines can also undergo dehydrogenation reactions leading to the formation of imines, which are frequently unstable in water and hydrolyze to ketones or, in cases where one of the R groups is an H, to aldehydes:

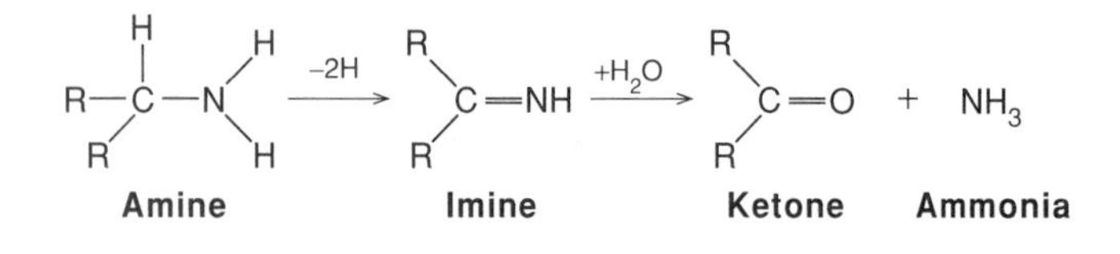

Table 1.2
Types of Covalent Linkages Most Commonly Found in Biomolecules

	H	C	O	N	P	S
H						
C	—C—H	—C—C— C=C				
O	—O—H	—C—O— C=O				
N	N—H	—C—N= C=N—	—			
P	—	—	P—O P=O	—		
S	—S—H	—C—S—	S—O— S=O	—	—	—S—S—

Table 1.3
Most Common Valences Displayed by Atoms in Covalent Linkages

Element	Valence
H	+1
C	−4 to +4
O	−2
P	+5
N	−3
S	+6, −2, −1

Aldehydes and ketones both may be reduced to alcohols by hydrogenation (see the alcohol dehydrogenation reaction). Aldehydes may react with either water or alcohol to form aldehyde hydrates, or hemiacetals, respectively. Reaction of an aldehyde with two molecules of alcohol leads to acetal formation.

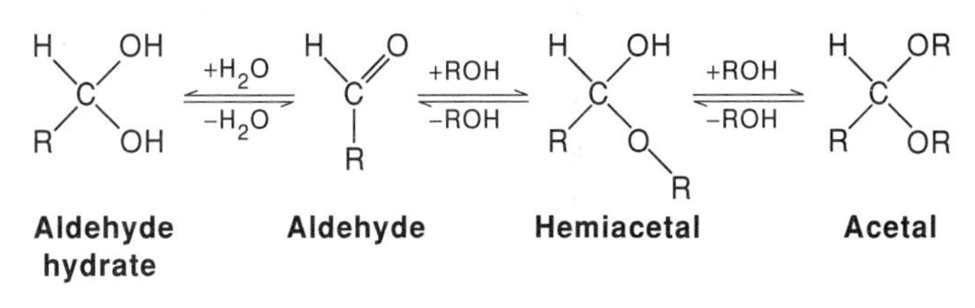

Figure 1.14
Different functional groups found in biomolecules. This figure includes the major functional groups. Other functional groups are found in minor amounts.

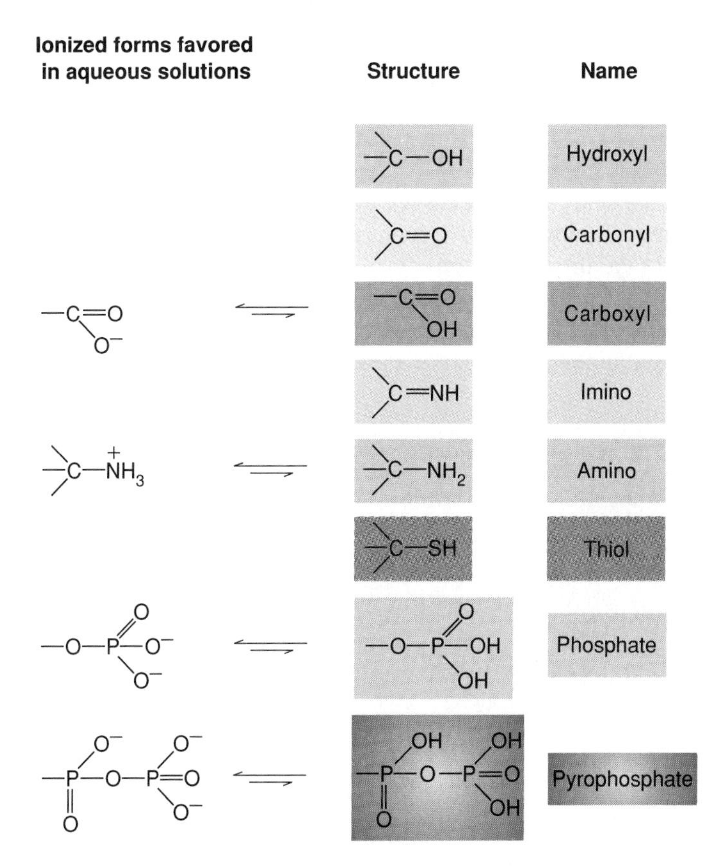

Dehydrogenation of an aldehyde hydrate leads to carboxylic acid formation.

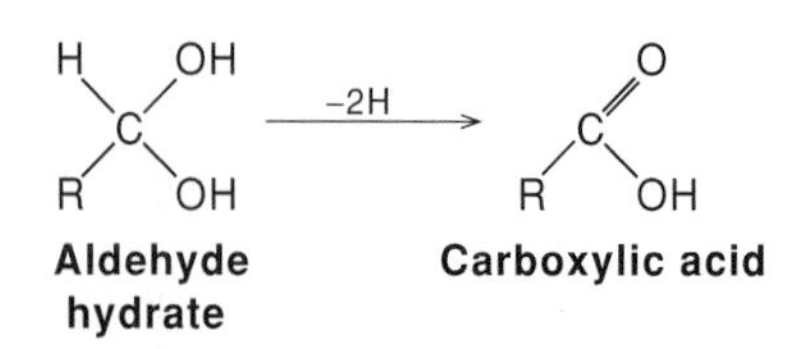

Aldehydes and ketones may also isomerize to the enol form as long as the adjacent carbon atom contains at least one H atom. In the reaction a hydrogen migrates and the double bond shifts.

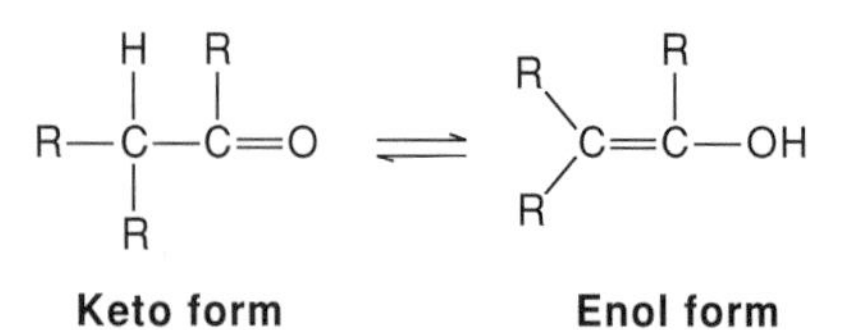

Pyrophosphates may hydrolyze to inorganic phosphoric acid (phosphate) and an organophosphoric acid.

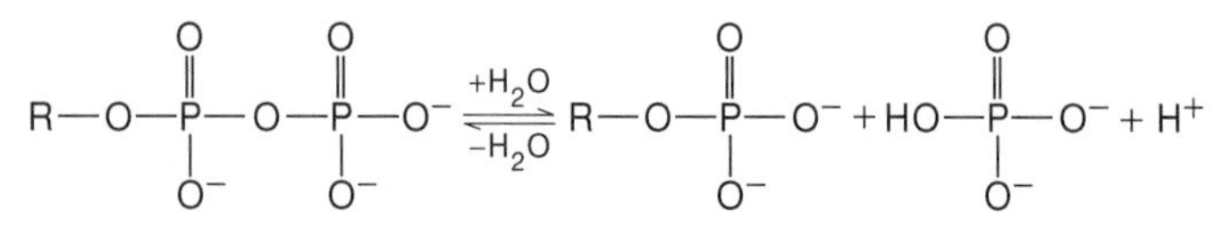

Hydrolysis reactions of this sort yield considerable energy, which can be utilized in biosynthesis.

All of the functional groups that we have described are electrostatically neutral in organic solvents. However, in water many of these functional groups either lose or gain protons to become charged species. Such ionization reactions are very important in biochemical systems because they frequently influence solubility and reactivity.

Carboxylic and phosphoric acids lose one or more protons in water to become negatively charged. The ionized forms are stabilized by resonance as shown:

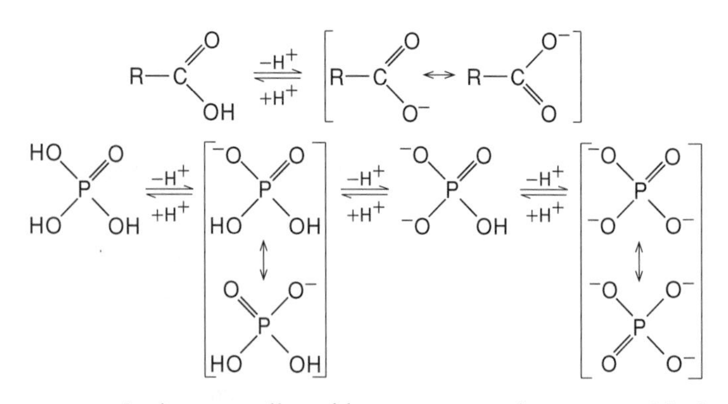

Amines usually add a proton to become positively charged.

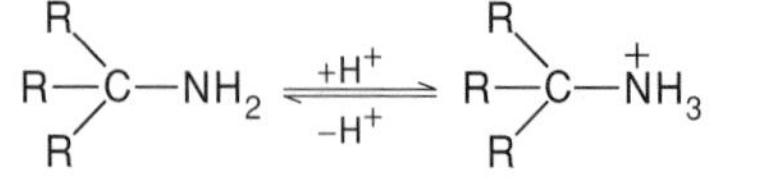

Near neutrality (10^{-7} M H^+), where most biochemical systems function, the carboxyl group exists mainly in the negatively charged form, phosphoric acid exists mainly in the diionized form, and amino groups exist mainly in the positively charged form. This fact has interesting consequences for amino acids, since they contain one amino group and one carboxyl group. The amino acids are usually neutral overall, even though they contain two charged groups, one resulting from the deprotonation of the carboxyl group and the other resulting from the protonation of the amino group. Amino acids existing as dipolar ions are called zwitterions.

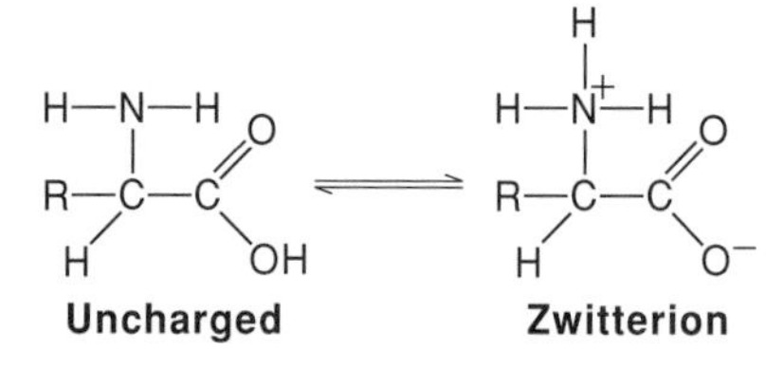

These are some of the more important reactions involving covalent bond breakage or formation in biochemistry. By now two things should be apparent about biochemical reactions: (1) as stated at the outset, the number of reactions in biochemistry is much more limited than in ordinary chemistry; (2) as far as the reactants and products are concerned, biochemical reactions can be understood in the same terms as ordinary chemical reactions.

Conditions for Biochemical Reactions Differ from Those for Ordinary Chemical Reactions

Although biochemical reactions resemble ordinary chemical reactions, they differ in some important ways. Chemical reactions are frequently carried out in nonaqueous solvents, using elevated temperatures and pressures, acids or bases, or other harsh reagents, conditions that would destroy the functional organization of a living cell. Biochemical reactions usually take place under very mild conditions in aqueous solution. However, many chemical reactions do not proceed at reasonable rates under such conditions. Biochemical reactions proceed at substantial rates because of the very special nature of the enzyme catalysts that accelerate them.

Enzymes are structurally complex, highly specific catalysts; each enzyme usually catalyzes only one type of reaction. The enzyme surface binds the interacting molecules, or substrates, so that they are favorably disposed to react with one another (fig. 1.15). The specificity of enzyme catalysis also has a selective effect, so that only one of several potential reactions will take place. For example, a simple amino acid can be utilized in the synthesis of any of the four major classes of macromolecules or can be simply secreted as waste product (fig. 1.16). The fate of the amino acid is determined as much by the presence of specific enzymes as by its reactive functional groups.

Figure 1.15

The structure of the complex formed between the enzyme lysozyme and its substrate. The crevice that forms the site for substrate binding (the active site) runs horizontally across the enzyme molecule. The individual hexose sugars of the hexasaccharide substrate are shown in a darker color and labeled A–F.

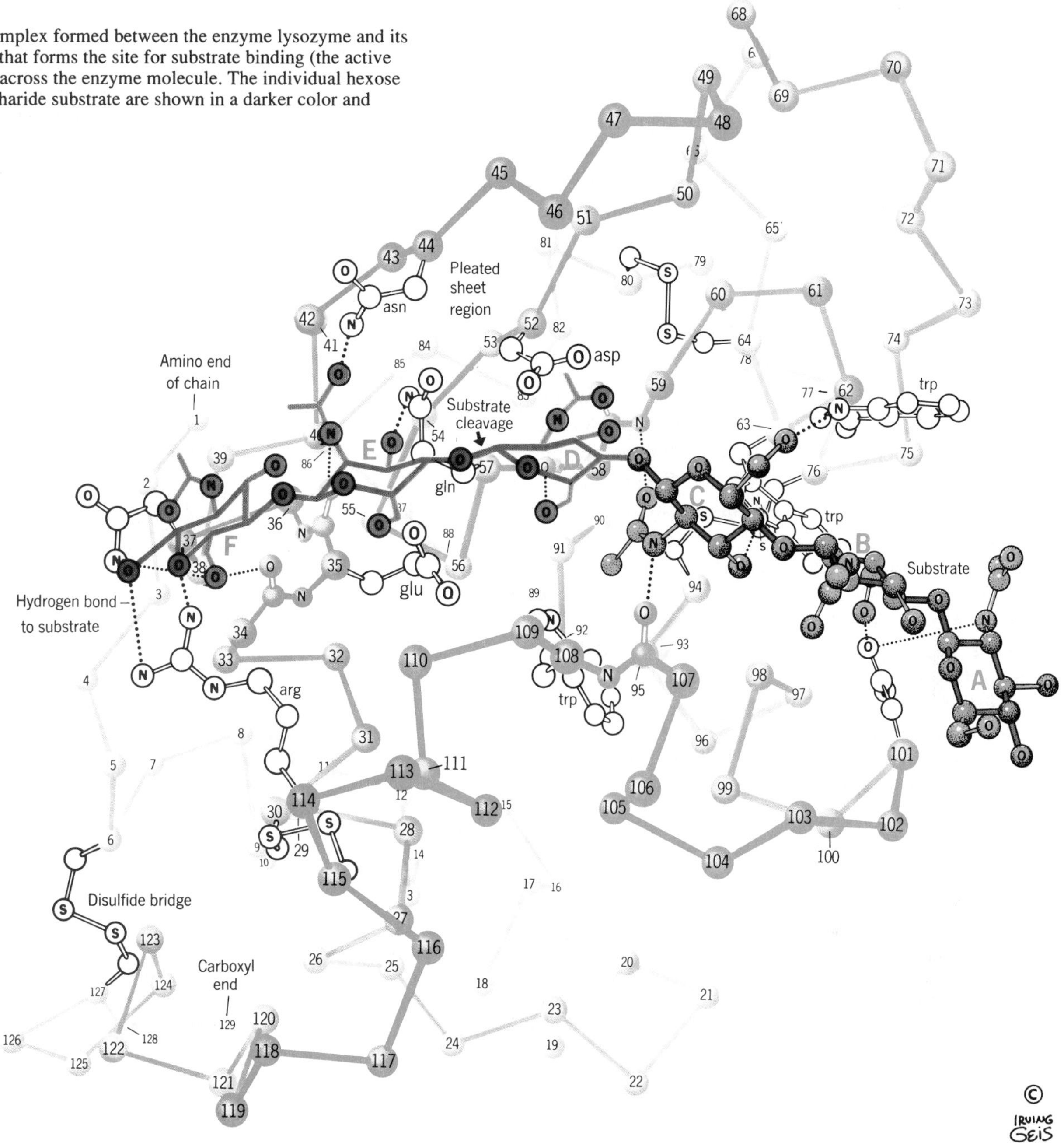

Figure 1.16

The different fates of an amino acid. Depending on which enzymes are present and active and on the needs of the organism, an amino acid can be metabolized in different ways. Each of these conversions involves one or more steps, and usually each step requires a specific enzyme.

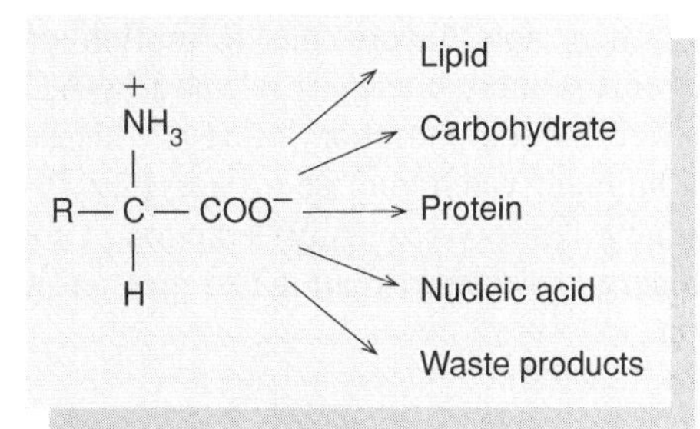

Many Biochemical Reactions Require Energy

An appreciable amount of energy is needed to build a cell. Even maintaining a cell in a steady nongrowing state requires energy input. Chemical energy is needed to drive many biochemical reactions, to do mechanical work, and for transport of substances across the plasma membrane. The ultimate source of energy that drives a cell's reactions is sunlight (fig. 1.17). Light energy is converted into chemical energy in the chloroplasts of plant cells or in the photosynthetic grana of certain microorganisms. The main form of chemical energy produced in the chloroplast is a nucleotide containing three phosphoric acid groups attached in sequence, adenosine triphosphate or ATP (fig. 1.18). Organisms that cannot harness the light rays of the sun themselves to make ATP are able to make ATP from the breakdown of organic nutrients originating from plants or other organisms.

Many nutrients consist of partly degraded macromolecules or various other small molecules that after absorption must be converted into a form suitable for the production of ATP. One of the simplest and yet most effective substances useful in ATP synthesis is the six-carbon sugar glucose. Degradation of a molecule of glucose can produce 38 molecules of ATP by the following overall reaction, which involves many enzymes:

$$38\ H^+ + C_6H_{12}O_6 + 6\ O_2 + 38\ ADP + 38\ P_i^* \rightarrow 6\ CO_2 + 38\ ATP + 44\ H_2O$$

Two characteristics of this equation should be noted. First, the glucose is degraded by oxidation to CO_2 and H_2O. A substantial fraction of the energy released by this complete oxidation of glucose is used in the production of ATP. Second, ATP is being synthesized not from small molecular precursors but simply by the addition of a single phosphate (P_i) to adenosine diphosphate (ADP). Considerable energy is released when ATP is hydrolyzed to ADP and P_i, and the ADP can be reutilized many hundreds of times. These are two of the reasons that ATP is so effective as an energy source. A third reason is that ATP is quite stable in water; it does not lose its terminal phosphate readily except in enzyme-catalyzed reactions.

Most biochemical reactions fall into one of two classes: degradative or synthetic. Degradative, or catabolic, reactions result in the breakdown of organic compounds to simpler substances. Synthetic, or anabolic, reactions lead to the assembly of biomolecules from simpler molecules. Most anabolic processes require energy to drive them. This energy is usually supplied by coupling the energy-requiring biosynthetic reactions to energy-releasing catabolic reactions. Most frequently the energy-releasing reaction involves ATP hydrolysis, either directly or indirectly.

As an example of coupling the hydrolysis of ATP to an energy-requiring reaction, let us examine the use of glucose in an anabolic reaction, the synthesis of a polysaccharide containing glucose as the monomer. The first step in glucose utilization involves its conversion to glucose-6-phosphate. The reaction of glucose with P_i to form glucose-6-phosphate is energetically unfavorable. Another way of saying this is that the equilibrium for the reaction favors the reactants, not the products.

$$\text{Glucose} + P_i \leftrightarrows \text{glucose-6-phosphate}$$

Thus, even in the presence of a catalyst for this reaction, very little glucose-6-phosphate would be formed. In order to make glucose-6-phosphate efficiently, its formation from glucose is coupled to the energetically favorable reaction of ATP breakdown:

$$ATP \rightleftharpoons ADP + P_i$$

When these two reactions are biochemically coupled, the net reaction is

$$\text{Glucose} + ATP \rightleftharpoons \text{glucose-6-phosphate} + ADP$$

The net reaction now favors the formation of glucose-6-phosphate, because more energy is released by ATP hydrolysis than is required for the phosphorylation of glucose. In the cell this is what happens when glucose-6-phosphate is synthesized. The unfavorable (energy-requiring) reaction is coupled to the favorable (energy-releasing) reaction to give an overall favorable reaction.

*P_i is the inorganic phosphate ion, which has the structural formula

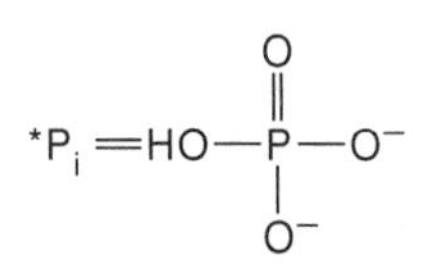

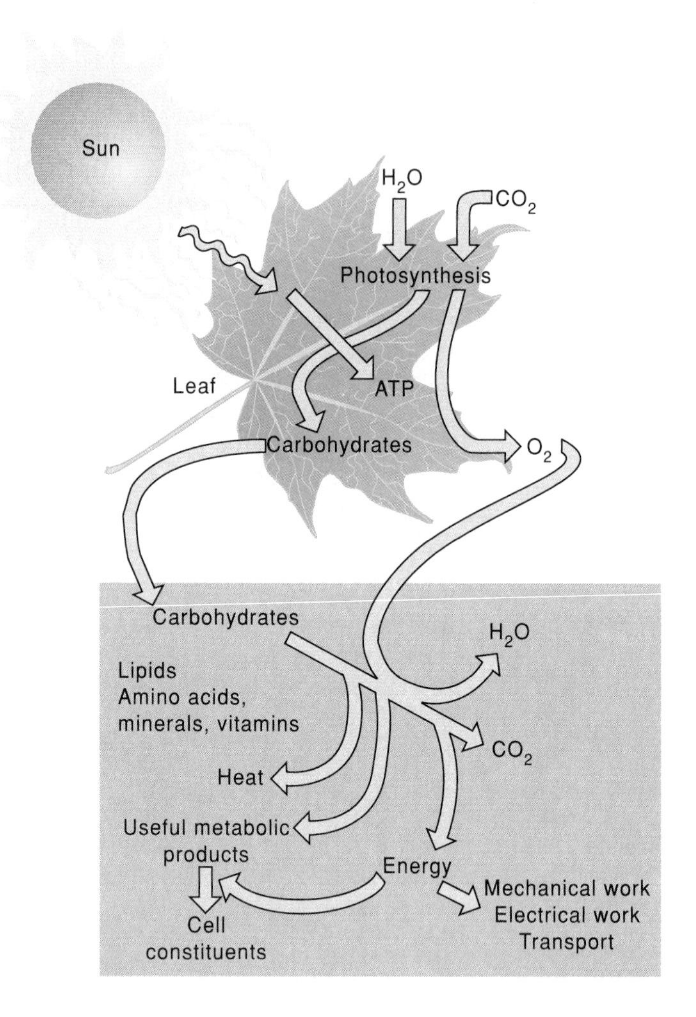

Figure 1.17

Flow of energy in the biosphere. The sun's rays are the ultimate source of energy. These rays are absorbed and converted into chemical energy (ATP) in the chloroplasts. The chemical energy is used to make carbohydrates from carbon dioxide and water. The energy stored in the carbohydrates is then used, directly or indirectly, to drive all the energy-requiring processes in the biosphere.

Figure 1.18

The structures of ATP and ADP and their interconversion. The two compounds differ by a single phosphate group.

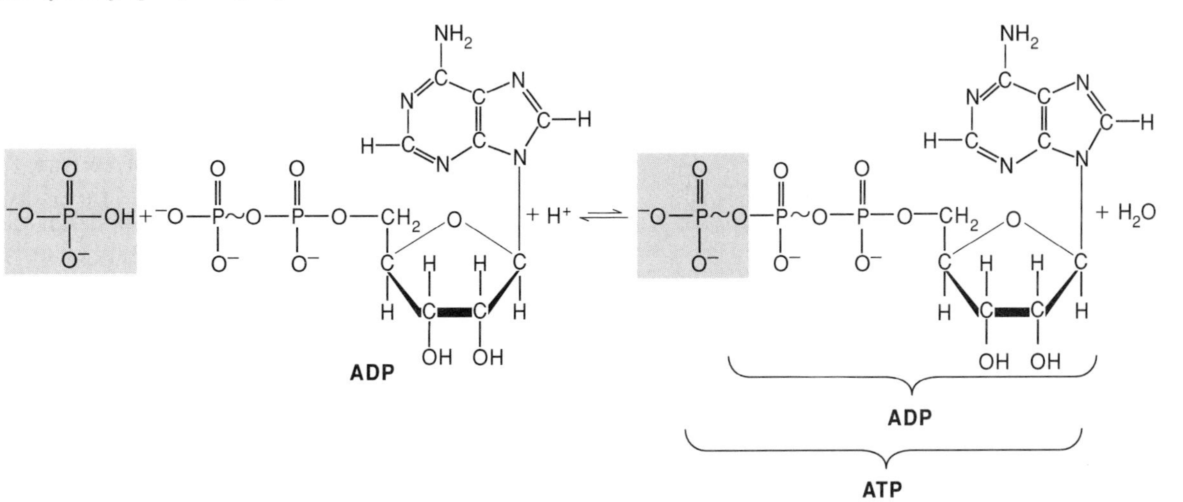

Biochemical Reactions Are Localized in the Cell

Biochemical reactions are organized so that different reactions occur in different parts of the cell. This organization is most apparent in eukaryotes, where membrane-bounded structures are visible proof for the localization of different biochemical processes. For example, the synthesis of DNA and RNA takes place in the nucleus of a eukaryotic cell. The RNA is subsequently transported across the nuclear membrane to the cytoplasm, where it takes part in protein synthesis. Proteins made in the cytoplasm are used in all parts of the cell. A limited amount of protein synthesis also occurs in chloroplasts and mitochondria. Proteins made in these organelles are used exclusively in organelle-related functions. Most ATP synthesis occurs in chloroplasts and mitochondria. A host of reactions that transport nutrients and metabolites occur in the plasma membrane and the membranes of various organelles. The localization of functionally related reactions in different parts of the cell concentrates reactants and products at sites where they can be most efficiently utilized.

Biochemical Reactions Are Organized into Pathways

Most biochemical reactions are integrated into multistep pathways utilizing several enzymes. For example, the breakdown of glucose into CO_2 and H_2O controls a series of reactions that begins in the cytosol and continues to completion in the mitochondrion. A complex series of reactions like this is referred to as a biochemical pathway (fig. 1.19). Synthetic reactions, such as the biosynthesis of amino acids in the bacterium *Escherichia coli,* are similarly organized into pathways (fig. 1.20). Frequently pathways have branchpoints. For example, the synthesis of the amino acids threonine and lysine starts with

Figure 1.19

Summary diagram of the breakdown of glucose to carbon dioxide and water in a eukaryotic cell. As depicted here, the process starts with the absorption of glucose at the plasma membrane and its conversion into glucose-6-phosphate. In the cytosol, this six-carbon compound is then broken down by a sequence of enzyme-catalyzed reactions into two molecules of the three-carbon compound pyruvate. After absorption by the mitochondrion, pyruvate is broken down to carbon dioxide and water by a sequence of reactions that requires molecular oxygen.

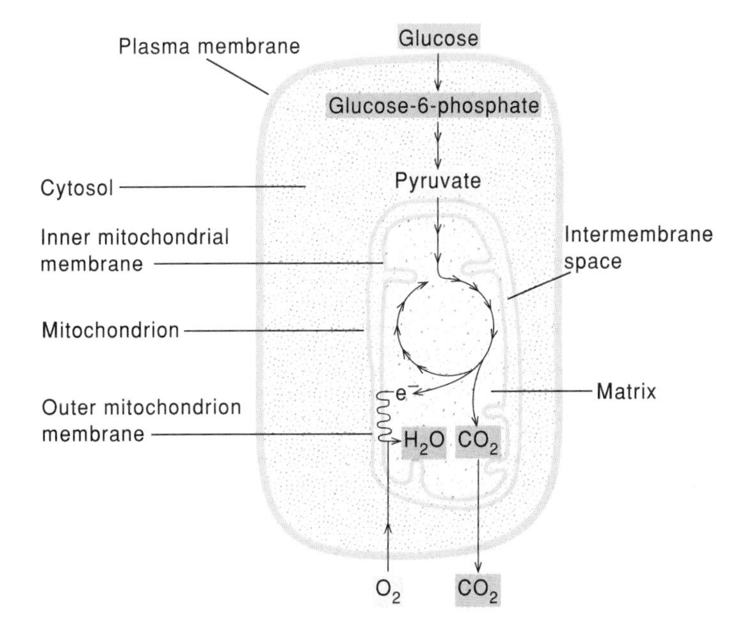

oxaloacetate. After three steps, a branchpoint is reached with the formation of the organic compound aspartic-β-semialdehyde. One branch of this pathway leads to the synthesis of the amino acid lysine, and another branch leads to the synthesis of the amino acids methionine, threonine, and isoleucine.

Figure 1.20

Synthesis of various amino acids from oxaloacetate. Each arrow represents a discrete biochemical step requiring a unique enzyme. Thus aspartic acid is produced in one step from oxaloacetate, whereas isoleucine is produced in five steps from threonine.

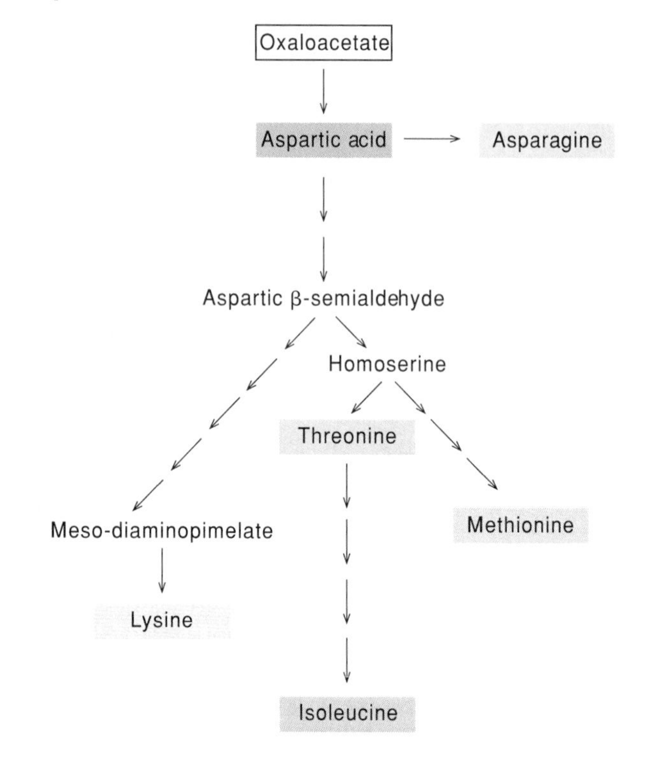

To understand the role of each biochemical reaction we must identify its position in a pathway and also consider how that pathway interacts with others.

Biochemical Reactions Are Regulated

Hundreds of biochemical reactions take place even in the cells of relatively simple microorganisms. Living systems have evolved a sophisticated hierarchy of controls that permits them to maintain a stable intracellular environment. These controls insure that substances required for maintenance and growth are produced in adequate amounts but without huge excesses. Biochemical controls have developed in such a way that the cell can make adjustments in response to a changing external environment. Adjustments are needed because the temperature, ionic strength, acid concentration, and concentration of nutrients present in the external environment vary over much wider limits than could be tolerated inside the cell.

The rate of intracellular reactions is a function of the availability of substrates and enzymes. Enzyme activity is controlled at two different levels. First and foremost, the rate of a catalyzed reaction is regulated by the amount of the catalyzing enzyme that is present in the cell. Control of enzyme amounts is usually accomplished by regulating the rate of enzyme synthesis; in some cases the rate of enzyme degradation is also regulated. We can think of controls that regulate the total amount of enzyme present as coarse controls. They define the limits of possible enzyme activity as being anywhere from 0 to 100% of the full activity of the enzyme. Fine controls that act directly on enzymes are also present. Only certain special enzymes, called regulatory enzymes, are susceptible to this second type of regulation. Regulatory enzymes usually occupy key points in biochemical pathways, and their state of activity frequently is decisive in determining the utilization of the pathway.

A simple example will serve to illustrate how these two types of controls work. *E. coli* can synthesize all of the amino acids required for protein synthesis. Histidine is one of these amino acids; its synthesis starts from the sugar phosphate compound phosphoribosylpyrophosphate (PRPP) and requires ten enzymes. Each enzyme catalyzes one reaction in a ten-step pathway. The synthesis of all ten enzymes is regulated by the end product of the pathway, histidine. If there is sufficient histidine for protein synthesis, then the cell ceases to make the enzymes for this pathway. The shutdown is triggered by controls that sense the level of histidine in the cell and as a result turn off synthesis of the messenger RNA required to make the enzymes. The histidine pathway also provides an example of a regulatory enzyme. The activity of the first enzyme in the pathway is directly inhibited by histidine. Thus, when there is sufficient histidine present in the cell, the activity of the first enzyme in the pathway is inhibited, and no more material or energy is funneled into the synthesis of unneeded histidine. Finally, if there is abundant histidine available from the external environment, the extracellular histidine is absorbed into the cell and the synthesis of the enzymes of the histidine pathway is brought to a halt. Bacterial cells grown in a histidine-rich growth medium for several generations contain only trace amounts of these enzymes.

The underlying principle in regulation is to maintain a favorable intracellular environment in the most economical manner. The cell makes products in the amounts that are needed. Each pathway is regulated in a somewhat different way, assuring that biochemical energy and substrates are efficiently utilized.

Organisms Are Biochemically Dependent on One Another

As we stated at the beginning of this chapter, 3–4 billion years ago the first self-replicating molecules appeared on earth. These entities had to have the capacity for extracting nutrients from the chemical compounds that existed in prebiotic times. We have some general notions about what types of substances were present at that time. One of the most important substances that was not present at that time in significant amounts was molecular oxygen, O_2. Currently this form of oxygen is required by all forms of life visible to the naked eye.

The O_2 that is used by most organisms is ultimately converted by them into CO_2. Oxygen is utilized at a rapid rate and it would soon disappear if it were not for special classes of photosynthetic organisms that are constantly producing more O_2 by the oxidation of water.

The oxygen story is an example of the dependence of one class of organisms on another for certain chemicals. A similar situation exists with the elements carbon and nitrogen, which must be converted from gaseous forms, CO_2 and N_2, to organic forms utilizable by most organisms. Reduced carbon compounds are constantly being lost by oxidation to gaseous CO_2. The supply of organic carbon compounds required by all forms of life is replenished by photosynthetic organisms; these include most plants and certain microorganisms. Similarly, nitrogen in organic molecules is constantly being lost to the atmosphere in the form of gaseous nitrogen. The reactions required for the conversion of nitrogen to a reduced form more usable to the majority of organisms occurs in only a limited number of microorganisms; yet without these nitrogen-fixing organisms life as we know it would soon vanish.

As we ascend the evolutionary tree, we find increasingly complex multicellular forms. Such organisms generally require more complex nutrients, which must ultimately be supplied to them by simpler living forms. Bacteria like *E. coli* can make all of their own amino acids from a reduced form of nitrogen, such as NH_3, and a reduced form of carbon, such as glucose. Humans, on the other hand, must receive most of their amino acids as nutrients. Humans and other complex organisms have gained new biochemical capacities, which permit them to synthesize the components associated with highly specialized differentiated tissues. At the same time, they have lost many of the biochemical systems required to survive on simpler nutrients.

Many biochemical reactions of great importance take place in only a limited number of organisms. This fact increases the complexity of the study of biochemistry. We must learn many reactions; we must also be aware of the biochemical potentials of different organisms. This is the only way we can understand the biochemical interdependency of organisms.

Information for the Synthesis of Proteins Is Carried by the DNA

DNA contains the genetic information transmitted to each daughter cell when cells divide. The DNA usually exists in the form of nucleoprotein (DNA-protein) complexes called chromosomes. A prokaryotic cell contains a single chromosome. Prior to cell division this chromosome duplicates and segregates so that an identical complement of DNA goes to each of two newly formed daughter cells.

Eukaryotic cells are more complex than prokaryotic cells and usually contain more DNA, which is partitioned between several chromosomes. In both prokaryotes and eukaryotes, almost all cells of the same organism contain the same number of chromosomes. In eukaryotes most of the chromosomes are localized in the nucleus. Thus the DNA is isolated from the main body of the cytoplasm—a unique feature of eukaryotes and the primary distinction between prokaryotes and eukaryotes. Some organelles, notably the mitochondria and the chloroplasts, also contain a single circular chromosome.

Eukaryotic chromosomes are detectable by light microscopy at the stage just prior to cell duplication. At this stage, called mitosis, chromosomes appear as elongated refractile structures that can be seen to segregate in equal numbers and types to each of the daughter cells before cell division (fig. 1.21). Each chromosome carries specific hereditary (genetic) information necessary for the synthesis of specific compounds essential for cell maintenance, growth, and replication. Each chromosome contains a single very long DNA molecule composed of 10^6 or more nucleotides in a specific sequence. The sequence of nucleotides in the chromosomal DNA determines the sequence of amino acids in the protein polypeptide chains of the organism. The relationship between base sequences and resultant amino acids is known as the genetic code. Each grouping of three bases, called a triplet, represents a specific amino acid and is called a codon. The genetic code ensures that the organism's characteristics will be reflected by the sequence of nucleotides in its DNA. When chromosomes replicate, the DNA replicates precisely, so that the same nucleotide sequence is passed along to each of the daughter cells resulting from mitosis and cell division.

The DNA does not transfer its genetic information directly to protein. Rather, this information passes through an intermediary, the messenger RNA (mRNA). The mRNA is made on a DNA template in the nucleus of a eukaryotic cell and then passes into the cytoplasm, where it serves in turn as a template for the synthesis of the polypeptide chain. The overall process of information transfer from DNA to mRNA (transcription) and from mRNA to protein (translation) is depicted in figure 1.22.

Biochemical Systems Have Been Evolving for Almost Four Billion Years

Biochemical systems are conservative and opportunistic. That is because they tend to evolve one step at a time, using a readily accessible route that leads to an advantage. To appreciate biochemical systems today it is useful to have some understanding of how they came to be.

The earth was formed by a process of accretion about 4.6 billion years ago. Initially it was a molten mass lacking the gravitational pull to retain its gases at the prevalent elevated temperatures. And yet, within a mere 700 million years of the planet's birth, as calculated from the isotopic record of sediments, cellular life almost certainly existed. What raw materials were available to bring about this miraculous turn of events?

Figure 1.21

Mitosis and cell division in eukaryotes. After DNA duplication has occurred, mitosis is the process by which quantitatively and qualitatively identical DNA is delivered to daughter cells formed by cell division. Mitosis is traditionally divided into a series of stages characterized by the appearance and movement of the DNA-bearing structures, the chromosomes.

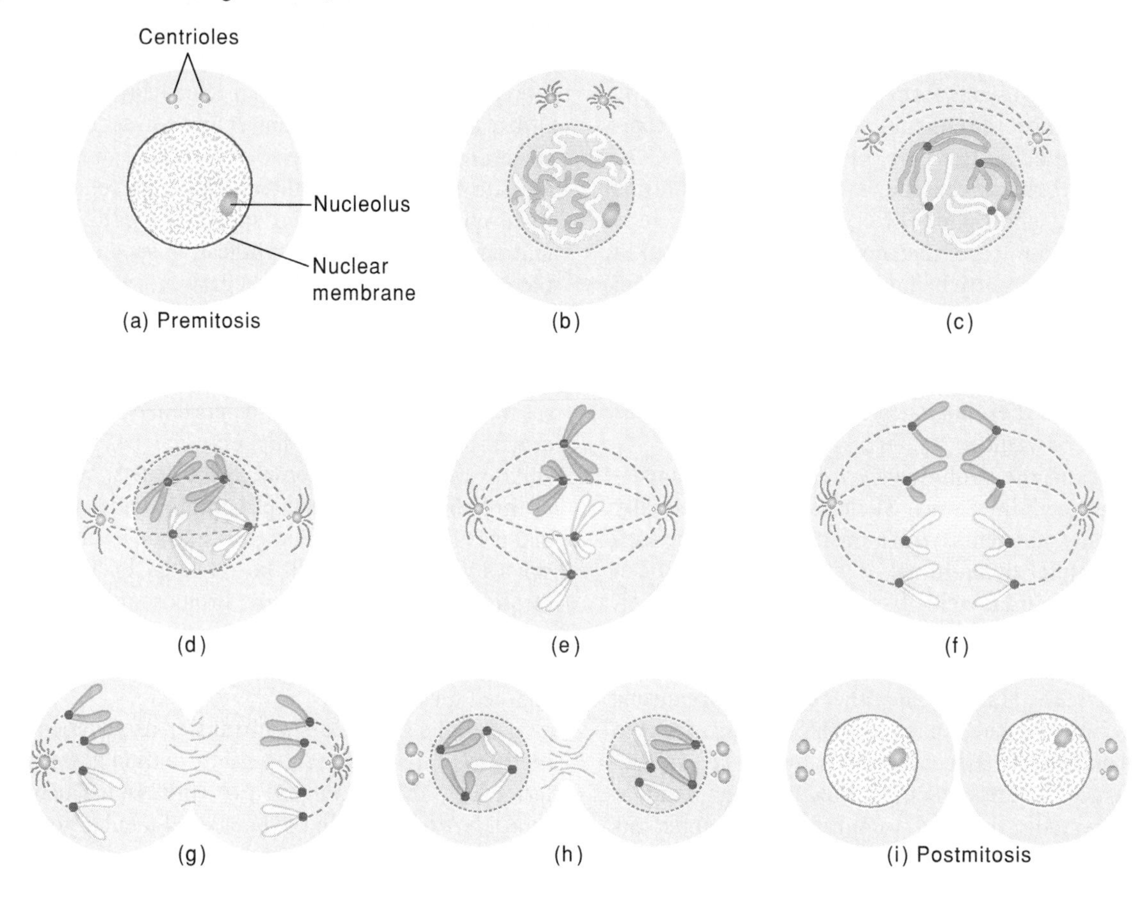

What were the sources of energy used to drive the necessary reactions? Where did the important reactions take place? Was it in the atmosphere, in the oceans, on dry land, or all three?

Table 1.4 shows a distribution of the major elements found in the earth's crust, the ocean water, and the human body. The composition of the human body, which is reasonably representative of living organisms, differs appreciably from that of the earth's crust. The four most abundant elements in the human body are hydrogen, carbon, nitrogen, and oxygen. Of these, only oxygen belongs to the highest-abundance class of elements making up the earth's crust. Nevertheless, the elements needed to make living things were present in sufficient quantities in the earth's crust and its primitive oceans and the atmosphere.

The original water and air associated with the newly formed planet were lost because of the high temperatures. As the earth cooled, water and various gases on the surface and in the atmosphere were produced by an outgassing process. Today the earth is only about 0.5% by weight water, but because of water's low density, most of it is present on the earth's surface, where it has a major impact on the environment. Water cycles through the gaseous form in the atmosphere and the liquid form in the oceans and bodies of fresh water.

Geologic evidence indicates that appreciable amounts of the total water mass have always been present as liquid water. Thus the temperature of the planet has for the most part been between 0 and 100° C, a range that is conducive to the formation of biomolecules and the origin and propagation of life. The earth has also been kind to living things in other ways. The buffering action of various clays and minerals is believed to have maintained the pH level of the oceans between 8.0 and 8.5, which is close to the pH inside living cells.

Unlike the temperature and pH of the surface water, the composition of the atmosphere has changed drastically since the origin of life. In fact, the processes taking place in living things are primarily responsible for these changes. Today's atmosphere, which is about one millionth of the mass of the earth

Figure 1.22

Transfer of information from DNA to protein. The nucleotide sequence in DNA specifies the sequence of amino acids in a polypeptide. DNA usually exists as a two-chain structure. The information contained in the nucleotide sequence of only one of the DNA chains is used to specify the nucleotide sequence of the messenger RNA molecule (mRNA). This sequence information is used in polypeptide synthesis. A three-nucleotide sequence in the messenger RNA molecule codes for a specific amino acid in the polypeptide chain.

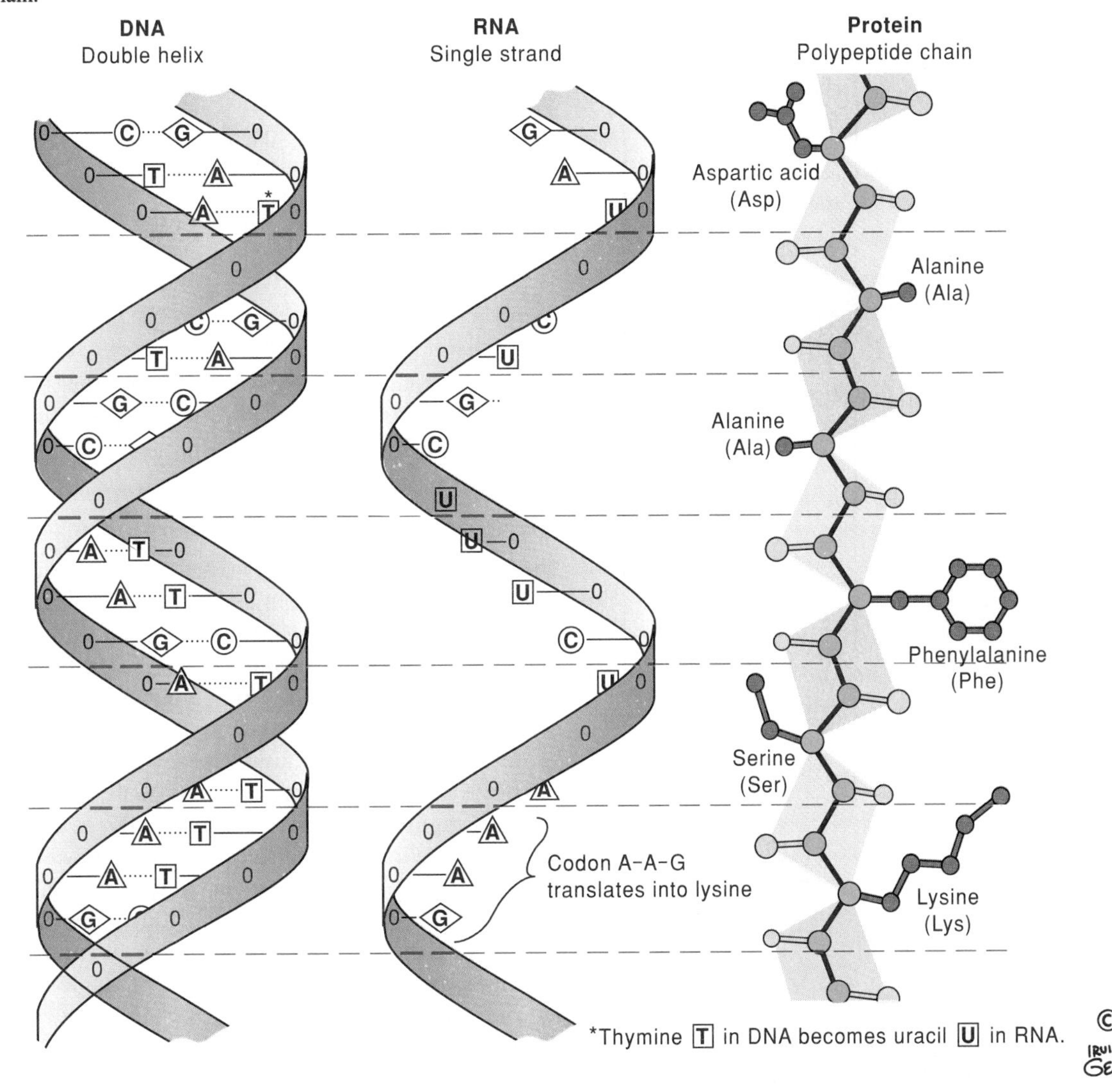

itself, is mainly composed of nitrogen (78%) and oxygen (21%). Most of the remaining atmosphere is argon (0.9%), water (variable up to 4%), and carbon dioxide (0.034%). The gases that made up the primitive atmosphere were quite different; especially conspicuous was the absence of gaseous oxygen. Most of the oxygen in the present atmosphere is due to the oxidation of water by photosynthetic organisms. The main forms of carbon and nitrogen in the primitive atmosphere were probably CO_2 and N_2 as they are today. In addition, and probably of great significance to the origin of life, small amounts of the more reduced forms of carbon and H_2 gas were present. Thus the primitive earth probably had a weakly reducing atmosphere as contrasted with today's highly oxidizing atmosphere. This situation was most favorable to the origin of life, as organic compounds that enter the biomass tend to be in a reduced state and they are readily oxidized in the presence of gaseous oxygen. It is generally believed that the first organics formed in this primitive atmosphere and then rained down to a more concentrated form to make larger bioorganic molecules.

The origin of life probably occurred in three phases (fig. 1.23): (1) The earliest phase was a period of chemical evolution during which the compounds needed for the nucleation of life must have been formed. These compounds include the most important class of biological macromolecules, the nucleic acids. In

Table 1.4
Distribution of the 24 Elements Used in Biological Systems[a]

Element	Atomic No.	Earth's Crust	Ocean	Human Body
Hydrogen (H)	1	2,882	66,200	60,562
Carbon (C)	6	56	1.4	10,680
Nitrogen (N)	7	7	<1	2,440
Oxygen (O)	8	60,425	33,100	25,670
Fluorine (F)	9	77	<1	<1
Sodium (Na)	11	2,554	290	75
Magnesium (Mg)	12	1,784	34	11
Silicon (Si)	14	20,475	<1	<1
Phosphorus (P)	15	79	<1	130
Sulfur (S)	16	33	17	130
Chlorine (Cl)	17	11	340	33
Potassium (K)	19	1,374	6	37
Calcium (Ca)	20	1,878	6	230
Vanadium (V)	23	4	<1	<1
Chromium (Cr)	24	8	<1	<1
Manganese (Mn)	25	37	<1	<1
Iron (Fe)	26	1,858	<1	<1
Cobalt (Co)	27	1	<1	<1
Nickel (Ni)	28	3	<1	<1
Copper (Cu)	29	1	<1	<1
Zinc (Zu)	30	2	<1	<1
Selenium (Se)	34	<1	<1	<1
Molybdenum (Mo)	42	<1	<1	<1
Iodine (I)	53	<1	<1	<1

[a]Amounts are given in atoms per 100,000.

this phase of evolution, the synthesis of nucleic acids was "noninstructed." (2) As soon as some nucleic acids were present, physical forces between them must have led to an "instructed" synthesis, in which the already formed molecules served as templates for the synthesis of new polymers. It seems likely that "feedback loops" selected out certain nucleic acids for preferential synthesis. At some point during this period of instructed synthesis more nucleic acids and possibly protein macromolecules were formed. The products of this phase of molecular self-organization must sooner or later have begun to resemble the complex organized units that we observe in the self-reproducing biosynthetic cycles of living cells. (3) In the final phase of the origin of life we find the beginnings of the divergent process of biological evolution, the development of the simplest single-celled organisms, and their differentiation into complex multicellular beings.

Figure 1.23

The origin of life probably occurred in three overlapping phases: In Phase I, chemical evolution, involved the noninstructed synthesis of biological macromolecules. In Phase II, biological macromolecules self-organized into systems that could reproduce. In Phase III, organisms evolved from simple genetic systems to complex multicellular organisms. The arrow pointing from left to right emphasizes the unidirectional nature of the overall process.

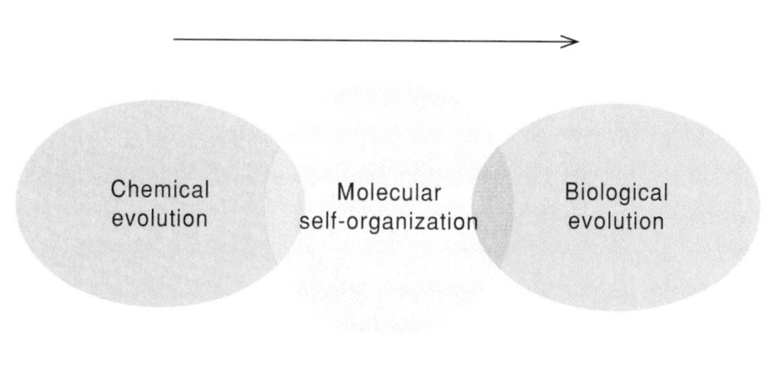

All Living Systems Are Related through a Common Evolution

The classical view of evolution, based on morphological differences among organisms, can be diagrammed as a branching tree in which all existing organisms are shown at the tips of the branches (fig. 1.24). An evolutionary tree starts from the simple ancestral prokaryotic cell, which branches off in three main directions into the archaebacteria, the eubacteria, and the eukaryotes. Each of these kingdoms has continued to branch in elaborate ways; we have shown only some of the main branch points in figure 1.24. Prokaryotes for the most part have remained as relatively undifferentiated single-celled organisms containing a single chromosome. By contrast, eukaryotes have changed dramatically. Although the organisms on many branches of the eukaryotic part of the evolutionary tree have remained as relatively undifferentiated single-cell forms, significant numbers of eukaryotic organisms have evolved into multicellular forms in which the individual cells of the total organism have differentiated to serve different functions. This process has given rise to plants, animals, and fungi.

A somewhat different view of biological evolution has arisen from a comparison of nucleic acid sequences in different organisms. The best-known sequences for such studies have come from the 16S ribosomal RNAs in different organisms. This comparative study has provided us with the evolutionary pattern shown in figure 1.25, where distances are proportional to sequence differences. A most striking characteristic of this unrooted tree is the distinctness of the three primary kingdoms as evidenced by the large sequence distances that separate each of

Figure 1.24

Classical evolutionary tree. All living forms have a common origin, believed to be the ancestral prokaryote. Through a process of evolution some of these prokaryotes changed into other organisms with different characteristics. The evolutionary tree indicates the main pathways of evolution.

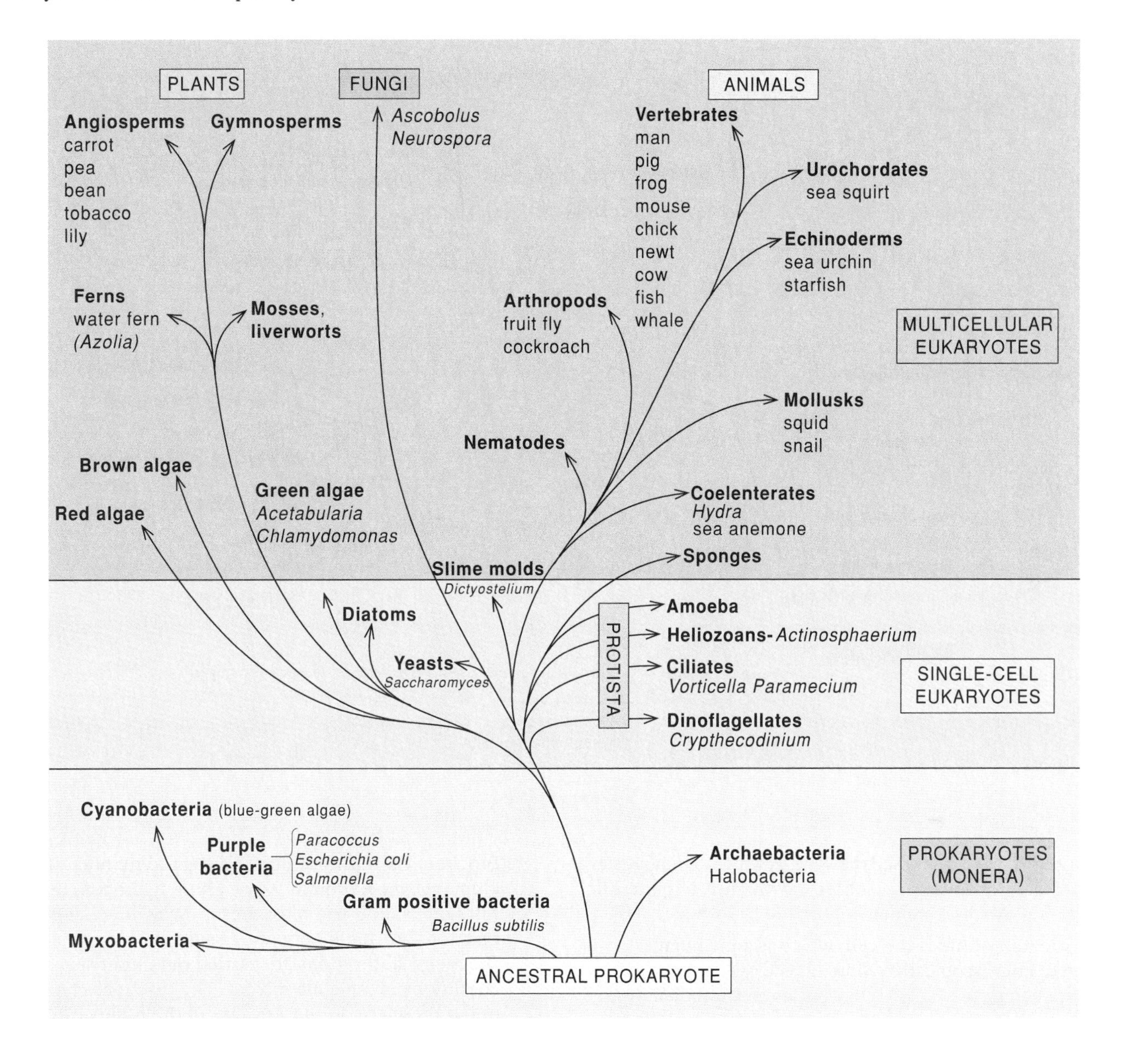

the kingdoms from the others. Although we do not know the position of the root of the tree, it seems likely that the archaebacteria are closer to the common ancestor of all three kingdoms than are the eubacteria or the prokaryotes.

The organisms most familiar to us, the multicellular plants and animals, occupy a shallow domain within the eukaryotic line of descent. True, the developmental programs of the multicellular forms have generated an incredible diversity in form and function. Nevertheless, both bacteria and unicellular eukaryotes span far greater evolutionary histories. This fact is reflected in the greater biochemical diversity that we find in the unicellular microorganisms.

Figure 1.25

An evolutionary tree can be constructed by comparing the complete sequences of 21 different 16S and 16S-like ribosomal RNAs (rRNAs). The scale bar represents the number of accumulated nucleotide differences (mutations) per sequence position in the rRNAs of the various organisms. (Source: N. R. Pace, G. J. Olsen, and C. R. Woese, in *Cell* 45:325, 1986. Copyright © 1986 Cell Press, Cambridge, Mass.)

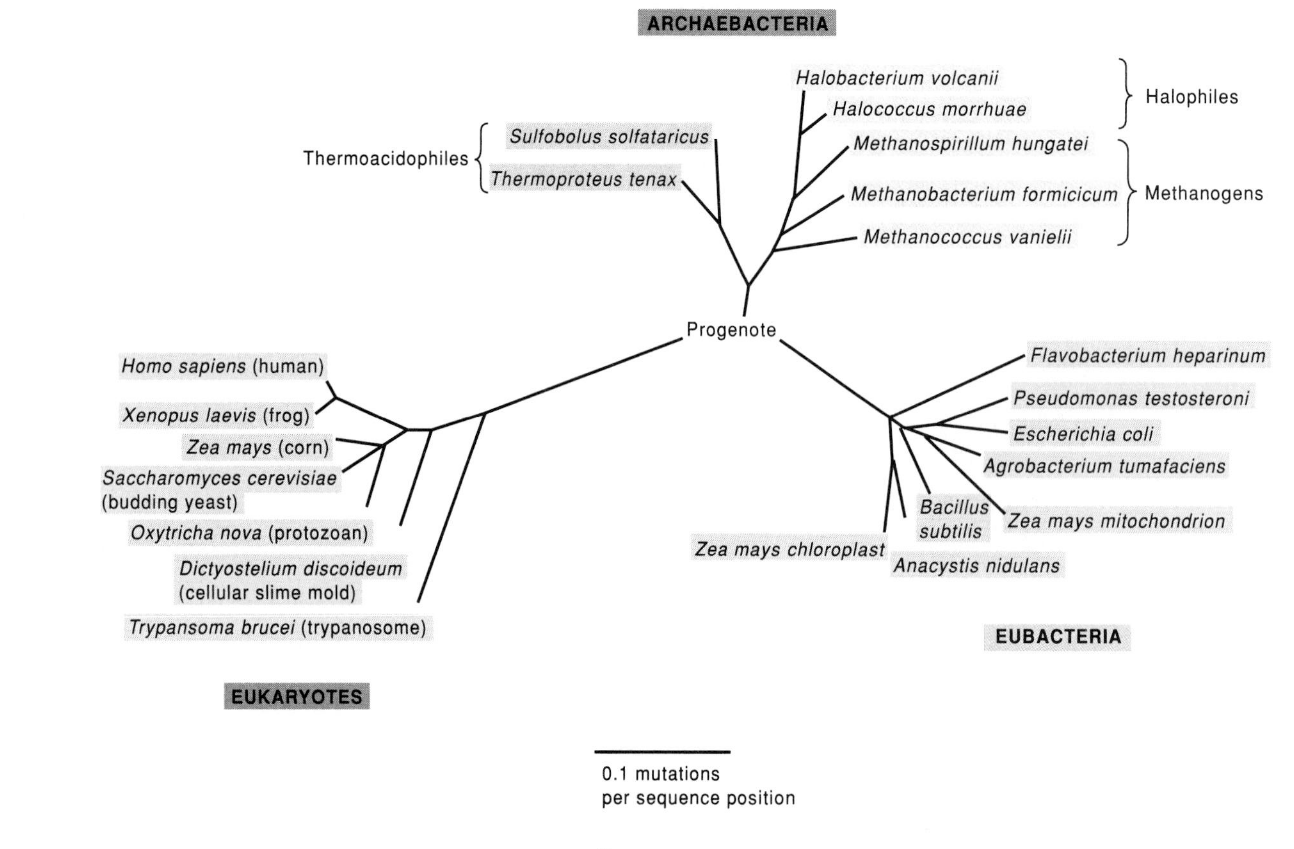

Summary

In this chapter we have discussed the ways in which biochemistry parallels ordinary chemistry and those in which it is quite different. The chief points to remember are the following.

1. The basic unit of life is the cell, which is a membrane-enclosed, microscopically visible object.
2. Cells are composed of small molecules, macromolecules, and organelles. The most prominent small molecule is water, which constitutes 70% by weight of the cell. Other small molecules are present only in quite small amounts; they are precursors or breakdown products of macromolecules or coenzymes. There are four types of macromolecules: lipids, carbohydrates, proteins, and nucleic acids.
3. Noncovalent intermolecular forces largely determine the folded structure adopted by a macromolecule, particularly the relative affinity of different groupings on the macromolecule for water. In general, hydrophobic groupings are buried within the folded macromolecular structure, while hydrophilic groupings are located on the surface, where they can interact with water.
4. Biochemical reactions utilize a limited number of elements, most prominently carbon, hydrogen, oxygen, nitrogen, sulfur, and phosphorus. Many biochemical reactions are simple organic reactions.
5. Biochemical reactions are carried out under very mild conditions in aqueous solvent. The reactions can proceed under these conditions because of the highly efficient nature of protein enzyme catalysts.
6. Biochemical reactions frequently require energy. The most common source of chemical energy used is adenosine triphosphate (ATP). The splitting of a phosphate from the ATP molecule can provide the energy needed to make an otherwise unfavorable reaction go in the desired direction.
7. Biochemical reactions of different types are localized to different parts in the cell.
8. Biochemical reactions are frequently organized into multistep pathways.
9. Biochemical reactions are regulated according to need by controlling the amount and activity of enzymes in the system.

10. Most organisms depend on other organisms for their survival. Frequently this is because a given organism cannot make all of the compounds needed for its growth and survival.
11. The specific properties of any protein are due to the specific sequence of amino acids in its polypeptide chains. This sequence is determined by the genetic information carried by the sequence of DNA nucleotides. DNA transfers the information to messenger RNA, which serves as the template for protein synthesis.
12. Shortly after the earth was formed and had cooled to a reasonable temperature, chemical processes produced compounds that would be used in the development of living cells. Nucleic acids are the most important compounds for living cells, and it is believed that they played the central role in the origin of life.
13. Evolutionary trees based on morphology or biochemical differences indicate that all living systems are related through a common evolution.

Selected Readings

Becker, W. M., *The World of the Cell.* Menlo Park, Calif.: Benjamin/Cummings, 1986. A very readable cell biology book that could be referred to while taking biochemistry.

de Duve, C., *Blueprint For A Cell.* Burlington, N.C.: California Biological Supply Co., 1991. A short book on the origin of life that contains an excellent reference list.

Dickerson, R. E., Chemical evolution and the origin of life. *Sci. Am* 239(3):70–86, 1978.

Doolittle, R. F., The genealogy of some recently evolved vertebrate proteins. *Trends Biochem. Sci.* 10:233–237, 1985.

Kimura, M., The neutral theory of molecular evolution. *Sci. Am.* 241(5):98–126, 1979.

Schopf, J. W., The evolution of the earliest cells. *Sci. Am.* 229(3):10–138, 1978. An authoritative account from a foremost geologist.

Stillinger, F. H., Water revisited. *Science* 209:451–457, 1980. A reminder of the central importance of water to the origin of life.

Wilson, A. C., The molecular basis of evolution. *Sci. Am.* 253(4):164–173, 1985.

Thermodynamics in Biochemistry

The primary usefulness of thermodynamics to biochemists lies in predicting whether particular chemical reactions could occur spontaneously. A simple illustration is to predict what compounds could possibly serve as energy sources for an organism. You are aware from everyday experience that oxidation of organic molecules by molecular oxygen releases energy. For example, wood or coal burns with a large output of heat. Similarly, organisms can obtain energy by oxidizing carbohydrates, fats, or proteins. Some organisms oxidize hydrocarbons, some oxidize reduced forms of sulfur, and others oxidize iron. But no organisms live by oxidizing molecular nitrogen, and the explanation lies in thermodynamics. The reaction cannot occur spontaneously. This example illustrates the importance of thermodynamics in controlling all life. Because organisms live by extracting chemical energy from their surroundings, thermodynamics is not an esoteric subject. It is a matter of life or death.

We say that thermodynamics determines whether a process "could" occur, because thermodynamics tells us only whether the process is possible, not whether it actually will occur in a finite period of time. The rate at which a thermodynamically possible reaction occurs depends on the detailed mechanism of the process. For a biochemical process to occur rapidly, appropriate enzymes must be available. The distinction between spontaneity and speed is more critical for biochemists than it is for chemists, because if a chemical reaction does not proceed rapidly a chemist can change the pressure or temperature or increase the concentration of the reactants. A living organism is under more rigid constraints: It must function at a fixed temperature and pressure and within a limited range of concentrations of reactants.

In this chapter we will elaborate on the thermodynamic quantities that we introduced in chapter 1. We will then expand the discussion to show how the concept of free energy is used in predicting biochemical pathways, and we will explore the central role of ATP in providing energy for biochemical reactions.

Thermodynamic Quantities

The properties of a substance can be classified as either intensive or extensive. Intensive properties, which include density, pressure, temperature, and concentration, do not depend on the amount of the material. Extensive properties, such as volume and weight, do depend on the amount. Most of the thermodynamic properties we will be discussing are extensive properties. These include energy (E), enthalpy (H), entropy (S), and free energy (G).

Energy, enthalpy, entropy, and free energy are all properties of the state of a substance. This means that they do not depend on how the substance was made or how it reached a particular state. In a chemical reaction, it is the difference between the initial and final states that is important; the pathway that is taken to get from the initial state to the final state has no bearing on whether the overall reaction releases or consumes energy (fig. 2.1).

Figure 2.1

The change in energy of a system depends only on the initial and final states, not on the path by which the system gets from one state to the other. This diagram illustrates the conversion of a phosphate ester of glucose (glucose-6-phosphate) to free glucose and inorganic phosphate ion (P_i) by two different pathways. Although the two routes proceed through intermediate compounds that differ in energy (A, B, C, and D), the overall energy change (ΔE) is the same. However, the work done and the amount of heat absorbed or released generally will not be the same for the two paths.

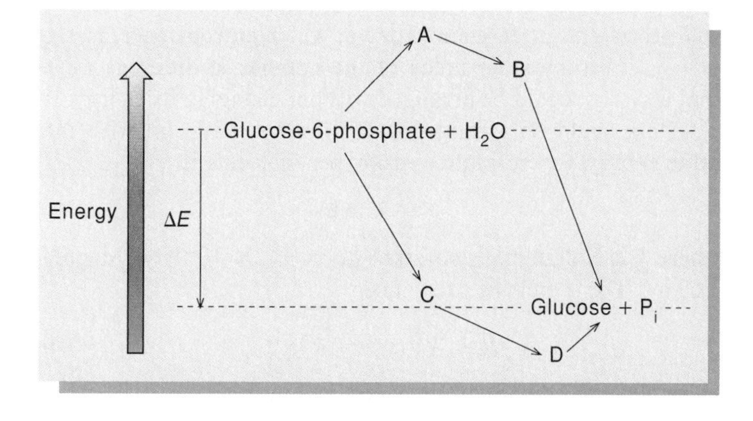

Figure 2.2

A system and its surroundings. Heat flow into the system is designated as a positive quantity (q), and work that the system does on the surroundings is designated as a positive quantity (w). The first law of thermodynamics relates q and w to changes in the energy of the system. Any change in the energy of the system (ΔE_{sys}) is balanced by an opposite change in the energy of the surroundings (ΔE_{sur}), so that the overall energy change (ΔE_{tot}) is zero.

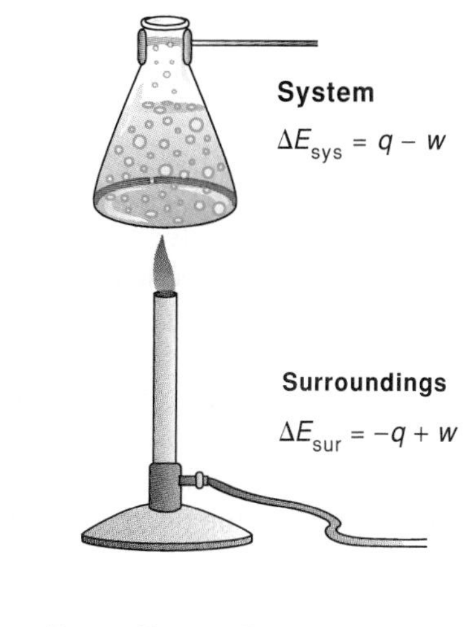

$$\Delta E_{tot} = \Delta E_{sys} + \Delta E_{sur} = 0$$

The First Law of Thermodynamics: Any Change in the Energy of a System Requires an Equal and Opposite Change in the Surroundings

Of the thermodynamic properties we have just listed, energy is probably the most familiar. Energy is the capacity to do work. The energy of a molecule includes the internal nuclear energies and the molecular electronic, translational, rotational, and vibrational energies. Electronic energies, which reflect the interactions among the electrons and nuclei, usually are much larger than the translational, rotational, and vibrational energies.

In biochemistry, we are concerned not so much with the absolute energies of molecules as with changes in energy that occur in the course of reactions. It is easier to evaluate a change in energy (ΔE) than to calculate the absolute energies of the reactants or products, because many of the terms that contribute to the total energy do not change much during a chemical reaction. Electronic energies usually dominate ΔE, as they do the total energies. Good estimates of the energy change resulting from a chemical reaction can usually be obtained by calculating the difference between the bond energies of the reactants and products (see table 4.2).

The first law of thermodynamics says that the total amount of energy in the universe is constant. Energy can undergo transformations from one form to another; for example, the chemical energy of a molecule can be transformed into thermal, electrical, or mechanical energy. But any change in the total energy of one part of the universe is matched by an equal and opposite change in another part, so that the overall energy of the universe remains constant.

To apply the first law to a chemical reaction, we must take into account all the energy changes that occur. This means including any changes in the surroundings, as well as in the system of interest (fig. 2.2). The system might be a reaction occurring in a test tube or a living cell; the surroundings are all the rest of the universe. In practical terms, however, we usually need to consider only the immediate surroundings, because only these are likely to be influenced by what happens in the system. According to the first law, the overall energy remains constant even though energy in some form may flow from the system to its surroundings or from the surroundings to the system.

If the amount of matter in a system is constant, there are only two means by which the system can gain or lose energy: transfer of heat or performance of work. The energy of a system increases if the system absorbs heat from its surroundings or if the surroundings do work on the system; it decreases if the system gives off heat to the surroundings or does work. A common way of stating the first law of thermodynamics is to equate the change in energy of the system (ΔE) to the difference between the heat absorbed by the system (q) and the work done by the system (w):

$$\Delta E = q - w \tag{1}$$

Both q and w depend on the path of the reaction, but their difference, ΔE, is independent of the path and therefore defines a state function.

The relative energy of a compound can be measured in a bomb calorimeter, a device in which the compound is thoroughly combusted and the resulting heat is measured. Energies (and enthalpies, discussed next) are usually given in units of kilocalories per mole or kilojoules per mole. One kilojoule (kJ) is the amount of energy needed to apply 1 newton of force over 1 km; 1 kilocalorie (kcal) is the heat needed to raise the temperature of 1 kg of water from 14.5 to 15.5° C. One kcal is equivalent to 4.184 kJ. Physicists generally prefer to use kilojoules; most chemists, including biochemists, prefer to use kilocalories.

It is customary to record energies of molecules with reference to the energies of their constituent elements. The standard energy of formation reported for a molecule, ΔE°_f, is equal to the energy change associated with the formation of the molecule from the elements in their standard states. The energy change in any other reaction can be obtained by subtracting the energies of formation of the reactants from the energies of formation of the products. The complex organic molecules found in cells have relatively weak chemical bonds, and thus higher energies, than H_2O, CO_2, and the other small molecules from which they are formed.

Enthalpy is a function of state that is closely related to energy but is usually more pertinent for describing the thermodynamics of chemical or biochemical reactions. The change in enthalpy (ΔH) is related to the change in energy (ΔE) by the expression

$$\Delta H = \Delta E + \Delta(PV) \tag{2}$$

where $\Delta(PV)$ is the change in the product of the pressure (P) and volume (V) of the system. ΔH is the amount of heat that is absorbed from the surroundings if a reaction occurs at constant pressure and no work is done other than the work of expansion or contraction of the system. (The work done when a system expands by ΔV against a constant pressure P is $P\,\Delta V$. This type of work is generally not very useful in biochemical systems.) In most biochemical reactions, there is little change in either pressure or volume, so the difference between ΔH and ΔE is relatively small.

In most chemical reactions that occur spontaneously, the enthalpy of the system decreases. If no work is done, the system gives off heat to the surroundings. But changes in enthalpy or energy do not provide a reliable way of determining whether a reaction can proceed spontaneously. For example, although LiCl and $(NH_4)_2SO_4$ both dissolve readily in water, the former process releases heat, whereas the latter absorbs heat. A mixture of solid $(NH_4)_2SO_4$ and water proceeds spontaneously to a state of higher enthalpy. This reaction is driven by an increase in the entropy of the system.

The Second Law of Thermodynamics: In Any Spontaneous Process the Total Entropy of the System and the Surroundings Increases

The second law of thermodynamics is that the universe inevitably proceeds from states that are more ordered to states that are more disordered. This phenomenon is measured by a thermodynamic function called entropy, which is denoted by the symbol S. A reaction in which entropy increases (ΔS is positive) will proceed in preference to one in which entropy decreases.

Entropy is an index of the number of different ways that a system could be arranged without changing its energy. If a system could be arranged in Ω different ways, all with the same energy, the absolute entropy per molecule is

$$S = k \ln \Omega \tag{3}$$

where k is Boltzmann's constant ($k = 3.4 \times 10^{-24}$ cal/degree Kelvin). For a mole of substance,

$$S = Nk \ln \Omega = R \ln \Omega \tag{4}$$

Here N is the number of molecules in a mole (6×10^{23}) and R is the gas constant ($R \approx 2$ cal/(degree K · mole). Quantitative values for entropies are usually given in entropy units (1 eu = 1 cal/degree K).

The underlying idea here is that the more ways a particular state could be obtained, the greater is the probability of finding a system in that state. A system that is highly disordered could be obtained in many different arrangements that are all energetically equivalent. Thus a state in which molecules are free to move about and rotate into many different orientations or conformations will be favored over a state in which motion is more restricted. The second law makes the remarkably general assertion that the total entropy change in any reaction that occurs spontaneously must be greater than zero. But note that this statement specifies the total entropy, which means that we must consider the entropy change in the surroundings as well as that in the system. The entropy of the system can decrease if the entropy of the surroundings increases by a greater amount.

The absolute entropies of small molecules can be calculated by statistical mechanical methods. Table 2.1 shows the results of such calculations for liquid propane. The largest contributions to the entropy come from the translational and rotational freedom of the molecule, and much smaller contributions from vibrations; electronic terms are insignificant. Although exact calculations of this type become intractable for large biological molecules, the relative sizes of the contributions from different types of motions are similar to those in small molecules. Thus entropy is associated primarily with translation and

Table 2.1
Contributions to the Entropy of Liquid Propane at 231 K

	kcal/(degree K · mole)
Translational entropy	36.04
Rotational entropy	23.38
Vibrational entropy	1.05
Electronic entropy	0.00
Total	60.47

rotation. This is very different from enthalpy, in which electronic terms are dominant and translational and rotational energies are comparatively small.

Statistical mechanical calculations show that the translational entropy of a molecule depends on $\frac{3}{2}R \ln M_r$ (plus some smaller terms), where R is the gas constant and M_r is the molecular weight. Suppose that a molecule, Y, undergoes a dimerization reaction so that its molecular weight doubles:

$$2\,\mathrm{Y} \rightarrow \mathrm{Y}_2$$

Intuitively, we expect dimerization to decrease the entropy because the two monomeric units can no longer move independently. We can calculate the effect quantitatively as a function of the molecular weight as follows. If M_r is the molecular weight of the monomer, then the change in translational entropy in going from the monomeric state to the dimer is approximately

$$\begin{aligned}\Delta S &\approx \frac{3}{2}R \ln 2M_r - 2\left(\frac{3}{2}R \ln M_r\right) \\ &= \frac{3}{2}R \ln 2 - \frac{3}{2}R \ln M_r \\ &= -\frac{3}{2}R \ln (M_r/2) \end{aligned} \tag{5}$$

The decrease of the translational entropy resulting from dimerization is a logarithmic function of the molecular weight.

Structural features that make molecules more rigid reduce rotational and vibrational contributions to entropy. Thus the formation of a double bond or ring decreases the entropy even when the molecular weight is unchanged. The formation of comparatively rigid macromolecular structures from flexible polypeptide or polynucleotide chains also requires an entropy decrease, although this can be offset by increases in the entropy of the surrounding water molecules (see chapter 4).

The entropy of a compound depends strongly on the physical state of the material. A gas has more translational and rotational freedom than a liquid, and a liquid has more freedom than a solid. As a result, entropy increases when a solid melts or a liquid vaporizes.

It can be shown that the increase in the entropy of a system that undergoes an isothermal, reversible process is

$$\Delta S = \frac{\Delta H}{T} \tag{6}$$

where T is the absolute temperature in degrees kelvin. An isothermal process is one that occurs at constant temperature. A reversible process is one that proceeds infinitely slowly through a series of intermediate states in which the system is always at equilibrium. For any real process occurring at a finite rate, the system is not strictly at equilibrium, and ΔS is larger than the value given by equation (6).

From equation (6), the entropy increase on vaporization or melting can be determined simply from the heat of vaporization divided by the boiling point, or the heat of fusion divided by the melting point. The entropy increase on vaporization of water is 26 eu/mole and that on melting of ice is 5.3 eu/mole. These values are consistent with our intuition that the increase in translational and rotational freedom is much greater in going from a liquid to a gas than in going from a solid to a liquid.

The entropy of a solution is increased by mixing of solvents, and it is decreased by interactions among the solvent molecules or interactions of solutes with the solvent. The mixing of two miscible liquids is a thermodynamically favorable process because it increases the number of positions that are available to the molecules. The entropy change on going from the unmixed liquids to the mixed state can be calculated from the expression

$$\begin{aligned}\Delta S &= n_a R \ln \frac{1}{X_a} + n_b R \ln \frac{1}{X_b} \\ &= -n_a R \ln X_a - n_b R \ln X_b \end{aligned} \tag{7}$$

where n_a and n_b are the number of moles of A and B that are mixed, and X_a and X_b are the corresponding mole fractions in the final solution. Because X_a and X_b are always less than 1, ln X_a and ln X_b will be negative. This means that the dilution of each component resulting from the mixing makes a positive contribution to the entropy. Equation (7) applies to the mixing of ideal solutions, in which there are no interactions among the molecules. Any intermolecular interactions will decrease the entropy by restricting the system's translational and rotational freedom.

Solvation, the interaction of a solute with the solvent, makes an important negative contribution to the entropy of a solution. Solvation can take the form of hydrogen bonding to donor or acceptor groups on the solute, or of a looser clustering of solvent molecules oriented around the solute (fig. 2.3). In general, the entropy of solvation by water becomes more negative with an increase in the charge or polarity of the solute. Small ions are solvated more strongly than large ions with the same charge, and anions are solvated more strongly than cations.

Figure 2.3

The entropy decrease resulting from solvation. When a salt is dissolved in water, the entropy of the dissociated cations and anions increases because of the increased possibilities for translation and rotation. But at the same time the movement of water molecules becomes restricted in the vicinity of the ions. The net effect is frequently a decrease in the entropy of the solution. Such a decrease in entropy can occur if the solution releases heat to the surroundings, because this will increase the entropy of the surroundings.

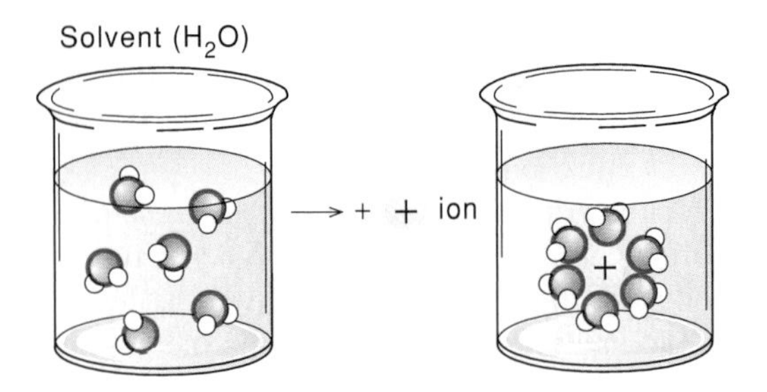

It is noteworthy that although enthalpy depends mainly on electronic interactions, whereas entropy depends mainly on translation and rotation, solvation affects both enthalpy and entropy. Enthalpies and entropies of solvation usually tend to oppose each other. For charged species, the more negative (favorable) the enthalpy of solvation, the more negative (unfavorable) the entropy of solvation.

From our earlier discussion, you might expect that the dissociation of a proton from a carboxylic acid, which increases the number of independent particles, would lead to an increase in entropy. However, this effect is more than counterbalanced by solvation effects. The charged anion and proton both "freeze out" many of the surrounding molecules of water (fig. 2.4). Thus the ionization of a weak acid decreases the number of mobile molecules, and so leads to a decrease in entropy. The entropy of ionization of a typical carboxylic acid in water is about −22 eu/mole. The entropy of dissociation of a proton from a quaternary ammonium group ($R{-}NH_3^+ \rightarrow R{-}NH_2 + H^+$) is usually smaller, because in this case the dissociation does not alter the number of charged species in the solution.

A different type of solvation effect occurs when an apolar molecule is added to water. The result is a decrease in entropy, but not because of favorable interactions between the molecule and the solvent. The water orients on the surface of the apolar molecule to form a relatively rigid cage held together by hydrogen bonds (see fig. 1.10). This effect plays important roles in governing the folding of proteins and determining the structure of biological membranes (see chapter 4).

Substantial changes in entropy can occur when a small molecule binds to a protein or other macromolecule. Of particular interest is binding of a substrate or inhibitor to an enzyme. It is instructive to compare the entropy changes here with those that accompany the binding of gas molecules to the surface of a solid catalyst. When gas molecules adsorb on a solid surface, there is a large decrease in entropy (fig. 2.5). The translational entropy of the gas disappears, and the thermodynamics of the adsorbed molecules becomes more like that of a solid. This negative change in entropy is an obstacle to the industrial use of solid catalysts for reactions of gases. To make the reaction proceed, the unfavorable entropy change must be overcome by increasing the pressure or tailoring the catalyst so that the enthalpy of interaction with the gas is strongly negative. In contrast, the binding of a substrate to an enzyme frequently has a positive ΔS. The reason is that water molecules are displaced when the substrate binds (fig. 2.5). Binding of an ester substrate to pepsin,

Figure 2.4

An ionization reaction often decreases the entropy of a solution, instead of increasing it as one might at first expect, because clustering of water molecules around the ions can result in a net decrease in the number of free particles.

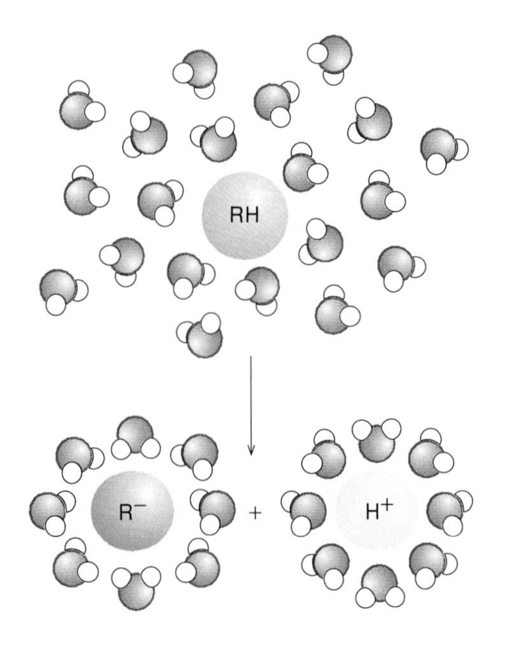

Figure 2.5

The entropy of binding of gas molecules to a solid catalyst is negative because of the restricted movements of the adsorbed molecules. By contrast, the entropy change on binding of a substrate to an enzyme is frequently positive. This effect arises because the restricted movement of the substrate is more than compensated for by the release of bound water from the enzyme and the substrate.

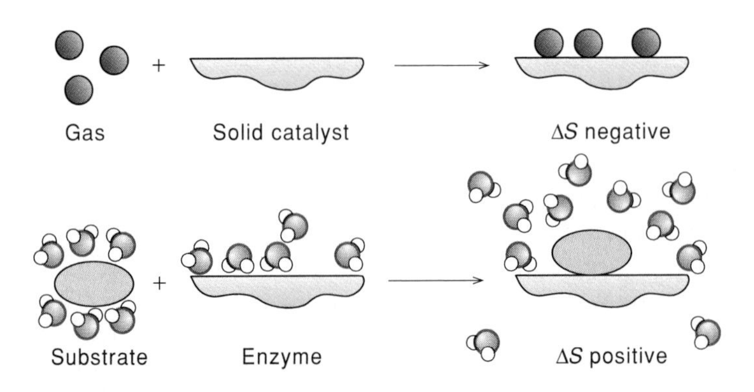

Table 2.2
Standard Enthalpy and Entropy Changes on Forming Complexes between Cadmium Ion and Methylamine or Ethylenediamine (en)

Reaction[a]	ΔH° (kcal/mole)	ΔS° (eu/mole)	$T\,\Delta S^\circ$ (kcal/mole)	ΔG° (kcal/mole)
$Cd^{2+} + 4\ CH_3NH_2 \rightarrow Cd(CH_3NH_2)_4^{2+}$	−13.7	−16.0	−4.77	−8.94
$Cd^{2+} + 2\ en \rightarrow Cd(en)_2^{2+}$	−13.5	−3.3	+0.98	−14.50

[a]en = ethylenediamine; temperature = 25° C.
Source: Data from Spike and Parry, "Thermodynamics of chelation I. The statistical factor in chelate ring formation" in *Journal American Chemical Society* 75:2726, 1953.

for example, produces an entropy increase of 20.6 eu/mole, and binding of urea to urease produces an entropy increase of 13.3 eu/mole. Because of the favorable entropy change, the binding of the substrate may occur spontaneously even if ΔH is unfavorable.

Another important entropy effect when an enzyme and substrate combine could be called the chelation effect. This effect is best understood by discussing a relatively simple case of metal chelation. Cadmium ion tends to be quadrivalent, so if there is one amino group in a ligand molecule, as in methylamine, the cadmium can bind to four molecules. If there are two amino groups, as in ethylenediamine, the cadmium will combine with two ligand molecules, as shown in table 2.2. Notice that the entropy change is much more unfavorable for the combination with four methylamines than for that with two ethylenediamines. Water molecules are released from the cadmium ion when the ligands bind, but the entropy increase from this release is about the same whether methylamine or ethylenediamine is added. The less favorable entropy change resulting from association with methylamine is due to the larger number of molecules that must attach to the cadmium in this case.

In general, the chelation effect means that a molecule with *n* points of attachment to another molecule will bind more strongly than *n* molecules with one point of attachment, even though the enthalpy change upon binding at each point is the same. The chelation effect is important in substrate binding to enzymes, where there typically are multiple points of attachment. Several weak interactions can produce an overall tight binding because of the additive contributions of the small, favorable enthalpy changes and the lack of a proportional decrease in entropy. This effect is even greater in the binding of two proteins or the binding of a protein to a nucleic acid, because there usually are many points of interaction between these large molecules.

The final column in table 2.2 indicates the changes in free energy accompanying the reactions of cadmium ion with the two amine compounds. Free energy is a function of both enthalpy and entropy; it provides the most useful criterion as to whether a reaction could proceed spontaneously, as explained in the next section.

Free Energy Provides the Most Useful Criterion for Spontaneity

We have seen that a system tends toward the lowest enthalpy and the highest entropy. The tendency for the enthalpy of a system to decrease can be explained simply by the second law. If no work is done, an enthalpy decrease means a transfer of heat from the system to the surroundings, and if the surroundings are at a lower temperature such a flow of heat will be driven by an increase in the entropy of the surroundings. But enthalpy changes do not afford a reliable rule for determining whether a reaction can proceed spontaneously, because the enthalpy of a system can increase if the entropy of the system also increases. The second law does provide a reliable rule, but its application is often difficult because it requires that we consider the entropy changes in the surroundings as well as the system.

A more convenient function for predicting the direction of a reaction was discovered by Josiah Gibbs. He was the first to appreciate that in reactions occurring at equilibrium and constant temperature, the change in entropy of the system is numerically equal to the change in enthalpy divided by the absolute temperature. This relationship is the one already presented in equation (6). The equation can be transposed to

$$\Delta H - T\,\Delta S = 0 \tag{8}$$

In search of a criterion for spontaneity, Gibbs proposed a new function called the free energy, defined by the equation

$$\Delta G = \Delta H - T\,\Delta S \tag{9}$$

Here ΔH and ΔS are the changes in enthalpy and entropy in the system alone, not including the surroundings. For a reaction occurring at equilibrium, such as the melting of ice at 0° C, the change in free energy is zero. For the same reaction occurring at a higher temperature, say 10° C, the term $T\,\Delta S$ is larger, making ΔG negative. Ice at 10° C melts spontaneously. In the reverse reaction, conversion of water to ice at 10° C, there would be a positive change in free energy. This process does not occur spontaneously. Gibbs proposed that a reaction can occur spontaneously if, and only if, ΔG is negative. If ΔG is zero, the system is in equilibrium and no net reaction will occur in either direction.

Free energy, like energy, enthalpy, and entropy, is a state function and an extensive property of a system. If the free energy change is favorable (negative), and a good pathway exists, a reaction will occur. If no pathway exists for the conversion, a catalyst may be added that provides an acceptable pathway. However, if the free energy change is unfavorable (positive), no catalyst can ever make the reaction proceed.

Applications of the Free Energy Function

The free energy function dominates most discussions of thermodynamics in biochemistry. Not only does the sign of ΔG determine the direction in which a reaction will proceed, but the magnitude of ΔG indicates just how far the reaction must proceed before the system comes to equilibrium. This is because the standard free energy change, $\Delta G°$, has a simple relationship to the equilibrium constant. We will elaborate on these points in the following sections. Despite its usefulness, however, many people find the free energy function difficult to grasp intuitively. The reason is that ΔG is a composite of enthalpic and entropic terms, which often make opposite contributions.

Values of Free Energy Are Known for Many Compounds

The standard free energy of formation of a compound, $\Delta G°_f$, is the difference between the free energy of the compound in its standard state and the total free energies of the elements of which the compound is composed, again when the elements are in their standard states. The standard states usually are chosen to be the states in which the elements or molecules are stable at 25° C and 1 atmosphere pressure. For oxygen and nitrogen, these are the gases O_2 and N_2; for solid elements such as carbon, they are the pure solids. For most solutes, the standard states are taken to be 1 M solutions. However, in biochemistry the standard state for hydrogen ion in solution is usually defined as a 10^{-7} M solution because this is close to the concentration in most systems of interest to biochemists.

Standard free energies of formation are known for thousands of compounds. They usually are given in units of kcal/mole or kJ/mole. The values for a few compounds of biological interest are collected in table 2.3. By subtracting the sum of the free energies of formation of the reactants from the sum of the free energies of formation of the products, it is possible to calculate the standard free energy change in any reaction for which all the free energies of formation are known.

From the values listed in table 2.3, we can calculate the standard free energy change for the reaction

$$\text{Oxaloacetate}^{2-} + H^+ (10^{-7}\ \text{M}) \rightarrow CO_2(g) + \text{pyruvate}^-$$

as

$$\Delta G° = -113.44 - 94.45 - (-9.87 - 190.62)$$
$$= -7.4 \text{ kcal/mole}$$

Table 2.3
Standard Free Energies of Formation of Some Compounds of Biological Interest

Substance	$\Delta G°_f$ (kcal/mole)	$\Delta G°_f$ (kJ/mole)
Lactate ions	−123.76	−516
Pyruvate ions	−113.44	−474
Succinate dianions	−164.97	−690
Glycerol (1 M)	−116.76	−488
Water	−56.69	−280
Acetate anions	−88.99	−369
Oxaloacetate dianions	−190.62	−797
Hydrogen ions (10^{-7} M)	−9.87[a]	−41
Carbon dioxide (gas)	−94.45	−394
Bicarbonate ions	−140.49	−587

[a]This is the value for hydrogen ions at a concentration of 10^{-7} M. The free energy of formation at unit activity (1 M) is 0.

The free energy change, when all the reactants and products are in their standard states (1M oxaloacetate dianion and pyruvate anion, 10^{-7}M hydrogen ion, and 1 atm CO_2), is −7.4 kcal/mole. The negative value of $\Delta G°$ means that the reaction would proceed spontaneously under these conditions. However, some of the concentrations are not very realistic. At pH 7, carbon dioxide would be present partly in the form of the bicarbonate anion, rather than as gaseous CO_2. To take this into account, we can add the standard free energy change for the reaction of CO_2 with water to give the bicarbonate anion plus a proton:

$$CO_2(g) + H_2O \rightarrow HCO_3^- + H^+$$

This calculation yields a correction of −140.49 − 9.87 − (−56.69 − 94.45) = 0.8 kcal/mole. The free energy change for the reaction of oxaloacetate to form pyruvate and 1M bicarbonate ions instead of CO_2 is −7.4 + 0.8 = −6.6 kcal/mole.

The preceding calculation illustrates the point that the standard free energy change for a reaction can be found by adding or subtracting the free energies of any other reactions that combine to give the desired reaction. Another example is the calculation of the standard free energy of hydrolysis of ATP at pH 7. This calculation can be done by combining the free energy change for the hydrolysis of glucose-6-phosphate with the free energy change for forming glucose-6-phosphate from glucose and ATP, as shown in table 2.4. We will return to these reactions in a later section.

Table 2.4
Calculating the Standard Free Energy of ATP Hydrolysis ($\Delta G^{\circ\prime}$) by Adding the Free Energies of Two Other Reactions

Reaction[a]	$\Delta G^{\circ\prime}$ (kcal/mole)
Glucose + $ATP^{4-} \rightleftharpoons$ glucose-6-phosphate^{2-} + ADP^{3-} + H^+	−5.4
Glucose-6-phosphate^{2-} + $H_2O \rightleftharpoons$ glucose + HPO_4^{2-}	−3.0
$ATP^{4-} + H_2O \rightleftharpoons ADP^{3-} + HPO_4^{2-} + H^+$	−8.4

[a]The values of $\Delta G^{\circ\prime}$ are for reactions at pH 7 in the absence of Mg^{2+}. In the presence of 10 mM Mg^{2+}, $\Delta G^{\circ\prime}$ for ATP hydrolysis is about −7.5 kcal/mole.

The Standard Free Energy Change in a Reaction Is Related Logarithmically to the Equilibrium Constant

In biochemistry we are most concerned with reactions occurring in aqueous solution. Suppose we have a chemical reaction with the stoichiometry

$$a\mathrm{A} + b\mathrm{B} \rightleftharpoons c\mathrm{C} + d\mathrm{D}$$

where *a, b, c,* and *d* refer to the moles of A, B, C, and D, respectively. The free energy change in the reaction is

$$\Delta G = G_{\text{final state}} - G_{\text{initial state}} \qquad (10)$$

If the reaction occurs at constant temperature and pressure, equation (10) can be expressed as the difference between the standard free energies of the products and reactants, ΔG°, plus a correction for the concentrations:

$$\Delta G = \Delta G^\circ + RT \ln \frac{[\mathrm{C}]^c[\mathrm{D}]^d}{[\mathrm{A}]^a[\mathrm{B}]^b} \qquad (11)$$

The last term in equation (11) is the correction for concentration and as such is an entropic contribution to ΔG. It is derived by using equation (7) to find the entropy changes associated with diluting the reactants and products from their standard states (1 M) to the actual concentrations in the solution. (In a rigorous treatment, we should use activities instead of concentrations in this formula, but for simplicity we will ignore the difference, keeping in mind that it can be substantial in some cases.) If the concentrations of the reactants exceed those of the products, so that the ratio $[\mathrm{C}]^c[\mathrm{D}]^d/[\mathrm{A}]^a[\mathrm{B}]^b$ is less than 1, the logarithm will be negative, making ΔG more negative than ΔG° and favoring the reaction in the forward direction. A concentration ratio greater than 1 will favor the reverse reaction.

When the reaction comes to equilibrium,

$$\frac{[\mathrm{C}]^c[\mathrm{D}]^d}{[\mathrm{A}]^a[\mathrm{B}]^b} = K_{eq} \qquad (12)$$

where K_{eq} is the equilibrium constant for the reaction. We also know that at equilibrium $\Delta G = 0$. Therefore, from equation (11),

$$\Delta G^\circ = -RT \ln K_{eq} \qquad (13)$$

Table 2.5
Relationship between ΔG° and K_{eq} (at 25° C)

ΔG° (kcal/mole)[a]	K_{eq}
−6.82	10^5
−5.46	10^4
−4.09	10^3
−2.73	10^2
−1.36	10
0	1
1.36	10^{-1}
2.73	10^{-2}
4.09	10^{-3}
5.46	10^{-4}
6.82	10^{-5}

[a]ΔG° values at 25° C are calculated from the equation

$$\Delta G^\circ = -RT \ln K_{eq} = -1.98 \times 298 \times 2.3 \log K_{eq} = -1364 \log K_{eq}$$

Thus the standard free energy change for a reaction can be used to obtain the equilibrium constant. Conversely, if we know K_{eq}, we can find ΔG°. Because of the logarithmic relationship, K_{eq} has a very steep dependence on ΔG° (table 2.5). A reaction that proceeds to 99% completion is, for most practical purposes, a quantitative reaction. It requires an equilibrium constant of 100, but a standard free energy change of only −2.7 kcal/mole, which is little more than half the standard free energy change for the formation of a hydrogen bond.

Equations (11) and (13) are two of the most important thermodynamic relationships for biochemists to remember. If the concentrations of reactants and products are at their equilibrium values, there is no change in free energy for the reactions going in either direction. Living cells, however, maintain some compounds at concentrations far from the equilibrium values, so that their reactions are associated with large changes in free energy. We will amplify on this point in chapter 12.

We have mentioned that biochemists usually define the standard state of protons as 10^{-7} M and report values of free energy and equilibrium constants for solutions at pH 7. These values are designated by a prime and written as $\Delta G^{\circ\prime}$, $\Delta G'$ and K'_{eq}. *Unprimed symbols are used to designate values based on a standard state of 1* M for protons (pH 0). For a reaction that releases one proton, the relationship between K'_{eq} and K_{eq} is $K'_{eq} = 10^7 K_{eq}$. In evaluating the standard free energies $\Delta G^{\circ\prime}$ and ΔG°, it is critical to use the equilibrium constants K'_{eq} and K_{eq}, respectively, because these can be very different quantities.

Free Energy Is the Maximum Energy Available for Useful Work

The free energy change gives a quantitative measure of the maximum amount of useful work that could be obtained from a reaction that occurs at constant temperature and pressure. By "useful" work, we mean work other than the unavoidable work of expansion or contraction against the fixed pressure of the surroundings. If ΔG is zero, the system is at equilibrium, which means that we could not obtain any useful work from the process. If ΔG is less than zero, the process could yield useful work as the system proceeds spontaneously toward equilibrium. If ΔG is greater than zero, the process is headed away from equilibrium, and we would have to perform work on the system in order to drive it in this direction. The farther the reactants are from equilibrium, the larger is the value of $-\Delta G$, and the larger is the amount of work that we might obtain from the reaction. However, $-\Delta G$ gives only the maximum amount of useful work. Remember that work and heat, unlike ΔG, ΔH, and ΔS, are not functions of state. The amount of work that is actually obtained depends on the path that the process takes, and it can be zero even if $-\Delta G$ is large.

Biological Systems Perform Various Kinds of Work

To sustain and propagate life requires that cells do various types of work. This work takes three major forms, related to three broad categories of cellular activities:

1. Mechanical work: changes in location or orientation. Mechanical work is done whenever an organism, cell, or subcellular structure moves against the force of gravity or friction. As examples, consider the contracting muscles that propel a runner up a hill, the swimming of a flagellated protozoan in a pond, the migration of chromosomes toward the opposite poles of the mitotic spindle, and the movement of a ribosome along a strand of messenger RNA.
2. Concentration and electrical work: movements of molecules and ions across membranes. Concentration work, the movement of a molecule or ion across a membrane against a prevailing concentration gradient, establishes the localized concentrations of specific materials on which most essential life processes depend. Concentration work is sometimes referred to as osmotic work. Examples include the uptake of amino acids from the blood by muscle cells, pumping of sodium ions out of a marine microorganism, and movement of nitrate from the soil into the cells of a plant root. Electrical work is required to move a charged species across a membrane against an electrical potential gradient. Although the most dramatic example of this is the generation of large potential differences in the electric organ of the electric eel, electrical work is done by almost all types of cells. It underlies the mechanisms of excitation of nerve and muscle cells and the conduction of impulses along axons.
3. Synthetic work: changes in chemical bonds. Synthetic work is necessary for the formation of the complex organic molecules of which cells are composed. As we have seen, these are in general molecules of higher enthalpy and lower entropy than the simple molecules that are available to organisms from their environment, so that free energy must be expended in their synthesis. Synthetic work is most obvious during periods of growth of an organism, but it also occurs in nongrowing, mature organisms, which must continuously repair and replace existing structures. The continuous expenditure of energy to elaborate and maintain ordered structures that were created out of less-ordered raw materials is one of the most characteristic properties of living cells.

Favorable Reactions Can Drive Unfavorable Reactions

In table 2.4, we made use of the principle that the free energies of all the components of a solution are additive. In general, if the free energy changes associated with two reactions A $\rightleftharpoons$ B and C $\rightleftharpoons$ D are ΔG_{AB} and ΔG_{CD}, the free energy change accompanying the combined process A + C $\rightleftharpoons$ B + D is simply $\Delta G_{AB} + \Delta G_{CD}$. It is this principle that allows living organisms to synthesize complex molecules with high enthalpies and low entropies. Thermodynamically unfavorable reactions can be driven by coupling them to favorable processes.

From equation (13) you can see that whereas free energy changes combine additively, equilibrium constants combine multiplicatively. If the equilibrium constants for the reactions A $\rightleftharpoons$ B and C $\rightleftharpoons$ D are K_{AB} and K_{CD}, the equilibrium constant for A + C $\rightleftharpoons$ B + D is $K_{AB}K_{CD}$.

There are numerous ways to achieve a coupling of favorable and unfavorable reactions. As an example, let's return to the formation of glucose-6-phosphate and water from glucose and inorganic phosphate ion (P_i):

$$\text{Glucose} + P_i \rightleftharpoons \text{glucose-6-phosphate} \quad (14)$$

This reaction has an unfavorable $\Delta G^{\circ\prime}$ of +3.0 kcal/mole at 298 K (K_{eq} is 0.0062); the reaction will not occur spontaneously. On the other hand, the reaction

$$\text{ATP} + H_2O \rightleftharpoons \text{ADP} + P_i + H^+ \quad (15)$$

Figure 2.6

The formation of glucose-6-phosphate (G-6-P) has a positive $\Delta G^{\circ\prime}$; the hydrolysis of ATP and ADP has a negative $\Delta G^{\circ\prime}$. If the two reactions are combined, the overall $\Delta G^{\circ\prime}$ is negative.

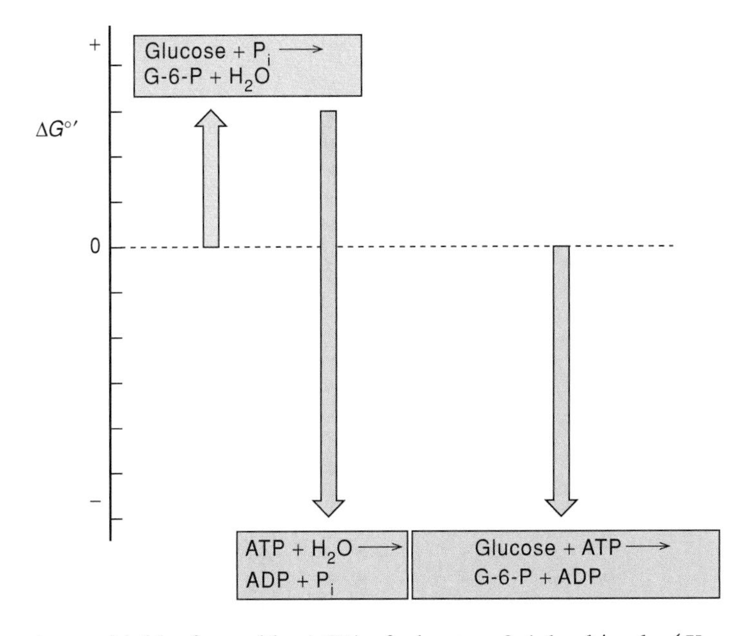

has a highly favorable $\Delta G^{\circ\prime}$ of about -8.4 kcal/mole ($K_{eq} \approx 1.35 \times 10^6$). If reactions (14) and (15) are combined to give the reaction

$$\text{Glucose} + \text{ATP} \rightleftarrows \text{glucose-6-phosphate} + \text{ADP} \qquad \textbf{(16)}$$

the overall $\Delta G^{\circ\prime}$ is $3.0 - 8.4$, or -5.4 kcal/mole, and the overall equilibrium constant is 8.7×10^3. The combined reaction is thermodynamically favorable (fig. 2.6). A cell could use this reaction to synthesize glucose-6-phosphate, provided that it has a source of ATP.

To take advantage of such a thermodynamic combination of favorable and unfavorable processes, a cell must have a catalytic mechanism for actually linking the two reactions. The breakdown of ATP to ADP and P_i, reaction (15), would be fruitless if it occurred independently of reaction (13). In many cells, the combined reaction (16) is catalyzed by an enzyme that facilitates the transfer of phosphate from ATP directly to glucose (hexokinase). This is a common motif in biosynthetic processes. But the coupling mechanism does not have to be so direct.

Another common mechanism for coupling an unfavorable reaction to a favorable one is simply to arrange for one of the reactions to precede or follow the other. If the free energy changes for the reactions $A \rightleftarrows B$ and $B \rightleftarrows C$ are ΔG_{AB} and ΔG_{BC}, the free energy change for $A \rightleftarrows C$ is $\Delta G_{AB} + \Delta G_{BC}$. As an example, consider the following sequence of reactions:

$$\text{Acetyl-CoA} + \text{oxaloacetate} \xrightarrow{1} \text{citryl-CoA} \xrightarrow{2} \text{citrate} + \text{coenzyme A} \qquad \textbf{(17)}$$

The first step has a $\Delta G^{\circ\prime}$ of -0.05 kcal/mole, which is close to zero; it will not occur to any great extent unless the concentrations of acetyl-CoA and oxaloacetate are greater than the concentration of citryl-CoA. The second step, however, has a highly favorable $\Delta G^{\circ\prime}$ of -8.4 kcal/mole. When the two steps are combined, $\Delta G^{\circ\prime}$ for the overall reaction is about -8.3 kcal/mole and the equilibrium constant lies far in the forward direction. These two reactions are catalyzed by the enzyme citrate synthase, by a mechanism that insures that they always occur together.

In the case of reactions that occur sequentially, with one step pulling or pushing the other, it is not necessary for the two steps to be catalyzed by the same enzyme, although this can help to speed up the overall sequence. For example, in many organisms the reactions shown in (17) are preceded by a reaction in which malate is converted to oxaloacetate:

$$\text{Malate}^{2-} + \text{NAD}^+ \rightarrow \text{oxaloacetate}^{2-} + \text{NADH} + \text{H}^+ \qquad \textbf{(18)}$$

This step is catalyzed by a separate enzyme, malate dehydrogenase. It has a very unfavorable $\Delta G^{\circ\prime}$ of about $+7$ kcal/mole. When this reaction is followed by reaction (17), the combined $\Delta G^{\circ\prime}$ is about -1.3 kcal/mole, so the equilibrium constant for the overall sequence is favorable. You can view this as simply an illustration of the principle of mass action: (17) pulls (18) along by removing one of the products.

This last point deserves additional emphasis. It is important to keep in mind that what determines whether or not a process will occur spontaneously is ΔG, not ΔG°. As we saw in equation (11), the actual free energy change depends on the concentrations of the reactants and products. The hydrolysis of ATP (reaction 15), for example, has a $\Delta G^{\circ\prime}$ of about -8.4 kcal/mole, but the actual ΔG in the cytosol of living cells is typically more negative than this by between 5 and 6 kcal/mole because the concentration ratio, $[ADP][P_i]/[ATP]$, is much less than 1.

ATP as the Main Carrier of Free Energy in Biochemical Systems

Virtually all living organisms use ATP for transferring free energy between energy-producing and energy-consuming systems. Processes that proceed with large negative changes in free energy, such as the oxidative degradation of carbohydrates or fatty acids, are used to drive the formation of ATP, and the hydrolysis of ATP is used to drive biosynthetic reactions and other processes that require increases in free energy. In the human body, about 2.3 kg of ATP is formed and consumed every day in the course of these reactions.

The Hydrolysis of ATP Yields a Large Amount of Free Energy

Adenosine triphosphate (ATP) can be hydrolyzed in two different ways, as shown in figure 2.7. Hydrolysis of the linkage between the β and γ phosphate groups yields ADP and P_i. Hydrolysis between the α and β phosphates gives adenosine monophosphate (AMP) and pyrophosphate ion ($HP_2O_7^{3-}$). The standard free energy change ($\Delta G^{\circ\prime}$) is about -8.4 kcal/mole in either case. The formation of AMP and pyrophosphate, however, can be pulled forward by hydrolysis of the pyrophosphate

Figure 2.7

Alternative routes of ATP hydrolysis. The charged species shown are the main ones present at physiological pH and ionic strength. The phosphate groups of ATP are referred to as α, ß, and γ as indicated. Under physiological conditions, ATP and ADP also bind Mg^{2+} (not shown).

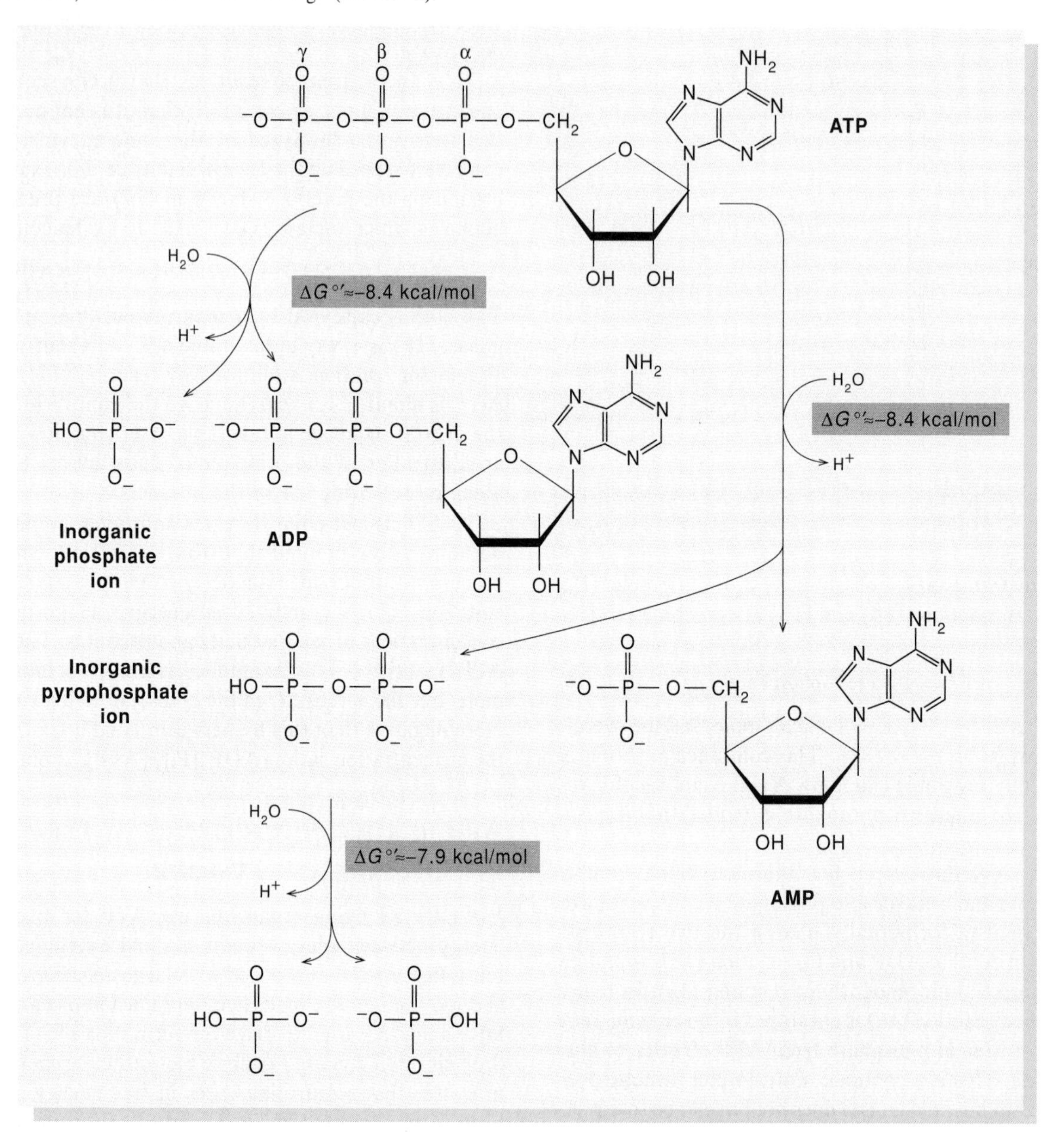

to give two equivalents of P_i. This secondary reaction has a $\Delta G^{\circ\prime}$ of about -7.9 kcal/mole, and is catalyzed by pyrophosphatase enzymes present in most types of cells. The overall $\Delta G^{\circ\prime}$ of about -16.3 kcal/mole for breakdown of ATP to AMP and 2 P_i makes this process effectively irreversible ($K_{eq} \approx 1 \times 10^{12}$). Cells commonly use this series of reactions in biosynthetic processes in which a reversal would be intolerable, such as in the synthesis of nucleic acids. The simpler hydrolysis to produce ADP and P_i allows some reversibility but has the advantage that less free energy is needed to resynthesize ATP from ADP than from AMP.

The standard free energies for the hydrolysis of ATP, ADP, or pyrophosphate depend on the pH, on the ionic strength, and also on the concentration of Mg^{2+}, which binds to both the reactants and the products and is required as a cosubstrate by most enzymes that use ATP. At pH 7, the principal ionic forms of ATP, ADP, and P_i have net charges of -4, -3, and -2, respectively (see fig. 2.7). Because the hydrolysis of ATP^{4-} to give $ADP^{3-} + HPO_3^{2-}$ is accompanied by release of a proton, raising the pH makes the hydrolysis more favorable. Increasing the Mg^{2+} concentration makes the hydrolysis less favorable.

$\Delta G°'$ decreases in magnitude from -8.4 kcal/mole in the absence of Mg^{2+} to about -7.7 kcal/mole in the presence of 1 mM Mg^{2+}, and to about -7.5 kcal/mole at 10 mM. The value -7.5 kcal/mole is probably close to the $\Delta G°'$ under typical physiological conditions, and will be used elsewhere in this text. But remember that the low ratio of [ADP][P_i] to [ATP] can make ΔG for hydrolysis of ATP in living cells substantially more negative than the $\Delta G°'$. In resting cells, the enzymatic reactions that synthesize ATP usually are more than adequate to keep up with the processes that consume it.

Why is $\Delta G°'$ for the hydrolysis of ATP so negative? The answer involves several different factors. At pH 7 in the presence of 10 mM Mg^{2+}, $\Delta H°'$ and $-T\Delta S°'$ for ATP hydrolysis are both negative and the two terms make similar contributions to the overall value of $\Delta G°'$. The favorable entropy change results partly from the fact that the proton released in the reaction is diluted into a solution with a very low proton concentration. In addition, solvent water molecules probably are less highly ordered around ADP and P_i than they are around ATP. The negative $\Delta H°'$ results partly from the fact that the negatively charged oxygen atoms in ATP tend to repel each other. The phosphoric anhydride bond in ATP also is weakened by competition between the phosphorus atoms, which both tend to pull electrons away from the bridging oxygen. Another consideration is that the major products of ATP hydrolysis at pH 7 (ADP^{3-} and $HOPO_3{}^{2-}$) have a larger total number of resonance forms than the reactant ATP^{4-} does (fig. 2.8). The increased resonance lowers the energy of ADP^{3-} and $HOPO_3{}^{2-}$ relative to ATP^{4-}. The magnitudes of all of these effects vary with the pH, because protonation of the oxygen atoms in ATP, ADP, or P_i relieves some of the repulsive electrostatic interactions, decreases the contributions that some of the resonance forms make to the structure, and decreases the ordering of nearby water molecules. At very low pH, when ATP, ADP, and P_i are all fully protonated, $\Delta G°$ probably becomes slightly positive, favoring ATP formation rather than hydrolysis.

In addition to ATP and other nucleoside triphosphates, cells use a variety of other organic phosphate compounds in energy metabolism. These include acetyl phosphate, glycerate-1,3-bisphosphate, phosphoenolpyruvate, phosphocreatine, and phosphoarginine. Figure 2.9 shows the structures of these compounds and some simpler phosphate esters. The compounds are ranked in the figure in order of their standard free energies of hydrolysis, with the materials having the most negative values of $\Delta G°'$ at the top. Given equal concentrations of reactants and products, any compound in the figure could be synthesized, in principle, at the expense of any of the compounds above it. Thus ADP could be phosphorylated to ATP by phosphocreatine, glycerate-1,3-bisphosphate, or phosphoenolpyruvate, and ATP could be used to convert glucose to glucose-6-phosphate or glycerol to glycerol phosphate. The reverse processes will occur only if the ratio of reactants to products is sufficiently high, but this is not necessarily an unusual circumstance. In cells of some tissues, the reaction

$$\text{Creatine} + \text{ATP} \rightleftharpoons \text{creatine phosphate} + \text{ADP} + \text{H}^+ \quad \textbf{(19)}$$

occurs readily in either direction in response to changes in the concentrations of the reactants and products. The same is true of the reaction

$$\text{Glycerate-3-phosphate} + \text{ATP} \rightleftharpoons \text{glycerate-1,3-bisphosphate} + \text{ADP} + \text{H}^+ \quad \textbf{(20)}$$

Figure 2.8

The phosphate groups of ATP, ADP, and P_i can be written in a variety of resonance forms. This figure shows the major resonance forms of the β phosphate group in ATP and of the same group in ADP. The hydrolysis of ATP to ADP results in an increase in the number of resonance forms available to this group. This increase contributes to the negative $\Delta H°'$ of hydrolysis of ATP. At physiological pH there are also favorable contributions to $\Delta G°'$ from the release of a proton and from the disordering of water around the polyphosphate chain.

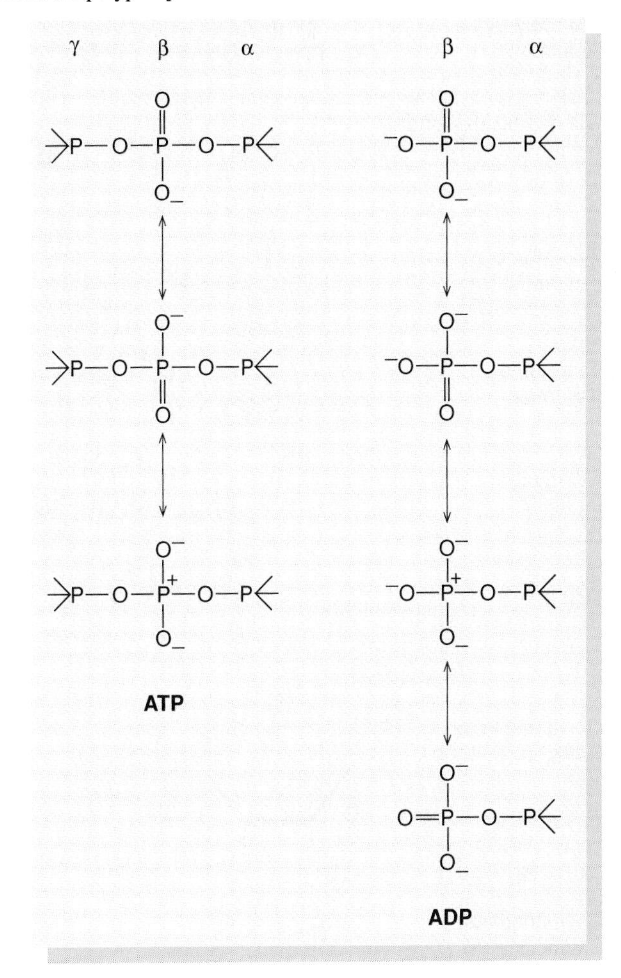

Figure 2.9

Standard free energies of hydrolysis for some common phosphorylated compounds. The $\Delta G^{\circ\prime}$ value refers to hydrolysis of the phosphate group indicated by the symbol Ⓟ. Note that ATP occupies an intermediate position among these compounds. Given equal concentrations of the reactants and products, ADP can accept a phosphate group from any of the compounds above it, and ATP can donate a phosphate to the unphosphorylated forms of the compounds below it.

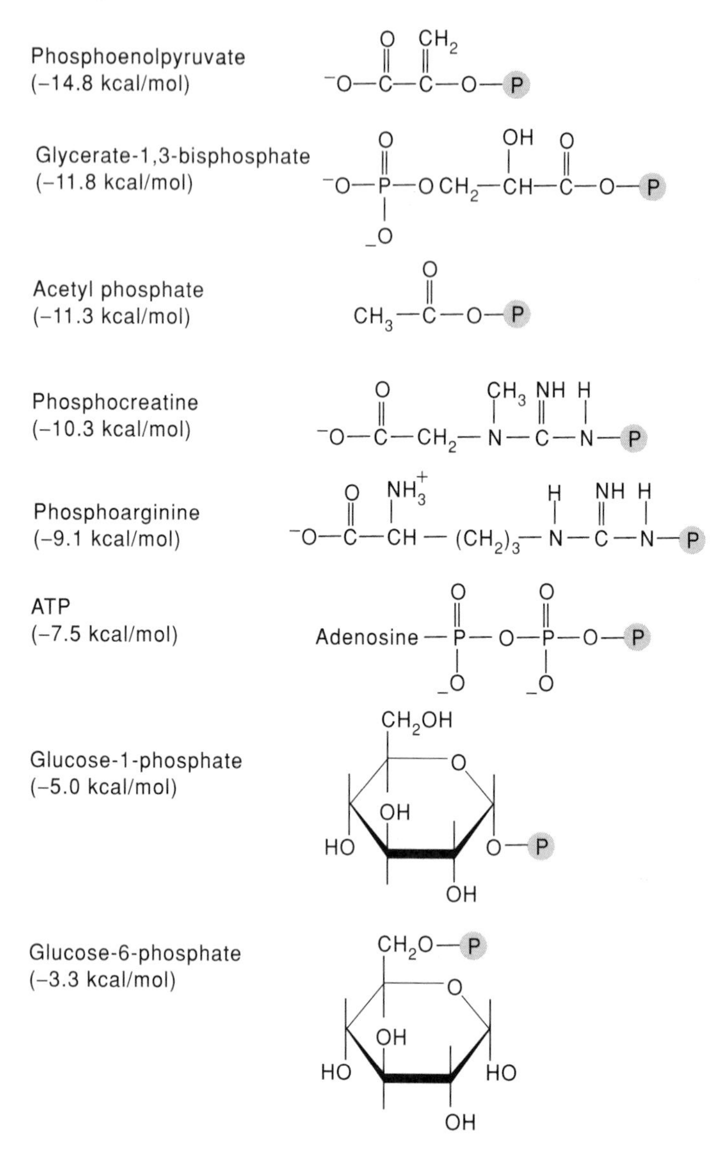

Most of the phosphorylated materials shown in figure 2.9 participate in only a few biochemical reactions, and many of them are formed only in certain types of cells or organisms. ATP, in contrast, is formed by all living things, and it participates in literally hundreds of different enzymatic reactions. What is it about ATP that has made this molecule such a universal currency of free energy in biology? Several considerations are relevant here. First, $\Delta G^{\circ\prime}$ for hydrolysis of ATP to ADP is large enough so that this reaction releases a substantial amount of free energy, enough to drive many of the reactions that are important for biosynthetic pathways. At the same time, the $\Delta G^{\circ\prime}$ is small enough so that ATP itself can be synthesized readily at the expense of available nutrients. We touched on this point above in discussing the relative merits of hydrolysis at the α-β or β-γ positions in ATP.

The second consideration returns us to the distinction between spontaneity and speed. Although the hydrolysis of ATP has a large negative $\Delta G^{\circ\prime}$, it also is critical that ATP is a relatively stable compound in aqueous solution. It does not hydrolyze rapidly under physiological conditions of pH and temperature. The hydrolysis, though far downhill thermodynamically, is slowed by a substantial activation barrier. But the barrier must be of such a nature that it can easily be overcome enzymatically. This feature allows the free energy of hydrolysis to be channeled quickly and selectively into reactions where it is needed, but to be conserved when energy is not in demand.

Third, the products of the hydrolysis of ATP provide opportunities for coupling to a wide variety of chemical reactions. We have seen how the phosphate group is incorporated into glucose-6-phosphate, and later chapters will provide many illustrations of similar enzymatic processes. We also will see reactions in which the pyrophosphate group or the adenylyl group (AMP) is incorporated into the products. Such a broad array of reactions could not be driven by a molecule that decomposed to release an inert material such as N_2.

Finally, the adenine and ribosyl groups of ATP, ADP, and AMP provide additional structural features that allow these molecules to bind to enzymes, and thus to participate in regulating enzymatic activities. This may be part of the reason that no known organisms base their energy-transfer reactions entirely on inorganic pyrophosphate or other polyphosphate compounds without a nucleoside moiety.

Summary

In this chapter we have discussed some principles of thermodynamics as they relate to biochemical reactions. The following points are of greatest importance.

1. Thermodynamics is useful in biochemistry for predicting whether a given reaction could occur and, if so, how much work a cell could obtain from the process.
2. The thermodynamic quantities energy, enthalpy, entropy, and free energy are properties of the state of a system. Changes in these quantities depend only on the difference between the initial and final states, not on the mechanism whereby the system goes from one state to the other.
3. Energy is the capacity to do work.
4. The first law of thermodynamics says that energy cannot be created or destroyed in a chemical reaction. If the energy of a system increases, the surroundings must lose an equivalent amount of energy, either by the transfer of heat or by the performance of work.
5. The energy of a molecule includes translational, rotational, and vibrational energy, as well as electronic and nuclear energy. Electronic terms usually account for most of the change in energy, ΔE, in a chemical reaction.
6. The change in enthalpy, ΔH, is given by the expression $\Delta H = \Delta E + \Delta(PV)$. For most biochemical reactions, ΔE and ΔH are nearly equal. The organic molecules found in cells generally have much higher enthalpies than the simpler molecules from which they are built.
7. In most reactions that proceed spontaneously, the enthalpy of the system decreases. If no work is done, the system gives off heat to the surroundings. But in some spontaneous reactions, heat is absorbed in the absence of work and the enthalpy of the system increases. Such reactions invariably show an increase in the entropy of the system.
8. Entropy is a measure of the order in a system: Systems that are highly ordered have low entropies. The entropy of a molecule depends mainly on translational and rotational freedom. Biological macromolecules generally have much lower entropies than their building blocks.
9. The second law of thermodynamics states that there must be an overall increase in the entropy of the system and its surroundings in any process that occurs spontaneously. An isolated system proceeds spontaneously to states of increasingly greater entropy (greater disorder).
10. The change in the free energy of a system is defined as $\Delta G = \Delta H - T\Delta S$, where T is the absolute temperature. A reaction at constant pressure and temperature can occur spontaneously if, and only if, ΔG is negative. The maximal amount of useful work that can be obtained from a reaction is equal to $-\Delta G$.
11. For a reaction in solution, ΔG depends on the standard free energy change ($\Delta G°$) and on the concentrations of the reactants and products. The standard free energy change is related to the equilibrium constant by the expression $\Delta G° = -RT \ln K_{eq}$. Increasing the concentration of the reactants relative to the concentration of the products makes ΔG more negative.
12. Reactions that are thermodynamically unfavorable can be coupled to favorable reactions. The coupling of the reactions may be direct or sequential, as in a biochemical pathway.
13. ATP is the main coupling agent for free energy in living cells. The free energy provided by the hydrolysis of ATP is used to drive many reactions that would not occur spontaneously by themselves.
14. Several features make ATP particularly well suited for its role. First, hydrolysis of ATP to ADP and P_i or to AMP and PP_i releases a considerable amount of free energy. Second, ATP does not hydrolyze rapidly by itself, but it can be hydrolyzed readily in enzymatically catalyzed reactions. This difference allows the free energy of hydrolysis to be channeled into reactions where it is needed, but to be conserved when energy is not in demand. Third, the products of the hydrolysis of ATP provide opportunities for coupling to a wide variety of chemical reactions. Finally, the adenine and ribosyl groups of ATP, ADP, and AMP provide additional structural features that allow these molecules to bind to a large number of enzymes, and thus to participate in regulating enzymatic activities.

Selected Readings

Alberty, R. A., and F. Daniels, *Physical Chemistry,* 5th ed. New York: Wiley, 1975.

Cantor, C. R., and P. R. Schimmel, *Biophysical Chemistry.* San Francisco: Freeman, 1980.

Ingraham, L. L., and A. B. Pardee, Free Energy and Entropy in Metabolism. In *Metabolic Pathways,* Vol. 1 (D. M. Greenberg, ed.). New York: Academic Press, 1967.

Tinoco, I., Jr., K. Sauer, and J. C. Wang, *Physical Chemistry, Principles and Applications in Biological Sciences,* 2d ed. Englewood Cliffs, N.J.: Prentice-Hall, 1985.

Van Holde, K. E., *Physical Biochemistry,* 2d ed. Englewood Cliffs, N.J.: Prentice-Hall, 1985.

Problems

1. How do conditions in the cell limit the manipulation of reaction conditions?
2. What is meant by a state function? Why is enthalpy a state function?
3. What is the basic difference between intensive and extensive thermodynamic parameters?
4. Why can we equate energy and enthalpy for most biochemical reactions?
5. A reaction mechanism cannot be defined by free energy considerations. Explain.
6. Transfer of a hydrophobic molecule (e.g., a hydrophobic amino acid side chain) from an aqueous to a nonaqueous environment is entropically favorable. Explain.
7. As we will see in chapter 14, oxaloacetate is formed by oxidation of malate. The reaction

   $$\text{L-malate} + NAD^+ \rightarrow \text{Oxaloacetate} + NADH + H^+$$

   has a $\Delta G°'$ of $+7.0$ kcal/mole. Suggest reasons that the reaction proceeds in the direction of oxaloacetate production in the cell.
8. Cite three factors that make ATP ideally suited to transfer energy in biochemical systems.
9. You wish to measure the $\Delta G°'$ for the hydrolysis of ATP,

   $$ATP \rightarrow ADP + P_i$$

   but the equilibrium for the hydrolysis lies so far toward products that analysis of the ATP concentration at equilibrium is neither practical nor accurate. However, you have the following data that will allow calculation of the value indirectly.

   $$\text{Creatine phosphate} + ADP \rightarrow ATP + \text{creatine} \quad K'_{eq} = 59.5 \qquad \textbf{(P1)}$$

   $$\text{Creatine} + P_i \rightarrow \text{creatine phosphate} \quad \Delta G°' = +10.5 \text{ kcal/mole} \qquad \textbf{(P2)}$$

   Assume that $2.3RT = 1.36$ kcal/mole.
   (a) Calculate the value of $\Delta G°'$ for reaction (P1).
   (b) Calculate the $\Delta G°'$ for hydrolysis of ATP.
10. As we will see in a later chapter, mitochondria establish a proton gradient across the inner mitochondrial membrane. Translocation of protons from the matrix (or inner surface of the inner membrane) to the outer surface of the inner membrane establishes the gradient. Translocation of the protons is an endergonic process. Explain. (Information in chapter 15 is necessary to answer this problem.)
11. What is the minimum number of moles of protons that must be translocated across a membrane to drive phosphorylation of $ADP \rightarrow ATP$ if $\Delta\psi$ is -150 mV and ΔpH is 0.5? Assume that phosphorylation of ADP requires 11 kcal/mole.
12. The hydrolysis of lactose (D-galactosyl-β (1,4) D-glucose) to D-galactose and D-glucose occurs with a $\Delta G°'$ of -4.0 kcal/mole.
    (a) Calculate K'_{eq} for the hydrolytic reaction.
    (b) What are the $\Delta G°'$ and K'_{eq} for the synthesis of lactose from D-galactose and D-glucose?
    (c) Lactose is synthesized in the cell from UDP-galactose plus D-glucose and is catalyzed by lactose synthase. Given that $\Delta G°'$ of hydrolysis of UDP-galactose is -7.3 kcal/mole, calculate for $\Delta G°'$ and K'_{eq} for the reaction

    $$\text{UDP-galactose} + \text{D-glucose} \rightarrow \text{Lactose} + \text{UDP}$$

13. Dihydrolipoamide dehydrogenase catalyzes the reaction

    $$\text{Dihydrolipoamide} + NAD^+ \rightleftharpoons \text{lipoamide} + NADH + H^+$$

    $\Delta G°'$ is $+1.38$ kcal/mole. Calculate the steady-state ratio of lipoamide/dihydrolipoamide if the ratio of $NADH/NAD^+$ is (a) 1:10 and (b) 10:1.
14. For each of the following reactions, calculate $\Delta G°'$ and indicate whether the reaction is thermodynamically favorable as written.
    (a) Glycerate-1,3-bisphosphate + creatine → phosphocreatine + 3-phosphoglycerate
    (b) Glucose-6-phosphate → glucose-1-phosphate
    (c) Phosphoenolpyruvate + ADP → pyruvate + ATP
    (d) Glycerol phosphate + ADP → glycerol + ATP
15. Although ATP is an important phosphate donor, in biosynthetic reactions the AMP portion of the molecule is often transferred to an acceptor with the release of pyrophosphate. Such a transfer occurs as an intermediate step in the reaction.

    $$R{-}COO^- + ATP + CoASH \rightarrow R{-}CO{-}SCoA + AMP + \text{pyrophosphate} \qquad \textbf{(P3)}$$

    Inorganic pyrophosphatases hydrolyze the pyrophosphate, yielding two molecules of inorganic phosphate. Assume that the hydrolysis of $R{-}CO{-}SCoA$ to $R{-}COO^- +$ CoASH proceeds with $\Delta G°'$ of -10 kcal/mole and that hydrolysis of a pyrophosphate anhydride bond yields -7.5 kcal/mole. Calculate the K'_{eq} for reaction (P3) in the presence and in the absence of inorganic pyrophosphatase. What role do pyrophosphatases play in biosynthetic reactions dependent on adenylate transfer?

Structure and Function of Major Components of the Cell

2

PART

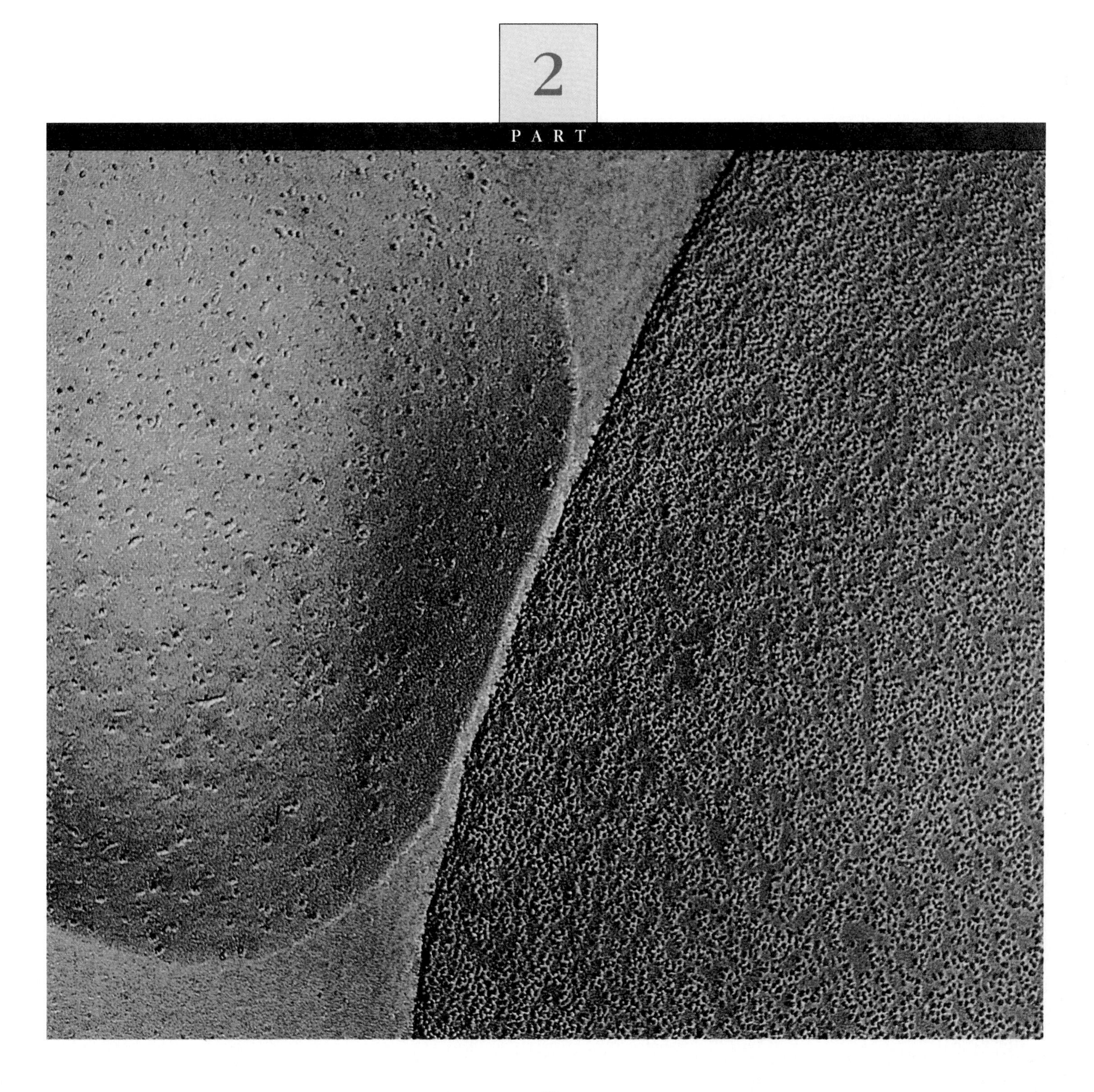

Scanning electron micrograph of human erythrocyte plasma membrane prepared by freeze fracture (120,000X). The micrograph has been color-enhanced to match the normal yellow tinge of blood plasma. When embedded in ice, membranes usually fracture between the monolayer leaflets of the lipid bilayer. The outer fracture face shows a large number of protrusions (presumably proteins); there is a relative lack of such particles on the inner fracture face. (© Harold H. Edwards/Visuals Unlimited.)

The major components of the living cell, other than water, are organic polymers: nucleic acids, proteins, carbohydrates, and lipids. Part 2 describes the structures and functions of three of these molecular species and their building blocks. Nucleic acid structure and function is treated later, with other closely related topics (see chapter 25). However, some instructors may prefer that you turn to these aspects of nucleic acids after the chapter on carbohydrates, treating that material as part of this section.

Proteins are composed of one or more polypeptides, and each polypeptide is composed of many amino acids regularly linked by peptide bonds into long chains. The polypeptide chains aggregate into long fibrous molecules that provide structural support, or into compact globular molecules important in metabolism (chapters 3–5).

Carbohydrates form both linear and branched-chain structures that are sometimes complexed with proteins (chapter 6). In their pure form, carbohydrate polymers with straight- and branched-chain structures store readily usable chemical energy. Linear and cross-linked polysaccharides also form cell walls in bacteria and plants. Short straight- or branched-chain oligosaccharides, linked to asparagine or serine side chains in proteins, usually enhance the recognition properties of proteins. This feature facilitates transport of proteins to an appropriate location in the cell or extracellular milieu and increases their specificity once they get there.

Lipids (chapter 7) are even more varied in their roles than carbohydrates. Some lipids are stored in fat deposits, where they can be called upon to satisfy the energy requirements of the cell when more accessible sources of biochemical energy are exhausted. Other lipids are used in membrane construction. Finally, the members of a select class of complex lipids function as hormones.

3

CHAPTER

The Building Blocks of Proteins: Amino Acids, Peptides, and Polypeptides

n the middle of the nineteenth century, the Dutch chemist Geradus Mulder extracted a substance common to animal tissues and the juices of plants, which he believed to be "without doubt the most important of all substances of the organic kingdom, and without it life on our planet would probably not exist." At the suggestion of the famous Swedish chemist Berzelius, Mulder named this substance protein (from the Greek *proteios,* meaning "of first importance"), and assigned to it a specific chemical formula ($C_{40}H_{62}N_{10}O_{12}$). Although he was wrong about the chemistry of proteins, he was right about their being indispensable to living organisms. The term "protein" endures.

Proteins are the most abundant of cellular components. They include enzymes, antibodies, hormones, transport molecules, and even components for the cytoskeleton of the cell itself. Proteins are also informational macromolecules, the ultimate heirs of the genetic information encoded in the sequence of nucleotide bases within the chromosomes. Structurally and functionally, they are the most diverse and dynamic of molecules and play key roles in nearly every biological process. Proteins are complex macromolecules with exquisite specificity; each is a specialized player in the orchestrated activity of the cell. Together they tear down and build up molecules, extract energy, repel invaders, act as delivery systems, and even synthesize the genetic apparatus itself.

In the first three chapters of part 2 we will discuss the basic structural and chemical properties of proteins. In this chapter we will concentrate on the structural and chemical properties of amino acids, peptides, and polypeptides—the building blocks of proteins. From our presentation you will learn the following:

1. Certain acidic and basic properties are common to all amino acids found in proteins except for the amino acid proline.
2. Side chains give amino acids their individuality. These side chains serve a variety of structural and functional roles.
3. The alpha carboxyl group of one amino acid can react with the alpha amino group of another amino acid to form a dipeptide.
4. Many amino acids, reacting in a similar way, can become linked to form a linear polypeptide chain.
5. The amino acid sequence in a polypeptide can be determined by a process of partial breakdown into manageable fragments, followed by stepwise analysis proceeding from one end of the chain to the other.
6. Polypeptide chains with a prespecified sequence can be synthesized by well-established chemical methods.

Amino Acids

Every protein molecule can be viewed as a polymer of amino acids. There are twenty common amino acids. Figure 3.1*a* shows the structure of a single amino acid. At the center is a tetrahedral carbon atom called the alpha (α) carbon (C_α). It is covalently bonded on one side to an amino group (NH_2) and on the other side to a carboxyl group (COOH). A third bond is always hydrogen, and the fourth bond is to a variable side chain (R). In neutral solution (pH 7), the carboxyl group loses a proton and the amino group gains one. Thus an amino acid in solution, while neutral overall, is a double charged species called a zwitterion (fig. 3.1*b*).

Figure 3.1

Amino acid anatomy. (*a*) Uncharged amino acid. (*b*) Doubly charged zwitterion.

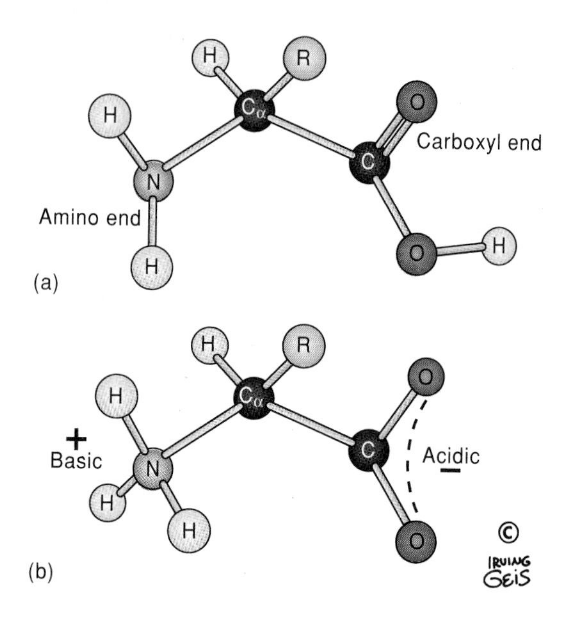

The structures of the twenty amino acids commonly found in proteins are listed in table 3.1. All of these amino acids except proline have an α ammonium ion ($-N^+H_3$) attached to the α carbon. In proline one of the N—H linkages is replaced by an N—C linkage forming part of a cyclic structure. Various ways of classifying amino acids according to their R groups have been proposed. In table 3.1 we have divided the amino acids into three categories. The first category contains eight amino acids with relatively apolar R groups; the second category contains seven amino acids with uncharged polar R groups; and the third category contains five amino acids with R groups that normally exist in the charged state.

Amino acids are often abbreviated by three-letter symbols; when this proves to be too cumbersome (as in certain kinds of charts and figures), one-letter symbols are used. Both designations are given in table 3.1, together with the molecular weight (M_r) of each amino acid.

In addition to the twenty commonly occurring α-amino acids, a variety of other amino acids are found in minor amounts in proteins and in nonprotein compounds. The unusual amino acids found in proteins result from modification of the common amino acids. In a few cases these amino acids are incorporated directly into the polypeptide chains during synthesis (see box 29A for selenocysteine). Most frequently the amino acid is modified after incorporation (see box 18B for the modification of proline to hydroxyproline). The unusual amino acids found in nonprotein compounds are extremely varied in type and are formed by a number of different metabolic pathways (see chapter 18).

Amino Acids Have Both Acid and Base Properties

The charge properties of amino acids are very important in determining the reactivity of certain amino acid side chains and in the properties they confer on proteins. The charge properties of amino acids in aqueous solution may best be considered under the general treatment of acid–base ionization theory. We will find this treatment useful at other points in the text as well.

Recall that water can be considered a weak acid (or a weak base) because it dissociates into a proton and a hydroxide ion, according to the equilibrium

$$H_2O \rightleftharpoons H^+ + OH^- \tag{1}$$

The equilibrium expression for this reaction is

$$K_{eq} = \frac{[H^+][OH^-]}{[H_2O]} \tag{2}$$

Because water dissociates to such a small extent, the concentration of undissociated water is high and does not vary significantly for chemical reactions in aqueous solution. Therefore, the denominator in this equation is effectively constant, with a value of 55.5. The constant K_w for the dissociation of water is redefined by the expression

$$K_w = [H^+][OH^-] = 10^{-14}\,(\text{mole/liter})^2 \tag{3}$$

at 25° C.

In pure water we expect equal amounts of H^+ ("hydrogen ion") and OH^- ("hydroxide ion"). From equation (3) we can calculate the concentration of H^+ or OH^- in pure water to be 10^{-7} M. Therefore, a solution with a H^+ concentration of 10^{-7} M is defined as neutral. A H^+ concentration greater than 10^{-7} M indicates an acidic solution; a H^+ concentration less than 10^{-7} M indicates a basic solution. Rather than deal with exponentials, it is convenient to express the H^+ concentration on a pH scale, the term pH being defined by the equation

$$\text{pH} = \log(1/[H^+]) = -\log[H^+] \tag{4}$$

According to this definition a neutral solution has a pH of 7. Other values of pH and corresponding H^+ and OH^- concentrations are given in table 3.2.

The most common equilibria that biochemists encounter are those of acids and bases. The dissociation of an acid may be written as

$$HA \rightleftharpoons H^+ + A^- \tag{5}$$

The equilibrium constant for this reaction is called the acid dissociation constant, K_a, written as

$$K_a = \frac{[H^+][A^-]}{[HA]} \tag{6}$$

Table 3.1
Structure of the Twenty Amino Acids Found in Proteins

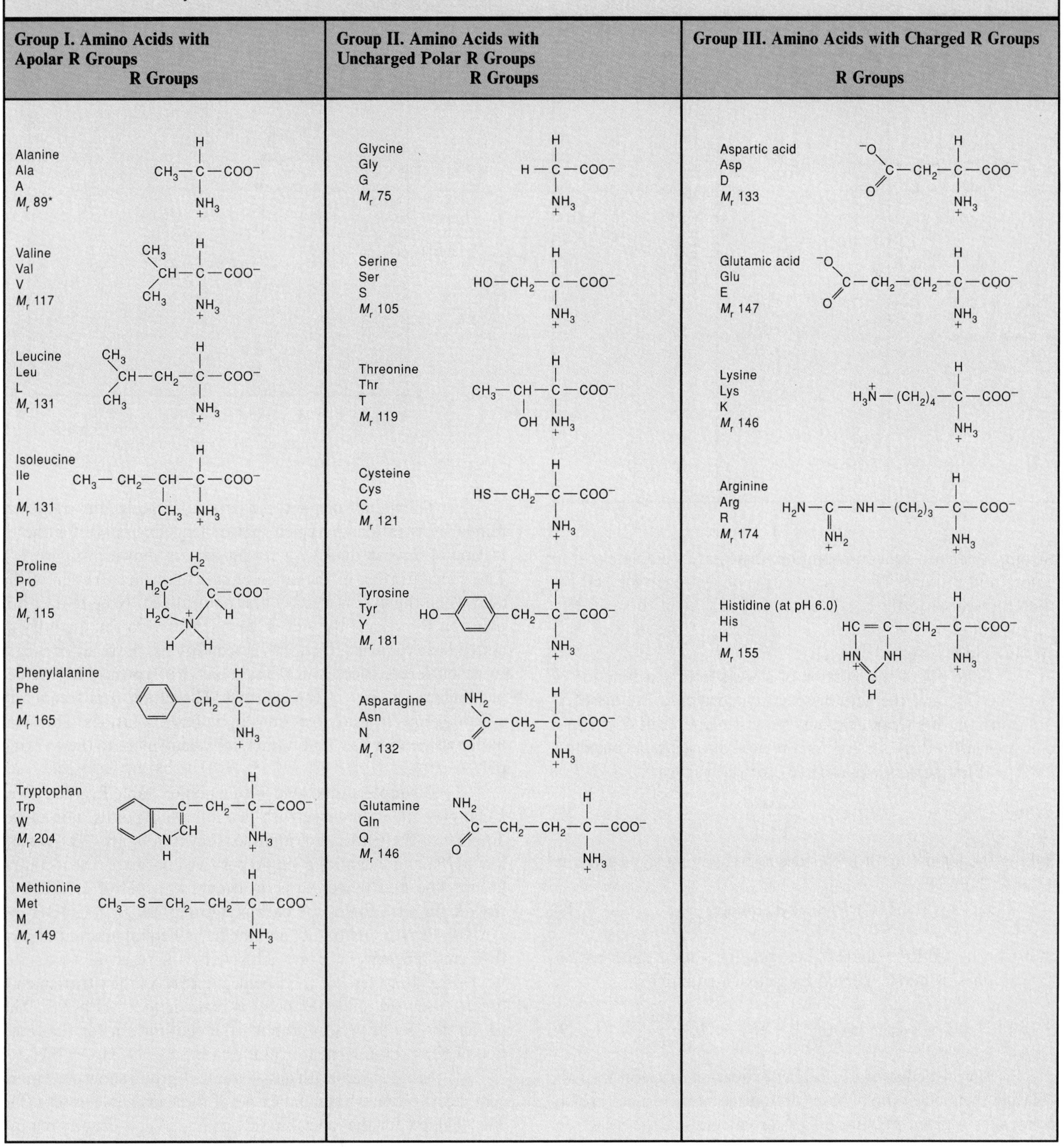

[a]Molecular weights in this text are expressed in units of grams per mole.

Table 3.2
The pH Scale

pH	$[H^+]$	$[OH^-]$
0	10^0	10^{-14}
1	10^{-1}	10^{-13}
2	10^{-2}	10^{-12}
3	10^{-3}	10^{-11}
4	10^{-4}	10^{-10}
5	10^{-5}	10^{-9}
6	10^{-6}	10^{-8}
7	10^{-7}	10^{-7}
8	10^{-8}	10^{-6}
9	10^{-9}	10^{-5}
10	10^{-10}	10^{-4}
11	10^{-11}	10^{-3}
12	10^{-12}	10^{-2}
13	10^{-13}	10^{-1}
14	10^{-14}	10^0

Strong acids in aqueous solution dissociate completely into anions and protons. The concentration of hydrogen ion $[H^+]$ is therefore equal to the total concentration C_{HA} of the acid HA that is added to the solution. Thus the pH of the solution of a strong acid is simply $-\log C_{HA}$.

The pH of the solution of a weak acid is a function of both the C_{HA} and the acid dissociation constant. The dissociation constant of a weak acid may be written in terms of the species present in the equation for the acid dissociation constant.

First solving equation (6) for $[H^+]$ gives

$$[H^+] = \frac{K_a[HA]}{[A^-]} \quad (7)$$

Taking the logarithm of both sides and changing signs gives us

$$-\log[H^+] = -\log K_a + \log\frac{[A^-]}{[HA]} \quad (8)$$

Substituting pH for $-\log[H^+]$ and pK_a for $-\log K_a$ in equation (8), we obtain the Henderson-Hasselbach equation:

$$\text{pH} = \text{p}K_a + \log\left[\frac{A^-}{HA}\right] = \text{p}K_a + \log\left[\frac{\text{base}}{\text{acid}}\right] \quad (9)$$

The Henderson-Hasselbach equation is useful for calculating the molar ratio of base (proton acceptor) to acid (proton donor) for a given pH and pK or for calculating the pK given the ratio of base (proton acceptor) to acid (proton donor). It can be seen that when the concentration of anion or base is equal to the concentration of undissociated acid (i.e., when the acid is half neutralized), the pH of the solution is equal to the pK of the acid.

Figure 3.2

The dependence of pH on the equivalents of base added to a typical weak acid. Note that at the pK_a, $[A^-] = [HA]$.

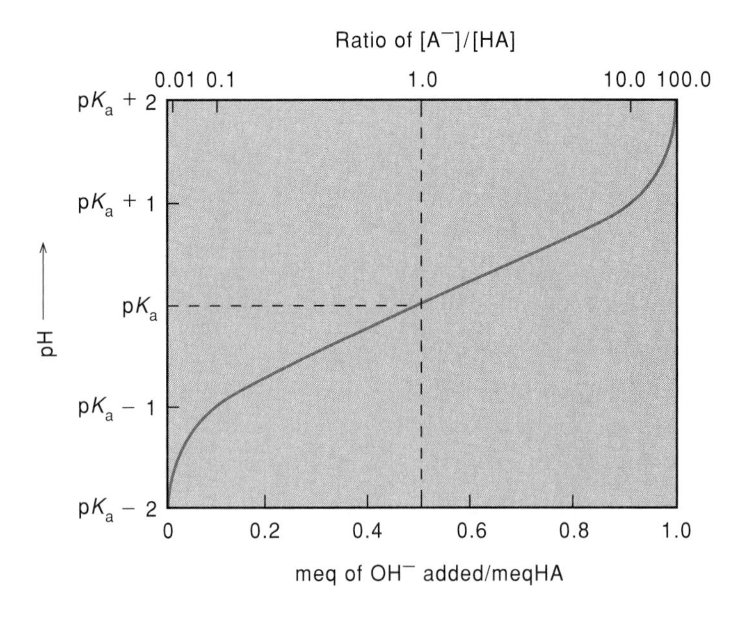

The values of pK for a particular molecule are determined by titration. A typical pH dependence curve for the titration of a weak acid by a strong base is shown in figure 3.2. The concentration of the anion equals the concentration of the acid when the acid is exactly half neutralized. Note that at this point on the curve, the pH is least sensitive to the quantity of added base (or acid). Under these conditions, the solution is said to be buffered. Biochemical reactions are typically highly dependent on the pH of the solution. Therefore, it is frequently advantageous to study reactions in buffered solutions. The ideal buffer is one that has a pK numerically equivalent to the working pH.

A simple amino acid with a nonionizable R group gives a complex titration curve with two inflection points. For an example, see the titration of alanine, shown in figure 3.3. At very low pH, alanine carries a single positive charge on the α-amino group. The first inflection point occurs at a pH of 2.3. This is the pK for titration of the carboxyl group, pK_1 ($-COOH \rightarrow -COO^-$). At a pH of 6.0, alanine has an equal amount of positive and negative charge. This value is referred to as the isoelectric point (pI) or the isoelectric pH. As the titration continues, a second inflection point is reached at a pH of 9.7. The pK at this point, pK_2, represents the equilibrium for the titration of the proton from the amino group ($-N^+H_3 \rightarrow NH_2$).

Amino acids with an ionizable R group show even more complex titration curves, indicative of three ionizable groups (fig. 3.4). The pK for the ionizable side chain, pK_R, is usually readily distinguishable from the pK values for the ionizable α-carboxyl and α-amino groups, pK_1 and pK_2, respectively, as the latter have numerical values close to the comparable pK values of alanine (see fig. 3.3 and table 3.3). Note that the only ionizable R group with a pK_R in the vicinity of 7, where most biological systems

Figure 3.3

Titration curve of alanine. The predominant ionic species at each cardinal point in the titration is indicated.

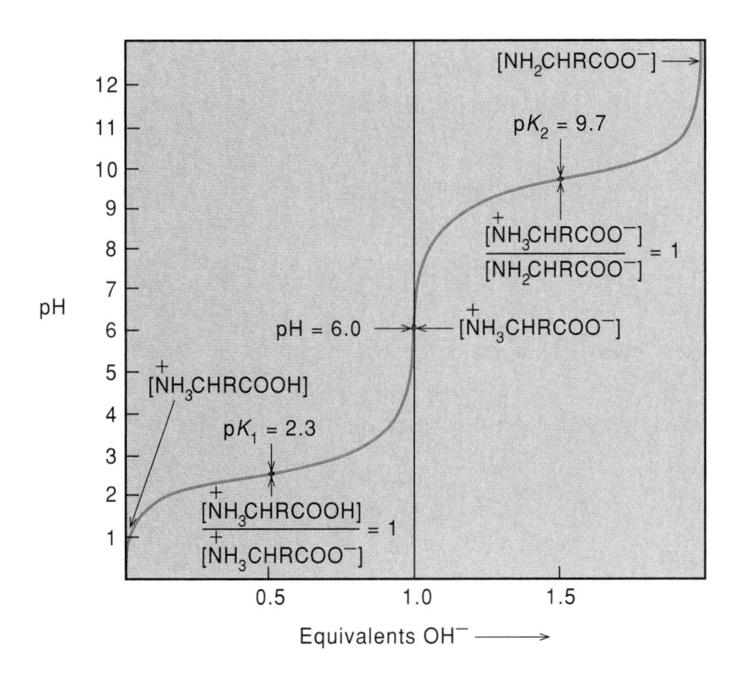

Figure 3.4

Titration curves of glutamic acid, lysine, and histidine. In each case, the pK of the R group is designated pK_R.

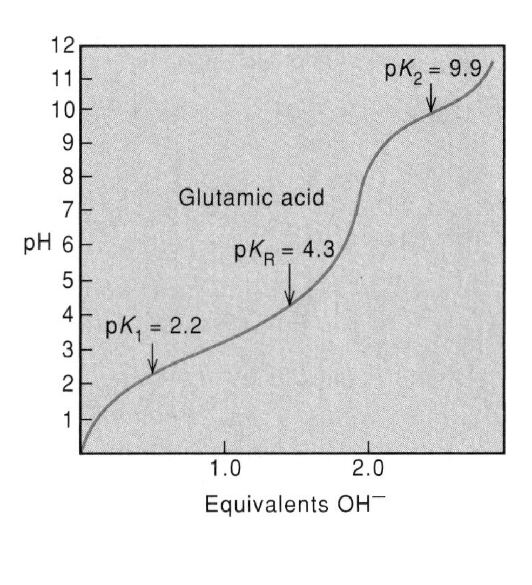

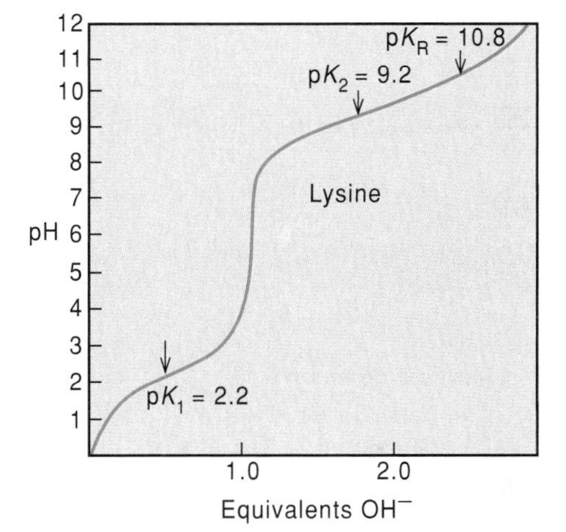

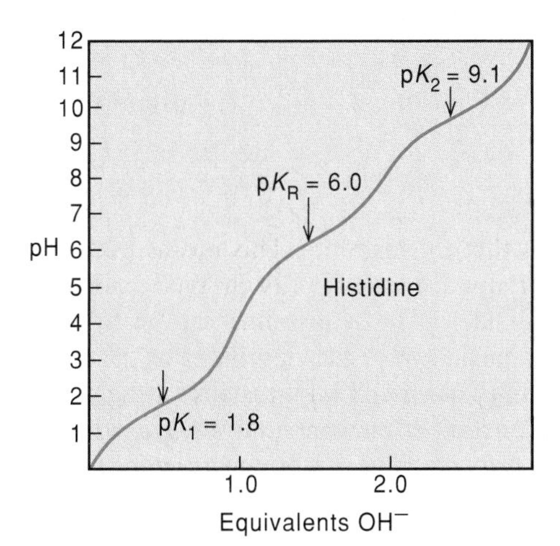

Table 3.3

Values of pK for the Ionizable Groups of the Twenty Amino Acids Commonly Found in Proteins

Amino Acid	pK_1 (α —COOH)	pK_2 (α —NH_3^+)	pK_R (R Group)
Alanine	2.35	9.87	—
Arginine	1.82	8.99	12.48
Asparagine	2.1	8.84	—
Aspartic acid	1.99	9.90	3.90
Cysteine	1.92	10.78	8.33
Glutamic acid	2.10	9.47	4.07
Glutamine	2.17	9.13	—
Glycine	2.35	9.78	—
Histidine	1.80	9.33	6.04
Isoleucine	2.32	9.76	—
Leucine	2.33	9.74	—
Lysine	2.16	9.18	10.79
Methionine	2.13	9.28	—
Phenylalanine	2.16	9.18	—
Proline	1.95	10.65	—
Serine	2.19	9.21	~13
Threonine	2.09	9.10	~13
Tryptophan	2.43	9.44	—
Tyrosine	2.20	9.11	10.13
Valine	2.29	9.74	—

Figure 3.5

Equilibrium between charged and uncharged forms of amino acid side chains.

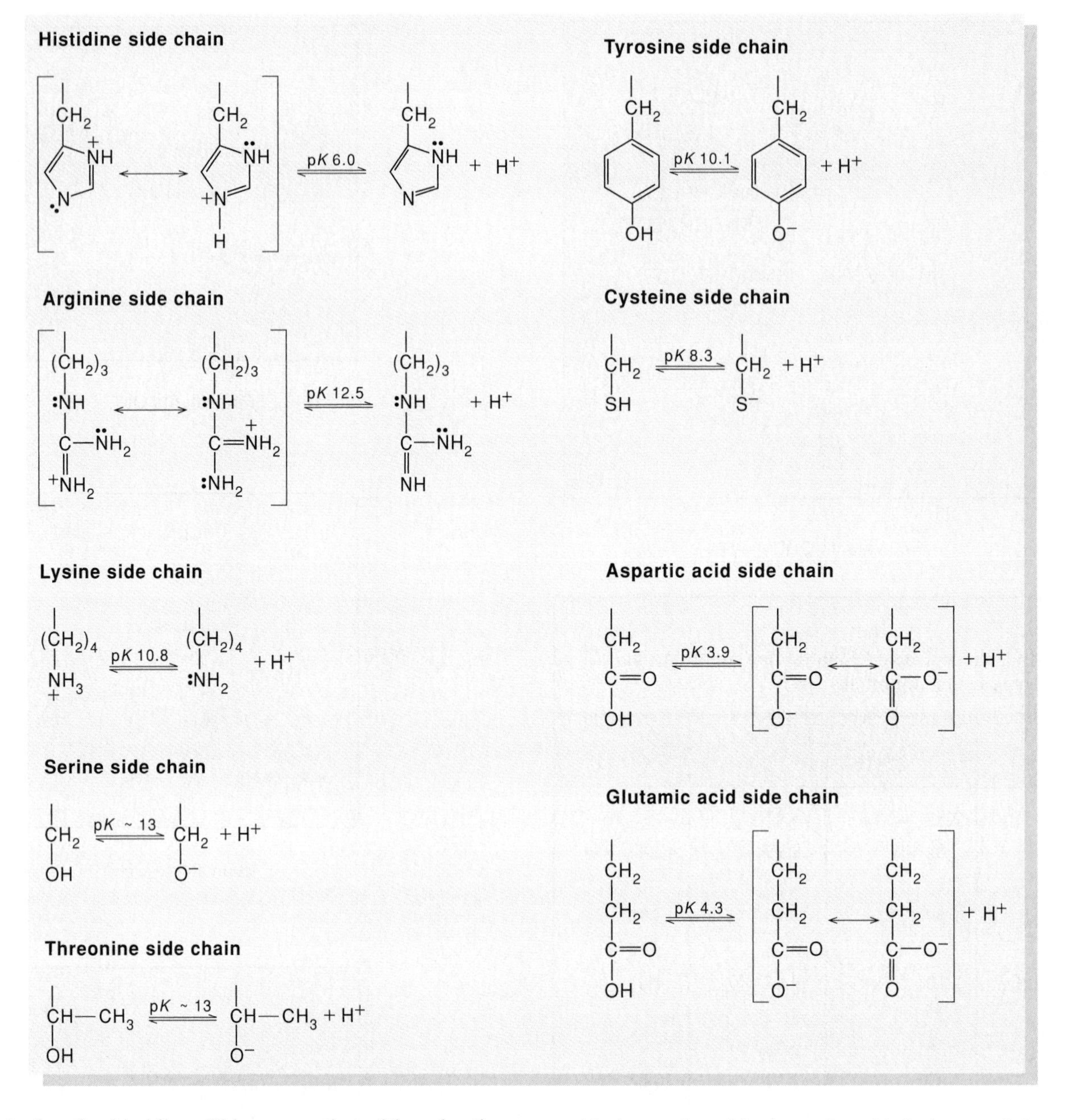

function, is that for histidine. This means that although other ionizable groups are usually fully charged under biological conditions, the side chain of histidine can be fully charged, uncharged, or partially charged, depending on the precise situation. This variability has major implications for the way the histidine side chain functions in enzyme catalysis. The side chain can serve as either a proton donor or a proton acceptor (see discussion in chapter 9).

An additional point should be noted from table 3.3. Whereas the amino acid side chains (R groups) that are normally charged at physiological pH are restricted to five amino acids (aspartic acid, glutamic acid, lysine, arginine, and sometimes histidine), a number of potentially ionizable R groups are part of other amino acids. These include cysteine, serine, threonine, and tyrosine. The ionization reactions for all of the potentially ionizable side chains are indicated in figure 3.5.

The acidic and basic groups within a protein can be titrated just like free amino acids to determine their number and their pK_a values. A titration curve for β-lactoglobulin is shown in figure 3.6. This protein contains 94 potentially ionizable groups. The protein is positively charged at low pH and negatively charged at high pH. At intermediate pH values a

Figure 3.6

Titration curve of ß-lactoglobulin. At very low values of pH (<2) all ionizable groups are protonated. At a pH of about 7.2 (indicated by horizontal bar) 51 groups (mostly the glutamic and aspartic amino acids and some of the histidines) have lost their protons. At pH 12 most of the remaining ionizable groups (mostly lysine and arginine amino acids and some histidines) have lost their protons as well. (Source: R. H. Haschenmeyer and A. E. V. Haschenmeyer, *A Guide to Study by Physical and Chemical Methods.* Copyright ©1973, John Wiley & Sons, Inc., New York, N.Y.)

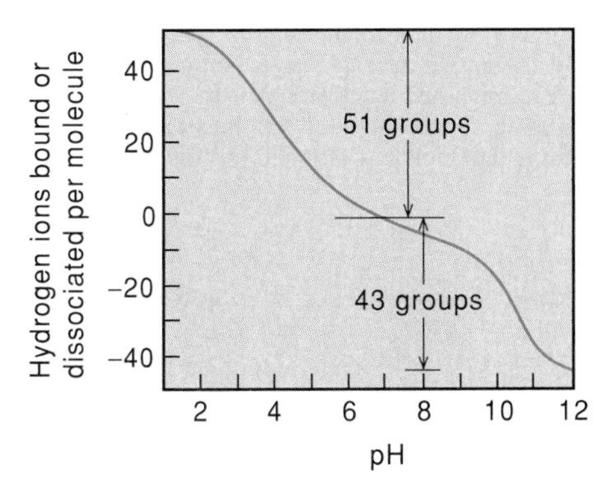

Figure 3.7

Ultraviolet absorption spectra of tryptophan (Trp), tyrosine (Tyr), and phenylalanine (Phe) at pH 6. The molar absorptivity is reflected in the extinction coefficient, with the concentration of the absorbing species expressed in moles per liter. (Source: D. B. Wetlaufer, *Advances in Protein Chemistry.* 17:303-390, 1962. Copyright ©1962 Academic Press Inc., San Diego, Calif.)

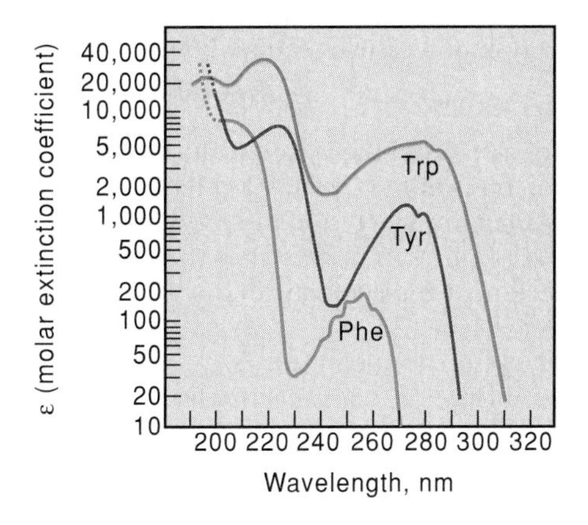

point is found where the sum of the positive side-chain charges exactly equals the sum of the negative charges, so that the net charge on the protein is zero. This value, as we have noted, is the isoelectric point (pI) of the protein; for β-lactoglobulin the pI is about 5.2. The isoelectric point is not an invariant quantity. The binding of charged species present in the solution could raise or lower the pI, depending on their charge.

Aromatic Amino Acids Absorb Light in the Near-Ultraviolet

The aromatic amino acids phenylalanine, tyrosine, and tryptophan all possess absorption maxima in the near-ultraviolet (fig. 3.7). These absorption bands arise from the interaction of radiation with electrons in the aromatic rings. The near-ultraviolet absorption properties of proteins are determined solely by their content of these three aromatic amino acids. In solution, UV absorption can be quantified with the help of a conventional spectrophotometer and used as a measure of the concentration of proteins (see box 3A).

All Amino Acids Except Glycine Show Asymmetry

One of the most striking and significant properties of amino acids is their chirality or handedness. The word "chiral" is related to the Greek word meaning hand. Just as the right hand is related to the left hand by a mirror image, so, in general, a naturally occurring amino acid is related to a stereoisomer by its mirror image. This observation is true of nineteen out of the twenty amino acids; the one exception is glycine.

The chirality of amino acids stems from the chiral or asymmetric center, the α-carbon atom. The α-carbon atom is a chiral center if it is connected to four different substituents. Thus

Figure 3.8

The covalent structure of alanine, showing the three-dimensional structure of the Ⓛ and Ⓓ stereoisomeric forms.

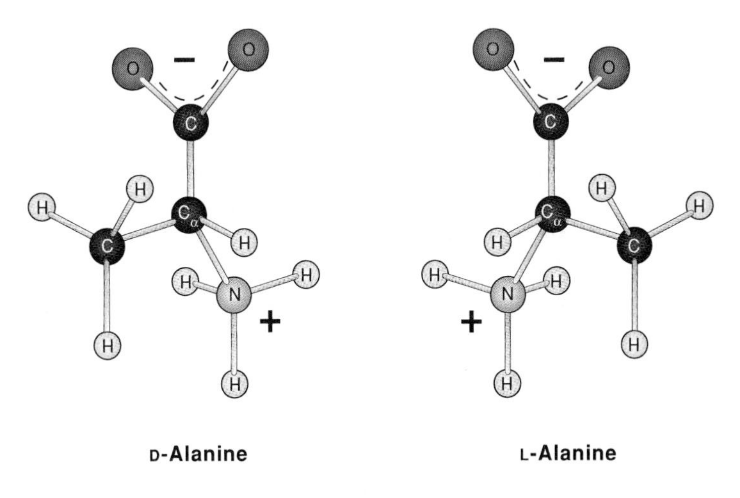

glycine has no chiral center. Two of the amino acids, isoleucine and threonine, possess additional chiral centers because each has one additional asymmetric carbon. You should be able to locate these carbons by simple inspection.

Two structures that constitute a stereoisomeric pair are referred to as enantiomers. The two enantiomers for alanine are illustrated in figure 3.8. These two isomers are called L-alanine and D-alanine, according to the way in which the substituents are arranged about the asymmetric carbon atom. The naming by L and D (for "dextrorotatory" and "levorotatory"; see chapter 6) refers to a convention established by Emil Fischer many years ago. According to this convention all amino acids found in proteins are of the L form. Some D-amino acids are found in bacterial cell walls and certain antibiotics.

Measurement of Ultraviolet Absorption in Solution

The general quantitative relationship that governs all absorption processes is called the Beer-Lambert law:

$$I = I_0\,10^{-\epsilon cd}$$

where I_0 is the intensity of the incident radiation, I is the intensity of the radiation transmitted through a cell of thickness d (in centimeters) that contains a solution of concentration c (expressed either in moles per liter or in grams per 100 ml), and ϵ is the extinction coefficient, a characteristic of the substance being investigated (see fig. 3.7).

Light absorption is measured by a spectrophotometer as shown in the illustration. The spectrophotometer usually is capable of directly recording the absorbance A, which is related to I and I_0 by the equation

$$A = \log_{10}(I_0/I)$$

Hence $A = \epsilon cd$, and A is a direct measure of concentration. We can see from figure 3.7 that the ϵ values are largest for tryptophan and smallest for phenylalanine.

Since protein absorption maxima in the near-ultraviolet (240–300 nm) are determined by the content of the aromatic amino acids and their respective values, most proteins have absorption maxima in the 280-nm region. By contrast, absorption in the far-ultraviolet (around 190 nm) is shown by all polypeptides regardless of their aromatic amino acid content. The reason is that absorption in this region is due primarily to the peptide linkage.

Figure 1

Schematic diagram of a spectrophotometer for measuring light absorption. Laboratory instruments for making measurements are much more complex than this, but they all contain the same basic components: a light source, a monochromator, a sample, and a detector. λ is the wavelength of the light, I_0 and I are the incident light intensity and the transmitted light intensity, respectively, and d is the thickness of the absorbing solution.

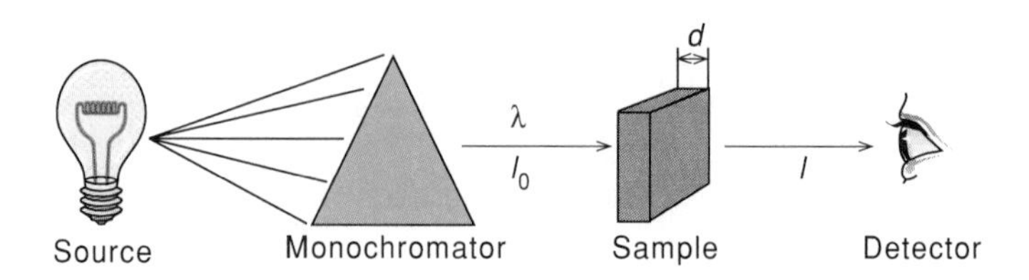

Another convention for referring to configurations is called the R, S convention. As the R, S convention is not as popular for amino acids or sugars as it is for other types of biomolecules, such as lipids, we will not discuss this notation until chapter 11 (see box 11A).

Peptides and Polypeptides

Amino acids can link together by a covalent peptide bond between the α-carboxyl end of one amino acid and the α-amino end of another. Formally, this bond is formed by the loss of a water molecule, as shown in figure 3.9. The peptide bond has partial double-bond character owing to resonance effects; as a result, the C—N peptide linkage and all of the atoms directly connected to C and N lie in a planar configuration called the amide plane. In the following chapter we will see that this amide plane, by limiting the number of orientations available to the polypeptide chain, plays a major role in determining the three-dimensional structures of proteins.

Any number of amino acids can be joined by successive peptide linkages, forming a polypeptide chain. The polypeptide chain, like the dipeptide, has a directional sense. One end, called the N-terminal or amino-terminal end, has a free α-amino group, whereas the other end, the C-terminal or carboxyl-terminal end, has a free α-carboxyl group. The sequence of main-chain atoms from the N-terminal end to the C-terminal end is C_α—C—N—C_α, etc., and in the opposite direction it is C_α—N—C—C_α, etc. Short polypeptide chains, up to a length of about 20 amino acids, are called peptides or oligopeptides if they are fragments of whole polypeptide chains. A small protein molecule may contain a polypeptide chain of only 50 amino acids; a large protein may contain chains of 3,000 amino acids or more. One of the larger single polypeptide chains is that of the muscle protein myosin, which consists of approximately 1,750 amino acid residues. Figure 3.10 shows a section of a polypeptide chain as a linear array with α carbons and planar amides alternating as repeating units of the main chain. Different side chains are attached to each α carbon.

In addition to the covalent peptide bonds formed between adjacent amino acids within a polypeptide chain, covalent disulfide bonds can be formed within the same polypeptide chain or between different polypeptide chains (fig. 3.11). Such disulfide linkages have an important stabilizing influence on the structures formed by many proteins (see chapter 4).

Figure 3.9

Formation of a dipeptide from two amino acids. (*a*) Two amino acids. (*b*) A peptide bond (CO—NH) links amino acids by joining the α-carboxyl group of one with the α-amino group of another. A water molecule is lost in the reaction. It is conventional to draw dipeptides and polypeptides so that their free amino terminal is to the left and their free carboxyl terminal is to the right. The amide plane refers to six atoms that lie in the same plane.

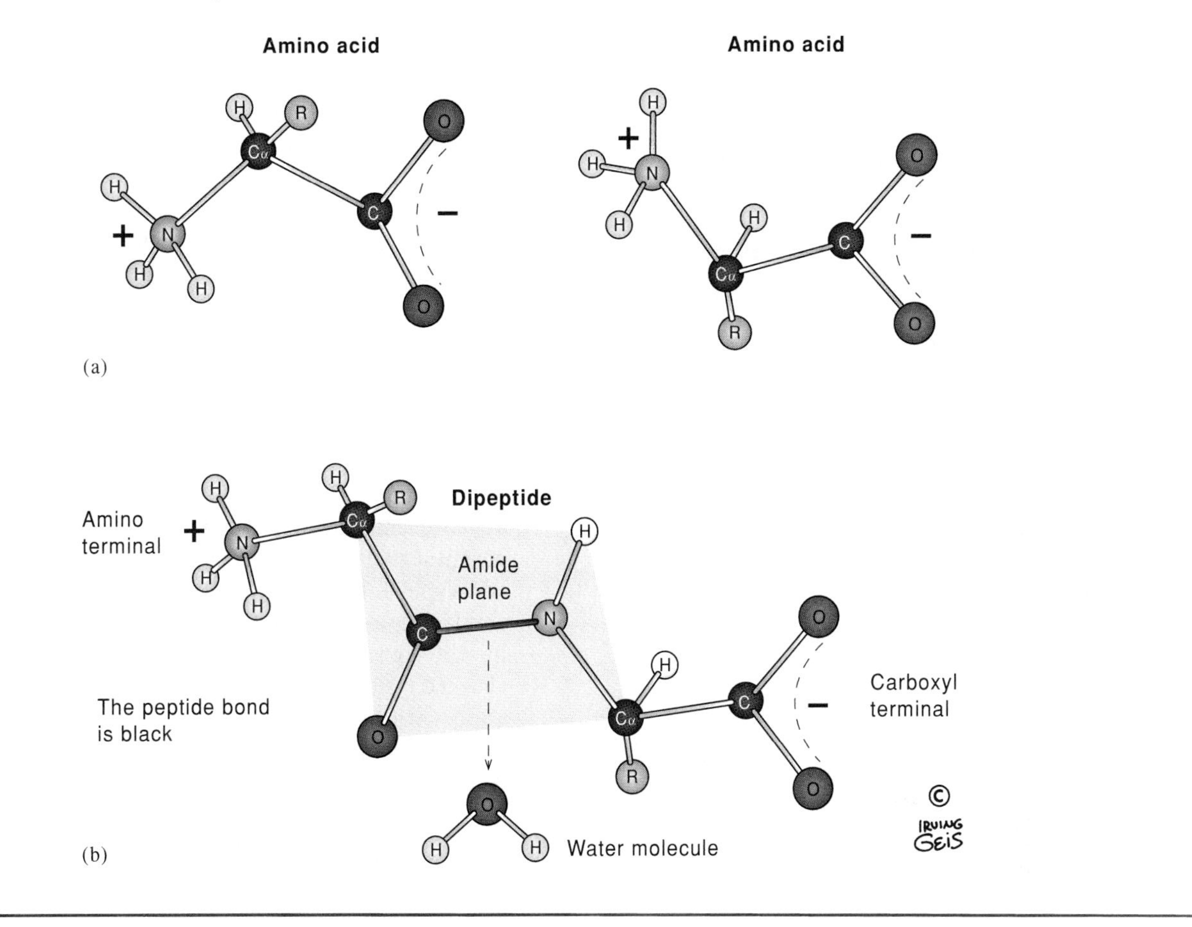

Figure 3.10

A polypeptide chain, with the backbone shown in color and the amino acid side chains in outline.

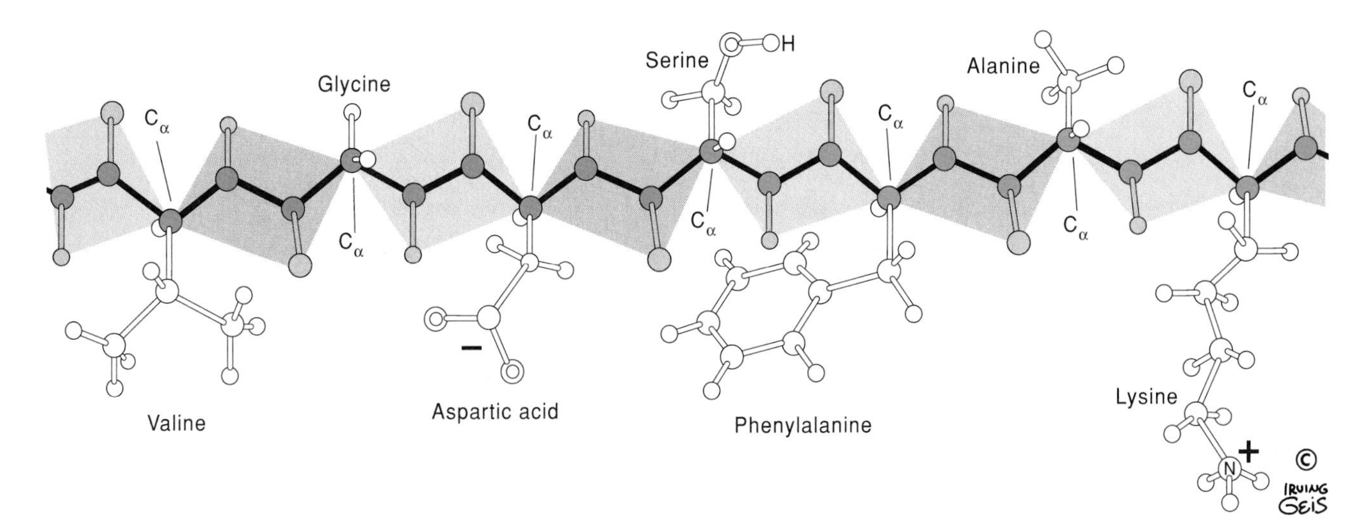

Determination of Amino Acid Composition of Proteins

Each protein is uniquely characterized by its amino acid composition and sequence. A protein's amino acid composition is defined simply as the number of each type of amino acid composing the polypeptide chain. In order to discover a protein's amino acid composition it is necessary to (1) break down the polypeptide chain into its constituent amino acids, (2) separate the resulting free amino acids according to type, and (3) measure the quantities of each amino acid.

Figure 3.11

Disulfide bonds can form between two cysteines. The cysteines can exist in the cytosol as free amino acids (as shown), in which case they give rise to cystine, or they can be on polypeptide chains. In the latter instance, they can be on the same polypeptide chains or different polypeptide chains. In either case the formation of covalent disulfide bonds stabilizes structural relationships.

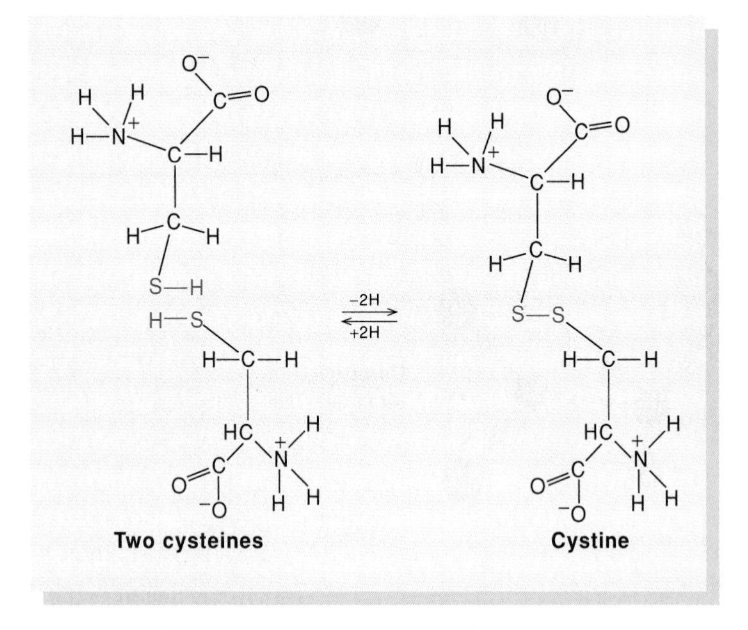

Cleavage of the peptide bonds is usually achieved by boiling the protein in 6-*N* HCl; this treatment causes hydrolysis of the peptide bonds and the consequent release of free amino acids (fig. 3.12). Although acid hydrolysis is the most frequently used means of breaking a protein into its constituent amino acids, it results in the partial destruction of the indole ring of tryptophan. Consequently, the amount of tryptophan in the protein must be estimated by an alternative method (e.g., spectroscopic absorption) when using acid hydrolysis. In addition, acid hydrolysis results in the loss of ammonia from the side-chain amide groups of glutamine and asparagine, with the consequent production of glutamic and aspartic acids (fig. 3.13). Therefore, estimates of amino acid composition based on acid hydrolysis show glutamine and glutamic acid combined and measured as glutamic acid. Similarly, asparagine and aspartic acid are combined and measured as aspartic acid.

Separation of amino acids for quantitative analytical purposes is usually achieved by ion-exchange chromatography. The general efficacy of chromatographic techniques is based on a difference in affinity between each compound to be separated and an immobile phase or resin. The resin consists of some relatively chemically inert polymer that has weakly basic side-chain constituents that are positively charged at pH 7. If we were to add some of this resin to a solution containing free aspartic acid and lysine at pH 7, the negatively charged aspartic acid would have a higher affinity for the resin than would the positively charged lysine. If we were then to pump a solution of these two amino acids through a column containing such a positively charged resin, the progress of the aspartic acid through the column would be retarded relative to the lysine, owing to the greater affinity of the aspartic acid for the resin (fig. 3.14).

Ion-exchange resins have been developed that have differential binding affinities for all the naturally occurring amino acids. Such resins are effective in separating a solution of amino

Figure 3.12

Acid hydrolysis of a protein or polypeptide to yield amino acids.

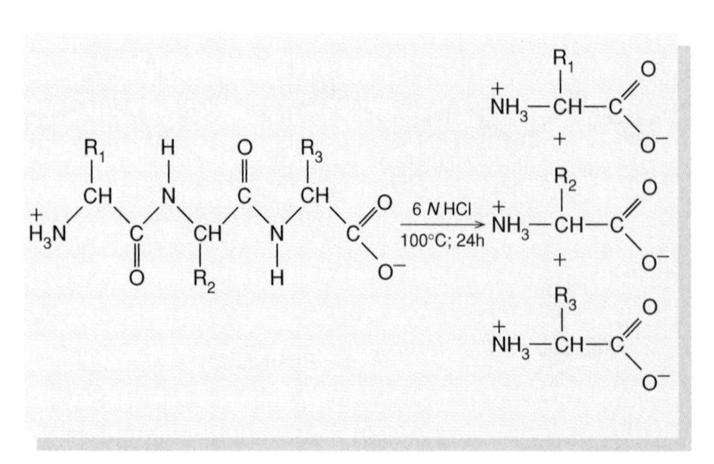

Figure 3.13

Acid hydrolysis of protein converts glutamine to glutamic acid. A similar reaction occurs for asparagine, which possesses an identical side-chain functional group.

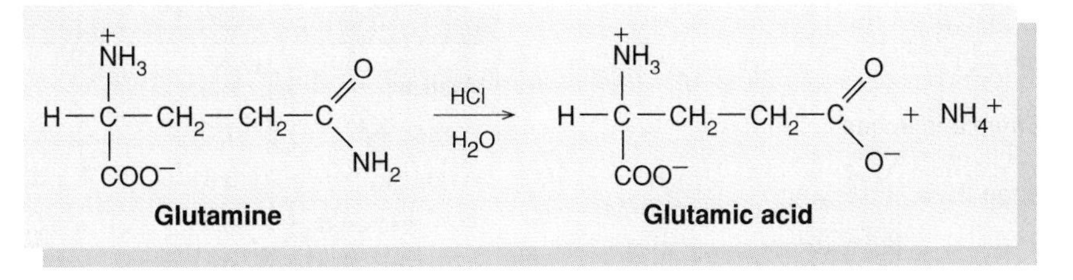

Figure 3.14

Migration of aspartic acid ⊖ and lysine ⊕ through a column with a higher affinity for aspartic acid. Views show the column at successively increasing time intervals after starting the elution.

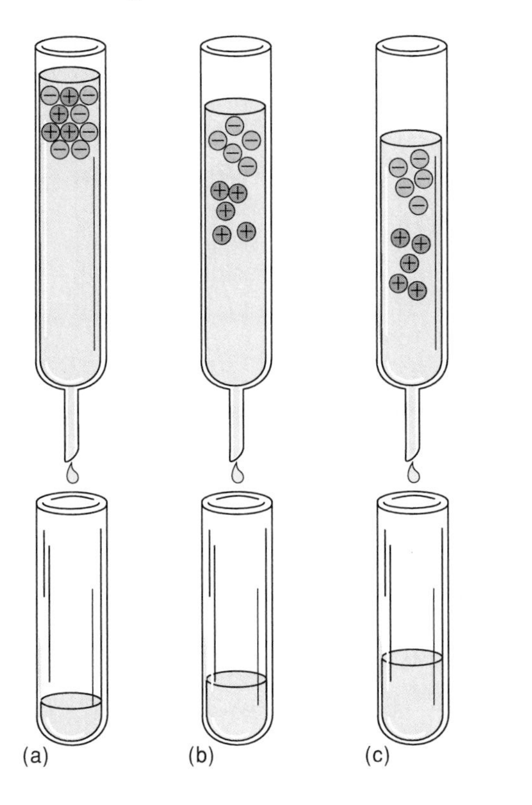

acids into its components. We must emphasize that the details of the forces responsible for the differential binding of amino acids to an ion-exchange resin are quite complicated and depend additionally on side-chain polarity, on subtle differences in the p*K* values of α-amino and α-carboxyl groups, on solvation effects, and on other factors. To enhance the separation properties of the column, such separation techniques frequently exploit changes in the pH of the solution buffer (eluting buffer) used to remove the compounds of interest. For example, a column might initially be run with the eluting buffer at a pH that results in some amino acids being so strongly bound to the resin that they are essentially immobile. However, after the separation and elution of the less strongly bound amino acids, the pH of the eluting buffer can be appropriately shifted to lessen the charge difference between the resin and the strongly bound amino acids. These amino acids can then be eluted and separated according to the newly established pattern of resin-binding affinities.

Quantitative determination of the separated amino acids is achieved by their reaction with ninhydrin to produce a colored reaction product. This product is measured spectrophotometrically. As shown in figure 3.15, the ninhydrin reaction abstracts an amino group from each amino acid, so that the amount of colored product formed is proportional to the amount of amino acid initially present.

Quantitative measurements of amino acid composition are usually carried out on an amino acid analyzer, a device that automates the previously described operations. As illustrated in

Figure 3.15

Reaction of ninhydrin with an amino acid yields a colored complex. The ninhydrin reaction permits qualitative location of amino acids in chromatography and quantitative assay of separated amino acids.

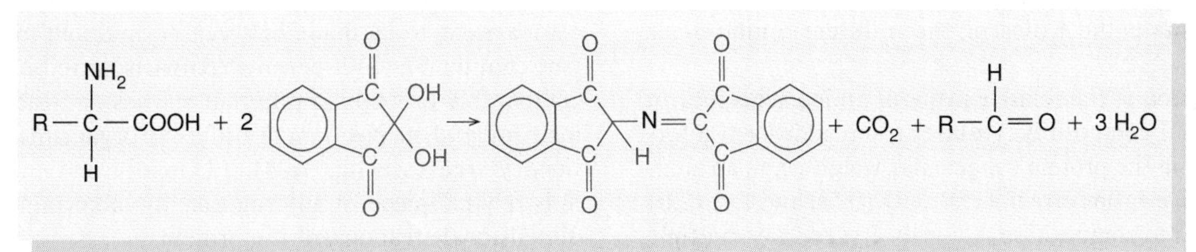

Figure 3.16

Schematic diagram of an amino acid analyzer. The amino acids are passed through an ion-exchange column and thereby separated. Eluted fractions are mixed and reacted with ninhydrin. The intensity of the resulting colored product is measured in a spectrophotometer and the results are displayed on a recording chart.

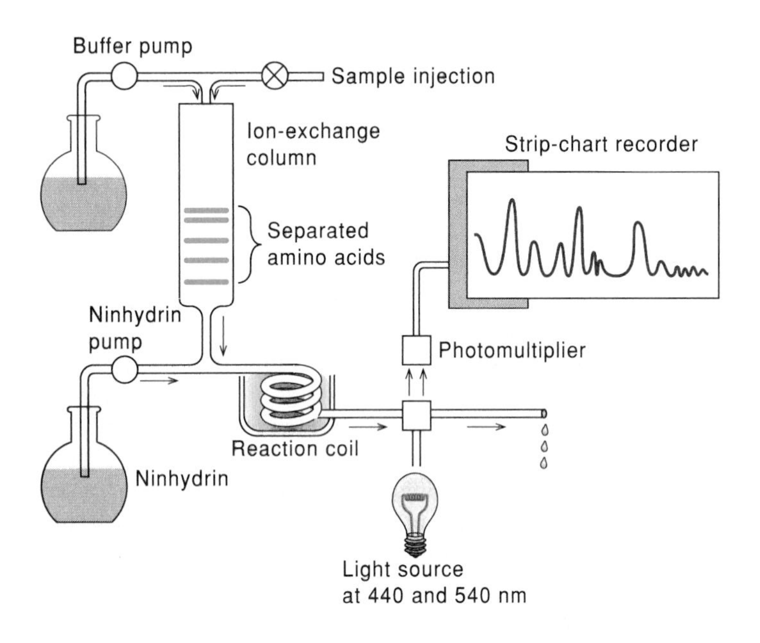

figure 3.16, the amino acid analyzer consists of an ion-exchange column through which the appropriate eluting buffer is pumped after the amino acids are introduced at the top of the column. As the separated amino acids emerge, they are mixed with ninhydrin solution and passed through a heated coil of tubing to allow the formation of the colored ninhydrin reaction product. The separated ninhydrin reaction products then pass through a cell that measures their optical absorbance at 540 and 440 nm and plots the results on a strip-chart recorder. The absorbance is measured at two wavelengths because proline, which is substituted at its amino group, forms a different ninhydrin reaction product, with an absorption maximum that is correspondingly different from that of the remaining amino acids.

Usually the amino acid analyzer is first standardized by running through it a sample containing known quantities of amino acids, in order to account for any differences in their ninhydrin reaction properties. In this way it is possible to directly relate the amount of amino acid present to the amount of colored product formed, as measured by the area under the "peak" produced on the strip-chart recorder (see fig. 3.16). Similarly, the amino acid hydrolysate of a protein of unknown composition can be run through the analyzer, and the relative peak areas can be used to estimate the ratios of the different amino acids present.

Conversion of the relative ratios of amino acids into an estimate of actual composition requires some additional information concerning the protein's molecular weight; e.g., an analysis giving relative ratios of Ala (1.0), Gly (0.5), and Lys (2.0) could correspond to composition Ala_2-Gly-Lys_4 or any multiple thereof. The required information is usually available, and in any case, an estimation of composition based on a minimum molecular weight of the protein is always possible. Results for three proteins are shown in table 3.4.

Table 3.4
Amino Acid Content of Proteins (in percent)

Constituent	Insulin (Bovine)	Ribonuclease (Bovine)	Cytochrome (Equine)
Alanine	4.6	7.7	3.5
Amide NH_3	1.7	2.1	1.1
Arginine	3.1	4.9	2.7
Aspartic acid	6.7	15.0	7.6
Cysteine	0	0	1.7
Cystine	12.2	7.0	0
Glutamic acid	17.9	12.4	13.0
Glycine	5.2	1.6	5.6
Histidine	5.4	4.2	3.4
Isoleucine	2.3	2.7	5.4
Leucine	13.5	2.0	5.6
Lysine	2.6	10.5	19.7
Methionine	0	4.0	2.1
Phenylalanine	8.6	3.5	4.5
Proline	2.1	3.9	3.3
Serine	5.3	11.4	0
Threonine	2.0	8.9	8.4
Tryptophan	0	0	1.5
Tyrosine	12.6	7.6	4.9
Valine	9.7	7.5	2.4

Determination of Amino Acid Sequence of Proteins

The most important properties of a protein are determined by the sequence of amino acids in the polypeptide chain. This sequence is called the primary structure of the protein. We know the sequences for thousands of peptides and proteins, largely through the use of methods developed in Fred Sanger's laboratory and first used to determine the sequence of the peptide hormone insulin in 1953. Knowledge of the amino acid sequence is extremely useful in a number of ways: (1) it permits comparisons to be made between normal and mutant proteins (see chapter 5); (2) it permits comparisons to be made between comparable proteins in different species and thereby has been instrumental in positioning different organisms on the evolutionary tree (see fig. 1.24); (3) finally and most important, it is a vital piece of information for determining the three-dimensional structure of the protein.

Figure 3.17

Steps involved in the sequence determination of the B chain of insulin. Amino acids are represented here by their single-letter codes (see table 3.1).

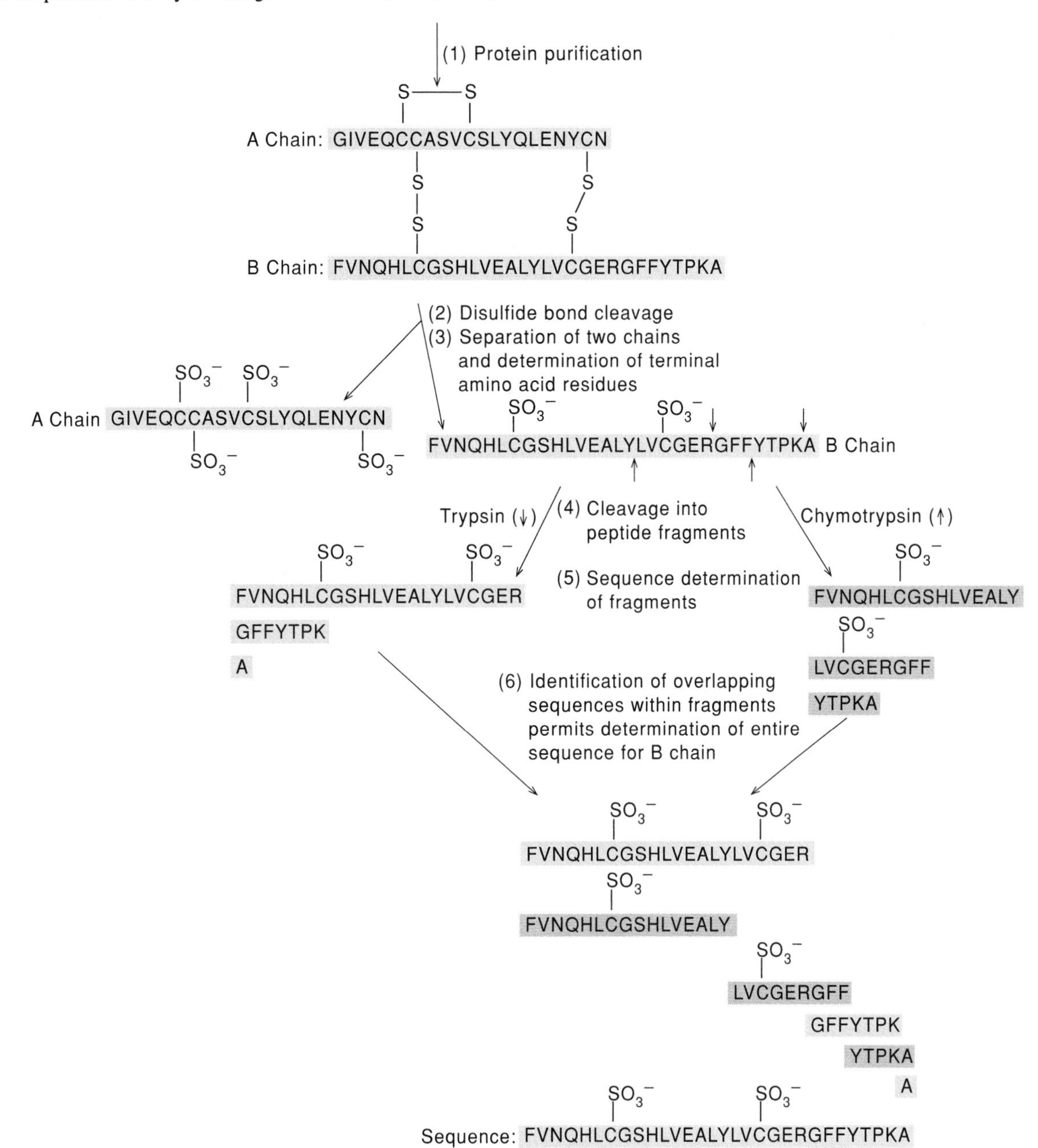

Determining the order of amino acids involves the sequential removal and identification of successive amino acid residues from one or the other free terminal of the polypeptide chain. However, in practice it is extremely difficult to get the required specific cleavage reaction of the desired products to proceed with 100% yield. This obstacle becomes significant when sequencing long polypeptides, because the fraction of the total material of minimum polypeptide chain length becomes constantly smaller as the successive removal of terminal residues continues. Conversely, the amino acid released from the polypeptide chain becomes increasingly contaminated with amino acids released from previously unreacted chains.

Because of this fundamental chemical limitation, the polypeptide chain must be broken down into sequences short enough for the chemistry to produce reliable results. The short sequences are then reassembled to obtain the overall sequence. The steps actually involved in protein sequencing (fig. 3.17) are (1) purification of the protein; (2) cleavage of all disulfide

Figure 3.18

Disulfide cleavage reactions. Prior to sequence analysis inter- and intrachain disulfide linkages are irreversibly cleaved by one of the two procedures shown.

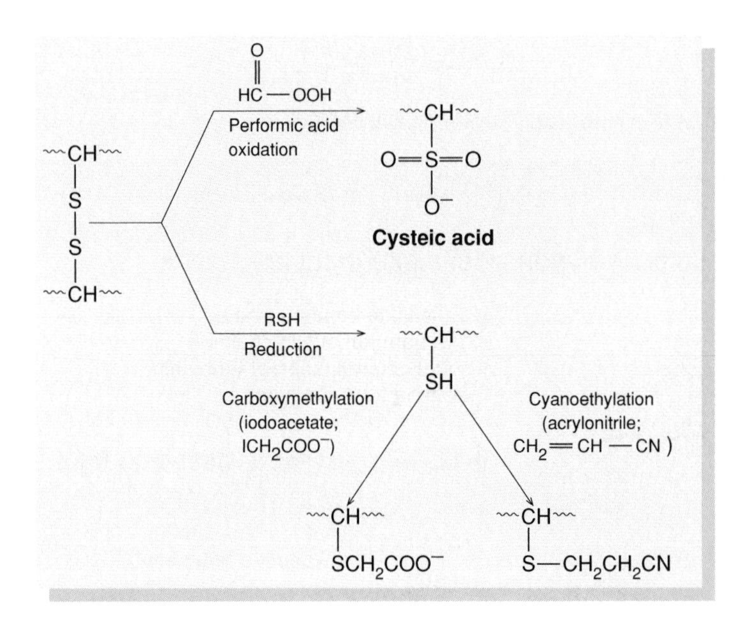

Figure 3.19

Polypeptide chain end-group analysis. (*a*) Amino-terminal group identification. A more sensitive method, the dansyl chloride method, is described in box 3B. (*b*) Carboxyl-terminal group indentification. Identification of this amino acid is considerably more difficult.

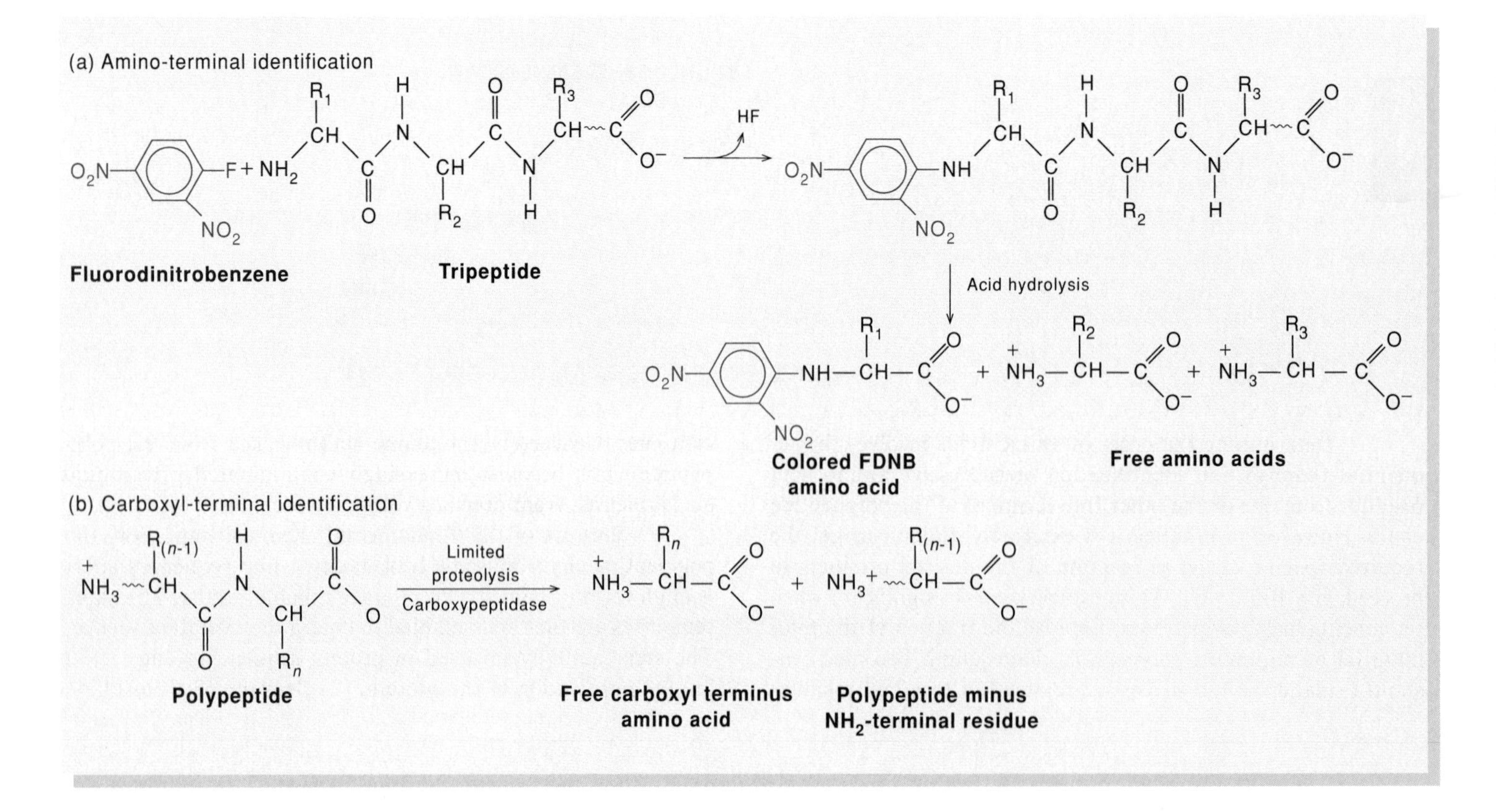

Figure 3.20

Site of action of some endopeptidases used for polypeptide chain cleavage prior to sequence analysis. Of the four different enzymes used, trypsin is used most frequently because of its high specificity.

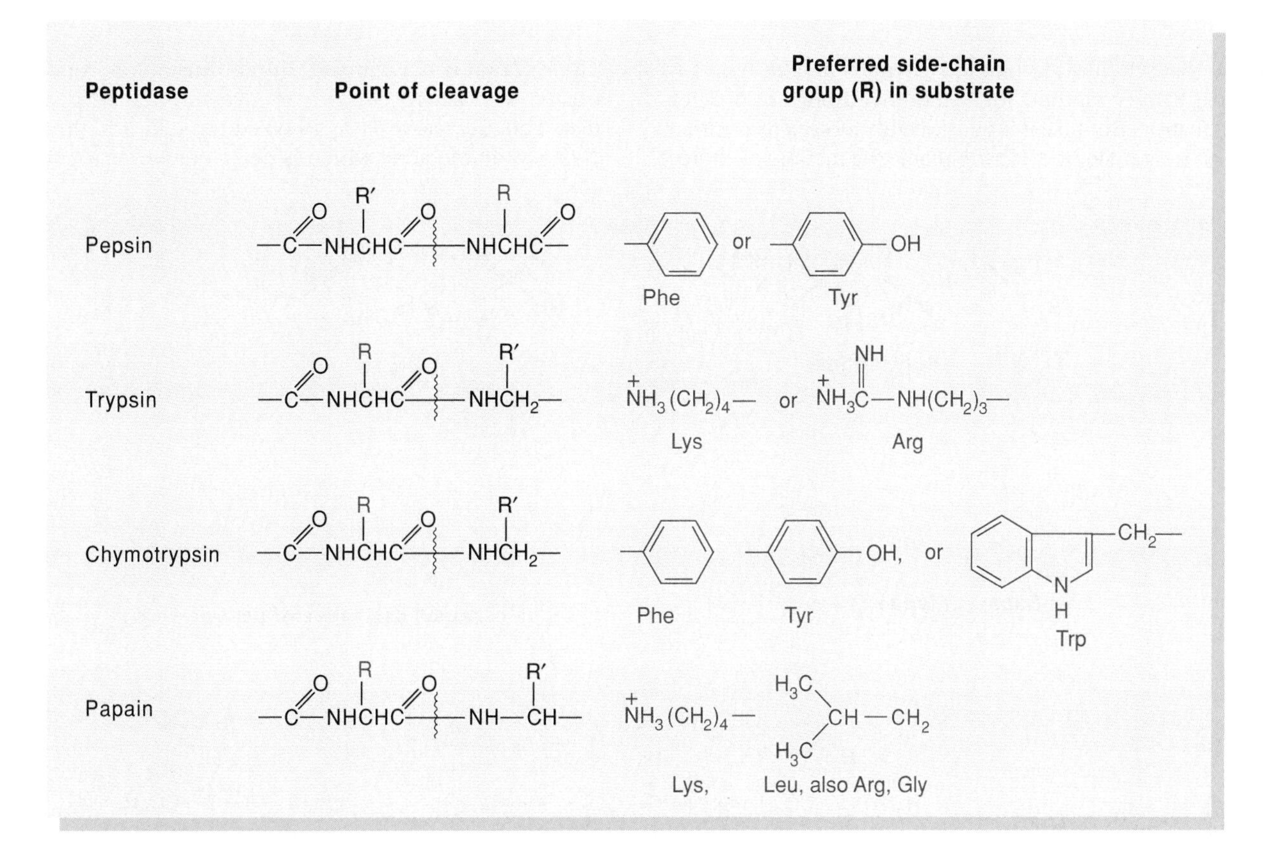

bonds; (3) determination of the terminal amino acid residues; (4) specific cleavage of the polypeptide chain into small fragments in at least two different ways; (5) independent separation and sequence determination of peptides produced by the different cleavage methods; and (6) reassembly of the individual peptides with appropriate overlaps to determine the overall sequence.

The first step, protein purification, will be discussed in chapter 5. Once the protein is pure, sequence analysis can begin, with cleavage of the disulfide bonds. Cleavage is achieved by oxidizing the disulfide linkages with performic acid (fig. 3.18). Sometimes this step results in the production of two or more polypeptide chains, in which case the individual chains must be separated.

The third step is to determine the polypeptide-chain end groups. If the polypeptide chains are pure, then only one N-terminal and one C-terminal group should be detected. The amino-terminal amino acid can be identified by reaction with fluorodinitrobenzene (FDNB) (fig. 3.19). Subsequent acid hydrolysis releases a colored DNP-labeled amino-terminal amino acid, which can be identified by its characteristic migration rate on thin-layer chromatography or paper electrophoresis. A more sensitive method of end-group determination involves the use of dansyl chloride (see box 3B).

Chemical methods for carboxyl end-group determination are considerably less satisfactory. Treatment of the peptide with anhydrous hydrazine at 100° C results in conversion of all the amino acid residues to amino acid hydrazides except for the carboxyl-terminal residue, which remains as the free amino acid and can be isolated and identified chromatographically. Alternatively, the polypeptide can be subjected to limited breakdown (proteolysis) with the enzyme carboxypeptidase. This results in release of the carboxyl-terminal amino acid as the major free amino acid reaction product. The amino acid type can then be identified chromatographically.

Step 4 involves breaking down the polypeptide chain into shorter, well-defined fragments for subsequent sequence analysis. Fragmentation can be achieved by the use of endopeptidases, which are enzymes that catalyze polypeptide-chain cleavage at specific sites in the protein. Figure 3.20 shows the specificity of four endopeptidases commonly used for this purpose. Another specific chemical method for polypeptide-chain

The Dansyl Chloride Method for N-Terminal Amino Acid Determination

The dansyl chloride method provides an alternative to the Sanger method for N-terminal amino acid determination. Because it is considerably more sensitive than the Sanger method, it has become the method of choice. The reaction is diagrammed in the illustration. A polypeptide is treated with dansyl chloride to give an *N*-dansyl peptide derivative. This derivative is hydrolyzed to yield a highly fluorescent *N*-dansyl-amino acid, which is detected chromatographically.

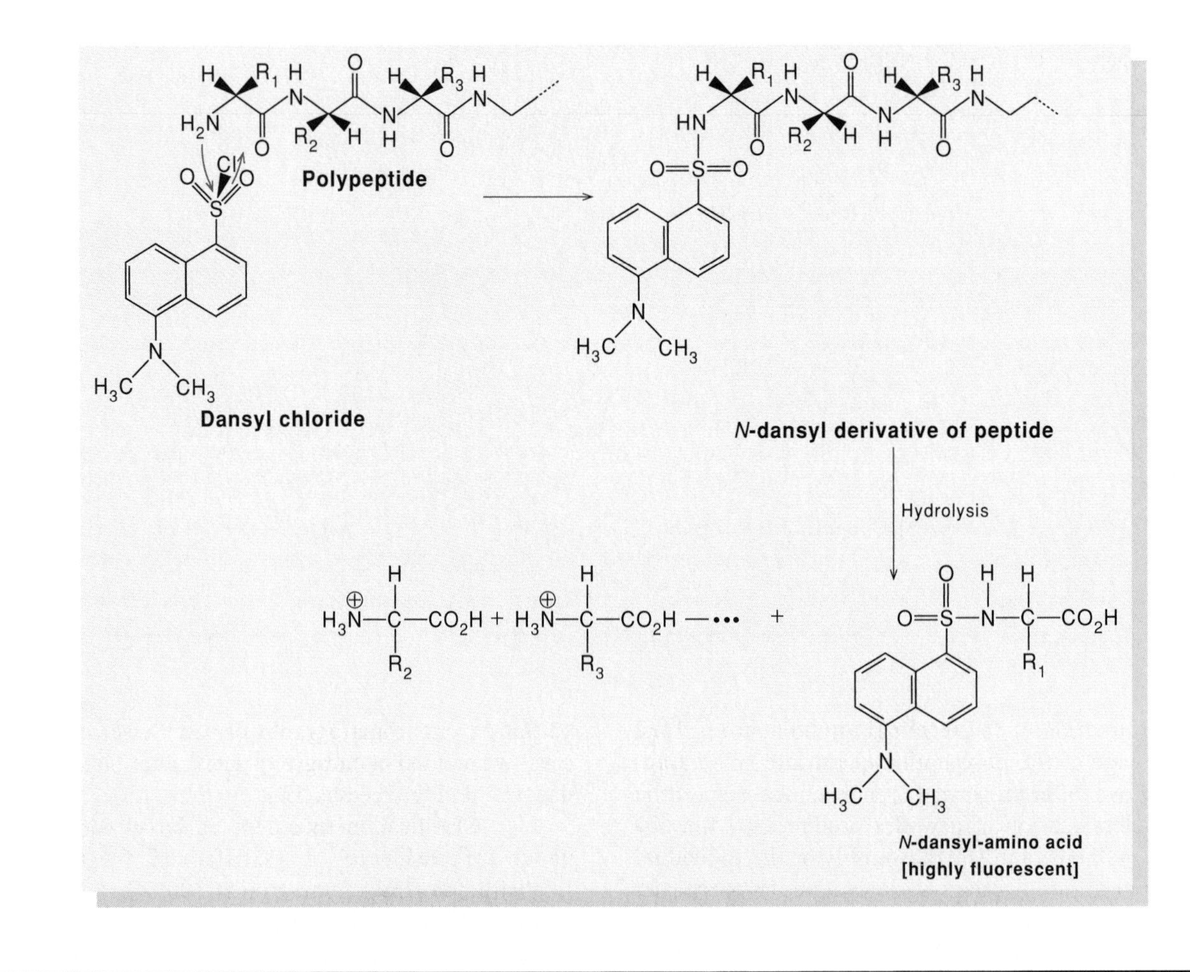

cleavage involves reaction with cyanogen bromide. This reaction cleaves specifically at the methionine residues, with the accompanying conversion of free carboxyl-terminal methionine to homoserine lactone (fig. 3.21). Although this methionine reaction product differs from the twenty naturally occurring amino acids, it is nevertheless readily identified by subsequent conversion to homoserine.

Peptides resulting from cleavage of the intact protein are generally separated by ion-exchange chromatographic methods. The isolated peptides may then be analyzed (step 5) to determine both their amino acid composition and their sequence. Sequence determination involves the stepwise removal and identification of successive amino acids from the polypeptide amino terminal by means of the Edman degradation (fig. 3.22). This process is carried out by reacting the free amino-terminal group with phenylisothiocyanate to form a peptidyl

Figure 3.21

The cleavage of polypeptide chains at methionine residues by cyanogen bromide. The cleavage reaction is accompanied by the conversion of the newly formed free carboxyl-terminal methionine to homoserine lactone.

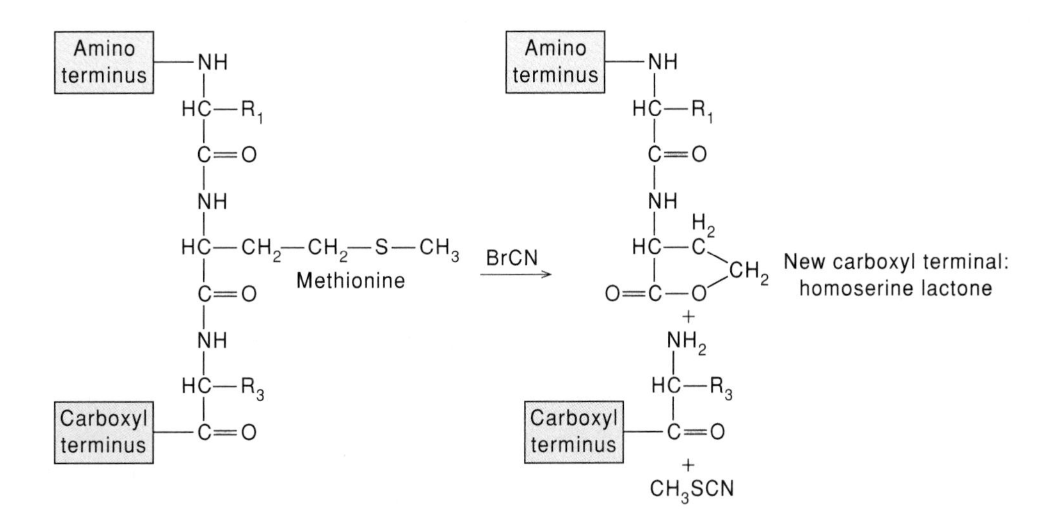

Figure 3.22

The Edman degradation method for polypeptide sequence determination. The sequence is determined one amino acid at a time, starting from the amino-terminal end of the polypeptide. First the polypeptide is reacted with phenylisothiocyanate to form a polypeptidyl phenylthiocarbamoyl derivative. Gentle hydrolysis releases the amino-terminal amino acid as a phenylthiohydantoin (PTH), which can be separated and detected spectrophotometrically. The remaining intact polypeptide, shortened by one amino acid, is then ready for further cycles of this procedure. A more sensitive reagent, dimethylaminoazobenzene isothiocyanate, can be used in place of phenylisothiocyanate. The chemistry is the same.

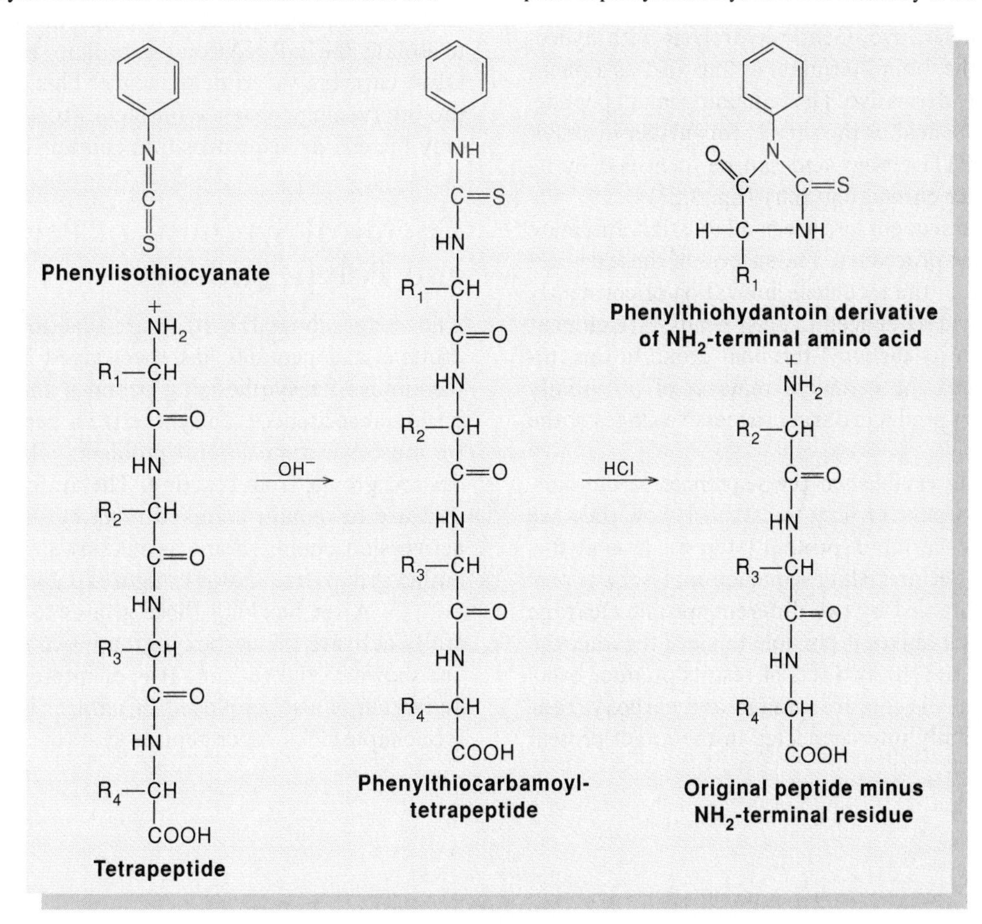

Figure 3.23

Thin-layer chromatography of amino acid–phenylthiohydantoin derivatives on silica gel plates. (*a*) Separation is done in a 98:2 mixture of chloroform and ethanol. (*b*) This is followed by further separation using an 88:2:10 mixture of chloroform, ethanol, and methanol. More sophisticated procedures, using column chromatography, give superior resolution and improved sensitivity. Automated sequencers always use such procedures. A general description of the use of columns is given in chapter 5.

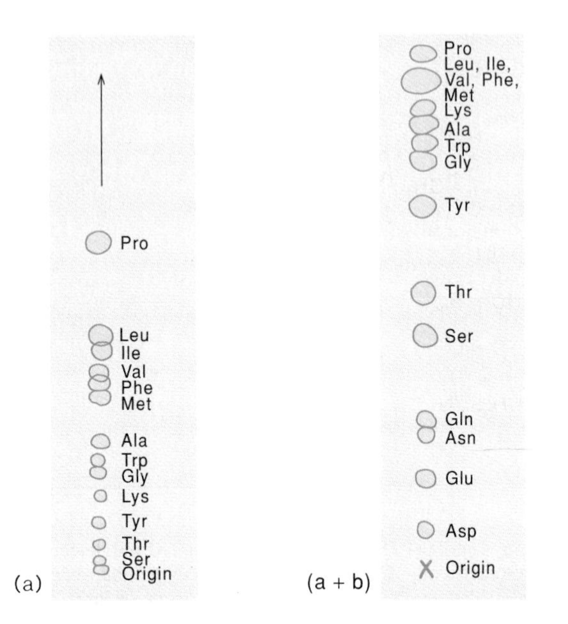

phenylthiocarbamyl derivative. Gentle hydrolysis with hydrochloric acid releases the amino-terminal amino acid as a phenylthiohydantoin (PTH) derivative. The remaining intact peptide, shortened by one amino acid, is then ready for further cycles of this procedure. The PTH-amino acid can be identified by its properties on thin-layer chromatography (fig. 3.23).

Devices called sequenators are available that automate the Edman degradation procedure. The success of these devices depends in large part on the technical innovation of covalently linking the peptide to be sequenced to glass beads. Attachment of the peptide through its carboxyl-terminal group to this immobile phase facilitates the complete removal of potentially contaminating reaction products during successive stages of the degradation.

Finally, having established the sequences of the individual peptides, it is necessary only to establish how they are connected together in the intact protein (step 6). It is at this stage that we see why the preceding sequence analysis was performed on peptides obtained by two different specific cleavage methods. This approach makes it possible to piece together the overall sequence, because the two sets of results produce overlapping sequences. That is, the free amino and carboxyl residues of peptides originally interconnected in the intact protein and liberated by one specific cleavage method will recur in the internal sequences of the peptides liberated by a second specific method.

Once the protein's primary sequence has been determined, the location of disulfide bonds in the intact protein can be established by repeating a specific enzymatic cleavage on another sample of the same protein in which the disulfide bonds have not previously been cleaved. Separation of the resulting peptides will show the appearance of one new peptide and the disappearance of two other peptides, when compared with the enzymatic digestion product of the material whose disulfide bonds have first been chemically cleaved. In fact, these difference techniques are generally useful in the detection of sites of mutations in protein molecules of previously known sequence, since a single substitution will generally affect the chromatographic properties of only a single peptide released during proteolytic digestion.

Great progress has been made in recent years in devising procedures for sequencing the DNA that encodes for proteins (see chapter 27). Knowing the sequence of coding triplets in DNA allows us to read off the amino acid sequence of the corresponding protein. Nevertheless, such studies have produced the remarkable observation that some eukaryotic DNA sequences coding for proteins are not continuous, but instead contain untranslated intervening DNA sequences. Although these results have profound implications for protein evolution, they obviously confound the general applicability of DNA-sequencing methods for the purposes of protein primary-structure determination. In cases like this, the usual solution has been to isolate the mRNA for the protein and use this to make a DNA carrying the same sequence. This procedure circumvents the intervening-sequence problem because the mRNA carries only the coding sequences (see chapter 27).

Chemical Synthesis of Peptides and Polypeptides

Knowledge about the structure-function interrelationships in proteins and peptides has encouraged biochemists to develop techniques for synthesizing peptides and proteins with predetermined sequences. To synthesize a peptide in the laboratory, we must overcome several problems related to preventing undesired groups from reacting. The amino and carboxyl groups that are to remain unlinked must be blocked; so must all reactive side chains. Some protecting groups for carboxyl and amino groups are shown in figures 3.24 and 3.25, respectively.

After blocking those groups to be protected, we generally activate the carboxyl group; two methods for doing this are shown in figure 3.26. It is of interest that carboxyl-group activation is also employed in natural biosynthesis in the cell (see chapter 29). After peptide synthesis, the protecting groups

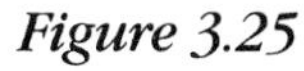

Figure 3.24

Carboxyl protecting groups used in peptide synthesis. The symbol ▲ in the amino acid structure on the left, stands for one of the protecting groups (middle) leading to the named compound indicated on the right. The protecting group prevents the carboxyl group from participating in subsequent reactions involved in peptide synthesis.

Amino acid	Carboxyl-terminal protecting group ▲	Esters formed
$RCH(NH_2)C(=O)$—▲	CH_3CH_2O—	Ethyl
	O_2N—C_6H_4—CH_2—O—	Nitrobenzyl
	H_3C—$C(CH_3)_2$—O—	*t*-Butyl

Figure 3.25

Amino protecting groups used in peptide synthesis.

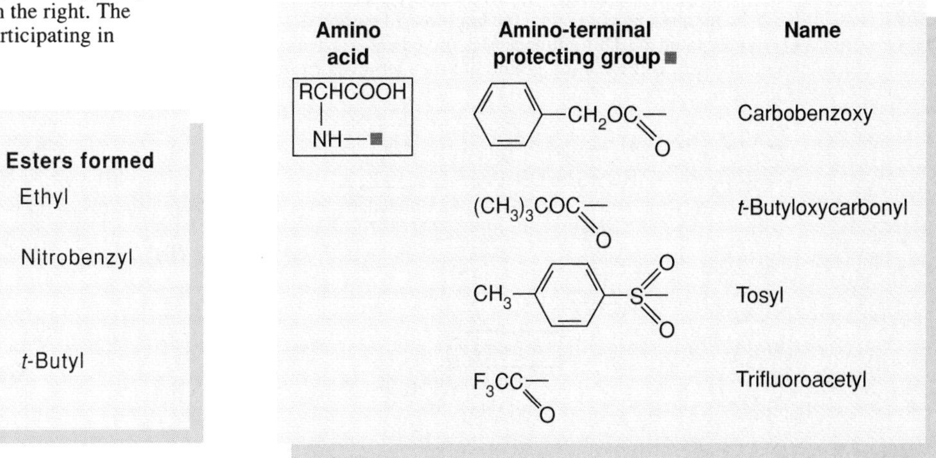

Figure 3.26

Different ways of activating the carboxyl group for peptide synthesis. Activated amino acids will react spontaneously with most α-amino acids as illustrated in figure 3.27.

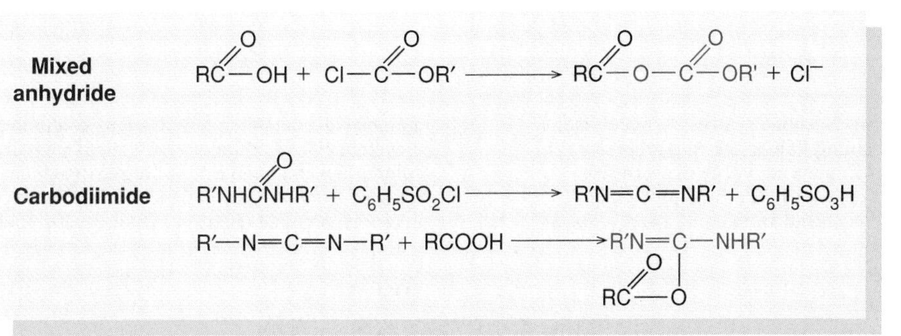

must be removed by a mild method. The overall process—comprising protection, activation, coupling, and unblocking—is shown in figure 3.27.

An important variation of the usual methods of peptide synthesis involves attaching a protected (*t*-butyloxycarbonyl group) amino acid to a solid polystyrene resin; removal of the amino protecting group; condensation with a second protected amino acid; and so on. In the last step, the finished peptide is cleaved from the resin. This method (outlined in figure 3.28) has the advantage that cumbersome purification between steps, often resulting in serious losses, is replaced by mere washing of the insoluble resin. Since each reaction is essentially quantitative, very long peptides, and even proteins, can be synthesized by this method. Indeed, Li synthesized a 39-amino-acid protein hormone, adrenocorticotropic hormone, by this method, and Merrifield synthesized bovine pancreatic ribonuclease, which contains 129 amino acids in a single polypeptide chain. A number of variants of ribonuclease that contain one or more changes in amino acid sequence also have been made by this method. The importance of the Merrifield process was underscored by the awarding of a Nobel Prize to Merrifield in 1984.

Figure 3.27

Schematic diagram illustrating the chemical method for peptide synthesis. First the amino acids to be linked are selected. The carboxyl group and the amino group that are to be excluded from peptide synthesis are protected (steps 1 and 1′). Next the amino acid containing the unprotected carboxyl group is carboxyl-activated (step 2). This amino acid is mixed and reacted with the other amino acid (step 3). Protecting groups are then removed from the product (step 4).

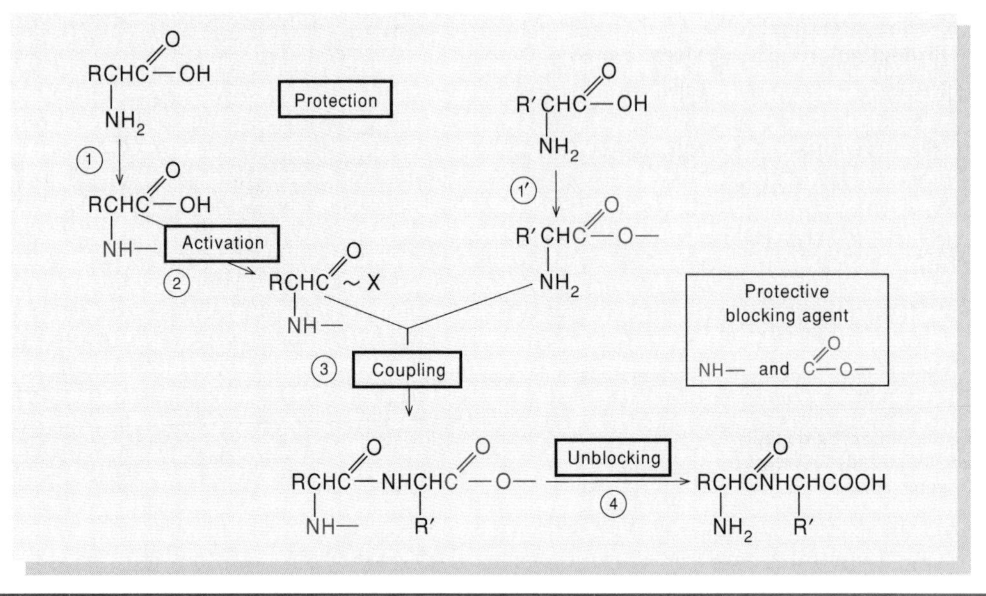

Figure 3.28

Merrifield procedure for solid-state dipeptide synthesis. (1) Polymer is activated. (2) Amino acid containing BOC protecting group is carboxyl-linked to polymer. This amino acid will be the carboxyl-terminal amino acid in the final peptide. (3) The BOC protecting group is removed from the polymer-linked amino acid. (4) A second amino acid, containing a BOC on its α-amino group and a dicyclohexylcarbodiimide (DCC) activated group, is reacted with the column-bound amino acid to form a dipeptide. (5) The dipeptide is released from the polymer and the BOC protecting group by adding hydrogen bromide (HBr) in trifluoroacetic acid.

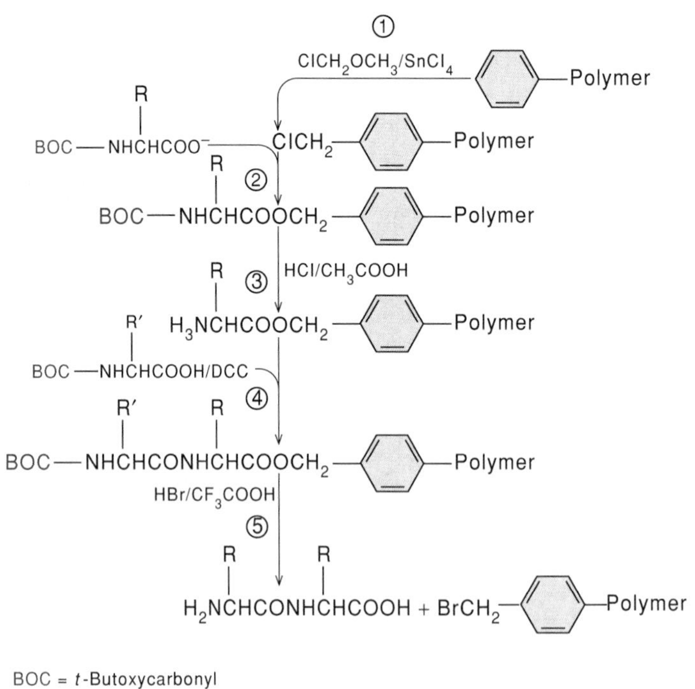

Summary

In this chapter we have dealt with some of the fundamental properties of amino acids and polypeptide chains. The following points are especially important.

1. Nineteen of the twenty amino acids commonly found in proteins have a carboxyl group and an amino group attached to an α-carbon atom; they differ in the side chain attached to the same α carbon.
2. All amino acids have acidic and basic properties. The ratio of base to acid form at any given pH can be calculated from the pK with the help of the Henderson-Hasselbach equation.
3. All amino acids except glycine are asymmetric and therefore can exist in at least two different stereoisomeric forms.
4. Peptides are formed from amino acids by the reaction of the α-amino group from one amino acid with the α-carboxyl group of another amino acid.
5. Polypeptide formation involves a repetition of the process involved in peptide synthesis.
6. The amino acid composition of proteins can be discovered by first breaking down the protein into its component amino acids and then separating the amino acids in the mixture for quantitative estimation.
7. The amino acid sequences of proteins can be discovered by breaking down the protein into polypeptide chains and then partially degrading the polypeptide chains. For each polypeptide chain fragment, the sequence is determined by stepwise removal of amino acids from the amino-terminal end of the polypeptide chain. Two different methods of forming polypeptide chain fragments are used so as to produce a map of overlapping fragments, from which the sequence of undegraded polypeptide chains in the proteins can be deduced.
8. Polypeptide chains with a predetermined amino acid sequence can be synthesized by chemical methods involving carboxyl-group activation.

Selected Readings

Barrett, G. C. (ed.). *Chemistry and Biochemistry of Amino Acids.* New York: Chapman and Hall, 1985. A recent and authoritative volume on this classical subject.

Gray, W. R., End group analysis using dansyl chloride. *Methods in Enzymology* 25:121–138, 1972. This volume of *Methods in Enzymology* contains several chapters on end-group analysis.

Hunkapiller, M. W., J. E. Strickler, and K. J. Wilson, Contemporary methodology for protein structure determination. *Science* 226:304–311, 1984.

Kent, S. B. H., Chemical synthesis of peptides and proteins. *Ann. Rev. Biochem.* 57:957–989, 1988. Comprehensive and up-to-date.

Merrifield, B., Solid phase synthesis. *Science* 232: 341–347, 1986.

Sanger, R., Sequences, sequences and sequences. *Ann. Rev. Biochem.* 57:1–28, 1988.

Problems

1. (a) A 10-mM solution of a weak monocarboxylic acid has a pH of 3.00. Calculate K_a and pK_a for this carboxylic acid.
   (b) You add 0.06 g NaOH (M_r = 40) to 1,000 ml of the acid solution in part (a). Calculate the final pH, assuming no volume change.
2. Given the pK_a values in the text, predict how the titration curves for glutamic acid and glutamine would differ.
3. You have 50 ml of 10-mM fully protonated histidine. How many millimoles of base must be added to bring the histidine solution to a pH that is equivalent to the pI?
4. Calculate the isoelectric point for histidine, aspartic acid, and arginine. Calculate the fractional charge for each ionizable group on aspartate at pH equal to pI. Do the results verify the isoelectric point of aspartic acid?
5. Which of the naturally occurring amino acid side chains are charged at pH 2? pH 7? pH 12? (Consider only those amino acids whose side chains have >10% charge at the pH indicated.)
6. Amino acids are sometimes used as buffers. Indicate the appropriate pH value(s) of a buffer containing aspartic acid, histidine, and serine.
7. Ten ml of a 10-mM solution of lysine was adjusted to pH 11.20. Draw the structures of the principal ionized forms present in solution. Use the pK_a values shown in table 3.3 and calculate the concentration of each principal form.
8. For the tripeptide shown below, the numbers in parentheses are the pK_a values of the ionizable groups.

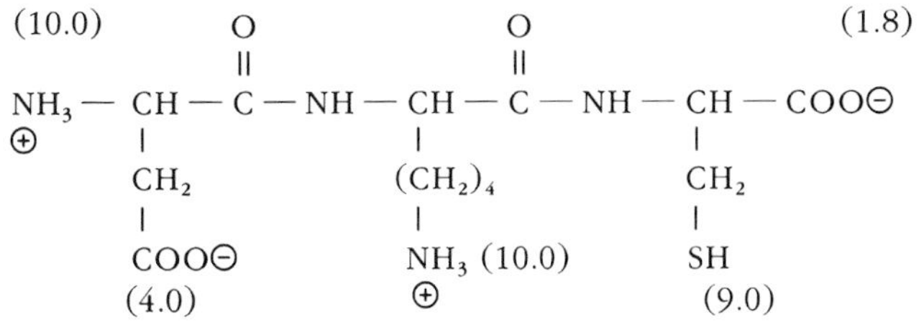

```
(10.0)        O               O                     (1.8)
              ‖               ‖
NH3 — CH  —   C — NH — CH  —  C — NH — CH — COO⊖
⊕     |                |                |
      CH2             (CH2)4            CH2
      |                |                |
      COO⊖            NH3 (10.0)        SH
      (4.0)            ⊕                (9.0)
```

   (a) Estimate the net charge at pH 1 and pH 14.
   (b) Estimate the isoelectric pH.
9. Polyhistidine is insoluble in water at pH 7.8 but is soluble at pH 5.5. Explain the observation. Would you expect the polymer to be soluble at pH 10?

10. Protamines are basic proteins whose sulfate salts are often used as a precipitating agent added to crude cellular extracts. What types of biomolecules would you expect to be precipitated by protamine sulfate at pH 7? What is the physical basis for the precipitation?
11. A mixture of alanine, glutamic acid, and arginine was chromatographed on a weakly basic ion-exchange column (positively charged) at pH 6.1. Predict the order of elution of the amino acids from the ion-exchange column. Are the amino acids separated from each other? Explain.
Suppose you have a weakly acidic ion-exchange column (negatively charged), also at pH 6.1. Predict the order of elution of the amino acids from this column. Propose a strategy to separate the amino acids using one or both columns. Explain your rationale. (Assume only ionic interactions between the amino acids and the ion exchange resin.)
12. For the following peptide sequences, determine the products resulting from the following treatments:
(a) Trypsin digestion. (b) Treatment of the peptide with succinic anhydride, then trypsin. (c) Reaction with ethyleneimine followed by trypsin. (d) Chymotrypsin.
(e) Cyanogen bromide (CNBr).

Ala-Glu-Lys-Phe-Val-Cys-Tyr-Met-Gly-Phe

13. You have a peptide that is a potent inhibitor of nerve conduction and you wish to obtain its primary sequence. Amino acid analysis reveals the composition to be Ala (5); Lys; Phe. Reaction of the intact peptide with FDNB releases free DNP-alanine on acid hydrolysis. (ϵ-DNP-lysine but not α-DNP-lysine is also found.) Trypsin digestion gives a tripeptide (composition Lys, Ala_2) and a tetrapeptide (composition Ala_3, Phe). Chymotryptic digestion of the intact peptide releases a hexapeptide and free alanine. Derive the peptide sequence.
14. Performic acid oxidation followed by acid hydrolysis of a decapeptide yielded the following amino acid composition: Ala(1); Asp(2); cysteic acid(2); Gly(2); methionine sulfone(1); Phe(1); Val(1). In addition, one mole of ammonium ion was detected per mole of peptide hydrolyzed. The following results were obtained.
(i) Carboxypeptidase A released Asn, then Ala.
(ii) Two cycles of sequential Edman degradation released the phenylthiohydantoins of Asp and Val, in that order.
(iii) Chymotrypsin released two peptides whose compositions were determined after acid hydrolysis: CHT A (Ala, Asp, Cys, Gly, Met); and CHT B (Asp, Cys, Gly, Phe, Val).
(iv) Treatment of the original peptide with 2-bromoethylamine and then with trypsin released three peptides whose compositions were determined after acid hydrolysis: T-1 (Ala, Asp);
T-2 (Asp, modified Cys, Gly, Val); and T-3 (modified Cys, Gly, Met, Phe).
Deduce the primary sequence of the peptide. If there is a portion of the sequence whose assignment is ambiguous or if there is a disulfide bond, so indicate. If there is an ambiguity, what other single cleavage might you use to assign an unambiguous sequence?
15. You have isolated from a rare fungus an octapeptide that prevents baldness and you wish to determine its amino acid sequence. The amino acid composition is Lys_2, Asp, Tyr, Phe, Gly, Ser, Ala. Reaction of the intact peptide with FDNB yields DNP-alanine plus 2 moles ϵ-DNP-lysine upon acid hydrolysis. Cleavage with trypsin yields peptides whose compositions are: (Lys, Ala, Ser) and (Gly, Phe, Lys) plus a dipeptide. Reaction with chymotrypsin releases free aspartic acid, a tetrapeptide with the composition (Lys, Ser, Phe, Ala), and a tripeptide whose composition following acid hydrolysis is (Gly, Lys, Tyr). What is the sequence?

4

CHAPTER

The Three-Dimensional Structure of Proteins

The enormous structural diversity of proteins begins with the amino acid sequences of polypeptide chains. Each protein consists of one or more unique polypeptide chains, and each of these polypeptide chains is folded into a three-dimensional structure. The final folded arrangement of the polypeptide chain in the protein is referred to as its conformation. Most proteins exist in unique conformations exquisitely suited to their function. It is the availability of a wide variety of conformations that permits proteins as a group to perform a broader range of functions than any other class of biomolecules.

In this chapter we will deal primarily with the structural properties of proteins, and in the following chapter we will consider the functional diversity of proteins. Traditionally proteins have been divided into two groups: fibrous and globular. Fibrous proteins aggregate to form highly elongated structures having the shape of fibers or sheets. Each protein unit going into these aggregated structures is built from a repeating structural motif, and for that reason the molecules have a basically simple structure that is relatively easy to analyze. By contrast, globular proteins are more complex, containing one or more polypeptide chains folded back on themselves many times to give an approximately spherical shape. We will consider the relatively simple fibrous proteins first.

Fibrous Protein Structures

Two names stand out above all others in the history of the discovery of fibrous protein structure. These are the names Linus Pauling and Robert Corey. Before they ventured into the world of proteins, Pauling and Corey spent a great deal of time studying the simple crystals formed by amino acids and low-molecular-weight peptides. From their crystallographic investigations, Pauling and Corey formulated two rules that describe the ways in which amino acids and peptides interact with one another to form noncovalently bonded crystalline structures. These rules remain central to our understanding of how amino acids interact with one another in protein polypeptide chains.

The first rule was that the peptide C—N linkage and the four atoms to which the C and the N atoms are immediately linked always form a planar structure, as though the C—N linkage were a double bond rather than a single bond as normally written. Pauling reasoned that the C—N linkage is a res-

Figure 4.1

Resonance and the planar structure of the peptide bond. (*a*) Two major hybrids contribute to the structure of the peptide bond. In structure 1 the C—N bond is a single bond with no overlap between the nitrogen lone electron pair and the carbonyl carbon. The carboxyl carbon is sp^2- hybridized and is therefore planar, while the nitrogen is sp^3-hybridized and pyramidal. By contrast, in structure 2 there is a double bond between the amide nitrogen and the carbonyl carbon; also, the nitrogen atom bears a charge of +1 and the carbonyl oxygen bears a charge of −1. Both the carboxyl carbon and the amide nitrogen are sp^2 hybridized, both are planar, and all six atoms lie in the same plane. (*b*) The structure of the peptide bond is a compromise between the two resonating hybrids, structures 1 and 2. (*c*) Dimensions of the peptide bond and surrounding linkages. The C—N bond length of 1.325 Å is significantly less than the length of a single C—N bond, 1.47 Å.

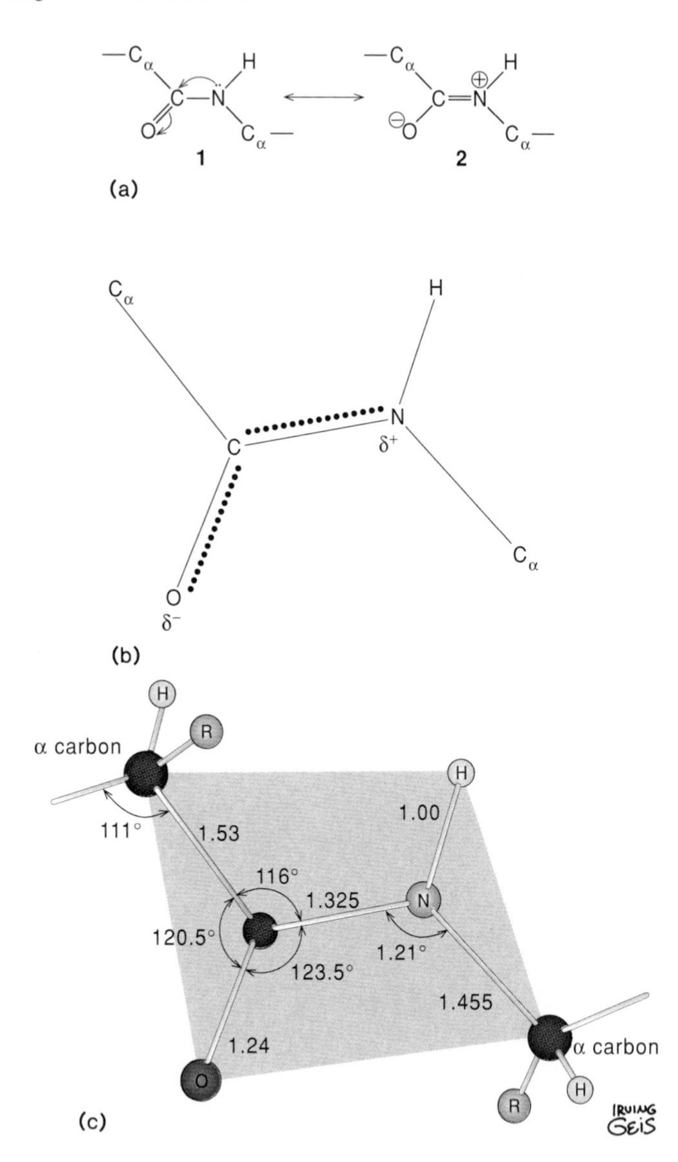

Figure 4.2

Basic dimensions of a dipeptide. The conformational degrees of freedom of a polypeptide chain are restricted to rotations about the single-bond connections between the adjacent planar transpeptide groups to C_α, i.e., the C_α—C_2 and C_α—N_1 single bonds. The corresponding rotations are represented by ψ and ϕ, respectively, which have values of 180° for the fully extended configuration shown. (Reprinted with permission from R. E. Dickerson and I. Geis, *The Structure and Action of Proteins,* Benjamin/Cummings, Menlo Park, Calif., 1969.)

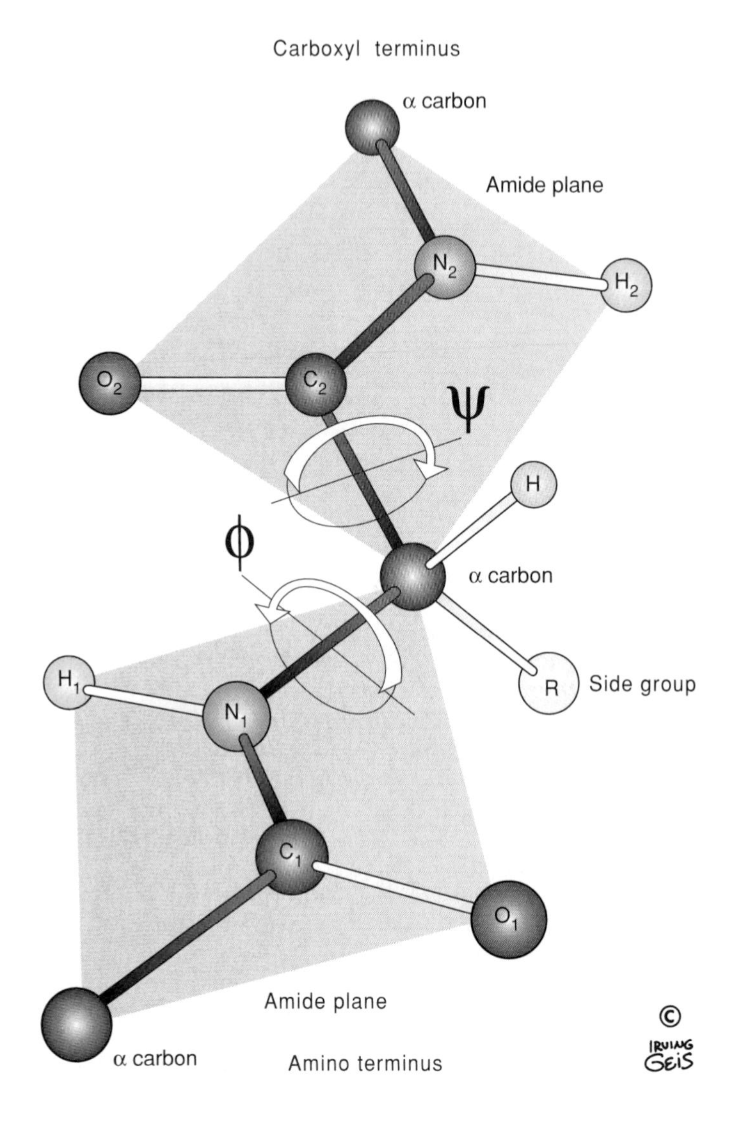

onating structure with partial double-bond character (fig. 4.1), so that it locks the peptide grouping into a planar conformation. This characteristic is extremely important because it greatly reduces the flexibility in the polypeptide chain. The only flexibility remaining in the polypeptide backbone results from rotation about the α carbon that joins two peptide planar groups (fig. 4.2).

The second rule that Pauling and Corey formulated was that peptide carbonyl and NH groups always form the maximum number of hydrogen bonds. Recall that hydrogen bonds are formed by a partially unshielded proton from one molecule and a nitrogen or oxygen atom possessing a lone pair of electrons that originates from another molecule (see fig. 1.7). The attraction between the two groups is strongest along the lone-pair orbital axis of the hydrogen bond acceptor group. As a rule the angle between the N or O acceptor and the N—H or

Figure 4.3

Major hydrogen-bond donor and acceptor groups found in proteins. Note that the angle between the O acceptor and the N—H donor is 180° in the hydrogen-bond complex.

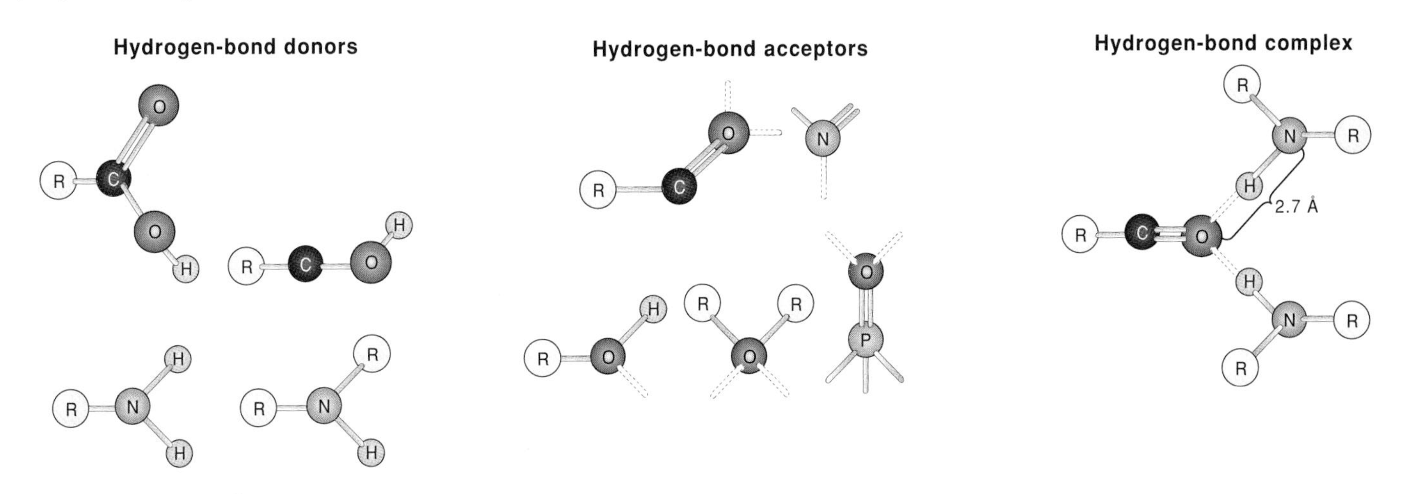

O—H donor is close to 180° (fig. 4.3). Thus a hydrogen bond brings two interacting groups close together and orients them in a certain way. This second rule of Pauling and Corey further limits the number of conformations available to polypeptide chains.

With these rules in their arsenal, Pauling and Corey examined the x-ray diffraction patterns of a number of fibrous proteins. Despite the vast number of fibrous proteins that exist in nature, the majority of them give diffraction patterns that fall into one of three types: the α pattern (box 4A), the β pattern, and the collagen pattern. Fiber diffraction gives information about the repeating units of a protein structure only, but since fibrous proteins are arranged as simple repetitious units, this was just the information that was needed. The first type of diffraction pattern observed for a subgroup of the keratins was consistent with a helical arrangement of the polypeptide chains and an advance per turn along the helix axis of 5.4 Å. (The advance per turn is called the pitch of the helix.) If the fibers are tilted by about 30° with respect to the x-ray beam, they show a prominent diffraction spot indicative of an advance per amino acid along the helix axis of 1.5 Å.

The α-Keratins Contain Helically Arranged Polypeptide Chains Held Together by Intramolecular Hydrogen Bonds

From diffraction data, their two rules, and a set of precisely constructed space-filling molecular models, Pauling and Corey went to work testing the structural possibilities. Their models were designed so that covalently linked atoms were accurately spaced according to well-known dimensions for such groups (table 4.1). Moreover, the individual atoms were made of a size so that nonbonded atoms could get no closer to one another than their van der Waals radii would normally allow (see table 4.1). Recall from chemistry that the average separation between two nonbonded atoms is known as their van der Waals separation. There is a very high repulsion if nonbonded atoms get closer than this, and there is a rapid fall-off in favorable interaction if the distance is larger. With their carefully constructed "toys" and a boyish enthusiasm, the two learned scientists tried to arrange the polypeptide chains so as to maximize the number of peptide hydrogen bonds in a way that was consistent with the x-ray diffraction data. By trial and error they came to the conclusion that the most acceptable structure was a polypeptide chain arranged in a right-handed helical coil. This structure, known as the alpha (α) helix, has a rigid, regularly repeating backbone structure with an advance per amino acid residue along the helix axis of 1.5 Å, corresponding to the prominent spot in the diffraction pattern. The diffraction spots resulting from the helix itself could be explained by a spacing between adjacent turns of the polypeptide backbone along the helix axis of 5.4 Å.

In figure 4.4 we see three different ways of representing the structure of the α helix. In (*a*) we see a ball-and-stick model in which the planar groups are highlighted and the hydrogen bonds are represented by dashed lines. In (*b*) and (*c*) we see a space-filling model and a wire model, respectively, representing the same structure. The latter two models are shown in both side views and top view. The space-filling model, similar to the one used by Pauling and Corey, reveals that the α helix is a very tightly packed structure with no unfilled cavities, whether one looks at a profile or down the helix. The wire model (*c*) illustrates the helical structure best. But for purposes of discussion the ball-and-stick model (*a*) is the most suitable. Careful inspection shows that the polypeptide backbone follows the path of a right-handed helical spring in which each residue's carbonyl group forms a hydrogen bond with the amide NH group of the residue four amino acids further along the polypeptide chain. All residues in the α helix have nearly identical conformations, so they lead to a regular structure in which each 360° of helical turn incorporates approximately 3.6 amino acid residues and rises 5.4 Å along the helix axis direction. This arrangement gives the observed advance per amino acid residue along the helix axis ($5.4/3.6 = 1.5$ Å).

X-Ray Diffraction: Applications to Fibrous Proteins*

-ray diffraction played a major role in discovery of the structure of fibrous proteins. In most cases the fibers under study are oriented in two dimensions by stretching. In this box we illustrate how the technique is used to study the α form of the synthetic polypeptide poly-L-alanine.

A stretched fiber containing many poly-L-alanine molecules is suspended vertically and exposed to a collimated monochromatic beam of CuK_α x-rays, as shown in figure 1(*a*) of the illustration. Only a small percentage of the x-ray beam is diffracted, while most of the beam travels through the specimen with no change in direction. A photographic film is held in back of the specimen. A hole in the center of the film allows the incident undiffracted beam to pass through.

Coherent diffraction occurs only in certain directions specified by Bragg's law: $2d \sin \theta = n\lambda$, where d is the distance between identical repeating structural elements, θ is the angle between the incident beam and the regularly spaced diffracting planes, λ is the wavelength of x-rays used, and n is the order of diffraction, which may equal any integer but is usually strongest for $n = 1$. For small θ, $\sin \theta \approx \theta$ and $d \approx 1/\theta$, so that a spot far out on the photographic film is indicative of a repeating element of small dimension.

Figure 1(*b*) shows the diffraction pattern that is obtained when the fiber axis is normal to the beam. Note the strong off-vertical reflection at 5.4 Å (arrow). A different diffraction pattern (*c*) is obtained when the fiber axis is inclined to the beam at 31°. Note the strong reflection at 1.5 Å in the upper part of the diagram (arrow).

*Source: C. H. Bamford et al., *Synthetic Polypeptides.* Copyright © 1956, Academic Press, Orlando, Fla.

Figure 1

(*a*) Experimental arrangement for obtaining x-ray diffraction pattern shown in (*b*). (*b*) Diffraction pattern of the α form of a cluster of poly-L-alanine molecules oriented vertically. (*c*) Diffraction pattern of the same fiber bundle with the fiber axis inclined to the beam at 31°. (*b* and *c* Brown & Trotter, 1956.)

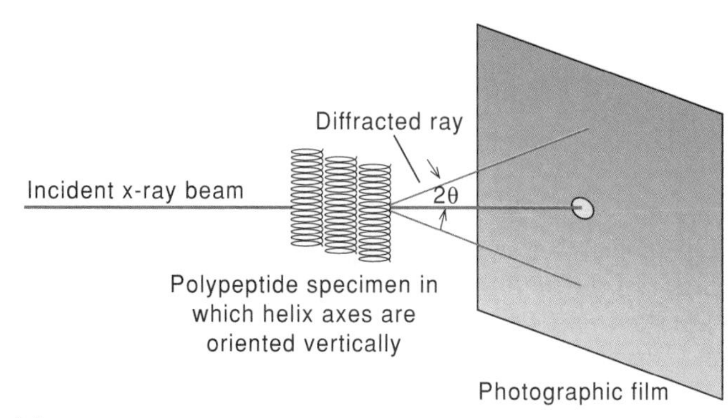

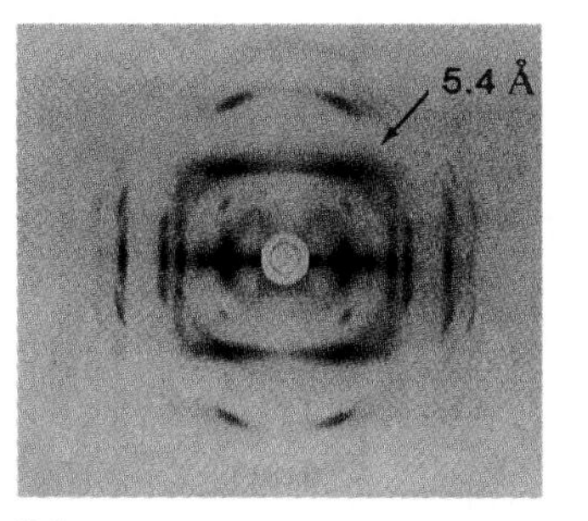

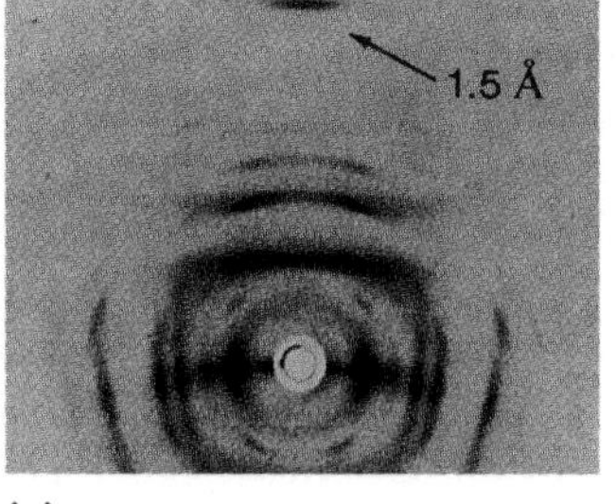

Table 4.1
Radii for Covalently Bonded and Nonbonded Atoms

Covalent Bond Radii (in Å)				Van der Waals Radii (in Å)			
Element	*Single Bond*	*Double Bond*	*Triple Bond*	*Element*			
Hydrogen	0.30			Hydrogen	1.2		
Carbon	0.77	0.67	0.60	Carbon	2.0		
Nitrogen	0.70			Nitrogen	1.5		
Oxygen	0.66			Oxygen	1.4		
Phosphorus	1.10			Phosphorus	1.9		
Sulfur	1.04			Sulfur	1.8		

Figure 4.4

Three ways of projecting the α-helix. (*a*) This simple ball-and-stick model highlights the planar peptides. The interpeptide hydrogen bonds are shown by dashed lines and the amino acid side chains are indicated by R groups. Approximately two turnings of the helix are shown. There are about 3.6 residues per turn. In (*b*) and (*c*) we see identical projections of a side view using space-filling and wire models, respectively. The space-filling models use van der Waals radii for the atoms.

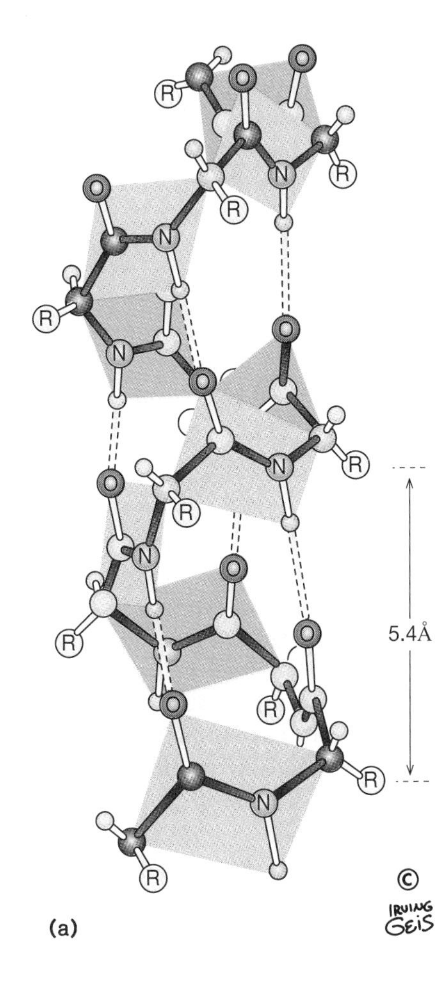

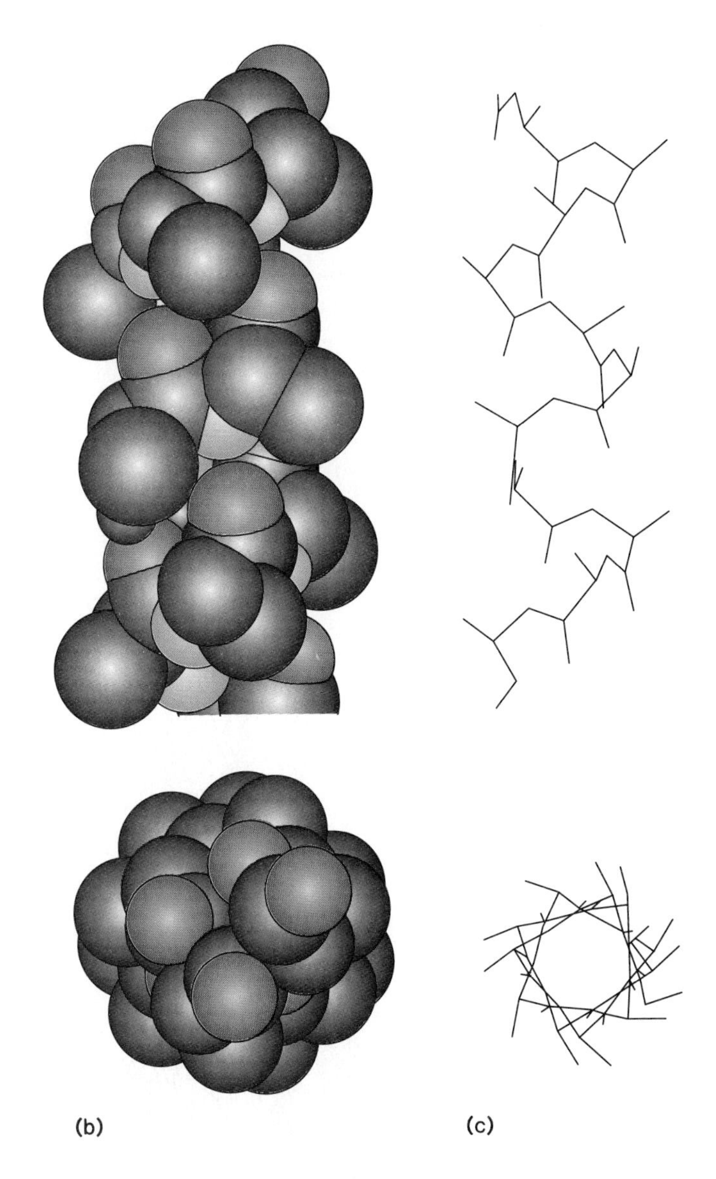

Although alternative helical arrangements having different hydrogen-bonding patterns and different geometries are conformationally possible, the α helix is by far the most commonly observed helical arrangement found for polypeptide chains in proteins. The special stability of the helix is probably related not only to the formation of stable hydrogen bonds between all the carbonyl and NH groups, but also to the tight packing achieved in folding the chain to form the structure.

Fibrous proteins in which the α helix is a major structural component are found in hair, scales, horns, hooves, wool, beaks, nails, and claws; these proteins are referred to as α-keratins. When individual α helices aggregate in side-by-side fashion to produce such proteins, they usually form long cables in which the individual helices are spirally twisted so that the resulting cable has an overall left-handed twist (fig. 4.5). The formation of such a cable appears to result from optimization of packing among the amino acid side-chain residues between helices. In figure 4.6 we see how the side-chain residues of an α helix are arranged in a spiral fashion so that residues falling on the same side of a helix generally do not lie along a line parallel to the helix axis. The packing together of helices is consequently optimized when the helices interact at an angle of about 18°. Obviously, if the α helices involved in such a packing interaction were straight, they would soon become separated. However, with the left-twisted cable structure characteristic of the keratins, their packing interaction can be preserved. The coiled-coil character of such fibers consequently represents a trade-off between some local deformations that coil the α helix and the optimization of extended side-chain packing interactions in the cable as a whole.

Figure 4.5

The assembly of hair α-keratin from one α helix to a protofibril, to a microfibril, and finally, to a single hair.

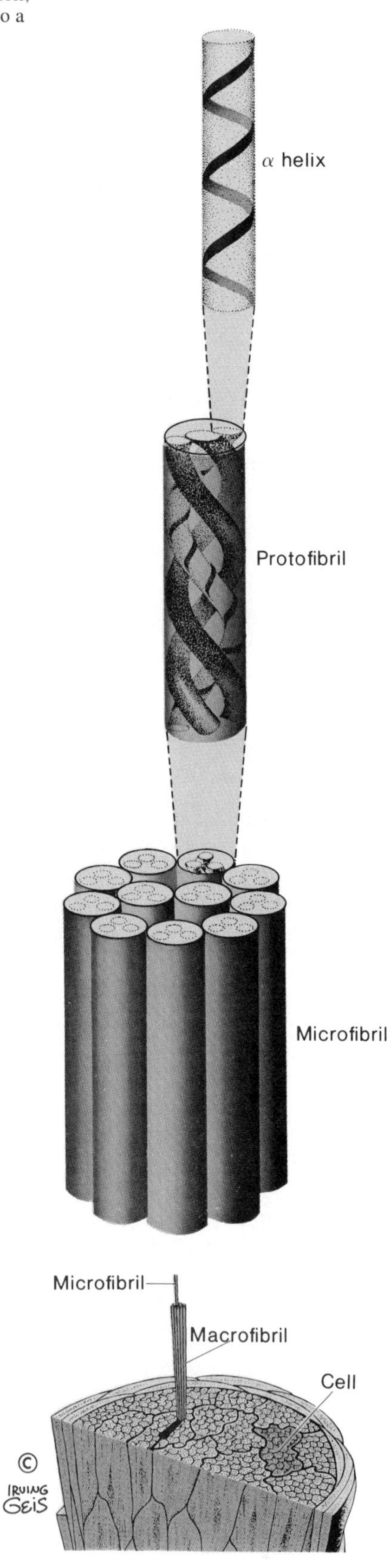

Figure 4.6

Coiling of α helices in α-keratins. Residues on the same side of an α helix form rows that are tilted relative to the helix axis. Packing helices together in fibers is optimized when the individual helices wrap around each other so that rows of residues pack together along the fiber axis. Helices in coiled coil (*c*) are oriented in parallel.

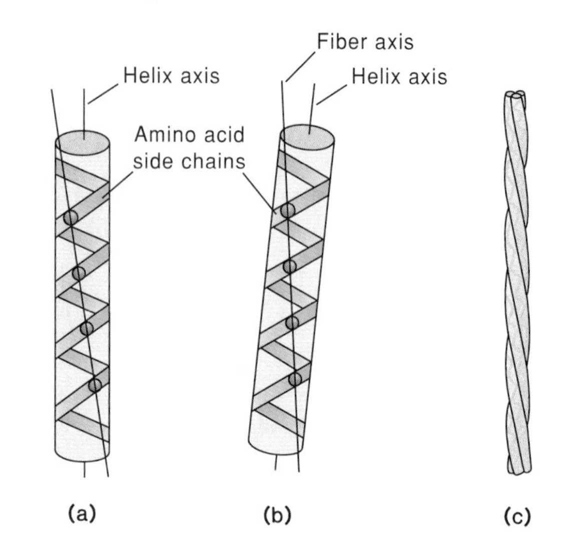

The springiness of hair and wool fibers results from the tendency of the α-helical cables to untwist when stretched and spring back when the external force is removed. In many forms of keratin, the individual α helices or fibers are covalently linked by disulfide bonds formed between cysteine residues of adjacent polypeptide chains. In addition to giving added strength to the fibers, the pattern of these covalent interactions serves to influence and fix the extent of curliness in the hair fiber as a whole. Chemical reactions and mechanical processes involving reductive cleavage, reorganization, and reoxidation of these interhelix disulfide bonds form the basis of the "permanent wave."

The β-Keratins Form Sheetlike Structures with Extended Polypeptide Chains

Pauling and Corey noticed that certain fibrous proteins give radically different diffraction patterns and behave macroscopically as sheets rather than fibers. The diffraction patterns yielded by fibrous proteins from different sources revealed a repeat along the direction of the extended polypeptide chain at intervals of either 13.0 or 14.0 Å. The similarity of intervals suggested two closely related structures. Pauling and Corey interpreted these repeating patterns as resulting from extended polypeptide chains lying side-by-side in either a parallel or an antiparallel fashion (fig. 4.7). Although both structures exist in nature, the antiparallel sheet is more common in fibrous proteins. In such proteins the structure is known as the antiparallel β-pleated sheet (fig. 4.8). Regular hydrogen bonds form between the peptide backbone amide NH and carbonyl oxygen groups of adjacent chains. The β sheet can be easily extended into a multistranded structure simply by adding successive chains in the appropriate direction to the sheet.

Figure 4.7

Two forms of the ß-sheet structure: (*a*) the antiparallel and (*b*) the parallel ß sheet. The advance per two amino acid residues is indicated for each structure.

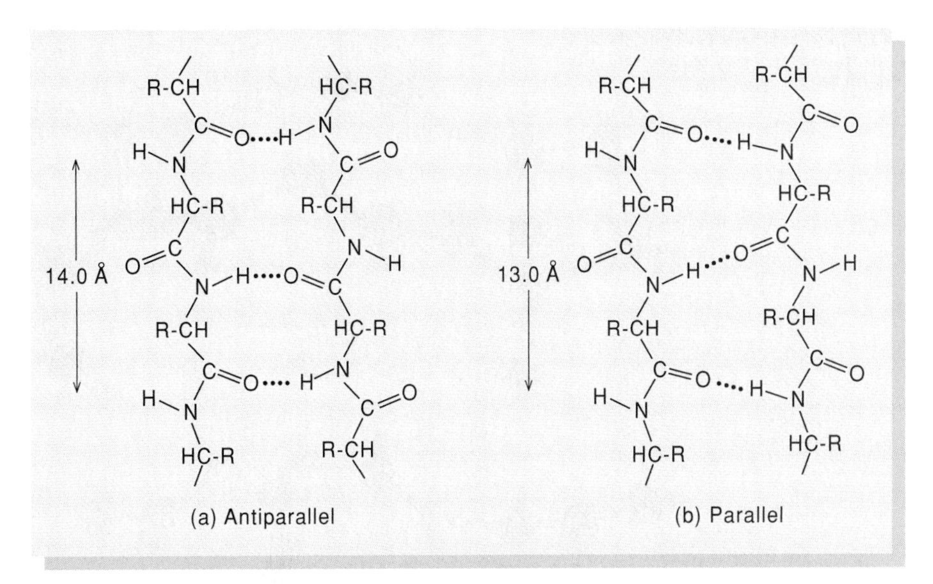

Figure 4.8

The antiparallel ß sheet. This structure is composed of two or more polypeptide chains in the fully extended form, with hydrogen bonds formed between the chains. Hydrogen bonds are shown as dashed lines.

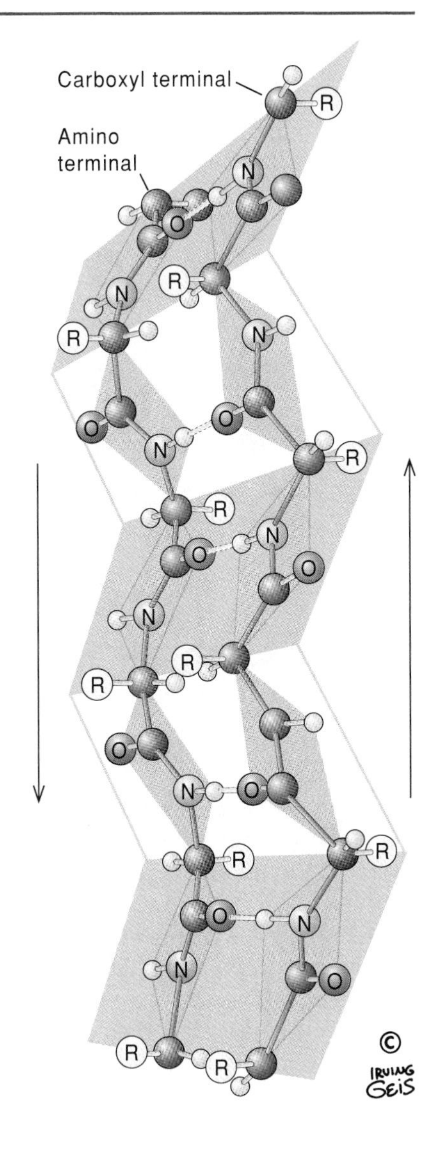

Parallel and antiparallel β-pleated sheets are both composed of polypeptide chains that have conformations pointing alternate R groups to opposite sides of the sheet, but have their peptide planes nearly in the sheet plane to allow good interchain hydrogen bonding. Nevertheless, the chain conformation that produces the best interchain hydrogen bonding in parallel sheets is slightly less extended than that for the antiparallel arrangement. As a result, the parallel sheet has both a shorter advance per amino acid residue, 6.5 Å (versus 7.0 Å for the antiparallel structure), and a more pronounced pleat.

The best-known β-sheet structure in nature is silk, a variety of fibrous proteins produced by certain insects. Silks are composed of stacked antiparallel β-pleated sheets (fig. 4.9). Sequence analysis of silk proteins shows them to be largely composed of glycine, serine, and alanine, where every alternate residue is glycine. Since the side-chain groups of a flat antiparallel sheet point alternately upward and downward from the plane of the sheet, all the glycine residues are arranged on one surface of each sheet and all the substituted amino acids on the other. Two or more such sheets can consequently be intimately packed together to form an arrangement in which two adjacent glycine-substituted or alanine-substituted sheet surfaces interlock with each other (see fig. 4.9*b*). Owing to both the extended conformations of the polypeptide chains in the β sheets and the interlocking of the side chains between sheets, silk is a mechanically rigid material that resists stretching.

Collagen Forms a Unique Triple-Stranded Structure with Interstrand Hydrogen Bonds

Collagen is a particularly rigid and inextensible protein that serves as a major constituent of tendons and many connective tissues. In the electron microscope it can be seen that collagen fibrils have a distinctive banded pattern with a periodicity of

Figure 4.9

The three-dimensional architecture of silk (*a*). The side chains of one sheet nestle quite efficiently between those of neighboring sheets (*b*).

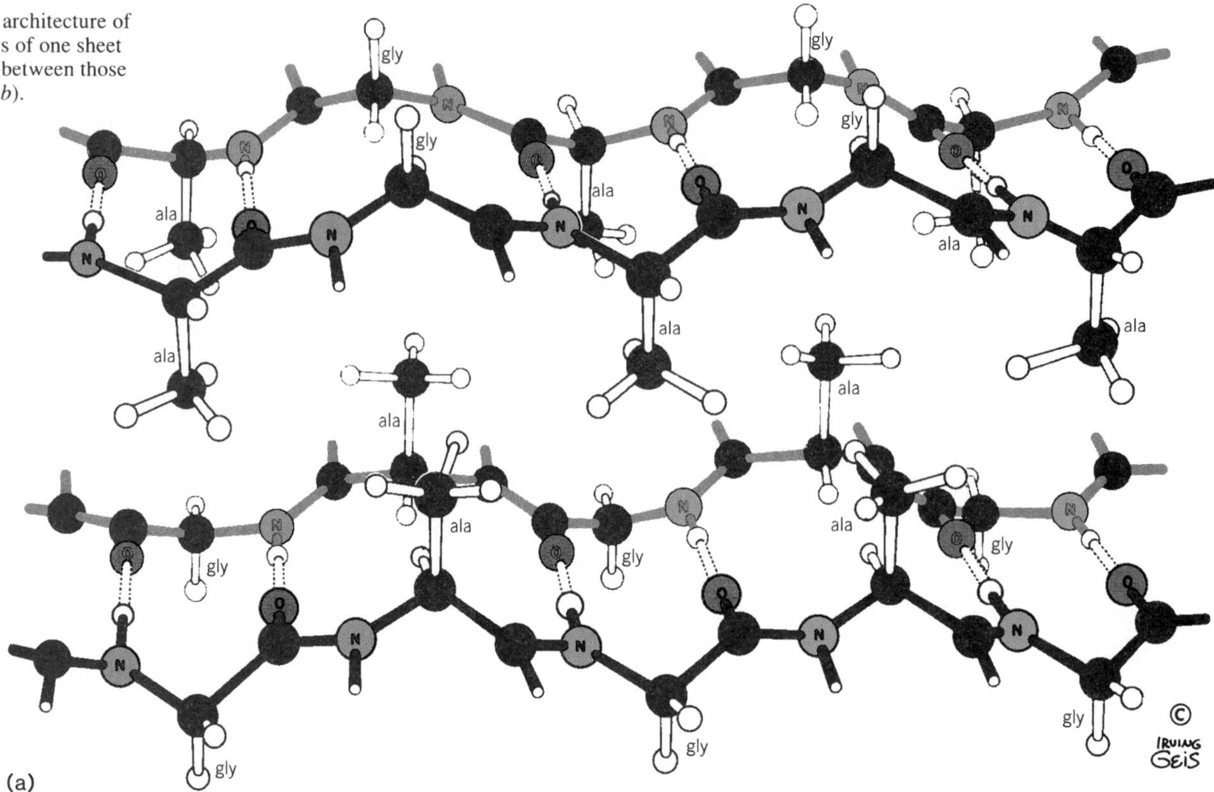

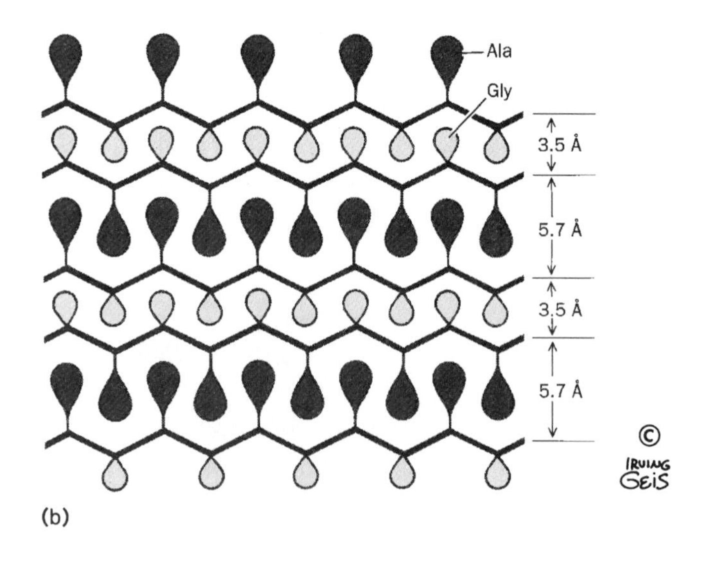

680 Å (fig. 4.10). These fibrils are of varying thickness, depending on the source and the mode of preparation. Individual fibrils are composed of collagen molecules 3,000 Å long that aggregate in a staggered side-by-side fashion (fig. 4.11).

Analysis of collagen indicated a most unusual amino acid composition in which glycine, proline, and hydroxyproline are the dominant amino acids. Further characterization of the polypeptide chains showed that these amino acids are arranged in a repetitious tripeptide sequence, Gly-X-Y, where X is frequently a proline and Y is frequently a hydroxyproline. This unusual amino acid sequence and the unique diffraction pattern of collagen suggested that collagen is a totally different type of fibrous protein. Pauling attempted to determine its structure by his molecular model approach but failed. At a meeting in Cambridge in 1958 where the correct structure was presented, Pauling suggested, more or less in jest, that the structure he had derived might be more stable than the one found in nature. The correct structure was in fact found by Ramachandran, whose interpretation was supported by Crick and Rich.

The repeating proline residue in collagen excluded the possibility that the polypeptide chains could adopt either an α-helical or a β-sheet conformation. Instead, individual collagen polypeptide chains assume a left-handed helical conformation and aggregate into three-stranded cables with a right-handed twist (fig. 4.12). When viewed down the polypeptide chain axis (fig. 4.13*b*), the successive side-chain groups can be seen to point toward the corners of an equilateral triangle. The glycine at every third residue is required because there is no room for any other amino acid inside the triple helix, where the glycine R groups are located. The three collagen chains do not form hydrogen bonds among residues of the same chain. Instead, the collagen chains within each three-stranded cable form interchain hydrogen bonds. This arrangement produces a highly interlocked fibrous structure that is admirably suited to its biological role of providing rigid connections between muscles and bones as well as structural reinforcement for skin and connective tissues.

Figure 4.10

An electron micrograph of collagen fibrils from skin. (Courtesy of Jerome Gross, Massachusetts General Hospital.)

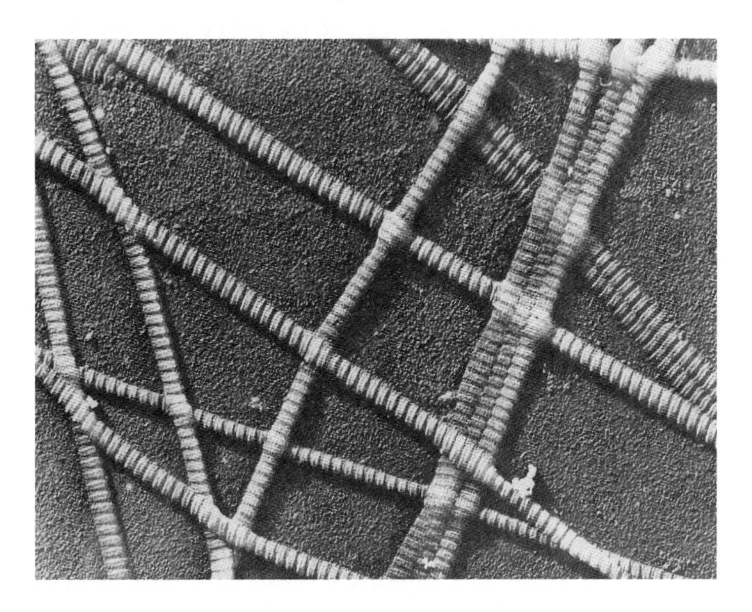

Figure 4.11

The banded appearance of collagen fibrils in the electron microscope arises from the schematically represented staggered arrangement of collagen molecules (*above*) that results in a periodically indented surface. *D*, the distance between cross striations, is ~680 Å so that the length of a 3,000-Å-long collagen molecule is 4.4*D*. (©Michael C. Webb/Visuals Unlimited.)

Collagen molecule

Packing of molecules

Hole zone 0.6*D*

Overlap zone 0.4*D*

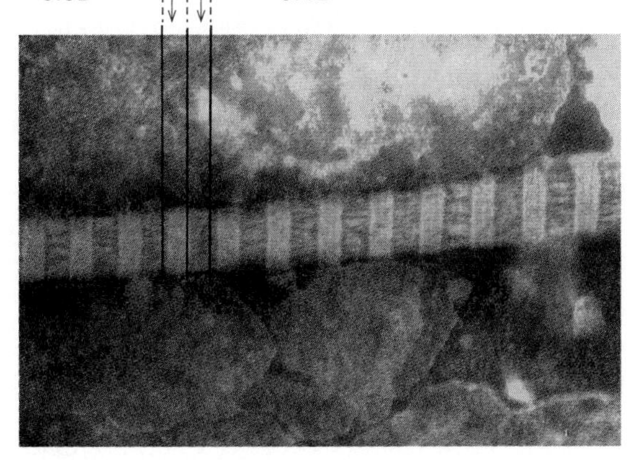

Figure 4.12

The triple helix of collagen.

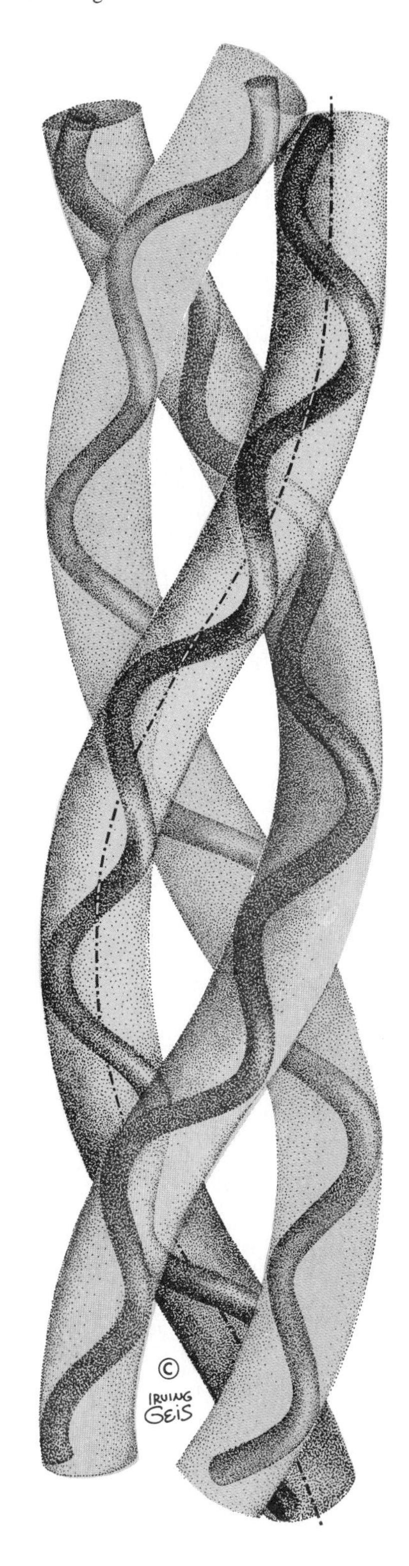

Figure 4.13

The basic coiled-coil structure of collagen. Three left-handed single-chain helices wrap around one another with a right-handed twist. (*a*) Ball-and-stick single-collagen chain. (*b*) View from top of helix axis. Note that glycines are all on the inside. In this structure the C=O and N—H groups of glycine protrude approximately perpendicularly to the helix axis so as to form interchain hydrogen bonds.

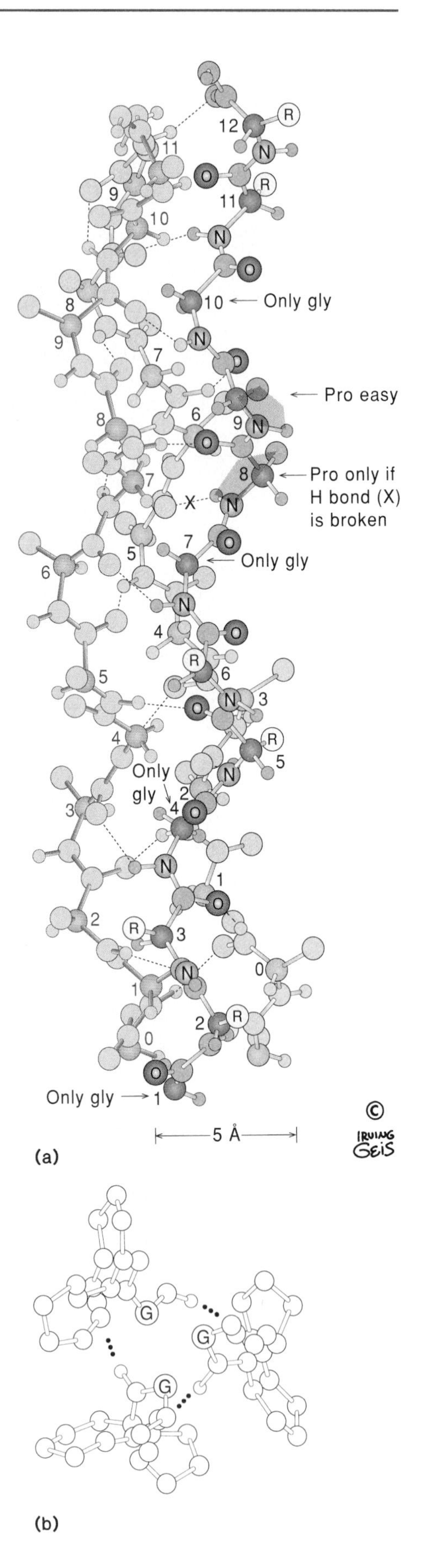

Although there exist in living organisms additional types of fibrous proteins, as well as polysaccharide-based structural motifs, we have focused here on those three arrangements whose structural properties are currently the best understood and the most widely distributed. Two of these, the α-keratins and the silks, incorporate polypeptide secondary structures that also commonly occur in globular proteins. Collagen, in contrast, is a protein that evolution has developed to play a more specialized role.

The Use of Ramachandran Plots to Predict Sterically Permissible Structures

In figure 4.2 we saw a ball-and-stick model of a short section of a polypeptide chain. As we noted, many geometric features of this structure are fixed as a result of bonded interactions between adjacent atoms. The bond lengths and bond angles are nearly constant for all proteins. Additionally, the backbone peptide bond has substantial double-bond character so that all the atoms of the peptide bond, together with the connected α-carbon atoms (conventionally labeled C_α), lie in a common plane with the carbonyl oxygen and amide hydrogen in the *trans* configuration. Consequently, the only adjustable geometric features of the polypeptide-chain backbone involve rotations about the single covalent bonds that connect each residue's C_α to the adjacent planar peptide groups.

Rotations about the C_α—N bond are labeled with the Greek letter ϕ (phi), and rotations about the C_α-carbonyl carbon are labeled ψ (psi) (fig. 4.14). All possible conformations of a

Figure 4.14

The conformation corresponding to $\phi = 0°$, $\psi = 0°$. This conformation is disallowed by the steric overlap between the H and O atoms of adjacent peptide planes. Rotation of both ϕ and ψ by 180° gives the fully extended conformation seen in figure 4.2. Curved arrows for ϕ and ψ indicate positive variations in angle.

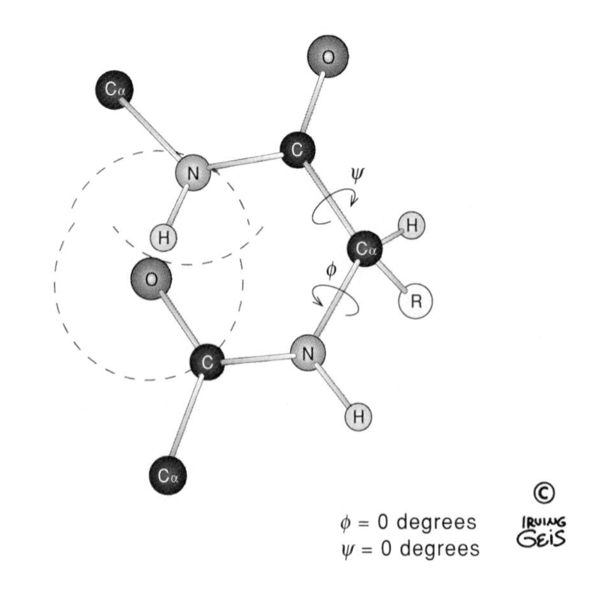

polypeptide chain can be described in terms of their ϕ,ψ conformational angles, a description that automatically takes account of the fixed geometric features of the polypeptide backbone. Thus any polypeptide conformation can be represented as a point on a plot of ϕ versus ψ, where ϕ and ψ have values that range from $-180°$ to $+180°$. By convention, the formation corresponding to $\phi = 0$, $\psi = 0$ is one in which both peptide planes that are connected to a common C_α atom lie in the same plane, as shown in figure 4.14. Positive variations in ϕ correspond to clockwise rotations of the preceding peptide about the C_α—N_1 bond when viewed from C_α toward N_1 (see fig. 4.2). Positive variations in ψ correspond to clockwise rotations of the succeeding peptide about the C_α—C_2 bond when viewed from C_α toward C_2 (again, see fig. 4.2).

Experiments with models that approximate the polypeptide atoms as hard spheres, with appropriate van der Waals radii, quickly reveal that many ϕ,ψ angular combinations are impossible because of steric collisions between atoms along the backbone or between backbone atoms and the side-chain R groups. For example, it is clear that the $\phi = 0$, $\psi = 0$ conformation shown in figure 4.14 is impossible. The reason is that this conformation results in noncovalently bonded interatomic contacts that are considerably less than the sum of the van der Waals radii of the atoms involved. In fact, of all the possible ϕ,ψ combinations, only a relatively restricted number of conformations are sterically allowed. The Ramachandran plot (fig. 4.15) shows explicitly how the accessible regions of ϕ,ψ space

Figure 4.15

Ramachandran plot, showing which atomic collisions (using a hard-sphere approximation) produce the restrictions of the main-chain angles ϕ and ψ. The cross-hatched regions are allowed for all residues, and each boundary of a prohibited region is labeled with the atoms that collide in that conformation. Additional shaded regions are for glycine residues only. The numbering scheme for amide atoms used in the derivation diagram is given in figure 4.2. Each boundary of a prohibited region is labeled with the atoms that collide in that conformation. For an explanation of the various labeled structures, see the text.

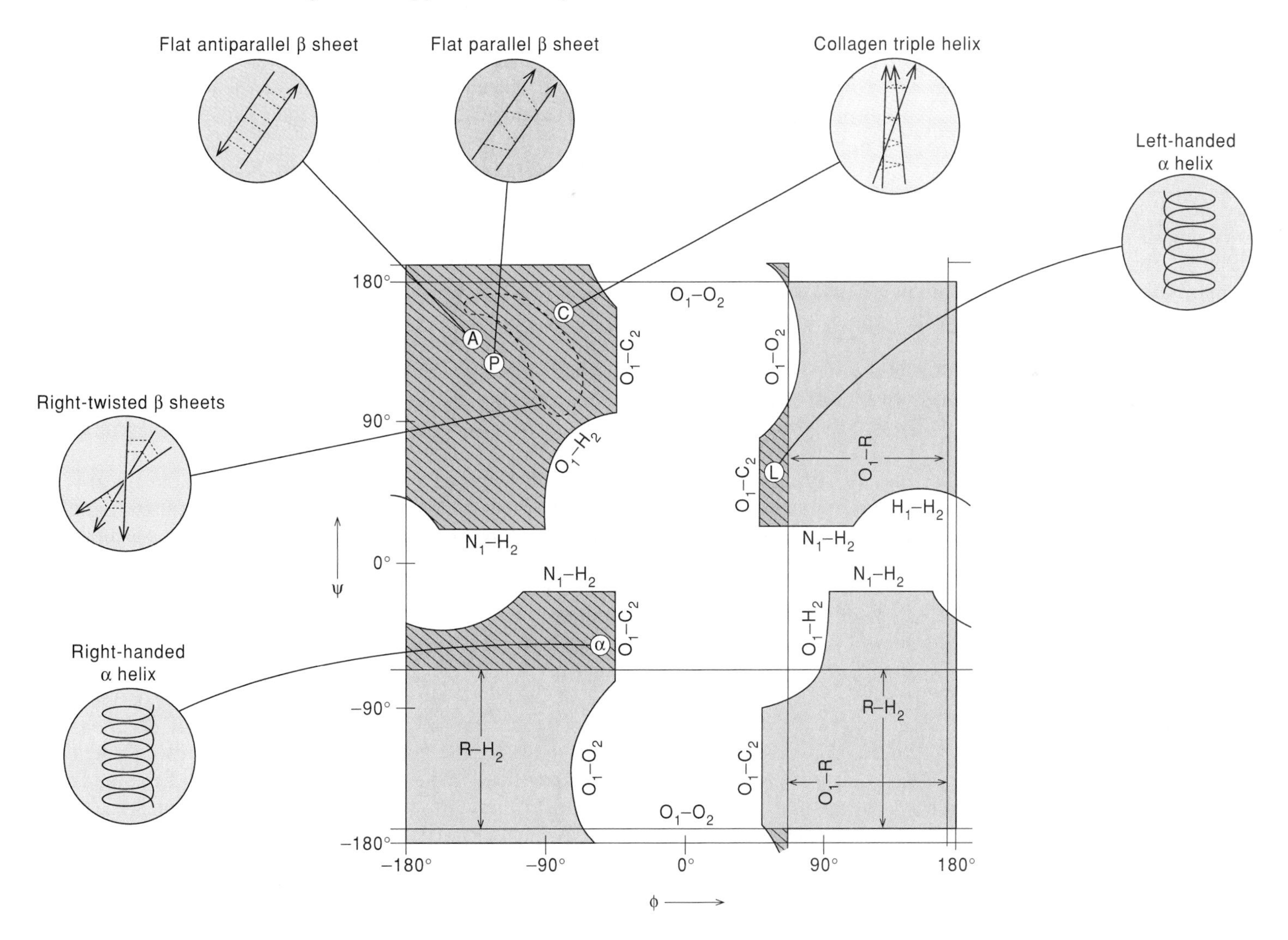

Figure 4.16

Ramachandran plot of main-chain angles ϕ and ψ, experimentally determined for approximately 1,000 nonglycine residues in eight proteins whose structures have been refined at high resolution (chosen to be representative of all categories of tertiary structure).

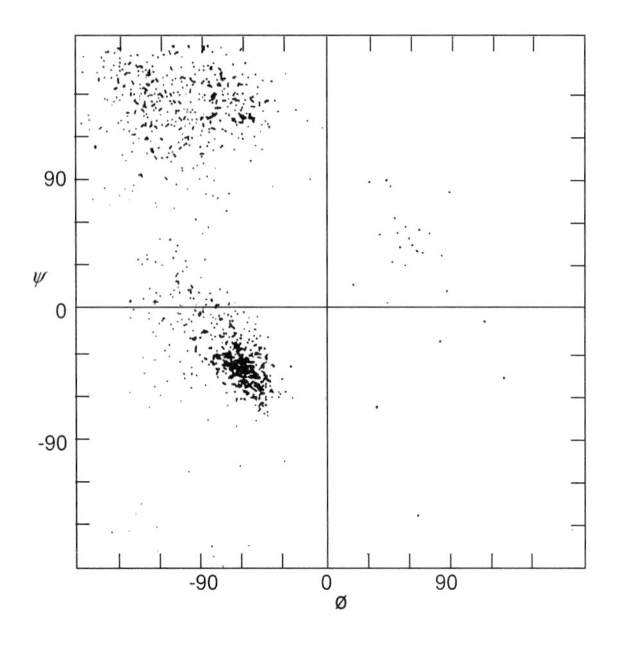

are limited by steric interactions among the polypeptide backbone and side-chain groups, assuming that the atomic groups behave as rigid spheres having appropriate van der Waals radii. In reality, the atoms in molecules do not behave as rigid spheres, so real proteins span a slightly greater range of values than suggested by this plot.

Figure 4.16 shows the distribution of some observed conformational values for proteins whose three-dimensional structures are known from crystallography. The great majority of these lie within the bounds defined by allowable steric interactions. The exceptional residues are usually glycines. Glycine frequently can assume conformations that are sterically hindered in other amino acids because its R group, a hydrogen atom, is considerably smaller than the CH_2 or CH_3 groups connected to C_α in all other amino acids.

In summary, owing to the basic geometric properties of the polypeptide chain, the sterically allowed conformations are greatly restricted by the occurrence of unfavorable steric interactions between various atomic groups. As a result, fibrous proteins with regularly repeating structures can be defined by single values for the coordinates ϕ and ψ. Proteins that have less regular structures would require more than a single set of coordinates but nevertheless they would be expected to obey the same set of limits established by the Ramachandran plot.

Analysis of Globular Proteins

It is not surprising that repeating structures with long-range order were the first protein structures to be understood. The demands on the available technology were minimal. Much more sophisticated technology was required to interpret the diffraction patterns of most proteins that have less long-range repetition. A host of crystallographers struggled over these structures, but the outstanding pioneers were Jonathan Kendrew and Max Perutz. They led their research teams to discovery of the structures of myoglobin and hemoglobin, the first globular-protein structures to be deciphered.

Today, enormous advances in protein chemistry and computer technology have systematized the necessary research and greatly reduced the amount of work and time required to investigate a protein structure. Accurate structure determinations have now been made on over 500 different proteins. It appears that detailed three-dimensional structures of proteins also may be deduced either by a combination of electron diffraction and low-dose imaging in the electron microscope or by high-resolution two-dimensional nuclear magnetic resonance (NMR) spectroscopy. Nevertheless, essentially all the information summarized here comes from the results of x-ray crystallography. From this wealth of data, patterns of structure are becoming apparent that suggest, among other things, that the overall folding arrangements of proteins may some day be predictable from the amino acid sequences of the polypeptide chain.

Three-Dimensional Protein Structure Can Be Analyzed by X-Ray Diffraction of Protein Crystals

The quality of the information potentially available through x-ray diffraction depends on the degree or extent of order of the protein molecules in a given sample. A sample under investigation usually consists of a hydrated purified protein. If the individual protein molecules are packed in a random order relative to one another, the distances between atoms or groups of atoms can be derived, but we cannot tell how these spacings are ordered in three dimensions.

If the protein is fibrous, it is often possible to learn its two-dimensional order. For example, when a fibrous sample consisting of long α helices is stretched, the helix axes become oriented in the direction of stretching. The resulting fibrous bundle gives a characteristic x-ray diffraction pattern indicating an ordered molecular arrangement along the helix axis. As we have seen, the pitch of the α helix (5.4 Å) and the advance per residue along the helix axis (1.5 Å) were detected by this technique. But the orientation of specific atoms in the helix cannot be determined from diffraction patterns of stretched fibers because of the lack of three-dimensional order. To deduce the correct three-dimensional structure for the α helix (and the β sheet), it was necessary to work with molecular models. (Currently, computer programs are available for such purposes.) By

Figure 4.17

Schematic diagram of the procedures followed for image reconstruction in light microscopy (top) and x-ray crystallography (bottom).

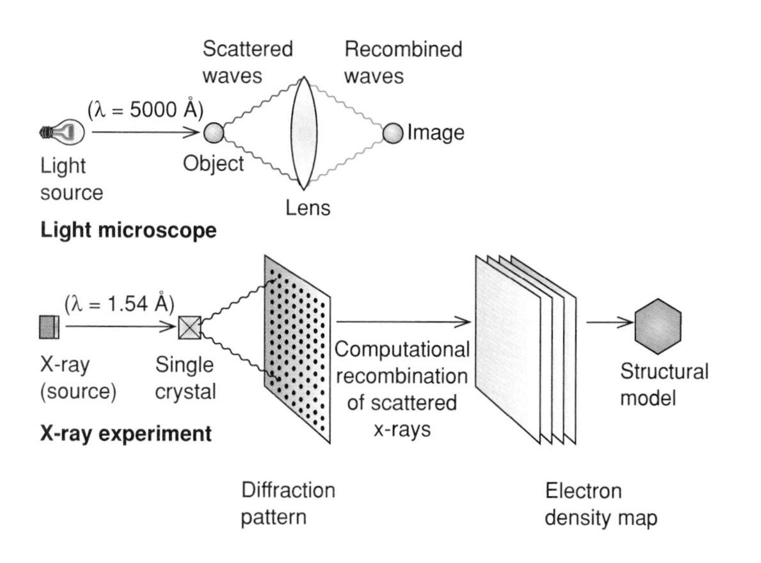

trial and error, model structures were built that were consistent with the steric limitations of the polypeptide chain that had the helix pitch and the advance per residue indicated by the diffraction pattern of stretched fibers.

The most information about a protein's structure is obtained from ordered three-dimensional protein crystals; this is the main interest of x-ray crystallographers. The goal in x-ray crystallography is to obtain a three-dimensional image of a protein molecule in its native state at a sufficient level of detail to locate its individual constituent atoms. The way this is done can most easily be appreciated by considering the more familiar problem of how we obtain a magnified image of an object in a conventional light microscope. In a visible-light microscope, light from a point source is projected on the object we wish to examine. When the light waves hit the object, they are scattered so that each small part of the object essentially serves as a new source of light waves. The important point is that the light waves scattered from the object contain information about its structure. The scattered waves are collected and recombined by a lens to produce a magnified image of the object (fig. 4.17).

Given this picture, we might ask what prevents us from simply putting a protein molecule in place of our object and viewing its magnified image. The basic problem here is one of resolution. The resolution, or extent of detail, that can be recovered from any imaging system depends on the wavelength of light incident upon the object. Specifically, the best resolution obtainable equals $\lambda/2$, or one-half the wavelength of the incident light. Since λ lies in the range of 4,000–7,000 Å for visible light, a visible-light microscope clearly does not have the resolving power to distinguish the atomic structural detail of molecules. What we need is a form of incident radiation with a wavelength comparable to interatomic distances. X-rays emitted from excited metal atoms, with wavelengths in the range of one to a few angstroms, would be most suitable.

However, simply replacing a visible-light source with an x-ray source does not solve all the problems. For example, to get a three-dimensional view of a protein, some provision must be made for looking at it from all possible angles, an obvious impossibility when dealing with a single molecule. Furthermore, when x-rays interact with proteins, very few of the rays are scattered. Most x-rays pass through the protein, but a relatively large number of them interact destructively with the protein, so that a single molecule would be destroyed before scattering enough x-rays to form a useful image. Both these problems are overcome by replacing a single protein molecule with an ordered three-dimensional array of many molecules, which scatters x-rays essentially as if it were one molecule. This ordered array of protein molecules forms a single crystal, so the general technique is called protein x-ray crystallography.

The problems do not end here, because although the protein crystal readily scatters incident x-rays, there are no lens materials available that can recombine the scattered x-rays to produce an image. Instead, the best that can be done is to directly collect the scattered x-rays in the form of a diffraction pattern. Although recording the diffraction pattern results in loss of some important information, experimental techniques have been developed for recovering the lost information. Eventually the scattered waves can be mathematically recombined in a computational analog of a lens. By collecting the diffraction pattern of the crystal in many orientations, it is possible to construct a three-dimensional image of the protein molecule (see box 4B).

4B

BOX

Interpretation of Diffraction Patterns from Protein Crystals

Crystals suitable for protein x-ray studies may be grown by a variety of techniques, which generally depend on solvent perturbation methods for rendering proteins insoluble in a structurally intact state. The trick is to induce the molecules to associate with each other in a specific fashion to produce a three-dimensionally ordered array. A typical protein crystal useful for diffraction work is about 0.5 mm on a side and contains about 10^{12} protein molecules (an array 10^4 molecules long along each crystal edge). Note especially that, because protein crystals are from 20 to 70% solvent by volume, crystalline protein is in an environment that is not substantially different from free solution.

The x-ray radiation usually employed for protein crystallographic studies is derived from the bombardment of a copper target with high-voltage (50 kV) electrons, producing characteristic copper x-rays with $\lambda = 1.54$ Å. Figure 1 shows, in schematic fashion, the x-ray diffraction pattern from a protein crystal. Several features about this pattern bear explanation. First, as you can see, the diffraction pattern consists of a regular lattice of spots of different intensities. The spots are due to destructive interference of waves scattered from the repeating unit of the crystal. For the crystal whose diffraction pattern is shown, the repeating unit (or crystal unit cell) contains four symmetrically arranged protein molecules. Corresponding symmetrical features appear in the spot-intensity pattern. Further, the lattice spacing of the diffraction spots is inversely proportional to the actual dimensions of the crystal's repeating unit or unit cell. Consequently, both the crystal's unit-cell dimensions and general molecule packing arrangement can be derived from inspection of the crystal's diffraction pattern.

Information concerning the detailed structural features of the protein is contained in the intensities of the diffraction spots. All the atoms in the protein structure make individual contributions to the intensity of each diffraction spot. Therefore, to deduce the three-dimensional structure, all the spots must be measured, either by scanning the x-ray films with a densitometer or by measuring the diffraction spots individually with a scintillation counter.

Initial studies of a protein's tertiary structure are generally carried out at low resolution, that is, using intensity data near the origin (center) of the diffraction pattern. Diffraction data near the origin reflect large-scale structural features of the molecule, while those nearer the edge correspond to progressively more detailed features. Figure 2 provides examples of electron-density maps calculated at different resolutions to show how various levels of structural detail appear at different degrees of resolution.

A powerful aspect of protein crystallography is that once the native structure is known, various cofactors or enzyme substrate

Figure 1

Schematic view of an x-ray diffraction pattern. The spacing of the spots is reciprocally related to the dimensions of the repeating unit cell of the crystal. The symmetry of the spots (e.g., the mirror planes in the sample shown) and the pattern of missing spots (alternating spots along the mirror axes) give information on how molecules are arranged in the unit cell. Information concerning the structure of the molecule is contained in the intensities of the spots. Spots closest to the center of the film arise from large-scale or low-resolution structural features of the molecule, while those farther out correspond to progressively more detailed features. Circles show 5-Å and 3-Å regions of resolution. Mirror axes are labeled *m*. Spacing of vertically oriented spots, *b, and horizontally oriented spots, *a**, are reciprocally related to *b* and *a*, the dimensions of the unit cell.**

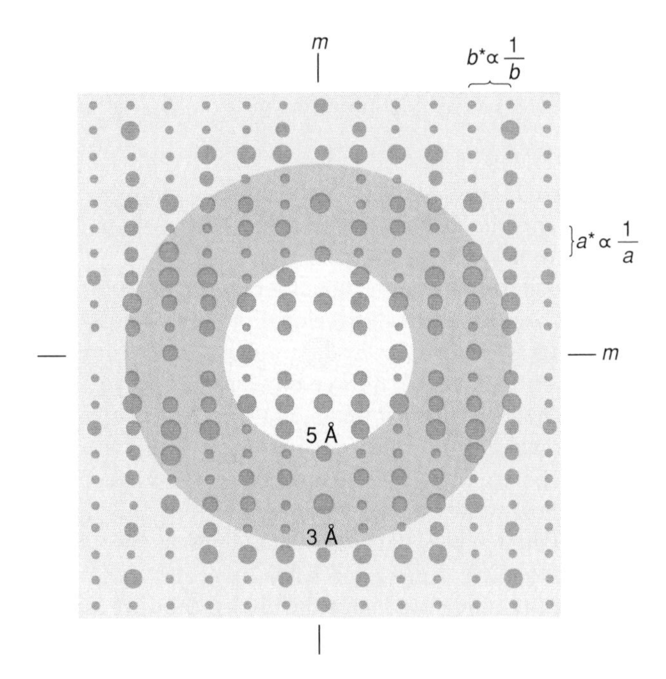

The Hierarchy of Globular Protein Structure

In our analysis of fibrous protein structures, we said little about the importance of interaction with water in determining the final folded structures of the proteins. This is because most of the side chains in fibrous proteins are routinely exposed to water except when they interact with each other to form multimolecular aggregates. Then the difference in their affinity for other, like side chains and for water becomes a major issue. In the case of globular proteins, the interaction of amino acid side chains is a major issue almost from the start, since globular proteins have a large number of their amino acid side chains buried in the interior of their folded structures. Hence in our analysis of

Figure 2

Views of crystallographic electron-density maps, showing how the structural detail revealed depends on the resolution of the data used to compute the maps. The actual molecular structure is inserted in its true position in the electron-density maps.

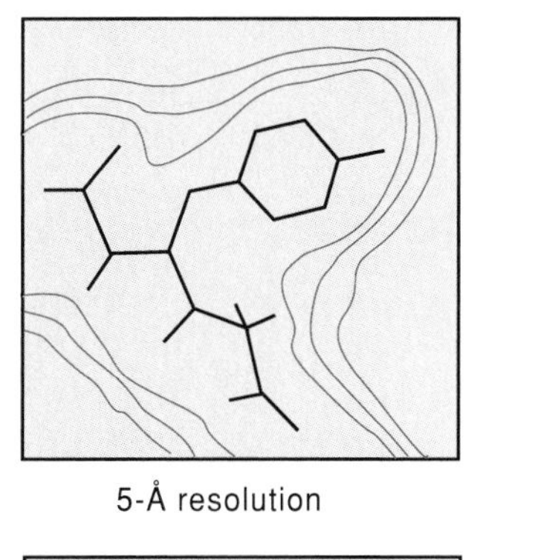

5-Å resolution

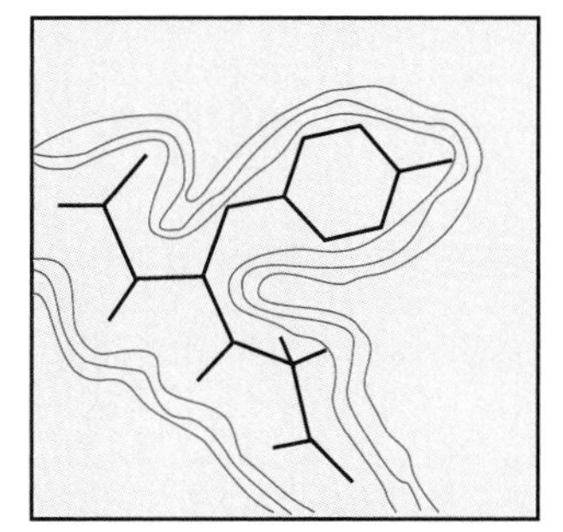

3-Å resolution

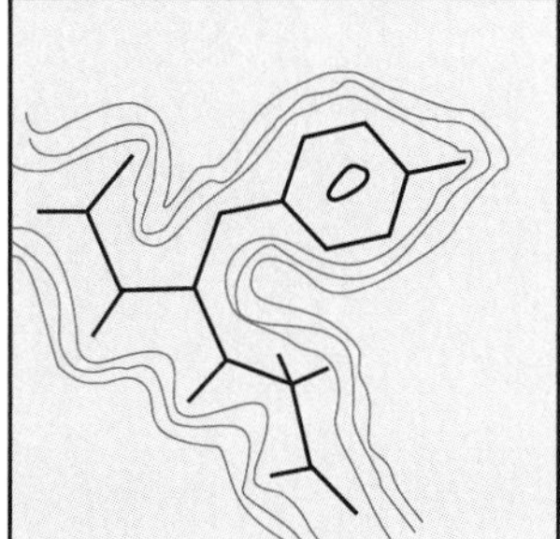

2-Å resolution

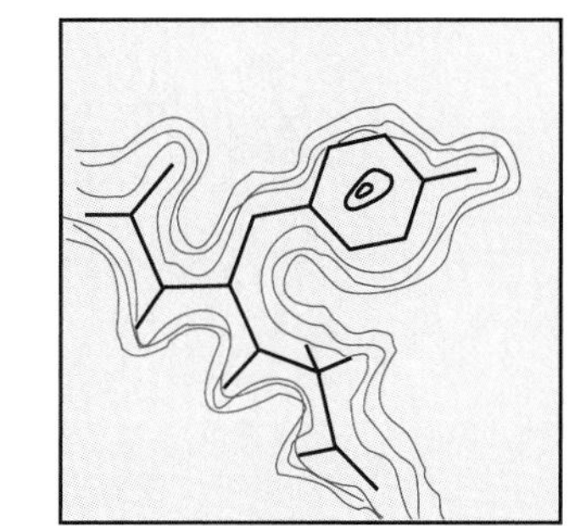

1.5-Å resolution

analogs can be bound to the molecule in the crystal. By simply measuring the diffraction intensities, we can compute a new map that allows direct and explicit examination of the structural interactions between the native protein and its substrate or cofactor molecules. Detailed analysis of these interactions has provided much of the foundation for our current understanding of many protein catalytic and functional properties.

the higher orders of structure of globular proteins we must be aware, not only of those structural considerations that were important in the analysis of fibrous proteins, but also of those considerations, first raised in chapter 1, that relate to the interaction of amino acid side chains with water.

We may think of the structure of globular proteins at four levels (fig. 4.18). Steric interactions restrict accessible conformations and reflect features of the protein's amino acid sequence, or primary structure. The requirements for hydrogen-bond preservation in the folded structure result in the cooperative formation of regular structural regions in proteins. This situation arises principally because of the regularly repeating geometry of the hydrogen-bonding groups of the polypeptide backbone, which leads to the formation of regular hydrogen-bonded secondary structures. Association between elements of secondary structure in turn results in the formation of structural domains, whose properties are determined both by chiral properties of the polypeptide chain and packing requirements that effectively minimize the molecule's hydrophobic surface area. Further association of domains results in the formation of the protein's tertiary structure, or overall spatial arrangement of the polypeptide chain in three dimensions. Finally, fully folded protein subunits can pack together to form quaternary structures, which may either serve a structural role or provide a structural basis for modification of the protein's functional properties.

Because globular proteins have nonrepeating structures, it is essential to have a means for displaying the entire three-dimensional structure with sufficient detail and yet not too much detail, so that the overall structural design can be appreciated (see box 4C).

Primary Structure Determines Tertiary Structure

Throughout the discussion of protein structures we have assumed that structures form because they represent the most stable way of arranging the polypeptide chains. The first direct support for this notion for a globular protein came from the studies of Cris Anfinsen. What Anfinsen did was to completely unfold pancreatic ribonuclease, an enzyme containing 124 amino acid residues with four disulfide bridges, and then to find conditions under which it could be refolded into its original native structure. First, the enzyme was denatured in a solution containing the hydrogen-bond-breaking reagent urea and β-mercaptoethanol, a thiol reagent that reduces disulfides to sulfhydryls, thus cleaving the covalent cross-links (fig. 4.19). These conditions have been used since as a general means of denaturing proteins, by completely disrupting the conformation without giving rise to coagulation or precipitation. The reduced, denatured ribonuclease was enzymatically inactive because its native structure had been destroyed. Renaturation was carried out by removing the urea, a step that caused the protein to refold. Finally the mercaptoethanol was removed to permit the air oxidation of the reduced disulfides back to disulfide cross-links. The result of this series of manipulations was an almost complete recovery of the enzymatic activity of the original ribonuclease molecule.

Thus it appears that the information for folding to the native conformation is embodied in its amino acid sequence, for of the many disulfide-paired ribonuclease isomers that are possible, only one was formed in major yield. Further studies have indicated that a similar result is obtained with many other proteins.

Figure 4.18

Hierarchies of protein structures.

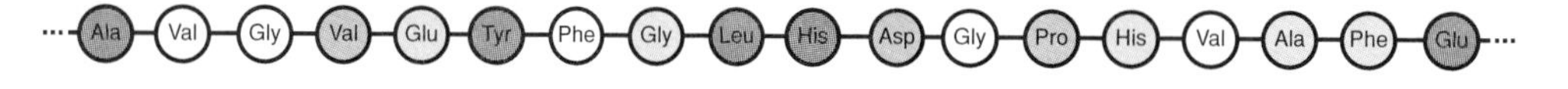

(a) Primary structure (amino acid sequence in the protein chain)

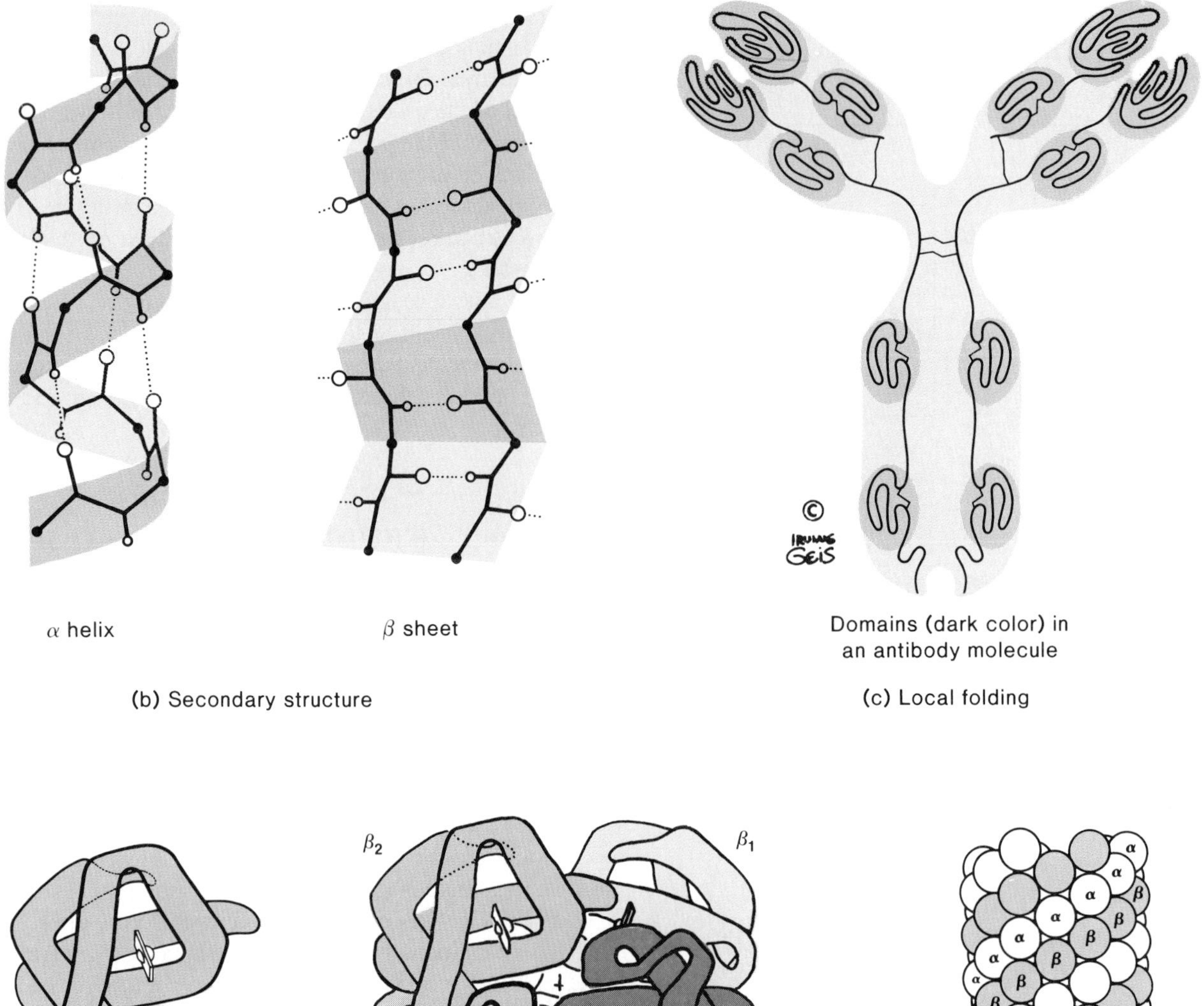

α helix

β sheet

(b) Secondary structure

Domains (dark color) in an antibody molecule

(c) Local folding

β

One complete protein chain (β chain of hemoglobin)

(d) Tertiary structure

The four separate chains of hemoglobin assembled into an oligomeric protein

(e) Quaternary structure

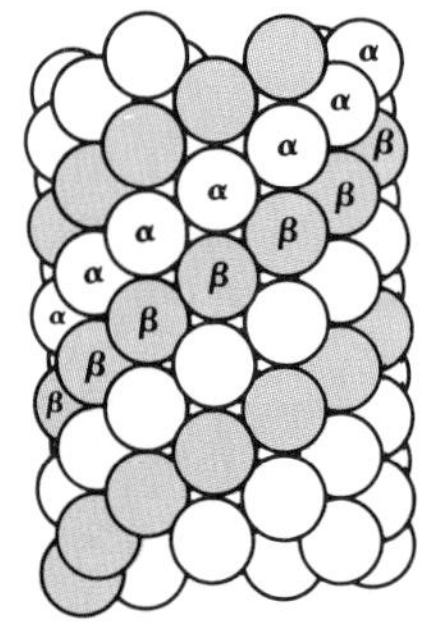

σ (white) and β (color) tubulin molecules in a microtubule

(f) Quaternary structure

Figure 4.19

Schematic representation of an experiment to demonstrate that the information for folding into a biologically active conformation is contained in the protein's amino acid sequence.

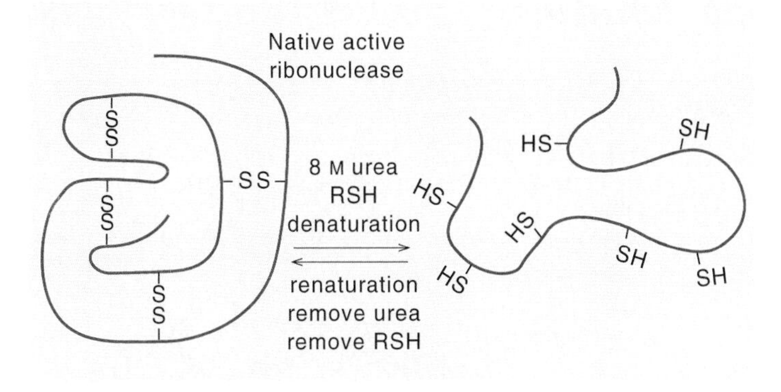

Despite the elegance and simplicity of the Anfinsen experiment, current indications are that there are groups of proteins that accelerate the folding process and that possibly return damaged or incorrectly folded proteins to their native state. These polypeptide-chain-binding proteins (PCB proteins) fall into two families: proteins that are related to the heat shock protein that has a subunit molecular weight of 70,000 (designated hsp70), and the GroEL family of proteins. The GroEL proteins are found in bacteria, mitochondria, and chloroplasts. The hsp70 proteins appear to be ubiquitous, occurring in bacteria, mitochondria, and the eukaryotic cytosol. They are especially abundant in the endoplasmic reticulum.

Hydrophobic Forces Play A Major Role in Determining Folding Patterns

The studies of Anfinsen not only show that folding can be a spontaneous process predetermined by the primary amino acid sequence, they suggest that folded structures are thermodynamically more stable. Thus the folding of globular proteins should be understandable in thermodynamic terms. Although x-ray diffraction cannot be used for studying proteins in solution, there are other optical methods available for the purpose (see box 4D).

Hydrogen bonding and van der Waals forces are of great importance in determining the secondary structures formed by fibrous proteins. To understand the complex folded structures found in globular proteins, additional types of interactions between amino acid side chains and water must also be considered. The so-called hydrophobic forces that lead to the interaction of hydrophobic groups in proteins are the hardest type of noncovalent interactions to appreciate. Whereas H bonding and van der Waals forces are due primarily to enthalpic factors, hydrophobic forces relate primarily to entropic factors. Furthermore, the entropic factors mainly concern the solvent, not the solute.

As a first step to understanding the nature of these forces it is useful to consider the interaction between water and a small hydrophobic molecule like hexane. It is tempting to ascribe the water insolubility of hexane to van der Waals attractive forces between these small hydrophobic molecules. However, thermodynamic measurements indicate that this explanation is wrong. The hydrophobic molecule hexane has a small favorable enthalpy for solution in water; however, it is highly insoluble in water because of an unfavorable entropic factor. What is this mysterious factor? Owing to the weak enthalpic interactions between hexane and water, the water withdraws slightly in the region of the apolar hydrophobic molecule and forms a relatively rigid hydrogen-bonded network with itself (for example, see the clathrate structure illustrated in figure 1.10). The network effectively restricts the number of possible orientations of water molecules directly opposed to the dissolved hexane molecules. This ordering of water constitutes an energetically unfavorable entropic effect.

The same type of entropic effect plays a major role in directing the folding of globular proteins. About half of the amino acid side chains in proteins are hydrophobic (e.g., alanine, valine, isoleucine, leucine, and phenylalanine). For these side chains, entropic effects strongly favor internal locations, where they are free from contacts with water (e.g., see figure 1.12, which shows the location of polar and apolar side chains for cytochrome *c*).

The native folded state of a globular protein reflects a delicate balance between opposing energetic contributions of large magnitude. Whereas entropic factors favor the folding of hydrophobic side chains into the interior regions of globular proteins, enthalpic factors favor interaction of hydrophilic side chains on the surface of the protein, where they interact with water. In the limited number of cases for which data are available, it appears that the overall entropy of folding is slightly negative (unfavorable) and the overall enthalpy is also slightly negative (favorable). On balance, folding is opposed by the entropy but favored by the enthalpy change and occurs because the latter factor outweighs the former.

Secondary Valence Forces Are the Glue that Holds Polypeptide Chains Together

When forces involved in determining protein conformation are being considered, we can ignore covalent bond energies, despite their large magnitude (table 4.2). This is because covalent bonds (except for disulfide bonds) are not made or broken when polypeptide chains fold into their native three-dimensional conformations. The bonds that are affected (made or broken) on folding are, by and large, of the noncovalent type, involving secondary valence forces. Typically the intermolecular bond energies between noncovalently linked atoms range in value from 0.1 to 6 kcal/mol. The intermolecular forces between noncovalently linked atoms may be grouped into four categories: hydrophobic forces, electrostatic forces, van der Waals forces, and hydrogen

4C
BOX

Visualizing Molecular Structures

The primary data of protein crystallography yield a three-dimensional electron-density map, which must be interpreted in terms of a complex three-dimensional model of all atom positions in the protein. Such modeling is now generally done by computer graphics.

The difficulty with complete models is that even for relatively small proteins, their complexity is almost impossible to comprehend. Therefore, it is more common to show only selected parts of a molecule (such as only the polypeptide backbone) and to highlight particular features of interest.

Here, we show four different presentations for ribonuclease, in which each model is seen from the same perspective. Each method of presentation has its advantages. The space-filling model (figure 1) is excellent for displaying the volume occupied by molecular constituents and the shape of the outer surface. Figure 2 shows a stereo pair of a space-filling model from exactly the same perspective as figure 1. When the pair is viewed with stereo glasses, the three-dimensional illusion is striking. Figure 3 shows the polypeptide chain, with N and O atoms labeled and with dotted lines representing hydrogen bonds. In addition, all α-carbon positions are numbered. Figure 4 shows an abstraction of the polypeptide chain in which the β strands are characterized as flat arrows and the α helices as spiral ribbons. This simplified style has proved useful in classifying and comparing proteins according to their secondary- and tertiary-structure folding patterns.

Figure 1

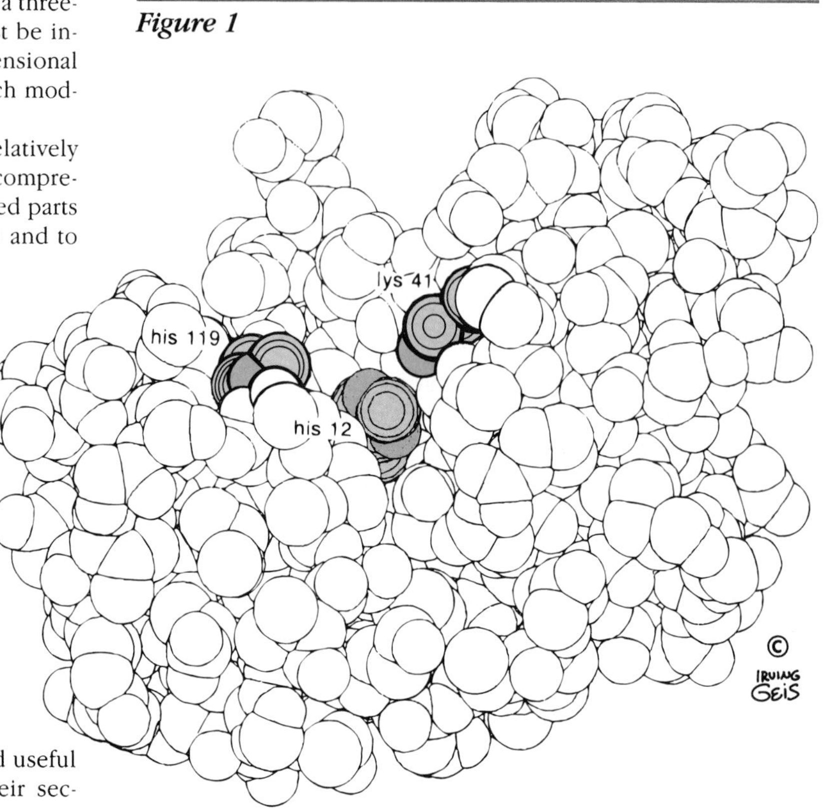

Figure 2

The 3-D illusion can be seen without stereo glasses by the following method. With your eyes about 10 inches from the page, stare at the drawings below as if you were looking straight ahead at a far-away object. A double image will form and the central pair should drift together and fuse. Then the illusion becomes apparent. Adjust the page, if necessary, so that a horizontal line is perfectly parallel with the eyes. As an aid to seeing one image with each eye, use two cardboard tubes from toilet-paper rolls. Close the right eye and focus the left eye on the left image. Do the same for the other eye. Now, with both eyes together, the 3-D illusion should appear.

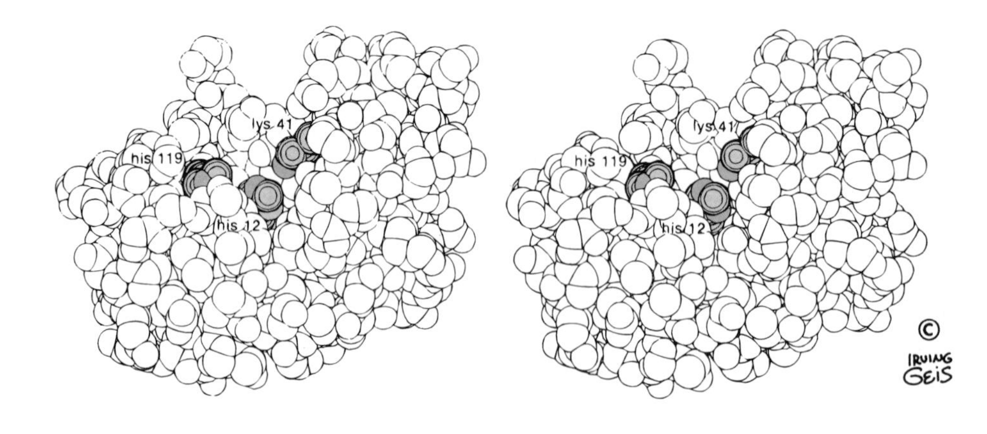

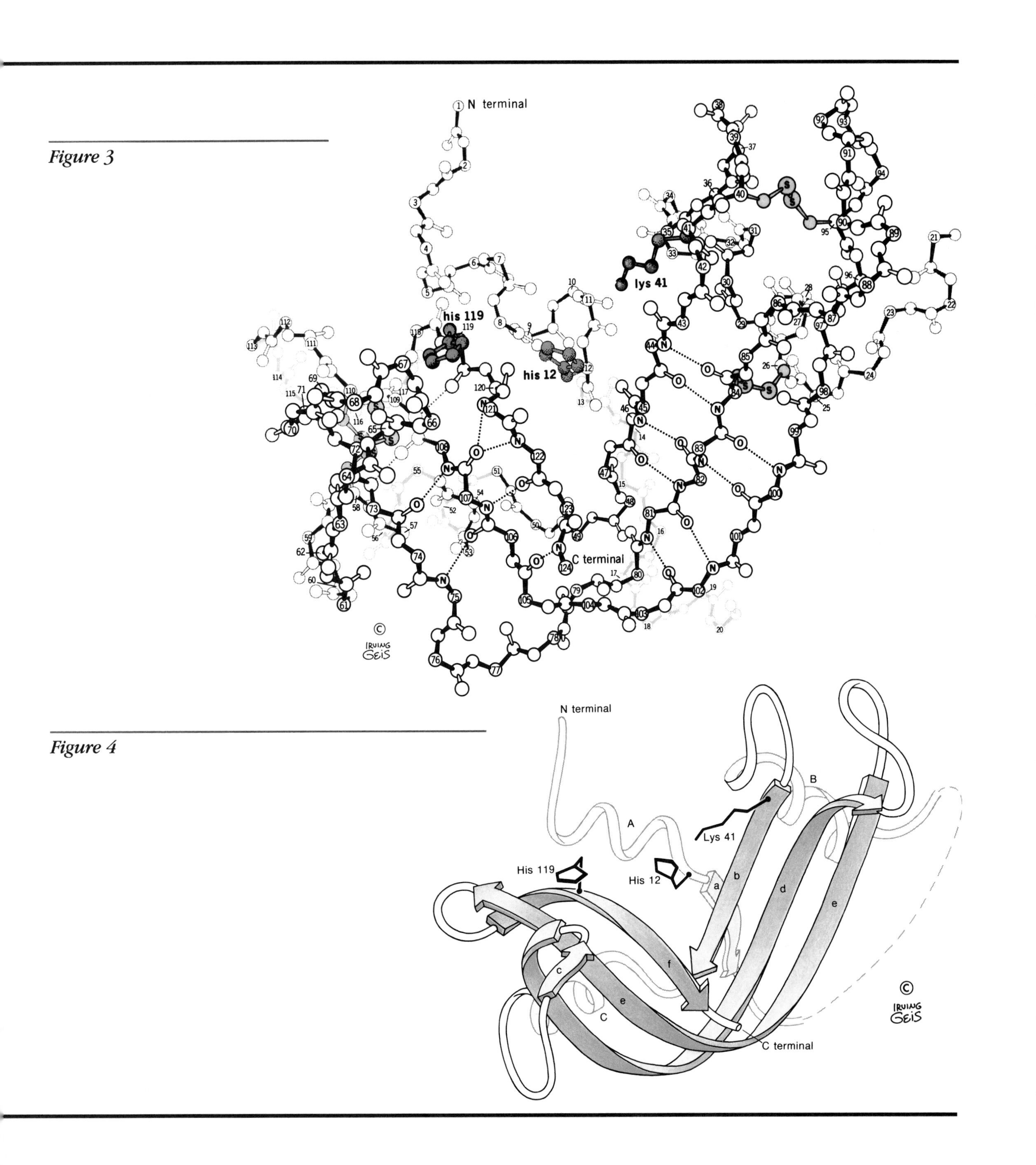

Figure 3

Figure 4

Radiation Techniques for Examining Protein Structure

The Anfinsen experiment raises the question, Must we always use x-ray diffraction to determine a protein's structure? In the case of ribonuclease the structure is known from x-ray diffraction, but at the time of Anfinsen's investigation it was not. He relied heavily on the fact that the reconstituted enzyme had the same activity as the native enzyme. This remains an acceptable approach as long as one is working with an enzyme that has a demonstrable catalytic activity, but what of the many proteins that do not?

Because of such limitations, x-ray diffraction is still the most powerful tool for determining protein structure. However, apart from the enormous amount of work it requires to use, x-ray diffraction suffers from two disadvantages: (1) it requires that a protein be available in the crystalline form, which is not always the case; and (2) there is no absolute assurance that a protein's structure in the crystalline form is the same as its structure in solution, which may be more like the environment in the cell.

Fortunately, there are at least a dozen other techniques, less fussy in their demands, that may be used to investigate protein structure either in the solid state or the solution state (see table 4D). The information obtained by these procedures is very extensive. As you can see from the table, it covers virtually every aspect of protein structure, with considerable overlap between what is yielded by different techniques. Excellent references explaining the use of other radiation techniques are given at the end of this chapter.

Table 4D
Radiation Techniques for Examining Protein Structure

Technique	Information Obtained
(a) X-ray diffraction	Detailed atomic structure
(b) Nuclear magnetic resonance spectrometry (NMR)	Structure of specific sites; ionization state of individual residues
(c) Electron paramagnetic resonance	Structure of specific sites; this includes structures of small molecules whether they be carbohydrates, lipids, small proteins or complexes between segments of DNA and proteins
(d) Spectrometry (EPR)	
(e) Optical absorption spectroscopy	Measurement of concentrations or rate of reactions
(f) Infrared absorption spectroscopy	Type and extent of secondary structure
(g) Light scattering	Molecular weight and size
(h) Ultraviolet absorption spectroscopy	Concentration and conformation
(i) Fluorescence	Proximity between specific sites
(j) Raman scattering	Structure of specific sites
(k) Optical rotatory dispersion (ORD)	Type and extent of secondary structure
(l) Circular dichroism (CD)	Type and extent of secondary structure
(m) Fluorescence polarization	Molecular weight, shape, flexibility and orientation of secondary structure units

bonds (H bonds). Hydrophobic forces have already been discussed in detail. We considered some aspects of the other types of secondary valence forces in the preceding section and in chapter 1; we will consider additional properties of these forces here.

Electrostatic Forces Electrostatic forces are of three types: charge–charge interactions, charge–dipole interactions, and dipole–dipole interactions. The energy of interaction between two charges Q_1 and Q_2 is proportional to the product of the charges and inversely proportional to the distance R between them (table 4.3):

$$\text{energy of interaction} \propto \frac{Q_1 Q_2}{R}$$

In solution this interaction is reduced by the dielectric constant of the surrounding medium:

$$\text{energy of interaction} \propto \frac{Q_1 Q_2}{\epsilon R}$$

If the two charges in question are buried within a protein, their interaction energy can be substantially increased because the dielectric constant in the regions inaccessible to water is much lower than the dielectric constant of water. The long-range nature of charge–charge interactions has led to the speculation that such forces can be important in accelerating interaction between proteins, between proteins and nucleic acids, and between proteins and small molecules such as coenzymes and substrates (see chapter 9).

Favorable charge–charge interactions between oppositely charged amino acids are less significant in determining protein folding than are the ion–dipole interactions between the

Table 4.2
Bond Energies between Some Atoms of Biological Interest

Energy Values for Single Bonds (kcal/mole)					
C—C	82	C—H	99	S—H	81
O—O	34	N—H	94	C—N	70
S—S	51	O—H	110	C—O	84
Energy Values for Multiple Bonds (kcal/mole)					
C=C	147	C=N	147	C=S	108
O=O	96	C=O	164	N≡N	226

Table 4.3
Dependence of Energy of Interaction on the Distance of Separation of the Interacting Species

Range of Interaction	Type of Interaction
$1/R$	Charge–charge
$1/R^2$	Charge–dipole
$1/R^3$	Dipole–dipole
$1/R^6$	Van der Waals (dipole–induced dipole) attractive forces
$1/R^{12}$	Van der Waals repulsive forces

charged groups of amino acid side chains and water. With very few exceptions, side chains containing charged groups as well as polar side chains are located on the protein surface at the protein–water interface.

Van der Waals Forces We referred to van der Waals forces earlier in the chapter, in connection with the models constructed by Pauling and Corey. Here we will take a more careful look.

Van der Waals interactions are of two types, one attractive and one repulsive. Attractive van der Waals forces involve interactions among induced instantaneous dipole moments that arise from fluctuations in the electron charge densities of neighboring nonbonded atoms. Such interactions amount to 0.1 to 0.2 kcal/mole; despite their small size, the large number of such interactions that occur when molecules come close together makes the interactions quite significant. As we will see, van der Waals forces favor close packing in folded protein structures.

Repulsive van der Waals interactions occur when noncovalently bonded atoms or molecules come very close together. An electron–electron repulsion arises when the charge clouds between two molecules begin to overlap. If two molecules are held together exclusively by van der Waals forces, their average separation will be governed by a balance between the van der Waals attractive and repulsive forces. This distance is known as the van der Waals separation. Some van der Waals radii for biologically important atoms are given in table 4.1. The van der Waals separation between two nonbonded atoms is given by the sum of their respective van der Waals radii.

Hydrogen Bonds We have discussed many characteristics of H bonds already. Nevertheless, there are still some points to be made. Evidence for a significant H bond comes from the observation of a decreased distance between the donor and acceptor groups making the H bond. Thus, from the van der Waals radii given in table 4.1, we can calculate that the distances between nonbonded H and O atoms and between nonbonded H and N atoms are 2.6 and 2.7 Å, respectively. When an H bond is present, the distances are usually reduced by about 0.8 Å. Some of the more important H-bond donors and acceptors are shown in figure 4.2.

Polypeptides carry a number of H-bond donor and acceptor groups, both in their backbone structure and in their side chains. Water also contains a hydroxyl donor group and an oxygen acceptor group for making H bonds (see fig. 1.8). Formation of the maximum number of H bonds between a polypeptide chain and water would require the complete unfolding of the polypeptide chain. However, it is not obvious that such an unfolding would result in a net energy gain. The reason is that water is a highly H-bonded structure, and for every H bond formed between water and protein, an H bond within the water structure itself must be broken. The strategy followed by most proteins is to maximize the number of intramolecular H bonds between the backbone peptide groups, but to keep most of the potential H-bond-forming side chains of the protein near the protein–water interface, where they can interact directly with water. It seems likely that such side-chain–water interactions will involve both H bonds, on the one hand, and charge–dipole interactions and dipole–dipole interactions on the other.

β Bends Are Useful for Building Compact Globular Proteins

Thus far we have described the geometry of protein secondary structures that resemble long rods or flat sheets. Obviously, to fold a polypeptide chain such as RNase to a compact globular form, there must be some way to change the direction of the polypeptide chain. Such folding might, for example, be required to connect adjacent ends of the polypeptide chains in an antiparallel β sheet. A commonly observed and particularly efficient way to do this is by formation of a tight loop in which a residue's carbonyl group forms a hydrogen bond with the amide NH group of the residue three positions farther along the polypeptide chain. The resulting so-called β bend (fig. 4.20) reverses the direction of the polypeptide chain.

Figure 4.20

The two major types of tight turn of ß bends (I and II). In type II, R_3 is generally glycine.

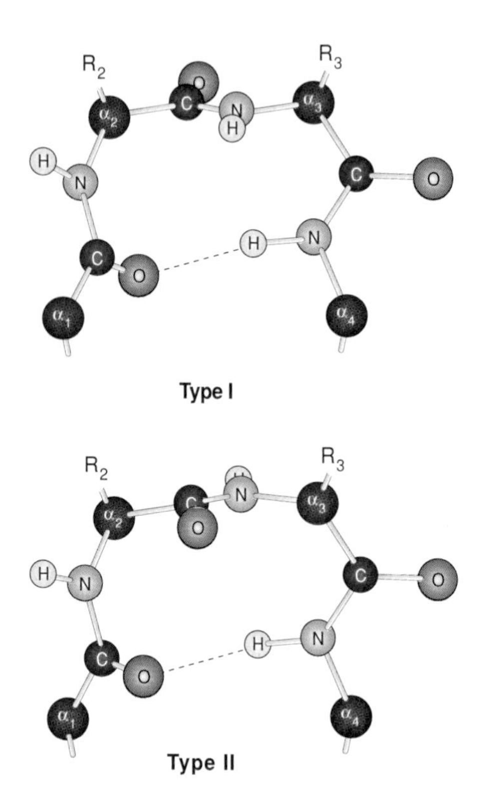

Figure 4.21

Lysozyme. In this and succeeding figures the polypeptide backbone is represented as a ribbon to allow the polypeptide-chain course to be followed easily.

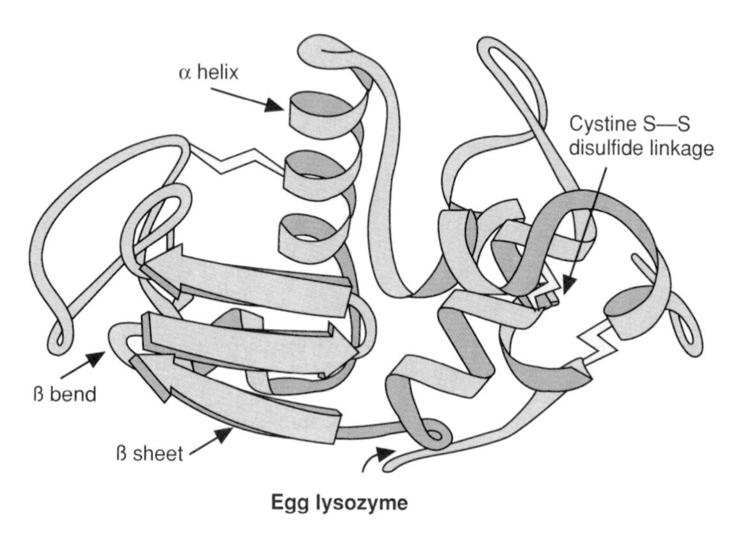

Several conformational variations of the β bend have been observed that are a function of the amino acid sequence in the bend. In particular, it has been observed that the amino acids glycine and proline occur frequently in β bends. Because of its small size, glycine is conformationally more flexible than other amino acids. It can therefore serve as a flexible hinge between regions of polypeptide chains whose steric interactions would otherwise keep them in more extended conformations. Proline, in contrast, is more conformationally restricted than other amino acids, since its cyclically bonded structure fixes its conformational degrees of freedom. In a sense, then, part of the geometry that results in bend formation is performed in proline-containing sequences. It appears probable that either situation might promote the formation of a bend during initial stages of protein folding and so cause structures such as antiparallel β sheets to assemble cooperatively, in a manner resembling the closure of a zipper.

Chirality and Packing Influence the Tertiary Structure of Globular Proteins

There is a seemingly endless array of different folding patterns that can be found for globular proteins. Despite this complexity there are rules that govern the folding process. We have already considered some basic rules that relate directly to thermodynamics. It will be helpful now to consider some additional rules relating to thermodynamic or kinetic aspects of the folding process.

Globular proteins, as their name implies, differ from fibrous proteins in that they generally have a more or less spherical shape. Nevertheless, three-dimensional structural studies of globular proteins show that they incorporate many of the secondary structural features that typify the fibrous proteins. Figure 4.21, for example, illustrates the first enzyme whose three-dimensional structure was determined, the 120-residue protein lysozyme. This protein has local regions of ordered α-helical and antiparallel β-sheet secondary structures, but it has, in addition, several extended regions incorporating β bends and extended polypeptide chains with a less regular conformation.

Among the approximately 500 proteins whose structures have been determined, there is a rich variety of alternative structural arrangements. Each different polypeptide sequence is associated with the formation of a unique tertiary structure. Careful comparisons have shown, however, that many proteins share some fundamental structural similarities. Furthermore, the similarities recur among proteins that show little similarity in sequence or function. This fact suggests that the recurring features have common physical origins. They appear, in fact, to arise from two different sorts of physical effects.

The first of these is the chiral effect, meaning the tendency for extended structural arrangements in proteins to be "handed," because their constituent polypeptide chains are composed of chiral L-amino acids. Chiral effects manifest themselves both in the manner in which regions of secondary structure are interconnected in globular proteins and in the geometric properties of globular protein sheets. The second effect of im-

Figure 4.22

The natural tendency for the polypeptide chain to twist in the right-hand direction produces structures with an overall right-handed connectivity. The structure represents a single fully extended polypeptide chain.

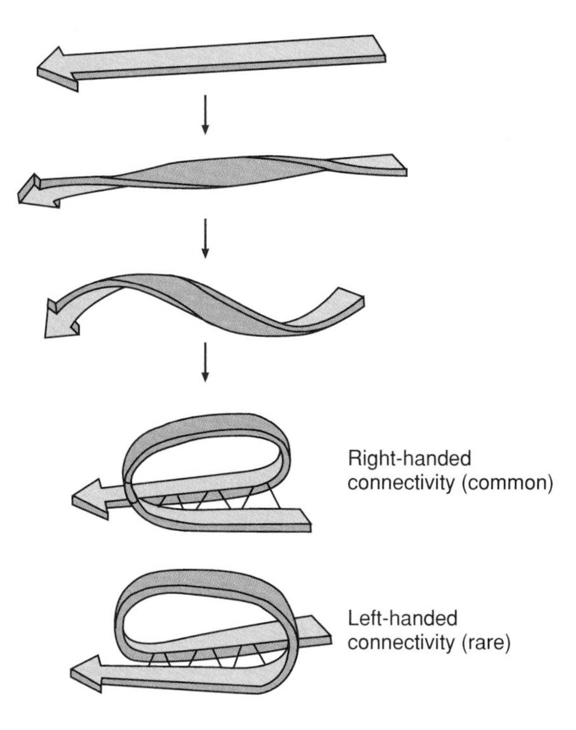

Figure 4.23

Three ways of making connections between ß strands. (*a*) A hairpin same-end connection is commonly found for ß strands in the antiparallel orientation. (*b*) A right-handed crossover connection is commonly found for ß strands in the parallel orientation. (*c*) A left-handed crossover connection is rarely found.

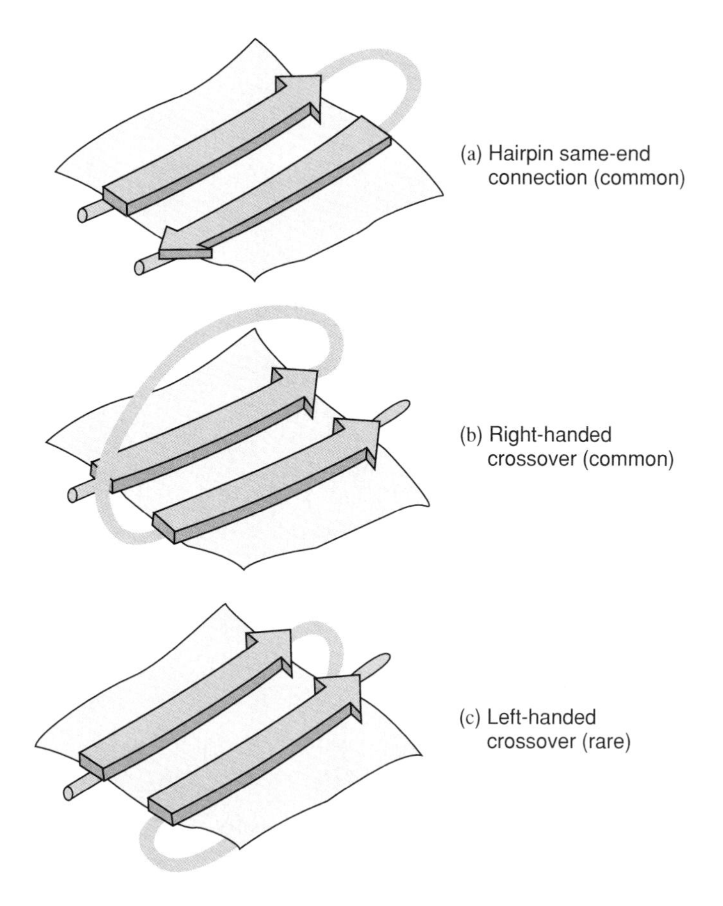

portance in tertiary structural organization relates to how secondary structural regions, such as α helices and β sheets, can most efficiently pack together so as to minimize the protein's solvent-accessible surface area.

How Chiral Properties Influence Conformation In the preceding discussion we noted that many potential polypeptide conformations are ruled out by unfavorable steric interactions. However, the frequent occurrence of structures such as α helices suggests that certain of these arrangements not only are allowed, but are particularly stable. As in the case of the α helix, the relative stability of a particular conformation is governed by the details of the interaction forces among the atoms comprising the polypeptide chain. Given the fact that proteins are composed primarily of chiral L-amino acids, it is not surprising that the most stable conformations of extended polypeptide chains are not straight. Instead, detailed conformational energy calculations have shown that extended polypeptide chains prefer to be slightly twisted in a right-handed sense when viewed down the polypeptide chain axis. Since the residues of straight, extended polypeptide chains alternate in position by 180° (e.g., see fig. 4.7), the cumulative effect of this tendency toward right twisting is to produce extended structures that are coiled in a right-handed direction (fig. 4.22).

The effects of this tendency for extended-chain structures to form right-handed coiled or twisted structures are revealed in two related but different structural features common to virtually all known proteins. The first is the kind of connection that occurs between the ends of parallel polypeptide strands that form β sheets in globular proteins. The connection from one β strand to the next can occur at the same end of the sheet in a simple hairpin turnaround only in antiparallel β sheets. In parallel β sheets a "crossover" connection to the other end of the sheet is required; theoretically, this can be either right-handed or left-handed (fig. 4.23). All crossover connections observed in protein structures are right-handed, irrespective of whether the strands they join are adjacent. This invariant pattern is most likely a result of the energetically favored nature of the right-handed crossover.

The second feature that reveals the influence of right-handed twisting is the geometry of globular protein parallel β sheets. The β sheets of globular proteins are always twisted in a right-handed sense when viewed along the polypeptide chain direction. The twisting behavior of β sheets is an important feature in protein structural architecture, since twisted β sheets frequently constitute the backbone of protein structures. Figure

Figure 4.24

A comparison of parallel ß-sheet structures forming the backbone structures in different enzymes (or parts of enzymes): (*a*) ß-barrel arrangement, (*b*) saddle shape.

ß-BARREL SHAPE

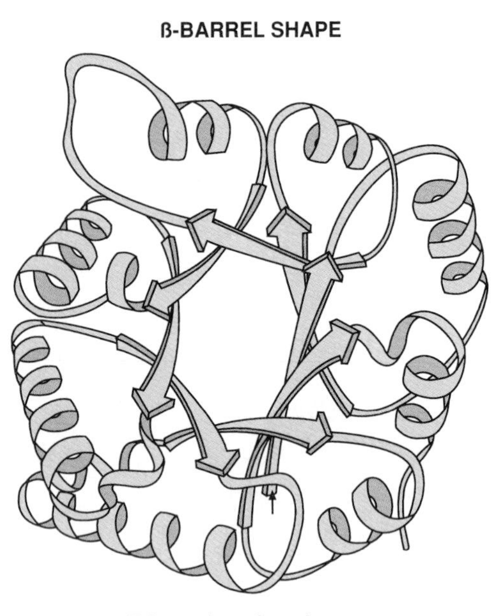

Triose phosphate isomerase

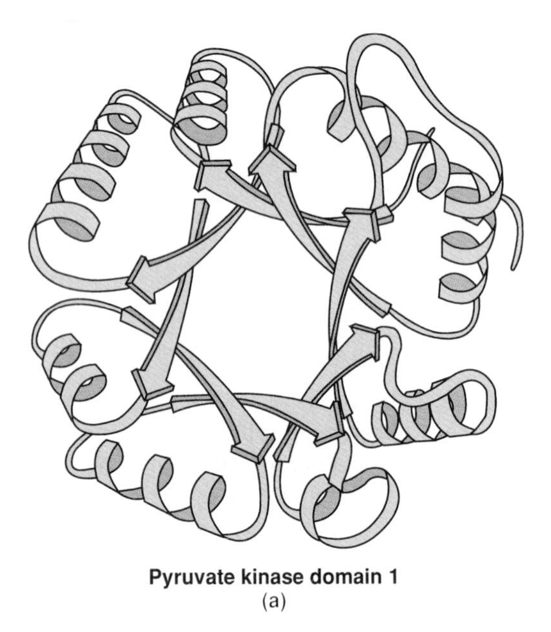

Pyruvate kinase domain 1

(a)

SADDLE SHAPE

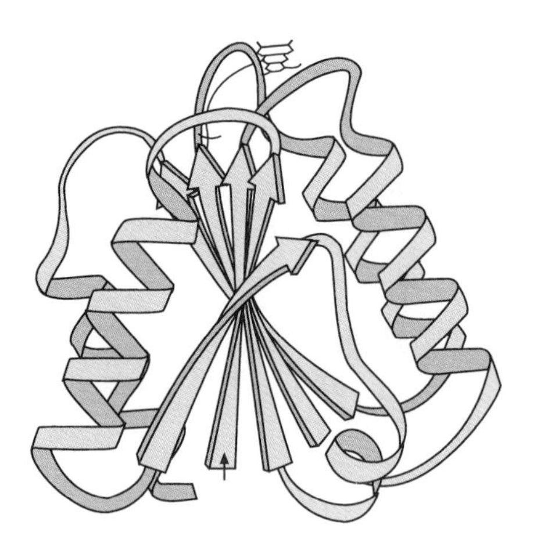

Flavodoxin

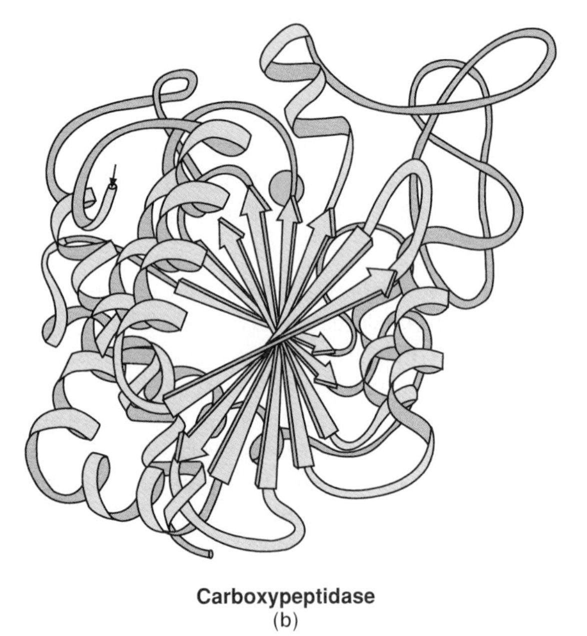

Carboxypeptidase

(b)

4.24 shows the polypeptide-chain folding of four proteins that incorporate twisted parallel or mixed β sheets. These include the exoprotease carboxypeptidase A, the electron-transport protein flavodoxin, and the glycolytic enzyme triose phosphate isomerase. Although these proteins all have right-twisted β sheets, it is clear that their overall geometries differ. That is, the β sheets in carboxypeptidase and flavodoxin are smoothly twisted to form saddle-shaped surfaces, while the β sheets in triose phosphate isomerase take the form of a cylinder or β barrel.

Within each overall type of parallel β sheet organization, the detailed hydrogen-bond pattern can be understood in terms of the forces acting within and between the polypeptide chains (fig. 4.25). In the case of the roughly rectangular sheets in carboxypeptidase and flavodoxin, the observed geometry reflects a competition between the tendency of the individual chains to twist and the tendency of the interchain hydrogen bonds to remain firm. Basically, the interchain hydrogen bonds tend to stretch when the sheet is twisted and so resist introduc-

Figure 4.25

Origin of ß-barrel and ß-sheet conformations. The observed geometry in ß-sheet structures represents a competition between the tendency of the individual chains to twist in a right-handed way and the tendency of the interchain hydrogen bonds to be preserved. Rectangularly arranged sheets give rise to the saddle shape. Staggered sheets give rise to the ß-barrel shape.

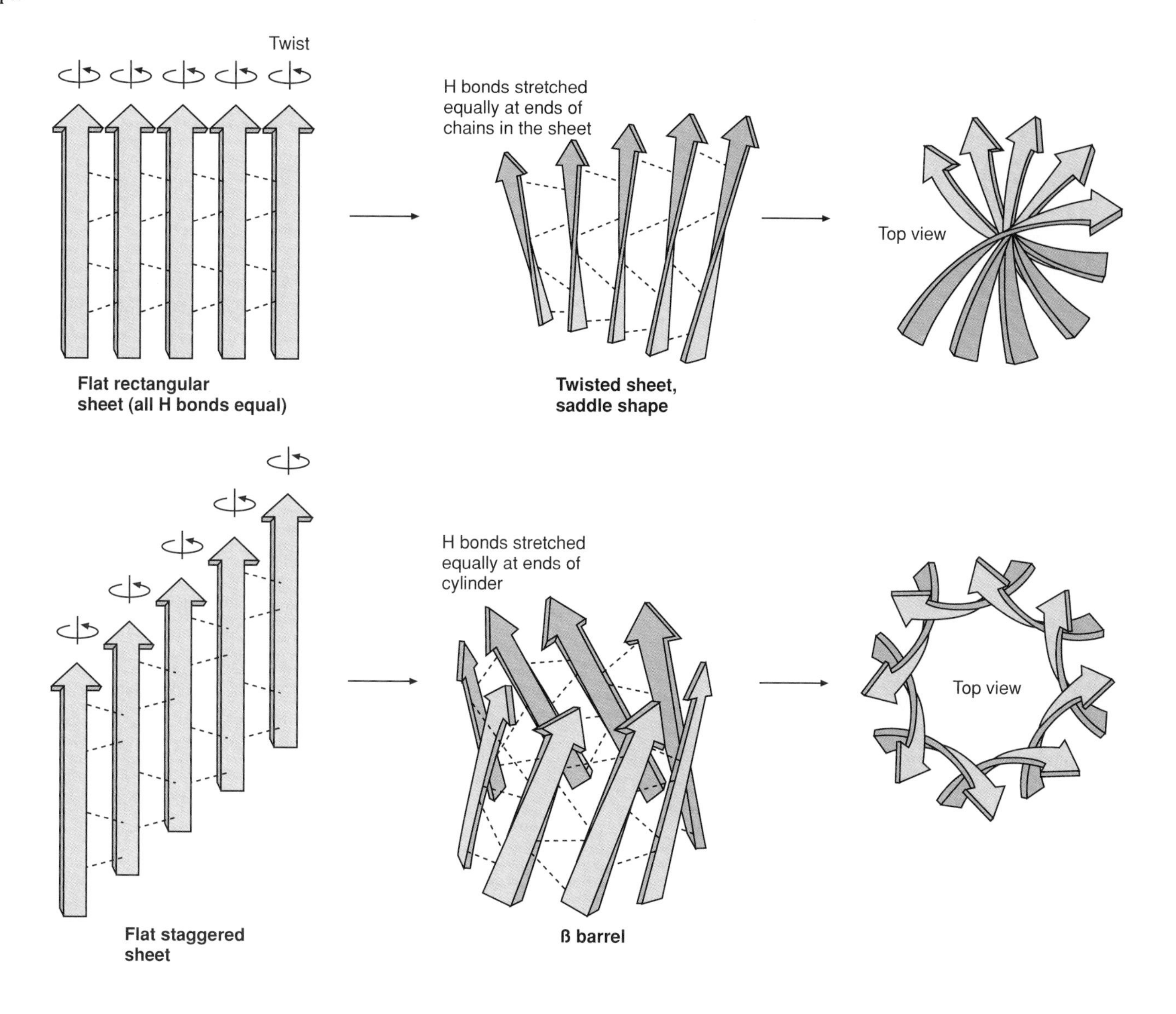

tion of twist into the sheet. The observed saddle-shaped geometry reflects the uniform distribution of these conflicting forces throughout the sheet, as shown in figure 4.25.

The β sheet that forms the barrel in triose phosphate isomerase has an hourglass-shaped surface with cylindrical curvature. Twisted β strands with a staggered hydrogen-bond pattern (see fig. 4.24*a*) automatically produce a cylindrical curvature. Conversely, twisted strands on a cylindrical surface necessitate a staggered hydrogen-bond pattern. Again, a compromise occurs between twisting and hydrogen bonding, leading to approximately straight chains with somewhat stretched hydrogen bonds at the top and bottom, which produce the hourglass shape. The differences in the geometries of rectangular and staggered plane sheets therefore result from differences in how adjacent sheet strands are hydrogen-bonded together. In either case, the operative forces are similar, and the final result reflects a compromise between chain twisting and preservation of good interchain hydrogen bonds.

Chiral preferences affect the connectivity as well as the sheet geometry in parallel β proteins. The right-handed crossover in parallel β barrels cannot go down the center, which is only large enough to accommodate the hydrophobic side chains.

Figure 4.26

Highly simplified sketches of (*a*) a singly wound parallel ß barrel, (*b*) a doubly wound ß sheet. Thin arrows next to the diagrams show the direction in which the chain is progressing from strand to strand in the sheet. The α and ß labels in the figure refer to regions of α helix and ß sheet.

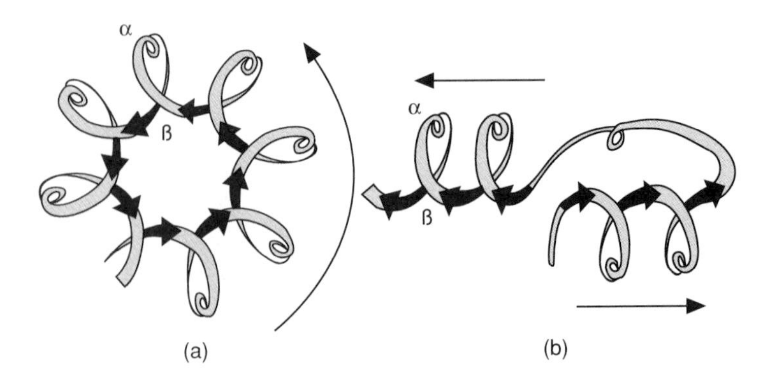

Figure 4.27

A ßαß loop. This arrangement forms the basis of many of the more extended structural arrangements, such as those shown in figure 4.24.

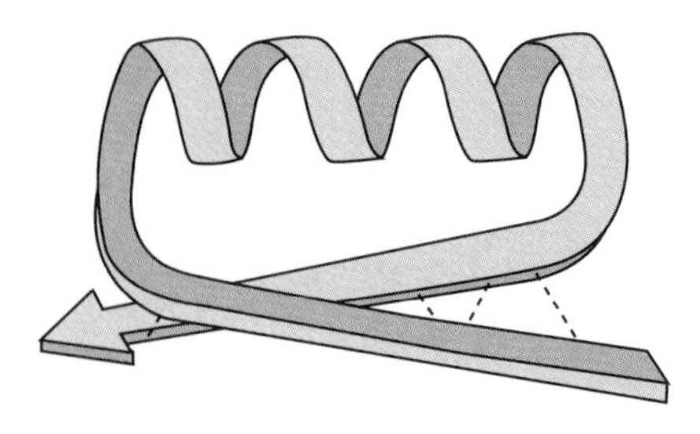

As a rule, the polypeptide backbone winds in a simple right-handed spiral around the barrel, moving over by one β strand at a time and packing helices or loops around the outside. Thus, although these structures tend to be large, their organization (fig. 4.26*a*) is very simple.

The saddle-shaped parallel β sheets, such as those in carboxypeptidase or flavodoxin (see fig. 4.24*b*), have a layer of helices and loops on each side. In order to accomplish this with right-handed crossover connections, the polypeptide chain must sometimes move along the sheet in one direction and sometimes in the other direction. The most common organizational pattern found in known protein structures starts in the middle of the sheet and winds toward one edge, with right-handed crossovers packing a layer of helices on one side of the sheet. Then the polypeptide chain returns to the middle of the sheet and winds out to the opposite edge, packing helices against the other side of the sheet (see fig. 4.26*b*). This pattern is often known as a nucleotide-binding domain, since most of these proteins bind a mononucleotide or dinucleotide cofactor in the middle of the C-terminal end of the β sheet.

Limits on Tertiary Structure Imposed by Efficiency of Packing between Secondary Structures An additional important factor in protein structural organization is the efficient packing together of secondary structural elements to form larger units. Figure 4.27 shows one of the most commonly observed arrangements, called a $\beta\alpha\beta$ loop. This is a special case of the right-handed crossover connection described in the previous section. A $\beta\alpha\beta$ loop is composed of a pair of adjacent hydrogen-bonded parallel β strands that are right-connected, with a stretch of α helix that packs tightly on the surface of the sheet. Generally, the β-sheet strands in a local $\beta\alpha\beta$ loop are part of a much larger β sheet. Structures such as triose phosphate isomerase (see fig. 4.24*a*) can be viewed as a series of overlapping $\beta\alpha\beta$ structural units. The overlapping produces a final structural arrangement in which an inner barrel, composed of β-sheet strands, packs tightly within an outer barrel composed of α helices.

Although tertiary structures composed of $\beta\alpha\beta$ loops are perhaps the most commonly observed pattern in known protein structures, other arrangements are found in proteins having either predominantly antiparallel β-sheet or predominantly α-helical conformations. The most common structural organization in antiparallel β proteins has two layers of β sheets (fig. 4.28). Other antiparallel β proteins have a single twisted β sheet covered on only one side by a layer of helices and loops (fig. 4.29).

No protein is stable as a single-layer structure, since it requires at least two layers to bury the hydrophobic core. Thus antiparallel β proteins are typically two-layer structures; antiparallel sheets are apparently quite stable when one side is exposed to solvent. Parallel sheets require at least one additional layer besides the sheet in order to make the crossover connections between the parallel β strands. In contrast to antiparallel β sheets, they apparently cannot tolerate solvent exposure on even one side and are always found as a structural "backbone" in protein interiors, with other layers of structure on both sides. Therefore, proteins with $\beta\alpha\beta$ loops generally have either three layers, as in the nucleotide-binding domains, or four layers, as in the parallel β barrels (see fig. 4.24). Usually the outer layers are formed of α helices, which must pack against one another and also against the surface formed by the β-sheet side chains.

The structural geometry of proteins that have only an α-helical secondary structure is also largely determined by requirements for efficient packing between the helices. We have already encountered one particularly stable interhelical packing arrangement in the discussion of fibrous proteins (see fig. 4.6). In the α-keratins, adjacent helices pack together with an interaction angle between helices of about 18°. This extended-helix interaction pattern forms the basis for a protein structural motif frequently seen in globular proteins, the 4-α-helical bundle. In this arrangement, four α helices, sequentially connected to their nearest neighbors, pack together to form an array with a roughly square cross section. Since each helix interacts with its neighbors at an angle of about 18°, the overall bundle has a left-handed twist (fig. 4.30). This commonly observed folding domain clearly represents a minimum accessible surface area arrange-

Figure 4.28

Examples of proteins containing ß-sheet domains.

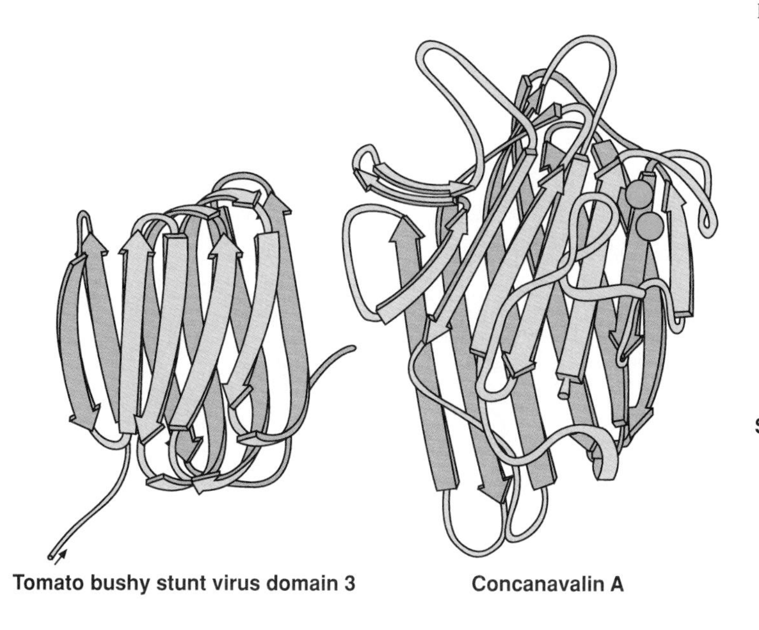

Figure 4.29

Examples of antiparallel ß proteins that are covered on only one side by larger helices and loops.

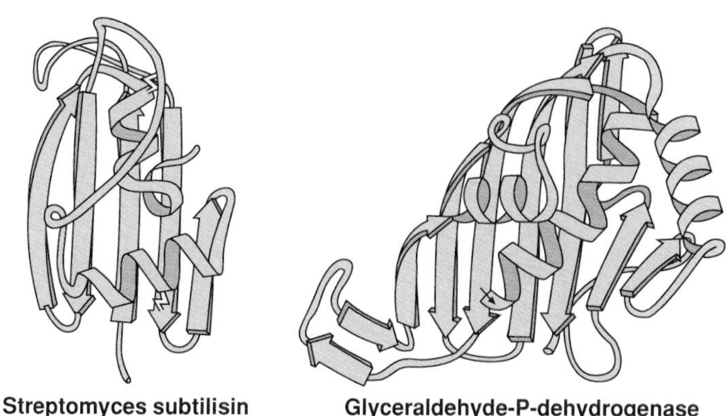

Figure 4.30

Examples of some proteins that share a common structural motif of four α helices.

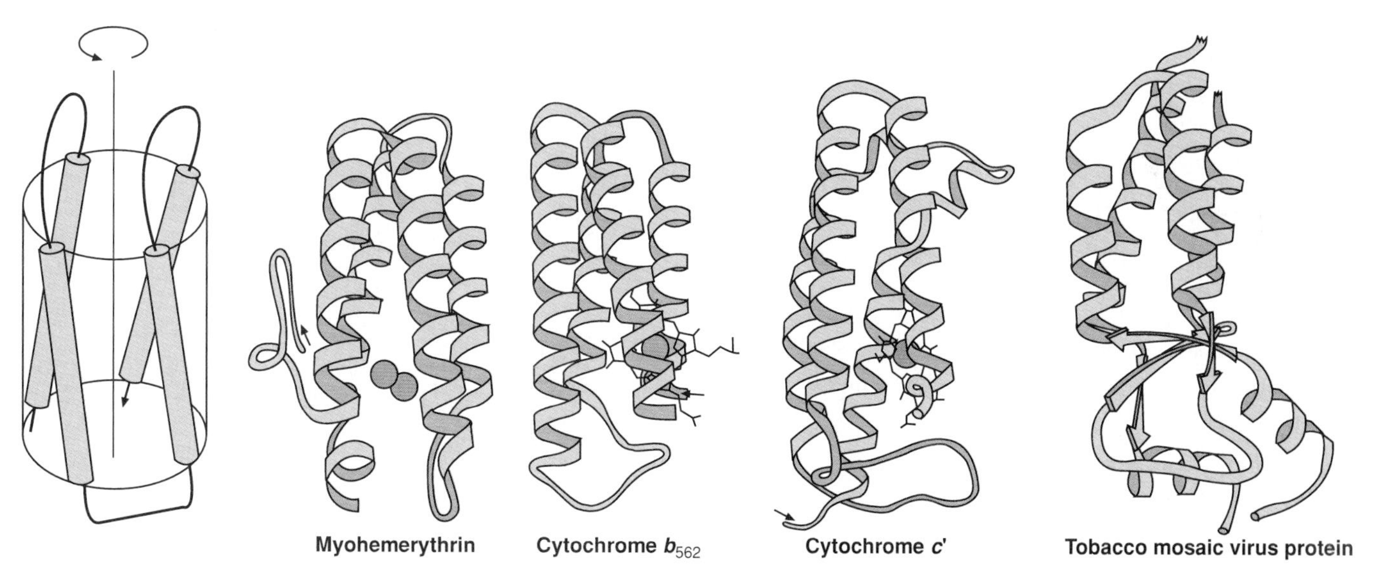

ment for four sequentially connected α helices of approximately equal length. Many α-helical proteins that lack these features have more complex and irregular geometries. However, even in these cases it appears that the relative orientations of adjacently packed helices reflect geometric restrictions that accompany close packing between helices.

In addition to the packing of elements of protein secondary structure, which is a dominant feature in most proteins, there are some cases, especially among the smallest structures, where the geometry and packing of disulfide bonds or nonpeptidyl groups are a dominant factor. Figure 4.31 shows examples of this sort, in which the secondary structures are short and irregular and cannot assume their native structures if the disulfides are broken (*a, b*), or if the nonpeptidyl groups are missing (*c*).

In summary, examination of a large number of protein tertiary structures has shown that they incorporate several different sorts of recurring structural arrangements. In general,

Figure 4.31

Examples of some small proteins or domains in which disulfide bonds. (*a, b*) or a porphyrin group (*c*) are a dominant factor holding the structure together. Porphyrins are depicted in figure 5.11 and, in greater detail, in figure 11.20.

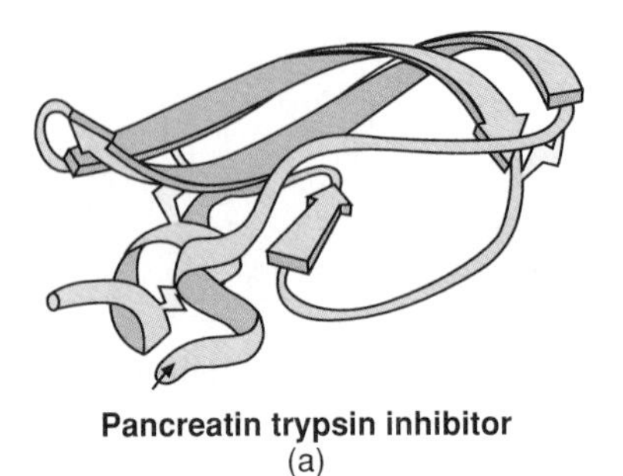

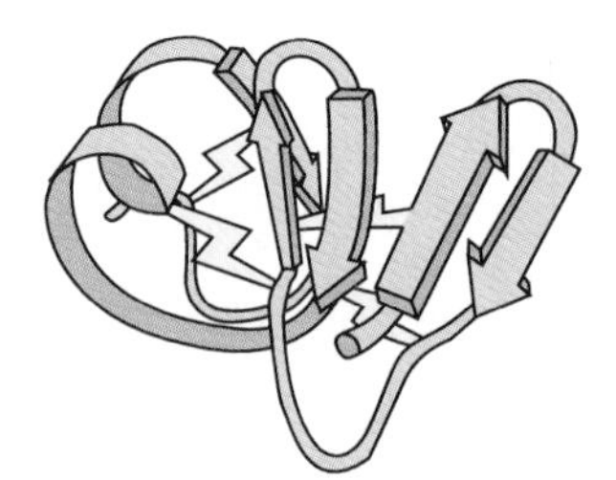

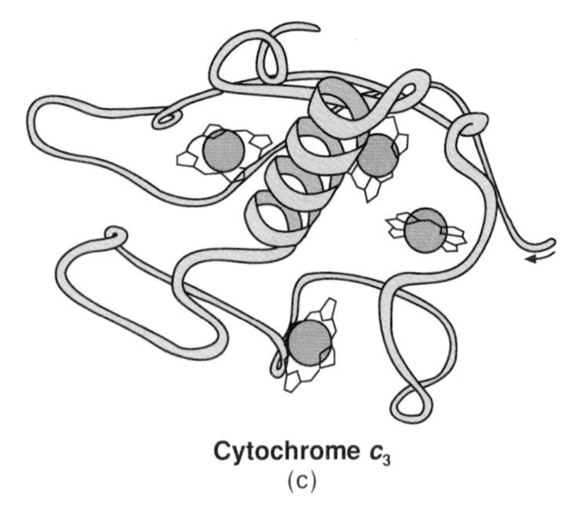

Figure 4.32

Papain, a protein in which the domains are very different from one another.

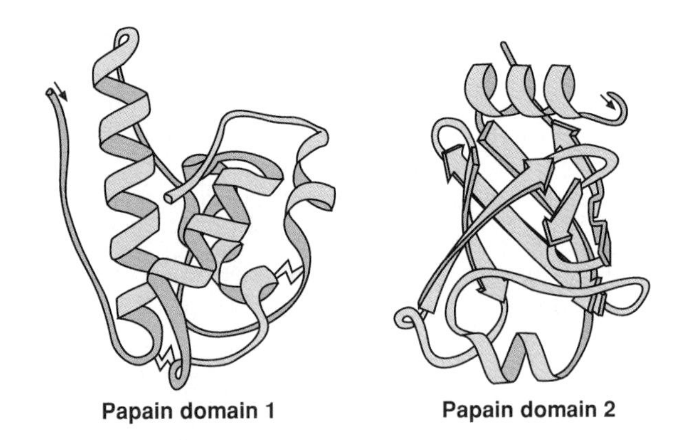

these arrangements owe their origins to physical effects, some of which predispose the most stable conformations of the polypeptide chain, and some of which govern the formation of intimately packed tertiary structures.

Domains as Functional Units of Tertiary Structure The patterns of tertiary structure described in this and previous sections frequently constitute the entire protein. However, within a single folded chain or subunit, contiguous portions of the polypeptide chain often fold into compact local units called domains, each of which might consist, for example, of a four-helix cluster or a barrel or an antiparallel β sheet. Sometimes the domains within a protein are very different from one another, as within the protease papain (fig. 4.32), but often they resemble each other very closely, as in rhodanase (fig. 4.33).

The separateness of two domains within a subunit varies all the way from independent globular domains joined only by a flexible length of polypeptide chain, to domains with tight and extensive contact and a smooth globular surface for the outside of the entire subunit, as in the proteolytic enzyme elastase (fig. 4.34). An intermediate level of domain separateness, characterized by a definite neck or cleft between the domains, is found in phosphoglycerate kinase (fig. 4.35).

Figure 4.33

Rhodanese domains 1 and 2 as an example of a protein with two domains that resemble each other extremely closely. Rhodanese is a liver enzyme that detoxifies cyanide by catalyzing the formation of thiocyanate from thiosulfate and cyanide.

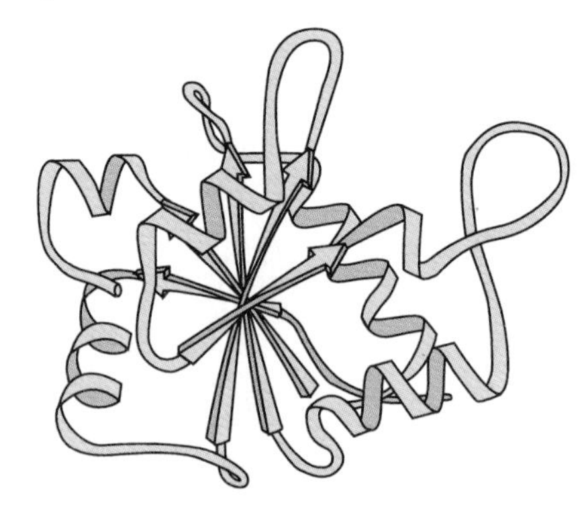

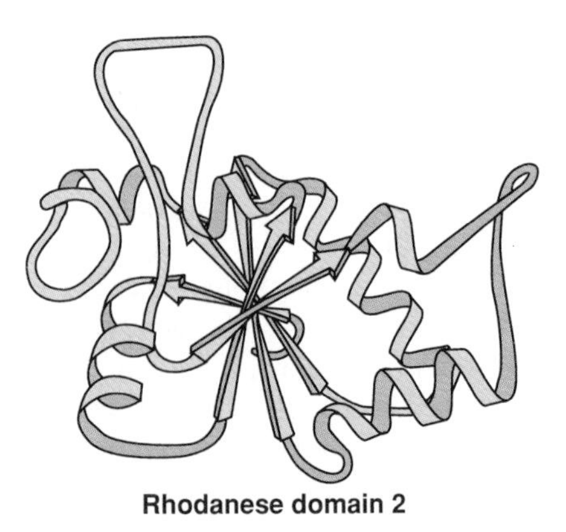

Figure 4.34

Schematic backbone drawing of the elastase molecule, showing the similar ß-barrel structures of the two domains.

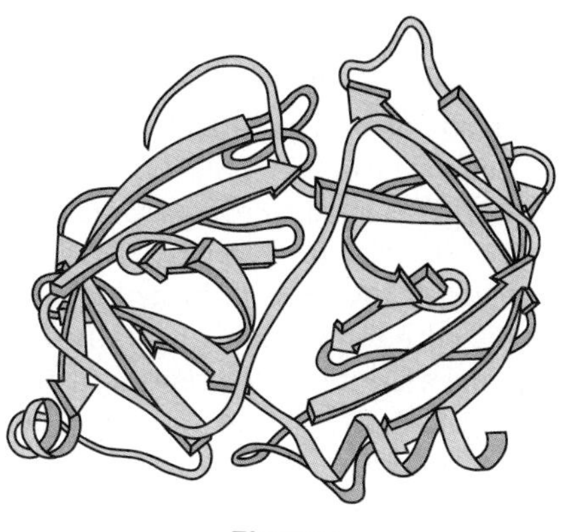

Figure 4.35

The dumbbell domain organization of phosphoglycerate kinase, with a relatively narrow neck between two well-separated domains.

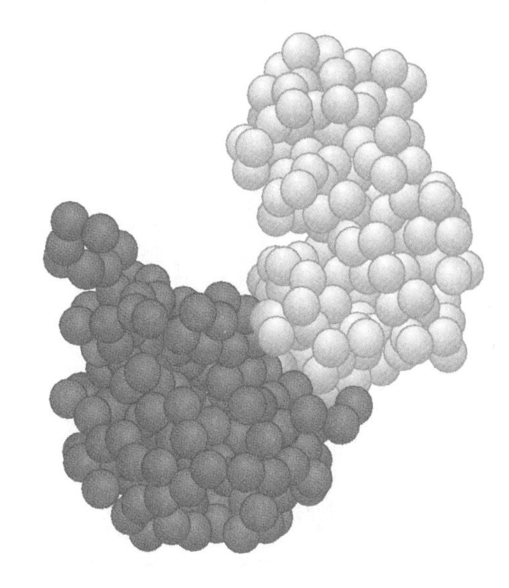

Domains as well as subunits can serve as modular bricks to aid in efficient assembly of the native conformation. Undoubtedly the existence of separate domains is important in simplifying the protein-folding process into separable, smaller steps, especially for very large proteins. There is no strict upper limit on folding size. Indeed, known domains vary in size all the way from about 40 residues to over 400.

Another important function of domains is to allow for movement. Completely flexible hinges would be impossible between subunits because they would simply fall apart. However, flexible hinges can exist between covalently linked domains. Limited flexibility between domains is often crucial to substrate binding, allosteric control (discussed in chapter 10), or assembly of large structures. In hexokinase, the two domains within the individual subunits hinge toward each other upon binding of the substrate glucose, enclosing it almost completely (fig. 4.36). In this manner glucose can be bound in an environment that excludes water as a competing substrate (see chapter 13 for further details on the hexokinase reaction).

Figure 4.36

Schematic representation of the change in conformation of the hexokinase enzyme on binding substrate. E and E′ are the inactive and active conformations of the enzyme, respectively. G is the sugar substrate. Regions of protein or substrate surface excluded from contact with solvent are indicated by a crinkled line. Figure 9.3 presents a more detailed view of the hexokinase molecule. (Source: W. S. Bennett and T. A. Steitz, "Glucose-induced conformational change in yeast hexokinase," *Proceedings. National Academy of Sciences USA*, 75:4848, 1978. Copyright ©1978 National Academy of Sciences, Washington, D.C.)

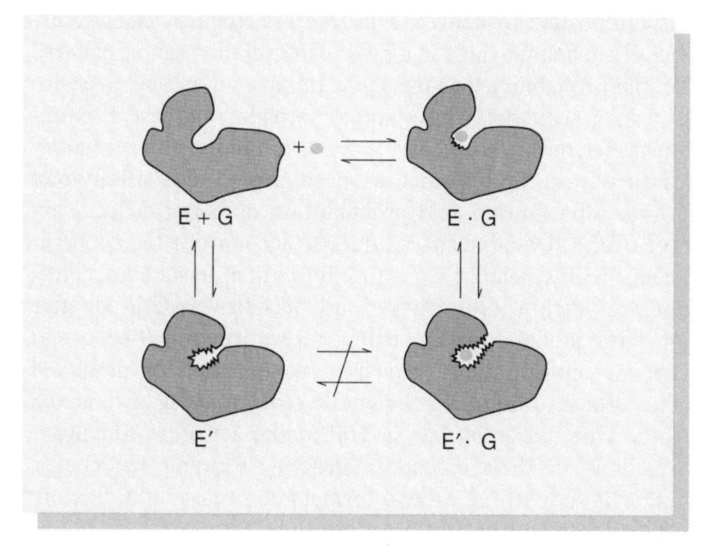

The Prediction of Secondary and Tertiary Structures

We began our discussion of globular protein tertiary structure by pointing out that the tertiary structure is determined by the primary structure, probably because in many cases the native tertiary structure is the most stable structure that can be formed from a given primary structure. If this is so, then it might be possible to predict a protein's structure from its primary sequence alone. Right now, however, x-ray diffraction and nuclear magnetic resonance (NMR) are yielding information on tertiary structures at such a rate that efforts in that direction may not be necessary—especially since most proteins are made of a limited number of domains, which tend to reappear in many proteins. Thus in the future we may be able to predict the structures of many proteins by combining the limited amount of structure information accumulated from x-ray diffraction or NMR studies with the vast data on amino acid sequence of proteins whose tertiary structure is unknown.

Figure 4.37

Relative probabilities that any given amino acid will occur in the α-helical, ß-sheet, or ß-hairpin-bend secondary structural conformations.

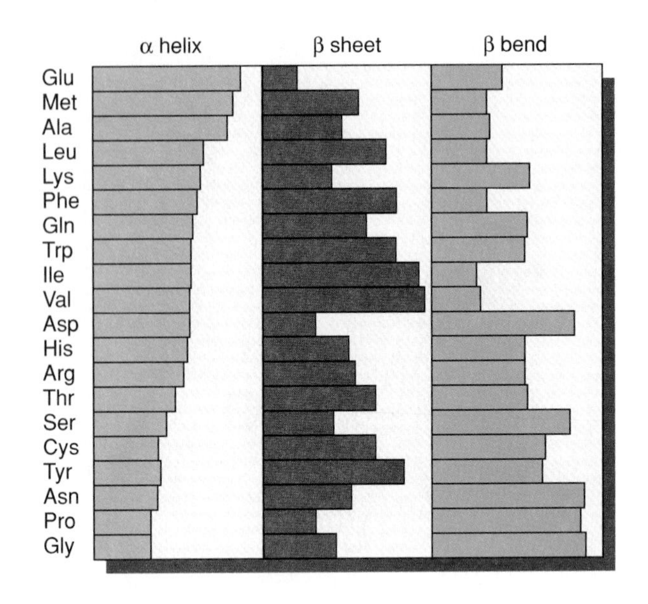

At this juncture it seems clear that various amino acids have tendencies to form different secondary structures. As shown in figure 4.37, glutamic acid, methionine, and alanine appear to be the strongest α-helix formers and valine, isoleucine, and tyrosine the most probable β-sheet formers, while proline, glycine, asparagine, aspartic acid, and serine occur most frequently in β-bend conformations. This type of information is of value in the prediction of secondary structural regions of proteins from their amino acid sequences. The observed frequencies of occurrence of each amino acid in a given conformation can be equated with the probability that the same amino acid will behave similarly in a sequence whose actual secondary structure is unknown. Therefore, in theory at least, to predict the secondary structure from the sequence we need only to sequentially plot the individual amino acid probabilities, or better, a local average over a few adjacent residues to account for the cooperative nature of secondary-structure formation. In such a scheme, sequences such as Gly-Pro-Ser and Ala-His-Ala-Glu-Ala give high joint probabilities for being, respectively, in β-bend and α-helical conformations. However, comparisons of predicted versus directly observed polypeptide conformations give mixed results. One reason is that several amino acids are somewhat ambiguous in their secondary-structure-forming tendencies; another is that strong β-bend formers do occasionally turn up in the middle of α helices.

Nevertheless, it is clear that the structural arrangement of a protein in its folded state must depend ultimately on its sequence. The information given in figure 4.37—that certain amino acids have general tendencies toward forming particular sorts of structures—is a basic point of departure. The attainment of a particular folding arrangement depends on details of both the short- and long-range interactions that uniquely characterize and stabilize each protein's structure.

Quaternary Structure Depends on the Interaction of Two or More Proteins or Protein Subunits

Although many globular proteins function as monomers, biological systems abound with examples of complex protein assemblies (table 4.4). This higher-order organization of globular subunits to form a functional aggregate is referred to as the quaternary structure of the protein. Protein quaternary structures can be classified into two fundamentally different types. The first involves the assembly of proteins (sometimes referred to as subunits because they constitute a part of the final structure) that are very different structures. Examples range from dimeric molecules that contain different molecular subunits to complex assemblies such as ribosomes, which contain twenty or more nonidentical protein subunits in addition to one or more RNA components. The organization of these sorts of quaternary structures depends on the specific nature of each interaction made between the different molecular subunits and their neighbors. Each intermolecular interaction generally occurs only once within a given aggregate arrangement, so that the overall complex structure has a highly irregular geometry. A widely used approach for determining the state of aggregation of proteins in solution is by sedimentation and diffusion analysis (see box 4E).

A second, commonly observed pattern of quaternary structure is typified by molecular aggregates composed of multiple copies of one or more different kinds of subunits. Owing to the recurrence of specific structural interactions between the subunits, such aggregates typically form regular geometric arrangements. Given that proteins are fundamentally asymmetrical objects (because they incorporate chiral L-amino acids), it is clear that the simplest pattern of quaternary structure involves formation of a linear aggregate. As illustrated in figure 4.38*a, b,* the formation of such an aggregate results from the repetition of one sort of specific structural interaction between adjacent subunits of the assembly. Structures of this type can be extended simply by the addition of successive subunits.

Somewhat more frequently observed than linear arrangements are helical arrangements of identical molecular subunits. As is the case for the amino acid residues in the α helix, in helical quaternary structures the individual subunits display different, local interactions with their nearest neighbors (see fig. 4.38*c, d*). However, the pattern of nearest-neighbor interactions is repeated for each subunit.

Helical molecular aggregates are frequently associated with self-assembling molecular structures. Some outstanding examples of such aggregates are the rodlike and filamentous viruses, in which the helical aggregation of the protein-coat subunits forms a cylindrical container for the virus's nucleic acid (fig. 4.39). Since the entire coat is assembled from multiple

Table 4.4
Molecular Weight and Subunit Composition of Selected Proteins

Protein	Molecular Weight	Number of Subunits	Function
Glucagon	3,300	1	Hormone
Insulin	11,466	2	Hormone
Cytochrome *c*	13,000	1	Electron transport
Ribonuclease A (pancreas)	13,700	1	Enzyme
Lysozyme (egg white)	13,900	1	Enzyme
Myoglobin	16,900	1	Oxygen storage
Chymotrypsin	21,600	1	Enzyme
Carbonic anhydrase	30,000	1	Enzyme
Rhodanese	33,000	1	Enzyme
Peroxidase (horseradish)	40,000	1	Enzyme
Hemoglobin	64,500	4	Oxygen transport
Concanavalin A	102,000	4	Unknown
Hexokinase (yeast)	102,000	2	Enzyme
Lactate dehydrogenase	140,000	4	Enzyme
Bacteriochlorophyll protein	150,000	3	Enzyme
Ceruloplasmin	151,000	8	Copper transport
Glycogen phosphorylase	194,000	2	Enzyme
Pyruvate dehydrogenase (*E. coli*)	260,000	4	Enzyme
Aspartate carbamoyltransferase	310,000	12	Enzyme
Phosphofructokinase (muscle)	340,000	4	Enzyme
Ferritin	440,000	24	Iron storage
Glutamine synthase (*E. coli*)	600,000	12	Enzyme
Satellite tobacco necrosis virus	1,300,000	60	Virus coat
Tobacco mosaic virus	40,000,000	2,130	Virus coat

Figure 4.38

Linear and helical quaternary aggregates of protein molecules. (*a*) A linear arrangement of hypothetical protein subunits (illustrated as simplified right shoes). The interactions in such a linear arrangement are all identical, and the structure lends itself to the formation of an indefinitely long linear structure whose subunits are related by translation in one dimension. (*b*) A helical arrangement in which equivalently interacting subunits are related by unit translations along the helix axis followed by a 180-degree rotation about the helix axis. (*c*) A helical arrangement in which subunits are related by unit translation plus a rotation of 360 degrees/*n* to give an *n*-fold helix. (*d*) An *n*-fold multiple-start helix, an arrangement in which subunits form different equivalent interactions with their nearest neighbors. Red arrows indicate two types of identical contact points.

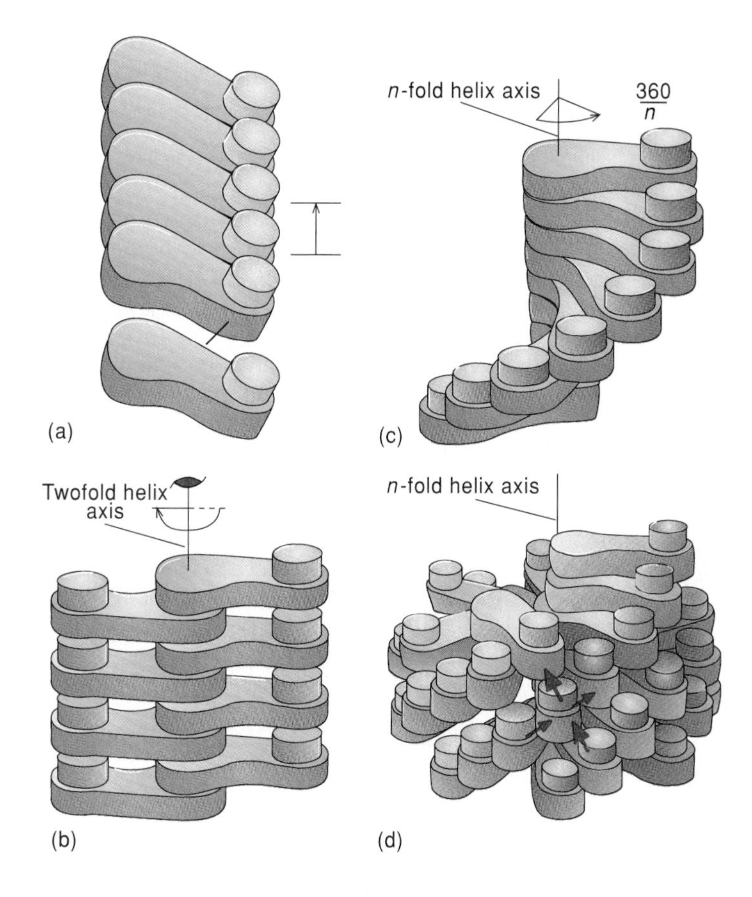

copies of the same protein, this arrangement represents a very efficient utilization of the information content of the virus nucleic acid.

Helical quaternary structures, then, are characterized by a repeating interaction that results in structures whose subunits are related both by some rise along and twist around a central axis. They are therefore similar to linear arrangements in that they are, at least potentially, indefinitely extendable. In fact, in helical viruses the length of the coat-protein structure is determined, not by the protein, but rather by the fixed length of the virus nucleic acid.

The Use of Sedimentation and Diffusion to Calculate Molecular Weights in Solution

In a centrifugal force field, protein molecules slowly migrate toward the bottom of a centrifuge tube at a rate that is proportional to their molecular weight (figure 1). The rate of sedimentation may be recorded by optical methods that do not interfere with the operation of the centrifuge. From this rate, we can obtain the sedimentation constant *s*. This constant equals the rate at which a molecule sediments, divided by the gravitational field (angular acceleration in a spinning rotor), as defined by the equation

$$s = \frac{dx/dt}{\omega^2 x} \tag{B1}$$

where dx/dt is the rate at which the particle travels at distance x from the center of rotation, ω is the angular velocity of the rotor in radians per second (hence $\omega^2 x$ is the angular acceleration), and t is the time of centrifugation in seconds. The sedimentation constant is usually given in Svedberg units (S); one S = 10^{-13} *s*.

From the sedimentation constant we can obtain the molecular weight, provided we have certain other information. This additional information includes the frictional coefficient (f) of the protein and the density of the protein. The coefficient f is bigger for larger proteins and, for proteins of the same molecular weight, it is larger for elongated, rodlike molecules. The density of the protein is important because of the buoyancy factor, $1 - \bar{v}_p\rho_s$, which takes into account the density difference between solvent (ρ_s) and the volume of water displaced per gram of protein ($\bar{v}_p$). The equation that relates s, M, and f is

$$s = \frac{M(1 - \bar{v}_p\rho_s)}{Nf} \tag{B2}$$

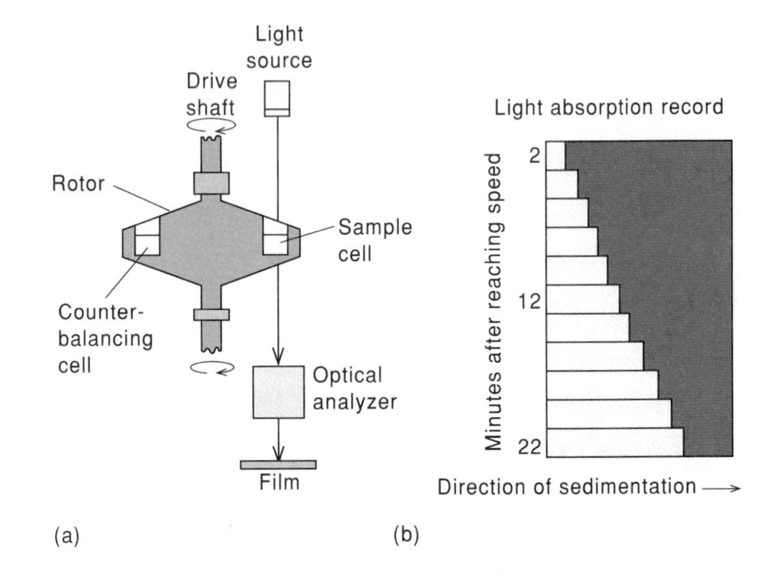

Figure 1

Apparatus for analytical ultracentrifugation. (*a*) The centrifuge rotor and method of making optical measurements. (*b*) The optical recordings as a function of centrifugation time. As the light-absorbing molecule sediments, the solution becomes transparent.

Table 4E
Physical Constants of Some Proteins

Protein	Molecular Weight	Diffusion Constant ($D \times 10^7$)	Sedimentation Constant (s)	pI[a] (Isoelectric)
Cytochrome *c* (bovine heart)	13,370	11.4	1.17	10.6
Myoglobin (horse heart)	16,900	11.3	2.04	7.0
Chymotrypsinogen (bovine pancreas)	23,240	9.5	2.54	9.5
β-Lactoglobulin (goat milk)	37,100	7.5	2.9	5.2
Serum albumin (human)	68,500	6.1	4.6	4.9
Hemoglobin (human)	64,500	6.9	4.5	6.9
Catalase (horse liver)	247,500	4.1	11.3	5.6
Urease (jack bean)	482,700	3.46	18.6	5.1
Fibrinogen (human)	339,700	1.98	7.6	5.5
Myosin (cod)	524,800	1.10	6.4	—
Tobacco mosaic virus	40,590,000	0.46	198	—

[a]pI = $-\log I$ = the isoelectric point, that is, the pH at which the protein carries no net charge (see box 5B).

where N is Avogadro's number. Thus in order to estimate the molecular weight from the sedimentation constant we must have a means of determining $\bar{v}_p$ and f.

For most proteins $\bar{v}_p$ is about 0.75 cc/g, so its value does not present much of a problem. The frictional coefficient, however, is a sensitive function of the shape, varying over a wide range, and we must usually know its value if we need a serious estimate. The value of f is usually found by working with the diffusion constant D, which is related to the frictional constant by

$$D = \frac{RT}{Nf} \text{ or } f = \frac{RT}{ND} \tag{B3}$$

where R is the gas constant and T is the absolute temperature. Substituting this expression for f in equation (B2) and transposing leads to

$$M_r = \frac{RTS}{D(1 - \bar{v}_p \rho_s)} \tag{B4}$$

The diffusion constant that we need if we want to use equation (B4) to calculate the molecular weight is usually obtained with the help of Fick's first law of diffusion:

$$\frac{dn}{dt} = -DA\left(\frac{dc}{dx}\right)_t \tag{B5}$$

This equation states that the amount dn of a substance crossing a given area A in time dt is proportional to the concentration gradient dc/dx across that area. The diffusion constant D, which is a function of both molecular weight and shape, is the proportionality constant. It can be measured by observing the spread of an initially sharp boundary between the protein solution and a solvent as the protein diffuses into the solvent layer. Once we know the value of the diffusion constant, we can combine the information with the sedimentation data and calculate the molecular weight of the protein.

Representative sedimentation and diffusion data, together with the calculated molecular weights, are presented in table 4E. The sedimentation constants are usually larger the greater the molecular weight. Likewise, the diffusion constants are usually inversely proportional to the molecular weights. Exceptions arise because of proteins with unusual shapes. For example, the globular protein urease and the rodlike protein myosin have similar molecular weights. Yet their sedimentation and diffusion constants both differ by about a factor of 3. Rapid diffusion can be a highly desirable property for an enzyme that must pass rapidly from one point to another. Such a protein would benefit from a globular shape, which is quite common for enzymes. By contrast, a rodlike protein could be advantageous to create a cytoskeletal boundary within the cytoplasm. Myosin is used for just such purposes in many cell types.

Figure 4.39

Tobacco mosaic virus structure. (*a*) Diagram of TMV structure, an example of a helical virus. The nucleocapsid (protein shell) is composed of a helical assembly of 2,130 identical protein subunits (protomers) with the RNA of the virus spiraling on the inside. (*b*) An electron micrograph of the negatively stained helical capsid (400,000×). In negative staining the virus is immersed in a pool of a heavy metal salt that is much more electron-dense than the virus. The result is that the darker portions of the figure that surround the virus appear more dense than those parts of the figure where the less dense nucleoprotein is located. (©Dennis Kunkel/Phototake.)

RNA
Protein
0
10 nm

(a)

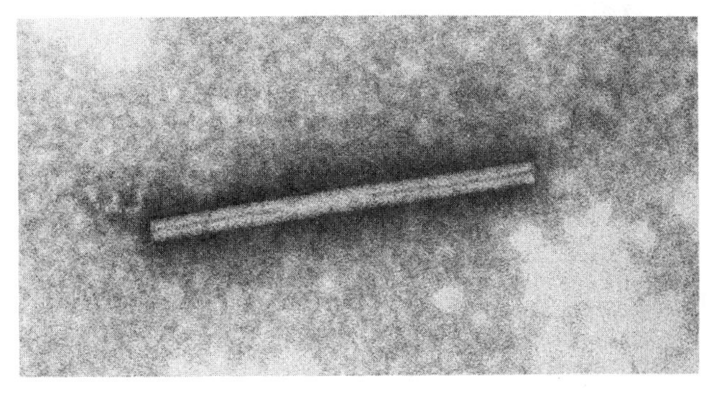

(b)

We can also imagine patterns of repeating molecular interactions that involve only twists between the subunits and so result in the formation of quaternary structures that are essentially like flat rings (fig. 4.40*a–e*). Such cyclically repeating interactions typically give rise to symmetrical molecular dimers and trimers. Larger aggregates, in contrast, most frequently do not form flat-ring structures, since flat rings would not provide enough total contact surface to stabilize an open, extended arrangement. Instead, they form arrangements that resemble geometric polyhedra. The formation of polyhedral aggregates

Figure 4.40

Quaternary structure with rotational and polyhedral symmetry. (*a*) Arrangements of two molecules related by twofold rotational symmetry to form symmetrical dimers. (*b*) Symmetrical trimers. In these arrangements, the intersubunit contacts are all identical. (*c*) The most common arrangement for tetrameric molecules (as in hemoglobin), where each subunit makes three different interactions with its neighbors: a "side-by-side" interaction, a "toe-to-toe" interaction, and a "heel-to-heel" interaction. (*d*) A common arrangement for hexameric molecules. (*e*) Octameric molecules. (*f*) A cubic quaternary structure with 24 subunits as found in some iron-storage proteins. (*g*) An icosahedral quaternary structure with 60 subunits, 3 to each triangular face. This pattern is frequently seen in viruses.

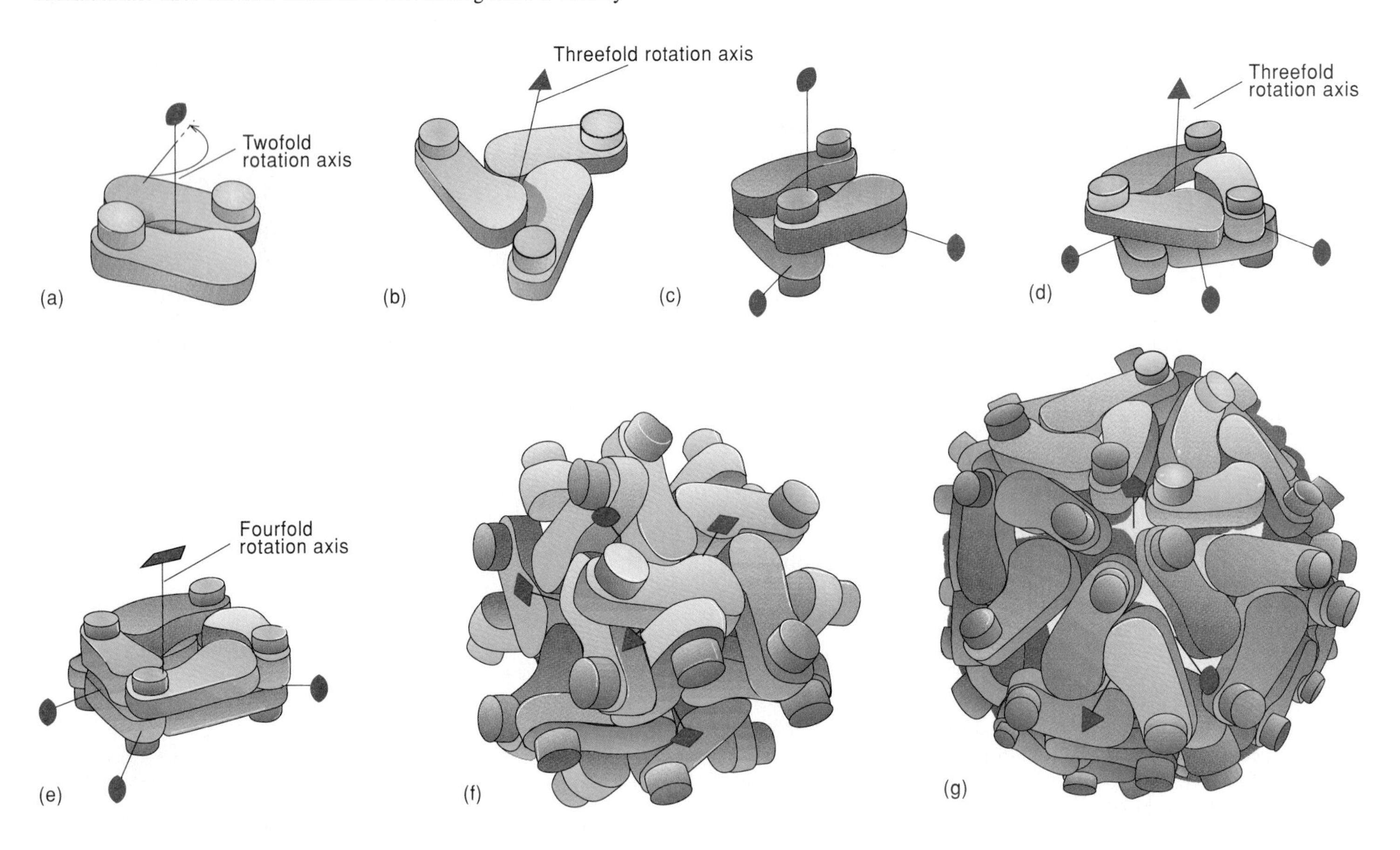

reflects the fact that the molecular subunits can have more than one type of intermolecular interaction. Figure 4.40*f, g* illustrates some of the types of structures observed. Such arrangements can be composed of identical subunits or of different types of subunits. One property that distinguishes the polyhedral and ring quaternary structures from linear and helical types is that they incorporate fixed numbers of subunit copies.

The structures of helical and polyhedral viruses demonstrate that quaternary structures play a central role in the self-assembly of very large biological structures from individual molecular subunits. Structural stabilization occurs when all the subunits interact in geometrically similar ways, i.e., essentially like the atoms in a salt crystal. Surprisingly, however, x-ray crystallography reveals that many quaternary interactions are not symmetrical or equivalent, even when they pertain to chemically identical subunits. The simplest sort of nonequivalence occurs at some dimer contacts, where individual side chains close to the twofold axis (which in such cases is only approximate) are forced to take up different positions in order to avoid overlapping. The departures from exact symmetry are usually local, in which case they probably have no functional consequences, but sometimes the nonequivalence extends to other parts of the subunit (e.g., in insulin and in malate dehydrogenase), where it can produce such effects as different binding constants for ligands. It is even easier, of course, for contacts between nonidentical subunits to be asymmetrical. For instance, in hemoglobin the contact between the two β chains is wider than that between the two α chains and produces the binding site for several important effector molecules (see chapter 5).

An even more extreme case of asymmetrical association occurs in the dimer of yeast hexokinase, where in place of the pure 180° rotation of a twofold axis, the subunits are related by a rotation of 156° plus a translation of 13.8 Å. Although this is basically a helical contact relationship, it cannot be extended past the dimer because if a third subunit were to bind by the same rule it would collide with the first one, as illustrated in figure 4.41. In hexokinase the asymmetrical association creates a significant conformational difference between two initially equivalent subunits so that the two active sites in the dimer have quite different functional properties.

Figure 4.41

A schematic drawing of a heterologous dimer interaction (lavender and green structures) in which infinite polymerization is sterically prevented. Addition of further subunits (pink and beige structures) to the free binding sites on the heterologous dimer is prevented by overlap of proteins. This arrangement of subunits is observed in the hexokinase dimer. (Adapted from a drawing obtained from T. A. Steitz.)

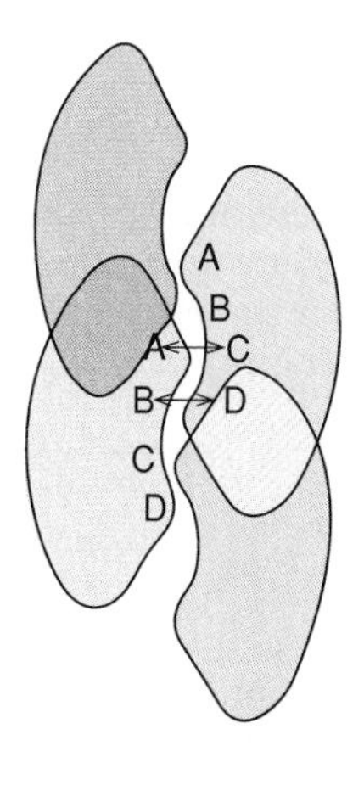

Yet another type of nonequivalent association occurs in icosahedral viruses, with only 60 symmetry-equivalent positions (see fig. 4.40*g*) but with more than 60 subunits. One way of reconciling this apparent contradiction is shown by the 180 subunits of tomato bushy stunt virus, which are placed so that five subunits are in contact around each fivefold axis, while six other subunits have a distinct but similar contact around each threefold axis (fig. 4.42). The versatility that permits such nonequivalent associations thus allows assembly of larger and more complex structures, and it may be even more common in biological structures that are too large to have been examined crystallographically.

The formation of a subunit aggregate can have extremely important functional results. In particular, as we will show later, the contact interactions provide a means of com-

Figure 4.42

The structure of the capsid (protein shell) for an icosahedral virus such as tomato bushy stunt virus. Pentons (P) are located at the twelve vertices of the icosahedron. Hexons (H), of which there are 20, form the edges and faces of the icosahedron. Each penton is composed of 5 protein subunits and each hexon is composed of six protein subunits. In all, the structure contains 180 protein subunits.

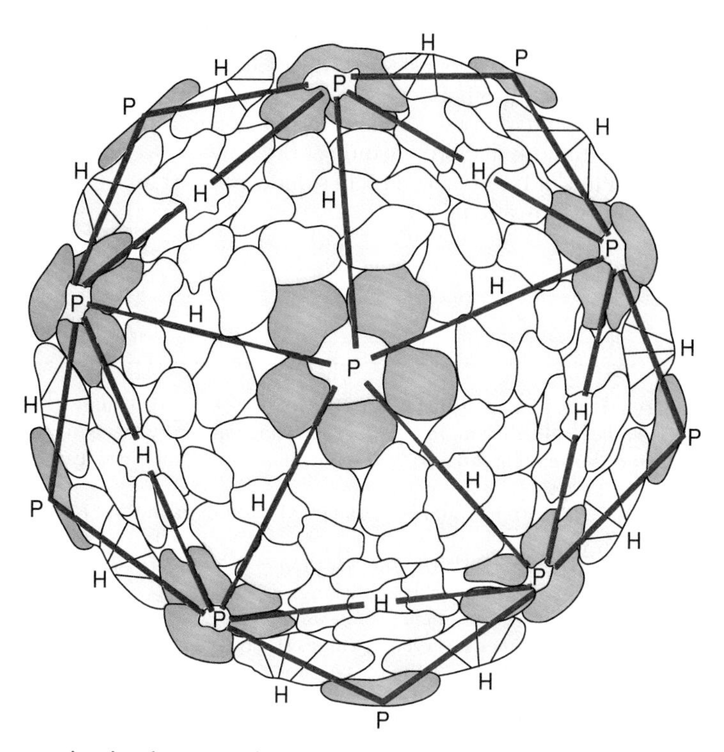

munication between the individual subunits (e.g., see chapters 5, 10, and 32). As a result, the interaction of a ligand or substrate molecule with one subunit of an aggregate can influence the course of subsequent events in other subunits of the aggregate. Such interactions, which form the basis for cooperativity in biochemical systems, are of great importance because they provide many of the control mechanisms for regulating biochemical processes.

Summary

In this chapter we have introduced the subject of the three-dimensional structures of proteins. This will be an important subject to keep in mind throughout the text. Our discussion focused on the following points.

1. Most proteins may be divided into two groups, fibrous and globular. Fibrous proteins usually serve structural roles. Globular proteins function as enzymes and in many other capacities.
2. The three most prominent groups of fibrous proteins are the α-keratins, the β-keratins, and collagen.
3. The α-keratins are composed of right-handed helical polypeptide chains in which all the peptide NH and carbonyl groups form intramolecular hydrogen bonds. When these helical coils interact they form left-handed coiled coils.
4. The β-keratins consist of extended polypeptide chains in which adjacent polypeptides are oriented in either a parallel or an antiparallel fashion. Sheets formed from such extended polypeptide chains may be stacked on top of one another.
5. Collagen fibrils are composed of extended polypeptide chains that are coiled in a left-handed manner. Three of these chains interact by hydrogen bonding and coil together into a right-handed cable. Collagen fibrils are composed of a staggered array of many such cables interacting in a side-by-side manner.
6. The structures of fibrous proteins are determined by the amino acid sequence, by the principle of forming the maximum number of hydrogen bonds, and by the steric limitations of the polypeptide chain, in which the peptide grouping is in a planar conformation.

7. X-ray diffraction provides data from which we can deduce the dimensions of the polypeptide chains in proteins. The use of x-ray techniques is, however, limited to molecules that can be oriented to achieve two- or three-dimensional order.
8. Fibrous proteins may achieve two-dimensional order, but they never achieve three-dimensional order. Therefore, the diffraction pattern of fibrous proteins gives information about the regularly repeating elements along the long axis of the fibers but tells us very little about the orientation of amino acid side chains.
9. Many globular proteins can be crystallized to achieve three-dimensional order. Study of the crystals of a globular protein can lead to a complete solution of its three-dimensional structure.
10. The forces that hold globular proteins together are the same as those that hold fibrous proteins together, but there is less emphasis on regularity and more emphasis on burying the hydrophobic regions in the interior of the protein.
11. The secondary structures found in the keratins recur in smaller patches in globular proteins. Such regions of secondary structure are folded into a seemingly endless array of tertiary structures.
12. Tertiary structures can be understood in terms of a limited number of domains. Chirality and packing considerations are major factors in determining the geometry of domains.
13. Quaternary structures are formed between nonidentical subunits to give irregular macromolecular complexes or between identical subunits to give geometrically regular structures.

Selected Readings

Anfinsen, B. C., Principles that govern the folding of protein chains. *Science* 181:223–230, 1973. Nobel Prize recounting by the man who showed that proteins fold spontaneously into their native structures.

Baron, M., D. G. Norman, and I. D. Campbell, Protein modules. *TIBS* 16:13–17, 1991.

Branden, Carl, and John Tooze, *Introduction to Protein Structure.* New York and London: Garland Publishing, 1991.

Cantor, C. R., and P. R. Schimmel, *Biophysical Chemistry,* vols. 1, 2, and 3. New York: Freeman, 1980. Includes several chapters (2, 5, 13, 17, 20, and 21) on the principles of protein folding and conformation.

Chothia, C., Principles that determine the structures of proteins. *Ann. Rev. Biochem.* 53:537–572, 1984.

Chothia, C., and A. Leak, Helix movements in proteins. *Trends Biochem. Sci.* 10:116–118, 1985.

Chothia, C., and A. V. Finkelstein, The classification and origins of protein folding patterns. *Ann. Rev. Biochem.* 59:1007–39, 1990.

Cohen, C., and D. A. D. Parry, α-Helical coiled coils—a widespread motif in protein. *Trends Biochem. Sci.* 11:245–248, 1986.

Creighton, T. E., *Proteins, Structures and Molecular Principles.* New York: Freeman, 1984. Very readable and reasonably comprehensive.

Dorit, R. L., L. Schoenbach, and W. Gilbert, How big is the universe of exons? *Science* 250:1377–1381, 1990. Predicts that there are between 1,000 and 7,000 different kinds of domains in all proteins found in nature.

Farber, G. K., and G. A. Petsko, The evolution of α/β barrel enzymes. *TIBS* 15:228–234, 1990.

Fasman, G. D., Protein conformation prediction. *TIBS* 14:295–299, 1989.

Fersht, A. R., The hydrogen bond in molecular recognition. *TIBS* 12:301–304, 1987.

Hogle, J. M., M. Chow, and D. J. Filman, The structure of polio virus. *Sci. Am.* 256(3):42–49, 1987.

Karplus, M., and J. A. McCannon, The dynamics of proteins. *Sci. Am.* 254(4):42–51, 1986. A reminder that proteins are not rigid inflexible structures.

Pauling, L., *The Nature of the Chemical Bond,* 3d ed. Ithaca: Cornell University Press, 1960. A classic on molecular structure.

Pauling, L., and R. B. Corey, Configurations of polypeptide chains with favored orientations around single bonds: two new pleated sheets. *Proc. Natl. Acad. Sci. USA* 37:729–740, 1953. Classic paper.

Pauling, L., R. B. Corey, and H. R Branson, The structure of proteins: two hydrogen-bonded helical configurations of the polypeptide chain. *Proc. Natl. Acad. Sci. USA* 27:205–211, 1951. Another classic paper.

Richardson, J. S., and D. C. Richardson, The *de novo* design of protein structures. *TIBS* 14:304–309, 1989.

Rose, C. D., A. R. Geselowizt, G. J. Lesser, R. H. Lee, and M. H. Zehfus, Hydrophobicity of amino acid residues in globular proteins. *Science* 229:834–838, 1985.

Rossman, M. G., and P. Argos, Protein folding. *Ann. Rev. Biochem.* 50:497–532, 1981.

Rossman, M. G., and J. E. Johnson, Icosahedral RNA virus structure. *Ann. Rev. Biochem.* 58:533–573, 1989.

Sali, A., J. P. Overington, M. S. Johnson, and T. L. Bundell, From comparisons of protein sequences and structures to protein modelling and design. *TIBS* 15:235–240, 1990.

Tonegawa, S., The molecules of the immune system. *Sci. Am.* 253(4):122–130, 1985.

Valegard, K., L. Liljas, K. Fridborg, and T. Unge, The three-dimensional structure of the bacterial virus MS2. *Nature* 345:36–41, 1990.

Wuthrich, K., Protein structure determination in solution by nuclear magnetic resonance spectroscopy. *Science* 243:45–50, 1989. The most effective technique for determining protein fine structure in cases where x-ray diffraction cannot be used.

Wright, P. E., What can two-dimensional NMR tell us about proteins? *TIBS* 14:255–259, 1989.

Yang, J. T., Protein secondary structure and circular dichroism: a practical guide. *Chemtracts, Biochemistry and Molecular Biology* 1:484–490, 1990.

Problems

1. The principal force driving the folding of some proteins is the movement of hydrophobic amino acid side chains out of an aqueous environment. Explain.
2. Outline the hierarchy of protein structural organization.
3. What is the role of loops or short segments of "random" structure in a protein whose structure is primarily α-helix?
4. What are some consequences of changing a hydrophilic residue to a hydrophobic residue on the surface of a globular protein? What are the consequences of changing an interior hydrophobic to a hydrophilic residue in the protein?
5. Some proteins are anchored to membranes by insertion of a segment of the N-terminal into the hydrophobic interior of the membrane. Predict (guess) the probable structure of the sequence (Met-Ala-(Leu-Phe-Ala)$_3$-(Leu-Met-Phe)$_3$-Pro-Asn-Gly-Met-Leu-Phe). Why would this sequence be likely to insert into a membrane?
6. Suppose that every other Leu residue in the peptide shown in problem 5 were changed to Asp. Would that necessarily alter the secondary structure? Explain whether insertion into the membrane would be altered.
7. Amino acid side chains coordinate to the metal cofactor in metalloproteins. Examples of these coordination ligands include Asp, Glu, His, and Cys. In most of the proteins studied, the side chains directly surrounding the ligand amino acid are highly conserved among homologous proteins isolated from different organisms, while nonconservative alterations in amino acid sequence are found at sites distant from the metal binding site. How do these observations fit the argument that biological structure dictates function?
8. Proteins that span biological membranes to provide ion-conducting channels or pores frequently have multiple α-helical segments aligned parallel to each other. Proteins that span the membrane with a single α-helical segment do not allow conduction of ions. Explain why ion conduction through a single α-helical segment does not occur.
9. An investigator purified a protein (protein X) from *E. coli.* She injected protein X into rabbits to generate antibodies that recognize and bind to protein X. Using an electrophoretic technique, she separated the proteins from a crude cell extract of *E. coli* and used the antibody to locate protein X on the gel. To her surprise, the antibody bound not only with protein X but also with a second, unrelated protein (protein Y). When proteins X and Y were sequenced, she found that the sequence of residues 67–78 in protein X and that of residues 120–131 in protein Y were identical. Help the investigator rationalize the data, recognizing that antigenic determinants (epitopes) of proteins are clusters of amino acids.
10. "Left- and right-handed α-helices of polyglycine are equally stable." Defend or refute the statement. (Consider glycine's chirality or lack thereof.)
11. Molecular weight analysis of a protein yields the following information:

Solvent	M_r
Dilute buffer	200,000
6 M Guanidinium chloride (GuHCl)	100,000
6 M GuHCl + 100 mM 2-mercaptoethanol	75,000 and 25,000

(Guanidinium chloride is a chaotropic (denaturing) reagent and 2-mercaptoethanol can reduce disulfide bonds.) What can you deduce about the protein's quaternary structure?

12. Using the Ramachandran diagram in the text, explain why polypeptides assume only a limited number of regular structures.
13. It might be argued that in protein structure, as in everyday life, it is a "right-handed world." Use examples of protein structure discussed in the chapter to support this contention.
14. Recombinant DNA technology allows the amino acid sequences in proteins to be altered. However, not all of these "genetically engineered" proteins yield stable, catalytically active proteins. Why?
15. Write all the quaternary forms possible for a hexameric protein composed of A and B type subunits. (Homohexamers are allowed.) What forces likely bind the subunits to each other?

Functional Diversity of Proteins

ow that we have described the chief types of protein structure, let's turn to the question of how these structures relate to the function for which they were designed. We will begin by introducing the proteins that occupy the different parts of the cell or extracellular environment. (The treatment will be brief at this point, since we will be discussing many of these proteins later in the text.) We will then examine two protein systems in some detail, to provide a perspective on how structure relates to function. Next, we will consider protein design from the evolutionary viewpoint. As we will see, small refinements arise from point mutations that lead to single amino acid changes, and grosser changes arise by a reshuffling of domains. Finally, recognizing that individual proteins cannot be characterized unless they are first isolated, we will conclude this chapter with a discussion of protein purification techniques.

Spatial Localization and Functional Diversity

The cell is a highly organized factory in which the constituent parts are assembled in different locations and specialized machinery exists for specific purposes. Thus single-cell organisms are compartmentalized so that specific reactions occur in unique locations. In multicellular organisms the localization of reactions is even greater. The workers in the biochemical factory of the organism are the proteins.

Proteins Are Directed to the Regions Where They Are Utilized

Our first consideration is how proteins get to their final destination, that is, the locations where they function. All proteins are made in the cytoplasm, but their final location depends on a variety of signals. We will give a brief overview of this subject here, reserving a consideration of the mechanisms for chapters 21 and 29.

All proteins are made on ribosomes. Except for a small number of ribosomes located inside the organelles themselves, the vast majority of proteins are made on ribosomes in the cytosol. Some of the ribosomes are freely floating in the cytosol and some are attached to the endoplasmic reticulum. The ribosomes that remain free account for the proteins that are targeted to locations in the cytosol, the nucleus, the peroxisomes, the mitochondria, and the chloroplasts (fig. 5.1). Ribosomes that are bound to the endoplasmic reticulum make proteins that are deposited in the lumen of the endoplasmic reticulum. From there

Figure 5.1

The different routes traveled by proteins during and after synthesis. In a typical eukaryotic cell, proteins are synthesized on free polysomes or in the endoplasmic reticulum on membrane-bound polysomes. Built into the proteins are amino acid sequences that determine in which of these two locations they will be synthesized. Arrows indicate location to which proteins are transported after synthesis. Some proteins synthesized on free polysomes remain in the cytosol; others become incorporated into mitochondria, chloroplasts (not shown), peroxisomes, or the nucleus. Some proteins synthesized on membrane-bound polysomes remain in the endoplasmic reticulum; others are transported to the Golgi. Some proteins transported to the Golgi remain there; others are transported to lysosomes, secretory vesicles, or the plasma membrane. Arrows indicate the directions of protein transport.

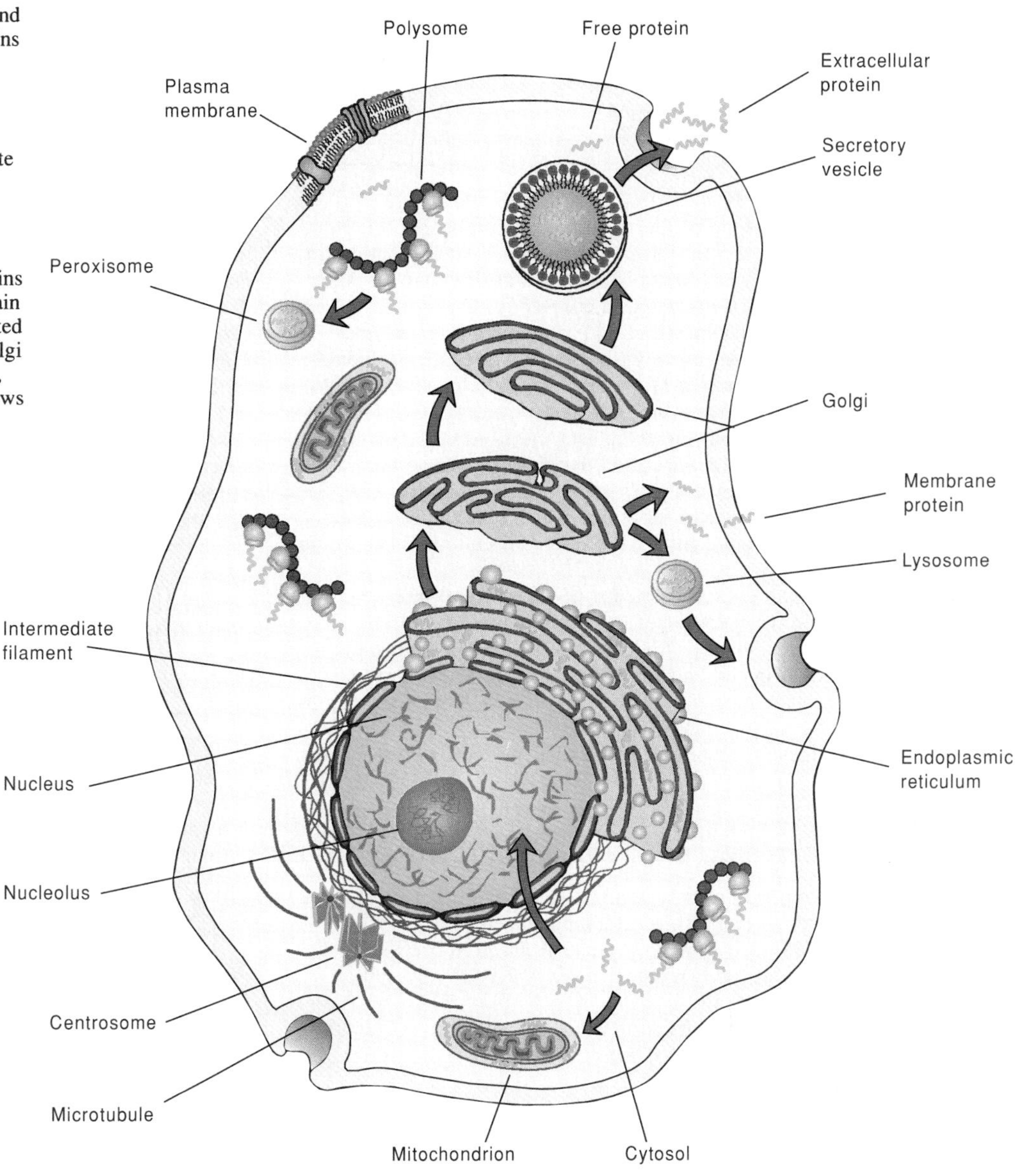

the newly synthesized proteins may be transferred to the Golgi apparatus while undergoing modifications of various sorts. At some point parts of the Golgi pinch off and the modified proteins that do not remain in the Golgi are transferred to specific locations such as the lysosomes, the plasma membrane, and the secretory granules. Those proteins targeted to the secretory granules are eventually exported.

Classification of Proteins According to Location Emphasizes Functionality

Because of the great structural and functional diversity of proteins, it is difficult to capture the important features or the whole range of them within any one classification scheme. For our present purposes we will classify proteins according to the locations they occupy when they are fully functional. This is a useful classification scheme because it emphasizes functional interrelatedness—proteins that go together work together. The structures and functions of many proteins found in different locations, both inside and outside the cell, are listed in table 5.1, which also notes points in the text where the various proteins are discussed in greater detail.

Protein Structure Is Suited to Protein Function

We have seen that highly elongated fibrous proteins are well suited for compartmentalization, for giving stable form to organellar and cellular structures, and for processes involving movement of the organism. Because of their generally low mobility, fibrous proteins are rarely associated with enzyme activity or used for transport purposes. For those functions globular proteins are more suitable. In this section we will consider two classical examples of protein assemblages that are ideally designed for the roles they play in the cell: hemoglobin and the skeletal muscle system.

Table 5.1
Some of the Main Proteins Found in Living Organisms

Protein/*Characteristics and Functions*

I. Main proteins of the cytoskeleton

Actin Bihelical filaments of aggregated globular monomers (monomer M_r = 42,000). Form cross-linked networks. In combination with myosin form actinomyosin, which functions in muscular contraction. Muscle is discussed in this chapter. Actin in the cytoskeleton is discussed in chapter 7.
Tubulin Hollow tube composed of 13 protofilaments; each protofilament contains an extensive linear aggregate of globular tubulin dimers (monomer M_r = 50,000). Tubulin exists mainly as single filaments emanating from the centrosome (see fig. 5.1) to locations throughout the cytoplasm. Involved in maintenance of cell shape. The mitotic apparatus that directs chromosomes to opposite poles of a dividing cell is composed of tubulin. The cilium is a special case of tubulin in which many microtubules interact to produce a complex apparatus for cell motility (see fig. 4.18).
Intermediate filaments Interrupted α-helical proteins interacting in a side-by-side twisted manner to form ropelike structures (monomer M_r = 40,000–75,000). Intermediate filaments are more abundant in cells subject to mechanical stress; they occupy locations near membranes where they appear to exert a protective function.
Spectrin Cytoskeletal protein particularly abundant in erythrocytes (see chapter 7).

II. Human plasma proteins

Albumin Osmotic regulation; transports acids and other substances (monomer M_r = 66,000). Most abundant serum protein.
α-Globulins A mixture of many proteins involved in transport and possibly other functions (monomer M_r = 20,000–400,000).
β-Globulins
Transferin Binds and transports iron (monomer M_r = 76,500).
β_2-Microglobulin Associated with the histocompatibility antigen (see chapter 33).
α-Globulins Antibodies (see chapter 33) (monomer M_r = 150,000).
Fibrinogen Circulating soluble protein, which, after proteolysis by thrombin, forms fibrin polymers of the blood clot (monomer M_r = 340,000) (see chapter 9).
Complement A mixture of about 11 proteins that work together to complement the immune system (see chapter 33) (monomer M_r = 80,000–200,000).

III. Some proteins of the extracellular matrix

Glycosaminoglycans Occupy large amounts of space forming hydrated gels (see chapters 6 and 21).
Proteoglycans Long glycosaminoglycans covalently linked to a core protein (see chapters 6 and 21).
Collagen (about 12 major types) The major proteins in the extracellular matrix (see chapters 4 and 29).
Types I–III Assemble into fibrils organized to meet the needs of the tissue.
Type IV Assembles into a laminar network.
Elastin Cross-linked random coil protein that gives elasticity to tissues.
Fibronectin A glycoprotein that helps to mediate cell-matrix adhesion (see chapter 7).
Integrins (several) Integral membrane protein that helps to bind cells to the extracellular matrix. Each protein usually consists of two different subunits.

Protein/*Characteristics and Functions*

IV. Digestive enzymes of the gastrointestinal tract

Amylase Degrades starch to disaccharides.
Pepsin Degrades proteins to large peptides.
Amylase As above.
Peptidases Split large peptides to small peptides.
Trypsin Degrades proteins to large peptides (see chapter 9).
Chymotrypsin Degrades proteins to large peptides (see chapter 9).
Lipase Degrades lipids into fatty acids and glycerol (see chapters 17, 22).
Ribonuclease Degrades RNA to oligonucleotides (see chapter 9).
Peptidases Degrade peptides to amino acids.
Disaccharidases Degrade disaccharides to monosaccharides.

V. Proteins of the cytosol

Many (between 300 and 1,000) Synthesis of most small molecules required by the cell. Synthesis of proteins, carbohydrates, and lipids (see chapters 11, 22, 23, 24).

VI. Proteins of the nucleus

Histones (5) Proteins that complex with DNA to make chromosomes (see chapter 26). There are five major histones.
Nucleic acid polymerizing enzymes (5 to 10) For DNA and RNA synthesis (see chapters 26 and 28). There are between 5 and 10 nucleic acid polymerases in different cells.

VII. Proteins of the mitochondria and the chloroplasts

Many (100 to 300) Proteins involved in energy production from metabolites or light (see chapters 14, 15, 16).

VIII. Proteins of the endoplasmic reticulum and the Golgi

Many (50 to 200) Enzymes involved in protein modification and in oligosaccharide and lipid synthesis (see chapters 17, 21, 22, 23).

IX. Proteins of the lysosomes and the peroxisomes

Many (30 to 100) Enzymes involved in a wide variety of degradation processes for removing undesired compounds. (See chapter 17 for role of peroxisomes in fatty acid degradation. See chapter 14 for role of peroxisomes or glyoxyosomes in utilization of C-2 carbon source.)

X. Proteins of the plasma membrane

Many (100 to 500) Proteins involved in transport across membranes and for transmission of important metabolic signals across the plasma membrane (see chapters 7, 24, 31).

Hemoglobin—An Allosteric Oxygen-Binding Protein

Hemoglobin is the best-known transport protein. Its chief function is to pick up oxygen in the lungs, where it is plentiful, and deliver it to tissues throughout the body. Figure 5.2 shows the central feature of hemoglobin (and myoglobin as well). This is a water-free pocket for the heme, with its central iron atom located where oxygen is bound. (Note: Heme is a complex of Fe^{2+} and protoporphyrin IX; its structure is presented in figure 5.11 and in atomic detail in chapters 11 and 15.) The hydrophobic character of the heme binding cavity is dictated by the apolar side chains that line it. This environment is particularly suited to binding the hydrophobic porphyrin ring and creates an environment where iron (Fe^{2+}) can bind oxygen reversibly without itself being oxidized to Fe^{3+}.

Hemoglobin consists of two α subunits, each with 141 amino acids, and two β subunits, each with 146 amino acids. Each subunit is capable of binding a single molecule of oxygen. In muscle cells, a reserve oxygen store is provided by the myoglobin molecule, which is similar in structure to hemoglobin except that it exists as a monomer. While the components of and hemoglobin are remarkably similar, their physiological responses are very different. On a weight basis, each molecule binds about the same amount of oxygen at high oxygen tensions (pressures). At low oxygen tensions, however, hemoglobin gives up its oxygen much more readily. These differences are reflected in the oxygen-binding curves of the purified proteins in aqueous solution (fig. 5.3).

The oxygen-binding curve for myoglobin (Mb) is hyperbolic in shape, as would be expected for simple one-to-one association of myoglobin and oxygen:

$$\text{Mb} + \text{O}_2 \rightleftarrows \text{MbO}_2$$

$$K_f = \frac{[\text{MbO}_2]}{[\text{Mb}][\text{O}_2]} = \text{equilibrium formation constant} \quad (7)$$

If y is the fraction of myoglobin molecules saturated, and if we express the oxygen concentration in terms of the partial pressure of oxygen $[O_2]$,* then

$$K_f = \frac{y}{[1 - y][\text{O}_2]} \text{ and } y = \frac{K_f[\text{O}_2]}{1 + K_f[\text{O}_2]} \quad (8)$$

This is the equation of a hyperbola, as shown in figure 5.3.

Hemoglobin (Hb) behaves differently. Its sigmoidal binding curve can be fitted by an association-constant expression with a greater-than-first-power dependence on the oxygen concentration:

$$K_f = \frac{[\text{HbO}_2]}{[\text{Hb}][\text{O}_2]^n} \text{ and } y = \frac{K_f\text{O}_2^n}{1 + K_f\text{O}_2^n} \quad (9)$$

Under physiological conditions the value of n is around 2.8, indicating that the binding of oxygen molecules to the four hemes in hemoglobin is not independent and that binding to any one heme is affected by the state of the other three. (A fuller discussion of the equations for multiple binding and the procedure for determining n from experimental data are presented in box 5A.) The first oxygen attaches itself with the lowest affinity, and successive oxygens are bound with a higher affinity. The exact value of n for hemoglobin is a function of the extent of oxygen binding as well as the presence of other factors discussed below.

*Partial pressure is usually indicated by a lower case p to the left. For simplicity the p has been omitted.

Figure 5.2

The heme pocket. The helices of hemoglobin (and myoglobin) form a hydrophobic pocket for the heme and provide an environment where the iron atom can be oxygenated when it reversibly binds oxygen. The chemical structure of heme is shown in figure 5.11 and is described in atomic detail in chapters 11 and 15.

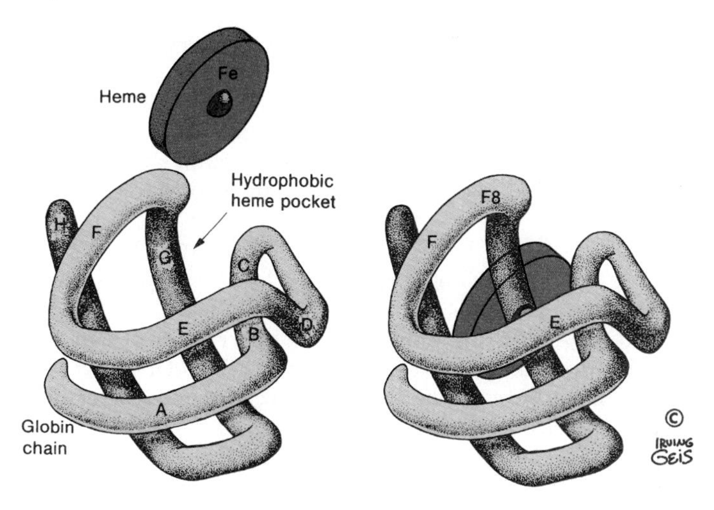

Figure 5.3

Equilibrium curves measure the affinity for oxygen of hemoglobin and of the simpler myoglobin molecule. Myoglobin, a protein of muscle, has just one polypeptide chain and resembles a single subunit of hemoglobin. The vertical axis gives the amount of oxygen bound to one of these proteins, expressed as a percentage of the total amount that can be bound. The horizontal axis measures the partial pressure of oxygen in a mixture of gases with which the solution is allowed to reach equilibrium. For myoglobin, the equilibrium curve is hyperbolic. Myoglobin absorbs oxygen readily, but becomes saturated at a low pressure. The hemoglobin curve is sigmoidal. Initially hemoglobin is reluctant to take up oxygen, but its affinity increases with oxygen uptake. At arterial oxygen pressure, both molecules are nearly saturated, but at venous pressure, myoglobin would give up only about 10% of its oxygen, whereas hemoglobin releases roughly half. At any partial pressure, myoglobin has a higher affinity than hemoglobin, which allows oxygen to be transferred from blood to muscle.

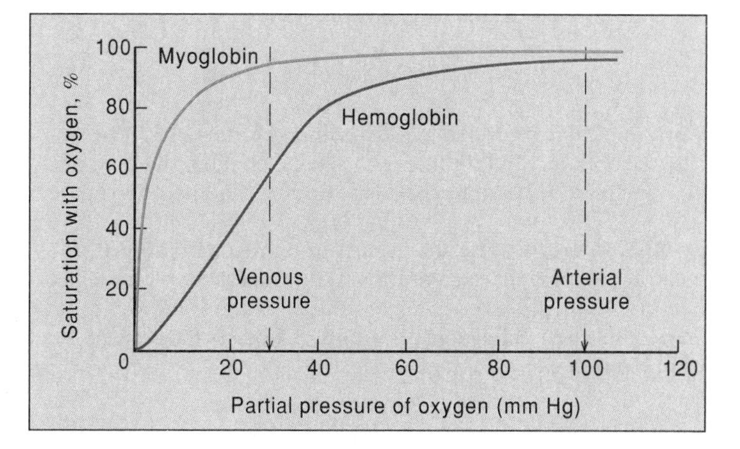

Theory of Multiple Binding and the Hill Plot

Binding studies require a measurement of the free ligand, B, in the presence of the macromolecule, A. From this we may determine the average binding number, called y. The value of y is equal to the ratio of the total number of B molecules bound to the total number of binding sites. If there is only one binding site per A molecule, the expression for y is rather simple:

$$y = \frac{[AB]}{[A] + [AB]} \tag{B1}$$

The value of [A] + [AB] is known from the total amount of A added to the solution. The value of [AB] is equal to the total amount of B added, minus the experimentally observed concentration of free B after A is added:

$$y = \frac{\text{total B} - \text{free B}}{\text{total A}} \tag{B2}$$

Let us first consider the simple binding of one ligand B to a protein molecule A with a formation constant K_f:

$$A + B \overset{K_f}{\rightleftharpoons} AB;\ K_f = \frac{[AB]}{[A][B]} \tag{B3}$$

From Equations (B2) and (B3) we can express y in terms of K_f, the formation constant, or K_d, the dissociation constant:

$$y = \frac{K_f[B]}{1 + K_f[B]} \tag{B4}$$

or, since $K_f = 1/K_d$,

$$y = \frac{[B]}{[B] + K_d} \tag{B5}$$

Taking the reciprocal of both sides, we obtain

$$\frac{1}{y} = 1 + K_d\left(\frac{1}{[B]}\right) \tag{B6}$$

By plotting the experimentally determined values of $1/y$ against $1/[B]$, we obtain a straight line with a slope equal to the dissociation constant K_d. The intercept on the ordinate should be 1 (figure 1).

When there is more than one site on A for binding B, the equations become more complex and in general the plots are not linear. If we consider the average binding number for n sites on a molecule, the total average binding number is the sum of the binding numbers for each of these sites.

$$y = \sum_{i=1}^{n} \frac{K_{fi}[B]}{1 + K_{fi}[B]} \tag{B7}$$

Figure 1

A binding plot to determine the dissociation constant K_d for the simple situation where there is one ligand binding site. y is the average binding number and [B] is the concentration of ligand.

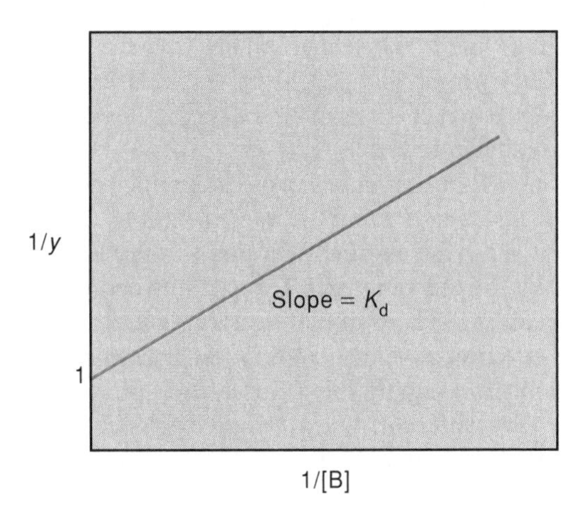

There are two situations in which the data for multiple binding may be treated rather simply. First, when all binding sites bind B with the same energy, all terms are equal and the solution to the sum is simply n times each term.

$$y = \frac{nK_f[B]}{1 + K_f[B]} \tag{B8}$$

Again, since $K_f = 1/K_d$,

$$y = \frac{n[B]}{[B] + K_d} \tag{B9}$$

and taking the reciprocal of both sides,

$$\frac{1}{y} = \frac{1}{n} + \frac{1}{n}K_d\left(\frac{1}{[B]}\right) \tag{B10}$$

If we plot $1/y$ against $1/[B]$, we find that the slope is equal to K_d/n and the intercept (when $1/[B]$ approaches 0) is $1/n$. Alternatively, we may plot the data as y versus $y/[B]$ (figure 2), in which case the slope is $-K_d$ and the intercept at $y/[B] = 0$ is n, since

$$y = n - \frac{yK_d}{[B]} \tag{B11}$$

Figure 2

A binding plot to determine the dissociation constant K_d and the number of binding sites n for the situation where there are n ligand binding sites per macromolecule with identical binding affinities.

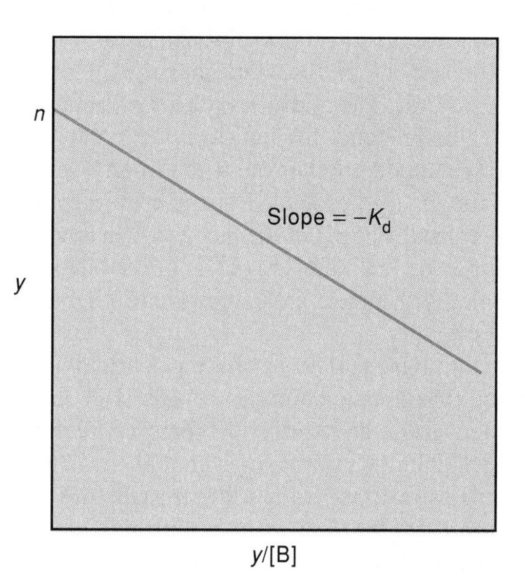

The second situation arises when the sites interact so strongly that only the fully saturated product AB_n is formed. The data may then be treated in the following way. The formation of only one major product means that the binding of the ligand molecule to the macromolecule greatly enhances the further binding of additional ligands such that at equilibrium, only three species exist in significant concentrations: A, B, and AB_n:

$$A + nB \rightleftharpoons AB_n$$

$$y = \frac{n[AB_n]}{[AB_n] + A} \tag{B12}$$

$$y = \frac{nK_f[B]^n}{1 + K_f[B]^n} \tag{B13}$$

$$\frac{1}{y} = \frac{1}{n} + \frac{1}{nK_f[B]^n} \tag{B14}$$

Figure 5.4

The structure of glycerate-2, 3-bisphosphate, an allosteric effector for hemoglobin oxygen release.

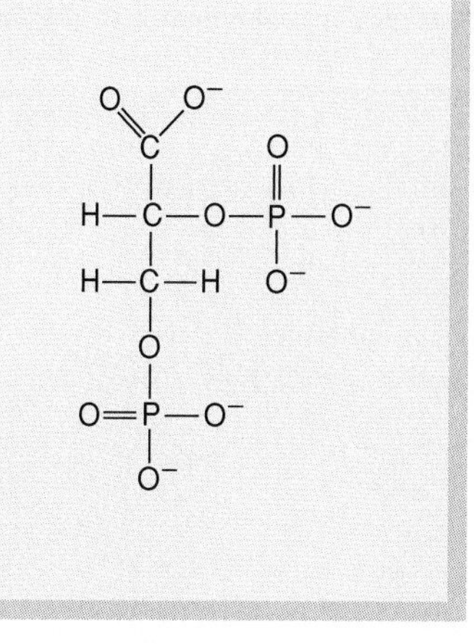

In general, a value of $n > 1$ indicates cooperative binding (or positive cooperativity) between small-molecule ligands, a value of $n < 1$ indicates anticooperative binding (or negative cooperativity), and a value of $n = 1$ indicates no cooperativity.

The cooperative binding of oxygen by hemoglobin is ideally suited to the conditions involved in oxygen transport. Thus in the lung, where the oxygen tension is relatively high, hemoglobin can become nearly saturated with oxygen, while in the tissues, where the oxygen tension is relatively low, hemoglobin can release about half its oxygen (see fig. 5.3). If myoglobin were used as the oxygen transporter, less than 10% of the oxygen would be released under similar conditions. The positive cooperativity associated with oxygen binding to hemoglobin is a special case of allostery in which the binding of "substrate" to one site stimulates the binding of "substrate" to another site on the same multisubunit protein. Although this is a special case of allostery, it is quite commonly observed in regulatory proteins (see chapter 10).

The Binding of Certain Factors to Hemoglobin Influences Oxygen Binding in a Negative Way

The combination of hemoglobin with oxygen depends not only on oxygen tension but also on pH, CO_2, and glycerate-2,3-bisphosphate (GBP or BPG). GBP (fig. 5.4) binds preferentially to the deoxygenated form of hemoglobin with a dissociation constant of about 10^{-5} M^{-1}. Its dissociation constant with HbO_2 is only about 10^{-3} M^{-1}. Since the concentrations of GBP and hemoglobin are both about 5 mM in the erythrocyte, we would expect most of the deoxy form to be complexed with GBP and most of the oxyhemoglobin to be free of GBP. The net effect of the GBP is to shift the oxygen-binding curve to higher oxygen tensions (fig. 5.5). This shift is not sufficient to lower the binding of oxygen at the high oxygen tensions in the capillaries of the

Figure 5.5

Oxygen binding curve for hemoglobin as a function of the partial pressure of oxygen. Two curves are shown, one in the absence and one in the presence of glycerate-2, 3-bisphosphate (GBP). GBP decreases the affinity between oxygen and hemoglobin, as shown by the displacement of the binding curve to high oxygen concentrations in its presence.

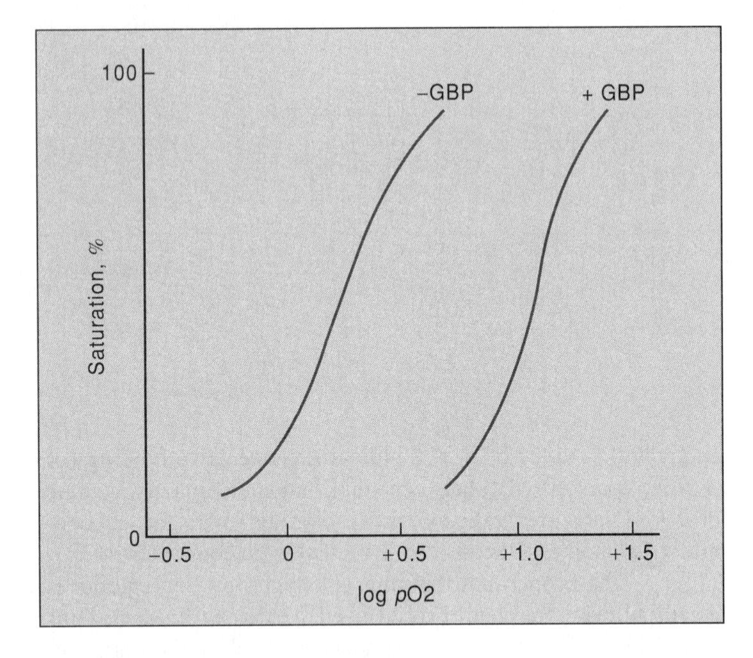

lungs, but it is sufficient to cause a substantially greater release of oxygen at the lower oxygen tensions that exist where it is needed, that is, in the rest of the body tissues.

The negative influences of H^+ and CO_2 on oxygen binding are causally related and together are known as the Bohr effect. The CO_2 that diffuses across the capillary wall from the tissue is largely in solution as CO_2 molecules, since hydration to form carbonic acid (H_2CO_3) is a slow reaction with a half-time of about 10 s. The substantial amount of CO_2 generated by the various decarboxylation reactions of intermediary metabolism diffuses from cells through interstitial fluid into the blood plasma, with only a small fraction becoming hydrated as carbonic acid. In the erythrocytes, however, hydration of CO_2 occurs very rapidly, catalyzed by carbonic anhydrase, which is concentrated in the erythrocytes. At the pH of blood (about 7.4) the carbonic acid largely dissociates into H^+ and HCO_3^-:

$$CO_2 + H_2O \overset{\text{Carbonic anhydrase}}{\rightleftharpoons} H_2CO_3 \rightleftharpoons H^+ + HCO_3^-$$

The production of protons through these two consecutive reactions explains why an increase in concentrations of CO_2 causes a lowering of pH (an increase in the H^+ concentration). The net result is a lowering of the affinity of hemoglobin for oxygen, since protons, like GBP, bind preferentially to deoxy hemoglobin. For example, at pH 7.6 and 40 mm Hg of oxygen tension, hemoglobin retains more than 80% of its oxygen; at pH 6.8 it retains only 45%. The negative effect of CO_2 on oxygen binding is mainly due, then, to the tendency of CO_2 to lower the pH (i.e., raise the H^+ concentration), and the negative effects of the protons on oxygen binding are qualitatively similar to the negative effects of GBP.

The Bohr effect is also closely related to the major role that hemoglobin plays in disposing of the CO_2 produced in tissues, and in controlling the blood pH. While oxygen is being delivered to the tissues in the venous blood, the CO_2 is being absorbed from the tissues (fig. 5.6). This process would stop very quickly if it were not for the erythrocytes and the hemoglobin. As we have just seen, the CO_2 that diffuses into the erythrocytes is rapidly converted into carbonic acid, which in turn dissociates into H^+ and HCO_3^-. The protons produced by this dissociation would lower the pH and reverse this dissociation if it were not for the buffering action of the hemoglobin. Loss of oxygen increases the acid dissociation constant of the hemoglobin so that it picks up the excess protons. This change serves two purposes; it helps to keep the pH constant and it enables the system to absorb more CO_2. The additional CO_2 is ultimately disposed of when the blood reaches the lung and the hemoglobin again becomes oxygenated. Another point to remember is that the majority of the HCO_3^- produced in the erythrocyte diffuses into the venous blood. The pH of the blood system is controlled within narrow limits by the buffering action of the bicarbonate and the hemoglobin, with minor assistance from other proteins in the bloodstream.

One must marvel at the way various factors work in concert so that hemoglobin can be useful in so many roles—oxygen deliverer, carbon dioxide remover, and pH stabilizer. From the explanation we have given it should be clear why the hemoglobin is confined to a cellular structure in the plasma rather than being present as a free plasma protein. Carbonic anhydrase and GBP are essential for efficient hemoglobin function, and their presence in the bloodstream at adequate concentrations would probably be unattainable or intolerable to the blood system. Furthermore, the protons released as a result of the carbonic anhydrase reaction are rapidly picked up by the deoxygenated hemoglobin before they have a chance to cause a potentially harmful lowering of the pH of the plasma.

X-Ray Diffraction Studies Reveal Two Conformations for Hemoglobin

X-ray diffraction studies on fully oxygenated hemoglobin and deoxygenated hemoglobin have shown that the molecule is capable of existing in two states, with significant differences in tertiary and quaternary structures (fig. 5.7). Further studies on partially oxygenated hemoglobin may indicate additional intermediate structures between these two extremes. Until these can be characterized in structural terms, the two-state model serves as a useful conceptual framework for explaining the allosteric mechanism of the hemoglobin system.

The hemoglobin tetramer is composed of two identical halves (dimers), with the $\alpha_1\beta_1$ subunits in one dimer and the $\alpha_2\beta_2$ subunits in the other. The subunits within the dimers are tightly held together; the dimers themselves are capable of motion with respect to one another (fig. 5.8). The interface between the movable dimers contains a network of salt bridges and hydrogen bonds when hemoglobin is in the deoxy conformation (fig. 5.9). The quaternary transformation that takes place on binding of oxygen causes the breakage of these bonds.

Figure 5.6

Transport of oxygen (O_2) and carbon dioxide (CO_2) in the circulatory system. In most tissues O_2 is released and CO_2 is withdrawn by the red blood cells; in the lungs these processes are reversed.

Lungs

CO_2

O_2

Other tissues

O_2

CO_2

Lungs

$$HCO_3^- + H^+ \xrightleftharpoons{\text{Carbonic anhydrase}} H_2CO_3 \rightleftharpoons CO_2 + H_2O$$

$$O_2 + HHb^+ \rightleftharpoons H^+ + HbO_2$$

$$HHb^+ + O_2 + HCO_3^- \rightleftharpoons HbO_2 + CO_2 + H_2O$$

Other tissues

$$H^+ + HbO_2 \rightleftharpoons HHb^+ + O_2$$

$$CO_2 + H_2O \xrightleftharpoons{\text{Carbonic anhydrase}} H_2CO_3 \rightleftharpoons H^+ + HCO_3^-$$

$$HbO_2 + CO_2 + H_2O \rightleftharpoons HHb^+ + O_2 + HCO_3^-$$

Figure 5.7

Three-dimensional structure of oxy- and deoxyhemoglobin as determined by x-ray crystallography. This is a view down the dyad (two fold) axis, with the ß chains on top. In the oxy-deoxy transformation (quaternary motion) $\alpha_1\beta_1$ and $\alpha_2\beta_2$ dimers move as units relative to each other. This allows glycerate-2, 3-bisphosphate to bind to the larger central cavity in the deoxy conformation. A close-up of the binding site is shown in figure 5.10.

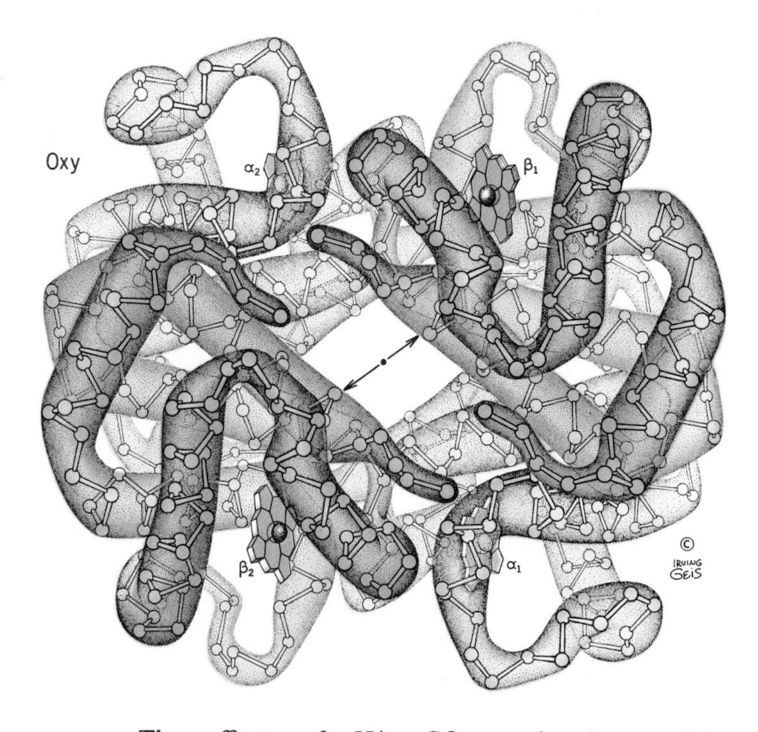

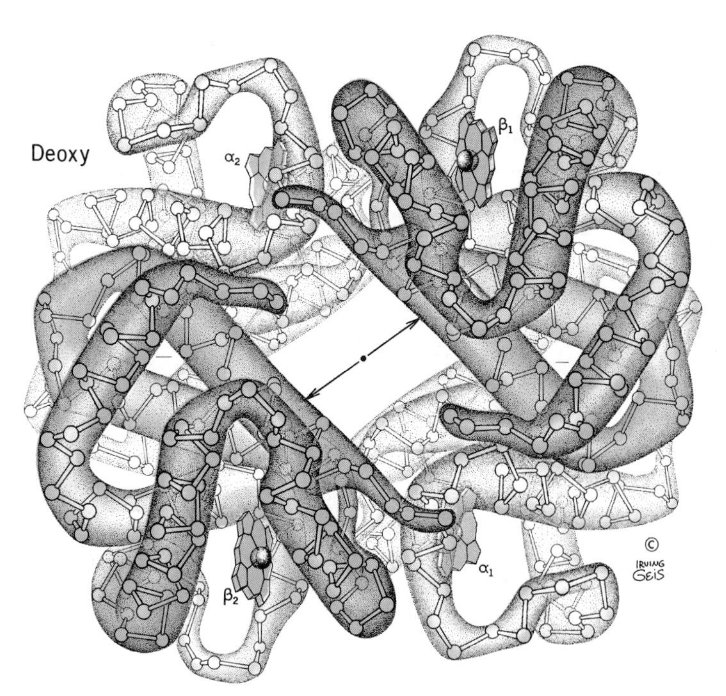

The effects of H^+, CO_2, and glycerate-2,3-bisphosphate on oxygen binding can be understood in terms of their stabilizing effect on the deoxy conformation. The decreased oxygen binding as the pH is lowered from 7.6 to 6.8 suggests the involvement of histidine side chains, because these are the only side-chain groups in proteins that have a pK in this pH range. Certain histidines in the charged form make salt linkages that contribute to the stability of the deoxy form (see fig. 5.9). As the pH is lowered, these histidines tend to become charged, which increases the stability of the deoxy form. Such a change should inhibit a structural transition to the oxy form and thereby lower the affinity of the protein for oxygen. Similarly, glycerate-2,3-bisphosphate binds most strongly to the deoxy form (fig. 5.10) and thereby discourages the transition to

Figure 5.8

The deoxy-to-oxy shift upon binding oxygen in one hemoglobin molecule. The projection shown in this figure is approximately perpendicular to the one shown in figure 5.7. The $\alpha_1\beta_1$ dimer moves as a unit relative to the $\alpha_2\beta_2$ dimer. The interface between the two dimers is crucial to the cooperativity effect in hemoglobin. The interface is not visible in this figure (see figure 5.9).

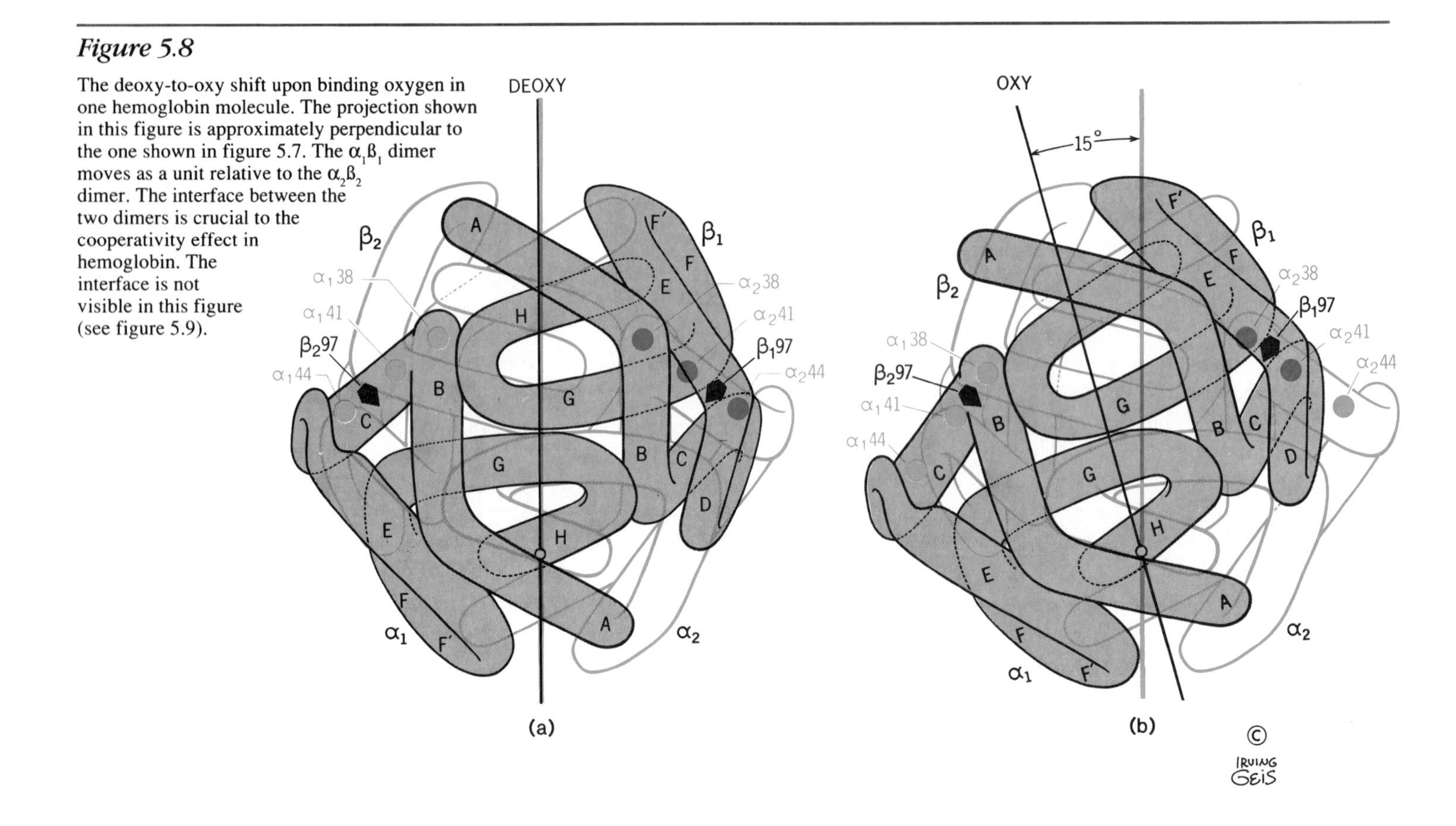

Figure 5.9

(*a*) The $\alpha_1\beta_2$ (and $\alpha_2\beta_1$) interface is shown schematically at the lower left and in detail (*b*). This is the regulatory zone of the hemoglobin molecule, which contains crucial hydrogen bonds and salt bridges. (*b*) All the hydrogen bonds and salt bridges shown here (dotted lines) exist only in the deoxy state, with the exception of $\alpha_1$41–$\beta_2$40 and $\alpha_1$94–$\beta_2$102, which exist only in the oxy state. Only the α carbons are shown in the backbone structure of the hemoglobin, except in the region of the interface.

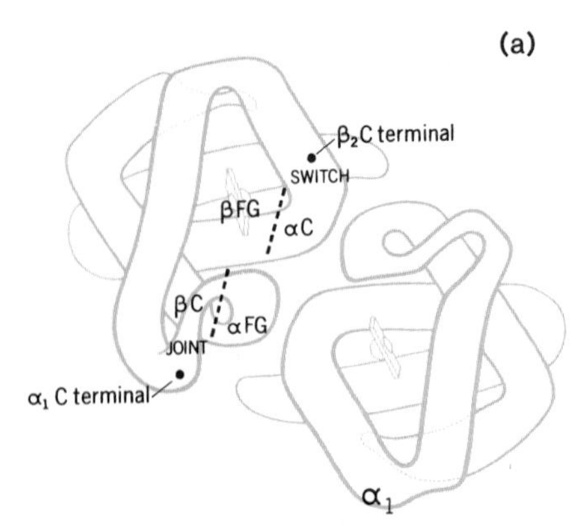

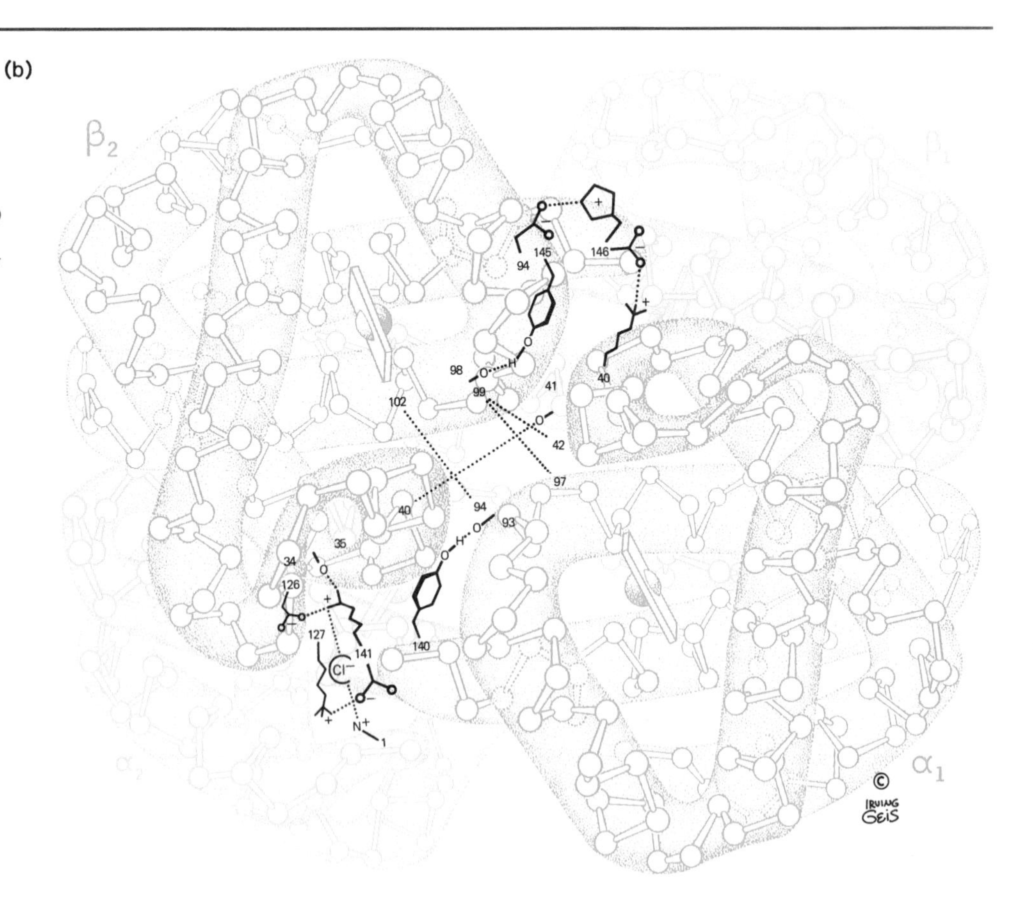

Figure 5.10

The binding of glycerate-2, 3-bisphosphate in the central cavity of deoxyhemoglobin between ß chains. The surrounding positively charged residues are the amino terminal, His 2, Lys 82, and His 143.

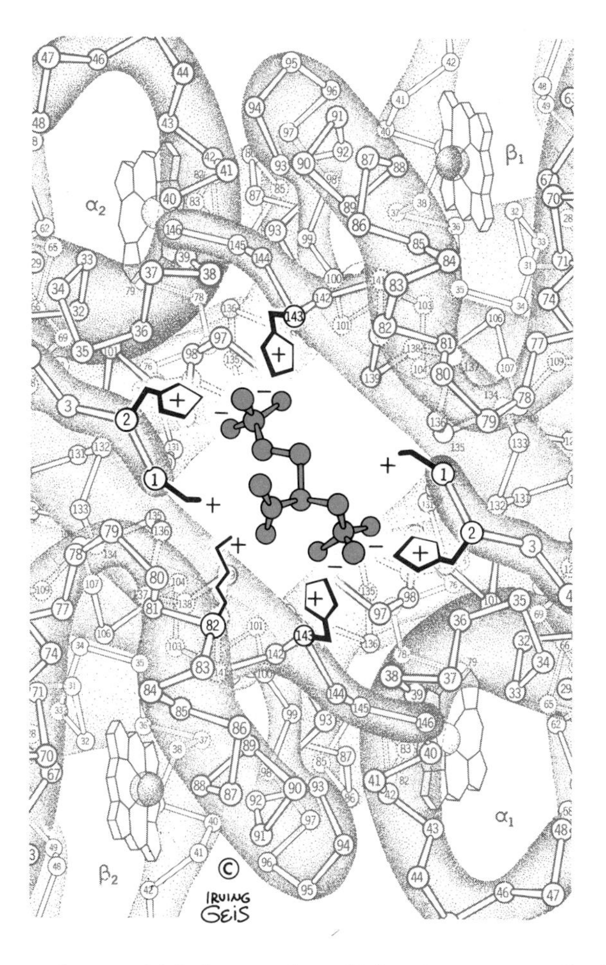

the oxy form, which lowers the affinity for oxygen. Carbon dioxide binds as bicarbonate to the α-amino groups in hemoglobin; this binding also favors the deoxy conformation. Binding of CO_2 is freely reversible, being favored by the high CO_2 tensions in the tissues. As a result hemoglobin becomes a carrier of CO_2 from tissues to the lungs, where it is discharged.

Changes in Conformation Are Initiated by Oxygen Binding

The oxygen binding at the heme group itself initiates the changes in tertiary and quaternary structure that are responsible for the cooperative effect seen on oxygen binding. The heme group contains an Fe^{2+} ion located near the center of a porphyrin ring. The Fe^{2+} makes four single bonds to the nitrogens in the heme ring, and a fifth bond to a histidine side chain of the F helix, F8 histidine (fig. 5.11). When oxygen is present it binds at the sixth coordination position of the iron on the other side of the heme. Movement of the iron upon oxygen binding (or release) pulls the F8 histidine and the F helix to which it is covalently attached (fig. 5.12). The tertiary-structure change in the F helix

Figure 5.11

A close-up view of the iron-porphyrin complex with the F helix in deoxyhemoglobin. Note that the iron atom is displaced slightly above the plane of the porphyrin.

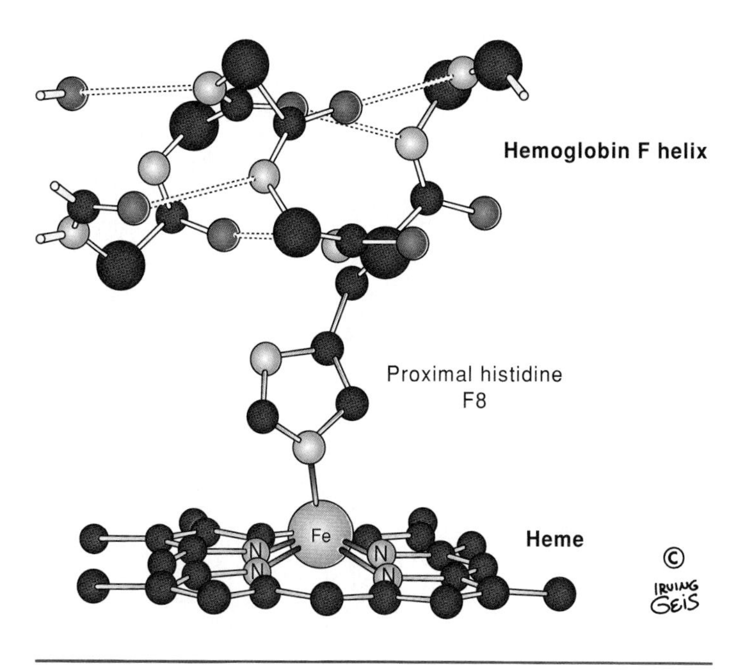

Figure 5.12

Downward movement of the iron atom and the complexed polypeptide chain on binding oxygen. The structure is shown before (black) and after (blue) binding oxygen. Movement of His F8 is transmitted to FG5 valine, straining and breaking the hydrogen bond to the penultimate tyrosine. Only the α chain is shown here.

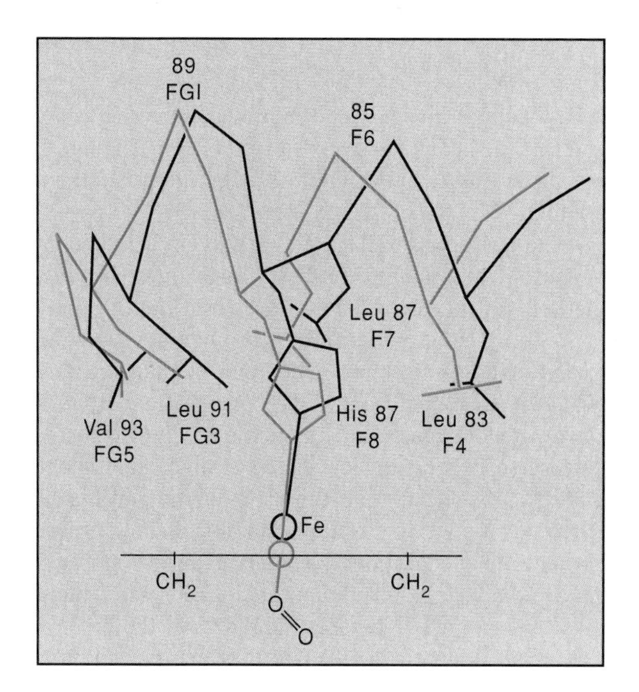

Figure 5.13

Invariant residues in the α and ß chains of mammalian hemoglobin. The colored dots, indicating the positions of invariant residues, line the heme pockets as well as the crucial $\alpha_1\beta_2$ interface. The invariant residues have been found in about 60 species. There are 43 invariant positions in the hemoglobin molecule.

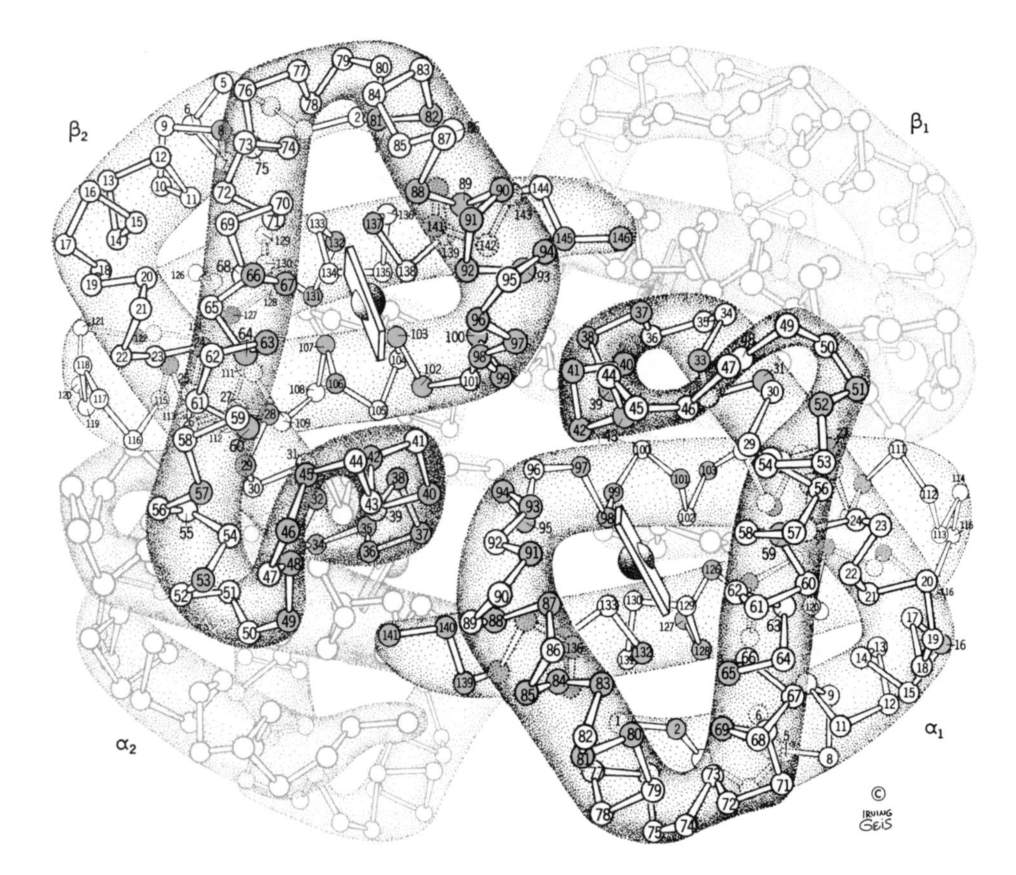

induces a strain in the rest of the protein that facilitates the conversion of the deoxy to the oxy structure. This change favors the binding of additional oxygen at other unoccupied sites in the tetramer.

The Fe^{2+} ion is well suited to its job in hemoglobin, not only because it has a natural affinity for oxygen, but also because it changes its electronic structure in a highly significant way in so doing. Fe^{2+} is normally paramagnetic, having four unpaired electrons in its outer *d* electronic orbitals. In this state it is too large to sit precisely in the plane of a porphyrin, as studies with model compounds have shown. Fe^{2+} is also paramagnetic when it is pentacoordinated in deoxyhemoglobin; as expected, it is displaced from the plane of the porphyrin by a few tenths of an angstrom unit. When O_2 binds, however, the Fe^{2+} becomes hexacoordinated and diamagnetic (no unpaired electrons). This change results in a major reorganization of its outer *d* orbitals, which decreases the radius of the Fe^{2+} so that it can move to an energetically more favorable position in the center of the porphyrin (see fig. 5.12).

The structural arguments advanced here to explain oxygen binding by hemoglobin are supported by amino acid sequences of α and β chains for a large number of hemoglobins from different species. Data from 60 species of α chains and 66 species of β chains reveal 43 invariant positions in the hemoglobin molecule. These invariants are plotted on a map of the hemoglobin structure in figure 5.13, where the invariant positions are shown by colored dots. In a sense, the dots provide a diagram of the working machinery of hemoglobin, for the invariant positions line the heme pockets where oxygen is bound and the crucial $\alpha_1\ \beta_2$ interface, which changes its orientation when oxygen binds. Electrostatic forces and hydrogen bonds stitch the interface together in the deoxy conformation when the molecule gives up its oxygen to the tissues. If there are changes, resulting from mutation at any of these positions, then we would expect trouble to develop. This is just what happens, as can be seen in figure 5.14, which shows the positions of pathological mutations in hemoglobin. Where there are changes in the heme pockets or in the $\alpha_1\ \beta_2$ interface, hemoglobin abnormalities occur; many of these are associated with serious diseases.

Figure 5.14

Positions of mutations in hemoglobin that produce a pathological condition. Comparison with figure 5.13 shows that these mutations, in general, show the same pattern as the distribution of the invariant positions. Dark circles indicate positions of abnormal residues, solid black dot indicates the valine β_6 mutation in sickle-cell anemia, heavy circles indicate M (Met) hemoglobin, and jagged perimeter indicates unstable hemoglobin. Dark color indicates increased oxygen affinity; light color indicates decreased oxygen affinity.

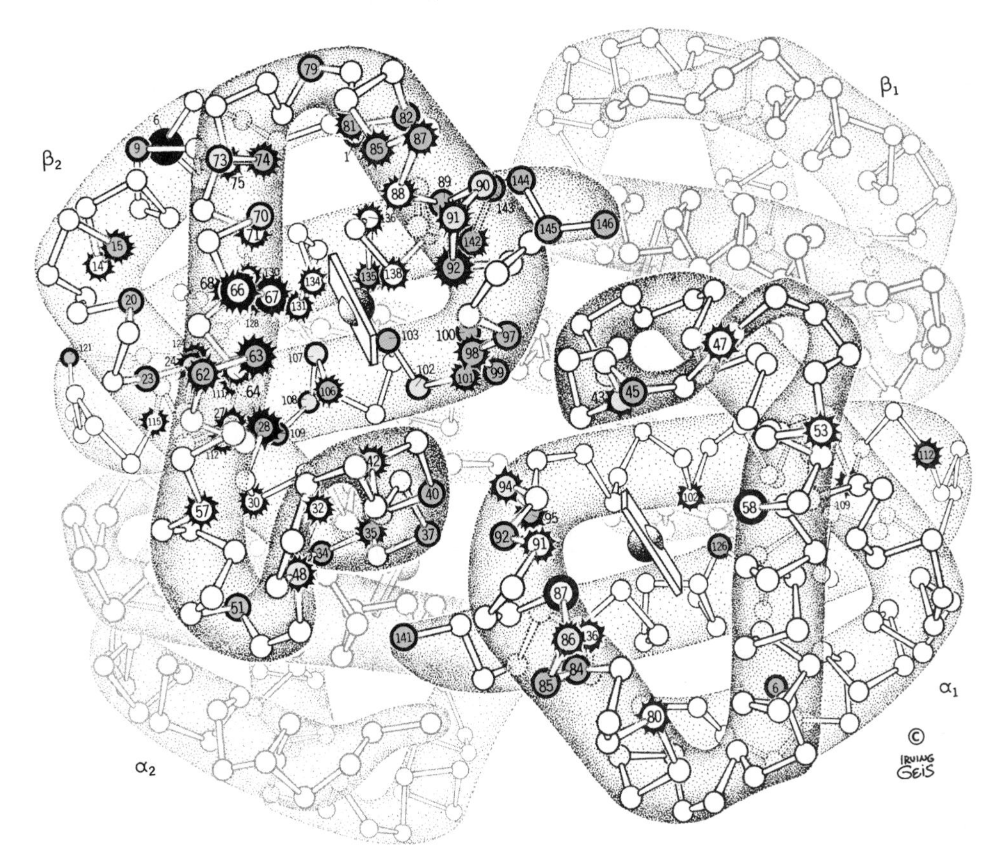

As a rule, invariant amino acids are the only critical loci for hemoglobin function. One striking exception occurs at position 6 of the β chain. A hydrophobic valine residue is substituted for glutamic acid (a charged side chain) with disastrous results. The specific consequence of the β_6 alteration is to cause hemoglobin tetramers to aggregate when they are in the deoxy state. This is because the β_6 valine fits into a hydrophobic pocket in an adjoining hemoglobin molecule. The aggregates form long fibers that stiffen the normally flexible red blood cell. The resulting distortion of the red cells leads to capillary occlusion, which prevents proper delivery of oxygen to the tissues. This pathological condition is known as sickle-cell anemia.

Two Models Have Been Proposed for the Way Hemoglobins and Other Allosteric Proteins Work

We have spent a good deal of time describing the function of hemoglobin because it is the best-understood regulatory protein and provides us with a model system for understanding in general terms how other allosteric proteins work. The first indication that hemoglobin was an allosteric protein came from the sigmoidal shape of its oxygen-binding curve (see fig. 5.3). Allosteric proteins are usually composed of two or more subunits. Different ligands may bind to quite different sites, or to quite similar sites as in the case of hemoglobin.

Two quite different models were proposed about twenty-five years ago to explain the unusual nature of the hemoglobin oxygen-binding curve. These models could also be used as a starting point for discussing other allosteric proteins (fig. 5.15). The first model, introduced by Monod, Wyman, and Changeux in 1965, is called the symmetry model. In this model hemoglobin can exist in only two conformations, one with all four of the subunits within a given tetramer in the low-affinity form and one with all four subunits in the high-affinity form (see fig. 5.15*a*). Also, in this model, the hemoglobin molecule is always symmetrical, i.e., all the subunits are either in one state or the other, and all of the binding sites have identical affinities. The binding of oxygen to one of the subunits favors the transition to the high-affinity form. The greater the number of oxygens binding to the tetramer, the more likely it is that the transition from the low-affinity form to the high-affinity form will occur.

Figure 5.15

Alternate models for hemoglobin allostery. (*a*) In the symmetry model hemoglobin can exist in only two states. (*b*) In the sequential model hemoglobin can exist in a number of different states. Only the subunit binding oxygen must be in the high-affinity form.

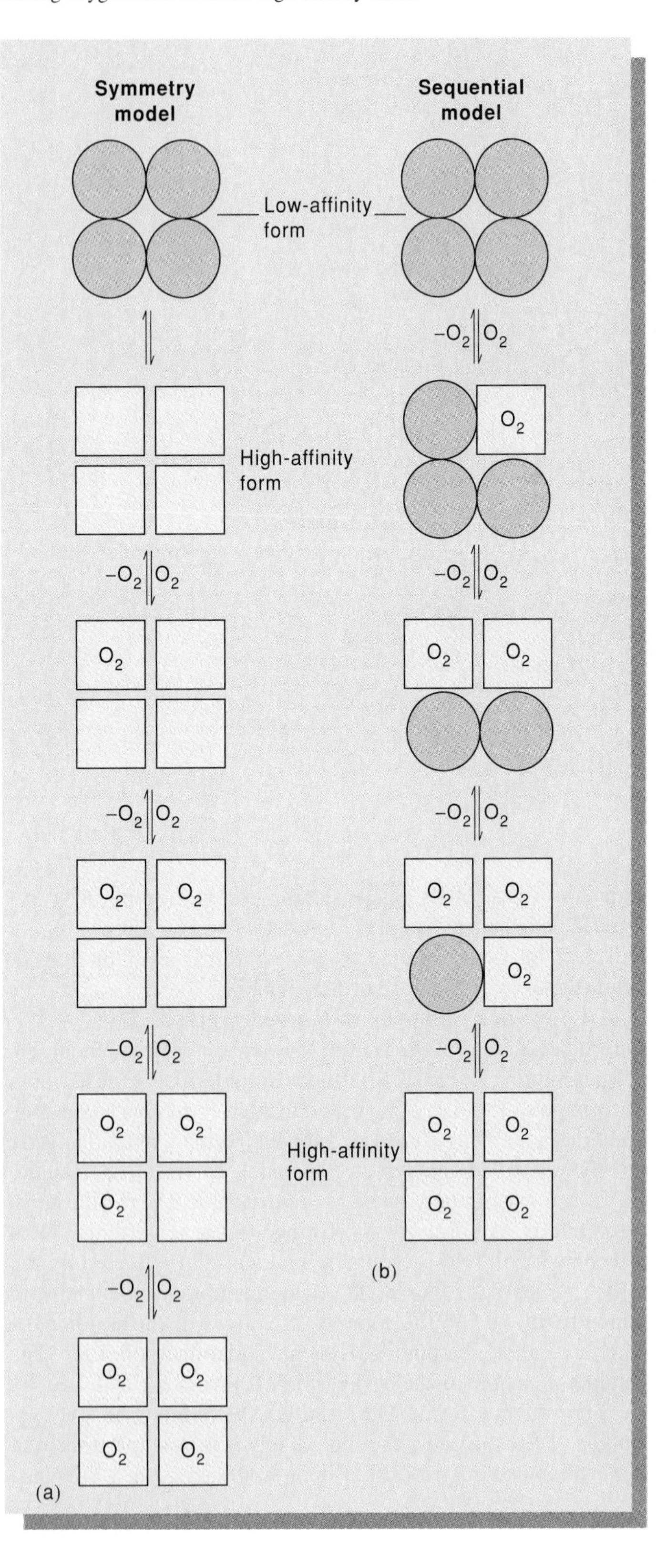

The second model, proposed by Koshland, Nemethy, and Filmer in 1966, is referred to as the sequential model (see fig. 5.15*b*). In this model the binding of an oxygen molecule to a given subunit causes that subunit to change its conformation to the high-affinity form. Because of its molecular contacts with its neighbors, the change increases the probability that another subunit in the same molecule will switch to the high-affinity form and bind a second oxygen more readily. The binding of the second oxygen has the same type of enhancing effect on the remaining unoccupied oxygen-binding sites.

Either of these models (or something in between) could account for the sigmoidal oxygen-binding curve of hemoglobin, and either is consistent with the fact that deoxygenated and fully oxygenated hemoglobin have different conformations. The only way to rigorously discriminate between these two models is to obtain structural information on partially oxygenated hemoglobin. This information is still lacking, so no final judgment can yet be made. Even when the situation is fully resolved for hemoglobin, there is no assurance that other allosteric proteins work in the same way.

Muscle—An Aggregate of Proteins Involved in Contraction

Vertebrate skeletal muscle represents a remarkable example of a supermolecular aggregate capable of undergoing a reversible reorganization. Voluntary muscle tissue is arranged into fibers that are surrounded by an electrically excitable membrane called the sarcolemma (fig. 5.16). Each fiber is composed of many myofibrils, which when viewed in the light microscope present a striated and banded appearance. As shown in figure 5.17, a myofibril exhibits a longitudinally repeating structure called the sarcomere. This 23,000-Å-long repeating unit is characterized by the appearance of several distinct bands, the less optically dense band being referred to as the I band and the more dense one as the A band. Furthermore, a dense line appears in the center of the I band, called the Z line; and a dense narrow band somewhat similar in appearance also occurs in the center of the A band, called the M line. Adjacent to the M line are regions of the A band that appear less dense than the remainder; these are referred to as the H zone.

Transverse sections of the sarcomere reveal that these patterns result from the interdigitation of two sets of filaments (fig. 5.17). For example, when a sarcomere is sectioned in the I band, a somewhat disordered arrangement of thin filaments (about 70 Å in diameter) is seen. In contrast, when sectioned in the H zone, a hexagonal array of thick filaments (about 150 Å in diameter) is apparent. The substantive observation is that a transverse section in the dense region of the A band shows a regularly packed array of interdigitating thick and thin filaments. This observation led Hugh Huxley to propose that the process of muscle contraction involves sliding the thick and thin filaments past each other (fig. 5.18).

Figure 5.16

The hierarchy of muscle organization. A voluntary muscle such as the bicep is a composite of many muscle fibers, connected to tendons at both ends. Each muscle fiber is composed of several myofibrils that are surrounded by an electrically excitable muscle fiber membrane (sarcolemma). Myofibrils exhibit longitudinally repeating structures called sarcomeres. The fine structure of the sarcomere is described in figure 5.17.

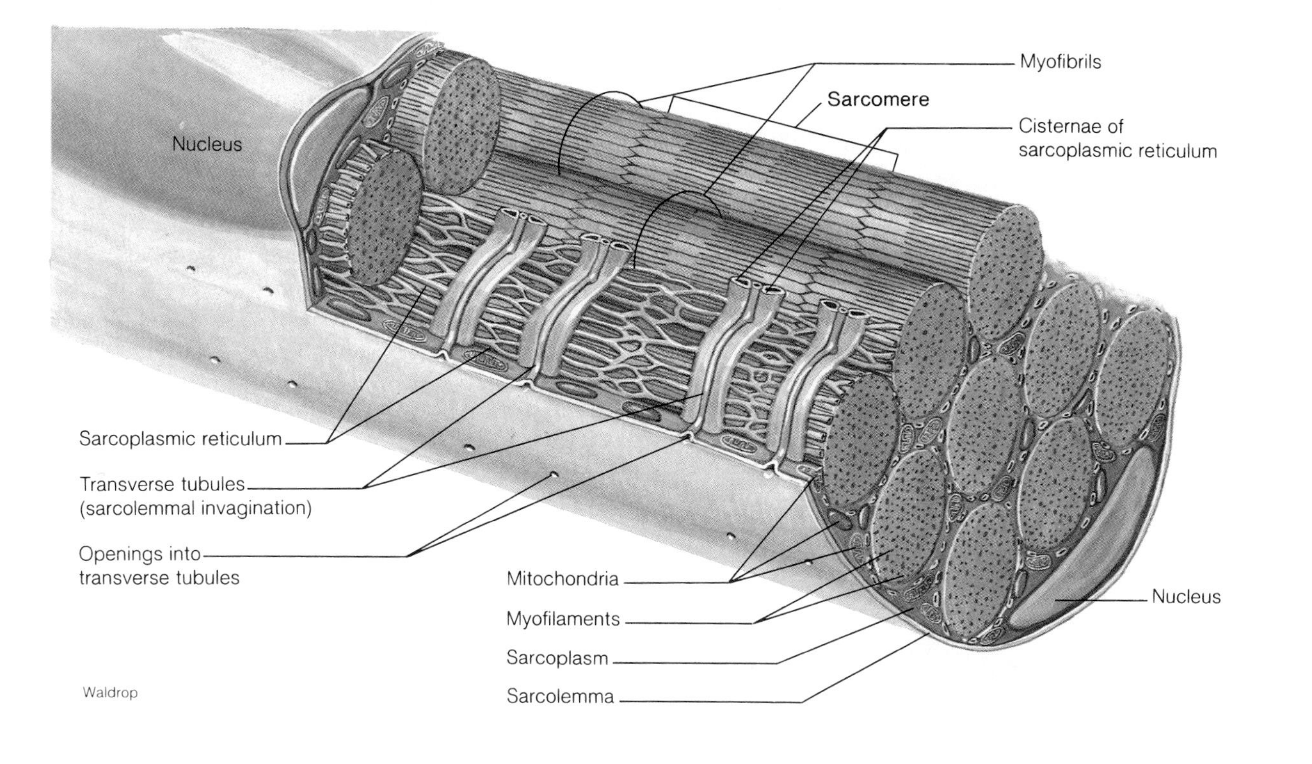

Figure 5.17

Electron micrograph of a striated muscle sarcomere showing the appearance of filamentous structures when cross-sectioned at the locations illustrated below. (Electron micrograph courtesy of Dr. Hugh Huxley, Brandeis University.)

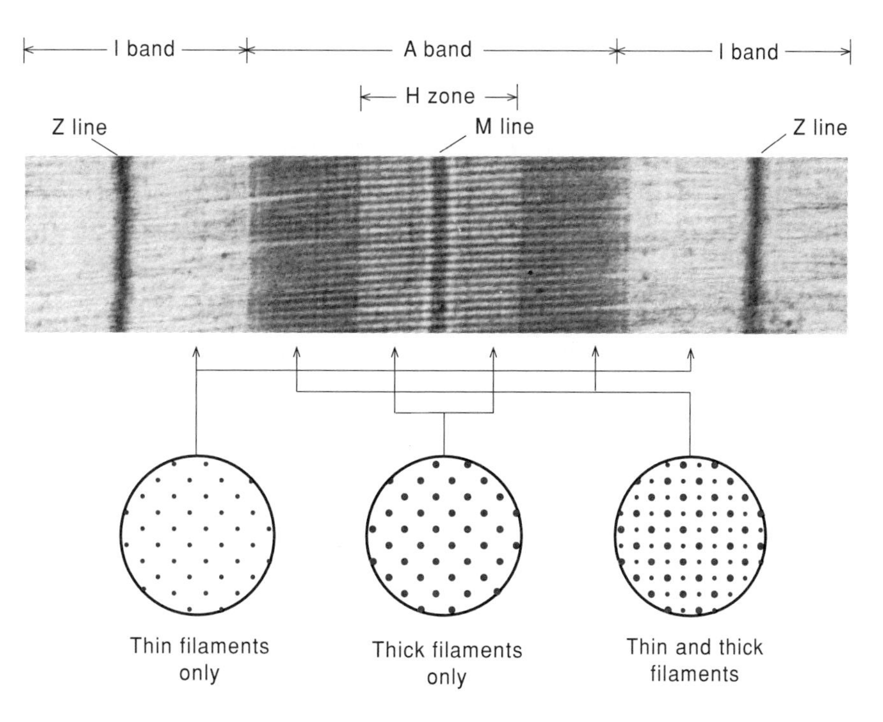

Figure 5.18

The sliding-filament model of muscle contraction. During contraction, the thick and thin filaments slide past each other so that the overall length of the sarcomere becomes shorter.

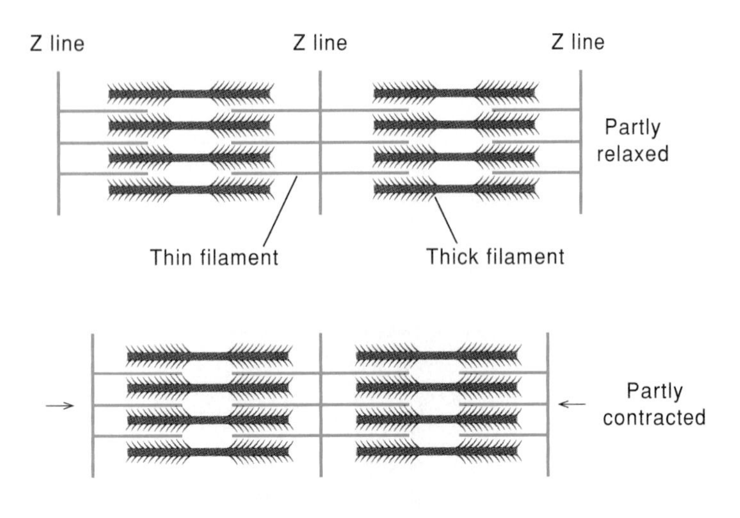

Subsequent analyses have shown that the thin filaments are composed of three proteins (fig. 5.19 and table 5.2). The main filamentous structure consists of an aggregate of globular actin molecules which takes on the form of a right-handed double helix. The individual actin molecules have a molecular weight of 42,000. Every turn of the actin helix incorporates 14 actin molecules and two molecules of the 360-Å-long filamentous protein tropomyosin (TM) that fit into the two grooves created by the actin double helix. The TM molecule is a dimer of two identical α-helical chains that wind around each other in a coiled coil. Each TM dimer spans seven actin monomers, and a succession of TM dimers extends the full length of the thin filament. Two molecules of troponin (TN) bind to the actin filament at each helical repeat. Troponin is a complex of three nonidentical subunits: TN-C, a calcium binding subunit; TN-T, a TM binding subunit; and TN-A, an "inhibitory" subunit. The TN-TM proteins form a regulatory complex whose properties we will discuss in a moment.

Figure 5.19

A molecular view of muscle structure. Formation of a thick filament involves the lateral aggregation of many myosin molecules to form the central bipolar structure shown in (*c*).

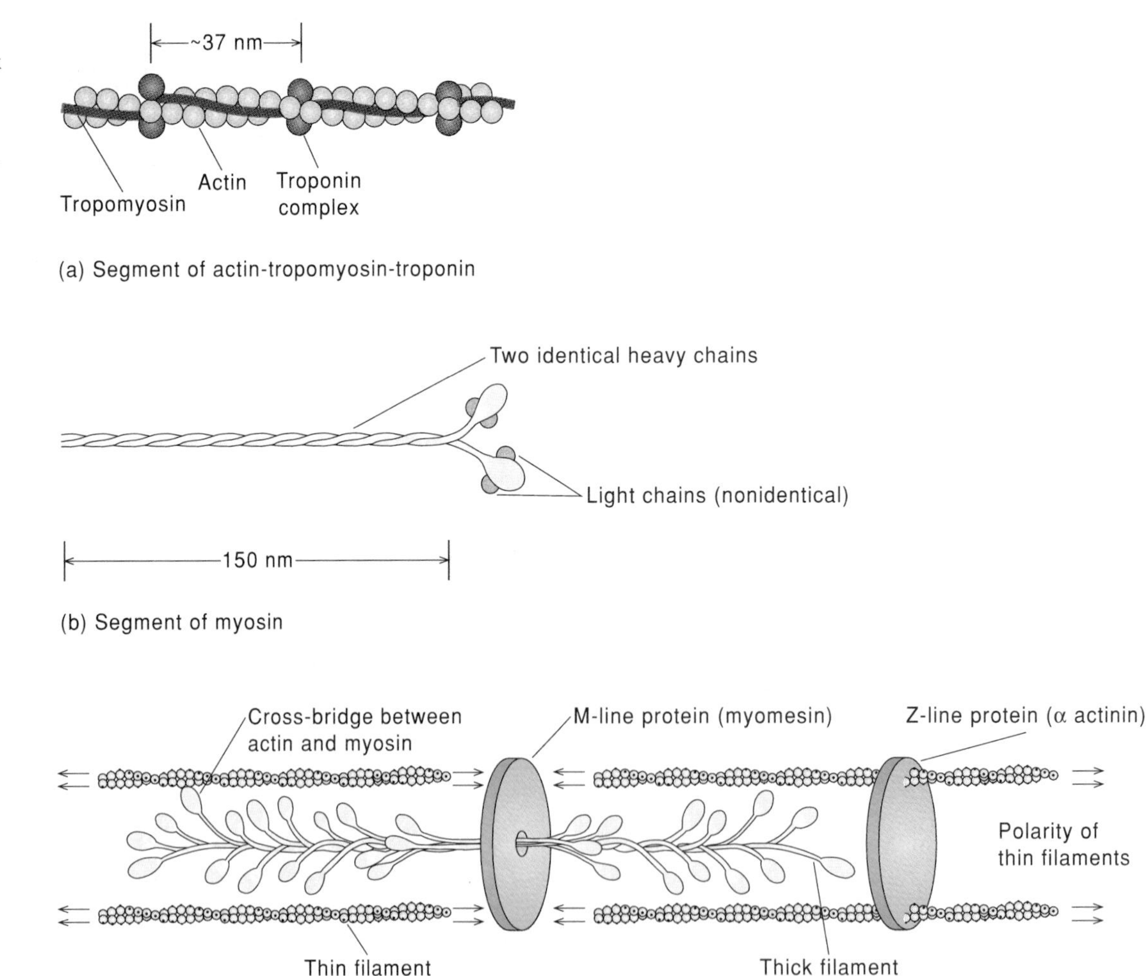

Thick filaments are composed of myosin, a large molecule containing two identical heavy chains (M_r = 223) and two light chains (M_r = 22 and 18). The structural organization of myosin is illustrated in figure 5.19. The molecule has two identical globular head regions that incorporate the light chains and a significant fraction of the heavy chains. The tails of the heavy chains form very long α helices that wrap around each other to form left-handed coiled coils. The individual myosin molecules can be cleaved into fragments by partial degradation with various proteases. By separation of such fragments, it has been demonstrated that the binding sites of myosin for actin and the ATPase activity of myosin are located in the globular head regions. The α-helical coiled coils form the backbone of the thick filament, while the remainder forms an arm that can provide a flexible extension or hinge for the globular head away from the body of the thick filament. The thick filament contains many myosin molecules oriented in a staggered bipolar fashion.

Granted that this is a marvelous piece of molecular architecture, but how does it actually work? The answer lies in the observation that actin cyclically binds the globular myosin head group to form cross-bridges in a reaction that depends on the myosin-catalyzed hydrolysis of adenosine triphosphate (ATP). The cyclic binding of actin to myosin is driven by the energy-releasing hydrolysis of ATP, catalyzed by the myosin head group in a manner that causes rearrangement of the actin-myosin cross-bridges. When muscle is completely relaxed, there is a minimum number of cross-bridges and the muscle is fully stretched. However, when the muscle is activated and under tension, it contracts and more cross-bridges are formed as the region of overlap between actin and myosin increases. At each stage of the contraction process it is essential to break the existing bridges with the help of ATP hydrolysis before new ones can be formed. It is important to realize that although ATP encourages more bridges to be formed, the ATP is required to break the bridges, not to form them. The breaking of bridges is required so that new and more numerous bridges can be formed. Thus cross-bridge formation is energetically favored.

A likely scenario for the contraction process is shown in figure 5.20. The ATP that binds to myosin causes the bridges to break or weaken. This bound ATP is rapidly hydrolyzed to ADP and P_i, but the hydrolysis products are not immediately released by the myosin. The P_i is released first, but its release requires effective contact with the actin. This is the regulated step in muscular contraction. Once the P_i has been released, a strong bridge forms between the myosin and the actin. This is followed by a structural change in the myosin that leads to the translocation of the myosin relative to the actin filament and finally to ADP dissociation. The translocation step is referred to as the power stroke in muscular contraction because it is at this point that the energy ultimately donated by the ATP is expended in the form of a complex structural change that is still not fully understood. Further contraction requires fresh ATP, followed by bridge dissociation and ATP hydrolysis. If conditions are right the P_i dissociates and the bridges reform, but this time they reform at points further along the actin. Cyclic repetition of this process results in a net increase in the number of actin-myosin bridges and further contraction of the sarcomere.

As we indicated earlier, the process of contraction is regulated or triggered by the TN-TM system. Since voluntary muscles are under the conscious control of the animal, one would expect a signal from the central nervous system to initiate the process of contraction. A nerve impulse communicated to the muscle causes a depolarization of the sarcolemma membrane that surrounds the muscle fibers. This in turn causes a release of Ca^{2+} from the endoplasmic reticulum in the cytoplasm of the muscle cell (fig. 5.21). The Ca^{2+} ions form a complex with the TN-C component of the troponin molecule (see table 5.2). This induces changes within the TN complex which overcomes the inhibitory effect of the TN-I subunit. Then, through TN-T, a signal is sent to TM that triggers the contraction event. The precise nature of this signal from troponin to tropomyosin is unclear; it appears to involve a movement of the tropomyosin on the actin surface that leads to an allosteric transition of the actin that encourages more favorable contact between the complementary binding sites on actin and myosin. As a result a strong bridge is formed and the P_i is released from the myosin. The remaining steps in muscular contraction have already been described. It is noteworthy that for some time after death, the muscle enters a state of rigor in which the muscle is fully contracted and the maximum number of bridges is formed. This is probably due to excessive neuronal firing and a considerable discharge of calcium from the sarcoplasmic reticulum. Normally the cytosolic Ca^{2+} concentration is restored to resting levels within 30 ms of receiving a signal and the myofibrils relax.

Table 5.2
Principal Proteins of Vertebrate Skeletal Muscle

Protein	M_r	Subunits	Function
Myosin	510,000	2 × 223,000 (heavy chains) 22,000–18,000 (light chains)	Major component of thick filaments
Actin	42,000	One type	Major component of thin filaments
Tropomyosin	64,000	2 × 32,000	Rodlike protein that binds along the length of actin filaments
Troponin	78,000	30,000 (TN-T) 30,000 (TN-I) 18,000 (TN-C)	Complex of three protein subunits involved in the regulation of muscle contraction

Figure 5.20

Steps in the contraction process. Since contraction is a cyclical process, the choice of a starting point is somewhat arbitrary Five frames are shown; the first two and the last two frames are identical to make the cyclical nature of the process clear. In the first frame (*a*), the myosin head groups contain the hydrolysis products of a single ATP molecule, ADP and P_i. A structural transition in the actin leads to contact between the actin and the myosin and the release of P_i. The release of the P_i is the rate-limiting step in muscular contraction. In the second frame (*b*), strong bridges form between actin and myosin. This is followed by a structural alteration in the myosin molecules and an effective translocation of the thick filament relative to the thin filament in (*c*). During this process the ADP is released. After the translocation step, the bridge structure is broken by the binding of ATP, which is rapidly hydrolyzed to ADP and P_i. Each thick filament has about 500 myosin heads and each head cycles about five times per second in the course of a rapid contraction.

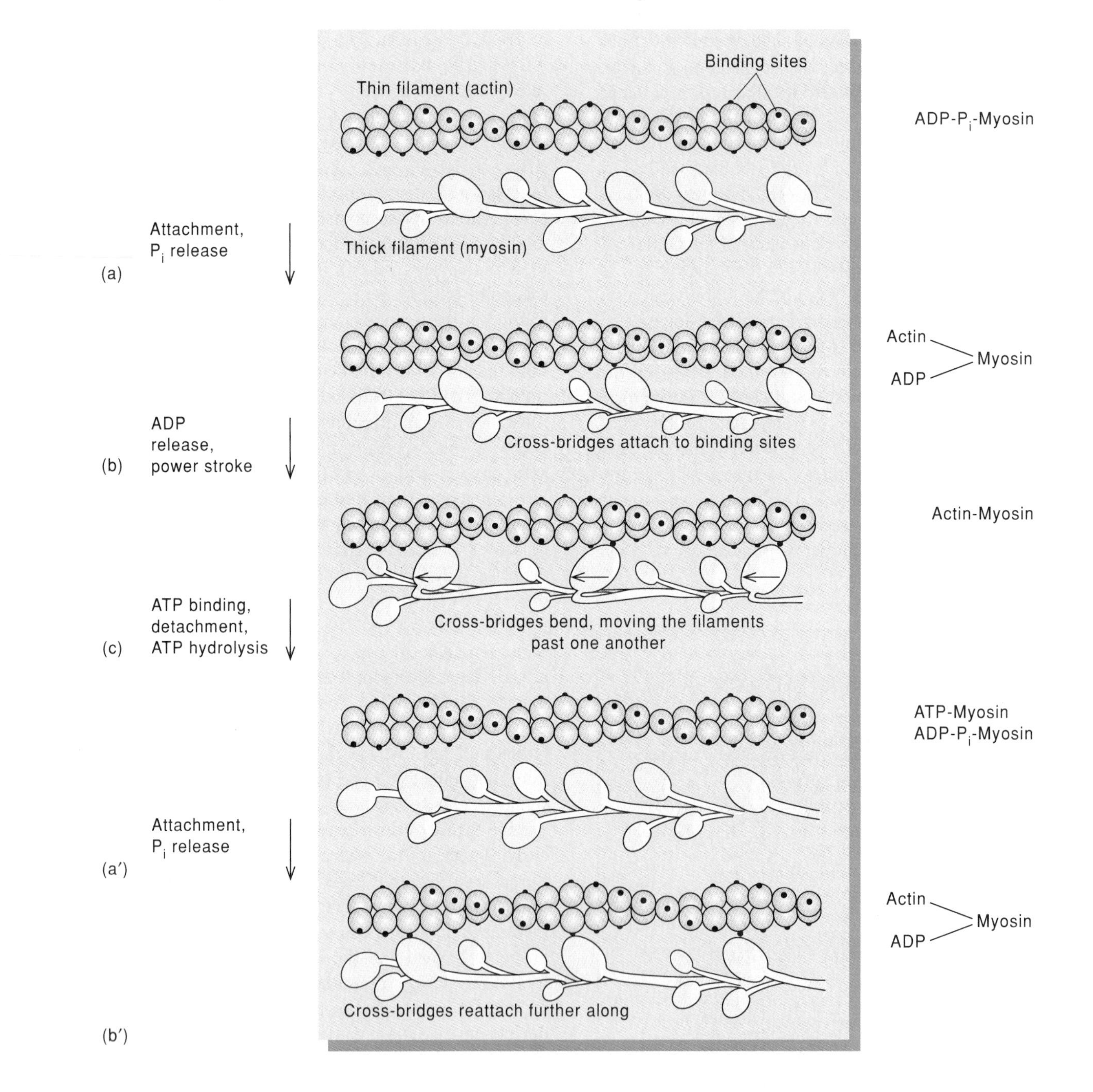

Protein Diversification as a Result of Evolutionary Pressures

We can gain further insight into protein function by asking the question, How do selective evolutionary pressures lead to the diversification of protein structure and function? The answer to this question is multifaceted.

Proteins that serve similar or identical biological functions in different living organisms are typically very similar in both their amino acid sequences and their tertiary structures. Such related families of molecules are generally assumed to reflect processes of divergent evolution. That is, they are thought to have evolved through gradual point-by-point modification of one ancestral molecule.

Figure 5.21

The effect of calcium on muscle contraction. Binding of calcium to the TN-TM-actin complex produces a shift in the location of TM, which produces an allosteric transition in actin. The allosteric transition in actin facilitates the release of P_i from myosin, which strengthens the interaction between actin and myosin.

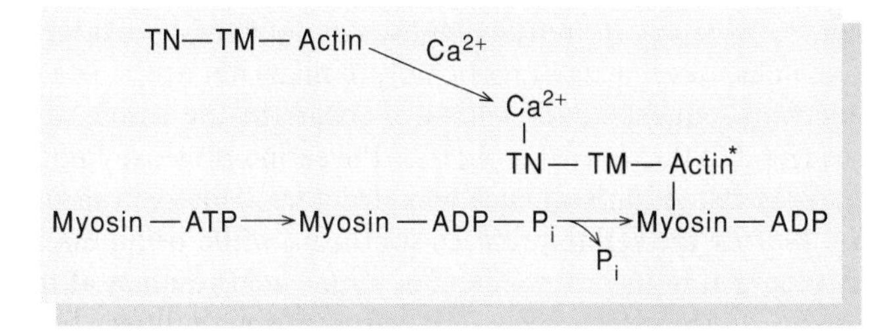

One of the most extensively studied protein families is the cytochrome *c* family. Cytochrome *c* proteins function as electron carriers in the mitochondrial electron-transport chains of all multicellular organisms (see chapter 15). Sequence comparisons of mitochondrial cytochrome *c* from organisms as diverse as humans and green plants reveal an extraordinary degree of sequence conservation. This fact suggests that the functional role of the molecule was highly refined by selective evolutionary pressures prior to the emergence of the first multicellular organisms. Some positions in the sequence are quite variable, whereas others are essentially invariant. Structural and chemical modification studies have shown that some of the invariant amino acid residues are associated with functionally important heme interactions, while others are important in governing the interactions of cytochrome *c* with its physiological oxidase and reductase.

If the sequence and structure of mitochondrial cytochrome *c* were indeed highly refined prior to the emergence of multicellular organisms, then its evolutionary precursors should still exist in prokaryotic organisms. In fact, in virtually all prokaryotic organisms that use oxidative or photosynthetic electron-transport chains to synthesize the high-energy intermediate adenosine triphosphate (ATP), we find molecules that are strikingly similar to mitochondrial cytochrome *c*. However, as might be expected, cytochrome *c* proteins from prokaryotes exhibit much more sequence diversity than the proteins typically found in higher organisms. In particular, the prokaryotic cytochrome *c* proteins often contain multiple amino acid insertions or deletions relative to mitochondrial cytochrome *c*. Nevertheless, from tertiary-structure determination of several prokaryotic proteins, we find that these molecules are all variations on a basic structural theme (fig. 5.22). Further, the prokaryotic molecules all show a strong conservation of those amino

Figure 5.22

Examples of structural diversification in prokaryotic cytochromes *c*. Three cytochromes are shown; cytochrome c_{550} from the denitrifying bacterium *P. denitrifican;* cytochrome *c* from mitochondria; cytochrome c_3 from the photosynthetic bacterium *P. rubrum*. The prokaryotic cytochromes contain more residues in their polypeptide chains (shaded) than does cytochrome *c*. Despite these variations there is a strong conservation of those amino acid residues that interact in functionally important ways with the protein's heme group.

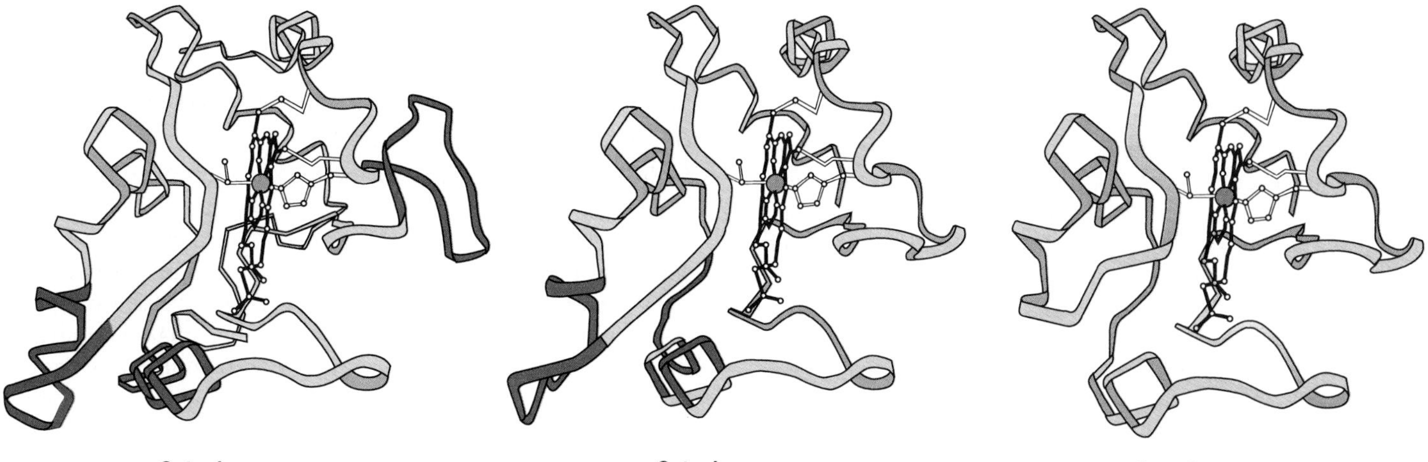

Figure 5.23

Schematic illustration of two serine proteases, elastase and subtilisin. Their molecules differ totally in sequence and tertiary structure but have catalytic sites that are nearly identical. The configuration of the active site for elastase is described in chapter 9.

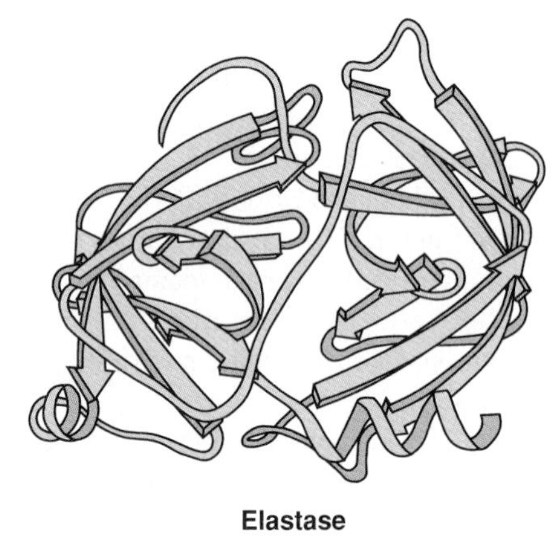

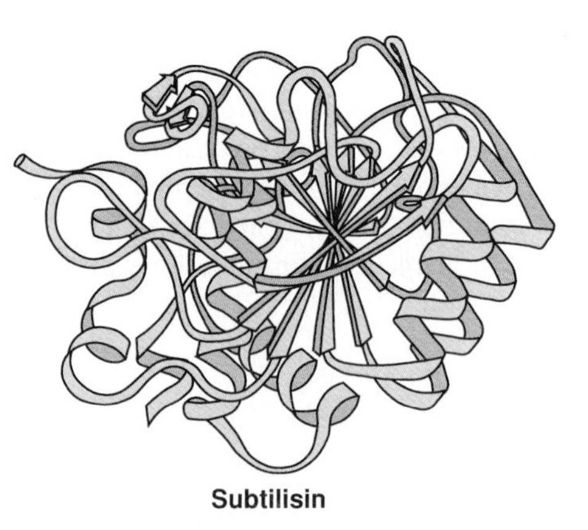

acid residues that interact in functionally important ways with the protein's heme group. The slight variations of the basic structural theme of cytochrome *c* that are observed appear to optimize the molecule's function in different organisms.

In addition to the cytochrome *c* proteins, several other families of proteins have been found to share similarities in amino acid sequence and tertiary structure. Again, the observed differences in sequence and structure among individual members reflect evolutionary pressures that modified a basic structural arrangement in order to diversify the functional properties of the molecules. Examples of such structurally and functionally related families include the oxygen-binding globins that we discussed earlier, the serine protease enzyme families (see chapter 9), and the dehydrogenases. Generally, related members within a given enzyme family catalyze chemically similar reactions but exhibit varying specificities for structurally different substrate molecules. For example, while all serine proteases catalyze the hydrolytic cleavage of peptide bonds, different members of this molecule family cleave polypeptides at different locations, depending on the nature of the amino acid side chains adjacent to the cleavage site (see chapter 9).

Although divergent evolutionary processes usually produce gradual changes in a given protein function, some changes are less gradual, as when a mutation occurs that radically alters protein function. Such mutations frequently result in the synthesis of functionally defective molecules and so constitute one cause of inheritable disease. Alternatively, amino acid substitutions in related proteins may result in the generation of new functions. The enzyme lysozyme, which binds and subsequently cleaves polysaccharide chains, and the protein α-lactalbumin, which transports sugars, are very similar in both sequence and structure. In this case, it appears that relatively slight modifications of a common ancestral precursor have resulted in selection for molecules with quite different functions.

Not all functionally related families of proteins arose by divergent evolution from a common ancestor. In some cases, proteins that have extensive functional or structural similarities appear to have arisen independently. An outstanding example of two molecules that are functionally similar, but are radically different in sequence and structure, occurs in the serine proteases (fig. 5.23). Many members of the serine protease family are closely related in sequence and structure. However, in ***Bacillus subtilis*** the serine protease subtilisin, while being essentially identical in its arrangement of amino acid residues at the active site to the other serine proteases, otherwise differs from them completely in sequence and tertiary structure. This situation presumably reflects convergent evolution on a particular active site arrangement required for the protein's catalytic function.

More frequently, proteins that differ completely in sequence and function have quite similar tertiary structures. In these cases, the observed structural similarities most probably reflect selection of a particularly stable structural arrangement. Examples of such structurally related molecules include those with similarly twisted β sheets (see fig. 4.24) and proteins organized as a bundle of four closely packed α helices (see fig. 4.30).

Gene Splicing Results in a Reshuffling of Domains in Proteins

In the preceding descriptions of protein diversification, we looked at evolutionary change that occurs as a consequence of the continuing selection of individual point mutations in the protein's encoding DNA. However, many proteins exhibit structural characteristics that suggest that they have resulted from processes of gene splicing. In particular, a surprisingly large fraction of known protein structures incorporate multiple copies of structurally similar domains. In many cases it appears that these

Figure 5.24

Pyruvate kinase domains 1, 2, and 3 as an example of a protein whose domains show no structural resemblance whatsoever.

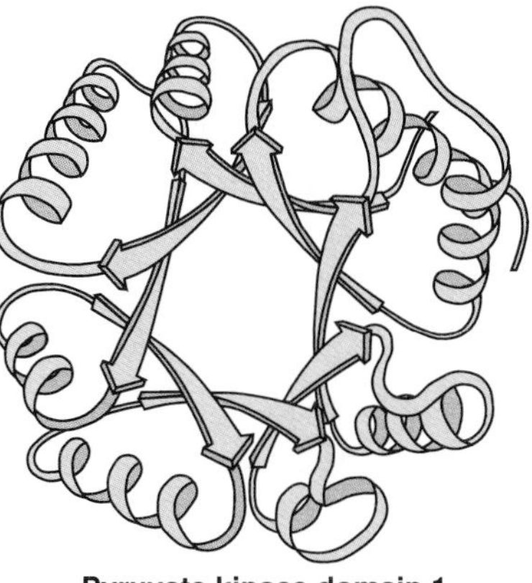

Pyruvate kinase domain 1

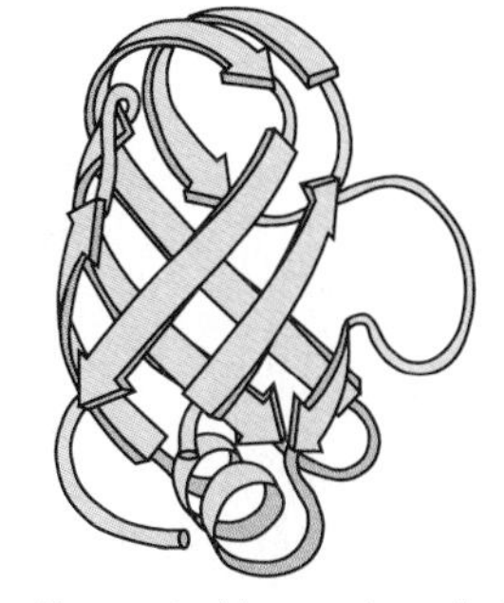

Pyruvate kinase domain 2

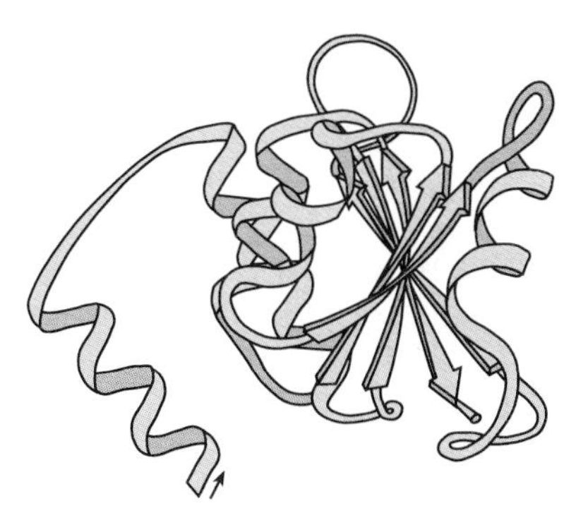

Pyruvate kinase domain 3

molecules have arisen by the splicing together of duplicate or multiple copies of a gene coding for a given structural domain, followed by the essentially independent fixation of mutations throughout the spliced genome. The eventual result is a protein composed of sequentially different but structurally similar repeating domains.

Additional evidence for the role of gene splicing in protein evolution comes from the observation that some large proteins are composed of several different structural domains, each of which may structurally resemble parts or the entirety of other known proteins. A good example is the glycolytic enzyme pyruvate kinase (fig. 5.24). This large protein is organized as three structural domains, two of which show convincing structural similarities to, respectively, the β barrel of triose phosphate isomerase (domain 1; see fig. 4.24) and a twisted β-sheet domain common to many dehydrogenases (domain 3; see fig. 4.24). The third domain of pyruvate kinase (domain 2) also has convincing similarity to a common structural type, the antiparallel β barrel.

Evolutionary Diversification Is Directly Involved in Antibody Formation

In most cases the fixation of new mutations is a relatively infrequent event, resulting in the gradual evolution of proteins such as cytochrome *c*. By contrast, one of the important biological defense mechanisms of higher organisms, the immune response, depends on the rapid generation of structurally novel molecules that can recognize and bind foreign substances that may be harmful to the organism. The molecules responsible for the initial recognition and binding of foreign substances are the immunoglobulins. These molecules are composed of two pairs of polypeptide chains of different length that are interconnected by covalent cysteine disulfide linkages (fig. 5.25). Sequence studies of various immunoglobulins have shown that both the heavy and light polypeptide chains contain repeating homologous sequences that are about 110 residues in length. Structural studies of the immunoglobulin molecule show that the sequentially homologous regions fold individually into similar structural domains, arranged as a bilayer of antiparallel sheets. The molecule in its entirety is formed of twelve similar structural domains, of which eight are formed by the two heavy chains and four by the two light chains.

Immunoglobulins that are specific for binding to different foreign substances vary greatly in the sequences found in the amino-terminal domains of both the heavy and light chains. It is these variable regions that form the binding sites between the immunoglobulin molecule and the foreign substances that trigger the immune response. The remarkable property of this system is that it can rapidly diversify the sequence of variable regions by mutation, gene splicing, and RNA splicing. The net result is that the organism can produce an enormous variety of antibodies from a quite limited amount of informational DNA originating in the germ-line tissue.

Protein Purification Procedures

To characterize the proteins that we have discussed so far, we must usually first isolate the protein under study from the complex mixture of proteins found in the organism. Once we have decided to purify a particular protein, we must weigh several factors. For example, how much material is needed? What level of purity is required? The starting material should be readily available and should contain the desired protein in relative abundance. If the protein is part of a larger structure, such as the nucleus, the mitochondria, or the ribosome, then it is advisable to isolate the large structure first from a crude cell extract.

Figure 5.25

The pattern of disulfide cross-links in immunoglobulin G. The subscripts L and H refer to light and heavy chains, respectively; C and V refer to regions of the sequence that are relatively constant or quite variable, respectively, in different IgG species. Each block of each sequence folds into an independent tertiary-structure domain, so that the final structure resembles a protein with 12 subunits, even though it is a single covalent entity.

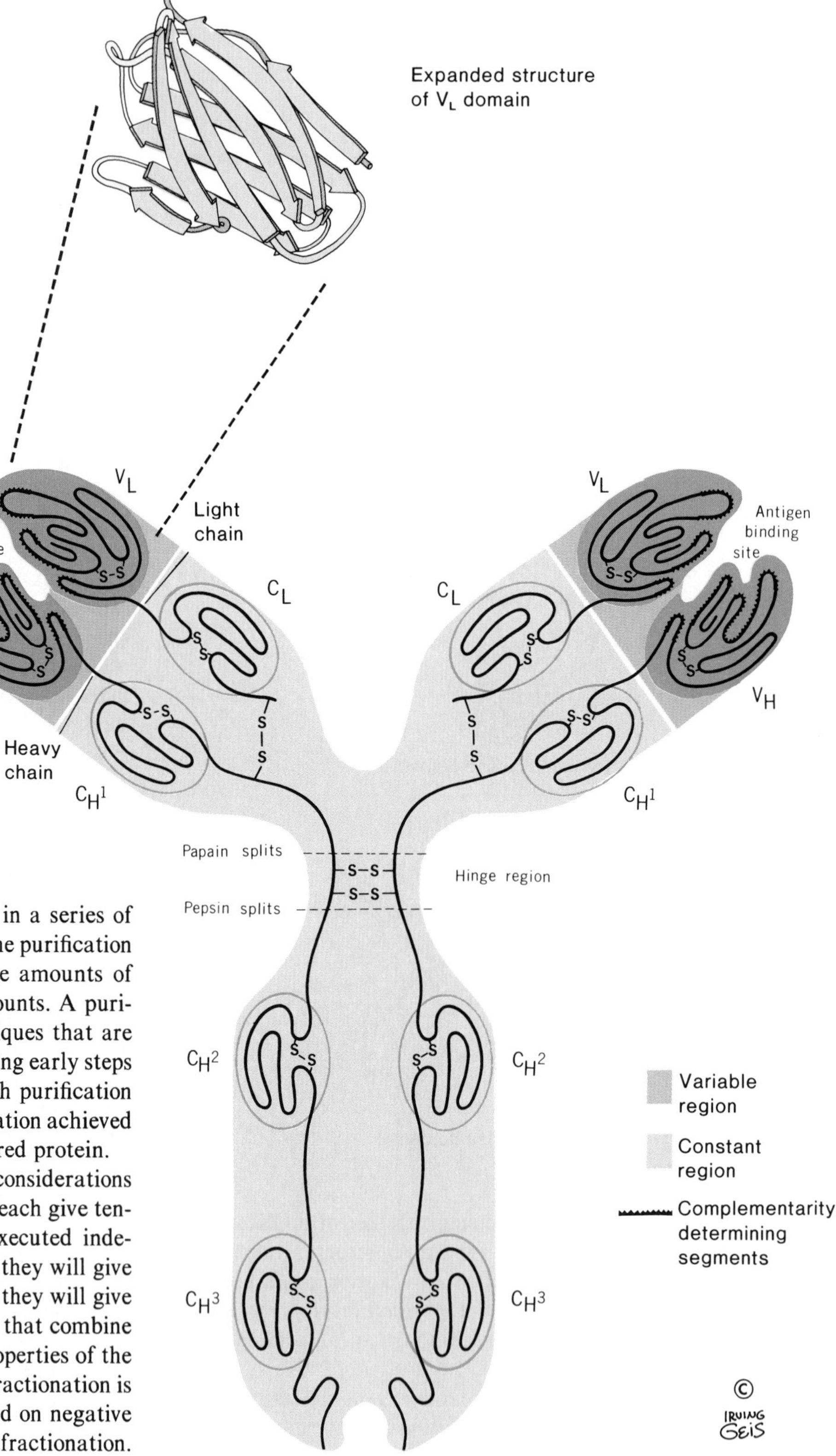

Purification must usually be performed in a series of steps, using different techniques at each step. Some purification techniques are more useful when handling large amounts of material, whereas others work best on small amounts. A purification procedure is arranged so that the techniques that are best for working with large amounts are used during early steps in the overall purification. The suitability of each purification step is evaluated in terms of the amount of purification achieved by that step and the percent recovery of the desired protein.

Combining techniques introduces new considerations and new problems. If two purification techniques each give tenfold enrichment for the desired protein when executed independently on a crude extract, this does not mean they will give 100-fold enrichment when combined. In general, they will give somewhat less. As a rule, purification techniques that combine most effectively usually are based on different properties of the protein. For example, a technique based on size fractionation is more effectively combined with a technique based on negative charge than with another technique based on size fractionation.

Table 5.3
Outline of Purification of UMP Synthase from Ehrlich Ascites Carcinoma

Fraction	Volume (ml)	Protein (mg)	OMPDase[a]			OPRTase[a]			Ratio of OMPDase to OPRTase
			Units[b]	Sp. Act.[c]	Percent Recovery	Units[b]	Sp. Act.[c]	Percent Recovery	
1. Streptomycin fraction	1040	11,700	40.4	0.0034		20.5	0.0018		2.0
2. Dialyzed $(NH_4)_2SO_4$ fraction	144	311	24.3	0.0078	60	8.7	0.0028	42	2.8
3. Affinity column eluate (concentrated)	0.475	0.51	4.0	7.8[d]	10	0.35	0.69	3.3	11.4

[a]OMPDase = OMP decarboxylase; OPRTase = orotate PRTase.
[b]Units refer to total amount of enzyme activity.
[c]Specific activity refers to the units of enzyme activity divided by the total protein.
[d]This value represents a 2,300-fold enrichment from fraction 1.

Throughout the purification we must have a convenient means of assaying for the desired protein, so we can know the extent to which it is being enriched relative to the other proteins in the starting material. In addition, a major concern in protein purification is stability. Once the protein is removed from its normal habitat, it becomes susceptible to a variety of denaturation and degradation reactions. Specific inhibitors are sometimes added to minimize attack by proteases on the desired protein. During purification it is usual to carry out all operations at 5° C or below. This temperature control minimizes protease degradation problems and decreases the chances of denaturation.

In their natural habitat, proteins are usually surrounded by other proteins and organic factors. When these are removed or diluted as during purification, the protein becomes surrounded by water on all sides. Proteins react differently to a pure aqueous environment; many are destabilized and rapidly denatured. A common remedial measure is to add 5 to 20% glycerol to the purification buffer. The organic surface of the glycerol is believed to simulate the environment of the protein in the intact cell. Two other ingredients that are most frequently added to purification buffers are mercaptoethanol and ethylenediamine tetraacetate (EDTA). The mercaptoethanol inhibits the oxidation of protein —SH groups, and the EDTA chelates divalent cations. The latter, even in trace amounts, can lead to aggregation problems or activate degradative enzymes.

The following two examples of purification show how various techniques can be effectively combined to produce purified proteins with a minimum of effort and loss of activity.

Purification of An Enzyme with Two Catalytic Activities

The last two steps in the biosynthesis of the mononucleotide uridine 5′-monophosphate (UMP) are catalyzed by (1) orotate phosphoribosyltransferase (OPRTase) and (2) orotate 5′-monophosphate (OMP) decarboxylase.

Mary Ellen Jones and her colleagues set out to purify the enzyme or enzymes involved in these two reactions. Their main goal was to determine whether the two reactions are carried out by one protein or more than one. Their findings indicated that the two reactions were both catalyzed by the same enzyme, consisting of a single polypeptide chain. To demonstrate this fact, it was necessary to monitor both enzyme activities at each step in the purification and show that both activities copurified. For this purpose, Jones used specific enzyme assays for both enzyme activities. All fractions were assayed for both enzymatic activities at each stage of the purification.

The main data associated with the purification are summarized in table 5.3. This table indicates the total protein obtained in each step, the number of enzyme units* for each enzyme, and the ratio of enzyme units to total protein, called the specific activity. In the absence of enzyme inactivation, the specific activity should be directly proportional to the enrichment. The percent recovery refers to the amount of enzyme activity in the indicated fraction, as compared with the amount present in fraction 1. This number is usually less than 100%. The apparent losses may reflect actual losses of enzyme during purification, or they may reflect inactivation (usually due to unknown causes) of the enzyme during purification.

The nine steps involved in the purification of UMP synthase from starting tissue are summarized in figure 5.26. All steps were carried out at 0–5° C. About 200 g of Ehrlich ascites cells, a mammalian tumor rich in the desired enzymes, was suspended in buffer and processed in a tissue homogenizer, which mechanically breaks down the tissue and the cell membranes (step 1). Then EDTA and an —SH reagent were added to this total cell lysate. Solid streptomycin sulfate was also added with stirring (step 2). Streptomycin sulfate aggregates nucleic acids so that they may be more easily removed by centrifugation. The

*Enzyme units are proportional to the amount of enzyme activity. The relationship between enzyme units and absolute amount of enzyme need not concern us here.

Figure 5.26

Outline of purification scheme for UMP synthase from Ehrlich ascites tumor cells of mice.

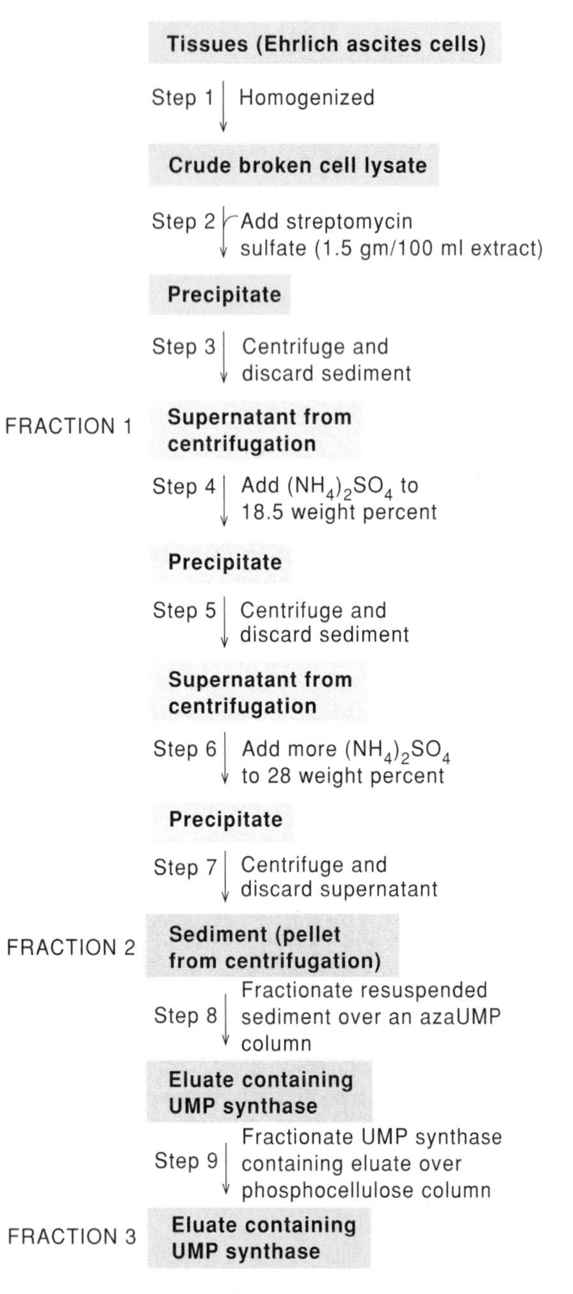

resulting slurry was subjected to high-speed centrifugation, and the resulting supernatant (see table 5.3, fraction 1) was carefully decanted for further processing (step 3).

Preliminary experiments had shown that the desired enzymes were in the 18.5–28% $(NH_4)_2SO_4$ fraction. (The reasoning behind use of $(NH_4)_2SO_4$ in purification is described in box 5B.) This knowledge served as the basis for the next three steps. First 239 g of solid $(NH_4)_2SO_4$ was added to 1040 ml of supernatant (step 4). The resulting precipitate was removed by centrifugation (step 5). Then an additional 120 g of $(NH_4)_2SO_4$ was added to the supernatant (step 6). The resulting slurry was centrifuged. This time the supernatant was discarded after centrifugation, leaving the sediment containing the enzyme activity for further processing (step 7). The sediment was resuspended in a dilute buffer for column chromatography (see table 5.3, fraction 2). Inspection of table 5.3 indicates only about a twofold increase in specific activity between fractions 1 and 2.

The main purification was achieved by two column steps, carried out in series. (Applications of column chromatography to protein purification are described in box 5C.) The first column was an affinity column containing an analog of UMP, 6-azauridine 5′-monophosphate (azaUMP), covalently attached to an agarose column support system. In dilute buffer, greater than 99% of the protein in fraction 2 is retained on this column. After thorough rinsing of the column with dilute buffer, 5×10^{-5} M azaUMP was added to the buffer. This addition resulted in the elution of the UMP synthase (step 8). Then the column eluant carrying the two enzyme activities associated with UMP synthase was resuspended in pure buffer and passed over a phosphocellulose column (step 9). Phosphocellulose was chosen because the negatively charged phosphate groups result in the retention of proteins by electrostatic attraction alone, but they also resemble the phosphate groups in the naturally occurring enzyme substrate, OMP. Thus the phosphocellulose column may be thought of as an ion-exchange column and an affinity column combined. In ordinary ion-exchange chromatography, the protein, after column loading, is eluted by increasing the ionic strength with a simple inorganic salt. In this example, Jones used a more specific method to elute the enzyme, which involved adding 10^{-5} M azaUMP and 2×10^{-5} M OMP to the original loading buffer. The addition does not substantially increase the ionic strength of the buffer. Therefore the only phosphocellulose-bound proteins that are likely to be eluted by this treatment are those with an especially high affinity for either of these nucleotides (azaUMP or OMP). This fact should greatly favor selective elution of those enzymes that carry specific sites for binding these nucleotides. The two column steps together resulted in an enzyme preparation (see table 5.3, fraction 3) that was approximately 2,300-fold purified over the starting material, as measured by the increase in specific activity.

The final product (see table 5.3, fraction 3) was examined by SDS gel electrophoresis and found to contain a single band with an estimated molecular weight of 51,000. (Gel electrophoresis is discussed in box 5D.) The denaturing conditions of an SDS gel would be expected to dissociate a multisubunit protein. Hence the SDS gel result indicated that the enzyme contains one type of polypeptide chain, but it does not tell us whether the enzyme contains one or more of these chains. Sedimentation analysis on a sucrose density gradient in a nondenaturing buffer indicated a single band with an estimated molecular weight of about 50,000. These two results taken together demonstrate that the enzyme in its native state contains a single polypeptide chain. The purity of the enzyme was also confirmed by isoelectric focusing and two-dimensional electrophoresis, with isoelectric focusing in the first direction followed by SDS gel electrophoresis in the second direction.

Methods of Protein Purification: Differential Precipitation with $(NH_4)_2SO_4$

Because proteins differ in their solubility, it is possible to separate them by differential precipitation. In this box we examine some of the factors that bear on the choice of solvents used in this procedure.

The solubility of a protein reflects a delicate balance between different energetic interactions, both internally within the protein and between the protein and the surrounding solvent. Consequently, the choice of solvent can affect both the solubility and the structure of a protein.

Proteins typically have charged amino acid side chains on their surfaces that undergo energetically favorable polar interactions with the surrounding water. The total charge on the protein is the sum of the side-chain charges. However, the actual charge on the weakly acidic and basic side-chain groups also depends on the solution pH. In fact, the acidic and basic groups within the protein can be titrated just like free amino acids (see fig. 3.6) to determine their number and their pK values.

Proteins tend to show a minimum solubility at their isoelectric pH. This fact is apparent for β-lactoglobulin in figure 1. The decrease in solubility at the isoelectric pH reflects the fact that the individual protein molecules, which would all have similar charges at pH values away from their isoelectric points, cease to repel each other. Instead, they coalesce into insoluble aggregates.

Proteins also show a variation in solubility that depends on the concentration of salts in the solution. These frequently complex effects may involve specific interactions between charged side chains and solution ions or, particularly at high salt concentrations, may reflect more comprehensive changes in the solvent properties. For the effect of salt concentration on the solubility of β-lactoglobulin, again see figure 1. Most globulins are sparingly soluble in pure water. The increase in solubility that occurs upon adding salts such as sodium chloride is often referred to as salting in.

The effects of four different salts on the solubility of hemoglobin at pH 7 can be seen in figure 2. All four salts produce the salting-in effect with this protein; two of them, sodium sulfate and ammonium sulfate, also produce a greatly decreased solubility of the protein at high salt concentrations. This result is called salting out and occurs with salts that effectively compete with the protein for available water molecules. At high salt concentrations, the protein molecules tend to associate with each other because protein–protein interactions become energetically more favorable than protein–solvent interactions.

Each protein has a characteristic salting-out point, and we can exploit this fact to make protein separations in crude extracts. For this purpose $(NH_4)_2SO_4$ is the most commonly used salt because it is very soluble and is generally effective at lower concentrations than many other salts.

Figure 1

Solubility of ß-lactoglobulin as a function of pH and ionic strength. The isoelectric pH (pI) for this protein is about 5.2. This corresponds to the point of minimum solubility.

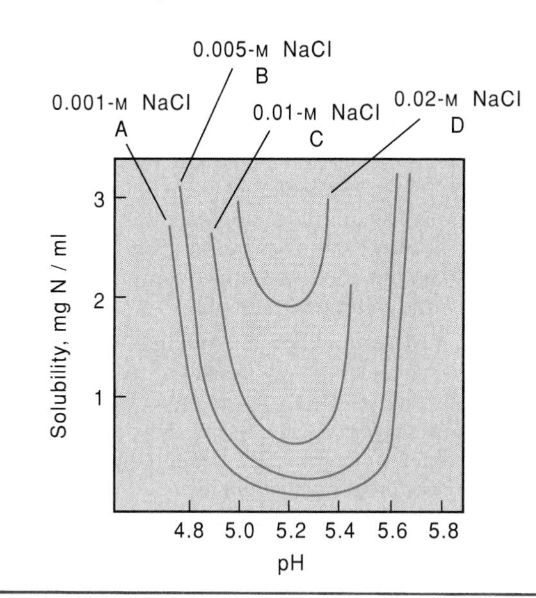

Figure 2

Solubility of horse carbon monoxide hemoglobin in different salt solutions. The addition of a moderate amount of salt (salting in) is required to solubilize this protein. At high concentrations, certain salts compete more favorably for solvent, decreasing the solubility of the protein and thus leading to its precipitation (salting out). (Source: E. J. Cohn and J. T. Edsall, *Proteins, Amino Acids, and Peptides as Ions and Dipolar Ions.* Copyright © 1942, Reinhold, New York, N.Y.)

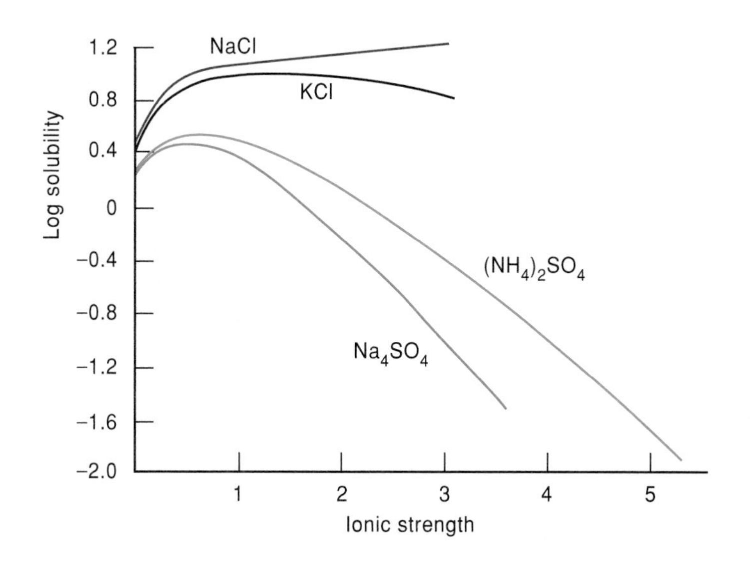

Methods of Protein Purification: Column Chromatography

Column procedures are the most effective and most varied of purification methods. Common to all column procedures is the use of a glass cylinder with an opening at the top and bottom. The cylinder is filled with a column of hydrated material, and the protein sample is applied to the top of the column. Then further buffer is passed through the column. In most column procedures, proteins bind differentially to the column material, and a change in the elution buffer causes them to be eluted differentially according to their degree of affinity for the column material. The eluant exiting from the bottom of the column is collected in equal-size fractions with the help of an automatic fraction collector (figure 1). Each fraction is analyzed, and fractions containing an appreciable amount of the desired protein are pooled for further purification.

Many variations on this basic procedure are in common use, as may be seen from the following examples.

1. In gel-exclusion chromatography a cross-linked dextran without any special attached functional groups is used for the column substrate (figure 2). Large molecules flow more rapidly through this type of column than small ones. The dextrans have different degrees of cross-linking, making them effective over different size ranges.
2. Ion-exchange chromatography makes use of the fact that proteins differ enormously in their affinity for positively or negatively charged columns. However, the cross-linked resins used in amino acid fractionation (see chapter 3) are rarely used for protein separations because proteins are too large to penetrate the resin beads. Instead, finely divided celluloses containing either positively or negatively charged groups are most commonly used (table 5C). The affinity of a protein for the column material is proportional to the salt concentration required to release the protein from the material. Typically, a column is loaded with protein solution at a low ionic strength so that most of the proteins bind to the column. After loading, elution is initiated by gradually increasing the salt concentration of the elution buffer. Proteins are eluted in the order of increasing affinity.
3. Finely divided celluloses may also be used in the column in conjunction with attached hydrophobic groups such as octyl alcohol. In this case, it is proteins with exposed hydrophobic centers that bind to the column with varying affinities. These proteins may be eluted in order of decreasing affinity for the column by increasing the level of free octyl alcohol in the eluting buffer.
4. Affinity chromatography makes use of chemical groups that have a special binding affinity for the proteins that are being sought. For example, many enzymes bind preferentially to cofactors such as adenosine triphosphate or pyridine nucleotides. Often, the separation of such enzymes from other proteins can be achieved on a column that has one of these cofactors attached. As the mixture passes through the column, proteins that have specific binding affinities for the cofactor bind to the column while other proteins pass through. The cofactor-binding proteins can then be eluted with a solution containing the same soluble cofactor.
5. High-performance liquid chromatography (HPLC) is not so much a new type of chromatography as a new way of applying old chromatographic techniques, using extremely high pressures. The same principles are involved, but the column materials usually consist of more finely divided particles made of physically stronger materials, which can withstand pressures of 5,000–10,000 psi without changing their structure. The column apparatus itself must also be designed to withstand high pressures.

Figure 1

Collecting fractions during column chromatography. Column material and elution procedure are chosen to effect optimal separation of the desired protein.

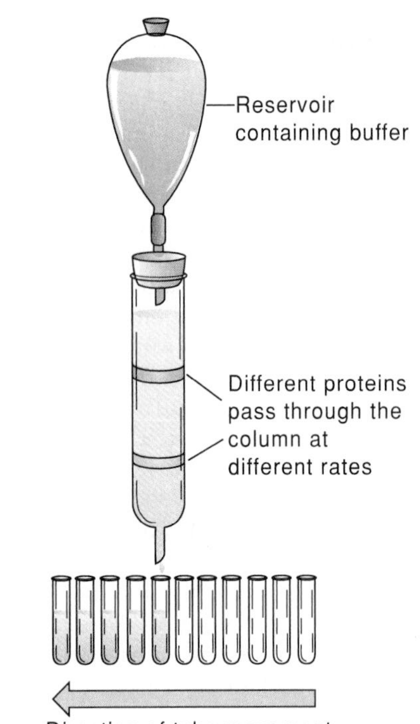

Figure 2

Polydextran column showing separation of small and large molecules. The column material is immersed in solvent, which penetrates the gel particles. A separation is initiated by layering a small sample containing different-size proteins on the top of the column. This sample is pushed through the column by opening the stopcock at the bottom and adding further solvent at the top to keep up with the flow. As shown, the small protein molecules can penetrate the gel particles but the big ones cannot. Therefore the big proteins move through the column much more rapidly, and a separation of the two proteins results.

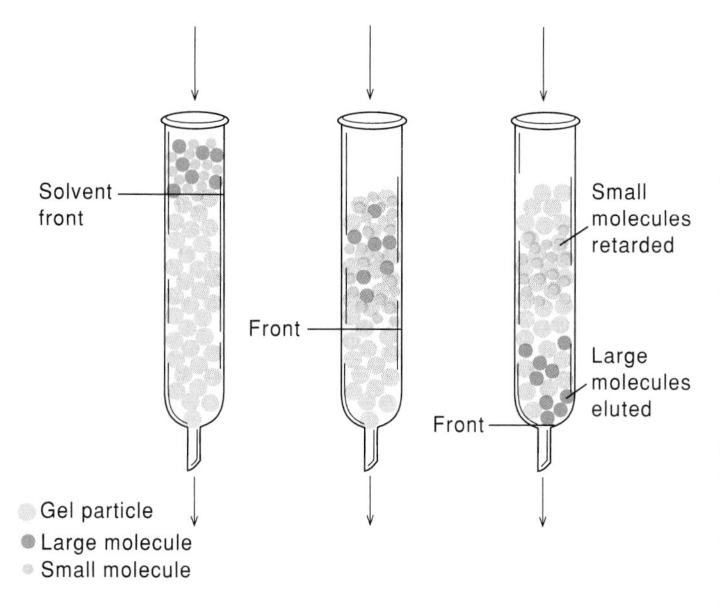

Table 5C

Some Column Materials for Ion-Exchange Chromatography of Proteins

Matrix	Functional Groups on Column
Phosphocellulose (PC)	$—PO_3^-$
Carboxymethyl cellulose (CMC)	$—CH_3—COO^-$
Diethylaminoethyl cellulose (DEAE)	$—(CH_2)_2—N(CH_2—CH_3)_2$

The conclusion drawn from these results was that both of the enzyme activities associated with UMP synthase—OMPDase and OPRTase—are contained in a single protein. The basis for the conclusion was that both enzyme activities are always present in the same fractions throughout the multistep purification. However, inspection of table 5.3 indicates a possible objection to this interpretation. In the columns showing percent recovery, it can be seen that substantial amounts of enzyme activity are lost for both enzymes during purification, but that considerably more activity is lost for the OPRTase. These losses could be due to actual loss of enzymes during purification or to some sort of inactivation of the enzyme sites. The preferential loss of OPRTase activity is emphasized by the last column in table 5.3, which gives the ratio of the two enzyme activities in the different fractions. Considered alone, these data could indicate that a separate catalytic unit of orotate PRTase is lost during purification. Jones thinks this is unlikely for two reasons: (1) the activity appears in no fractions other than with OMP decarboxylase during purification, and (2) the orotate PRTase activity is notably unstable. It appears, then, that both enzyme activities exist at distinct sites on a single protein. The greater loss in activity of one enzyme activity over the other can be attributed to a greater sensitivity of one reaction site over the other on the enzyme surface.

Purification of a Membrane-Bound Protein

The second purification procedure we will look at illustrates an unusual approach to the purification of a membrane-bound protein. The lactose carrier protein of *E. coli* is normally tightly bound to the plasma membrane. This protein is involved in the active transport of the dissaccharide lactose across the cytoplasmic membrane. When lactose carrier protein is present, the intracellular concentration of lactose can achieve levels 1,000-fold higher than those found in the external medium. Ron Kaback devised a simple yet elegant procedure for the purification of this protein.

Purification of the membrane-bound lactose carrier protein is a very different problem from the purification of the soluble OMP synthase. Both the approach to purification and the assays for the protein during purification are quite novel. The assay involves reconstituting a transport system with membranes that are free of lactose carrier protein, then adding the partially purified carrier protein and radioactively labeled lactose. The activity in this assay system is proportional to the transport of radioactive lactose across the membrane in the cell-free reconstituted system.

The results of the purification steps are tabulated in table 5.4, and the purification procedure is outlined in figure 5.27. In this procedure advantage was taken of the fact that the carrier protein in its native state is firmly bound to the cytoplasmic membrane. Thus the first step consisted of isolating these membranes from the rest of the cell constituents. Starting from the membrane fraction only 35-fold purification was required to achieve pure carrier protein. This rapid result was possible

5D
BOX

Methods of Protein Purification: Gel Electrophoresis

Gel electrophoresis is the best way to analyze mixtures and assess purity. Gel electrophoresis separates proteins according to their size and their charge. It is almost always performed in aqueous solution supported by a gel system. The gel is a loosely cross-linked network that functions to stabilize the protein boundaries between the protein and the solvent, both during and after electrophoresis, so that they may be stained or otherwise manipulated.

The most popular method of electrophoretic separation is called SDS gel electrophoresis. This method not only gives an index of protein purity but yields an estimate of the protein subunit molecular weights. The mixture of proteins to be characterized is first completely denatured by adding sodium dodecyl sulfate (a detergent) and mercaptoethanol and by briefly heating the mixture. Denaturation is caused by the association of the apolar tails of the SDS molecules with protein hydrophobic groups. Any cystine disulfide bonds are cleaved by a disulfide interchange reaction with mercaptoethanol.

The resulting unfolded polypeptide chains have relatively large numbers of SDS molecules bound to them. The success of the technique for molecular-weight estimation depends on two facts: (1) Each bound SDS molecule contributes one negative charge to the denatured protein complex, so that the charge of the protein in its native state is effectively masked by the more numerous charged groups of the associated detergent molecules. (2) The total number of detergent molecules bound is proportional to the polypeptide-chain length or, equivalently, the protein's molecular weight. As a result, the SDS-denatured protein molecules acquire net negative charges that are approximately proportional to their molecular weights (figure 1).

Another electrophoretic method frequently used for characterizing proteins is based on differences in their isoelectric points; this method is called isoelectric focusing. The apparatus usually consists of a narrow tube containing a gel and a mixture of ampholytes, which are small molecules with positive and negative charges. The ampholytes have a wide range of isoelectric points, and are allowed to distribute in the column under the influence of an electric field. This step creates a pH gradient from one end of the gel to the other, as each particular ampholyte comes to rest at a position coincident with its isoelectric point. At this stage, a solution of proteins is introduced into the gel. The proteins migrate in the electric field until each reaches a point at which the pH resulting from the ampholyte gradient exactly equals its own isoelectric point. Isoelectric focusing provides a way of both accurately determining a protein's isoelectric point and effecting separations among proteins whose isoelectric points may differ by as little as a few hundredths of a pH unit.

Figure 1

Gel electrophoresis for analyzing and sizing proteins. (*a*) Apparatus for slab-gel electrophoresis. Samples are layered in the little slots cut in the top of the gel slab. Buffer is carefully layered over the samples and a voltage is applied to the gel for a period of usually 1 to 4 h. (*b*) After this time the proteins have moved into the gel at a distance proportional to their electrophoretic mobility. The pattern shown indicates that different samples were layered in each slot. (*c*) Results obtained when a mixture of proteins was layered at the top of the gel in phosphate buffer, pH 7.2, containing 0.2% SDS. After electrophoresis the gel was removed from the apparatus and stained with Coomassie blue. The protein and its molecular weight are indicated next to each of the stained bands. (*d*) The logarithm of the molecular weight against the mobility (distance traveled) shows an approximately linear relationship. (Data of K. Weber and M. Osborn.)

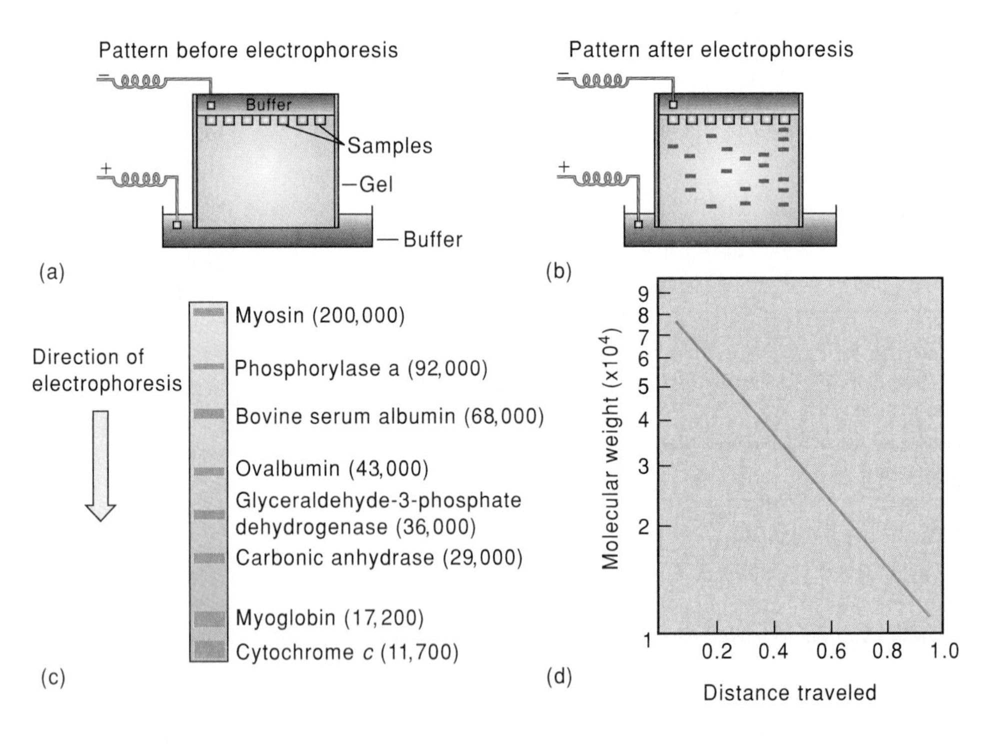

Table 5.4
Purification of the Lactose Carrier Protein

Fraction	Protein (mg)	Percent Recovery (Total Protein)	Percent Recovery (Carrier Protein)	Purification Factor
1. Membrane fraction	12.5	100	100	1.0
2. Urea-extracted membrane	5.6	45	76	1.7
3. Urea/cholate-extracted membrane	2.6	21	61	2.9
4. Octylglucoside extract	0.4	3.2	38	12
5. DEAE column peak	0.056	0.4	14	35

Figure 5.27

Outline of purification procedure for lactose carrier protein from *E. coli*.

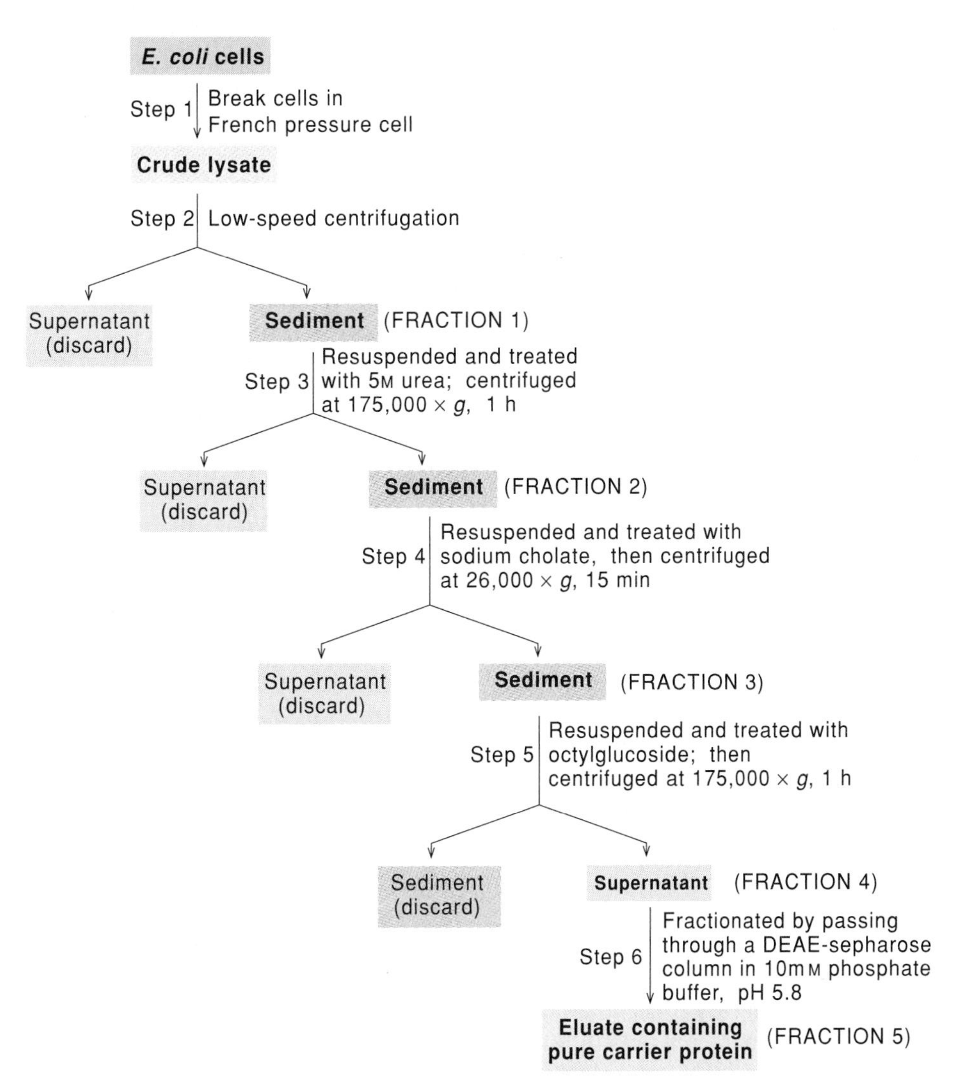

because a special strain of *E. coli,* containing about 100 times the normal carrier protein, was used as starting material for the purification. The high initial content was engineered by putting the carrier protein gene on a multicopy plasmid, which was then inserted into the cell—a procedure described in chapter 27.

Bacterial cells are much tougher than mammalian cells, requiring a more stringent procedure for cell disruption. In this case, the cells were placed in a so-called French pressure cell and bled through an orifice from very high pressures (~ 10,000 psi) to atmospheric pressure (step 1). Under these conditions the cells literally explode, fragmenting their membranes and releasing the cytoplasmic contents. The membranes were pelleted by a brief centrifugation, leaving a supernatant containing DNA, ribosomes, and cytoplasmic protein, which was removed by decantation (step 2). (See box 5E for differential centrifugation techniques.) The pelleted membrane fraction was next resuspended and extracted with 5 M urea and then reextracted with 6% sodium cholate. The pellet obtained by high-speed centrifugation from these two extractions still contained most (61%) of the carrier protein in the more rapidly sedimenting membrane fraction (fraction 3), although 79% of the total membrane-bound protein was released by these treatments (steps 3 and 4).

At this point, the carrier protein was released from a suspension of the membranes by addition of the hydrophobic reagent octylglucoside in the presence of *E. coli* phospholipid (step 5). It is believed that the octyl part of octylglucoside competes effectively with the membrane for binding to hydrophobic centers on the carrier protein. The *E. coli* phospholipid facilitates dissociation of the carrier protein from the membrane fraction. The solubilized octylglucoside-containing extract was fourfold enriched in carrier protein after a high-speed centrifugation to remove residual membrane and membrane-bound proteins.

Finally, the octylglucoside-containing extract was passed over a positively charged diethylaminoethyl (DEAE) sepharose column (sepharose is a form of cross-linked polydextran) in a buffer containing 10 mM potassium phosphate, 20 mM lactose, and 0.25 mg of washed *E. coli* lipid per milliliter. The carrier protein passed through this column as a symmetrical peak of protein (step 6). Most of the remaining protein in the extract adsorbed to the positively charged column. The fractions containing the bulk of the carrier protein activity were judged to be pure by SDS gel electrophoretic analysis. The purified protein contained a single polypeptide chain with an estimated molecular weight of 33,000.

Methods of Protein Purification: Differential Centrifugation

A typical crude broken cell preparation contains broken cell membranes, cellular organelles, and a large number of soluble proteins, all dispersed in an aqueous buffered solution. The membranes and the organelles can usually be separated from one another and from the soluble proteins by differential centrifugation. Differential centrifugation divides a sample into two fractions: the pelleted fraction, or sediment, and the supernatant fraction, that is, the fraction that does not sediment. The two fractions may then be separated by decantation.

Differential centrifugation involves the use of different speeds and different times of centrifugation (table 5E). For example, if the protein of interest were in the mitochondrial fraction, the crude lysate should be centrifuged first at 4000 × *g* for 10 min to remove cell membranes, nuclei, and (in the case of plant material) chloroplasts. The supernatant from this step contains, among other elements, the mitochondria, and would be decanted and recentrifuged at 15,000 × *g* for 20 min to obtain a sediment primarily containing mitochondria. If ribosomes instead of mitochondria were the goal, then the crude lysate would be centrifuged at 30,000 × *g* for 30 min and the resulting supernatant would be decanted and centrifuged at 100,000 × *g* for 180 min. If the soluble protein fraction were the goal, then the entire lysate would be centrifuged at 100,000 × *g* for 180 min and the resulting supernatant, containing the soluble protein, would be decanted for further processing.

Table 5E
Sedimentation Conditions for Different Cellular Fractions

Fraction Sedimented	Centrifugal Force (× *g*)	Time (min)
Cells (eukaryotic)	1,000	5
Chloroplasts; cell membranes; nuclei	4,000	10
Mitochondria; bacteria cells	15,000	20
Lysosomes; bacterial membranes	30,000	30
Ribosomes	100,000	180

The two purifications described here are as different as the two proteins involved. No two purifications are exactly alike, but the principles of purification, as stated at the outset, are quite similar. Fortunately, an almost endless variety of purification techniques exists. This variety is both helpful and challenging, as a great deal of knowledge and creativity are required to exploit it. In addition to professional expertise, the two most important things required to make a purification possible are an unambiguous assay for the protein in question and a means of stabilizing the protein during purification.

Summary

In this chapter we have considered the relationship between structural and functional properties of proteins and the means of purification so that individual proteins may be studied in isolation. The following points are the most important.

1. The polypeptide chains of proteins are synthesized on ribosomes. Ribosomes that remain free in the cytosol are associated with the synthesis of proteins that are targeted to locations in the cytosol, the nucleus, the peroxisomes, the mitochondria and the chloroplasts. Other ribosomes, which are attached to the endoplasmic reticulum, are engaged in the synthesis of proteins that are targeted to the endoplasmic reticulum, the Golgi, the lysosomes, the plasma membrane, and the secretory granules.
2. Proteins that cooperate in a particular biochemical process are usually found in the same location within the cell or extracellular milieu.
3. Each protein is carefully designed to carry out a specific function, as revealed by our discussion of hemoglobin and muscle.
4. Hemoglobin is a tetramer made of two almost identical subunits. The function of hemoglobin is threefold: to transport O_2 from the lungs to the tissues where it is consumed, to transport CO_2 from the tissues where it is produced to the lungs, where it is expelled, and to maintain the blood pH over a narrow range. The structure of hemoglobin is designed so that it will pick up the maximum amount of oxygen at high oxygen tensions in the lung tissue and will deliver the maximum amount of oxygen in the oxygen-consuming tissues.
5. Muscle is an aggregate of several different proteins involved in contraction. The main protein components of muscle are organized as overlapping filaments of two types: thin filaments, composed mainly of actin molecules, and thick filaments, composed of myosin molecules.
6. The process of muscular contraction entails a sliding of the two types of filaments past one another. In a fully contracted component of muscle tissue the actin and myosin filaments show a maximum overlap with one another. The contraction process involves the breakage and reformation of bridges between the actin and the myosin molecules in a reaction that requires the expenditure of ATP.
7. Protein diversification is the product of a long evolutionary process. It can be studied by comparing the structures of different proteins in the light of their functions. It can also be studied by comparing structurally related proteins with similar functions in different organisms, or by comparing structurally related domains between proteins that have different functions.
8. Antibodies provide the unique opportunity for studying diversification within a single organism, for antibodies with unique specificities are the result of an ongoing evolutionary process that takes place within the organism.
9. Whereas proteins must be studied *in vivo* in their normal milieu, to characterize them in great detail they must also be isolated in pure form. Protein purification is a complex art with a great variety of purification methods usually being applied in sequence for the purification of any protein.
10. Frequently the first step in protein purification is differential sedimentation of broken cell parts. In this way soluble proteins may be separated from organelle-sequestered proteins.
11. Column procedures are useful in fractionating proteins with different affinity properties and sizes. By the use of different column materials in conjunction with specific eluting solutions, highly purified protein preparations can be obtained.
12. Gel electrophoresis, which separates proteins according to their size and charge, can be used in purification or as a means of assaying the purity during a purification procedure.
13. The molecular weights of soluble proteins can be estimated by gel electrophoresis or can be rigorously determined by sedimentation and diffusion techniques.

Selected Readings

Akers, G. K., M. C. Doyle, D. Myers and M. A. Daugherty, Molecular code for cooperativity in hemoglobin. *Science* 255:54–63, 1992. A new model for hemoglobin allostery.

Allen, R. D., The microtubule as an intracellular engine. *Sci. Am.* 238(2):42–49, 1987.

Baldwin, J., Structure and cooperativity of haemoglobin. *Trends Biochem. Sci.* 5:224–228, 1980.

Berg, H. C., How bacteria swim. *Sci. Am.* 233(2):36–44, 1975.

Cantor, C. R., and P. Schimmel, *Biophysical Chemistry.* New York: Freeman, 1980. Especially see volume 2, entitled *Techniques for Study of Biophysical Structure and Function.*

Caplan, A. I., Cartilage. *Sci. Am.* 251(4):84–94, 1984.

Dickerson, R. E., and I. Geis, Hemoglobin. Menlo Park, Calif.: Benjamin/Cummings, 1983. A magnificent presentation of every facet of hemoglobin biochemistry and genetics.

Doolittle, R., Proteins. *Sci. Am.* 253(4):88–96, 1985. Overview emphasizing evolutionary considerations.

Eisenberg, D., and D. Crothers, *Physical Chemistry and Its Applications to the Life Sciences.* Menlo Park, Calif.: Benjamin/Cummings, 1979.

Eyre, D. R., M. A. Pdaz, and P. M. Gallop, Cross-linking in collagen and elastin. *Ann. Rev. Biochem.* 53:717–748, 1984.

Gething, M. J., and J. Sambrook, Protein folding in the cell. *Nature* 355:33–45, 1992.

Huxley, H. E., Sliding filaments and molecular motile systems. *J. Biological Chemistry* 265:8347–8352, 1990.

Hynes, R. O., Fibronectins. *Sci. Am.* 254(6):42–51, 1986.

Ingram, V. M., Gene mutation in human haemoglobin: the chemical difference between normal and sickle-cell haemoglobin. *Nature* 180:326–328, 1957. A classic paper.

Karplus, M., and J. A. McCammon, The dynamics of proteins. *Sci. Am.* 254(4):42–51, 1986. A reminder that proteins are dynamic structures.

Lawn, R. M., and G. A. Vehar, The molecular genetics of hemoglobin. *Sci. Am.* 254(3):48–65, 1986.

Martin, G. R., R. Timpl, P. K. Muller, and K. Kuhn, The genetically distinct collagens. *Trends Biochem. Sci.* 10:285–287, 1985. A brief, authoritative account of the most abundant protein found in vertebrates.

Methods in Enzymology. New York: Academic Press. A continuing series of over 250 volumes that discuss most methods at the professional level but are still understandable for students.

Pauling, L., H. A. Itano, S. J. Singer, and I. C. Wells, Sickle-cell anemia: a molecular disease. *Science* 110:543–548, 1949. A classic paper.

Pollard, T. D., and J. A. Cooper, Actin and actin-binding proteins. *Ann. Rev. Biochem.* 55:987–1036, 1986. Discusses structure and function.

Salemme, R., Structure and function of cytochromes *c. Ann. Rev. Biochem.* 46:299–329, 1977.

Scopes, R., *Protein Purification: Principles and Practice,* 2d ed. New York: Springer-Verlag, 1987. A recent general treatment of this subject.

Steinert, P. M., and D. R. Roop, Molecular and cellular biology of intermediate filaments. *Ann. Rev. Biochem.* 57:593–626, 1988.

Tonegawa, S., The molecules of the immune system. *Sci. Am.* 253(4):122–130, 1985.

Wilson, A. C., The molecular basis of evolution. *Sci. Am.* 253(4): 164–170, 1985.

Problems

Step	Volume (ml)	Total Protein (mg)	Total Units	Specific Activity (U/mg)	Yield (%)	Purification (*n*-fold)
Cell extract	2,800	70,000	2,700			
($(NH_4)_2SO_4$) fractionation	3,000	25,400	2,300			
Heat treatment	3,000	16,500	1,980			
DEAE chromatography	80	390	1,680			
CM-cellulose chromatography	50	47	1,350			
Bio-Gel A chromatography	7	35	1,120			

1. A method for the purification of 6-phosphogluconate dehydrogenase from *E. coli* is summarized in the table. For each step, calculate the specific activity, percentage yield, and degree of purification (*n*-fold). Indicate which step results in the greatest purification. Assume that the protein is pure after gel exclusion (Bio-Gel A) chromatography. What percentage of the initial crude cell extract protein was 6-phosphogluconate dehydrogenase?
2. Although used effectively in the 6-phosphogluconate dehydrogenase isolation procedure, heat treatment cannot be used in the isolation of all enzymes. Explain.
3. Assume that the isoelectric point (pI) of the 6-phosphogluconate dehydrogenase is 6. Explain why the buffer used in the DEAE cellulose chromatography must have a pH greater than 6 but less than 9 in order for the enzyme to bind to the DEAE.
4. Will the 6-phosphogluconate dehydrogenase bind to the CM-cellulose in the same buffer pH range used with the DEAE-cellulose? Explain. In what pH range might you expect the dehydrogenase to bind to CM-cellulose? Explain.
5. Examine the isolation procedure shown in problem 1 and explain why gel exclusion chromatography is used as the final step rather than as the step following the heat treatment.
6. A student isolated an enzyme from an anaerobe and subjected a sample of the protein to SDS-polyacrylamide gel electrophoresis. A single band was observed upon staining the gel for protein. His advisor was excited about the result, but suggested that the protein be subjected to electrophoresis under nondenaturing (native) conditions. Electrophoresis under nondenaturing conditions revealed two bands after the gel was stained for protein. Assuming the sample had not been mishandled, offer an explanation for the observations.
7. A salt-precipitated fraction of ribonuclease contained two contaminating protein bands in addition to the ribonuclease. Further studies showed that one contaminant had a molecular weight of about 13,000 (similar to ribonuclease) but an isoelectric point 4 pH units more acidic than the pI of ribonuclease. The second contaminant had an isoelectric point similar to ribonuclease but had a molecular weight of 75,000. Suggest an efficient protocol for the separation of the ribonuclease from the contaminating proteins.

8. You have a mixture of proteins with the following properties:
Protein 1: M_r 12,000, pI = 10
Protein 2: M_r 62,000, pI = 4
Protein 3: M_r 28,000, pI = 8
Protein 4: M_r 9,000, pI = 5

Predict the order of emergence of these proteins when a mixture of the four is chromatographed in the following systems.
   (a) DEAE-cellulose at pH 7, with a linear salt gradient elution.
   (b) CM-cellulose at pH 7, with a linear salt gradient elution.
   (c) A gel exclusion column with a fractionation range of 1,000–30,000 M_r, at pH 7.
9. The absorbance at 280 nm of an 0.5 mg ml^{-1} solution of protein A was 0.75. In addition, a shoulder on the absorbance spectrum at 288 nm was observed. An absorbance of 0.2 at 280 nm was measured for protein B in the same buffer and at the same concentration as protein A, but no spectral feature at 288 nm was observed. Amino acid analysis revealed that each protein contained approximately the same molar quantities of Tyr and Phe. Suggest a reason for the differences in absorbance at 280 nm and 288 nm.
10. The absorbance of a protein solution measured at 278 nm was 0.846. The protein content of that solution, calculated from quantitative amino acid analysis, was 460 μg/ml. Calculate the extinction coefficient of the protein in units of ml mg^{-1} cm^{-1}. Assume the molecular weight of the protein to be 42,000. Calculate the molar extinction coefficient.
11. You have isolated a manganese-containing pyrophosphatase and wish to determine its molecular weight. A Sephadex G-100 gel exclusion column was calibrated using proteins whose molecular weight had been established, with the following results.

Protein	Molecular Weight	Elution Volume (ml)
Serum albumin	69,000	84
Ovalbumin	43,000	92
Carbonic anhydrase	29,000	100
Chymotrypsinogen	23,000	104
Myoglobin	17,000	110
Blue dextran	2,000,000	60

The pyrophosphatase eluted from the calibrated column in 94 ml. Determine the molecular weight of the enzyme.
12. Individual proteins of known subunit molecular weight and the pyrophosphatase whose *native* molecular weight had been determined (problem 11) were denatured in buffer containing both SDS and 2-mercaptoethanol. A sample of each protein was electrophoretically separated by SDS-polyacrylamide gel electrophoresis. The marker dye migrated 12 cm. Staining the gel for protein revealed the following:

Protein	Molecular Weight (subunit)	Distance Migrated (cm)
Serum albumin	69,000	1.6
Catalase	60,000	2.2
Ovalbumin	43,000	3.5
Carbonic anhydrase	29,000	5.2
Myoglobin	17,000	7.4

The pyrophosphatase migrated 6.7 cm. However, when the 2-mercaptoethanol was omitted from the sample buffer, the pyrophosphatase migrated 3.8 cm. Determine the subunit molecular weight and comment on the quaternary structure of the enzyme.
13. A mutant form of alkaline phosphatase was focused to its isoelectric point and was found to differ from the native (nonmutant) alkaline phosphatase by an amount corresponding to one additional positive charge relative to the native enzyme. Assuming that the mutant enzyme's isoelectric point reflects the substitution of a single amino acid residue, what are the possibilities?
14. You wish to purify an ATP-binding enzyme from a crude extract that contains several contaminating proteins. In order to purify the enzyme rapidly and to the highest purity, you must consider some sophisticated strategies, among them affinity chromatography. Explain how affinity chromatography may be applied to this separation and explain the physical basis of the separation.
15. You have isolated a protein complex that sediments in the ultracentrifuge similarly to a hemoglobin marker. If the ultracentrifugation is repeated under identical conditions except that 2-M NaC1 is added to the dilute buffer, the protein sediments similarly to a myoglobin marker. What conclusion can you reach about the properties of the protein complex?
16. Predict the effect on O_2 transport of a mutant form of hemoglobin with a markedly decreased affinity for glycerate-2,3-bisphosphate.
17. The concentration of glycerate-2,3-bisphosphate (GBP) is approximately 5 mM in the erythrocyte. Calculate the concentration of hemoglobin in the erythrocyte and compare it with that of GBP. Assume that the hemoglobin content of blood is 14 gm/dl and that the erythrocytes occupy 45% of whole blood volume. For the calculations, assume that hemoglobin (M_r = 64,500) occupies 100% of the erythrocyte volume.

CHAPTER

Carbohydrates, Glycoproteins, and Cell Walls

arbohydrates are made from carbon skeletons richly laced with hydroxyl groups. The hydrophilic hydroxyl groups give carbohydrates the potential for strong interaction with water; they also give them functional groups to which various substituents can add. Finally, the hydroxyl groups provide the possibility of strong intra- or interchain interaction via hydrogen bonds. The simplest carbohydrates contain only carbon, hydrogen, and oxygen. Derivatives of carbohydrates contain nitrogen, phosphorus, and even sulfur compounds. Carbohydrates can interact with other macromolecules to form glycoproteins or glycolipids, and they are also an important component of nucleic acids.

Small carbohydrates such as glucose, fructose, and pyruvate occupy key roles in energy metabolism and supply carbon skeletons for the synthesis of other compounds. Polymeric carbohydrates are important as short-term energy-storage compounds and also as major structural compounds in plant and bacterial cell walls and in the extracellular matrix. Branched-chain polymeric carbohydrates covalently linked to proteins are used to give proteins certain surface characteristics that may be exploited to signal protein targeting and in a variety of cell–cell recognition processes.

Carbohydrates play so many roles that it is almost easier to discuss the things they can't do rather than the things they can do. There are two areas where they seem to be lacking in potential, at least by themselves. They cannot store genetic information and they cannot function as enzymes (so far as is known). In this chapter we will describe the structural and functional properties of carbohydrates.

Monosaccharides and Related Compounds

The simplest carbohydrates, sometimes referred to as monosaccharides or sugars, are either polyhydroxyaldehydes (aldoses) or polyhydroxyketones (ketoses). They can be derived from polyalcohols (polyols) by oxidation of one carbinol group to a carbonyl group. For example, the simple three-carbon triol glycerol can be converted either to the aldotriose, glyceraldehyde, or to the ketotriose, dihydroxyacetone, by loss of two hydrogens (fig. 6.1).

Since the middle carbon of glyceraldehyde is connected to four different substituents, it is a chiral center leading to two possible forms of glyceraldehyde. D-Glyceraldehyde is illustrated in figure 6.1 in the Fischer projection formula, in which the —OH group attached to the central carbon atom points to the right. If the central carbon were in the plane of the paper with tetrahedrally arranged substituents, the H and

Figure 6.1

Loss of two hydrogens by glycerol leads to the formation of glyceraldehyde or dihydroxyacetone, depending on whether the two hydrogens are lost from the end or middle position, respectively.

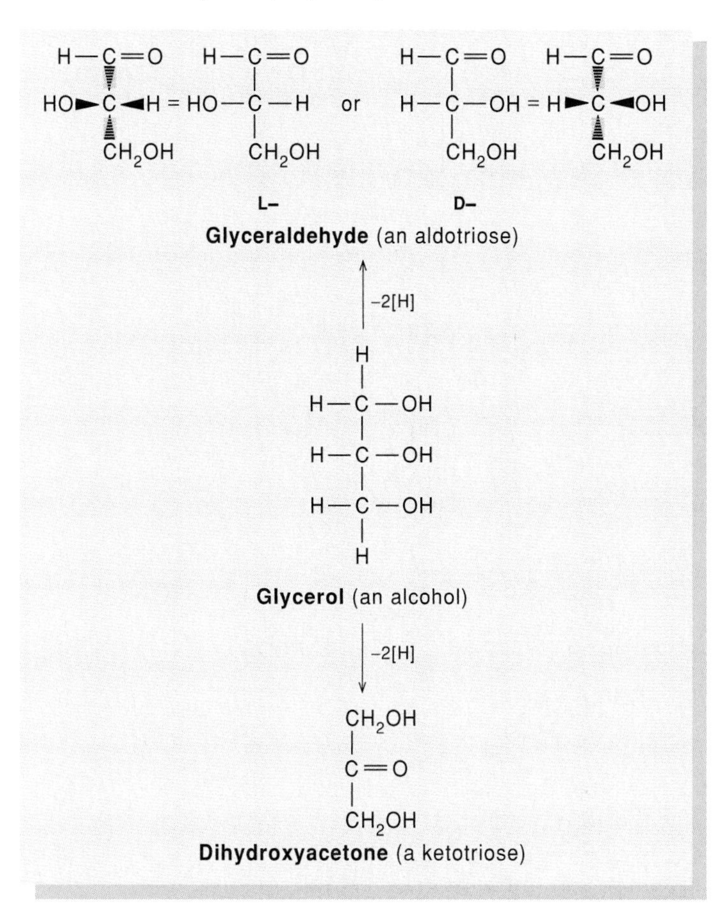

OH connected to it would project above the plane of the paper and the other two substituents would project below the plane of the paper. For larger sugars, Fischer projections are written with the most highly oxidized carbon, C-1, at the top.

A molecule such as glyceraldehyde, having one center of asymmetry (chiral center), is optically active, and the two forms of the molecule can be described as the dextrorotatory and levorotatory forms, according to the way in which they rotate plane-polarized light. The symbol *d* or (+) refers to dextrorotatory rotation, and the symbol *l* or (−) refers to levorotatory rotation. Rotation can be measured by making a solution of an optically active compound and measuring the rotation of plane-polarized light that passes through it (see box 6A). A mixture containing equal amounts of the two forms is optically inactive and is referred to as a racemic mixture.

The stereochemistry of sugars is based on configurational properties. The actual sign of rotation may still be indicated by the italic letters *d* and *l*, but the absolute configuration of the four different substituents around the asymmetric carbon atom is designated by the prefix symbols D and L. For the common sugars, the prefixes D and L refer to that center of asymmetry most remote from the aldehyde or ketone end of the molecule. By convention, all optically active centers are related to the asymmetric carbon of glyceraldehyde. Isomers stereochemically related to D-glyceraldehyde are designated D, and those related to L-glyceraldehyde are designated L. We may visualize the four-, five-, and six-carbon sugars as arising from the trioses through the stepwise condensation of formaldehyde to either glyceraldehyde or dihydroxyacetone. The biosynthesis of the sugars occurs by other means.

Families of Monosaccharides Are Structurally Related

A phosphorylated derivative of D-glyceraldehyde is an intermediate in the degradation of carbohydrates. This aldose is reversibly converted to its ketose isomer, dihydroxyacetone (see fig. 6.1), by an enzyme. The tetroses, pentoses, and hexoses related to D-glyceraldehyde are shown in figure 6.2. The ketoses (e.g., fructose) are similarly related to dihydroxyacetone (fig. 6.3).

Monosaccharides Cyclize to Form Hemiacetals

Aldehydes can add hydroxyl compounds to the carbonyl group. If a molecule of water is added, the product is an aldehyde hydrate, as shown in figure 6.4. If a molecule of alcohol is added, the product is a hemiacetal; the addition of a second alcohol results in an acetal. Sugars form intramolecular hemiacetals in cases where the resulting compound has a five- or six-membered ring.

Hemiacetal formation first became apparent from optical studies of D-glucose. The optical rotation of a freshly dissolved sample of D-glucose changes with time because there are two different hemiacetals that are convertible in solution (fig. 6.5). They are referred to as α-D-glucose, where $[\alpha] = 112°$, and β-D-glucose, where $[\alpha] = 19°$. A freshly prepared solution of either of these compounds will eventually approach an intermediate value that depends on the equilibrium between the two forms.

The convention for numbering hexoses is shown in the central structure of figure 6.5. The α designation for the D series indicates that the aldehyde or C-1 hydroxyl group is on the same side of the structure as the ring oxygen in the Fischer projection, and the β designation indicates that it is on the opposite side. For the L series the reverse is the case. When the sugar is dissolved in water, the hemiacetal is in equilibrium with the straight-chain hydrated form. The straight-chain form can produce either hemiacetal, α or β. Conversion of one isomer to another in solution is referred to as mutarotation. The different forms are referred to as anomers, and the anomeric carbon is the carbon that contains the reactive carbonyl. In the case of glucose this is the C-1 carbon. Equilibrium is reached without added catalyst in a few hours at room temperature. The open-chain form usually represents only a small fraction of the total (see figure 6.5 for actual percentages).

Figure 6.2

Configurational relationships of the D-aldoses. The most important sugars are starred. Note that in the two D series the configuration about the chiral center farthest from the carbonyl is the same.

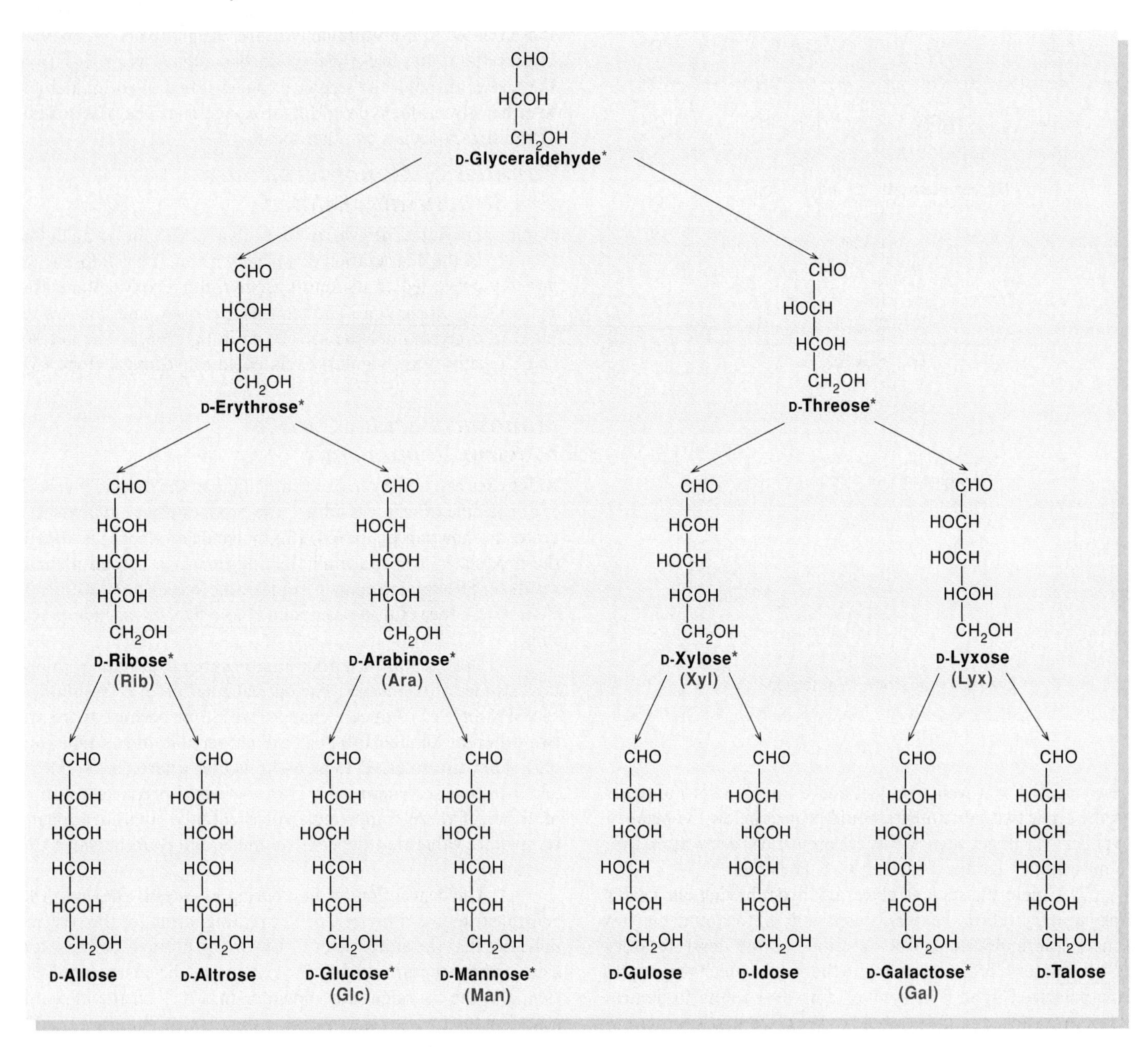

Figure 6.3

Configurational relationships of the D-ketoses. The most important sugars are starred.

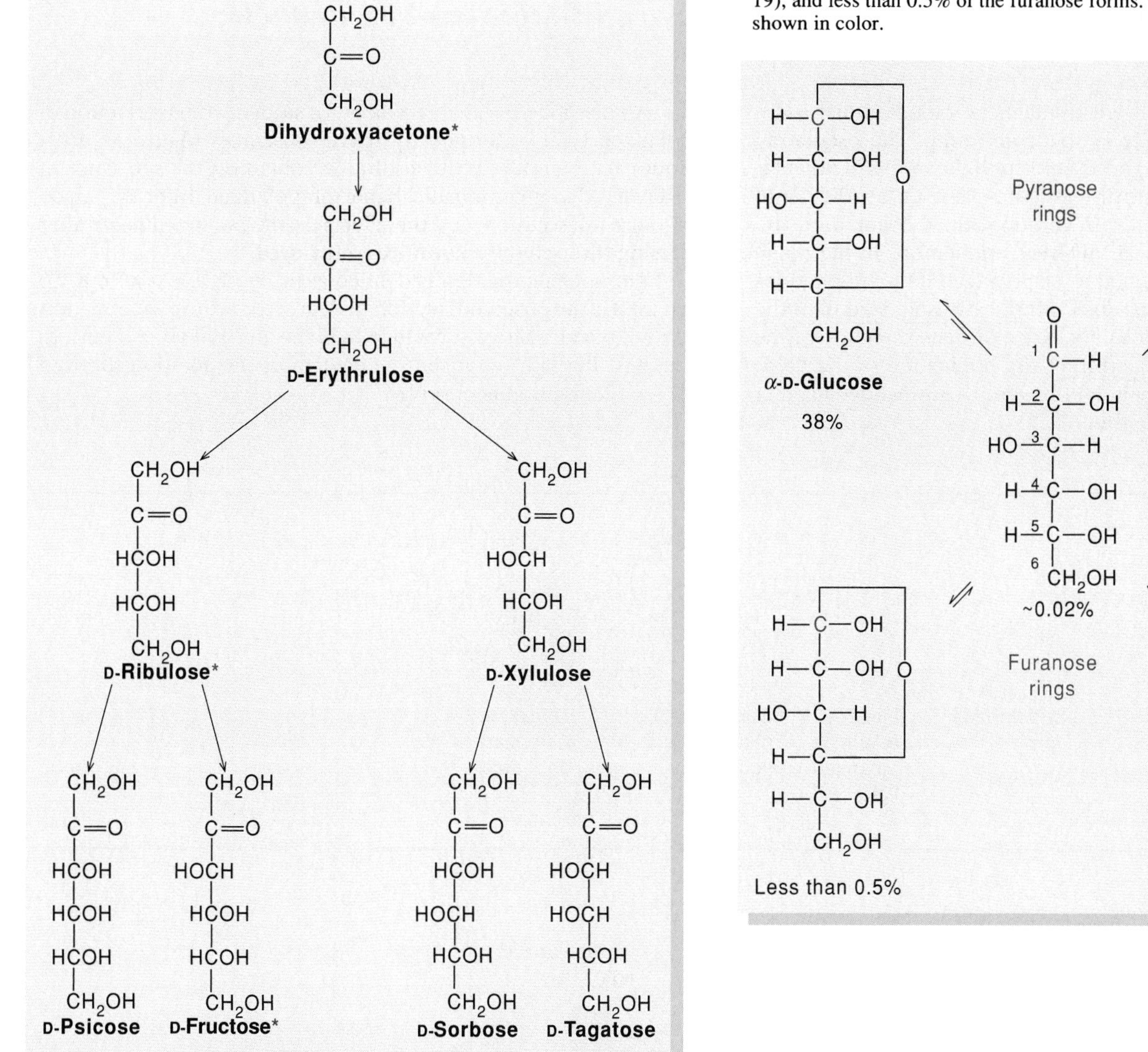

Figure 6.5

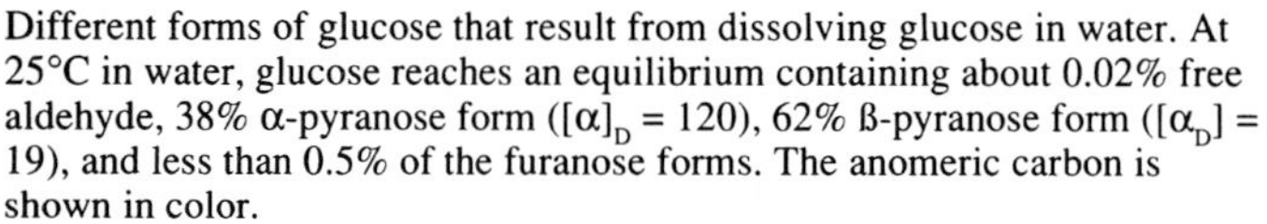

Different forms of glucose that result from dissolving glucose in water. At 25°C in water, glucose reaches an equilibrium containing about 0.02% free aldehyde, 38% α-pyranose form ($[\alpha]_D = 120$), 62% ß-pyranose form ($[\alpha_D] = 19$), and less than 0.5% of the furanose forms. The anomeric carbon is shown in color.

Figure 6.4

Aldehydes can add H_2O to form hydrates or can add alcohols to form hemiacetals and acetals.

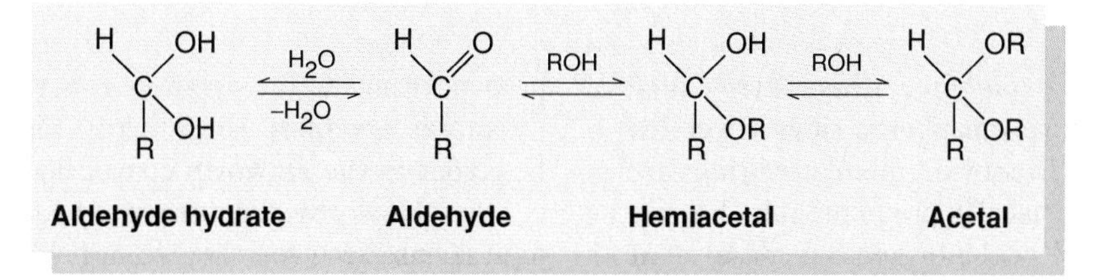

Methods for Structural Analysis: Polarized Light and Polarimetry

Light is a form of electromagnetic radiation that oscillates sinusoidally in space and time. The oscillating electric and magnetic fields of light are perpendicular to each other and are both in a plane perpendicular to the direction of the light ray. In an unpolarized beam, there are equally strong fields with all different orientations in the plane (figure 1). A light beam is said to be polarized if the orientations of the fields in the plane are fixed. In linearly polarized light the orientation of the fields does not vary along the direction of the beam. In circularly polarized light the orientation of the fields gradually changes in either a right-handed or left-handed manner along the direction of the beam.

A polarimeter is an instrument for studying the interaction of polarized light with optically active substances (figure 2). A cylindrical tube is filled with a solution containing the substance of interest. A monochromatic beam of polarized light is passed through the solution, and the effects on the polarized beam after passing through the solution are measured.

The specific rotation is defined as $[\alpha] = (100 \times A)/(c \times l)$, where A is the observed rotation in degrees, c is the concentration of the optically active substance in grams per 100 ml of solution, and l is the path length in decimeters of the solution through which the rotation is observed.

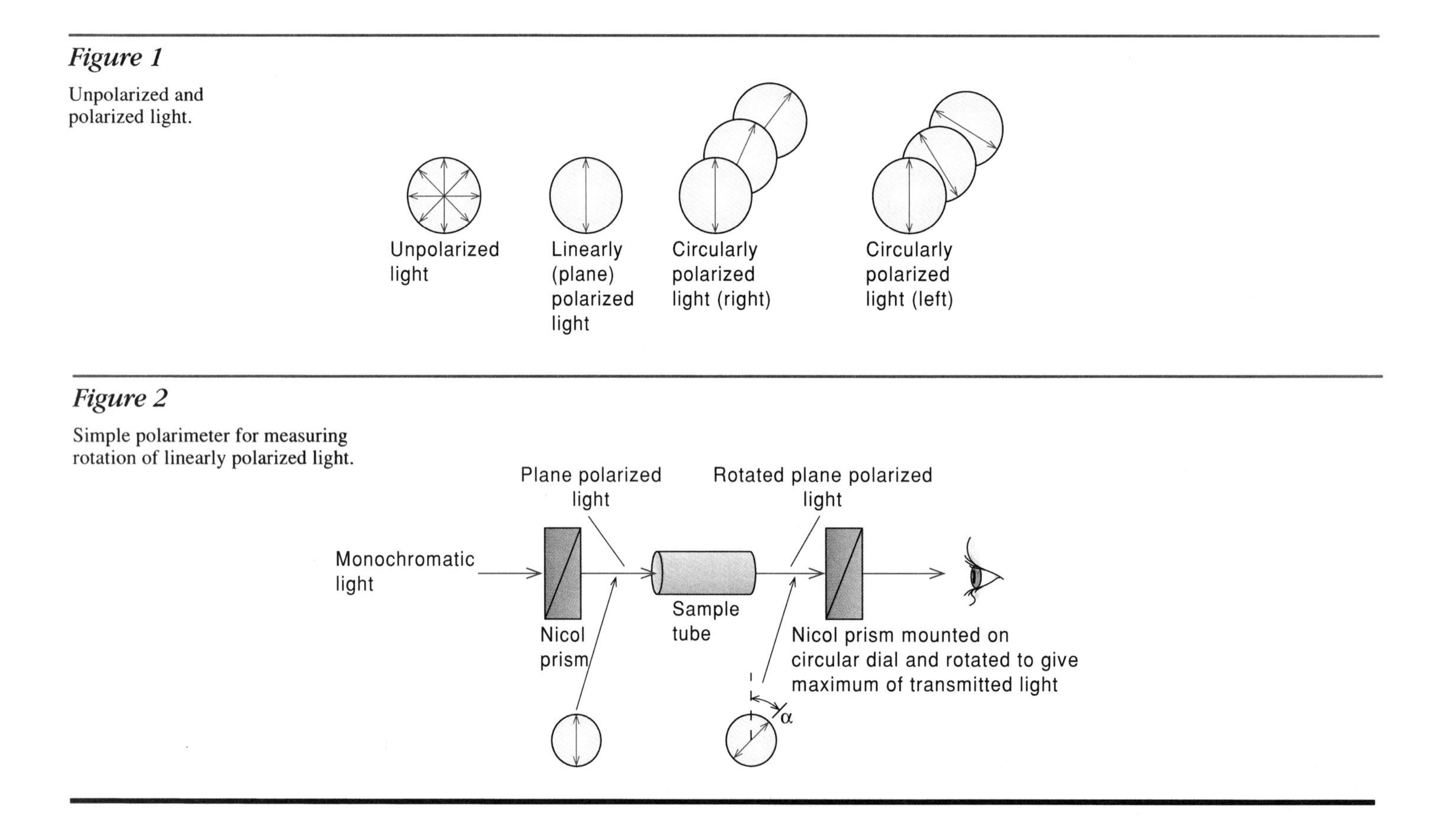

Figure 1

Unpolarized and polarized light.

Figure 2

Simple polarimeter for measuring rotation of linearly polarized light.

Hemiacetals with five-membered rings are called furanoses, and hemiacetals with six-membered rings are called pyranoses. In cases where either five- or six-membered rings are possible, the six-membered ring usually predominates. For example, for glucose less than 0.5% of the furanose forms exist at equilibrium (see fig. 6.5, bottom). The reason for the general preponderance of the pyranose is not known. Both furanoses and pyranoses are more realistically represented by pentagons or hexagons in the Haworth convention, as shown in figure 6.6.

Haworth structures are unambiguous in depicting configurations, but even they do not show correctly the spatial relationship of groups attached to rings. The normal valence angle

Figure 6.6

Comparison of the Fischer and Haworth projections for α- and ß-D-glucose. The Haworth projection is a step closer to reality.

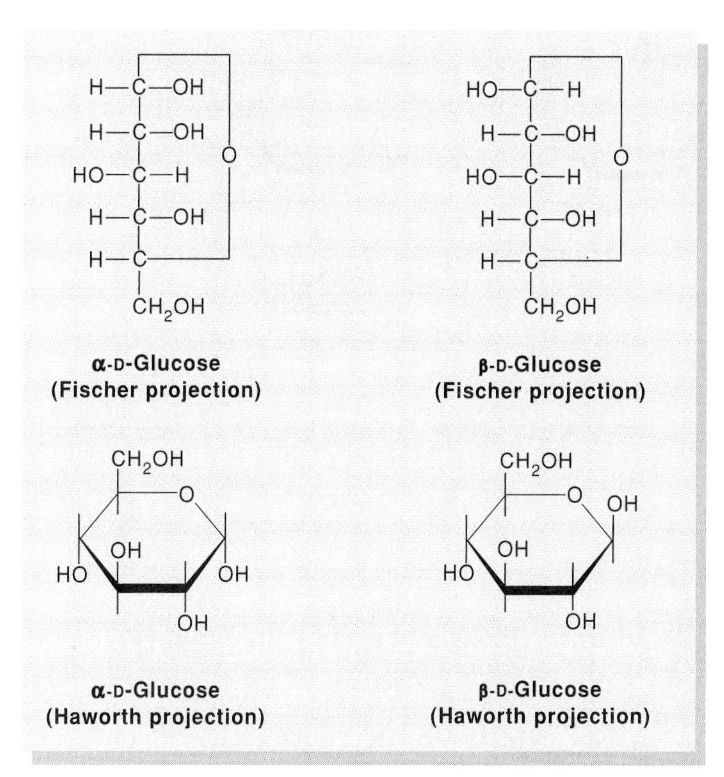

of saturated carbon (109°) prevents a stable planar arrangement for cyclohexane or the related pyranose molecule. The two most likely conformations* are the so-called chair and boat forms (fig. 6.7). Usually the chair form is considerably more stable than the boat form. The twelve substituent atomic groups of the ring carbons fall into two classes, those that are approximately perpendicular to the plane of the ring, i.e., axial, and those that are parallel to the plane of the ring, i.e., equatorial. As a rule, a substituent is at a lower energy state in the equatorial position because there is less chance of steric hindrance with other substituents. This fact becomes more important with larger substituents. The two anomers for the favored chair form of D-glucose are shown in figure 6.8. In sugar chemistry, formulating the conformations of aldohexoses is important in interpreting reactivity of hydroxyl groups and other sizable substituents. The most stable conformation for a particular aldohexose is the chair form, which places the maximum number of substituents larger than hydrogen in the equatorial position.

*When discussing sugars we make frequent use of the terms "configuration" and "conformation." These terms have different meanings. Two conformations of the same molecule are interconvertible without breakage of any chemical bonds; two configurations of the same molecule are not. For example, the chair and boat forms of α-D-glucose represent two different conformations of the same molecule. But α-D-glucose and β-D-glucose represent two different configurations of D-glucose.

Figure 6.7

Chair and boat forms for a generalized pyranose ring structure. Structures of this type are more realistic than the Haworth structure, as the carbon–carbon bond angle is correct. The chair form is usually favored over the boat form. The substituents are labeled "a" (axial) and "e" (equatorial). The axis of symmetry is labeled for both forms. Axial bonds are parallel to this axis of symmetry. Equatorial bonds are parallel to the nonadjacent sides of the rings. A large substituent generally prefers to be in an equatorial location.

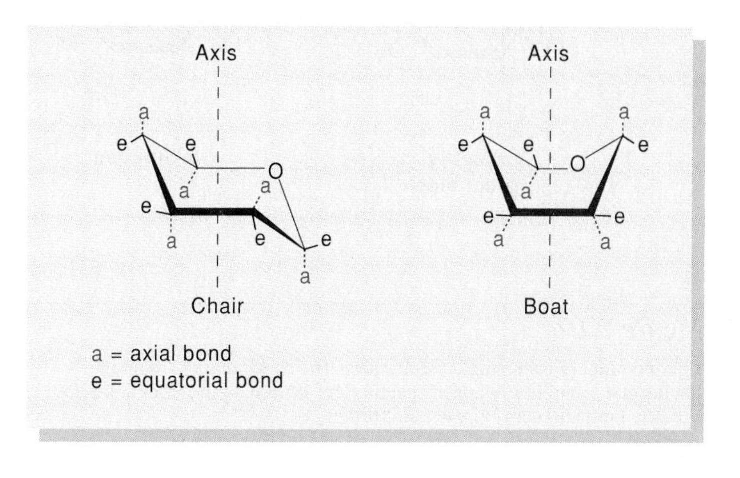

Figure 6.8

Chair configurations for the two anomers of D-glucose. Note that the largest substituent, —CH_2OH, is in an equatorial location in both structures. The differences between the two anomers are shown in color.

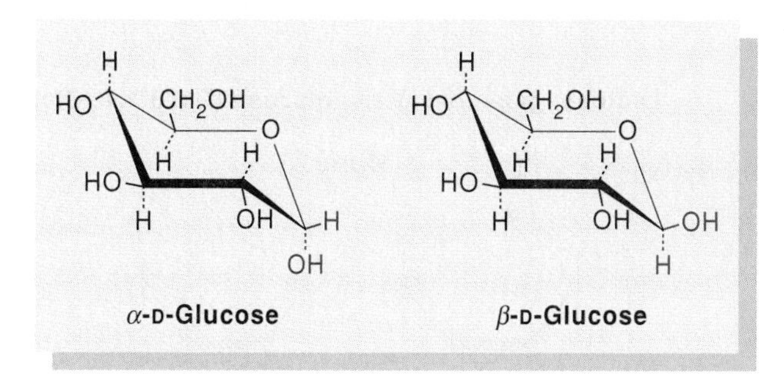

The furanose ring is nonplanar also and can exist in more than one conformation. The conformations for D-ribose (β-D-ribofuranose) and D-2-deoxyribose (β-D-2-deoxyribofuranose), the two pentoses found in all nucleic acids, will be discussed in chapter 25.

Monosaccharides Are Linked by Glycosidic Bonds

Warming glucose in methanol and acid produces a mixture of two new substances, α- and β-methylglucoside; the comparable derivatives of galactose are referred to as galactosides, and so on. Generally the bond between a sugar and an alcohol is referred to as a glycosidic bond, and the compound is known by

Figure 6.9

Formation of methyl glucosides. Glucosides (or glycosides) are quite stable in alkali but they hydrolyze readily in dilute acid.

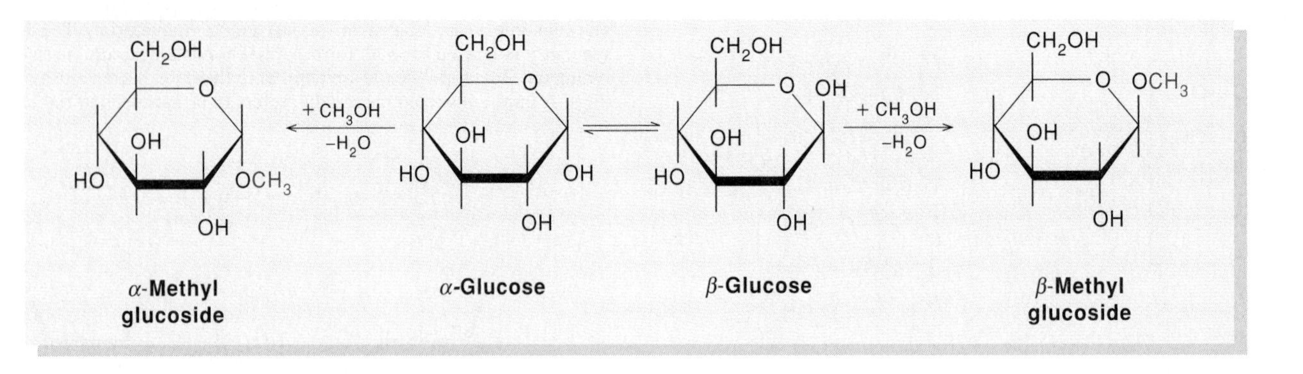

Figure 6.10

Four commonly occurring disaccharides. The configuration about the hemiacetal group has not been specified for lactose, maltose, or cellobiose because both anomers exist in equilibrium.

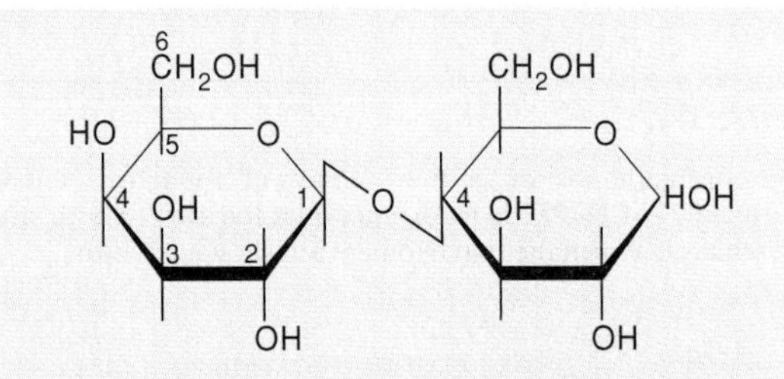

Lactose: galactose β(1,4)-glucose (Gal β(1,4)-Glc)

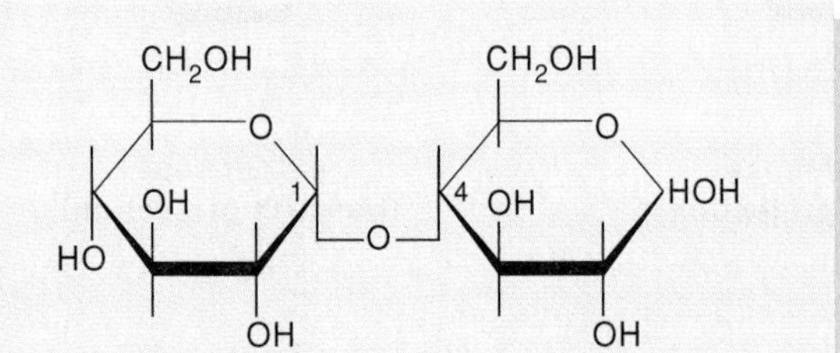

Maltose: glucose α(1,4)-glucose (Glc α(1,4)-Glc)

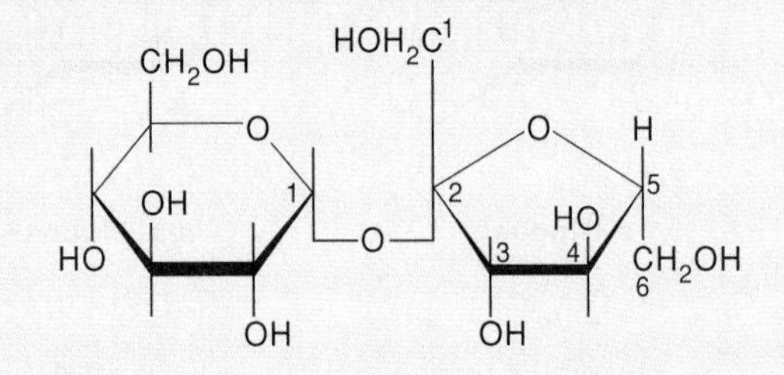

Sucrose: glucose α(1,2)-β-fructose (Glc α(1,2)-β-Fru)

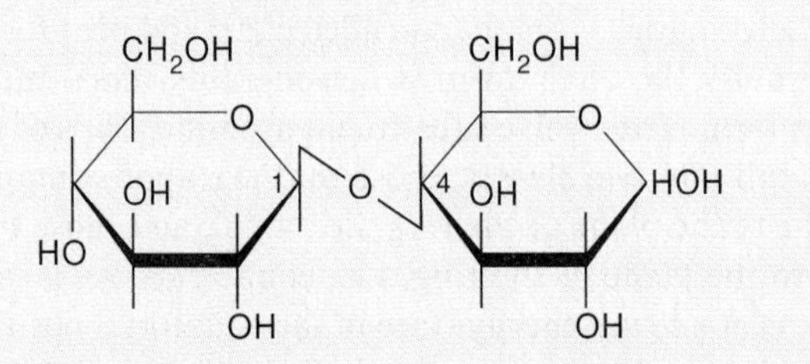

Cellobiose: glucose β(1,4) glucose (Glc β(1,4) Glc)

the generic name glycoside. The formation of a glycoside from a sugar and methanol by acid catalysis is identical to the formation of an acetal from an aldehyde and an alcohol (fig. 6.9). While the two forms of glucose in solution are in equilibrium through mutarotation, the corresponding glycosides are locked into one configuration. This is understandable, because mutarotation requires that, in the intermediate, the anomeric carbon adopt a carbonyl structure, which is not possible in a glycoside. A glycoside can be formed with aliphatic alcohols, phenols, and hydroxy carboxylic acids, as well as with another sugar.

Disaccharides and Polysaccharides

The most important glycosides are those formed with other sugars. Monosaccharides are glycosidically linked to form disaccharides (fig. 6.10). For instance, the disaccharide maltose contains a glycosidic bond between the C-1 of one glucose molecule and the C-4 of another glucose molecule. The compound is said to have an $\alpha(1,4)$ glycosidic linkage because the anomeric C-1 carbon of one sugar is connected to the C-4 of another sugar and the configuration about the anomeric carbon is α. Maltose possesses one potentially free aldehyde group and is

Figure 6.11

Some of the sugar building blocks found in polysaccharides.

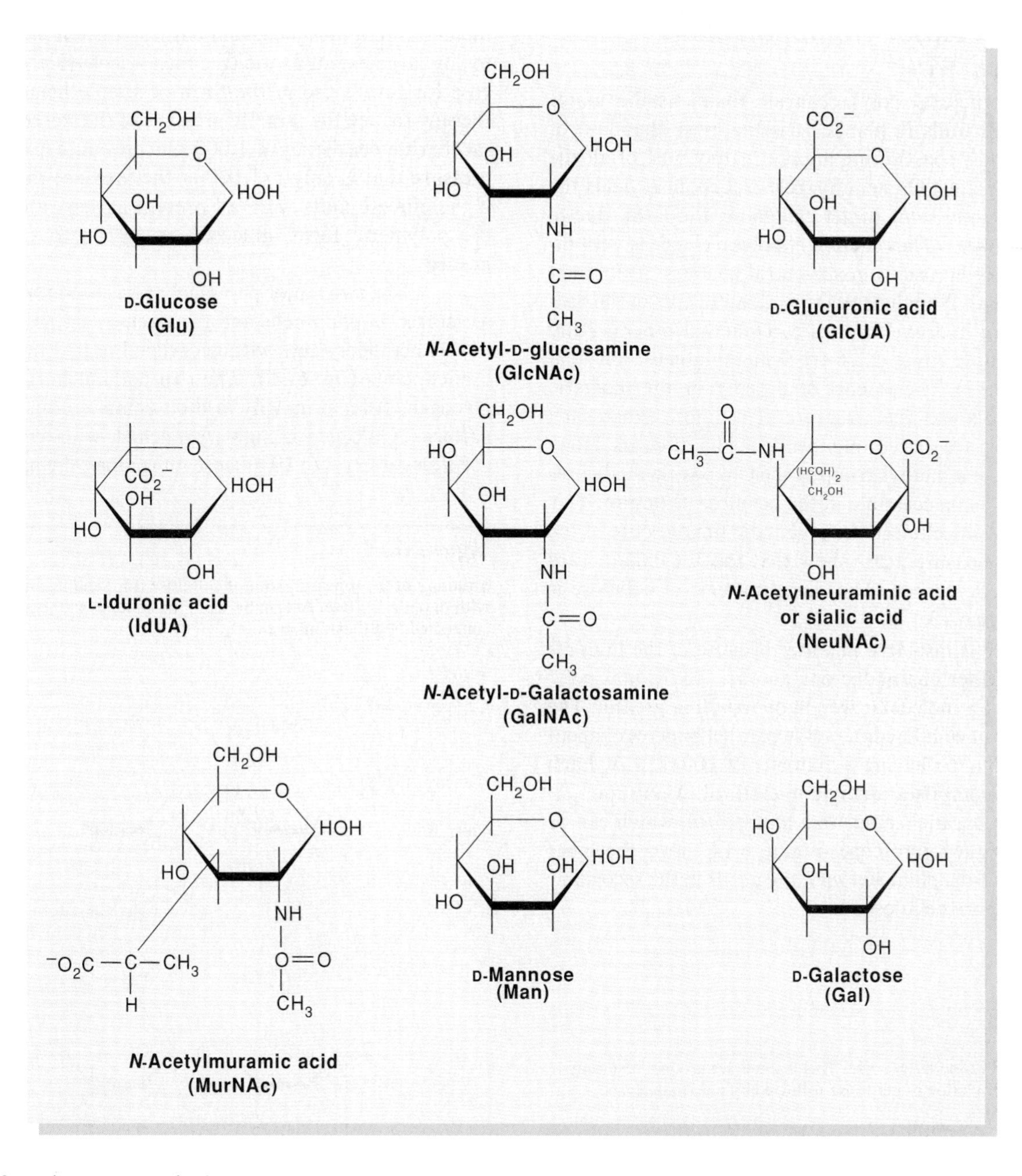

therefore referred to as a reducing sugar. The configuration about the hemiacetal hydroxyl group has not been specified in figure 6.10 because it can undergo mutarotation. Maltose is most familiar as a degradation product of starch.

The disaccharide cellobiose is identical with maltose except for having a $\beta(1,4)$ glycosidic linkage. Cellobiose is a degradation product of cellulose. Lactose is a disaccharide found exclusively in the milk of mammals. Lactose contains a $\beta(1,4)$ glycosidic linkage between galactose and glucose. Sucrose is found in abundance in sugar beets and sugar cane. On acid hydrolysis it yields equivalent amounts of D-glucose and D-fructose. Sucrose contains an $\alpha(1,2)\beta$ glycosidic linkage.

Most carbohydrates in nature exist as high-molecular-weight polymers called polysaccharides. Polysaccharides are composed of simple or derived sugars connected by glycosidic bonds. The most common building block used in polysaccharides is D-glucose. Other sugars that are used include D-mannose, D- and L-galactose, D-xylose, L-arabinose, D-glucuronic acid, D-galacturonic acid, D-mannuronic acid, D-glucosamine, D-galactosamine, and neuraminic acid. Some of these sugar building blocks are illustrated in figure 6.11; others are illustrated in figure 6.2. Polymers composed of a single type of building block are called homopolymers; those composed of more than one type are called heteropolymers.

Polysaccharides function in two quite distinct roles; some serve as a means for storage of chemical energy and others serve a structural function.

Cellulose Is a Major Homopolymer Found in Cell Walls

Cellulose is a structural polysaccharide found as the major component of cell walls in plants. It is the most abundant of organic compounds, constituting approximately 50% of all the carbon found in plants. On acid hydrolysis, cellulose yields the monomer glucose and some dimer cellobiose, the latter due to incomplete hydrolysis. The reaction of polysaccharides with dimethylsulfate is often useful in structural analysis. Treatment of a saccharide with dimethylsulfate in alkali results in the conversion of all free hydroxyl groups to *O*-methyl ethers. Fully methylated cellulose gives 2,3,6-tri-*O*-methylglucose on acid hydrolysis (fig. 6.12). The absence of a methyl on the anomeric C-1 says nothing about the structure of cellulose, since such methyl derivatives are susceptible to mild acid hydrolysis. However, the absence of a methyl group at the C-4 position indicates that this position is inaccessible in the cellulose structure. This fact suggests that in cellulose the glycosidic linkages are of the 1,4 type. Other measurements show that the 1,4 linkages are about the anomeric carbon. The repeating unit of cellulose is indicated in figure 6.13.

Cellulose is insoluble in water because of the high affinity of the polymer chains for one another. Individual polymeric chains have a molecular weight of 50,000 or greater. The molecular chains of cellulose interact in parallel bundles of about 2,000 chains, a bundle having a diameter of 100–250 Å. Each bundle of 2,000 comprises a single microfibril. Many microfibrils arranged in parallel comprise a macrofibril, which can be seen under the light microscope. Figure 6.14 shows the inner secondary walls of the plant *Valonia;* the fibrils in the secondary wall are almost pure cellulose.

Figure 6.12

Structure of 2,3,6-tri-*O*-methylglucose. This is the main product resulting from exhaustive methylation of cellulose followed by acid hydrolysis.

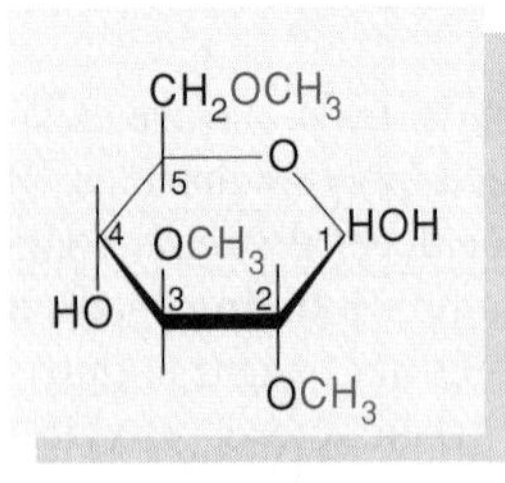

2,3,6-tri-*O*-methylglucose

Starch and Glycogen Are Major Energy-Storage Polysaccharides

Although glucose is the most important sugar involved in energy metabolism in most cells and tissues, it is not present in the cell to any large extent as the free monosaccharide. Cells store glucose for future use in the form of simple homopolymers, and thereby reduce the osmotic pressure of the stored sugar. A polysaccharide consisting of 1,000 glucose units exerts an osmotic pressure that is only 1/1,000 of the pressure that would result if the glucose units were all present as separate molecules. In the polymeric form, glucose can be stored compactly until needed.

The two major polysaccharides used for energy storage are starch in plant cells and glycogen in animal cells. Both are $\alpha(1,4)$ homopolymers with occasional $\alpha(1,6)$ linkages to make branchpoints (fig. 6.15). The two polysaccharides, starch and glycogen, differ primarily in their chain lengths and branching patterns. Glycogen is highly branched, with an $\alpha(1,6)$ linkage occurring every 8 to 10 glucose units along the backbone, giving

Figure 6.13

Structure of the repeating unit of cellulose (top) and its analysis by reaction with dimethylsulfate. As can be seen, the residues in the native structure are connected by ß(1,4) linkages.

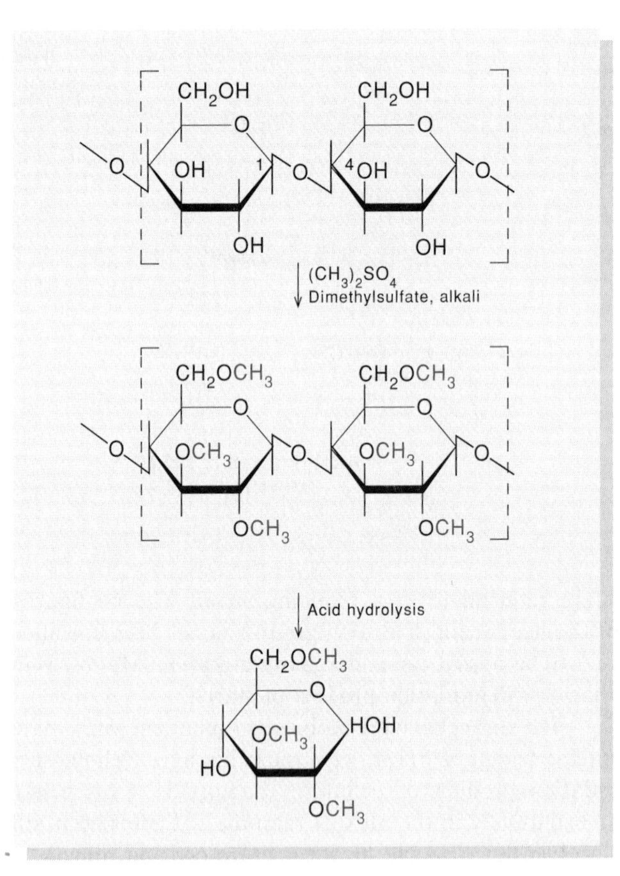

rise in each case to short side chains of about 8 to 12 glucose units each. Starch occurs both as unbranched amylose and as branched amylopectin. Like glycogen, amylopectin has $\alpha(1,6)$ branches, but these occur less frequently along the molecule (once every 12 to 25 glucose residues) and give rise to longer side chains (lengths of 20 to 25 glucose units are common). Starch deposits are usually about 10–30% amylose and 70–90% amylopectin.

Figure 6.14

Fibril arrangements in the cell wall of *Valonia* (12,000X). (Electron micrograph from A. Frey-Wyssling and K. Mühlethaler, *Ultrastructural Plant Cytology*, 1965, p. 298. Reprinted with permission from Elsevier Science Publishers.)

The Configurations of Glycogen and Cellulose Dictate Their Roles

It is a remarkable fact that the main energy-storage polysaccharides and the main structural polysaccharides found in nature both have a primary structure of (1,4)-linked polyglucose. Why should two such closely related compounds be used in totally different roles? A closer look at the stereochemistry of the α and β glycosidic linkage for polyglucose indicates why this is so.

Recall that D-glucose exists in the chair form of a pyranose ring (see fig. 6.8). The ring has a rigid character to it. We can think of it as a structural building block in a polysaccharide chain, just as we think of the rigid planar peptide grouping as a structural building block in a polypeptide chain. It is also possible to specify two torsional angles ϕ and ψ for rotation about the glycosidic C—O linkage (fig. 6.16). These angles are used extensively to discuss polypeptide configuration in proteins (see chapter 4). They have limited value in discussing polysaccharide structures, because much less information is available about such structures. Nevertheless, it is clear that only the $\beta(1,4)$-linked polyglucose has the capacity to form straight chains (see fig. 6.16*a*). A straight chain can be created by flipping each glucose unit by 180° relative to the previous one. This process should result in an almost fully extended polysaccharide chain, which is known to be characteristic of the structure of cellulose. Evidently this conformation is energetically favored. By contrast, $\alpha(1,4)$-linked units in a polyglucose cause a natural turning of the chain (see fig. 6.16*b*). Consistent with this fact is the observation that amylose adopts a coiled helical configuration. Indeed, one of the first helical structures

Figure 6.15

Structure of the storage polysaccharides glycogen and starch. The main chain is $\alpha(1,4)$-linked. Side chains are connected to the main chain by $\alpha(1,6)$ linkages.

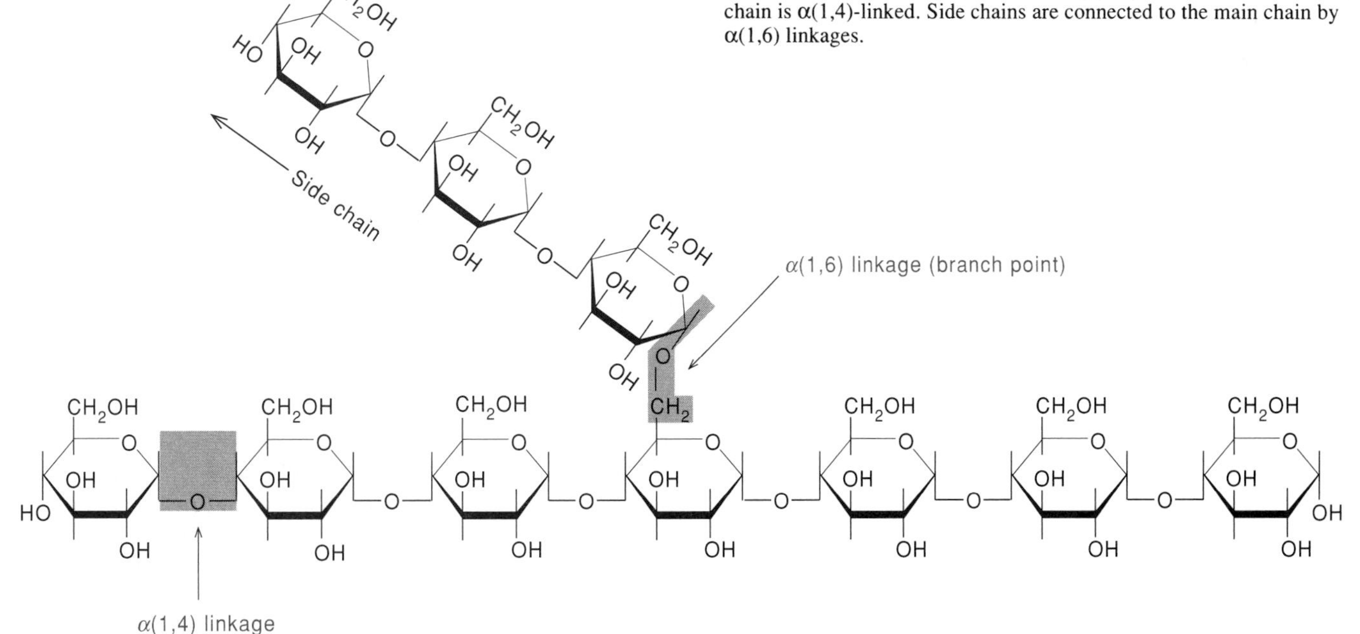

Figure 6.16

Energetically favored conformations of ß(1,4)-linked D-glucose (*a*) and α(1,4)-linked D-glucose (*b*). Note that in the ß(1,4) configuration in (*a*), alternating residues are flipped 180° relative to one another so that long straight chains result. In the α(1,4) configuration (*b*), the chain has a natural curvature.

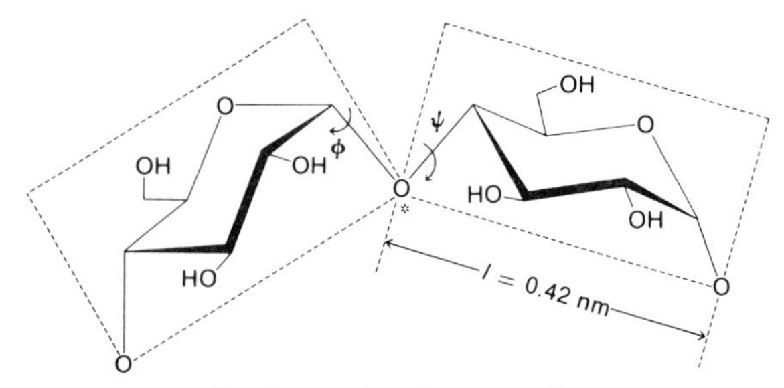

α(1,4)-linked D-glucose units

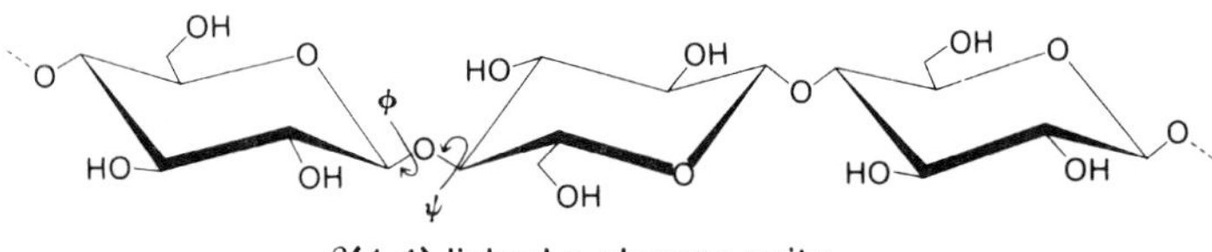

β(1,4)-linked D-glucose units

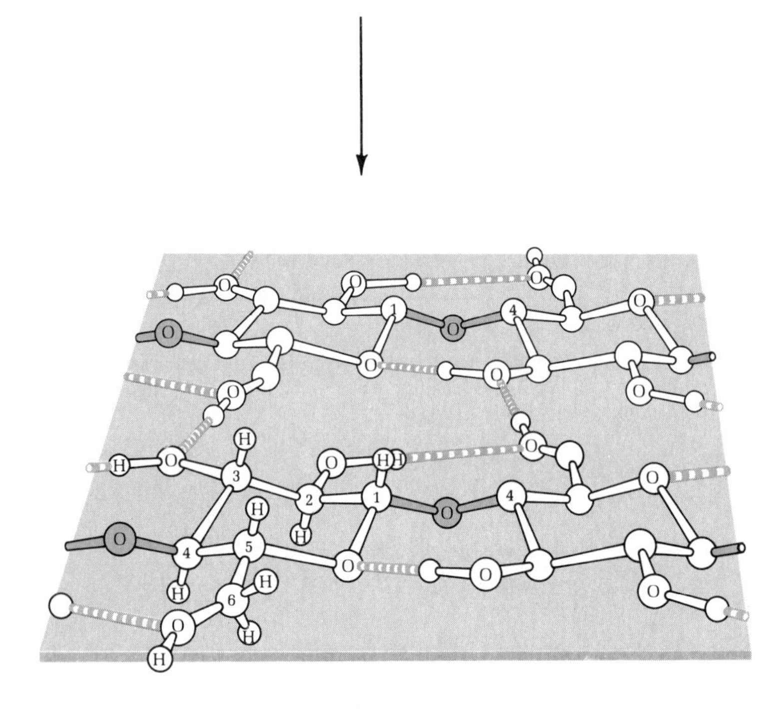

(a)

(b)

to be discovered (1943) was the left-handed helix of amylose wound around molecules of iodine (fig. 6.17). This structure is responsible for the characteristic blue color of the amylose-iodine complex.

The extended-chain form of polyglucose has been exploited in nature for structural purposes, leaving by default the coiled form for use as an energy-storage macromolecule. Correlated with this functional difference is the omnipresence of degrading enzymes for glycogen and starch and the very limited phylogenetic distribution of comparable enzymes for cellulose. Cellulose is degraded in the gastrointestinal tract of herbivores, such as the cow, or in insects, such as termites, by a protozoan that synthesizes the enzyme cellulase. Humans do not possess this enzyme and hence cannot degrade cellulose.

Figure 6.17

Structure of the helical complex of amylose with iodine (I_2). The amylose forms a left-handed helix with six glucosyl residues per turn and a pitch of 0.8 nm. The iodine molecules (I_2) fit inside the helix parallel to its long axis.

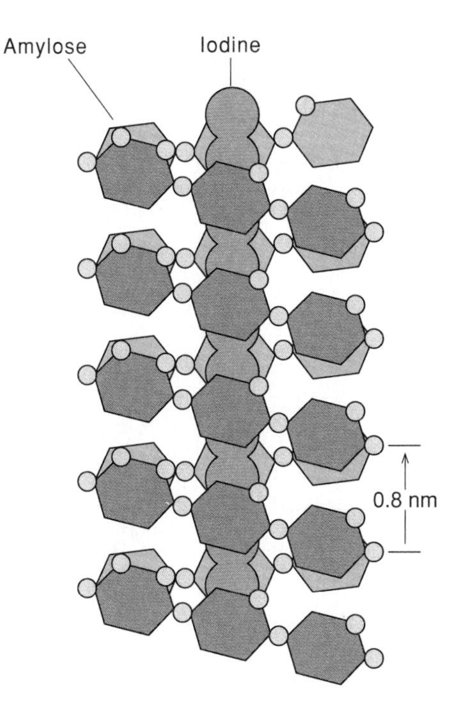

Figure 6.18

Structure of the repeating unit of chitin: a ß(1,4) homopolymer of *N*-acetyl-D-glucosamine.

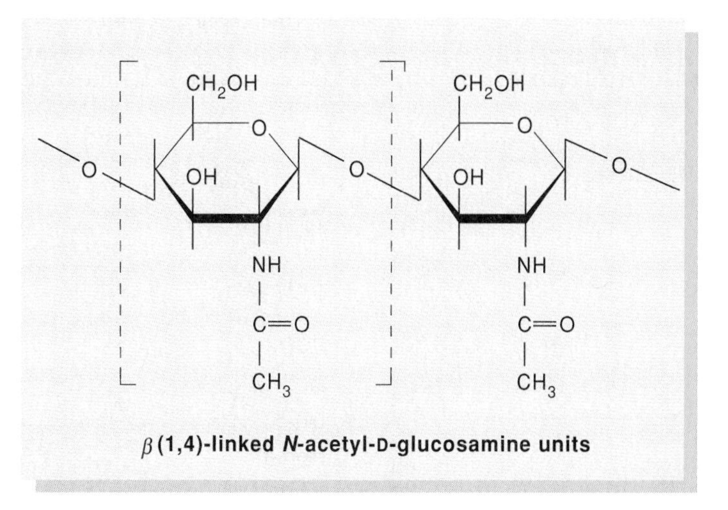

Figure 6.19

Structure of the repeating unit of hyaluronic acid: GlcUA ß(1,3)-GlcNAc ß(1,4).

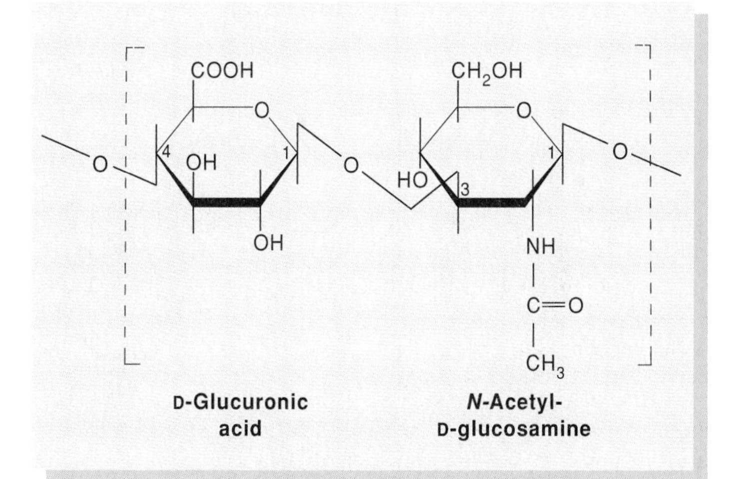

Chitin Contains a Different Building Block

Many polysaccharides consist of sugars other than glucose or different combinations of sugars. In some cases the new sugar is merely a derivative of glucose. Chitin is an example of a structural polysaccharide that uses a modified derivative of glucose. Chitin is found in the shells of crustaceans and insects and in the cell walls of fungi; it is a linear β(1,4) polymer of *N*-acetyl-D-glucosamine (fig. 6.18). A major rigid component of bacterial cell walls, the peptidoglycan, which we will discuss later, could be regarded as a substituted chitin.

Heteropolysaccharides Contain More than One Building Block

A large variety of carbohydrates contain either modified sugars, like the *N*-acetylglucosamine found in chitin, or two or more different sugars in straight-chain or branched-chain linkages.

Structurally, the simplest and best known of the heteropolysaccharides are the glycosaminoglycans. These are long, unbranched polysaccharide chains composed of repeating disaccharide subunits in which one of the two sugars is either *N*-acetylglucosamine or *N*-acetylgalactosamine (table 6.1). Glycosaminoglycans are highly negatively charged because of the presence of carboxyl or sulfate groups on many of the sugar residues. The high negative charge causes the polymeric chains to adopt a stretched or extended conformation. Their extended structure gives a high viscosity to the surrounding region, even in a dilute solution of the polysaccharide. Glycosaminoglycans are usually found in extracellular space in multicellular organisms, where they produce a viscous extracellular matrix that resists compression. Such an environment can be beneficial to the organism in various ways; it can provide a passageway for cell migration, supply lubrication between joints, or help maintain certain structural shapes such as the ball of the eye.

Hyaluronic acid is a copolymer of D-glucuronic acid and *N*-acetyl-D-glucosamine (fig. 6.19). It is much larger than other glycosaminoglycans, reaching molecular weights in excess of 10^6.

Table 6.1
Structure of Glycosaminoglycans

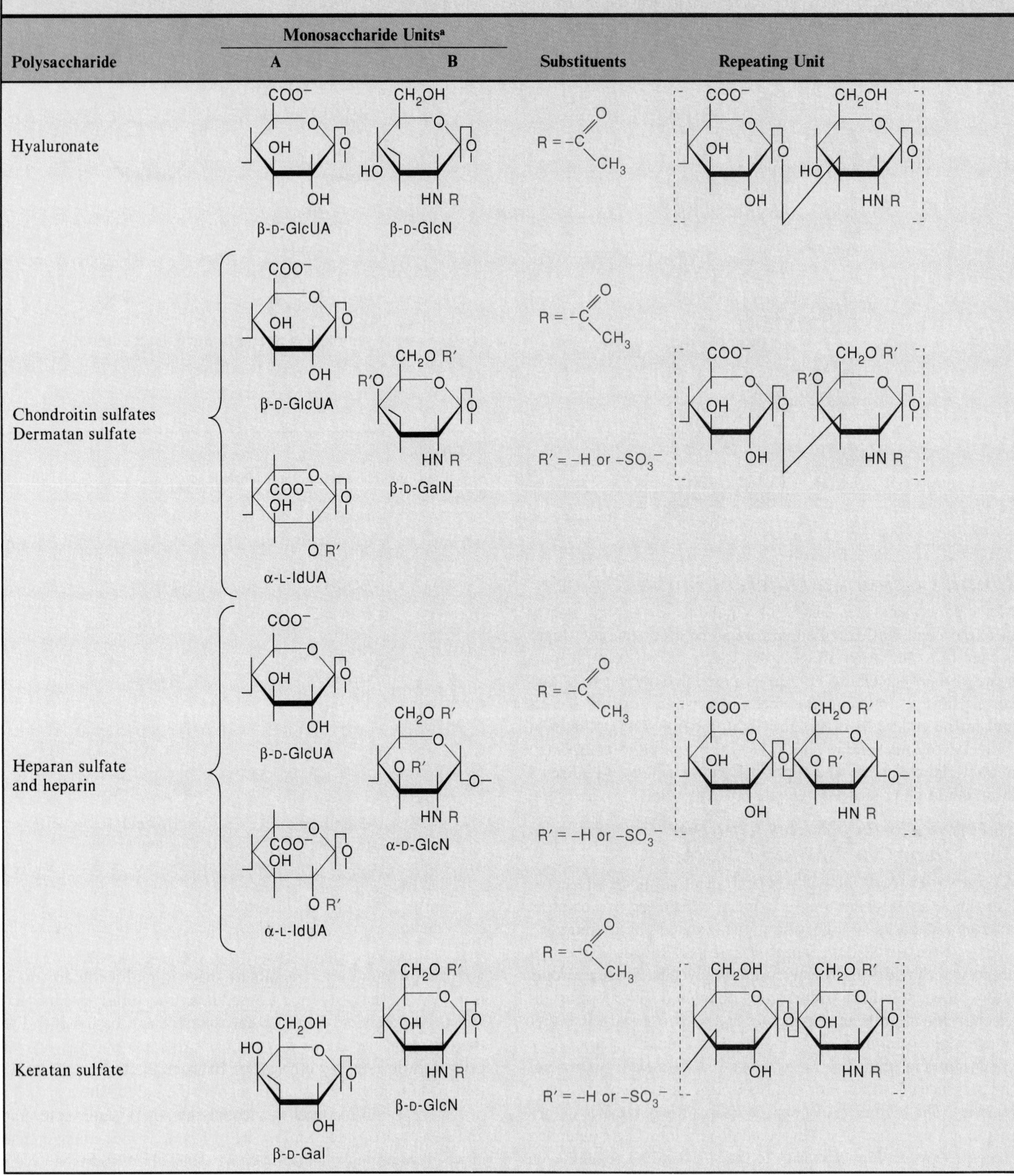

[a]The polysaccharides are depicted as linear polymers of alternating A and B monosaccharide units.
Abbreviations: GlcUA, glucuronic acid; IdUA, iduronic acid; GalN, galactosamine; GlcN, glucosamine; Gal, galactose.

Figure 6.20

(*a*) Structure of a typical proteoglycan. (*b*) The attachment site between a serine in the core protein and the glycosaminoglycan.

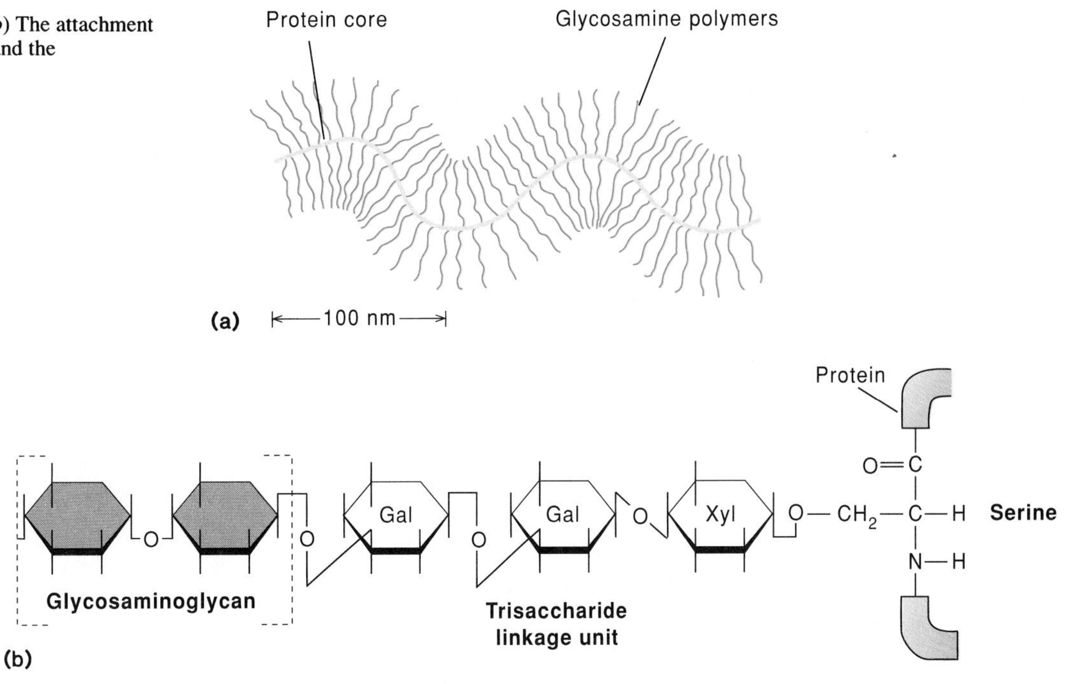

Proteoglycans

Most glycosaminoglycans are linked to a core protein as lateral extensions, forming a proteoglycan with a highly extended, brushlike structure (fig. 6.20). The linkage between the glycosaminoglycans and the core protein within the proteoglycan is mediated by a specific trisaccharide unit that is linked on one side to the repeating disaccharide unit of the glycosaminoglycan and on the other side to a serine hydroxyl group of the core protein (see fig. 6.20). In the extracellular matrix, hyaluronic acid and the other proteoglycans form aggregates with each other as well as with other macromolecular components, such as serum glycoproteins, growth factors, collagen, elastin, or the outer plasma membrane of cells. Hyaluronic acid, when complexed with other proteoglycans through their core proteins, produces very large complexes (fig. 6.21).

Thus far we have seen that glycosaminoglycans and their proteoglycans form highly extended aggregates that impart a rigid, gel-like structure to the extracellular matrix. It seems likely that many more specific reactions occur involving these compounds. However, it should be emphasized that relatively little is known about the organization or reactions of these molecules in the extracellular matrix. One exception is the case of the glycosaminoglycan heparin. We know that this compound functions as a highly specific anticoagulant. First it forms a complex with the plasma protein antithrombin III. This complex in turn inhibits the serine proteases of the blood clotting system.

Glycoproteins

Proteins are frequently adorned by straight-chain or branched oligosaccharides, in which case they are called glycoproteins. This type of modification can serve a variety of functions. It can stabilize the protein, facilitate its correct folding, be part of a lipid anchor for attaching the protein to a membrane, and provide the protein with surface characteristics that facilitate its recognition.

The carbohydrate moiety in a glycoprotein is attached to the polypeptide chain by a covalent connection to certain amino acid side chains. Two types of glycosidic linkages are commonly found: the O-glycosidic linkage involves attachment of the carbohydrate to the hydroxyl group of serine, threonine, or hydroxylysine; the N-glycosidic linkage involves attachment to the amide group of asparagine (fig. 6.22). Conformational constraints appear to be a major factor in determining the addition, although other factors also play a role. Thus, in the case of O-linked sugars, clusters of serines or threonines may be a preferred substrate. For N-glycosidic linkage, only asparagine residues of the sequence Asn-X-Ser(Thr) will accept carbohydrates. In this sequence X may be any amino acid except proline.

Glycoproteins are found in all cellular compartments and are also secreted from the cell. Those found in the cytoplasm have the simplest type of modification, consisting of a single *N*-acetylglucosamine that is O-glycosidically linked to Ser(Thr). Collagen, a secreted glycoprotein of the extracellular matrix, also has simple carbohydrates—the disaccharide Glcβ(1,2)Gal linked to hydroxylysine.

Figure 6.21

A hyaluronic acid–proteoglycan complex. The individual proteoglycans contain a core protein by which are linked various glycosaminoglycans as shown in figure 6.20. The core protein is noncovalently associated with a single hyaluronic acid molecule via two link proteins. Electron micrograph below shows a large aggregate (*A*) and a small aggregate (*B*) whose sizes are determined mainly by the length of the hyaluronate central filament of the individual apprepates. (From Buckwalter, J. A. and Rosenberg, L. Coli. Rel. Res. (1983) 3, 489–504.)

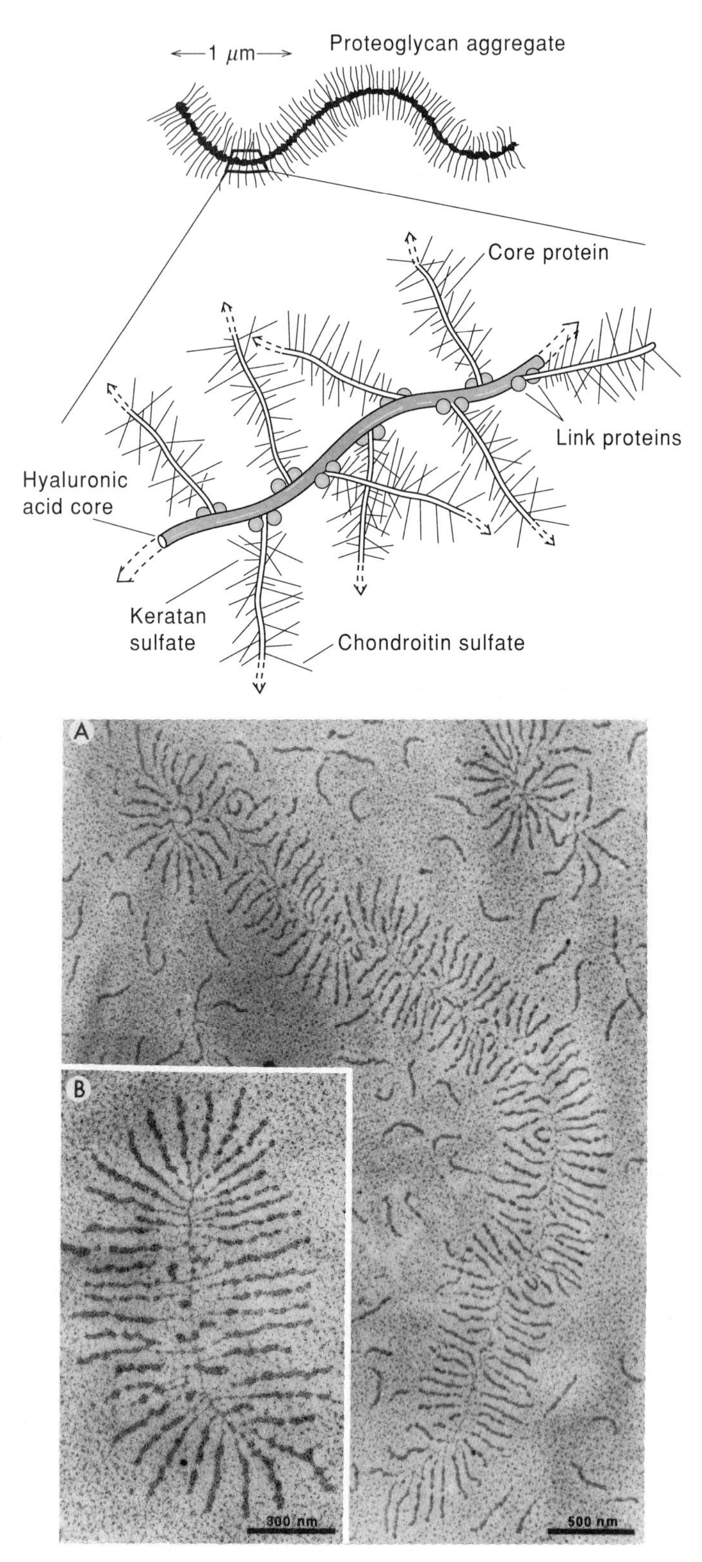

Figure 6.22

Linkages found between oligosaccharides and proteins in glycoproteins: (*a*) is an O-glycosidic linkage found in many glycoproteins and mucins; an analogous linkage occurs between *N*-acetylglucosamine and serine in glycoproteins; (*b*) is an N-glycosidic linkage found in many glycoproteins; and (*c*) is an O-glycosidic linkage found only in collagen.

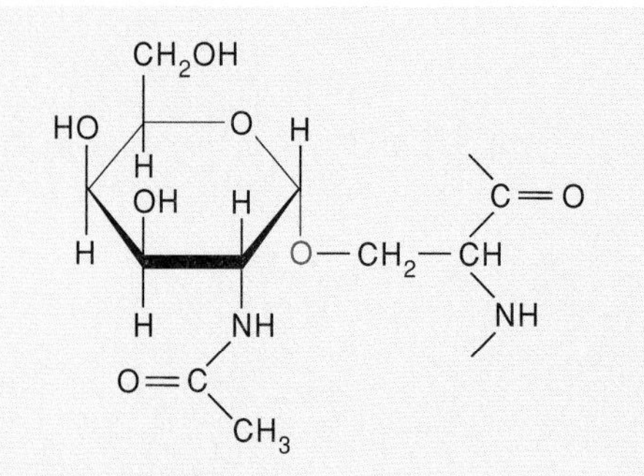

(a) The *N*-acetylgalactosamine-serine linkage

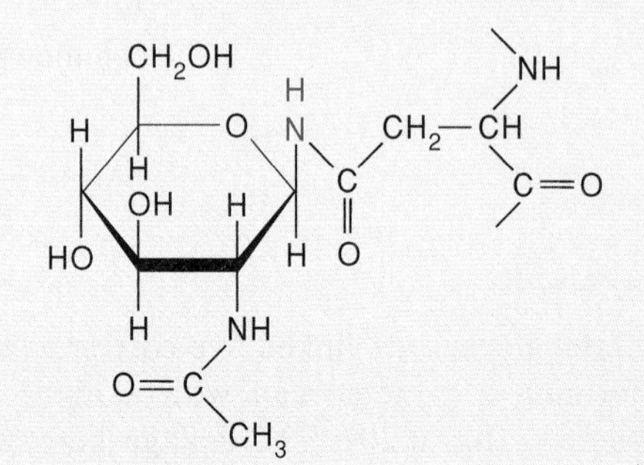

(b) The *N*-acetylglucosamine-asparagine linkage

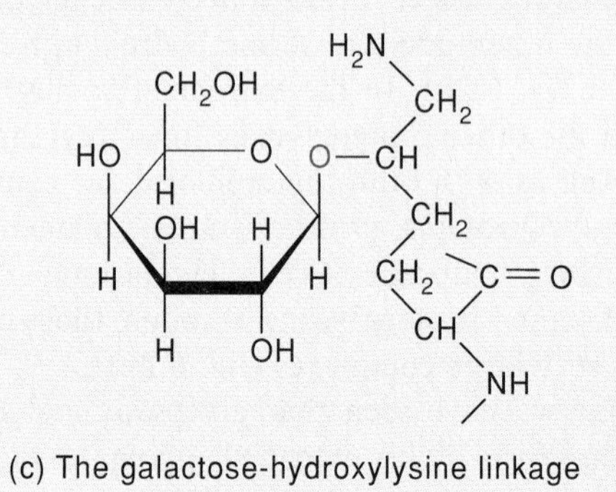

(c) The galactose-hydroxylysine linkage

Most secreted and membrane-bound glycoproteins have more complicated oligosaccharides containing between 4 and 30 sugar residues. A typical O-linked carbohydrate of a membrane glycoprotein is shown in figure 6.23. This structure can be further extended and branched by the addition of galactose, *N*-acetylgalactosamine, *N*-acetylglucosamine, fucose, or sialic acid in different linkages. Larger O-linked structures with branching occur on proteins with high Ser(Thr) content, forming mucins. These molecules are found in body fluids and carry the blood group determinants (see chapter 21).

Figure 6.23

Representative carbohydrate structures found on mammalian glycoproteins. (*a*) An O-glycosidically-linked carbohydrate from a red blood cell membrane glycoprotein; (*b*) an oligomannosyl N-linked carbohydrate; (*c*) a hybrid N-linked carbohydrate; and (*d*) a complex, lactosamine-containing carbohydrate. The latter three structures are found on many membrane and secreted glycoproteins of mammalian cells. Structure (*d*) may be much more complex, with the addition on the core mannose residues of extra branches of various lengths terminating with different sugars. It should be noted that all the N-linked carbohydrates (structures *b*, *c*, and *d*) have a common core of five sugars. (NeuNAc = sialic acid; Gal = galactose; GlcNAc = *N*-acetylglucosamine; Man = mannose; Fuc = fucose; Asn = asparagine; Ser = serine.)

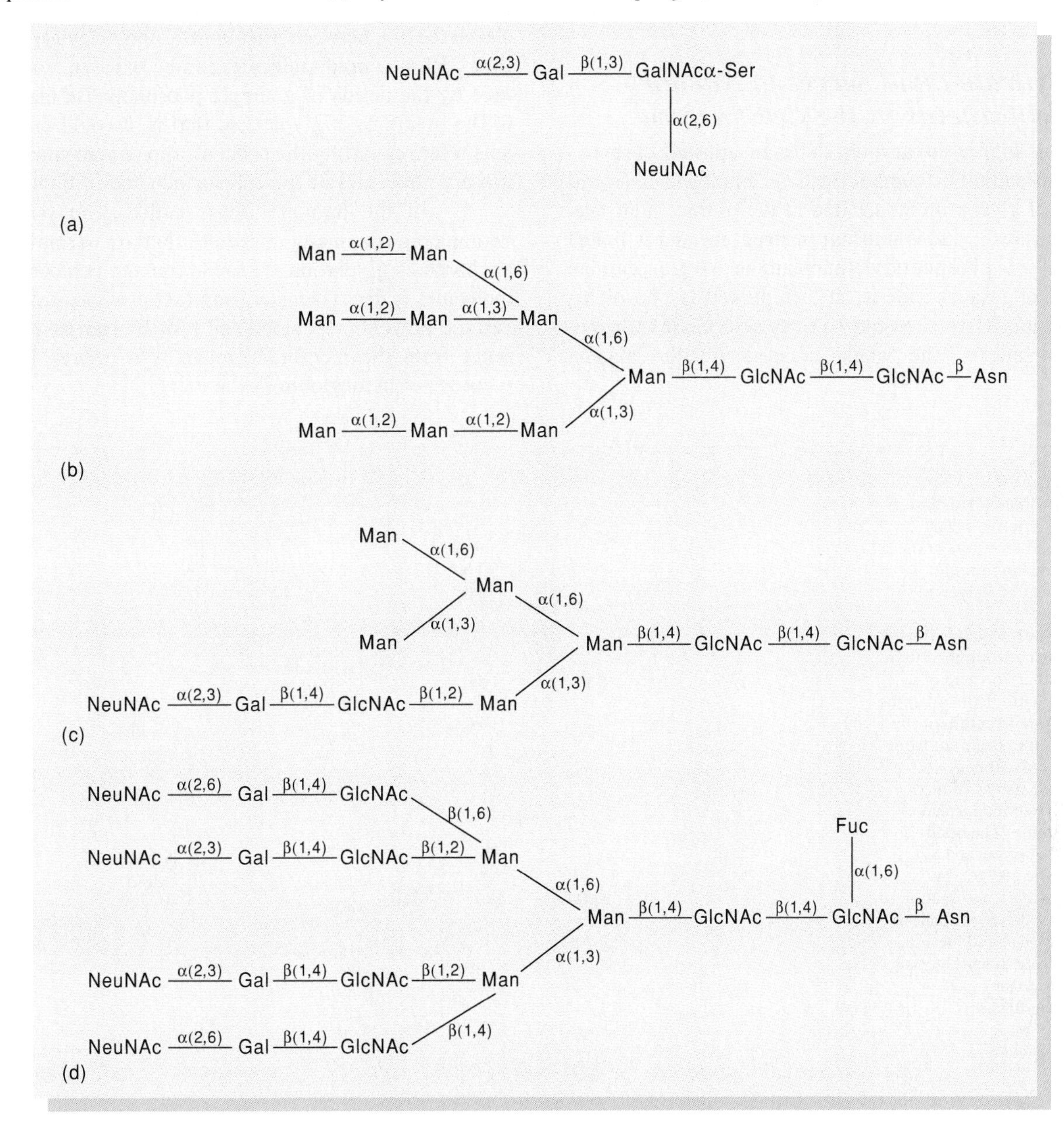

In contrast to the usual O-linked oligosaccharides of mammalian glycoproteins, N-linked carbohydrates always contain mannose and *N*-acetylglucosamine, and may also contain galactose, fucose, and sialic acid in combinations that vary as a result of differences in branching and linkage relationships. Three general types of structure form the basis of the variations in N-linked carbohydrates. These are termed oligomannosyl, hybrid, and lactosamine-containing (or complex) structures (see fig. 6.23). They all have a common core of five sugars attached to Asn because they are all synthesized by a common pathway. In fact, the wide variety of N-linked carbohydrates found in glycoproteins reflects intermediates of the biosynthetic pathway. The particular structures associated with a completed glycoprotein are a result of the conformation of that protein during biosynthesis, the availability of specific glycosyltransferase enzymes in the host cell, and the speed with which the glycoprotein travels through the secretory pathway along which glycosylation enzymes are located (see chapter 21).

Yeast glycoproteins carry only the oligomannosyl type of N-linked carbohydrates with mature structures that may contain hundreds of mannose residues. Thus far such structures have been found only in yeast. Yeast do not synthesize hybrid or lactosamine-containing structures, although the initial steps of N-linked carbohydrate biosynthesis appear to be the same in all eukaryotes (see chapter 21).

A Carbohydrate-Lipid Serves to Anchor Some Glycoproteins to the Cell Surface

Both yeast and higher eukaryotes share an unusual carbohydrate modification that is found at the C-terminal end of several diverse types of glycoproteins located at the extracellular surface. The oligosaccharide is unusual in structure and is linked to the protein via phosphatidylethanolamine. A glucosamine residue of the oligosaccharide is, in turn, linked to phosphatidylinositol, which is attached to two fatty acid chains (diacylglycerol) that anchor the whole molecule in the plasma membrane. This complex modification is termed a glycosyl-phosphatidylinositol (GPI) anchor, and proteins that carry such an anchor are said to be glypiated. The structure of a GPI anchor from a mammalian cell surface glycoprotein is shown in figure 6.24. The presence of this anchor in a glycoprotein may be ascertained by treatment with nitrous acid, which cleaves the glucosamine-*myo*-inositol linkage, or with phospholipases that cleave the link between *myo*-inositol and diacylglycerol. The fact that GPI-anchored molecules can be released from the cell surface by the action of a simple phospholipase may explain one of the functions of glypiation; that is, it could provide a mechanism for regulating the concentration of enzymes or other regulatory molecules at the cell surface and in tissue fluids.

In the human disease called paroxysmal noctornal hemoglobinuria, many molecules that are normally membrane-anchored by glypiation are found free in the blood. One of these molecules is decay-accelerating factor whose action at the cell surface prevents red blood cell lysis by complement. In its absence from the membrane much lysis occurs, leading to the presence of hemoglobin in the urine.

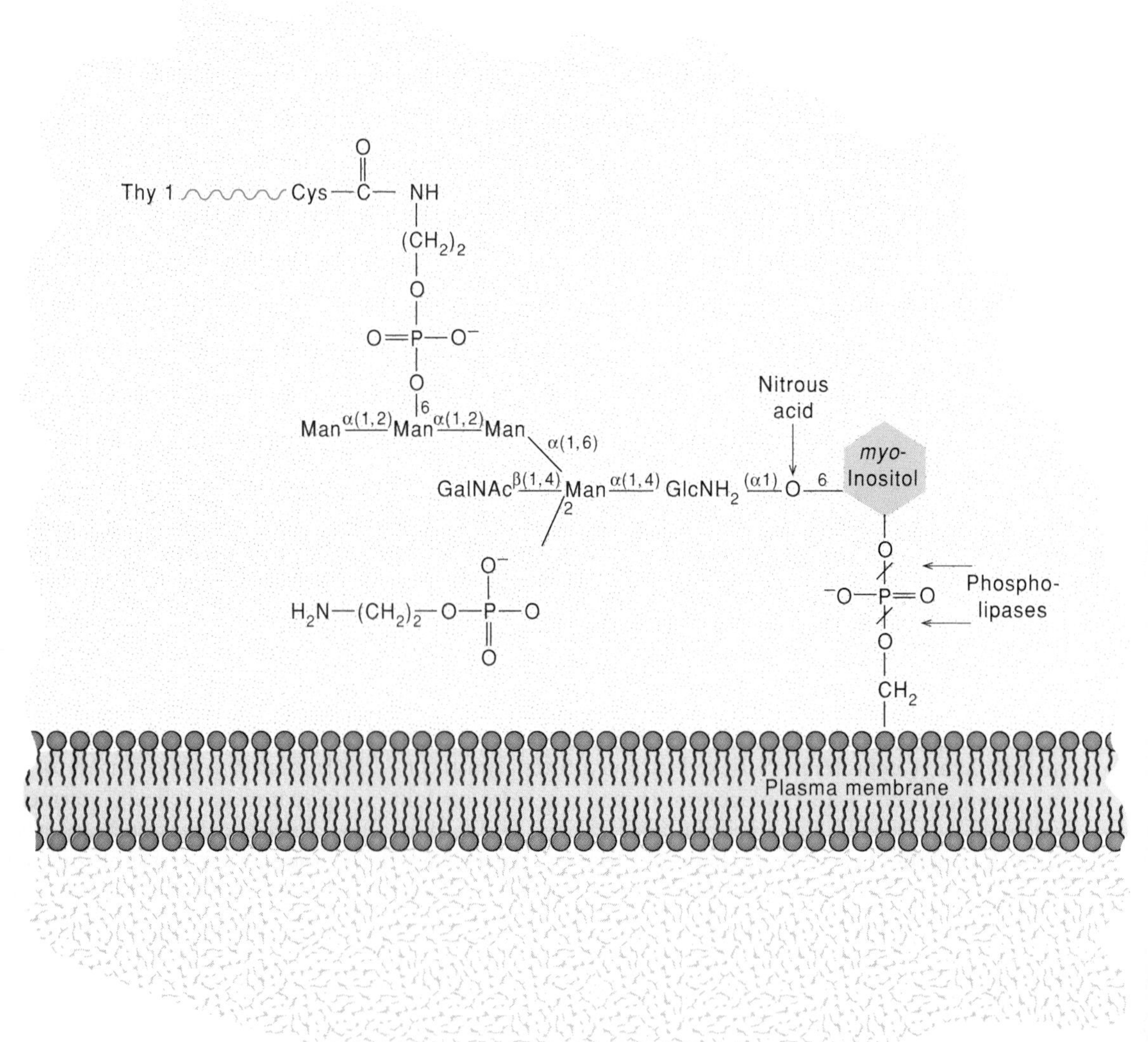

Figure 6.24

Glycophosphatidylinositol anchor of the membrane glycoprotein Thy 1. Many enzymes and receptors at the cell surface are anchored in the membrane via the diacylglycerol (DAG) portion of phosphatidylinositol, which is linked to a glycan that is in turn linked to the carboxyl-terminal amino acid of a protein via phosphatidylethanolamine. The oligosaccharide structure of the glycan region is quite different from that of O-linked or N-linked oligosaccharides. Most novel is the presence of glucosamine instead of *N*-acetylglucosamine. The bond between the glucosamine and *myo*-inositol can be cleaved by nitrous acid allowing identification of proteins that carry this modification. In addition, the glycan and protein can be released from diacylglycerol by cleavage with various phospholipase enzymes.

The GPI anchor serves to locate glycoproteins on the outer leaflet of the plasma membrane, where they are significantly more mobile than membrane proteins that span the bilayer. These glycoproteins are thus more accessible to other extracellular molecules and are more readily released by a phospholipase.

Carbohydrate Modification Is Important in Targeting Certain Enzymes to the Lysosomes

Soluble lysosomal hydrolases are targeted to lysosomes by a specific carbohydrate recognition marker that they acquire in the Golgi complex. Oligomannosyl carbohydrates on soluble lysosomal enzymes carry one or two phosphate residues at the 6-position of mannose (Man-6-P). These phosphorylated mannose residues are recognized by a glycoprotein called the Man-6-P receptor, which binds and transports lysosomal enzymes via several cellular compartments. In an acidic, prelysosomal compartment the binding between the Man-6-P receptor and the lysosomal enzyme is disrupted. The lysosomal enzyme continues in a vesicle destined to fuse to and thereby deliver its contents to the lysosome. Fibroblasts from patients with a lysosomal storage disease called I-cell disease cannot add the carbohydrate recognition marker and consequently their lysosomal hydrolases are largely secreted instead of being targeted to the lysosome. As a result, many molecules that are normally degraded by lysosomal hydrolases accumulate in the lysosomes. Morphologists have termed these dense lysosomes inclusion bodies, hence the name I-cell disease.

Thus far no other cases of protein targeting by carbohydrate modification are known, although their existence is a distinct possibility.

Carbohydrates of the Plasma Membrane

Carbohydrates appear prominently on the outside leaflet of the plasma membrane in the form of N- and O-linked glycoproteins, glycolipids, proteoglycans, and GPI-anchored proteins (fig. 6.25). In addition to being accessible for recognition by carbohydrate-binding proteins, cell-surface carbohydrates also appear to be important for cell shape and cell–cell interactions. Many infectious agents, such as bacteria, viruses, or parasites, recognize and bind to host cells via specific carbohydrate structures. For example, influenza virus binds to cells via specific types of sialic acids. Mutant mammalian cells with truncated carbohydrates are more rounded and aggregate with each other. Similar features are typical of red blood cells from patients with a rare blood disorder called HEMPAS (congenital dyserythropoetic anemia type II). In these patients, carbohydrates of red blood cell-surface glycoproteins are truncated, with the result that the red cells clump together. The removal of red cells from the circulation by this means is the cause of the anemia.

Figure 6.25

Diagram of a eukaryotic cell plasma membrane. Oligosaccharides are found on integral, transmembrane glycoproteins, glycolipids, glycophosphatidylinositol (GPI)-anchored glycoproteins, and glycoconjugates adsorbed at the cell surface. All carbohydrates on these molecules face the outside of the cell except O-linked *N*-acetylglucosamine, which may be found on the cytoplasmic portion of glycoproteins. Since many carbohydrates terminate in sialic acid, the external surface of the plasma membrane is negatively charged. The oligosaccharides of membrane glycoconjugates form a sugar coat, "glycocalyx," for the mammalian cell, which can be readily seen in the electron microscope by staining with ruthenium red or other sugar-binding dyes. Proteoglycans are found in the extracellular matrix and provide the support substance for cells in all tissues. Proteoglycan chains have also been found on certain membrane glycoproteins. Many membrane glycoproteins are receptors that bind other glycoproteins (e.g., growth factors), absorbing them to the cell surface.

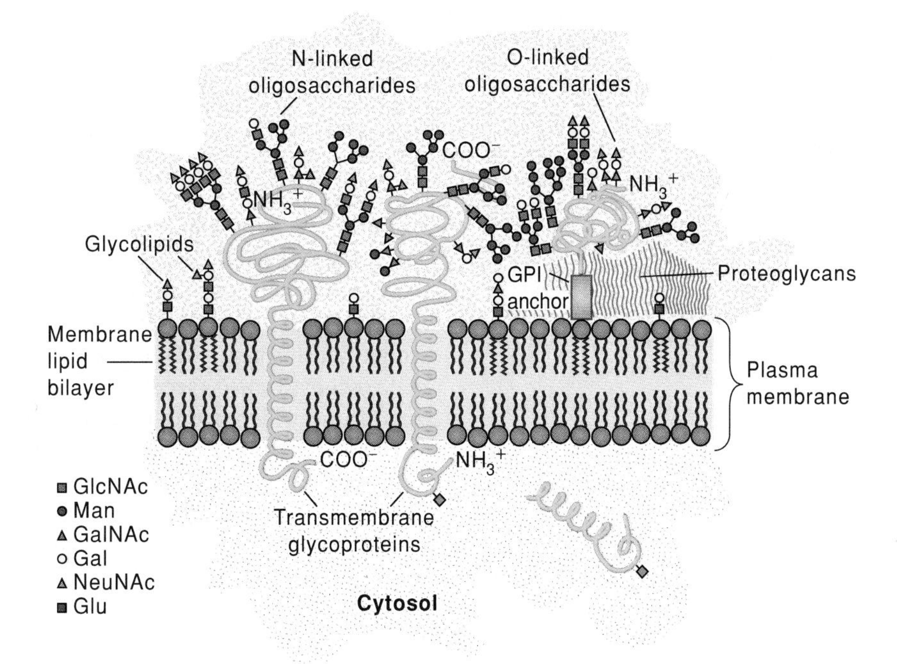

Figure 6.26

Separation of N-linked carbohydrates by lectin-affinity chromatography. Radiolabeled carbohydrates are released from a glycoprotein by an endoglycosidase (*N*-glycanase) that cleaves between the Asn of the protein and the first GlcNAc residue of the carbohydrate. They can then be separated into branched or biantennary (two branches) lactosamine-containing species or oligomannosyl species by affinity chromatography on concanavalin A-sepharose. Branched carbohydrates do not bind to the column (nor do O-linked oligosaccharides), biantennary carbohydrates bind and are eluted by 10-mM α-methylglucoside (α-MG), while oligomannosyl carbohydrates bind more strongly to the column and are eluted by increasing concentrations of α-methylmannoside (α-MM; 10 mM and 100 mM). Although several types of oligosaccharides are completely separated by this method, each species represented by a peak may include a mixture of related structures. For example, hybrid structures that may contain GlcNAc, Gal, and sialic acid attached to the $Man_5GlcNAc_2$ as shown are also eluted with α-MM(10).

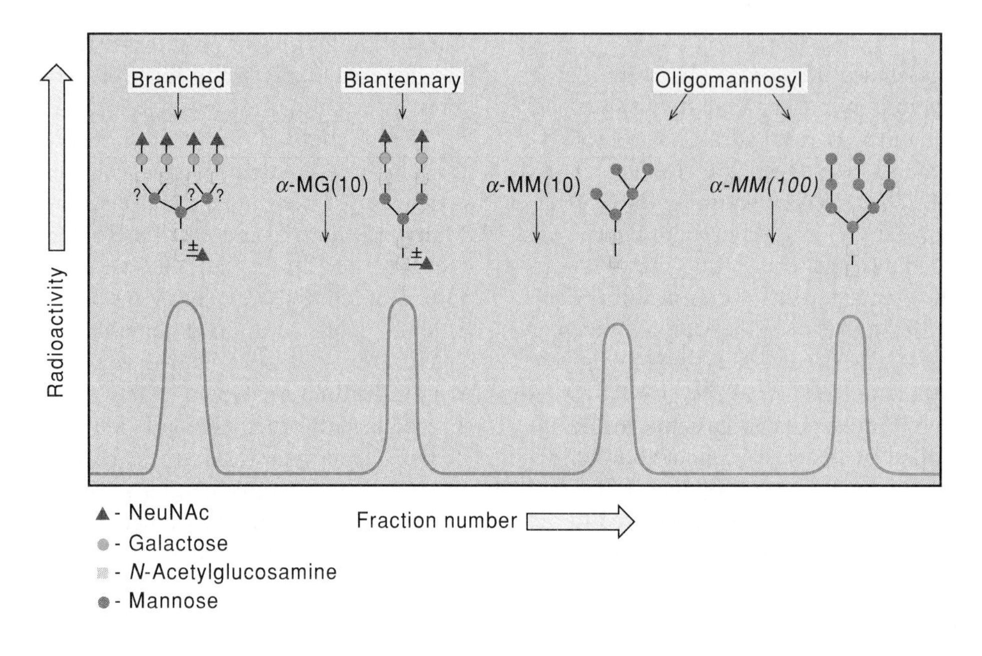

Structural Analysis of Carbohydrates

In order to correlate the structure of carbohydrates with their function, we must determine the precise arrangements of sugars in an oligosaccharide. For glycoproteins, often we must also determine the types of oligosaccharide structures at each glycosylation site. To do this, we first separate protease-generated fragments of the protein each containing one glycosylation site, then release the oligosaccharide from the peptide. Oligosaccharide moieties can be obtained free of protein by chemical or enzymic release from the protein backbone or by exhaustive proteolysis to give glycopeptides. Oligosaccharide mixtures can be separated on the basis of size or charge by standard chromatographic procedures and on the basis of structure by lectin-affinity chromatography (fig. 6.26). Lectins are proteins (or glycoproteins) that bind specific carbohydrate structures. For example, concanavalin A from the jack bean binds oligomannosyl N-linked carbohydrates but does not bind O-linked moieties or branched N-linked structures. Other lectins that are used in lectin-affinity chromatography include wheat germ agglutinin (binds sialic acid and *N*-acetylglucosamine), ricin (binds galactose), and lotus lectin (binds fucose).

We can determine the sequence of sugars in a pure oligosaccharide by sequential digestion with exoglycosidase enzymes (fig. 6.27). These are hydrolases that remove terminal, nonreducing sugars. They are highly specific for the particular sugar and its anomeric linkage (α or β). Some glycosidases are also specific for the type of linkage, cleaving a $\beta(1,2)$, for example, but not an $\alpha(1,4)$ linkage. Thus by sequential exoglycosidase digestion in combination with methods for detecting sugar removal (conventional or lectin-affinity chromatography), it is possible to determine the general structure of an oligosaccharide. Endoglycosidases are also most useful for carbohydrate structural analysis because they are able to recognize certain structural elements. Different endoglycosidases (D, F, or H) can be used to distinguish between complex, oligomannosyl, or hybrid structures as shown in figure 6.27*b*. The composition of an oligosaccharide can be determined after acid hydrolysis by gas chromatography or by separation of sugar oxyanions formed at high pH. The types of linkages between sugars can be determined by methylation analysis as described in figure 6.12. With combined information a complete structure can usually be deduced.

A much more rapid method of structural analysis is possible if $\geq 100\ \mu g$ of pure oligosaccharide is available. This is high field proton (^1H) nuclear magnetic resonance (NMR) spectroscopy. NMR spectroscopy is nondestructive and gives a spectrum in which one to three protons of each sugar—for example, the anomeric carbon (C-1) proton or the C-2 proton or methyl group protons—are resolved. The resolution improves dramatically with two-dimensional (2D) NMR, allowing all protons in small oligosaccharides to be resolved. An example of a partial NMR spectrum of an oligosaccharide is interpreted in box 6B.

Figure 6.27

Glycosidase enzymes used for structural analysis of oligosaccharides. (*a*) Exoglycosidase enzymes remove terminal, nonreducing sugars with specificity for the sugar and whether it is in α or ß linkage. Used sequentially and in conjunction with various separation techniques, exoglycosidases reveal the sequence of sugars in an oligosaccharide. Endoglycosidases may be used to remove the intact oligosaccharide from the protein backbone. (*b*) Different endoglycosidases vary in their specificity for cleaving N-linked structures. The enzyme that cleaves the GlcNAc-Asn bond is an amidase or peptide-*N*-glycosidase called PGNase F. It requires the glycosylated asparagine to be substituted on both sides (i.e., to be within a peptide) for its action and it leaves aspartic acid in the peptide after removal of the oligosaccharide. The enzyme that cleaves between the GlcNAc residue is an *N*-glycosidase called endoglycosidase F. It leaves one GlcNAc residue linked to the asparagine which need not be substituted for the carbohydrate to be a substrate.

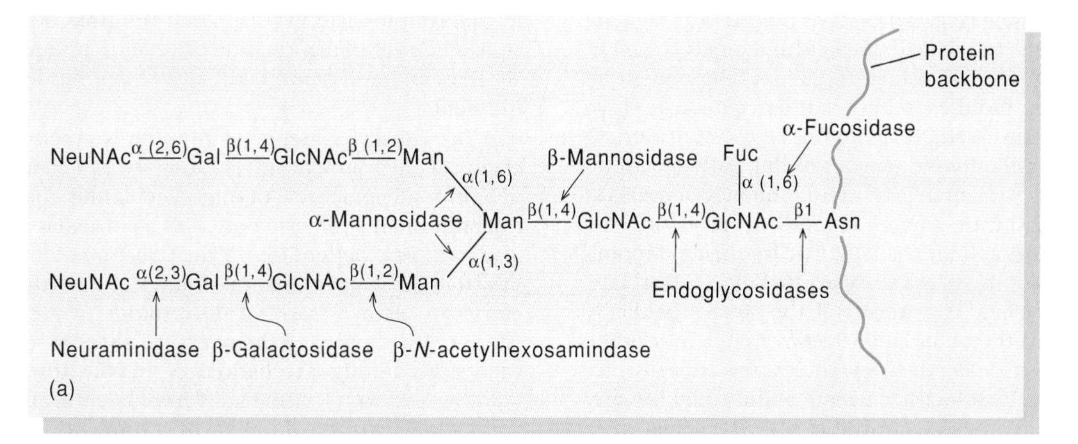

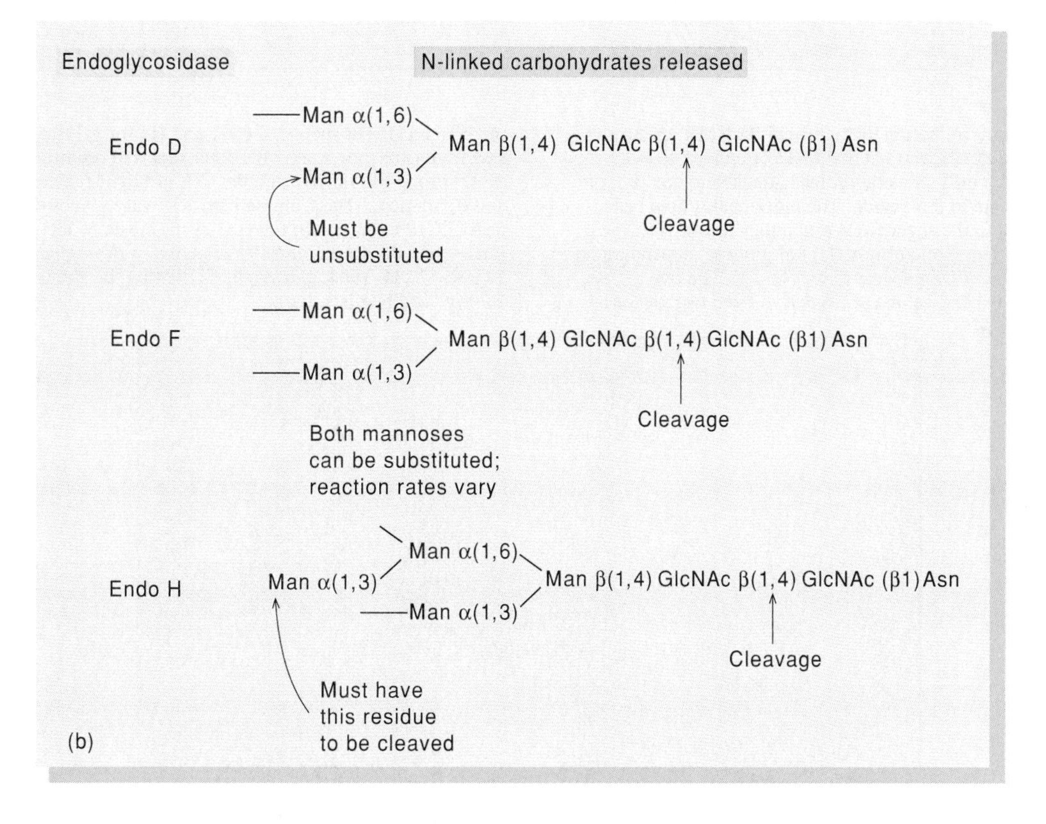

Methods for Structural Analysis: Nuclear Magnetic Resonance (NMR) Spectroscopy

Nuclear magnetic resonance (NMR) spectroscopy exploits the fact that when a spinning, paramagnetic, charged particle (e.g., a proton) is placed in a magnetic field, it aligns mainly with the field and precesses about the field with a frequency (the Lamar frequency) dependent on the particle properties and the strength of the magnetic field. To obtain an NMR spectrum, a sample of protons is placed in a strong magnetic field (generated by the magnet of the NMR spectrometer) and is irradiated with a range of radiofrequency energies at 90° to the main field. This treatment causes all the protons in the sample to absorb energy at their characteristic frequency, flipping their magnetic orientations 90° with respect to their original state. After the applied pulse field is switched off, the protons gradually relax to precess about the main field. Receiver coils in a probe surrounding the sample detect the frequencies of precessing protons as a set of oscillating electric currents, induced by the precessing magnetic vectors, which constitute the NMR signal. The magnitude of the induced voltage decays exponentially, giving rise to a free induction decay (FID). An FID is actually a mixture of sine waves arising from each of the chemical classes of protons in the sample. The FID is called the time domain of the NMR signal. Fourier transformation decodes the frequencies in the FID so they can be displayed in a plot of amplitude (amount) versus frequency.

The Lamar frequency of a proton is precisely dictated by its chemical environment and is expressed as a chemical shift in parts per million (ppm) Hz. For oligosaccharides, therefore, the chemical shift of an anomeric proton of a particular sugar (e.g., galactose) will vary depending on the structure of the oligosaccharide in which the sugar exists (see figure 1). Powerful NMR spectrometers can resolve many of the protons in a complex oligosaccharide. By identifying the chemical shifts of specific protons for each sugar in oligosaccharides of known structure (determined by other methods), data banks have been acquired which allow complete structures of unknown compounds to be deduced.

Figure 1

Oligosaccharide structural determination by high field ^{1}H-NMR spectroscopy. These traces are the partial spectra at 500 MHz of two related oligosaccharides from human milk. Several protons attached to one or more carbons of each sugar resonate in this region. The individual proton peaks are numbered to correspond to the sugar from which they are derived. Two peaks are obtained for the monomeric proton (H1) of glucose, depending on whether it derives from the α or ß anomeric form (1α, 1ß). The two resonances seen for fucose (5) and galactose (2) derive from two protons—the H1 and H5 for fucose, the H1 and H4 for galactose. Other regions of the spectrum (not shown) resolve other reported protons (e.g., protons of the *N*-acetyl group of GlcNAc and the CH_3 group of fucose). (Source: C. Campbell and P. Stanley, "The Chinese hamster ovary glycosylation mutants LEC11 and LEC12 express two novel GDP-fructose: *N*-acetylglucosaminide 3-α-L-fucosyltransferase enzymes" in *Journal of Biological Chemistry,* 259(18): 11208–11284, 1984. Copyright © 1984 by the American Society of Biological Chemists, Inc.)

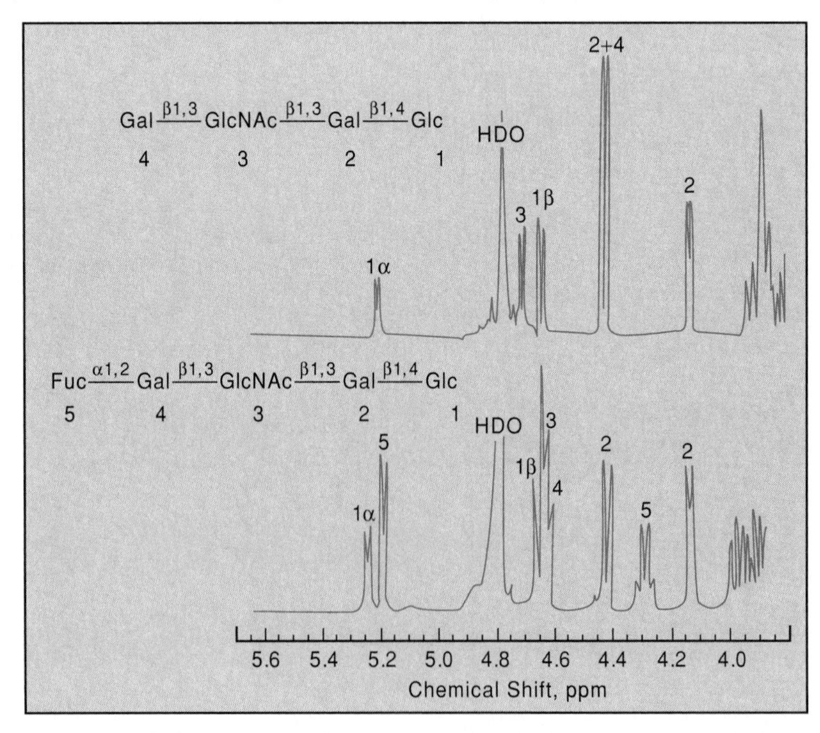

Figure 6.28

Diagram of a Gram negative cell envelope. The trimers of matrix protein of the outer membrane are associated with lipoprotein and with lipopolysaccharide (of variable polysaccharide length), and lipoprotein is covalently bound to peptidoglycan. Diagram also illustrates some general properties of membranes. Phospholipid molecules are illustrated with a circle for the polar groups and a line for each fatty acid acyl moiety. (Courtesy M. Inouye.)

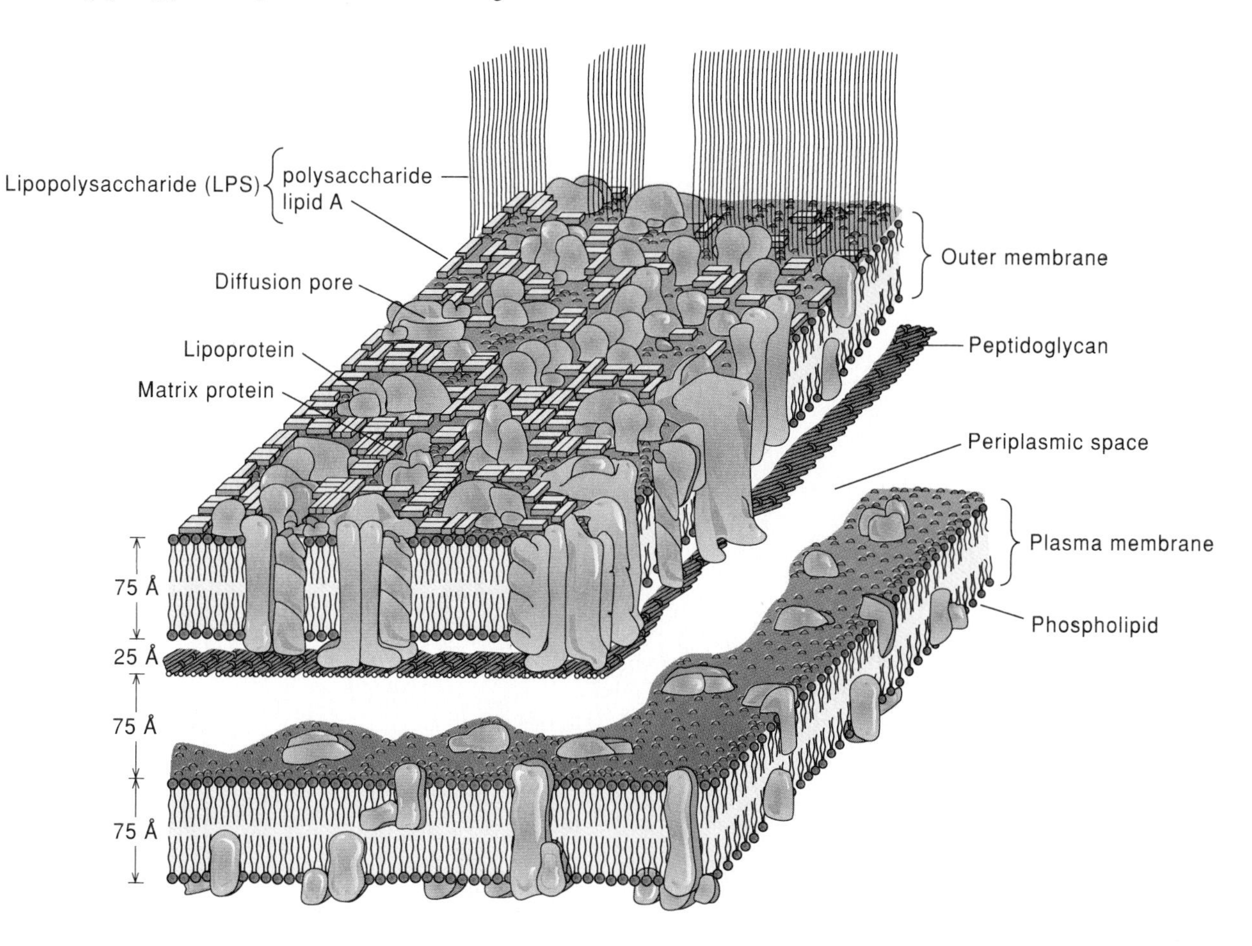

Peptidoglycans

As we noted in chapter 1, a unique feature of bacteria is the cell wall that surrounds the plasma membrane and provides the mechanical strength that enables bacteria to resist shear and osmotic shock. The cell wall is composed of a network of linear heteropolysaccharides cross-linked by peptides. A structure of this sort is called a peptidoglycan. Some bacterial cells (Gram negative cells) also possess an outer membrane composed of lipids, proteins, and polysaccharides. The main structural features of a Gram negative bacterial cell envelope, which is a composite of the two membranes and the cell wall, are illustrated in figure 6.28.

Here we will discuss the structural aspects of the cell wall, leaving a description of its biosynthesis to chapter 21. The peptidoglycan that constitutes the cell wall is a polymeric structure consisting of a heteropolysaccharide composed of amino sugars in one dimension, cross-linked through branched polypeptides in the other (fig. 6.29). The amino sugars alternate in the polymer, forming the glycan strands (see fig. 6.29). The carboxyl group of the lactic acid moiety of the acetylmuramic acid is substituted by a tetrapeptide, which in the Gram positive bacterium *Staphylococcus aureus* has the sequence L-alanyl-D-γ-glutamyl-L-lysyl-D-alanine. In this tetrapeptide the glutamyl residue is attached through its γ-COOH rather than its α-COOH. All the muramic acids are substituted in this way to form peptidoglycan strands. Variations of the same basic structure occur in all bacterial species. The peptidoglycan strands are further linked to each other by means of an interpeptide bridge. In *S. aureus,* this bridge is a pentaglycine chain that extends from the terminal carboxyl group of the D-alanine residue of one tetrapeptide to the ϵ-NH_2 group of the third amino acid, L-lysine, in another tetrapeptide (see fig. 6.29). The third dimension is probably built up by bridges extending in different planes. This gigantic macromolecule has the mechanical stability required for the cell wall.

Figure 6.29

Structure of the peptidoglycan of the cell wall of *Staphylococcus aureus*. (*a*) In this representation, X (*N*-acetylglucosamine) and Y (*N*-acetylmuramic acid) are the two sugars in the peptidoglycan. Light green circles represent the four amino acids of the tetrapeptide L-alanyl-D-glutamyl-L-lysyl-D-alanine. Dark green circles are pentaglycine bridges that interconnect peptidoglycan strands. The nascent peptidoglycan units bearing open pentaglycine chains are shown at the left of each strand. TA—P is the teichoic acid antigen of the organism, which is attached to the polysaccharide through a phosphodiester linkage. Teichoic acids are discussed in chapter 7 (see fig. 7.22). (*b*) The structure of X (*N*-acetylglucosamine) and Y (*N*-acetylmuramic acid), are connected by ß(1,4) linkages that alternate in the glycan strand. (*c*) The structure of a segment of the peptidoglycan before and after the final cross-linking reaction.

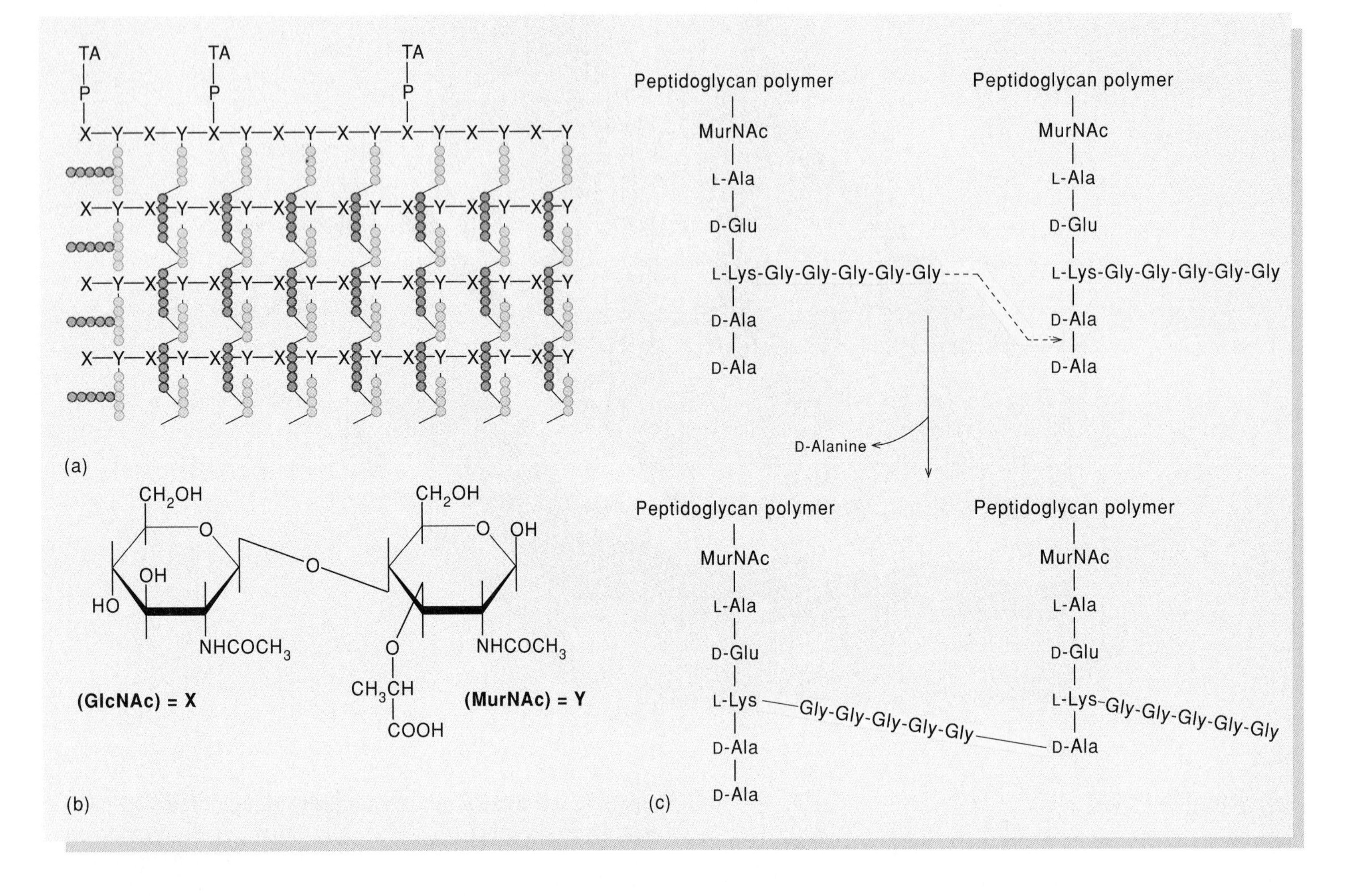

Summary

Carbohydrates are the most abundant organic substances. They are an important source of carbon compounds for all biomolecules and they also serve important energy and structural needs. In this chapter we have focused on the following points.

1. Carbohydrates may be divided into monosaccharides, oligosaccharides, and polysaccharides.
2. The monosaccharides are either polyhydroxyaldehydes or polyhydroxyketones. All monosaccharides are optically active because of the presence of one or more asymmetrical carbon atoms (chiral centers).
3. Straight-chain sugars in aqueous solution tend to form ring structures known as intramolecular hemiacetals, especially when the resultant is a five-membered (furanose) ring or a six-membered (pyranose) ring. Depending on which way the ring forms about the anomeric carbon, the structure is called an α or β hemiacetal.
4. For a given hemiacetal several conformations are possible. Usually the chair form is favored over the boat form because of the lower steric repulsion produced by side chains in the chair form. Polymer synthesis locks in a particular hemiacetal configuration.

5. Monosaccharides are linked by glycosidic bonds to form oligosaccharides and polysaccharides.
6. Polysaccharides function in two quite distinct roles: Some serve as a means for storage of chemical energy and others serve a structural function.
7. Polymers that use one type of building block (monomer) are called homopolymers, and those that use more than one are called heteropolymers.
8. Cellulose is the best known and most abundant structural polysaccharide. It is a homopolymer of glucose with $\beta(1,4)$ linkages between adjacent monomeric residues.
9. Starch and glycogen are energy-storage polysaccharides. They also are homopolymers of glucose but with $\alpha(1,4)$ linkages between adjacent residues. In addition, they contain branches with $\alpha(1,6)$ linkages.
10. Many other polysaccharides use different sugars as well as different combinations of sugars or modified sugars. Most of these function in a structural capacity.
11. Many heteropolysaccharides are linked to peptides (peptidoglycans) or proteins (proteoglycans or glycoproteins). Glycoproteins are mostly protein, with short, highly branched carbohydrate chains. By contrast, in proteoglycans the protein is the minor component.
12. Carbohydrates are also found linked to lipids in certain membrane structures.
13. Recent advances make it possible to determine structure/function relationships of oligosaccharides from a particular glycosylation site in a glycoconjugate.
14. The main component in bacterial cell walls is a peptidoglycan. The peptidoglycan is a polymeric structure made of a glycosaminoglycan extending in one dimension, cross-linked through oligopeptides extending in the other dimension. This cross-linked network completely surrounds the bacterial cell and provides it with the mechanical strength needed to resist osmotic shock and other mechanical stresses.

Selected Readings

Albersheim, P., and A. G. Darvill, Oligosaccharins. *Sci. Am.* 253(3):58–64, 1985.

Ferguson, M. A. J., and A. F. Williams, Cell surface anchoring of proteins via glycosylphosphatidylinositol structures. *Ann. Rev. Biochem.* 57:285–320, 1988.

Fransson, L.-A., Structure and function of cell-associated proteoglycans. *Trends Biochem. Sci.* 12:406–411, 1987. A recent account on a fast-moving subject.

Ginsberg, V., and P. Robbins, eds., *Biology of Carbohydrates.* New York: Wiley, 1984.

Hassell, J. R., J. H. Kimina, and L. Cantly, Proteoglycan core protein families. *Ann. Rev. Biochem.* 55:539–568, 1986.

Homans, S. W., M. A. J. Ferguson, R. A. Dwek, T. W. Rademacher, R. Anand, and A. F. Williams, Complete structure of the glycosylphosphatidylinositol membrane anchor of rat brain Thy-1 glycoprotein. *Nature* 333:269–272, 1988.

Lennarz, W. J., ed., *The Biochemistry of Glycoproteins and Proteoglycans.* New York, London: Plenum, 1980.

Lis, H. and N. Sharon, Lectins as molecules and as tools. *Ann. Rev. Biochem.* 55:35–67, 1986.

Maley, F., R. B. Trimble, A. L. Tarentino, and T. H. Plummer, Jr., Characterization of glycoproteins and their associated oligosaccharides through the use of endoglycosidases. *Anal. Biochem.* 180:195–204, 1989.

McNeil, M., A. G. Darvill, S. C. Fry, and P. Albersheim, Structure and function of the primary cell walls of plants. *Ann. Rev. Biochem.* 53:625–664, 1984.

Quiocho, F. A., Carbohydrate-binding proteins: tertiary structure and protein–sugar interactions. *Ann. Rev. Biochem.* 55:287–316, 1986.

Rademacher, T. W., R. B. Parekh, and R. A. Dwek, Glycobiology. *Ann. Rev. Biochem.* 57:785–838, 1988.

Ruoslahti, E., Structure and biology of proteoglycans. *Ann. Rev. Cell. Biol.* 4:229–255, 1988.

Sweeley, C. C., and H. Nuñez, Structural analysis of glycoconjugates by mass spectrometry and nuclear magnetic resonance spectroscopy. *Ann. Rev. Biochem.* 54:765–801, 1985.

Vliegenthart, J. F. G., L. Dorland, and H. van Halbeek, High-resolution, ^{1}H-nuclear magnetic resonance spectroscopy as a tool in the structural analysis of carbohydrates related to glycoproteins. *Adv. Carbohydr. Chem. Biochem.* 41:209–374, 1983.

von Figura, K., and A. Hasilik, Lysosomal enzymes and their receptors. *Ann. Rev. Biochem.* 55:167–193, 1986.

Problems

1. Compare the Haworth projections of D-glucose, D-mannose, and D-galactose. Indicate the differences between the structures.
2. Indicate the chiral carbons in D-glucose and determine the number of possible stereoisomers.
3. Is it possible that carbohydrates could have produced the diversity required to catalyze the myriad cellular reactions now relegated primarily to proteins?
4. In solution, D-glucose has a specific rotation of $[\alpha]^{20}$D = +52.7°. The specific rotation of pure β-D-glucose is +18.7° and that of pure α-D-glucose is +112.2°. Calculate the fraction of the α and β anomers in solution.
5. If either pure crystalline α- or β-D-glucose is dissolved in water, the final solution will contain the same fraction of α and β anomers as determined in problem 4. Explain chemically how this mutarotation occurs.
6. You are given two containers of polysaccharide by a colleague who has labeled one container as cellulose and one as glycogen. In his haste he may have mislabeled them and has asked you to verify which is which by methylation analysis. Indicate what product(s) you would expect from exhaustive methylation and mild acid hydrolysis of each polysaccharide and how the products could be used to differentiate the samples.

7. (a) Glycogen, starch, and cellulose are polymers of glucose. Suggest reasons, based on structure, that the physical form of each is appropriate to its role in nature. Why are the polymer forms of starch and glycogen much more desirable than an equivalent amount of free glucose in the cell?
   (b) Suggest how a cell might selectively synthesize starch but not cellulose.
8. Consider the packing of lipid triglycerides in adipocytes and of glycogen granules in the liver. Comment on the feasibility of using only glycogen, rather than lipid, as sole energy reserve. (See chapter 7, "Some fatty acids are stored as an energy reserve in triglycerides.")
9. Given the trisaccharide D-mannose-β(1,3)D-glucose-α(1,6)D-galactose, draw the structure of the trisaccharide using Haworth projections. Name and draw the structures of the products of exhaustive methylation with dimethyl sulfate methylation and mild acid hydrolysis of the trisaccharide.
10. Chemical degradation of glycosaminoglycans causes reduced viscosity of the synovial fluid and subsequent damage to joints. Explain.
11. A tetrasaccharide has the following composition: D-Man(2), D-Gal(1), D-Glu(1). The tetrasaccharide gave a positive reducing sugar test in which the glucose residue was oxidized. Exhaustive methylation and mild acid hydrolysis released 2,3,4,6-tetra-*O*-methylmannose, 2,3-di-*O*-methylgalactose, and 2,3,6-tri-*O*-methylglucose. Treatment of the tetrasaccharide with an α-mannosidase released mannose and a trisaccharide that yielded the following methylation products: 2,3,4,6-tetra-*O*-methylmannose, 2,3,6-tri-*O*-methylgalactose, and 2,3,6-tri-*O*-methylglucose. Deduce the sequence and specificity of anomeric linkages and indicate any ambiguity.
12. Which enzymes can be used to ascertain the presence of sialic acid, galactose, *N*-acetylglucosamine, and mannose in a complex carbohydrate?
13. How might a cell synthesizing an N-linked glycoprotein specifically glycosylate only two of ten asparagine residues in the protein?
14. What structural features of oligosaccharides complicate the determination of their sequence as compared with the sequence determination of proteins?

Lipids and Membranes

Lipids are biological molecules that are soluble in organic solvents. They have four major biological functions: (1) in all cells, the major structural elements of membranes are composed of lipids; (2) certain lipids, the triacylglycerols, serve as efficient reserves for the storage of energy; (3) many of the vitamins and hormones found in animals are lipids or derivatives of lipids; and (4) the bile acids help to solubilize the other lipid classes during digestion.

In this chapter, we will focus on the structures and functions of the major membrane lipids. From the structures of these compounds, we will see how they form the structural scaffolding of biological membranes and how proteins embedded in, or associated with, these membranes are arranged so as to carry out specific and essential functions in all cells. Lipid biosynthesis, membrane biogenesis, and the structures and roles of other classes of lipids are considered in part V (chapters 22 and 23), while one major function of biological membranes, transmembrane transport, is discussed in more detail in part VII (chapter 32). Nothing in those chapters depends on the intervening material, so if you prefer you can proceed to them directly after reading this chapter.

Fatty Acids

Compounds with the structural formula $CH_3(CH_2)_nCOOH$ that contain no carbon–carbon double bonds are known as saturated fatty acids. The two most abundant saturated fatty acids are palmitic and stearic acids (table 7.1). Some other saturated fatty acids present in smaller quantities in mammalian tissues are also shown in table 7.1. The sphingolipids, which we will consider later, contain longer-chain fatty acids (n = 20–24), as well as palmitic and stearic acids. Some tissues also contain short-chain fatty acids, such as decanoic acid (n = 10), found in milk.

Fatty acids with double bonds in the aliphatic chain are called unsaturated fatty acids. Monounsaturated fatty acids have one double bond, while polyunsaturated fatty acids contain more than one double bond. The double bonds in naturally occurring fatty acids are *cis*. The double bonds in polyunsaturated fatty acids are always separated by one methylene group.

Mammalian tissues contain all of the unsaturated fatty acids listed in table 7.1 with the exception of vaccenic acid, which is present in *E. coli* and other bacteria. However, *E. coli* and most other bacteria do not contain polyunsaturated fatty acids. While oleic acid is the most common unsaturated fatty acid in mammals, two other unsaturated fatty acids, linoleic and linolenic acids, are not synthesized by mammals and are therefore

Table 7.1
Fatty Acids

Common Name	Systematic Name	Structure	Abbreviation[a]
Saturated Fatty Acids			
Myristic acid	*n*-Tetradecanoic acid	$CH_3(CH_2)_{12}COOH$	14:0
Palmitic acid	*n*-Hexadecanoic acid	$CH_3(CH_2)_{12}CH_2CH_2COOH$	16:0
Stearic acid	*n*-Octadecanoic acid	$CH_3(CH_2)_{12}CH_2CH_2CH_2CH_2COOH$	18:0
Arachidic acid	*n*-Eicosanoic acid	$CH_3(CH_2)_{12}CH_2CH_2CH_2CH_2CH_2CH_2COOH$	20:0
Behenic acid	*n*-Docosanoic acid	$CH_3(CH_2)_{12}CH_2CH_2CH_2CH_2CH_2CH_2CH_2CH_2COOH$	22:0
Lignoceric acid	*n*-Tetracosanoic acid	$CH_3(CH_2)_{12}CH_2CH_2CH_2CH_2CH_2CH_2CH_2CH_2CH_2CH_2COOH$	24:0
Cerotic acid	*n*-Hexacosanoic acid	$CH_3(CH_2)_{12}CH_2CH_2CH_2CH_2CH_2CH_2CH_2CH_2CH_2CH_2CH_2CH_2COOH$	26:0
Unsaturated Fatty Acids			
Palmitoleic acid	*cis*-9-Hexadecenoic acid	$CH_3(CH_2)_5\overset{H}{C}=\overset{H}{C}(CH_2)_7COOH$	$16:1^{\Delta 9}$
Oleic acid	*cis*-9-Octadecenoic acid	$CH_3(CH_2)_7\overset{H}{C}=\overset{H}{C}(CH_2)_7COOH$	$18:1^{\Delta 9}$
Vaccenic acid	*cis*-11-Octadecenoic acid	$CH_3(CH_2)_5\overset{H}{C}=\overset{H}{C}(CH_2)_9COOH$	$18:1^{\Delta 11}$
Linoleic acid	*cis,cis*-9,12-Octadecadienoic acid	$CH_3(CH_2)_4\overset{H}{C}=\overset{H}{C}-CH_2-\overset{H}{C}=\overset{H}{C}(CH_2)_7COOH$	$18:2^{\Delta 9,12}$
α-Linolenic acid	All-*cis*-9,12,15-Octadecatrienoic acid	$CH_3CH_2\overset{H}{C}=\overset{H}{C}-CH_2-\overset{H}{C}=\overset{H}{C}-CH_2-\overset{H}{C}=\overset{H}{C}(CH_2)_7COOH$	$18:3^{\Delta 9,12,15}$
Arachidonic acid	All-*cis*-5,8,11,14-Eicosatetraenoic acid	$CH_3(CH_2)_3-\left(CH_2-\overset{H}{C}=\overset{H}{C}\right)_4-(CH_2)_3COOH$	$20:4^{\Delta 5,8,11,14}$
	All-*cis*-4,7,10,13,16,19-Docosahexaenoic acid	$CH_3\left(CH_2\overset{H}{C}=\overset{H}{C}\right)_6-(CH_2)_2COOH$	$22:6^{\Delta 4,7,10,13,16,19}$
Some Unusual Fatty Acids			
	2,4,6,8-Tetramethyl decanoic acid	$CH_3CH_2\left(\overset{CH_3}{CH}-CH_2\right)_3-\overset{CH_3}{CH}-COOH$	
Lactobacillic acid		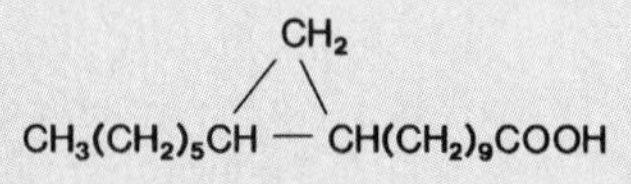 $CH_3(CH_2)_5\overbrace{CH-CH}^{CH_2}(CH_2)_9COOH$	
An α-mycolic acid		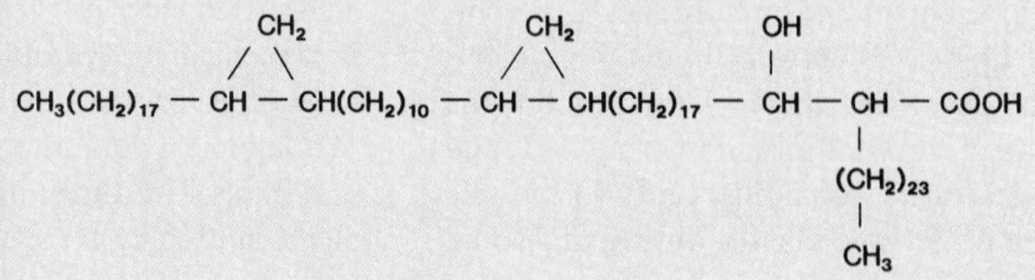 $CH_3(CH_2)_{17}-\overbrace{CH-CH}^{CH_2}(CH_2)_{10}-\overbrace{CH-CH}^{CH_2}(CH_2)_{17}-\overset{OH}{CH}-\underset{\underset{CH_3}{(CH_2)_{23}}}{CH}-COOH$	

[a] In these abbreviations the number to the left of the colon is the number of carbon atoms, and the number to the right is the number of double bonds. Δ9 signifies that there are 8 carbons between carboxyl group and double bond.

Figure 7.1

Space-filling and conformational models of (*a*) stearic and (*b*) linolenic acids. Each of these fatty acids has 18 carbon atoms, but the three double bonds in linolenic acid create a more rigid, curved molecule that interferes with tight packing in membrane structures.

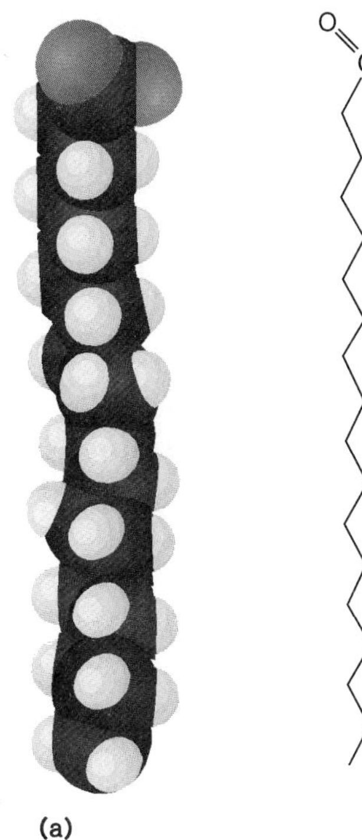

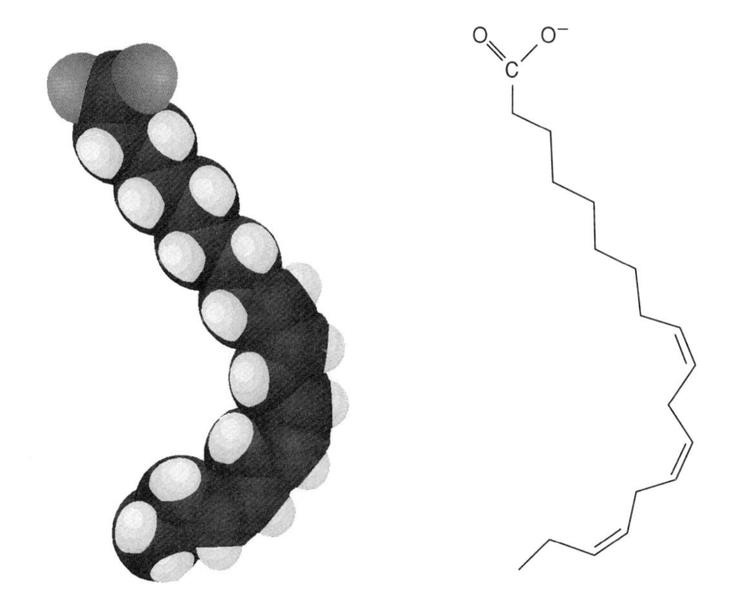

important dietary requirements. Like vitamins, these two fatty acids are required for growth and good health, and hence are called essential fatty acids. Plants are able to synthesize linoleic and linolenic acids and are the source of these fatty acids in our diet.

In addition to the commonly occurring fatty acids, many structural variations have evolved. There are over 100 other fatty acids found in various organisms, often associated with specialized functions. For instance, branched-chain fatty acids are found in many different tissues. The uropygial gland of the duck produces such a fatty acid (2,4,6,8-tetramethyldecanoic acid). The duck uses the fatty acids secreted by this gland to preen its feathers and thereby ensure that water continues to "run off its back." Other examples are fatty acids with a cyclopropane ring in the alkyl chain, found in many bacteria. The bacterium that causes tuberculosis, *Mycobacterium tuberculosis,* produces a family of complex fatty acids known as mycolic acids, which contain cyclopropane rings. One class of these is the α-mycolic acids (an example is given in table 7.1), and many structurally related α-mycolic acids are found in the mycobacteria and other related organisms (nocardiae and corynebacteria). These compounds appear to have a structural function in the outer part of the bacterial cell wall. There is much evidence to suggest that a major drug used in the treatment of tuberculosis, Isoniazid, functions by inhibiting an early reaction of α-mycolic acid biosynthesis.

Fatty acids are usually found as components of complex lipids, and only rarely as unesterified (free) fatty acids. Nevertheless, the pK_a for dissociation of the acid proton is around 4.7. Therefore, at pH 7.0, the fatty acid exists primarily in the dissociated form ($RCOO^-$):

$$CH_3(CH_2)_nCOOH \rightleftharpoons CH_3(CH_2)_nCOO^- + H^+$$

Because it exists as an anion at neutral pH, a fatty acid is not easily extracted from an aqueous medium by organic solvents such as hexane. However, if the pH is lowered by the addition of HCl or another strong acid, the fatty acid becomes protonated and is easily extracted by organic solvents.

Another property of fatty acids that you should note is the variation in their physical form at room temperature. If n equals 8 or less, the fatty acid is a liquid, whereas if n equals 10 or more, the fatty acid is a solid. If a fatty acid has a double bond, it has a lower melting point than the saturated fatty acid with the same number of carbons. *Cis*-unsaturated fatty acids are more condensed in length than the corresponding saturated fatty acids, and also contain one or more inflexible kinks (fig. 7.1). These properties explain the lowered melting temperatures of unsaturated fatty acids as well as the fact that they pack less tightly within membranes than saturated fatty acids, a point to which we will return later.

Table 7.2
Neutral Glycerides[a]

Common Name	Systematic Name	Structure
Triglyceride	1,2,3-Triacyl-*sn*-glycerol	O ‖ O CH_2OCR ‖ | R′ — COCH O | ‖ CH_2OCR'
Diglyceride	1,2-Diacyl-*sn*-glycerol	O ‖ O CH_2OCR ‖ | R′ — COCH | CH_2OH
Monoglyceride	1-Monoacyl-*sn*-glycerol	O ‖ CH_2OCR | HOCH | CH_2OH

[a]Because the substituents esterified to the first and third carbons of these glycerol derivatives are usually different, the second carbon atom is asymmetric. In naming and numbering these compounds, a special convention has been adopted: The prefix *sn*- (for *s*tereospecifically *n*umbered) immediately precedes "glycerol" and differentiates the naming of the compound from other approaches, such as the R S system described in chapter 11. The glycerol derivative is drawn in a Fischer projection with the secondary hydroxyl to the left of the central carbon, and the carbons are numbered 1, 2, and 3 from the top to the bottom. The prefix *rac*- (for *rac*emo) precedes the name if the compound is an equal mixture of antipodes. If the configuration is unknown or not specified, *x*- precedes the name.

Fatty acids are commonly analyzed by gas chromatography of the methyl esters. These esters are formed by esterification of the fatty acids with methanol (R represents hydrogen (H) or any group to which the fatty acid is esterified):

$$CH_3(CH_2)_nC(=O){-}OR + CH_3OH \xrightarrow[\text{HCl or } BF_3]{} CH_3(CH_2)_nC(=O)OCH_3 + ROH$$

Some Fatty Acids Are Stored as an Energy Reserve in Triacylglycerols

Fatty acids serve two major roles: they are major components of the triacylglycerols (table 7.2 and fig. 7.2) and of most of the complex lipids present in membranes. Triacylglycerols are the major uncharged glycerol derivatives found in animals and they are stored as an energy reserve. Monoacylglycerols and diacylglycerols are metabolites of triacylglycerols and of phospholipids (see chapters 17 and 22), and are normally present in cells in very small quantities.

Although triacylglycerols are found in the liver and intestine, they are primarily found in adipose tissue (fat), which functions as a storage depot for this lipid. The specialized cell in adipose tissue is called the adipocyte. Its cytoplasm is full of lipid vacuoles that are almost exclusively triacylglycerols (fig. 7.3) and that serve as an energy reserve for mammals. At times when the diet or glycogen reserves are insufficient to supply the body's need for energy, the fuel stored as fatty acyl components of the triacylglycerols is mobilized and transported to other tissues in the body. A second important function of adipose tissue is insulation of the body from cold. This function is most obvious in such cold-water mammals as the arctic (Beluga) whales, which have vast stores of fat (blubber).

Figure 7.2

Space-filling and conformational models of a triacylglycerol. Note the uncharged nature of this molecule, which serves as a means of storing fatty acids for future energy needs.

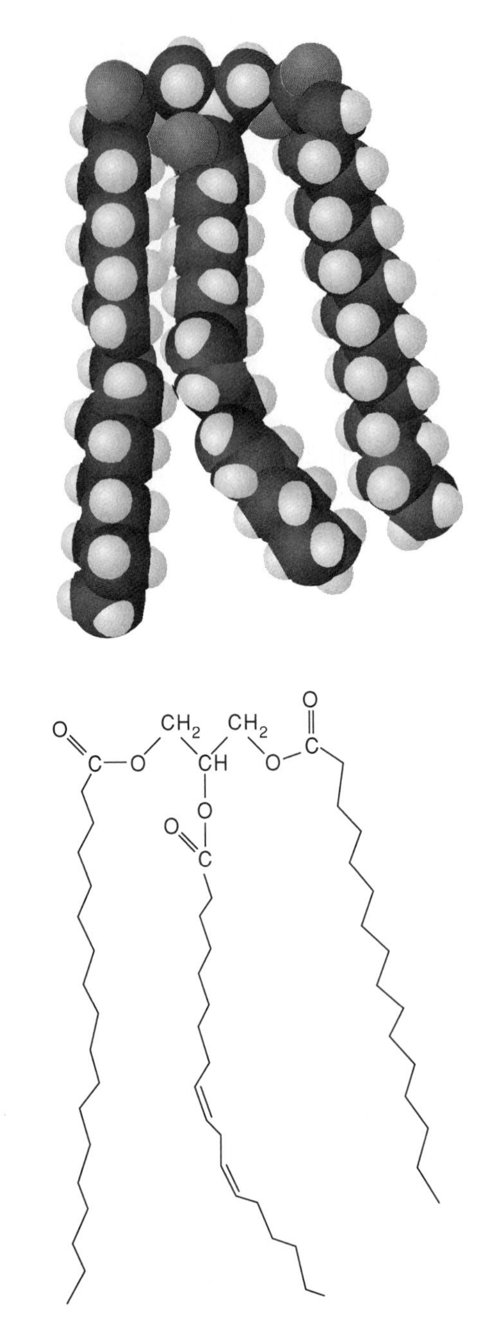

Triacylglycerols are structurally related to lipids that are found in membranes. They differ in one major respect; they are neutral, whereas most lipids found in membranes are charged at one end. This lack of charge has a profound effect on the role they can play; it suits them for compact storage in vesicles but makes them totally unsuitable for structural components in membranes.

Figure 7.3

Scanning electron micrograph of white adipocytes from rat adipose tissue (600X). (Courtesy of Dr. A. Angel and Dr. M. J. Hollenberg of the University of Toronto.)

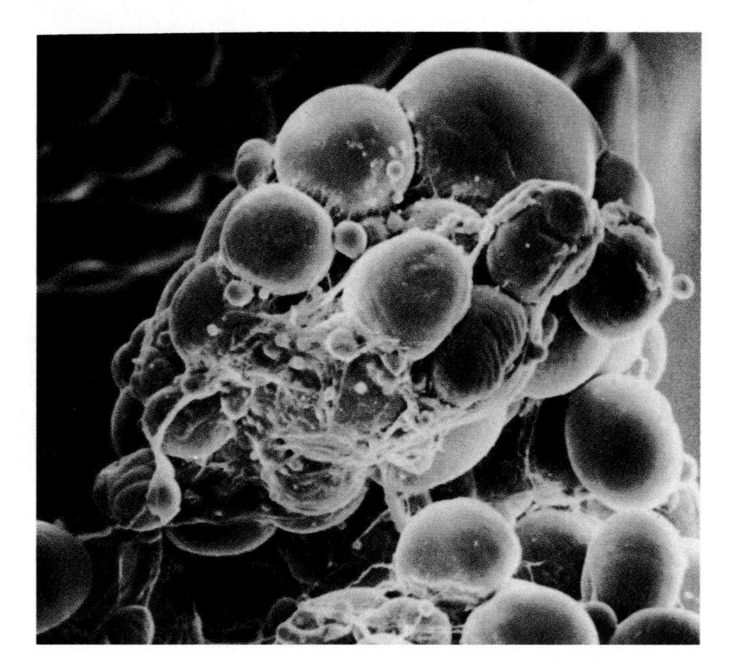

Membrane Lipids

The major lipids found in biological membranes include the phospholipids (phosphoglycerides and sphingomyelin), the glycosphingolipids, and cholesterol. Phosphoglycerides, quantitatively the most important structure group, contain, in addition to phosphate, a glycerol backbone and esterified fatty acids and alcohols. While the phosphoglycerides are the predominant lipids in biological membranes, cholesterol and the sphingolipids are important components of some cellular membranes, especially in eukaryotic cells. The special importance of phospholipids to living organisms is underscored by the nearly complete lack of genetic defects in the metabolism of these lipids in humans. Presumably, any such defects are lethal at early stages of development and therefore are never observed. Isolation and analysis of phospholipids are described in box 7A.

7A BOX

Isolation and Analysis of Phospholipids

Lipids are extracted from cells or tissues with organic solvents (for example, $CHCl_3$). Phospholipids are resolved from uncharged lipids (neutral lipids) by adsorption column chromatography (usually silicic acid) or by adsorption thin-layer chromatography. The various classes of phospholipids are usually separated on a small scale by thin-layer chromatography, as shown in the illustration. Large-scale preparations usually involve column chromatography or high-pressure liquid chromatography.

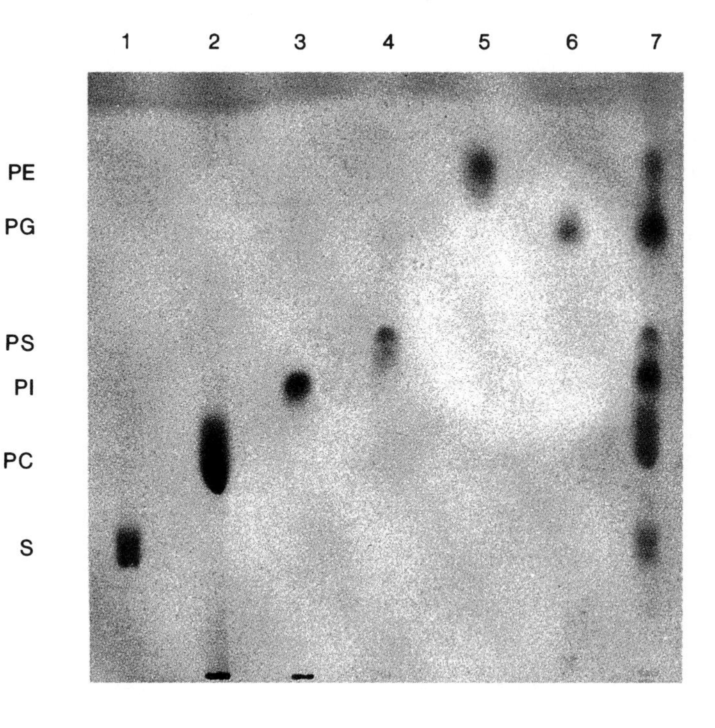

Figure 1

Thin-layer chromatography of the major phospholipids. Lanes 1 to 6 contain 0.1 mg of each lipid. Lane 7 is a mixture of the six phospholipids.
PE = phosphatidylethanolamine; PG = phosphatidylglycerol;
PS = phosphatidylserine; PI = phosphatidylinositol;
PC = phosphatidylcholine; S = sphingomyelin. Each compound was spotted on a silica gel G60 thin-layer plate that had been activated at 100°C 1 h before the analysis. The plate was developed in a solvent that contained $CHCl_3$:CH_3OH:CH_3COOH:H_2O (50:25:8:4 by volume). The compounds were visualized after spraying the plate with dilute sulfuric acid and heating in the oven.

Figure 7.4

The structure of a phosphatidic acid, a phosphoglyceride. The cluster of polar and charged oxygens gives phosphatidic acid its amphipathic properties. The fatty acids attached to carbons 1 and 2 are saturated and unsaturated, respectively.

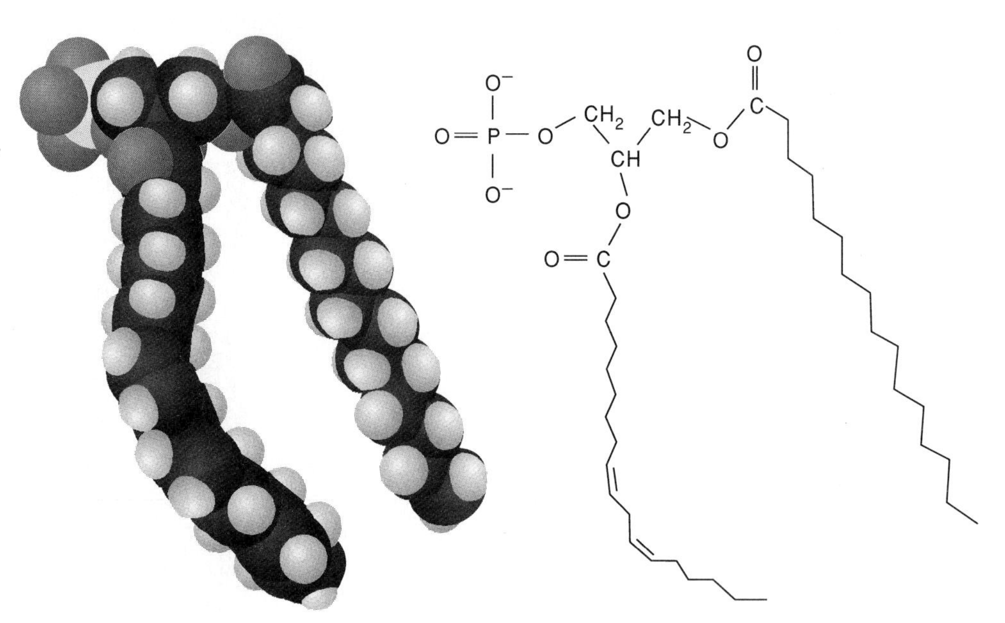

Figure 7.5

Phospholipids with alkyl or alkenyl ether substituents. The structures with alkenyl ether substituents are also called plasmalogens.

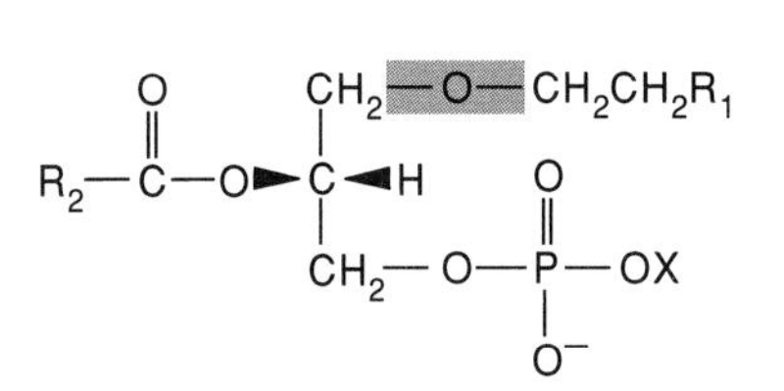

Phospholipid with an alkyl ether

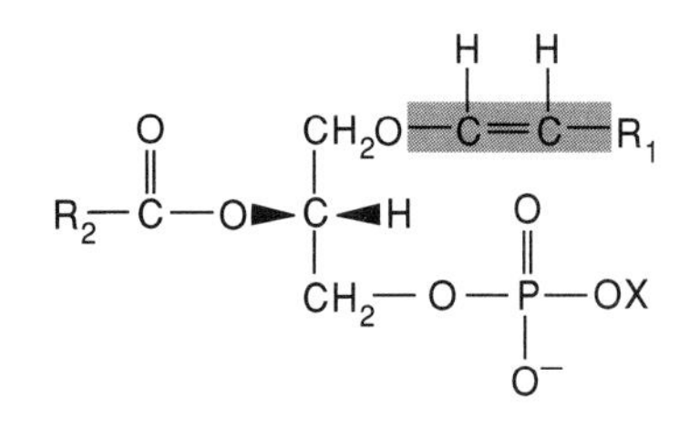

Phospholipid with an alkenyl ether

Phosphoglycerides Have a Glycerol-3-Phosphate Backbone

All phosphoglycerides have a glycerol-3-phosphate backbone, as shown in figure 7.4. The hydroxyls on carbons 1 and 2 are usually acylated with fatty acids, and in most phospholipids the fatty acid substituent at carbon 1 is saturated, while the one at carbon 2 is unsaturated. In some instances, the substituent on carbon 1 is an alkyl ether or an alkenyl ether as shown in figure 7.5.

Phosphoglycerides are classified according to the substituent (X) on the phosphate group (see table 7.3). If X is a hydrogen, the compound is called 3-*sn*-phosphatidic acid. If X—OH is choline, the lipid is called 3-*sn*-phosphatidylcholine (lecithin); this is the most abundant phospholipid in animal tissues (fig. 7.6). In addition to its role in membrane structure, phosphatidylcholine is an important structural component of the plasma lipoproteins and bile. The other major phosphoglycerides are listed in table 7.3.

An important subclass of phosphoglycerides is the lysophospholipids, in which one of the acyl substituents (usually from position 2) is missing, as shown in figure 7.7. If the acyl substituent at carbon 1 were removed, the acyl group from position 2 would spontaneously migrate to position 1. We can differentiate between deacylation at position 1 or 2 by analysis of the fatty acids derived from the lysophospholipid. If the fatty acid on the molecule is saturated, it is likely that the fatty acid from position 2 of the phospholipid has been cleaved. However, if the fatty acid on the lysophospholipid is mostly unsaturated, the fatty acid from position 1 has probably been cleaved and migration of the fatty acid from position 2 has occurred. The lysophospholipids are named simply by adding the prefix *lyso-* to the name of the original phospholipid (e.g., lysophosphatidylcholine). The lysophospholipids account for only 1–2% of the total phospholipids in animal cells.

Table 7.3
Major Classes of Phosphoglycerides

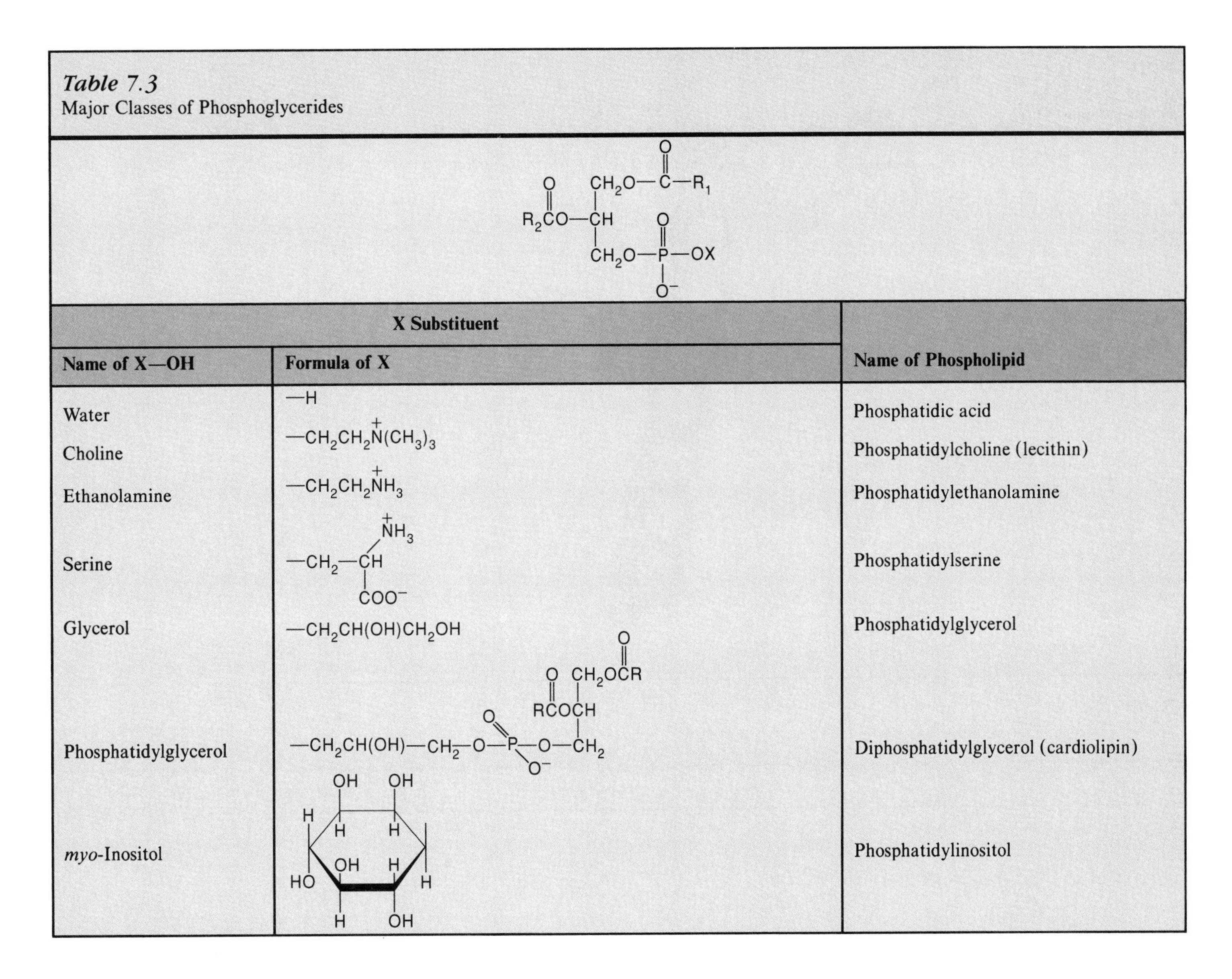

$$\begin{array}{l} \quad\quad\quad\quad\;\; CH_2O-\overset{O}{\overset{\|}{C}}-R_1 \\ R_2\overset{O}{\overset{\|}{C}}O-CH \\ \quad\quad\quad\quad\;\; CH_2O-\overset{O}{\overset{\|}{P}}(O^-)-OX \end{array}$$

X Substituent		
Name of X—OH	**Formula of X**	**Name of Phospholipid**
Water	$-H$	Phosphatidic acid
Choline	$-CH_2CH_2\overset{+}{N}(CH_3)_3$	Phosphatidylcholine (lecithin)
Ethanolamine	$-CH_2CH_2\overset{+}{N}H_3$	Phosphatidylethanolamine
Serine	$-CH_2-CH(\overset{+}{N}H_3)COO^-$	Phosphatidylserine
Glycerol	$-CH_2CH(OH)CH_2OH$	Phosphatidylglycerol
Phosphatidylglycerol	$-CH_2CH(OH)-CH_2-O-P(=O)(O^-)-O-CH_2-CH(OCOR)-CH_2OCOR$	Diphosphatidylglycerol (cardiolipin)
myo-Inositol		Phosphatidylinositol

Sphingolipids Contain a Long-Chain, Hydroxylated Secondary Amine

In addition to the phosphoglycerides, the sphingolipids are commonly found in eukaryotic cell membranes. Sphingomyelin contains phosphate and therefore is also a phospholipid, while the glycosphingolipids generally lack phosphate but contain sugar or oligosaccharide residues. The common structural feature of sphingolipids is a long-chain, hydroxylated secondary amine. There are three major long-chain bases (table 7.4) that contain 18 carbons and a number of other bases that differ in chain length, number of double bonds, or branching of the alkyl chain. Sphingosine (4-sphingenine) is quantitatively the most important long-chain base (usually 90% or more) in animal cells, whereas phytosphingosine (4-hydroxysphinganine) is characteristically found in plant tissues. Most bacteria, including *E. coli,* do not contain sphingolipids, whereas yeast cells do.

Figure 7.6
Structure of a phosphatidylcholine (lecithin).

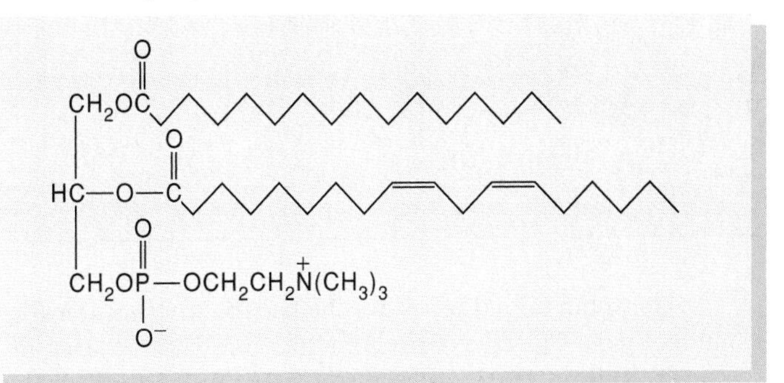

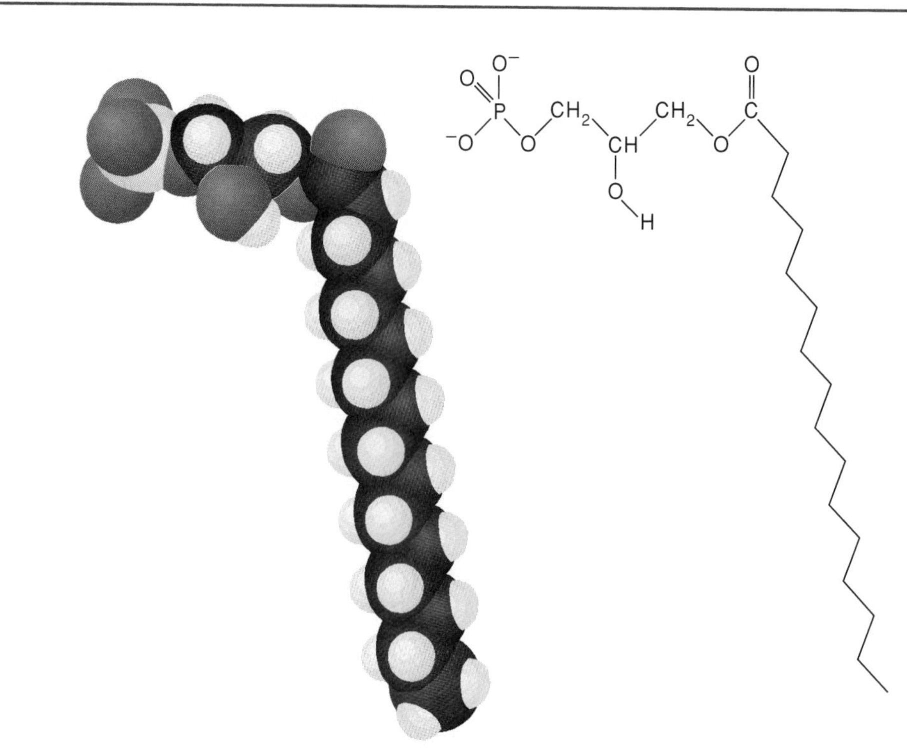

Figure 7.7

The basic structure for lysophospholipids. These phospholipids usually lack a substituent at position 2.

Table 7.4
Three Important Sphingolipid Bases

Structure	Systematic Name	Common Name
$CH_3(CH_2)_{12}C(H)=C(H)-C(H)(OH)-C(H)(\overset{+}{N}H_3)-CH_2OH$	4-Sphingenine	Sphingosine
$CH_3(CH_2)_{12}CH_2CH_2-C(H)(OH)-C(H)(\overset{+}{N}H_3)-CH_2OH$	Sphinganine	Dihydrosphingosine
$CH_3(CH_2)_{12}CH_2C(H)(OH)-C(H)(OH)-C(H)(\overset{+}{N}H_3)-CH_2OH$	4-Hydroxysphinganine	Phytosphingosine

The sphingolipid bases are the backbone structures for all sphingolipids. The free bases are toxic to cells and therefore are present only in trace quantities. The sphingolipid base is acylated on the amine with a fatty acid to give ceramide (fig. 7.8), which is common to all the sphingolipids. The fatty acid substituents are mainly C_{16}, C_{18}, C_{22}, or C_{24}, saturated or monounsaturated. In many cases, the acyl group also may contain an α-hydroxyl residue. Ceramide is further modified on the primary hydroxyl group to give the final sphingolipid structures; modification with phosphocholine gives sphingomyelin (fig. 7.9), while modification with carbohydrate gives the class

Figure 7.8

Structure of ceramide with sphingosine as the long-chain base. Sphingosine (shown in color) is very toxic to cells and is usually found only in trace amounts.

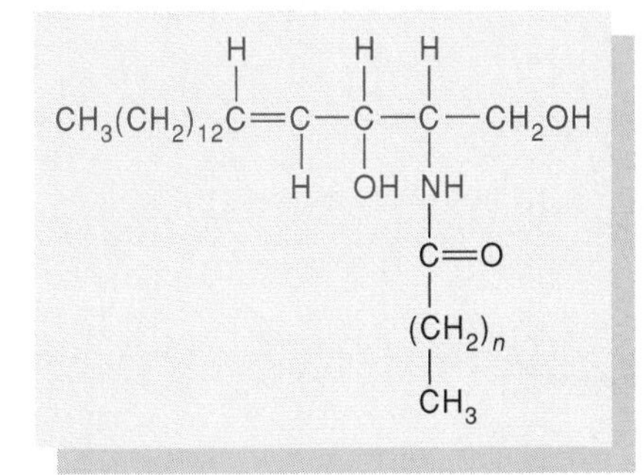

Figure 7.10

Structure of α-*N*-acetylneuraminic acid (sialic acid). Gangliosides are a subclass of glycosphingolipids that always contain one or more molecules of *N*-acetylneuraminic acid. This compound is also commonly found as a terminal residue of glycoproteins (see chapter 6).

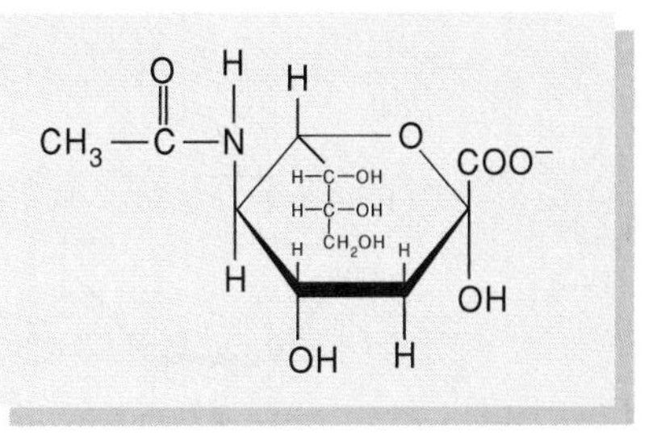

Figure 7.9

Structure of a sphingomyelin.

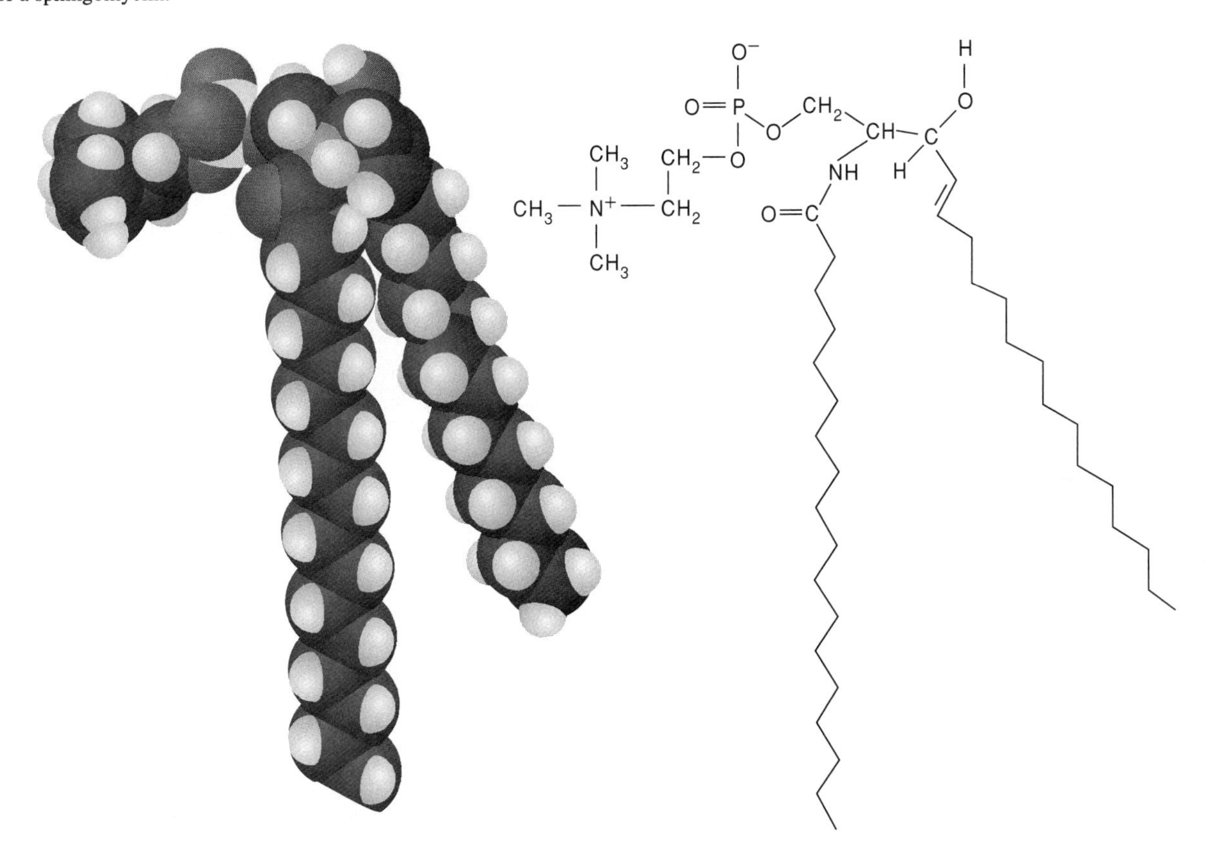

called glycosphingolipids. The carbohydrates most often associated with the glycosphingolipids are glucose, galactose, *N*-acetylglucosamine, and *N*-acetylgalactosamine. There is a subdivision of the glycosphingolipids called gangliosides, and these also contain one or more molecules of *N*-acetylneuraminic acid (sialic acid) (fig. 7.10) in addition to other carbohydrates. As the name implies, the gangliosides were first isolated from nerve tissue; subsequently, however, they were found in most other animal tissues. The structures of two important glycosphingolipids, globoside and GM_2, are shown in figures 7.11 and 7.12. Over 50 separate classes of glycosphingolipids have been identified on the basis of differences in the structure of the oligosaccharide.

Figure 7.11

Structure of globoside, a glycosphingolipid.

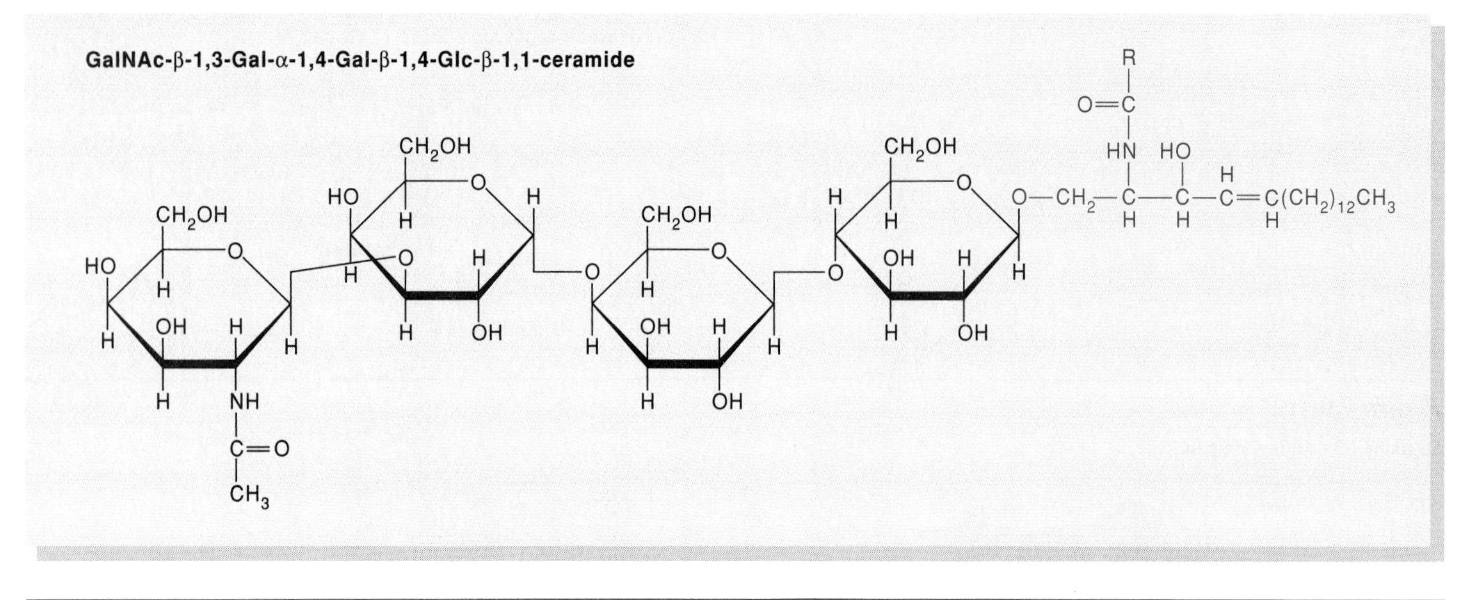

Figure 7.12

Structure of Tay-Sachs ganglioside (GM_2).

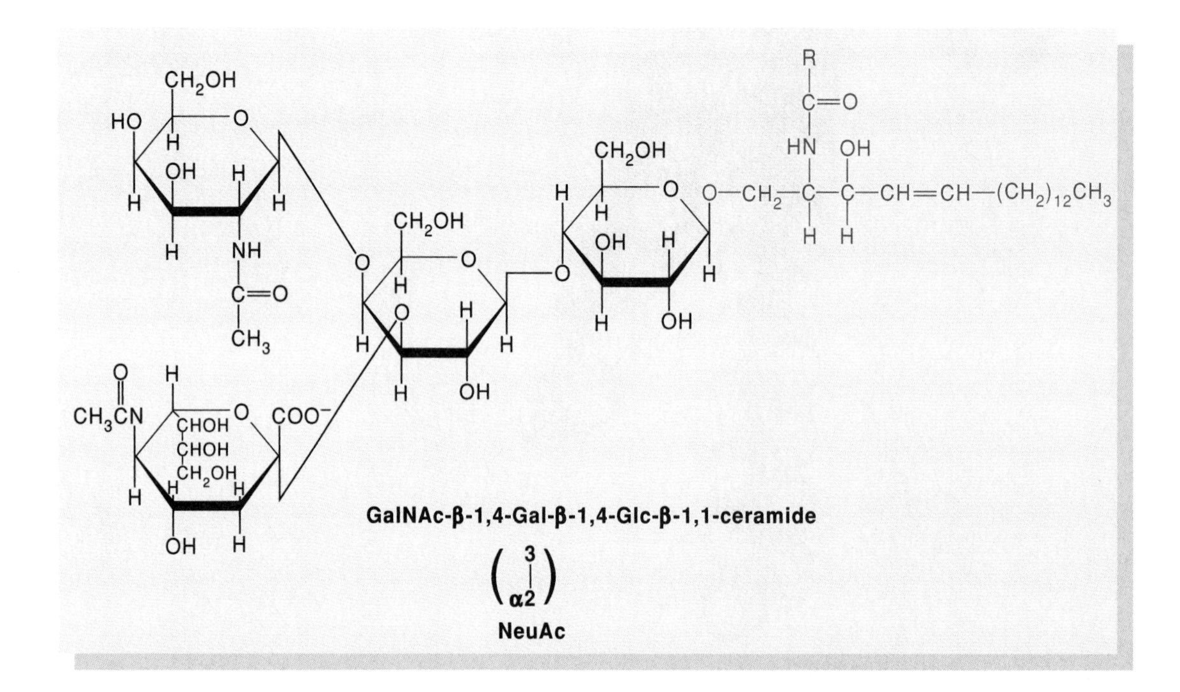

Sphingolipids are important components of the myelin sheath, a multilayered membranous structure that protects and insulates nerve fibers (see chapter 35). The lipids in human myelin contain 5% sphingomyelin (the original source of this lipid, as the name implies) and 15% galactosylceramide (galactocerebroside) (fig. 7.13). In addition, a sulfate derivative, 3′-sulfate-galactosylceramide, makes up 5% of the myelin lipid. The sphingolipids also are found in blood plasma as components of lipoproteins, primarily low-density lipoproteins, which are discussed in chapter 23.

Cholesterol Is a Steroid Found in Eukaryotic Membranes

Cholesterol is the most prominent member of the steroid family of lipids. Probably best known for its association with cardiovascular disease, it is an important structural component in some eukaryotic membranes, but is generally absent in most bacterial membranes.

Figure 7.13

Structure of galactosylceramide. This lipid comprises 15% of human myelin lipid.

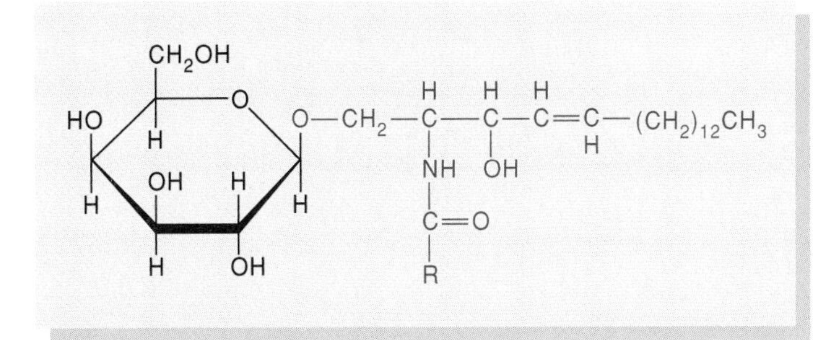

Figure 7.14

Structures of phenanthrene and perhydrocyclopentanophenanthrene.

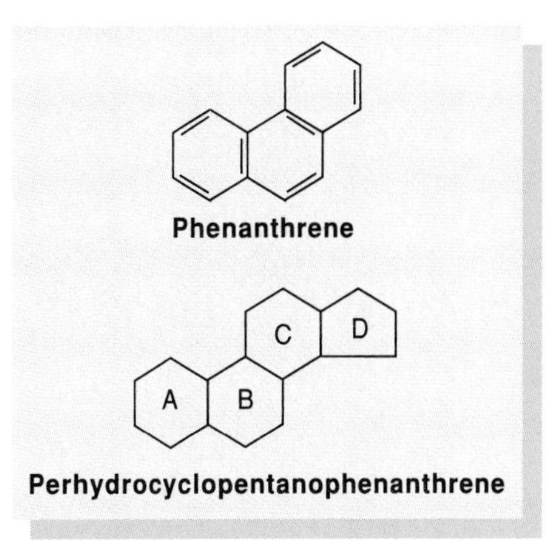

Figure 7.15

Structure of cholesterol, in three different views. The conventional projection is shown at the top center. The more realistic space-filling and conformational models are shown at the lower left and the lower right, respectively.

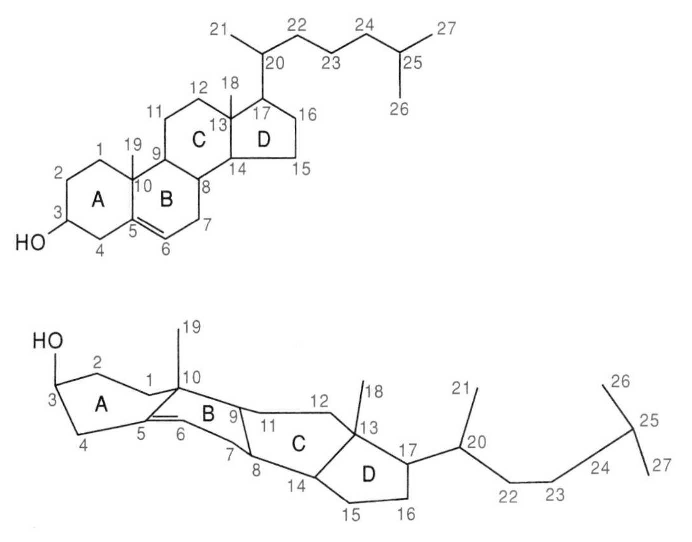

Cholesterol, like other steroids, is a derivative of the tetracyclic hydrocarbon perhydrocyclopentanophenanthrene (fig. 7.14). The four rings are identified by the first four letters of the alphabet, and the carbons are numbered in the sequence shown in figure 7.15. In addition to the basic ring structure, cholesterol contains a hydroxyl group at C-3, an aliphatic chain at C-17, methyl groups at C-10 and C-13, and a Δ^5 double bond.

The cyclohexane rings of steroids can adopt either the chair or boat conformation. The chair conformation is more stable and is the preferred conformation of steroids. The conformations of cholestanol and coprostanol, the two saturated derivatives of cholesterol, are shown in figure 7.16. The A and B rings can be joined in a *trans* configuration, as in cholestanol, or in a *cis* configuration, as in coprostanol. As you can see, the spatial orientation of the A and B rings of these two stereoisomers is very different.

Although it is important that we recognize the three-dimensional structure of steroids, such structural representations are too cumbersome for most uses in biochemistry. Hence a configurational convention for steroids has been adopted in which structural formulas are more easily drawn and recognized. The substituents of the steroid rings are related to the CH_3 group at position 10, which by definition projects above the plane of the rings. This methyl, which is said to be a β substituent, is indicated in the structural formulas by a solid line (—). Similarly, other groups that are above the plane of the rings are referred to as β. Those substituents below the plane of the rings are called α and are indicated in structural formulas by a dashed line (---). Examples of α and β substituents are shown in figure 7.16.

Figure 7.16

Conformational and conventional structures of cholestanol and coprostanol. The A and B rings are joined in a *trans* configuration in cholestanol and in a *cis* configuration in coprostanol. The methyl at position 10 is located above the plane of the rings and is said to be in the ß orientation. Other substituents are labeled ß or α, depending on whether they are above or below the planes of the rings.

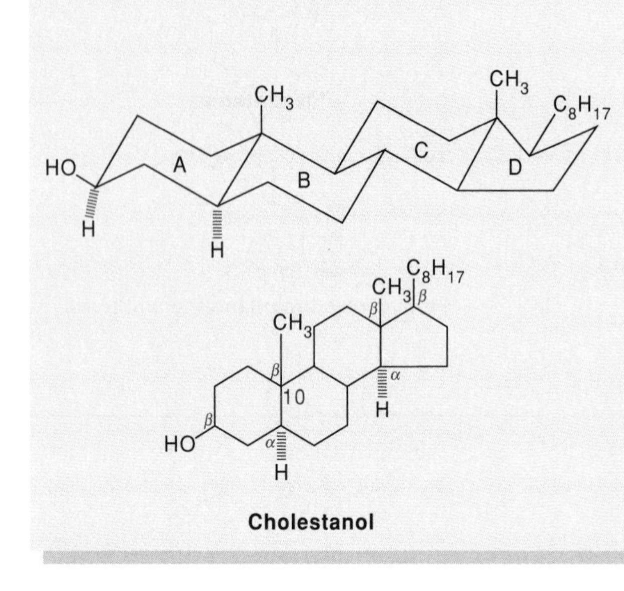

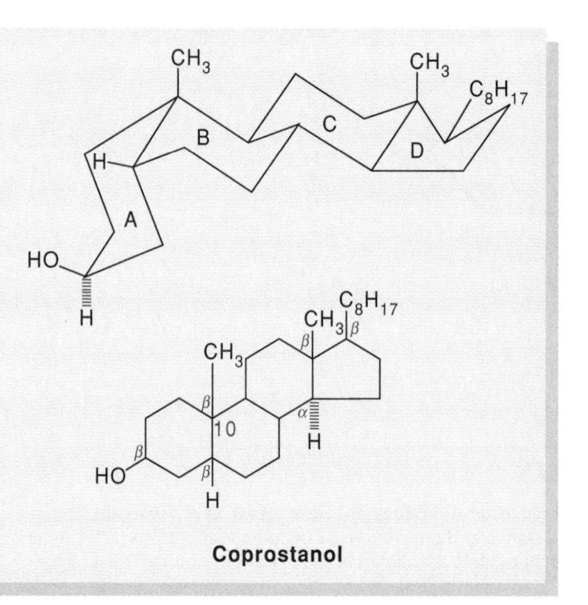

Figure 7.17

Structures formed by phospholipids in a cross section of aqueous solution. Each molecule is depicted schematically as a polar head group (●) attached to one or two fatty acyl hydrocarbon chains. The monolayer at the air–water interface is the first to form. When this interface has become saturated, further phospholipid forms bilayer vesicles, or in the case of lysophospholipids (one fatty acyl chain), micelles.

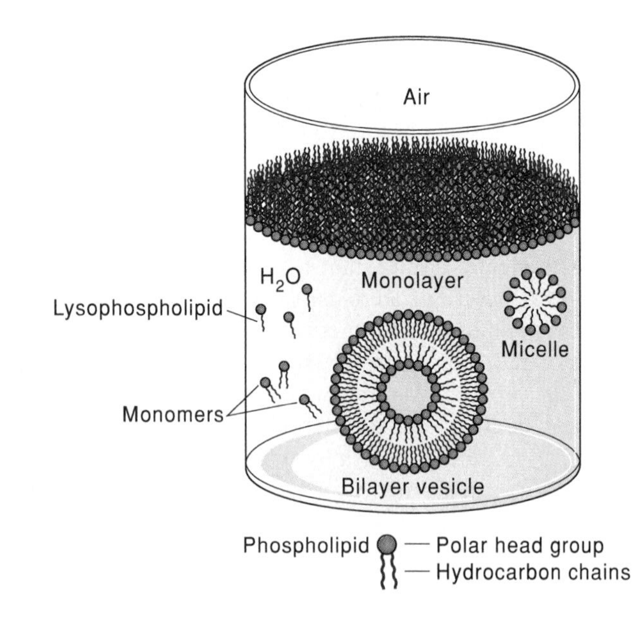

Membrane Lipids Are Amphipathic and Spontaneously Form Ordered Structures

Membrane lipids as a group are amphipathic ("having dual sympathy") molecules because they have both polar and nonpolar portions. The polar head groups of the phosphoglycerides and sphingolipids (e.g., the phosphate group plus the X-substituent of phosphoglycerides or the phosphocholine or carbohydrate of sphingolipids) prefer an aqueous environment (are hydrophilic), whereas the nonpolar (hydrophobic) acyl substituents ("tails") are excluded from aqueous environments. Even cholesterol has hydrophobic and hydrophilic "sides," although it differs altogether in chemical structure from the fatty-acid–containing lipids. It is this amphipathic property that causes most phospholipids to arrange spontaneously into ordered structures when suspended in an aqueous environment.

When phospholipid is added to water, very few lipid molecules exist freely in solution as monomers because of the large hydrophobic surface of the molecule. Instead, a "film" of phospholipid first forms on the water–air interface. Physical studies have shown that this film is a monolayer of phospholipid arranged such that the polar head groups are in contact with water, while the hydrocarbon tails extend up into the air phase (fig. 7.17). When more phospholipid is added to the solution, saturating the air–water interface, other assemblages of phospholipids are formed including bilayers, the structure preferred by most naturally occurring phospholipids, and micelles, which are usually formed in substantial quantity only by the lysophospholipids (see fig. 7.17). Both of these structures maximize hydrophobic interactions between the fatty acyl chains, effectively excluding water from their vicinity, and allow the polar head groups to interact with water molecules. Monolayers, bilayers, and micelles are the favored forms of the various phospholipids in aqueous solution because their formation results in an increase in entropy (positive ΔS), which is due to the fact that water molecules need not order themselves around the hydrophobic hydrocarbon tails of the phospholipid monomer (see fig. 1.10).

As shown in figure 7.17, phospholipid bilayers in aqueous solution are actually spherical "bubbles" or vesicles with water inside and out. This structure is favored over a planar bilayer because exposed hydrocarbon tails, which would occur around the periphery of a planar sheet of phospholipid, are not present. In vesicular structures, no hydrophobic groups need to be exposed to water molecules. Most naturally occurring phospholipids prefer to form vesicular bilayers instead of micelles in water solution because more efficient packing of the molecules can take place in the bilayer vesicle. Lysophospholipids, as well as free fatty acids and detergents (which we will discuss later), form micelles more readily than bilayers because of their geometry, which includes a smaller hydrophobic surface area relative to the diacyl-phospholipids.

The ability of phospholipids and glycolipids to spontaneously form these ordered structures is the basis for their role as the major components and structural determinants of biological membranes. Because lipid bilayers, with their hydrophobic interiors, are relatively impermeable to most hydrophilic molecules, some specific functions of membranes, including the transmembrane transport of hydrophilic molecules, are usually carried out instead by the other major membrane components, proteins. In the remainder of this chapter, we will consider the contributions of lipids and proteins to the structures and functions of biological membranes.

The Diversity of Biological Membranes

Typically, a biological membrane contains lipid, protein, and carbohydrate in ratios varying with the source of the membrane (table 7.5). Nearly always, the carbohydrate is covalently associated with protein (glycoproteins) or with lipid (glycolipids and lipopolysaccharides). Thus the membrane can be thought of as a lipid-protein matrix in which specific functions are carried out by proteins, while the permeability barrier and the structural integrity of the membrane are provided by lipids.

The one membrane structure common to all cells is the plasma membrane. This membrane encapsulates the cytoplasm and creates internal compartments in which essential functions are carried out. In addition to its role as a physical barrier that maintains the integrity of the cell, the plasma membrane provides functions necessary for the survival of a cell, including exclusion of harmful substances, acquisition of nutrients and energy sources, disposal of unusable and toxic materials, reproduction, locomotion, and interaction with components in the environment. All these functions require coordination both for short-range processes, such as sensation, and for long-range processes, such as growth and differentiation. If we look at a typical plasma membrane in the electron microscope, we see a trilaminar (three-layered) structure: two dark lines, representing the polar surfaces of the lipid bilayer, separated by a light region corresponding to the hydrophobic interior of the bilayer (fig. 7.18).

Table 7.5
Chemical Compositions of Some Cell Membranes

Membrane	Protein (%)	Lipid (%)	Carbohydrate (%)
Myelin	18	79	3
Human erythrocyte plasma membrane	49	43	8
Amoeba plasma membrane	54	42	4
Mycoplasma cell membrane	58	37	1.5
Halobacterium purple membrane	75	25	0

Source: G. Guidotti, "Membrane Proteins" in *Annual Review of Biochemistry,* 41:731, 1972. Copyright © 1972 Annual Reviews Inc., Palo Alto, Calif.

Figure 7.18

Cross section of microvilli of cat intestinal epithelial cells, showing the trilaminar (three-legged) structure of the cytoplasmic membranes (165,000X). (Courtesy of Dr. S. Ito.)

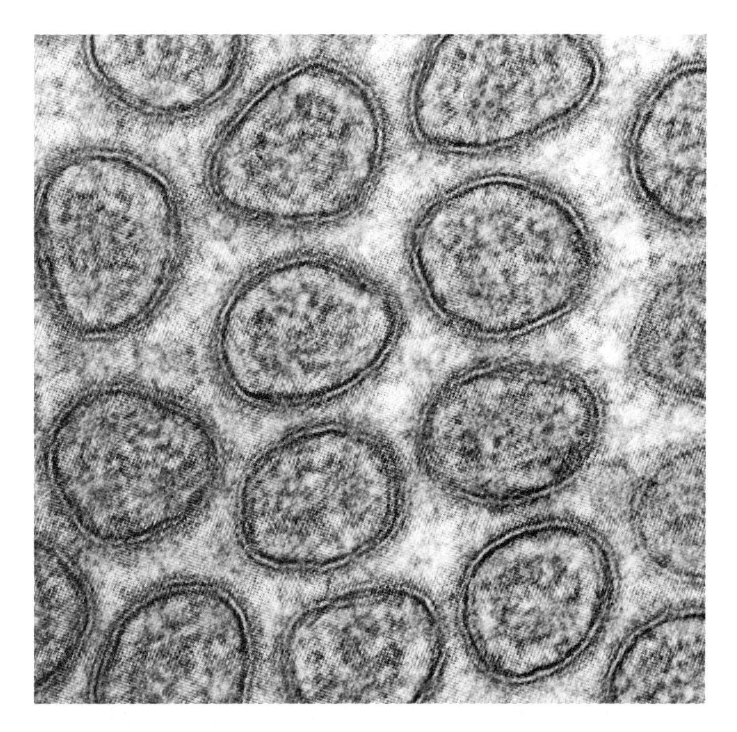

Eukaryotic cells contain numerous organelles of widely differing structure, each of which is specialized in its function—digestion (lysosomes), respiration (mitochondria), photosynthesis (chloroplasts), secretion (endoplasmic reticulum and Golgi apparatus), or nucleic acid biosynthesis (nucleus). Each organelle is surrounded by its own specialized membrane system, which has evolved to participate in its respective function (fig. 7.19). In contrast, prokaryotic cells (bacteria) typically have all functions integrated into the plasma membrane and lack specialized intracellular organelles. These differences in cell structure do not necessarily indicate different biochemical mechanisms, but merely the presence or absence of compartments specifically designed to fulfill separate functions. In the generally larger eukaryotic cells, each process is performed in a spatially isolated domain, whereas in the smaller prokaryotic cell the processes can operate for the most part within a single compartment.

From research done over the last few decades, a number of fundamental principles have emerged that appear to apply to most membrane systems that have been studied. For this reason, we will be able to examine aspects of membrane structure and function in systems as seemingly divergent as bacteria and mammalian mitochondria. In chapters 15 and 32 we will find that in both cases, the biosynthesis of ATP is a membrane-associated process that occurs by a similar mechanism. To take another example, certain bacterial membranes contain proteins that behave in artificial membrane systems much like nerve-cell membrane channels, which are partially responsible for propagation of the nerve impulse in vertebrates (chapter 35). Thus many mechanisms responsible for complex membrane phenomena are undoubtedly used repeatedly throughout the living kingdom.

Different Membrane Structures Can Be Separated According to Their Density

As with the structure of any biological entity, in order to study the structure of biological membranes we must first isolate them in a more or less intact form from the cell. In eukaryotic cells, this problem is complicated by the existence of several different membrane systems in addition to the plasma membrane, each surrounding a specific organelle. To separate membrane fractions, we must first disrupt the plasma membrane under conditions that leave subcellular organelles intact. One common procedure involves mild homogenization in a medium in which the osmolarity is below the normal physiologic value (hypotonic medium). Another method, nitrogen cavitation, involves forcing nitrogen gas into the cells under pressure and then rapidly releasing the pressure to "explode" the cell membrane.

Organelles can be isolated from disrupted cells by differential centrifugation, which separates them on the basis of their size (fig. 7.20). Ruptured plasma membrane fragments can be purified from the same mixture by equilibrium-density-gradient (isopycnic) centrifugation because of their low density (high lipid content) relative to intact organelles (table 7.6, column 4). This technique relies on centrifuging the sample into a preformed gradient of a solute, such as sucrose. When equilibrium is reached, each type of membrane or organelle is found in the region of the gradient corresponding to its own density (see fig. 7.20). Gradients of synthetic sucrose polymers (Ficoll) or colloidal silica particles (Percoll) also are used in these separations because of their inertness, ability to form stable gradients, and impermeability to biological membranes.

The properties of isolated rat liver organelles are summarized in table 7.6. The entries in column 2 provide some idea of the relative proportions of these organelles in the mammalian liver. For example, mitochondria represent 25% of the total cell protein, while lysosomes comprise about 2%. Interestingly, the plasma membrane, which completely surrounds the cell, also represents only 2% of the total protein. In addition to the soluble protein that is membrane-free, there is considerable soluble protein within the organelles, leaving somewhat less than 50% of the total cell protein in the membrane-associated form.

Figure 7.19

Electron micrograph of a cell from the rat pancreas, showing several different intracellular organelles. (PM = plasma membrane; NE = nuclear envelope; Nu = nucleolus; M = mitochondrion; ER = endoplasmic reticulum; Go = Golgi apparatus; arrows show pore complexes in the nuclear envelope; 24,000X.) (From S. L. Wolfe, *Biology of the Cell,* 2d ed., © Copyright 1981, Wadsworth Publishing Co.)

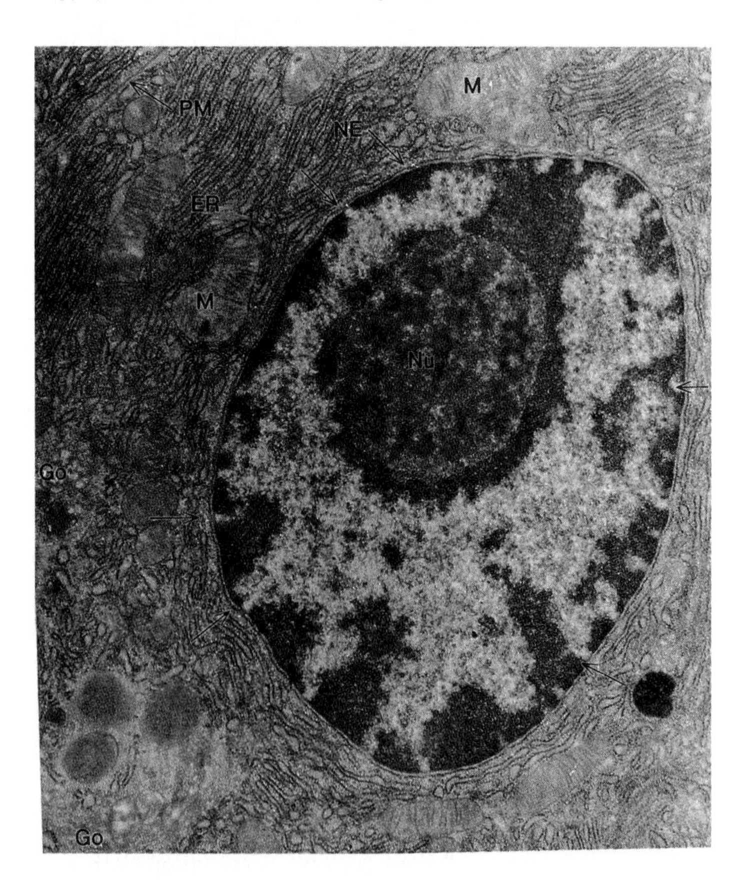

In column 3 of table 7.6, the relative sizes of the organelles are indicated. Note that the sedimentation behavior of an organelle in a sucrose density gradient (column 4) does not correlate with its size, but rather with its density, which is determined by its chemical composition (see fig. 7.20). Nucleic acid ($\rho \sim 1.7$) is more dense than protein ($\rho = 1.25$), and protein is more dense than lipid ($\rho \approx 0.9$–1.1). These facts account for the relatively high density of nuclei and the low density of the Golgi apparatus, which has a high lipid content.

Since each organelle has a specific function, it must also possess a unique complement of enzymes. This prediction is verified by the subcellular localization of numerous enzymes (see table 7.6, column 5), and these specific associations have greatly facilitated the assay and isolation of organelles from eukaryotic cells.

Once the subcellular organelles have been separated, their membranes can be isolated. For those organelles enclosed by a single membrane, treatment in hypotonic buffer (osmotic shock) followed by centrifugal separation of the membrane fragments from the intraorganellar soluble proteins allows us to study membrane composition. Nuclei and mitochondria, however, possess two membranes (inner and outer), and these must be separated before their individual chemical and physical properties can be studied. In these cases, selective solubilization of the outer membrane can be obtained by treatment with appropriate detergents (see later discussion), allowing purification of intact inner membranes. Procedures such as these have made it possible to analyze in detail the lipid and protein contents of organellar membranes as summarized in table 7.7. They have also provided experimental systems in which to study the structures and functions of each different membrane system.

Figure 7.20

Comparison of differential centrifugation (left), which separates on the basis of size, and isopycnic centrifugation (right), which separates on the basis of density. ρ is the density in grams per milliliter.

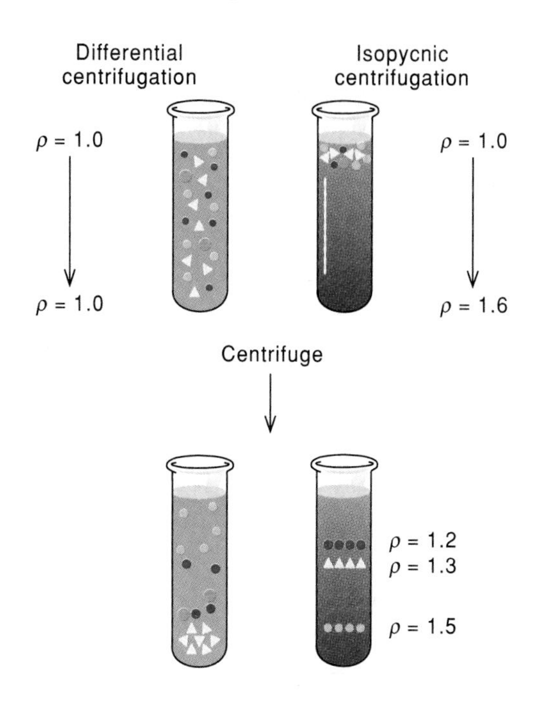

Table 7.6
Properties of Rat Liver Organelles

Organelle	Percent of Cell Protein	Diameter (μm)	Equilibrium Density in Sucrose (g/ml)	Organelle-Specific Enzyme Marker
Liver cell	100	20	1.20	—
Nuclei	15	5–10	1.32	DNA polymerase
Golgi apparatus	2	2	1.10	Glycosyl transferases
Mitochondria	25	1	1.20	Monoamine oxidase (outer membrane); cytochrome *c* (inner membrane)
Lysosomes	2	0.5	1.20	Acid phosphatase
Endoplasmic reticular vesicles	20	0.1	1.15	Cytochrome b_5 reductase and cytochrome b_5; glucose-6-phosphatase
Cytoplasmic membrane	2	—	1.15	Na^+-K^+ ATPase; viral receptors
Soluble protein	30	<0.01	—	—

Source: From M. H. Saier, Jr., and C. D. Stiles, *Molecular Dynamics in Biological Membranes,* Heidelberg Science Library, Vol. 22, Copyright © 1975, Springer Verlag, New York, N.Y. Reprinted with permission.

Table 7.7 Protein and Lipid Content of Organellar Membranes		
Membrane	**Approximate Protein/Lipid Ratio (wt/wt)**	**Approximate Cholesterol/Other Lipids (Molar Ratio)**
Golgi apparatus	0.7	0.08
Liver plasma membrane	1.0	0.40
Endoplasmic reticulum	1.0	0.06
Mitochondrial outer membrane	1.0	0.05
Mitochondrial inner membrane	3.0	0.03
Nuclear membrane	3.0	0.11
Lysosomal membrane	3.0	0.16

Source: M. H. Saier, Jr., and C. D. Stiles, ***Molecular Dynamics in Biological Membranes,*** Heidelberg Science Library, Vol. 22, Copyright © 1975, Springer Verlag, New York, N.Y.

Figure 7.21

Electron micrographs of sections through the surface layers of (*a*) a Gram positive and (*b*) a Gram negative bacterium. (cm = cytoplasmic membrane; om = outer membrane; pg = peptidoglycan; ta = teichoic acid.) Note the thick cell wall in (*a*), compared with the distinct inner and outer trilaminar membranes separated by a thin peptidoglycan layer in (*b*). (Courtesy of J. Stolz; 150,000X.)

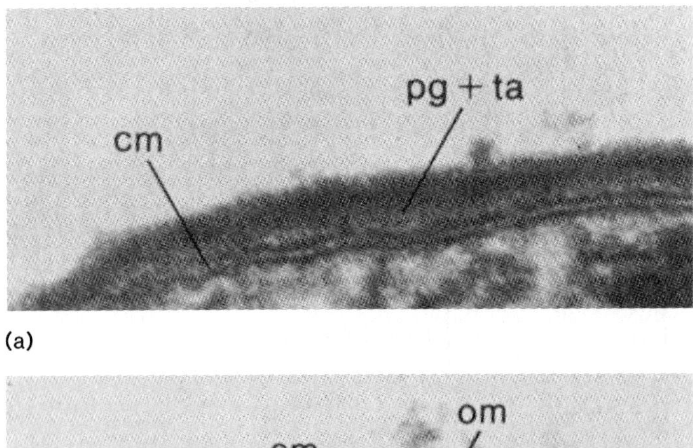

(a)

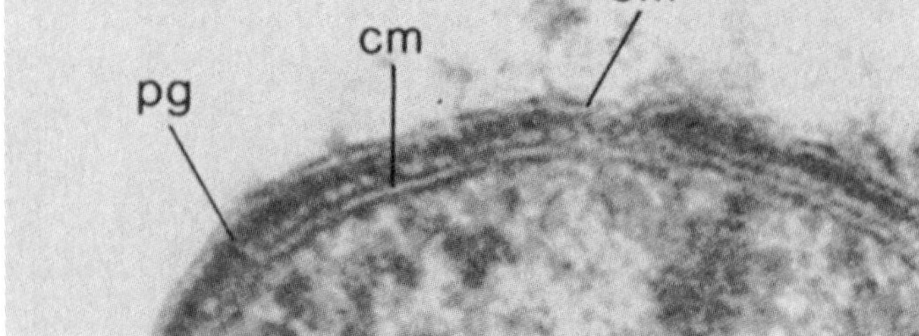

(b)

Bacteria Are Surrounded by One or Two Membranes and a Cell Wall

In contrast to animal cells, most prokaryotic cells are surrounded by a cell wall, which allows bacteria to live in a hypotonic environment without bursting and confers upon these cells their characteristic shape (rod, sphere, or spiral). In 1884, Christian Gram discovered that bacteria could be divided into those that retained a crystal violet-iodine dye complex after washing with alcohol (Gram positive) and those that did not (Gram negative). Even today, the Gram stain reaction is a useful tool in classifying bacteria, and this difference in staining has been found to correlate with a fundamental difference in cell-wall structure between Gram positive and Gram negative cells (fig. 7.21). Gram positive cells are surrounded by a cytoplasmic membrane and a thick cell wall consisting of a sugar–amino acid heteropolymer, or peptidoglycan (see chapter 6), and polyol phosphate polymers called teichoic acids (fig. 7.22*a*). Gram negative bacteria have a much thinner cell wall, consisting mainly of peptidoglycan and associated proteins; this cell wall is surrounded by a second, outer membrane composed of lipid, protein, and lipopolysaccharide (see fig. 7.22*b;* also see fig. 6.29). The biosynthesis of peptidoglycan and lipopolysaccharide is discussed in chapter 21. In Gram negative bacteria, the space between the inner and outer membranes, called periplasmic space, also contains proteins that have a variety of functions.

The two cell layers of Gram negative bacteria can be separated by treatment of the cells with lysozyme (which hydrolyzes peptidoglycan) and EDTA (which destabilizes the outer membrane) in isoosmotic sucrose solutions (fig. 7.23). Periplasmic proteins that are released by this first step can be separated by sedimenting the resulting spheroplasts, which have lost any nonspherical shape characteristic of the original cell because their cell wall has been digested. Then the spheroplasts can be treated with high-frequency sound (sonication) to rupture both the inner and the outer membranes, which quickly reseal into smaller spherical, closed vesicles (see fig. 7.23). Because of their higher carbohydrate content, outer membrane vesicles have a higher density than ones derived from the inner membrane and thus can be separated from them in a sucrose density gradient.

By these techniques, workers have discovered that electron-transport chains, ATP-synthesizing enzymes, many transport proteins, and other enzymes are located on the inner membrane of Gram negative bacteria, while the outer membrane harbors receptors for bacteriophage and bacteriocins, certain other transport proteins, and various phospholipases. The periplasmic space contains hydrolytic enzymes as well as nutrient-binding proteins involved in transmembrane transport and chemotaxis (chapter 32).

The organization of proteins in Gram positive bacteria is usually much simpler because these cells are surrounded by a single membrane. Thus soluble and cytoplasmic membrane proteins carry out functions similar to those of Gram negative bacteria, while macromolecular hydrolases are exported into the extracellular medium, where they function to scavenge nutrients from the environment.

Figure 7.22

Structures of some bacterial cell envelope constituents. (*a*) Some teichoic acids of Gram positive bacteria: (i) *Lactobacillus casei* (R = D-alanine); (ii) *Actinomyces antibioticus* (R = D-alanine); (iii) *Staphylococcus lactis* (R = D-alanine); (iv) *Bacillus subtilis* (R = glucose). Compounds (i)–(iii) are composed of repeating glycerol units, while (iv) is a ribitol teichoic acid to which D-alanine may be attached at either position 3 or 4 of the pentitol. (Adapted from R. Y. Stanier, E. A. Adelberg, and J. L. Ingraham, *The Microbial World,* 4th ed., Prentice-Hall, Englewood Cliffs, N.J., 1976. Used with permission.) (*b*) Schematic illustration of the structure of lipopolysaccharide in the outer membrane of *Salmonella typhimurium.* (EtN = ethanolamine; KDO = 2-keto-3-deoxyoctonic acid; Hep = L-glycero-D-mannoheptose; Abe = abequose; Man = mannose; Rha = rhamnose.) (Source: H. Nikaido, "Biosynthesis and assembly of lipopolysaccharide" in *Bacterial Membranes and Walls*, edited by L. Leive. Copyright © 1973 Dekker, New York, N.Y.)

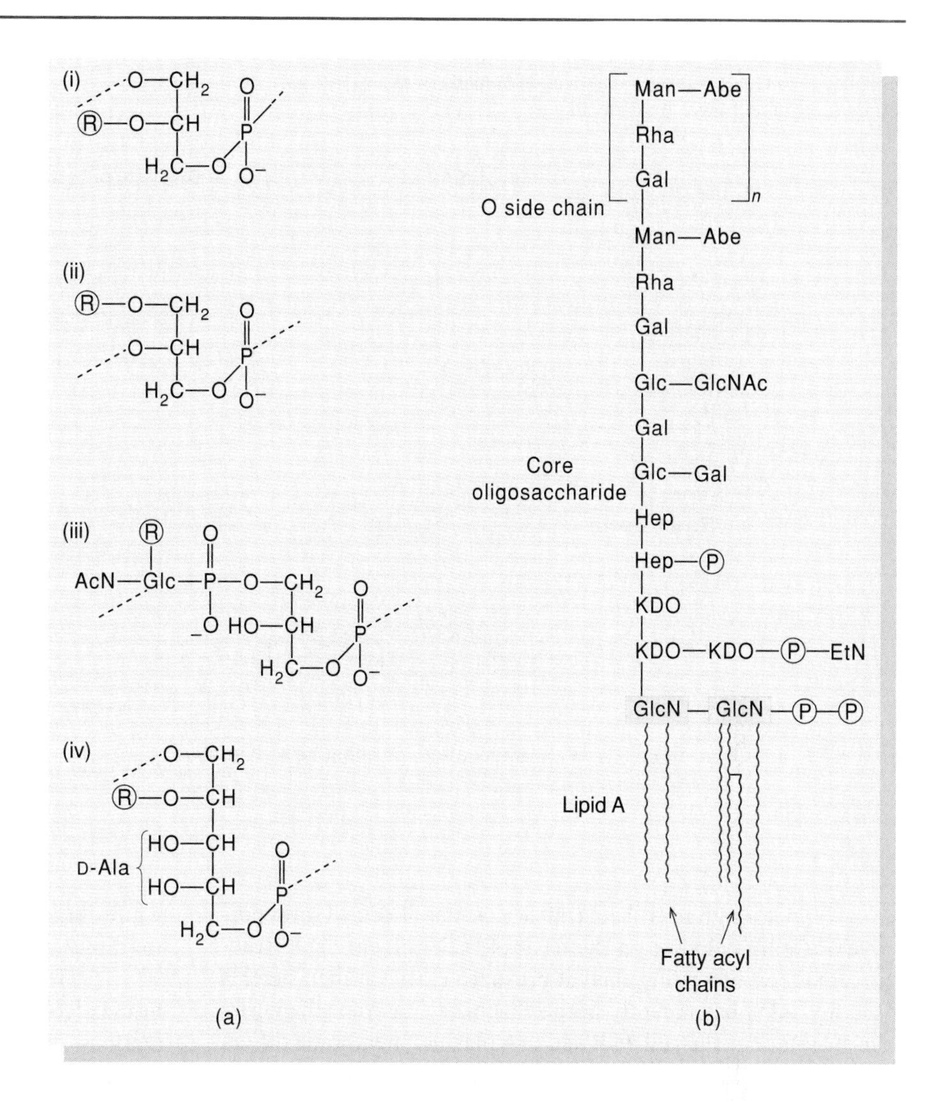

Figure 7.23

Separation of the periplasmic proteins and inner and outer membranes of a Gram negative bacterium. The periplasmic proteins are those proteins that are concentrated in the space between the inner and outer membranes. By treating intact cells with lysozyme and EDTA in an isotonic solution, the periplasmic proteins are released, leaving the spheroplasts. The proteins are removed by sedimentation; then the spheroplasts are suspended in a buffered saline solution and sonicated. This treatment ruptures the outer and inner membranes into smaller spherical closed vesicles. These may be separated by centrifugation on a sucrose density gradient. The outer membrane vesicles sediment to a lower point in the centrifuge tube because of their higher density.

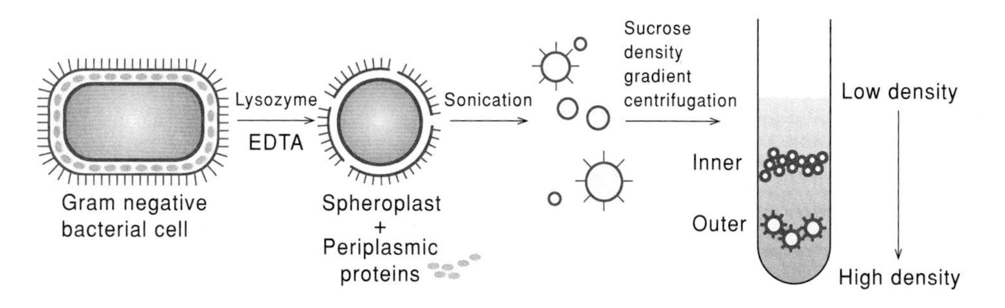

Figure 7.24

Structures of glycerol diethers and glycerol tetraethers, the major lipid components of archaebacterial membranes.

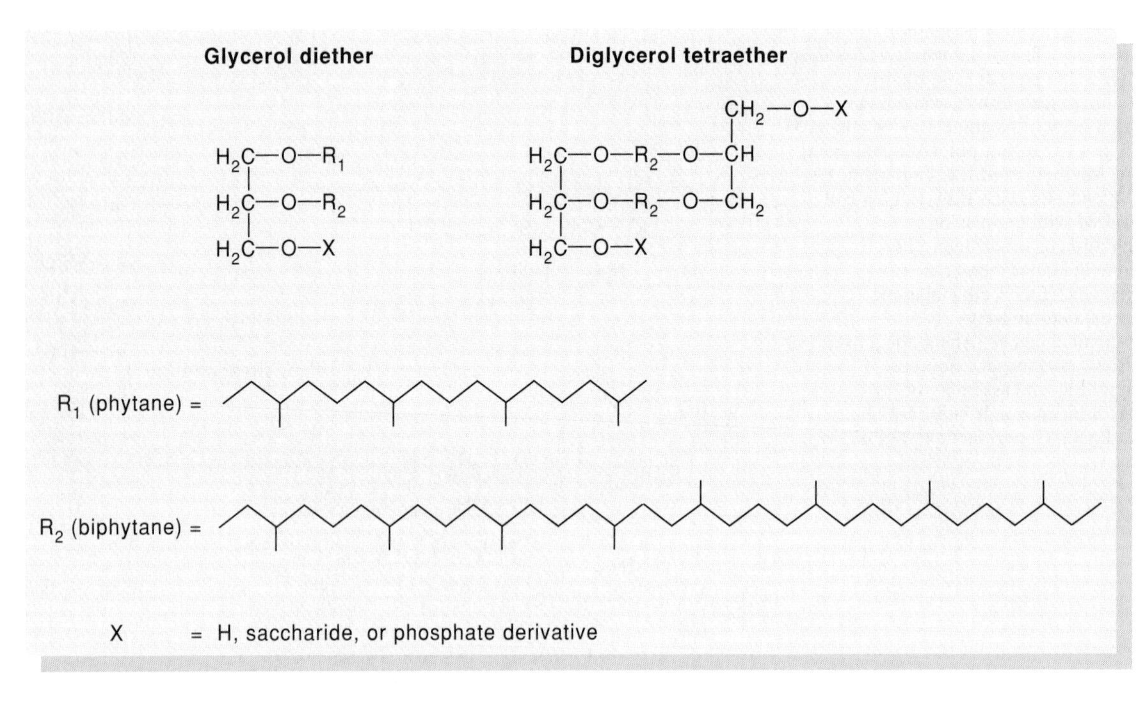

Although most bacteria can be classified as Gram positive or Gram negative on the basis of their cell envelope structure, another interesting class of prokaryotes, the archaebacteria (see fig. 1.25) has recently been recognized. In archaebacteria, the structure of the cell wall and cytoplasmic membrane differs considerably from that in most prokaryotes. The cell wall may consist of a peptidoglycanlike structure called pseudomurein, but more often it is composed of a complex polysaccharide or of glycoproteins or lipoproteins. The single membrane surrounding the archaebacterial cell is composed of unusual lipids that are mainly glycerol diethers or glycerol tetraethers. In these compounds, the first two carbon atoms of glycerol are attached to long-chain polyisoprenoid alcohols called phytane or biphytane (also see chapter 22) via ether linkages, while the X-substituent on the third carbon may be hydrogen, sugar, or a phosphate derivative (fig. 7.24).

Biological Membranes Contain a Complex Mixture of Lipids

Once isolated by any of the procedures just described, membrane fractions can be analyzed biochemically for their content of lipids and proteins. Lipids are usually extracted from proteins by treating the membranes with organic solvents such as chloroform and methanol. Phospholipids are resolved from uncharged (neutral) lipids by adsorption column chromatography (usually the adsorbant is silicic acid) or by adsorption thin-layer chromatography. The various classes of phospholipids are usually separated on a small scale by thin-layer chromatography, while large-scale preparations involve either column chromatography or high-pressure liquid chromatography (HPLC).

Table 7.8 compares the lipid compositions of membranes from a number of biological sources. Several generalizations can be drawn from the data in this table. First, phosphatidylcholine is the chief phospholipid found in membranes of animal cells, while phosphatidylethanolamine predominates in bacteria. Second, in addition to cholesterol, both sphingomyelin and glycolipids (except for lipopolysaccharides) are usually absent from prokaryotic membranes. Despite their widely varying lipid compositions, most of the membranes represented in the table have a characteristic trilaminar appearance in the electron microscope (see fig. 7.18). This similarity reflects the fact that the major structural components of biological membranes, the phospholipids, all form bilayer structures regardless of their exact composition.

Membrane Proteins

Although some characteristics of biological membranes can be explained by the properties of membrane lipids in aqueous solution, other characteristics, especially the ability to perform functions such as transport and enzymatic activities, depend on the presence of membrane-associated proteins. Therefore, in any model of membrane structure, we must also consider the properties of proteins found in biological membranes.

Table 7.8
Lipid Compositions of Membrane Preparations

Source	Lipid Composition (lipid percent)[a]									
	Cholesterol	**PC**	**SM**	**PE**	**PI**	**PS**	**PG**	**DPG**	**PA**	**Glycolipids**
Rat liver										
Cytoplasmic membrane	30.0	18	14.0	11	4.0	9.0	—	—	1.0	—
Endoplasmic reticulum (rough)	6.0	55	3.0	16	8.0	3.0	—	—	—	—
Endoplasmic reticulum (smooth)	10.0	55	12.0	21	6.7	—	—	1.9	—	—
Mitochondria (inner)	3.0	45	2.5	25	6.0	1.0	2.0	18.0	0.7	—
Mitochondria (outer)	5.0	50	5.0	23	13.0	2.0	2.5	3.5	1.3	—
Nuclear membrane	10.0	55	3.0	20	7.0	3.0	—	—	1.0	—
Golgi	7.5	40	10.0	15	6.0	3.5	—	—	—	—
Lysosomes	14.0	25	24.0	13	7.0	—	—	5.0	—	—
Rat brain										
Myelin	22.0	11	6.0	14	—	7.0	—	—	—	21
Synaptosome	20.0	24	3.5	20	2.0	8.0	—	—	1.0	—
Rat erythrocyte	24.0	31	8.5	15	2.2	7.0	—	—	0.1	3
Rat rod cell (outer segment)	3.0	41	—	37	2.0	13.0	—	—	—	—
E. coli cytoplasmic membrane	0	0	—	80	—	—	15.0	5.0	—	—
Bacillus megaterium cytoplasmic membrane	0	0	—	69	—	—	30.0	1.0	—	Trace

[a]PC = phosphatidylcholine; SM = sphingomyelin; PE = phosphatidylethanolamine; PI = phosphatidylinositol; PS = phosphatidylserine; PG = phosphatidylglycerol; DPG = diphosphatidylglycerol (cardiolipin); PA = phosphatidic acid.

Source: M. K. Jain and R. C. Wagner, ***Introduction to Biological Membranes.*** © 1980 John Wiley & Sons, Inc., New York; and from J. E. Rothman and E. P. Kennedy, "Asymmetrical distribution of phospholipids in the membrane of ***Bacillus megaterium***" in ***Journal of Molecular Biology.*** 110:603, 1977. Copyright © 1977, Academic Press Ltd., London England.

Membrane Proteins Are Integral or Peripheral

Membrane proteins are classified as peripheral or integral. Peripheral proteins are probably bound to the membrane as a result of specific interactions with exposed, hydrophilic portions of integral membrane proteins. As a consequence they can be dissociated from isolated membranes by agents that disrupt ionic or hydrogen bonds, such as high salt, EDTA (which chelates divalent cations), or urea. In contrast, integral membrane proteins appear to be deeply embedded in the membrane. They can be released from the membrane only by disrupting the hydrophobic interactions of membrane lipids with organic solvents or detergents. Significant hydrophobic interactions with membrane lipids and proteins probably are responsible for the interaction properties of integral membrane proteins. Figure 7.25 illustrates the differences between integral and peripheral membrane proteins.

To purify integral membrane proteins for study, the proteins first must be dissociated from the membrane matrix, a step that is usually accomplished with detergents. Detergents are amphipathic molecules that disrupt membranes by intercalation into the membrane matrix and solubilization of the component lipids and proteins. Examples of detergents that are

Figure 7.25

Schematic illustration of integral and peripheral membrane proteins. Peripheral membrane proteins may be removed by mild reagents without disruption of the membrane structure. Integral membrane proteins can be released only if the membranes are disrupted.

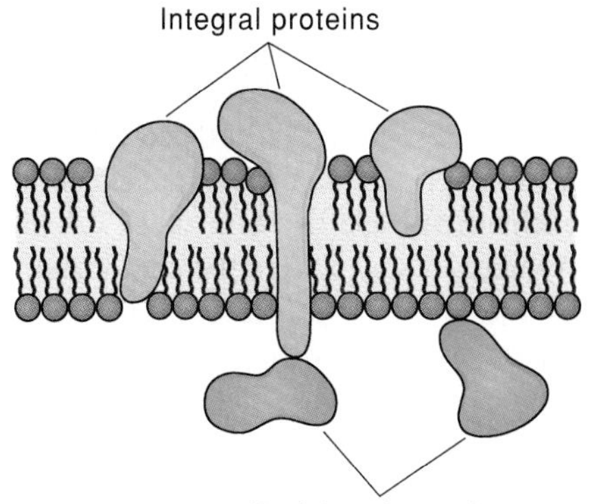

Figure 7.26

Structure of sodium deoxycholate, a steroid that is a naturally occurring ionic detergent.

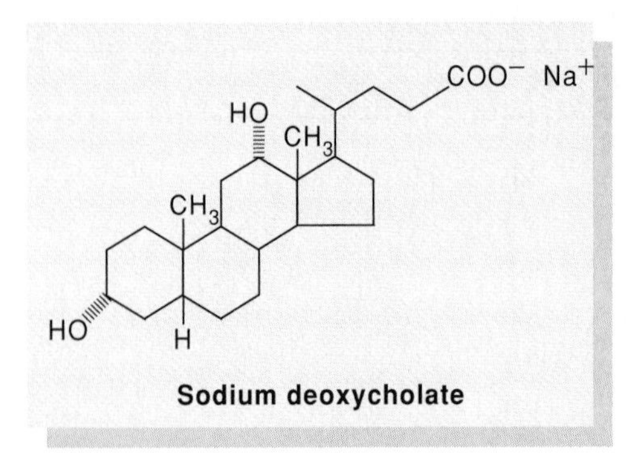

Figure 7.27

Two synthetically derived nonionic detergents. These detergents are useful for dissolving membranes.

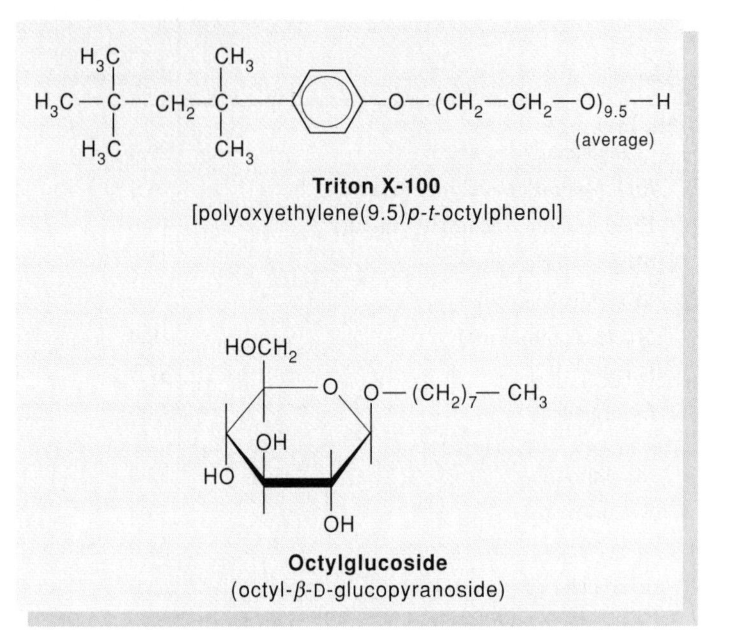

natural products include lysolecithin and sodium deoxycholate, a steroid bile salt (fig. 7.26). Both are ionic detergents. Synthetic detergents include Triton X-100 and octylglucoside (fig. 7.27), two nonionic detergents, and the ionic compounds cetyl trimethylammonium bromide and sodium dodecyl sulfate (SDS) (fig. 7.28), which is also an extremely effective protein denaturant.

Many detergents, including the nonionic ones and lysolecithin, dissolve membranes by forming detergent-lipid and detergent-lipid-protein mixed micelles (fig. 7.29). These detergents prefer to form micelles rather than bilayers because of their particular geometries. Thus an excess of these compounds, when added to a membrane suspension, will tend to shift the equilibrium of all amphipathic molecules in the mixture from bilayer to mixed micelle (see fig. 7.29).

Each detergent has a characteristic critical micellar concentration (CMC) above which it exists in aqueous solution almost entirely in a micellar form, and a characteristic hydrophilic-lipophilic balance (HLB), which is defined as the ratio of the molecular weight of the hydrophilic portion of the molecule to that of the hydrophobic portion. Both these properties appear to be important in determining the effectiveness of a particular detergent in dissolving membranes, although different membrane systems often respond differently to the same detergent.

Some of the properties of detergents commonly used in membrane biochemistry are listed in table 7.9. Because ionic detergents are more likely to alter the conformation of hydrophilic portions of membrane proteins, which are often responsible for catalytic activity, they are more likely to destroy biological function. This observation is in agreement with the fact that only ionic detergents bind to and denature soluble proteins. Little interaction is usually observed between most hydrophilic proteins and nonionic detergents.

Once dissociated from the membrane with detergent, individual membrane proteins can be isolated by a variety of separation techniques, provided an assay is available for the protein of interest. Ion-exchange chromatography (chapter 5) can be useful in separating solubilized membrane proteins if the solvent is a nonionic detergent, such as Triton X-100. Gel filtration and other procedures that separate proteins on the basis of size and shape (chapter 5) are not particularly useful because inclusion of integral membrane protein in detergent-lipid micelles, which can have molecular weights of 10^5 or greater, tends to mask individual size differences. Only with detergents possessing a high CMC, such as deoxycholate and octylglucoside, are separations based on size generally possible. These difficulties have led to the search for techniques that might be especially suited to membrane protein isolation. One such technique is hydrophobic interaction chromatography, which separates proteins on the basis of their relative hydrophobicities. This technique uses insoluble supports, such as agarose or polyacrylamide, to which hydrophobic alkyl or aryl groups have been covalently attached. Proteins bound to the resin by means of hydrophobic interactions can be eluted sequentially, according to the strength of this interaction, by gradual changes in hydrophobicity, ionic strength, pH, or temperature of the eluting buffer (fig. 7.30).

The molecular weights of membrane proteins and the complexity of a specific membrane can be estimated by the technique of polyacrylamide gel electrophoresis, after the membrane has been completely solubilized in the ionic detergent SDS (chapter 5). An example of this technique for examining proteins of the human erythrocyte membrane is shown in figure 7.31. In table 7.10 you will find a comparison of some physical and chemical properties of a number of integral membrane proteins that have been purified to apparent homogeneity. Some of

Figure 7.28

Two additional synthetically derived ionic detergents that can dissolve membranes. Sodium dodecyl sulfate is also an extremely effective protein denaturant and is commonly used in protein gel electrophoresis (see chapter 5).

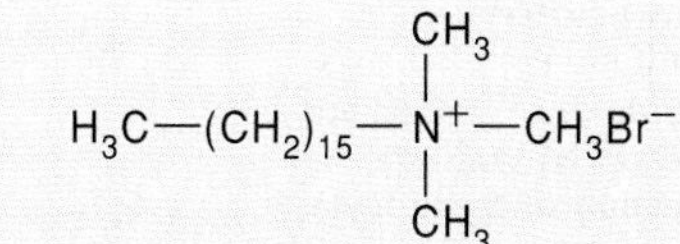

Cetyl trimethylammonium bromide

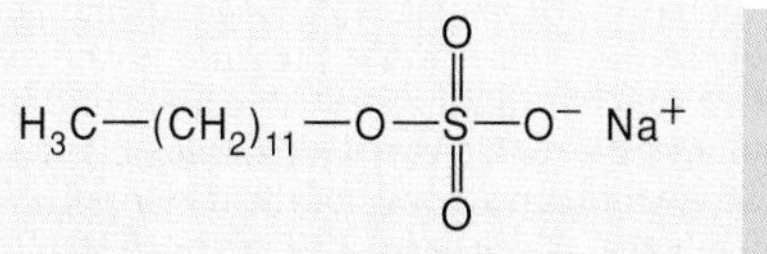

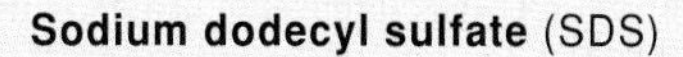

Figure 7.29

Detergent solubilization of biological membranes yields detergent-lipid-protein mixed micelles. The detergent disrupts the membrane and makes a complex with the hydrophobic portion of the integral membrane protein and the membrane lipid.

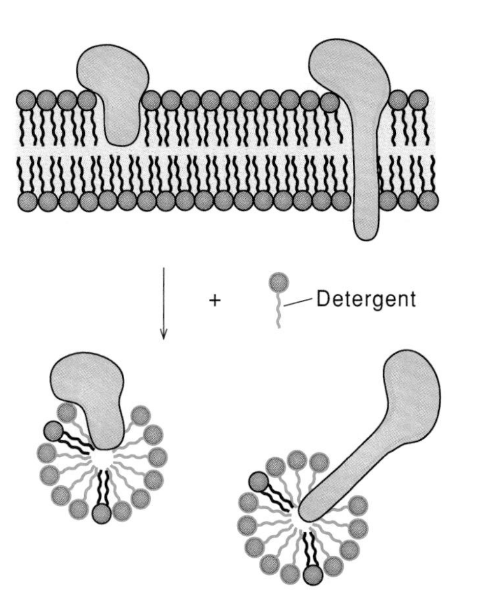

Table 7.9
Properties of Some Detergents

Detergent	Monomer M_r	CMC (mM)[a]	Micellar M_r (average)[a]
Nonionic			
Triton X-100	625	0.24	90,000
Octyl-β-D-glucoside	292	25	—
Ionic			
Deoxycholate (anion)	392	4	800
Lysolecithin (egg; mixture)	500–600	0.02–0.2	95,000
SDS	288	8	17,000

[a]Both CMC and micellar size are dependent on a number of factors, including ionic strength and pH. These numbers should therefore be used only for general comparisons, since each was determined under slightly different conditions.

Figure 7.30

Hydrophobic interaction chromatography. Proteins bound by hydrophobic interactions to insoluble alkyl agarose beads can be differentially removed by changes in ionic strength, hydrophobicity, pH, or temperature. General chromatographic procedures are discussed in chapter 5.

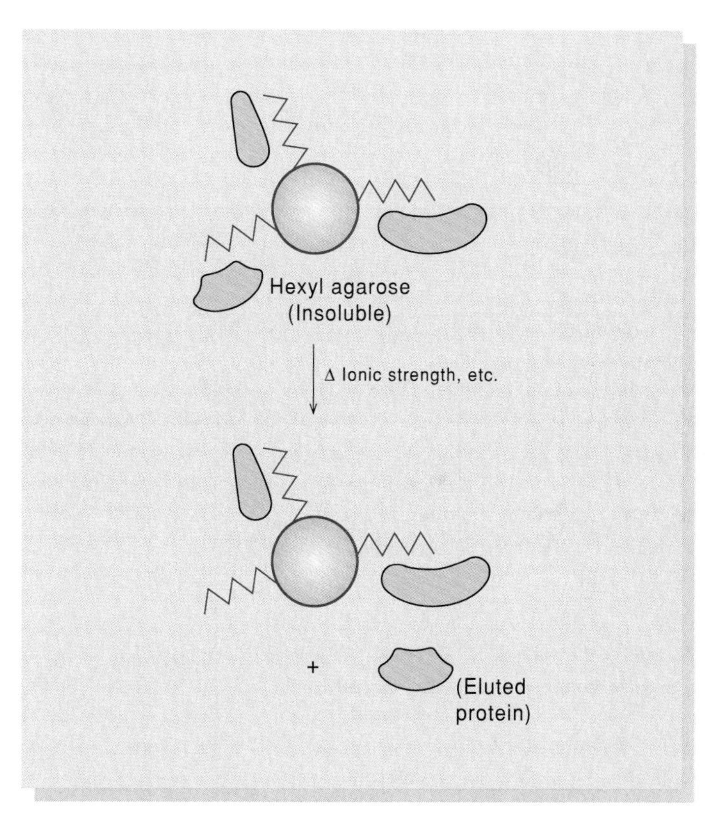

these proteins, such as bacteriorhodopsin from *H. halobium,* and lactose permease from *E. coli,* are highly enriched in hydrophobic amino acids. This property probably reflects the proportion of the polypeptide chain embedded in the hydrophobic portion of the membrane (see the following discussion). One property, however, that seems to be common to many isolated membrane proteins is that to maintain their biologically active structures they usually require a hydrophobic environment. Therefore, techniques that remove detergent and residual phospholipid from these proteins often lead to inactivation of any enzymatic or other biological activity they possess. This inactivation is often accompanied by aggregation or precipitation.

Figure 7.31

Proteins of the human erythrocyte membrane resolved by polyacrylamide gel electrophoresis in the presence of sodium dodecyl sulfate. Protein bands were stained with the dye Coomassie brilliant blue. The bands so stained are numbered 1 to 7. (After G. Fairbanks, T. L. Steck, and D. F. H. Wallach, Electrophoretic analysis of the major polypeptides of the human erythrocyte membrane, *Biochemistry* 10:2606, 1971.) PAS-1 to PAS-4 are sialoglycoproteins, which stain heavily with a reagent that detects carbohydrate. On this electrophoretogram, the anion channel (band 3) and glycophorin A (PAS-1) are not completely resolved. (From V. T. Marchesi, H. Furthmayr, and M. Tomita, "The red cell membrane"," Reproduced with permission from the Annual Review of Biochemistry, 45:667 © 1976 by Annual Reviews, Inc."

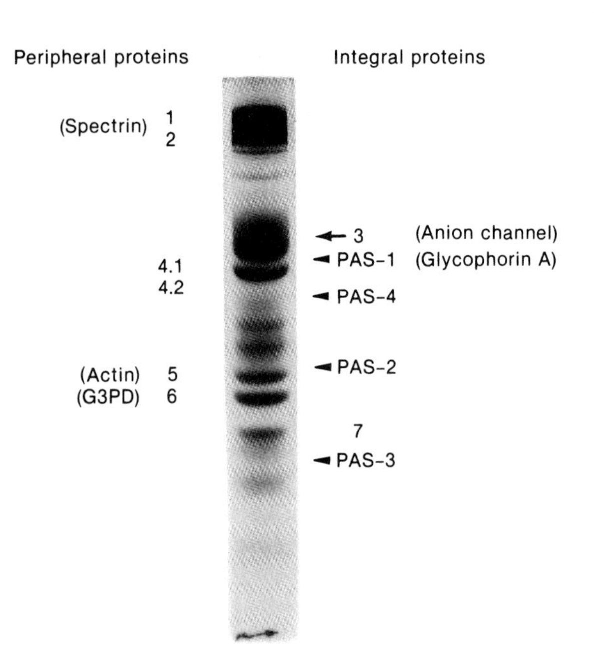

Table 7.10
Properties of Some Purified Membrane Proteins

Protein	Source	Monomer M_r	Subunit Structure	Percent Hydrophobic Amino Acids[a]	Covalent Carbohydrate
Cytochrome b_5	Liver endoplasmic reticulum	16,000	Dimer (0.4% deoxycholate)	40	−
Cytochrome b_5 reductase	Liver endoplasmic reticulum	43,000	Monomer (0.4% deoxycholate)	48	−
Anion-transport (band 3) protein	Human erythrocytes	95,000	Dimer	48	+
Glycophorin	Human erythrocytes	31,000	Dimer	38	++
Bacteriorhodopsin	*Halobacterium halobium*	27,000	Trimer	57	−
Lactose permease	*E. coli* plasma membrane	46,000[b]	Monomer or Dimer	59[b]	−
Mannitol permease	*E. coli* plasma membrane	68,000[c]	Dimer	46[c]	−
Porin	*E. coli* outer membrane	36,000	Trimer	34	−

[a]Mole percent of Pro, Ala, ½ Cys, Val, Met, Ile, Leu, Phe, and Trp.

[b]Deduced from the DNA sequence of the *lacY* gene. (D. E. Büchel, B. Gronenborn, and B. Müller-Hill, "Sequence of the lactose permease gene." *Nature* 283:541, 1980.)

[c]Deduced from the DNA sequence of the *mtlA* gene (C. A. Lee and M. H. Saier, Jr., "Mannitol-specific enzyme II of the bacterial phosphotransferase system III. The nucleotide sequence of the permease gene." *Journal of Biological Chemistry,* 258:10761, 1983.

Figure 7.32

Topography of glycophorin in the mammalian erythrocyte membrane. Carbohydrate residues (red hexagons) are all in the N-terminal domain on the outside of the cell and are attached mainly to the hydroxyl groups of serine and threonine residues of the protein. (Source: J. T. Segrest and L. D. Kohn, "Protein-lipid interactions of the membrane penetrating MN-glycoprotein from the human erythrocyte" in *Protides of the Biological Fluids,* 21st colloquium edited by H. Peeters. Copyright © 1973, Pergamon Press, New York, N.Y.)

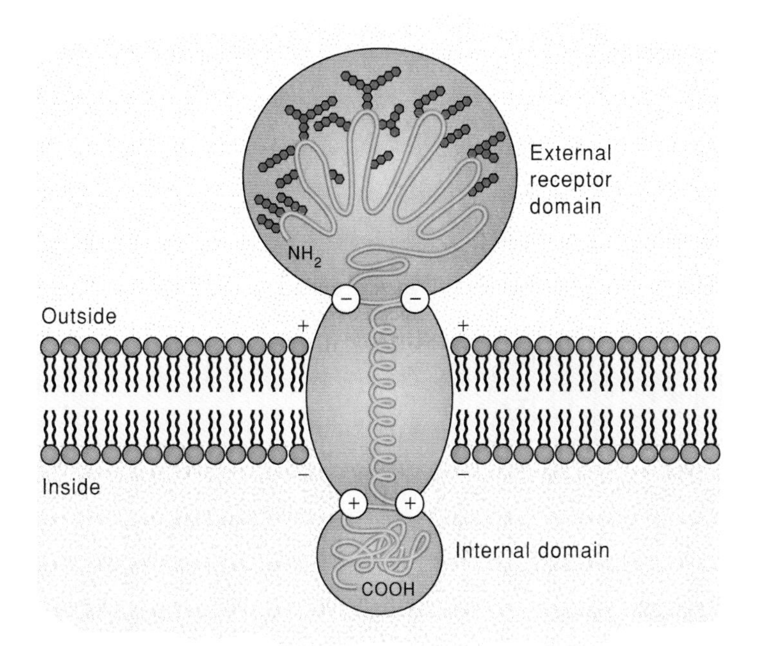

Figure 7.33

Possible disposition of the anion-channel (band 3) protein in the human erythrocyte membrane. Each identical subunit consists of 929-amino-acid residues, and the N-terminal methionine has been shown to be acetylated (NAc). Glycosyl residues are linked almost exclusively on an extracellular asparagine residue. The figure also shows the positions of cleavage by chymotrypsin and trypsin, giving rise to the fragments discussed in the text. (Source: D. Jay and L. Cantley, "Structural aspects of the red cell anion exchange protein" in *Annual Review of Biochemistry.* 55:511, 1986. Copyright © 1986 Annual Reviews Inc., Palo Alto, Calif.)

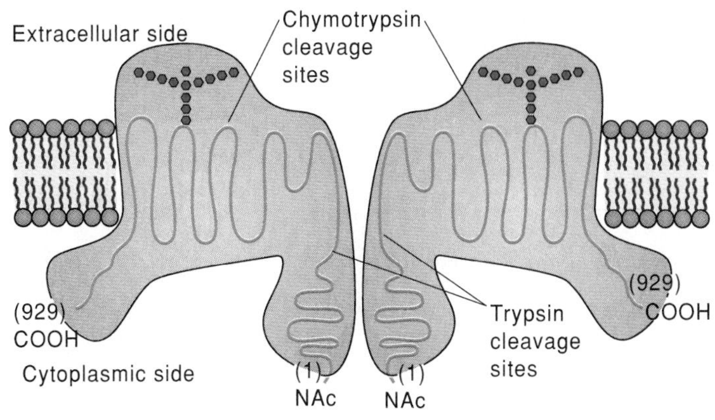

Integral Membrane Proteins Interact with Lipids through Their Hydrophobic Side Chains

In order to understand how proteins contribute to the structure of biological membranes, we need to examine their three-dimensional structures with respect to the membrane system in which they are found. A number of studies have focused on proteins that are major components of particular biological membrane systems. In this section we will describe some of the better-characterized integral membrane proteins.

The protein glycophorin has been isolated from the red blood cell membrane. It has a molecular weight of about 31,000 and contains a very large percentage of carbohydrate (60%). This carbohydrate bears the ABO- and MN-blood group antigenic specificities of the cell and is also the receptor for influenza virus. The probable structure of glycophorin is shown in figure 7.32. It is a 131-amino-acid polypeptide chain on which short carbohydrate chains are covalently attached to an asparagine residue (N-linked) or to serine or threonine residues (O-linked) that are present in the N-terminal region of the protein. This portion of the protein is highly polar, because of the presence of sugar and hydrophilic amino acid residues. The carboxyl end of the molecule is also rich in polar amino acids, but the central region of the protein contains a hydrophobic stretch of 19-amino-acid residues. Labeling experiments have shown that the carbohydrate-containing N terminal is localized on the external surface of the red blood cell and that this region comprises over 80% of the mass of glycophorin. Because the carboxyl terminal of the protein has been shown to be exposed to the cytoplasm, each molecule of glycophorin must span the membrane.

Other studies have shown that the hydrophobic portion of glycophorin has a high affinity for phospholipids and cholesterol, the two principal lipid components of the red blood cell membrane. Moreover, this portion of the molecule forms a very stable α helix in a hydrophobic environment. The length of the helix is about 40 Å, slightly less than the known width of biological membranes. These observations provide strong experimental evidence for the structural model in figure 7.32. Though much is known about the structure of glycophorin, its function is still unclear.

A second well-characterized integral membrane protein is the anion-channel, or band 3, protein of the human erythrocyte membrane. This protein plays an important role in CO_2 transport by enabling HCO_3^- to be exchanged for Cl^-. Labeling studies using intact cells and unsealed membrane ghosts (membranes with their normal contents removed) have shown that this 929-amino-acid glycoprotein also spans the membrane with the carbohydrate residues on the outside of the cell (fig. 7.33). In contrast to glycophorin, however, nearly half of the mass of the polypeptide, including the N terminal, is exposed to the cytoplasm of the cell. The C-terminal third of the molecule, including most of the amino acids to which sugar residues are attached, is at least partially available to labels from the outside of the cell. In addition, a 17,000-M_r segment, delineated by a

Figure 7.34

A model for the structure of bacteriorhodopsin. (*a*) Appearance of the membrane-exposed and membrane-removed helical sides; the hydrophobic residues are concentrated on the membrane-exposed side. (*b*) The way in which the helical units are believed to be clustered.

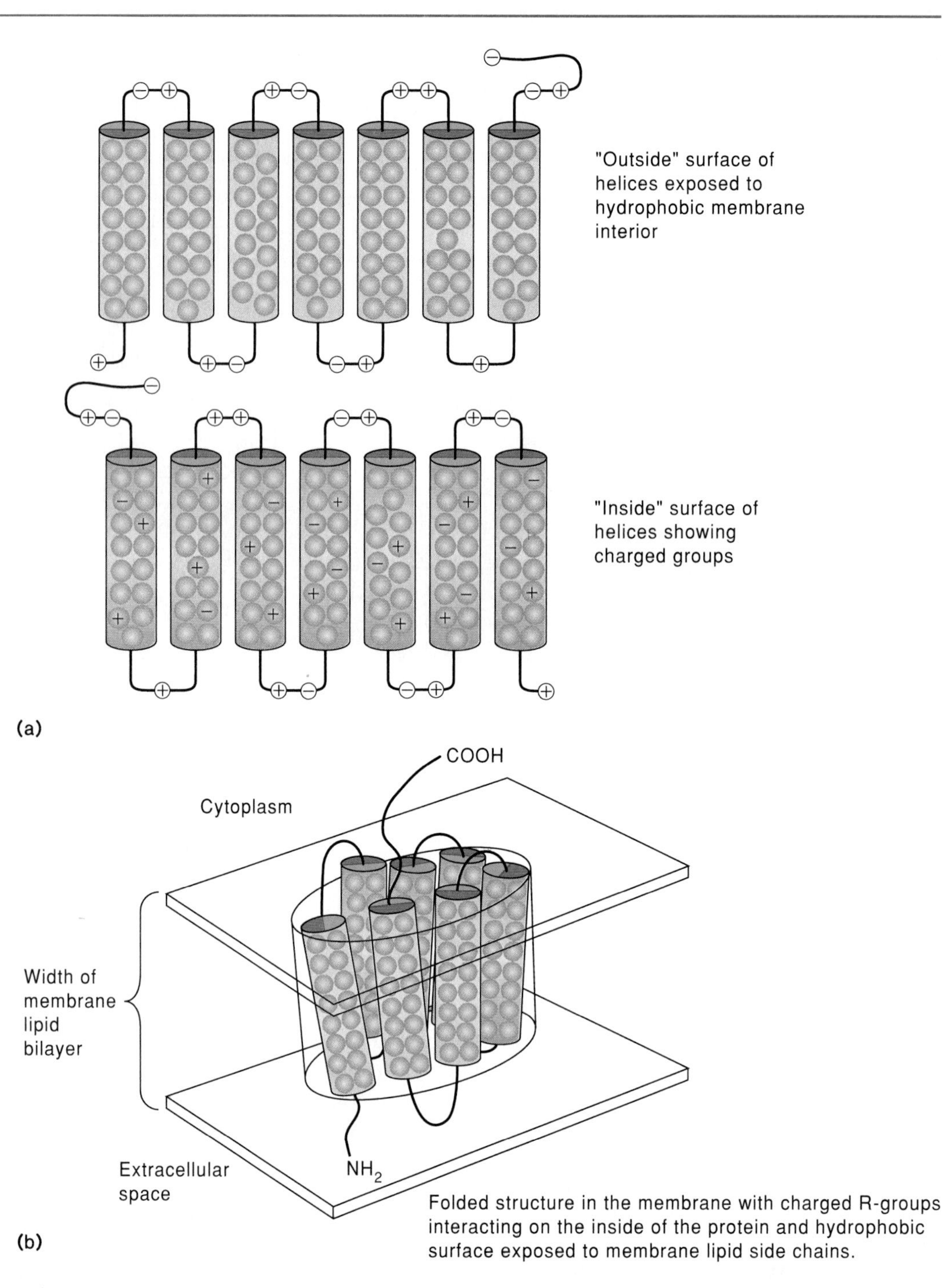

chymotrypsin-sensitive site on the outside of the cell and a tryptic cleavage site on the inside, has been shown to be tightly embedded in the membrane. Both of these latter domains, as compared with the completely water-soluble cytoplasmic domain, are rich in hydrophobic amino acids. The functional structure of the anion-channel protein appears to be a dimer of identical subunits. Recent studies suggest that within the hydrophobic portions of the molecule, the polypeptide chain of the anion channel traverses the membrane at least eight times, and possibly as many as twelve (see fig. 7.33).

In a few instances, it has been possible to isolate a biological membrane that contains only a single kind of protein. The so-called purple membrane from the halophilic archaebacterium *Halobacterium halobium* consists of lipid and one protein, bacteriorhodopsin. This protein functions as a proton pump in response to light and it is important to the organism for ATP synthesis under anaerobic conditions (chapter 32).

Because bacteriorhodopsin forms a highly ordered hexagonal lattice within the plane of the purple membrane, it has been possible to deduce its overall structure from electron micrographs and diffraction patterns. Initial studies showed that the 248-residue polypeptide chain comprising the molecule is organized as a bundle of seven α helices whose long axes are roughly perpendicular to the membrane surfaces (fig. 7.34).

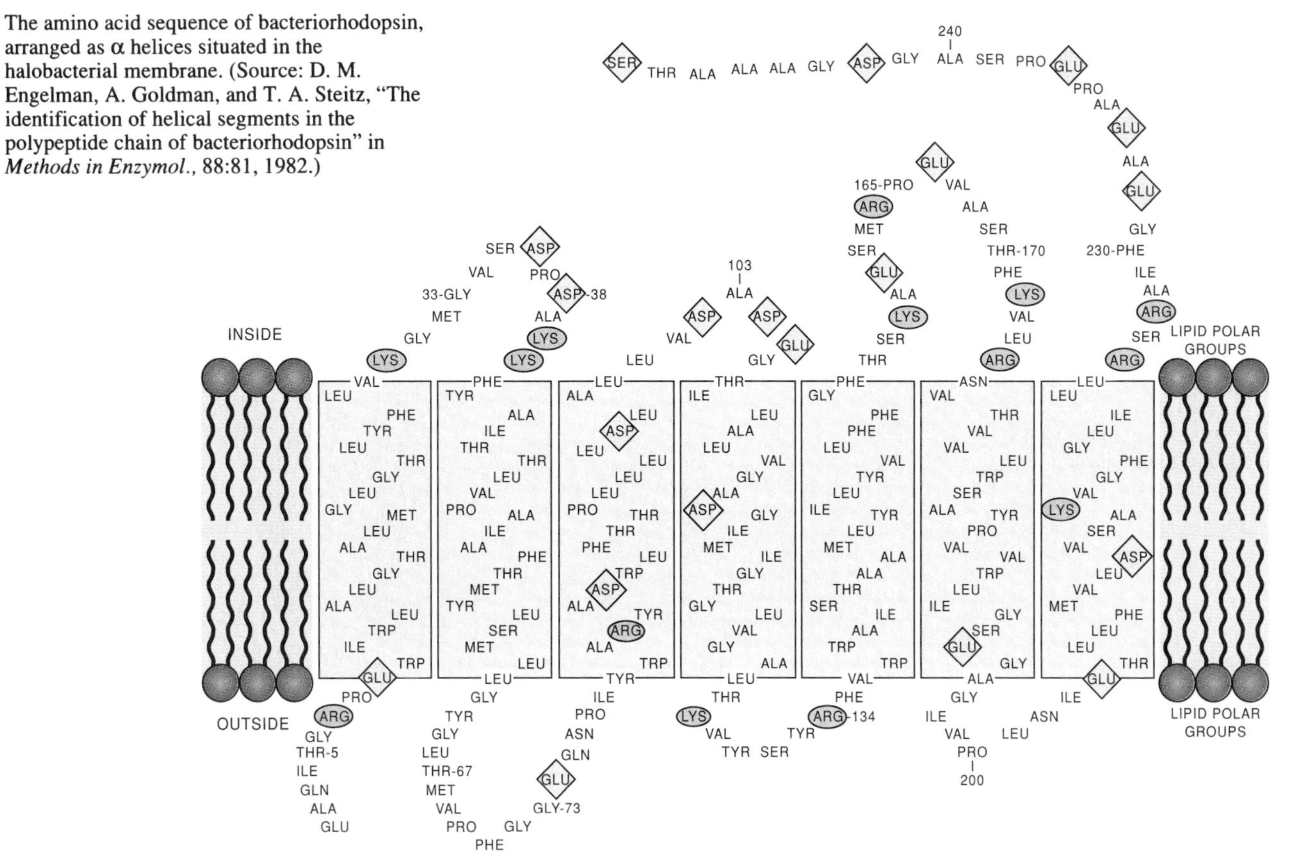

Figure 7.35

The amino acid sequence of bacteriorhodopsin, arranged as α helices situated in the halobacterial membrane. (Source: D. M. Engelman, A. Goldman, and T. A. Steitz, "The identification of helical segments in the polypeptide chain of bacteriorhodopsin" in *Methods in Enzymol.*, 88:81, 1982.)

From estimates of the number of residues required to make seven α helices long enough to each span the membrane (about 40 Å), it appears that virtually all of the polypeptide chain is required to form the helices, so that the interconnections between them must be quite short.

In more recent studies, the structure of bacteriorhodopsin has been determined to near-atomic resolution. These studies have led to several important suggestions concerning the structure of membrane proteins. First is the observation that the regions of structure within the membrane appear to be primarily α-helical. As outlined in chapter 4, α helices are secondary structures whose backbones are completely hydrogen-bonded, except at their ends. Consequently, they might be readily inserted into membranes, since there are no unsatisfied hydrogen-bonded groups whose stabilization would require the formation of hydrogen-bonded interactions with water. Second, the observed pattern of helices shows that only hydrophobic side chains are exposed to the hydrophobic interior of the membrane (figs. 7.34 and 7.35). The charged and polar amino acid side chains that do occur within the helices probably interact with each other or with water and are oriented toward the interior of the bundle of helices (see fig. 7.34). Put simply, the membrane protein bacteriorhodopsin is inside-out relative to globular proteins that exist in a polar water environment.

The first structure of a membrane protein at atomic resolution was determined by H. Michel, J. Diesenhofer, R. Huber, and co-workers. They were able to crystallize a protein comprising the photosynthetic reaction center (RC) from the purple sulfur bacterium *Rhodopseudomonas viridis* (also see chapter 16), and to determine its structure to a resolution of 2.3 Å using x-ray crystallography. This landmark achievement revealed for the first time many of the principles of membrane protein folding that previously had only been inferred for proteins such as bacteriorhodopsin.

The membrane-spanning RC complex consists of three protein subunits called L, M, and H, and associated with these are various cofactors for photosynthesis, including four molecules of bacteriochlorophyll, two of bacteriopheophytin, two quinone molecules, a nonheme iron, and a cytochrome molecule. The structure (fig. 7.36) clearly shows an array of roughly parallel α helices, five each from subunits L and M and one from subunit H, that are undoubtedly the eleven membrane-spanning regions of this protein. The amino acid residues comprising the helices are almost entirely hydrophobic, while the more hydrophilic regions of the protein form domains that are exposed on either side of the membrane-spanning region as shown in figure 7.36. We will consider the implications of this structure for the mechanism of photosynthesis in chapter 16.

Figure 7.36

The structure of the photosynthetic reaction center from purple bacteria. The presumptive transmembrane α helices are highlighted in color: *yellow (A–E)* = the five transmembrane helices of the M subunit; green (A–E) = the five transmembrane helices of the L subunit; purple (**A**) = the one transmembrane helix of the H subunit. Other cylindrical helices are also found on either hydrophilic surface in all of the subunits. Some of the cofactors associated with the reaction center are also shown. The solid lines connecting the α helices trace the backbone of the polypeptide chains of the three subunits. (Source: D. G. Rees, H. Komiya, T. O. Yeates, J. P. Allen and G. Feher, "The bacterial photosynthetic reaction center as a model for membrane proteins" *Annual Review of Biochemistry.* 58:607, 1989. Copyright © 1989 Annual Reviews Inc., Palo Alto, Calif.)

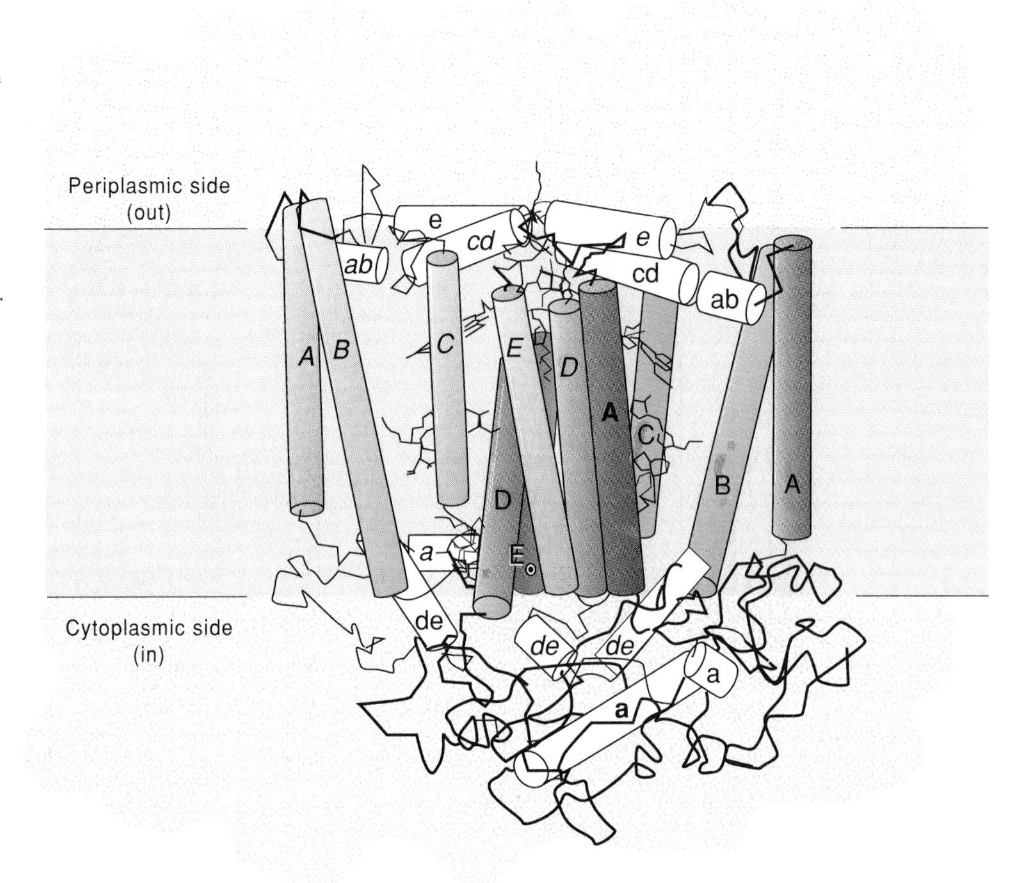

From the structures of integral membrane proteins that have been determined or inferred so far, a number of generalizations now appear valid:

1. A significant proportion of the polypeptide must interact with membrane lipids through hydrophobic interactions.
2. α helices are common intramembrane structures of polypeptides.
3. The main carbohydrate portions of plasma membrane glycoproteins are always found on the outside of the cell.
4. Hydrophilic regions of membrane proteins are exposed at the membrane surface, although the sizes and locations of these domains (inside or outside) are highly variable among different membrane proteins.

Given these ground rules, many workers have tried to predict the intramembrane structures of membrane proteins from their primary amino acid sequences alone, and they have had some success. For example, hydrophobic, membrane-spanning α helices can be predicted by looking for hydrophobic stretches of amino acids in the primary sequence that are long enough to span the membrane (about twenty residues). Computer algorithms have been developed to identify such regions in membrane proteins; these algorithms yield hydropathy plots, an example of which is shown in figure 7.37 for bacteriorhodopsin. In this case, the hydropathy plot reasonably accurately predicts the seven membrane-spanning helices inferred from the electron-diffraction structure.

Most Integral Membrane Proteins Can Move within the Membrane Structure

The currently accepted model of biological membrane structure, called the fluid mosaic model, was proposed by S. J. Singer and G. L. Nicolson in 1972 (fig. 7.38). In the fluid mosaic model, the essential structural repeating unit is the phospholipid molecule, in a bilayer arrangement with a thickness of about 50 Å. Integral membrane proteins are "dissolved" in the bilayer in a seemingly random fashion. Some proteins (such as certain mitochondrial cytochromes) are localized at one or the other of the two surfaces of the lipid bilayer; other proteins (such as glycophorin and the anion channel) span the membrane but with considerable mass extending into the aqueous surroundings; and still others (such as bacteriorhodopsin) are largely embedded in the hydrophobic matrix. A most important aspect of the fluid mosaic model is its dynamic character, with most components capable of relatively rapid lateral diffusion and of rotational motion about an axis perpendicular to the plane of the bilayer.

Figure 7.37

Example of a hydropathy plot for bacteriorhodopsin. The water-to-oil transfer free energy (ΔG) for successive 19-amino-acid segments is plotted as a function of the first amino acid in the sequence of a given segment. Regions below the line for $\Delta G = 0$ thus correspond to favorable partitioning of a region of the polypeptide into the lipid bilayer. Points A through G are the proposed start points of the transmembrane α helices of bacteriorhodopsin. This analysis was, in part, used to formulate the model shown in figure 7.35. (Source: D. M. Engelman, A. Goldman, and T. A. Steitz, "The identification of helical segments in the polypeptide chain of bacteriorhodopsin" in *Methods Enzymology* 88:81, 1982.)

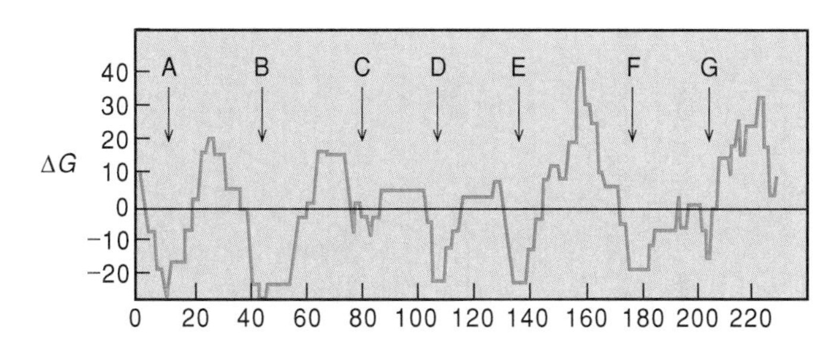

Figure 7.38

The fluid mosaic membrane as envisioned by Singer and Nicolson. The proteins within the bilayer have lateral mobility about an axis perpendicular to the plane of the membrane. Some rotational movement about an axis perpendicular to the plane of the membrane is also possible.

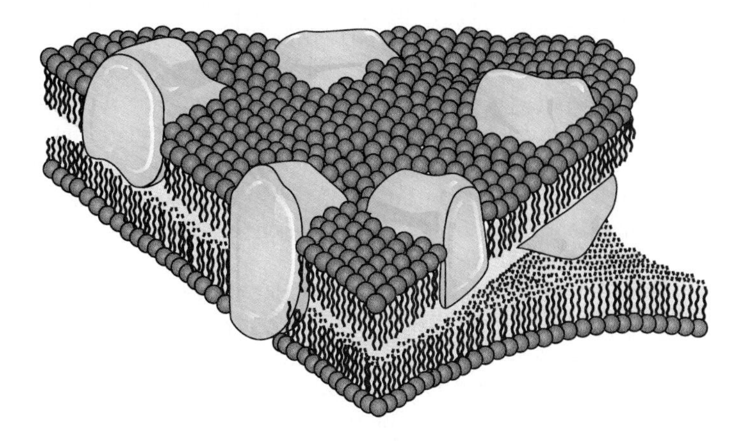

Rotation of lipids and proteins through the plane of the bilayer, however, is proposed to be a rare event. Figure 7.38 therefore depicts a hypothetical biological membrane at one point in space and time.

The abilities of membrane proteins and lipids to diffuse laterally, parallel to the plane of the bilayer, have been amply demonstrated by a variety of techniques. One of the more straightforward of these is the fluorescence photobleach recovery technique. In this method, fluorescently labeled membrane components are bleached (their fluorescence is destroyed) in a small area of the membrane by a short, intense flash of light. By monitoring this region as a function of time, using a fluorescence microscope, one can observe a reappearance of fluorescence (recovery), which is due to unbleached molecules diffusing into the treated area. If a specific membrane component is so labeled, its diffusion coefficient in the plane of the membrane can be calculated from the time course of fluorescence recovery. This technique has revealed lateral movements of both proteins and lipids in a number of biological membranes. For example, rhodopsin, a light-receptor protein of vertebrate retinal membranes (chapter 36), has proved to be highly mobile within the lipid bilayer. Other proteins vary enormously in their intramembrane diffusion coefficients (see the following discussion).

In an elegant experiment also designed to test protein mobility in biological membranes, L. Frye and M. Edidin fused a mouse cell and a human cell with the aid of a virus, called Sendai virus, that facilitates such fusion. Membrane proteins of the mouse cell were labeled with specific antibodies that fluoresced green, while similar labeling of the human cells was with red fluorescent antibody. If membrane proteins failed to diffuse, then the heterokaryon resulting from the fusion of these two cells should remain "half red" and "half green" even after long periods of incubation at physiological temperatures (37° C). In this experiment, however, significant intermixing of red and green labels was observed in less than 30 min, and complete mosaics of human mouse proteins were seen in all heterokaryons within 1 h. These observations provided direct evidence for the lateral mobility of integral membrane proteins and indirect evidence for the movement of the lipid components of the bilayer.

The specific orientations maintained by integral membrane proteins with respect to the bilayer suggest that rotation of these molecules through the plane of the bilayer (flip-flop or transverse motion) does not occur. Since most integral membrane proteins have at least some hydrophilic surface area exposed at one or both membrane faces, a transverse rotation would be an energetically unfavorable event, requiring transient interaction of these polar groups with the hydrophobic interior of the bilayer. Indeed, physical and chemical measurements have failed to detect such motions of proteins in biological membranes. Comparable measurements for phospholipids in model membranes have revealed that flip-flop of these molecules can occur, but that half-times for this process can be as long as days at physiological temperatures. These rates are many orders of magnitude smaller than those of lateral diffusion and again can be explained by the high energies required for these processes. As we will see, the asymmetric topography of biological membranes is essential to many membrane functions. This topography is maintained by the amphipathic nature of the membrane's lipid and protein constituents.

Temperature, Composition, and Extramembrane Interactions Affect Dynamic Properties

Diffusion rates in lipid bilayers are a function both of temperature and of the composition of the membrane being examined. Bilayers consisting of a single type of phospholipid typically show an abrupt change in physical properties over a characteristic and narrow temperature range (T_m). These temperature-dependent phase transitions are due to an organizational change in the fatty acyl side chains and can be detected by a variety of physical techniques. For example, differential scanning calorimetry measures heat absorption as a function of temperature and shows that over the temperature range of the phase transition, a relatively large amount of heat is absorbed per degree of temperature change compared with temperature ranges well above or below the transition (fig. 7.39). It is thought that the membranes pass from a state in which the fatty acyl chains are highly ordered (gel phase) to one in which they are far more mobile (liquid crystalline phase). This process is illustrated in figure 7.39. It is accompanied by increased rotational motion about the carbon–carbon bonds of the hydrocarbon chains of the phospholipids, allowing them to assume more random, disordered conformations.

In contrast to pure phospholipid bilayers, membranes isolated from cells usually undergo such phase transitions over a much broader (~10° C) temperature range. In some instances, distinct transitions cannot even be distinguished, because of the heterogeneity of lipids found in most biological membranes and because integral membrane proteins may decrease the mobility of lipids in their immediate vicinity. In general, lipids bearing short or unsaturated fatty acyl chains undergo phase transitions at lower temperatures than those containing long-chain saturated fatty acids. This is because short hydrocarbon chains have a smaller surface area with which to undergo hydrophobic interactions that stabilize the gel state and because *cis* unsaturation introduces "kinks" in the fatty acyl chain (see fig. 7.1) that also lead to more disorder in the bilayer. These effects of fatty acid composition on phase transition temperatures of model bilayers are apparent from the data presented in table 7.11, and in part explain the fact that broad phase transitions are a general characteristic of cellular membranes.

From the data in table 7.11, it can further be seen that the midtransition temperature of a pure phospholipid suspension also depends on the nature of the polar head group. Thus the dipalmitoyl esters of phosphatidic acid and phosphatidylethanolamine "melt" at temperatures about 20° C higher than the same derivatives of phosphatidylcholine and phosphatidylglycerol. Divalent cations, such as Ca^{2+} and Mg^{2+}, also affect membrane fluidity, presumably because of ionic interactions with neighboring phosphoryl head groups, tending to "tie" phospholipid molecules together and decrease their mobility. This is undoubtedly one of the reasons that divalent cations are well-known stabilizers of biological membranes, and their removal often facilitates lysis of cells as well as dissociation of peripheral membrane proteins.

Finally, cholesterol is a well-known modulator of membrane structure in animal cells and in one type of bacteria, the mycoplasmas. In these cells cholesterol intercalates among the fatty acyl chains, with its polar hydroxyl group interacting with the polar head groups of membrane lipids. At low concentrations of cholesterol in phospholipid bilayers, separate domains, or patches, of cholesterol plus phospholipid and pure phospholipid appear to exist, with a resultant broadening of the phase transition profile compared with phospholipid alone. Its effects

Figure 7.39

(*Top*) Differential scanning calorimetry of various phospholipids dispersed in water. Heat absorption is indicated by a trough in the plot relating differential heat flow to temperature. The lowest point in the trough is the phase transition temperature (T_m): (*a*) dipalmitoyl phosphatidylethanolamine; (*b*) dimyristoyl lecithin; (*c*) dipalmitoyl lecithin; (*d*) egg lecithin (plus ethylene glycol to prevent freezing).
(*Top*) (Reproduced, with permission, from the *Annual Review of Biophysics and Bioengineering.*,Vol. 5, © 1976 by Annual Reviews, Inc.)
(*Bottom*) Molecular interpretation of the heat-absorbing reaction during the phase transition.

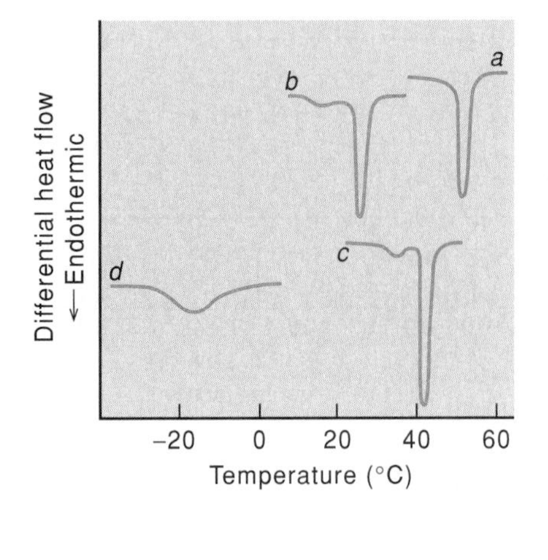

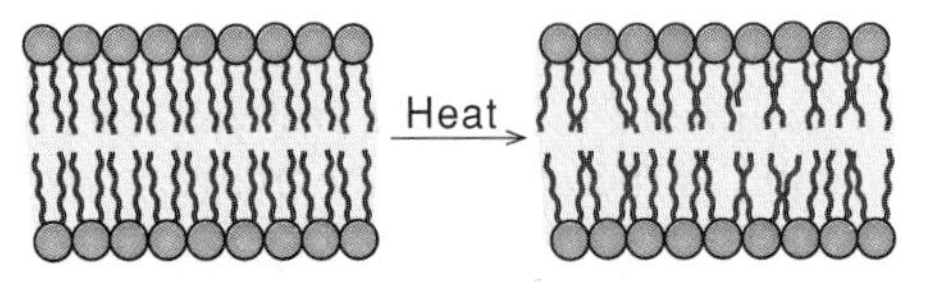

on various physical parameters of membranes depends on its proportion to other lipid components, as well as on the membrane system examined. However, a general property of membranes containing a high concentration of cholesterol appears to be an inhibition of processes dependent on a fluid environment. For example, the permeability of vesicles made from purified egg phosphatidylcholine to both water and glucose decreases when greater than 20 mol % cholesterol is incorporated into the vesicles. Intercalated cholesterol probably restricts the freedom of motion of the phospholipid hydrocarbon side chains, thereby decreasing the mobilities of membrane constituents.

Most cells maintain a lipid composition that allows for a relatively rapid lateral diffusion of many membrane components at normal growth temperatures. Membrane-associated processes, such as the vectorial reactions catalyzed by some transmembrane transport systems (chapter 32) and endo- and exocytosis, rely on a semifluid environment for their operation. Consequently, organisms have evolved intricate mechanisms to maintain this environment under a variety of conditions. One of the most remarkable of these is the ability of plant, animal, and bacterial cells to increase the proportion of membrane unsaturated fatty acids in response to a decrease in temperature. This ensures proper functioning of the membrane at the lower temperature. In bacteria, this modulation appears to be the result

Table 7.11
Midtransition Temperatures for Aqueous Suspensions of Phospholipids

Phospholipid[a]	T_m (°C)
Di-14:0 PC	24
Di-16:0 PC	41
Di-18:0 PC	58
Di-22:0 PC	75
Di-18:1 PC	~22
1-18:0, 2-18:1 PC	3
Di-14:0 PE	51
Di-16:0 PE	63
Di-14:0 PG	23
Di-16:0 PG	41
Di-16:0 PA	67

[a]Phospholipid abbreviations are as in table 7.8; additionally, Di-14:0, for example, refers to dimyristoyl (14 carbons, 0 double bonds).

Source: M. K. Jain and R. C. Wagner, *Introduction to Biological Membranes.* Copyright © 1980 John Wiley & Sons, Inc., New York, N.Y.

of temperature effects on the activities and/or the induction of synthesis of enzymes involved in the biosynthesis of phospholipids containing unsaturated fatty acyl chains. In plants and yeast, the increased solubility of O_2 at low temperatures apparently increases the proportion of unsaturated fatty acids because O_2 is a substrate of the desaturase enzyme that leads to their biosynthesis. Other factors that affect unsaturated fatty acid biosynthesis, and thus membrane fluidity, in various systems include light (in plants), nutrition, developmental stage, and aging.

Some Membrane Proteins Are Immobilized by External Forces

Despite the large body of evidence in favor of the lateral mobility of many membrane constituents at physiological temperatures, it is an oversimplification to view a biological membrane only as a random "sea" of lipids with proteins floating about aimlessly in them. Nearly all eukaryotic cells contain within their cytoplasm a cytoskeleton, made up in part of microtubules and microfilaments, which consist primarily of the proteins tubulin and actin (see table 5.1), respectively. Microfilaments, which are structurally similar to actin filaments of muscle cells, have been shown to form bundles just beneath the plasma membrane of many cells (fig. 7.40). These bundles are believed to have an important role in such processes as locomotion and phagocytosis, which involve local or general changes in the shape of the cell surface and thus in the plasma membrane. In many cases, direct or indirect association of microfilaments with the plasma membrane has been demonstrated. For example, the filamentous protein fibronectin, a peripheral, cell-surface glycoprotein in many animal cells, is believed to have a role in cell–cell and cell–substratum adhesion. A transmembrane association of fibronectin with cytoskeletal microfilaments has been deduced from immunofluorescent microscopy (fig. 7.41). Lateral diffusion of fibronectin has been shown to be at least 5,000 times slower than that of freely diffusible membrane proteins and

Figure 7.40

Membrane-associated cytoskeletal components of cultured mouse cells. (PM = plasma membrane, MF = microfilaments, MT = microtubules; 54,000X.) (From G. L. Nicolson, Transmembrane control of the receptors on normal and tumor cells. I. Cytoplasmic influence over cell surface components, *Biochem. Biophys. Acta* 457:57, 1976. Reprinted with permission from Elsevier Science Publishers.)

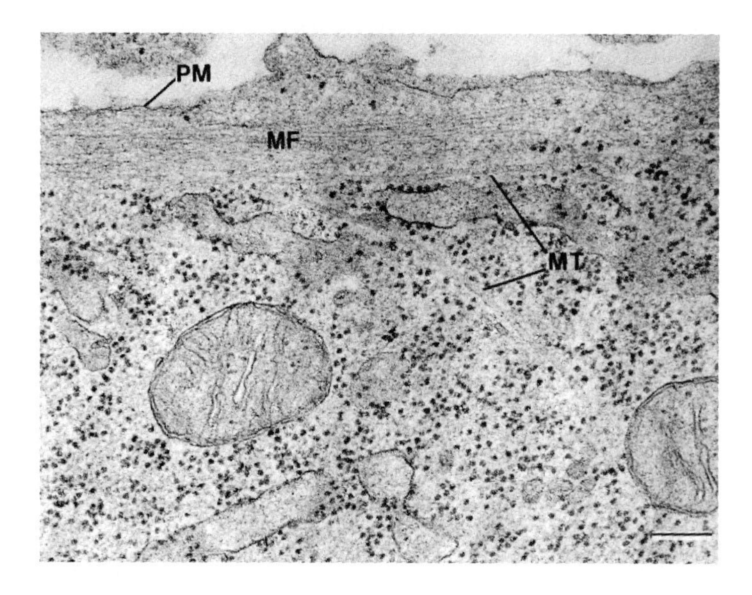

Figure 7.41

Colinearity of actin and fibronectin fibrils in cultured hamster fibroblast cells as seen by immunofluorescence in the light microscope. (*a*) The fluorescence is due to actin antibodies bound to intracellular microfilaments: These antibodies were produced in rabbits. Their location of binding is made visible by staining with fluorescent goat antirabbit antibodies. (*b*) The fluorescence is that of the fluorescent antifibronectin antibodies bound to extracellular fibronectin fibrils. The correspondence between the arrangement of actin and fibronectin filaments strongly suggests a transmembrane association between the two proteins. (From R. O. Hynes and A. T. Destree, "Relationships between fibronectin (LETS protein) and actin," *Cell* 15:875, 1978 © Cell Press.)

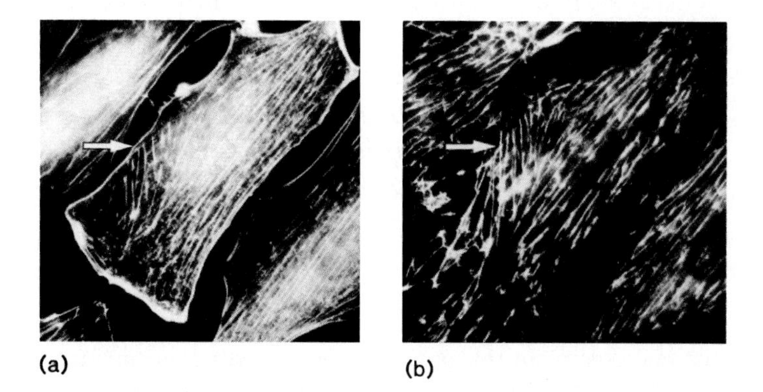

lipids. A number of transmembrane and membrane-associated proteins have been implicated in the attachment of extracellular fibronectin to intracellular actin filaments, including a transmembrane fibronectin receptor and the intracellular proteins talin, vinculin, and α-actinin. One possible arrangement of these proteins is depicted in figure 7.42.

A second well-characterized membrane cytoskeleton is that just beneath the plasma membrane of the erythrocyte. This scaffoldlike structure is believed to have a role in maintaining the biconcave shape of the red blood cell and in protecting it from the large shear forces it encounters in the peripheral circulation, while still conferring a degree of flexibility on the cell surface. The major structural protein of the red-cell membrane skeleton is called spectrin. It consists of α (260 kDa) and β (225 kDa) polypeptide-chain subunits. Each subunit folds into a rather extended structure consisting of 106-amino-acid repeating units, each of which in turn contains three α helices. An $\alpha\beta$ dimer of spectrin is formed when an α and a β subunit twist about one another in parallel; in the erythrocyte membrane, these dimers are joined end-to-end to produce an $\alpha_2\beta_2$ tetramer with a length of about 200 nm (fig. 7.43*a*). Just beneath the erythrocyte membrane, spectrin tetramers are indirectly attached to the membrane through the protein ankyrin, which cross-links spectrin with the anion-channel (band 3) protein, and through protein 4.1 (see fig. 7.31), which cross-links spectrin with glycophorin. In addition, each molecule of protein 4.1 also assists in the binding of several tetramers of spectrin to actin filaments from the cytoskeleton of the red cell. The result is a mesh-work of spectrin tetramers linked to both the cytoplasmic membrane and the cytoskeleton as depicted in figure 7.43*b*. As a consequence, the plasma membrane is structurally stabilized, and the mobilities of the anion channel and glycophorin are likely to be greatly reduced compared with those of other membrane proteins.

From the foregoing discussion, we can conclude that temperature, ionic environments, and composition all can affect the general physical state of a biological membrane, while local mobilities of membrane components can be influenced by protein–protein, lipid–protein, and lipid–lipid interactions.

Both Proteins and Lipids Are Asymmetrically Arranged in Biological Membranes

Physical chemical probes have provided ample evidence for the unidirectional, asymmetric orientation of membrane proteins with respect to the lipid bilayer. This asymmetry also has been more directly observed by the technique of freeze-fracture electron microscopy. In this procedure, whole cells or membranes are rapidly frozen, and the specimen is then struck with a sharp knife called a microtome. Very often, the fracture plane actually passes between the outer and inner monolayers (leaflets) of the membrane lipid bilayer because the relatively weak hydrophobic interactions between the fatty acyl chains of the two leaflets offer a "path of least resistance" (fig. 7.44). The inner surfaces of the two leaflets can then be viewed in the electron microscope after heavy-metal shadowing of the fractured specimen. Many biological membranes, when split in half by freeze-fracture, can be seen to have quite different morphologies of the inner and outer leaflet surfaces, indicating preferential adherence of some membrane proteins to one of the two monolayers (see fig. 7.44). This feature presumably reflects the asymmetric and unidirectional disposition of the polypeptides across the lipid bilayer. In a related method, freeze-etching electron microscopy, ice is sublimed away from the sample after freeze-fracture, to allow a view of the outer surfaces of the two leaflets. Both procedures are especially useful in examining membrane morphology, because they avoid the potentially destructive fixing, embedding, and staining steps of more conventional sample preparation procedures.

Proteins are not the only membrane constituents to show asymmetric orientations. Many lipids also have been shown to be unequally distributed between the inner and outer leaflets of many biological membranes. In 1972, M. S. Bretscher first showed that most of the phosphatidylethanolamine and all of the phosphatidylserine of human erythrocyte membranes are inaccessible to chemical modification from outside the cell. Both phosphoglycerides, however, could be modified in erythrocyte ghosts in which both the inner and outer leaflets were exposed

Figure 7.42

Schematic illustration of a so-called focal contact, showing how extracellular fibronectin is believed to be indirectly attached to the intracellular cytoskeleton through a transmembrane fibronectin receptor and several other peripheral membrane proteins. (Source: B. Alberts et al., *Molecular Biology of the Cell,* 2d ed. Copyright © 1989, Garland Publishing, New York, N.Y.)

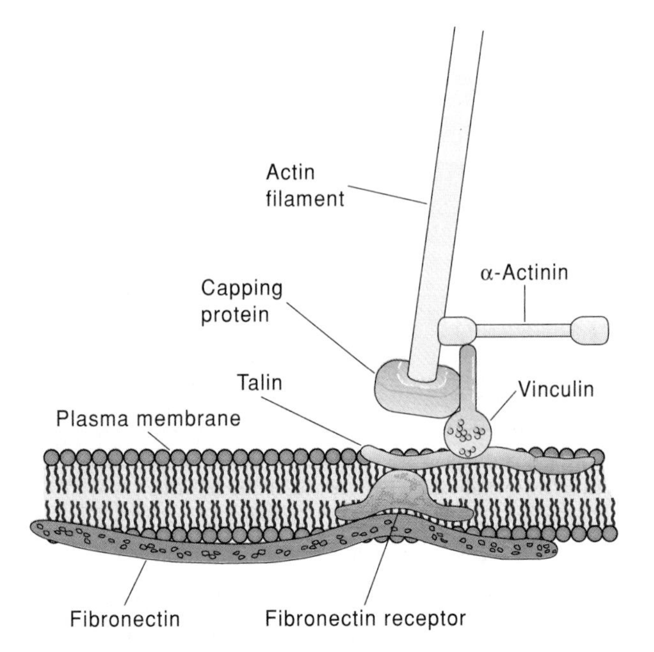

Figure 7.43

The structure of spectrin and the location of spectrin in the cytoskeleton. (*a*) An αβ dimer of spectrin. Both α and β subunits are extended structures consisting of end-to-end domains of 106 amino-acyl residues folded into three α helices (left); the subunits twist about one another loosely as shown. (*b*) The erythrocyte membrane skeleton. Spectrin tetramers ($\alpha_2\beta_2$), shown in yellow, are linked to the cytoplasmic domain of the anion channel (blue) by the protein ankyrin (red), and to glycophorin and actin filaments by protein 4.1. This structure lends structural stability to the red cell membrane while maintaining sufficient flexibility to allow erythrocytes to withstand substantial shear forces in the peripheral circulation.

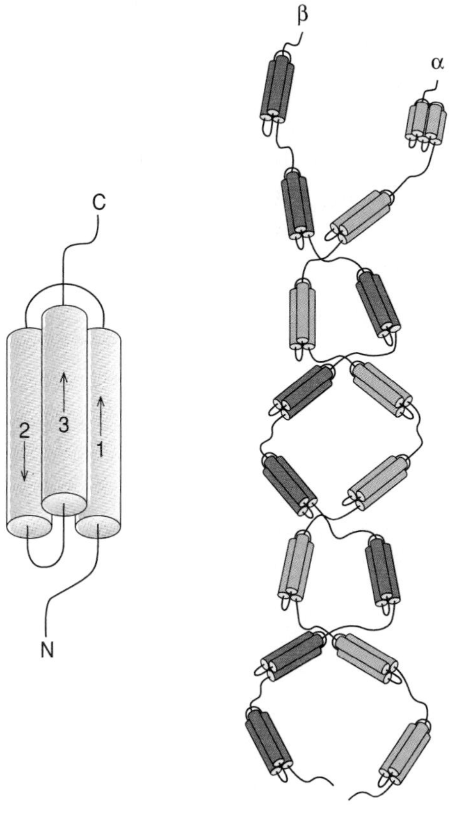

to the chemical reagent. These results suggested that lipid asymmetry also could exist in a biological membrane. Subsequent experiments using enzymes that degrade phospholipids (e.g., phospholipases and sphingomyelinase) showed that phosphatidylcholine and sphingomyelin were preferentially found in the outer leaflet of human red blood cell membranes (table 7.12). Thus, although total membrane lipid is equally distributed between outer and inner monolayers in the erythrocyte membrane, the lipid composition of each leaflet shows striking differences. Lipid asymmetry has now been found in membranes from a variety of biological sources; some examples are given in table 7.12.

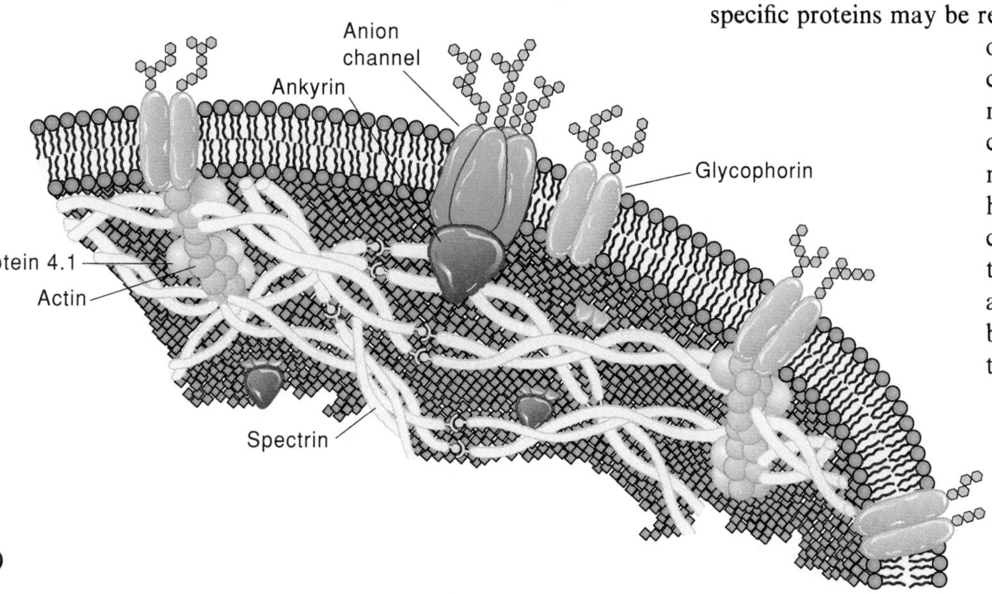

Given the asymmetric arrangements of proteins in membranes along with the nearly unidirectional nature of many membrane-associated processes (chapter 32), it is perhaps not surprising that certain lipids might be found predominantly on one side or the other of the bilayers. This could at least partially be a consequence of the fact that some integral membrane proteins appear to associate preferentially with certain lipids, and the orientation of the protein could therefore influence the orientation of bound lipid with respect to the bilayer. However, recent work strongly suggests that the asymmetric partitioning of phospholipids into the two bilayer leaflets is also due, at least in part, to the activities of specific phospholipid transporters, which are themselves integral membrane proteins. For example, a protein has been identified in membranes of erythrocytes from various sources that translocates phosphatidylserine from the outer to the inner membrane leaflet. This 31,000-dalton protein requires ATP for its activity, presumably to provide the energy necessary to catalyze this energetically unfavorable flip-flop. Such proteins may therefore be generally responsible for the generation of lipid asymmetry in biological membranes.

It is clear from energetic arguments how membrane protein and lipid asymmetry can be maintained, and, from functional requirements, why this asymmetry exists. Although specific proteins may be responsible for lipid asymmetry in biological membranes, a much more complex question is, How do integral membrane proteins adopt their specific asymmetric orientations in membranes? Although this question has not yet been fully answered, it is clearly related to the broader question of how biological membranes are assembled in the first place (membrane biogenesis). We will consider this topic in Part 5 (chapter 22).

Figure 7.44

Freeze-fracture electron microscopy. (*a*) When struck with a sharp knife, membranes embedded in ice usually fracture between the monolayer leaflets of the lipid bilayer. (*b*) Freeze-fracture electron micrograph of the plasma membrane of *Streptococcus faecalis,* showing a large number of protrusions (presumably proteins) on the outer fracture face and the relative lack of such particles on the inner fracture face (inset). (From H. C. Tsien and M. L. Higgins, "Effect of temperature on the distribution of membrane particles in *Streptococcus faecalis* as seen by the freeze-fracture technique," *J. Bacteriol.* 118:725, 1974. Reprinted with permission from American Society for Microbiology.)

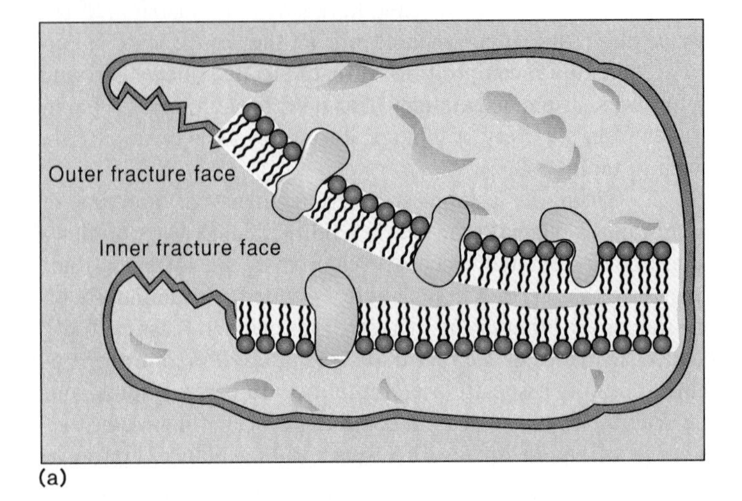

(a)

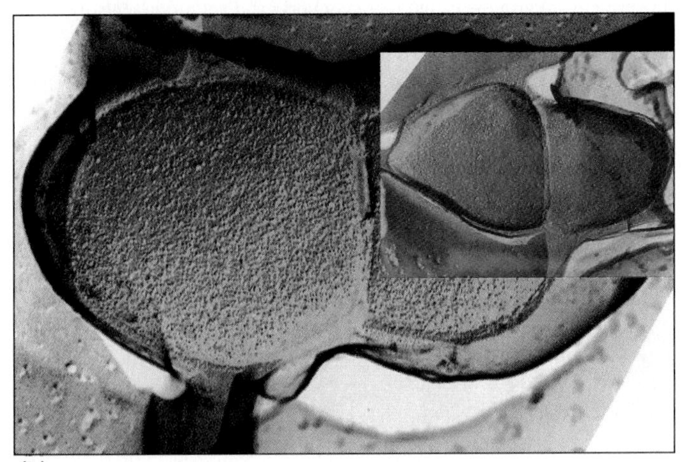
(b)

Table 7.12
Lipid Asymmetry in Biological Membranes

Membrane	Preferential Outside	Preferential Inside	Equal	Membrane	Preferential Outside	Preferential Inside	Equal
Various erythrocytes	PC, SM, glycolipids, cholesterol	PE, PS	—	*E. coli* outer membrane	—	PE	—
				Bacillus megaterium	—	PE	—
Rabbit sarcoplasmic reticulum	PE	PS	PC lyso PC	*Micrococcus lysodeikticus*	PG	PI	DPG
Mouse LM cell plasma membrane	SM	PE	PC	PC/PE artificial vesicles	PC	PE	—

Note: Abbreviations are as in table 7.8.

Source: J. A. F. Op den Kamp, "Lipid asymmetry in membranes" in *Annual Review of Biochemistry,* 48:47, 1979. Copyright © 1979 Annual Reviews, Inc. Palo Alto, Calif.

Summary

In this chapter we have focused on the structures and functions of lipids and membranes. The following points are central to this subject.

1. Lipids are biological molecules that are soluble in organic solvents. They comprise a wide diversity of structures and they serve four major biological functions: (a) some are major structural elements of membranes; (b) others are efficient reserves for the storage of energy; (c) many vitamins and hormones are lipids or lipid derivatives; and (d) the bile acids help to solubilize the other lipid classes during digestion.
2. Fatty acids are the major structural components of many membrane lipids and are the biological molecules in which energy can be stored most efficiently. They exhibit a large number of structural variations, primarily in chain length and in the number and location of double bonds. Triacylglycerols are the molecules in which fatty acids are stored for energy and are found mostly in adipose tissue.
3. Major lipids found in biological membranes include the phospholipids (phosphoglycerides and sphingomyelin), the glycosphingolipids, and cholesterol.
4. The phospholipids and glycosphingolipids contain a polar head group and a nonpolar tail portion composed of long-chain acyl or alkyl groups. The amphipathic nature of these

lipids allows them to form lipid bilayers spontaneously in an aqueous medium. Hence they play a major structural role in all biological membranes.

5. Cholesterol, a derivative of the tetracyclic hydrocarbon perhydrocyclopentanophenanthrene, is the most prominent member of the steroid family of lipids. Cholesterol is an important structural component of some eukaryotic membranes, but is generally absent from bacterial membranes.
6. Biological membranes are composed of lipids, proteins, and carbohydrates in various combinations.
7. Membrane lipids are complex mixtures that reflect the variety both of fatty acids and of polar head groups found in these molecules. The predominant lipid in eukaryotic membranes is phosphatidylcholine, while in bacteria it is phosphatidylethanolamine.
8. Membrane proteins are either peripheral or integral. Peripheral proteins may be dissociated from the membrane in a water-soluble form by relatively mild procedures that disrupt hydrogen or ionic bonds. Integral proteins can be brought into solution only by disrupting the bilayer structure, for example with detergents.
9. Physical and chemical analyses of purified integral membrane proteins show that they vary widely in structure, but that most of them require a hydrophobic environment for maintenance of their biologically active structures.
10. Some integral proteins, such as bacteriorhodopsin, have most of their mass buried within the lipid bilayer. Others have a large proportion of the polypeptide extending into the aqueous environment on either or both sides of the membrane. Examples of this second type are the anion-channel protein, glycophorin, and the photosynthetic reaction center of purple bacteria.
11. Many integral membrane proteins have been shown to span the lipid bilayer. The membrane-spanning regions of these proteins are often hydrophobic α helices.
12. If carbohydrates are covalently bound to a plasma membrane protein, they are always found on the external surface of the cell.
13. In 1972, Singer and Nicolson proposed the fluid mosaic model for membrane structure. In this model, the essential structural unit is a relatively impermeable lipid bilayer in which protein molecules are embedded. Both protein and lipid molecules are capable of rapid lateral diffusion. They are also capable, at physiological temperatures, of rotational motion about an axis perpendicular to the plane of the bilayer, but spontaneous rotation through the plane of the bilayer (transverse or flip-flop motion) is very rare. Thus in general the membrane maintains an asymmetric structure with respect to both protein orientation and the distributions of specific lipids between inner and outer leaflets. Biophysical techniques such as x-ray diffraction, and fluorescence spectroscopy, have confirmed these essential features of the fluid mosaic model.
14. Many factors have been shown to affect the general mobility of membrane components, including temperature, the fatty acid compositions of phospholipids and glycolipids, and the presence or absence of cholesterol. Local mobilities within a given membrane system are a function of protein–protein, lipid–protein, and lipid–lipid interactions.
15. In response to environmental changes, many cells can regulate the composition of their membranes to maintain the overall semifluid environment necessary for many membrane-associated functions.
16. The asymmetric orientation of integral membrane proteins with respect to the bilayer is necessary for their proper functioning. Similarly, the unequal partitioning of some membrane lipids between the two leaflets of the bilayer may also be necessary for the functions of some membrane systems. This partitioning may be influenced by both protein asymmetry and the activities of specific lipid transporters.

Selected Readings

Bennett, V., The membrane skeleton of human erythrocytes and its implications for more complex cells. *Ann. Rev. Biochem.* 54:273, 1985. A review of the organization of the red blood cell surface, including a discussion of the possible roles of the cytoskeleton.

Deuel, H. J., *The Lipids: Biochemistry,* vol. 3. New York: Interscience, 1957. A comprehensive and classical treatise on the biochemistry of lipids until the mid-1950s.

Diesenhofer, J., O. Epp, N. Miki, R. Huber, and H. Michel, X-ray structure analysis of a membrane protein complex. Electron density map at 3-Å resolution and a model of the chromophores of the photosynthetic reaction center from *Rhodopseudomonas viridis. J. Mol. Biol.* 180:385, 1984. First x-ray structure of an integral membrane protein.

Fleischer, S., and L. Packer (eds.), *Methods in Enzymology,* vols. 31 and 32. New York: Academic Press, 1974. Volume 31 of this invaluable series focuses on subcellular fractionation and membrane isolation techniques, including isopycnic and differential centrifugation. Volume 32 deals with the composition and characterization of various membranes and membrane components, and includes articles on model membrane systems.

Fleischer, S., and B. Fleischer (eds.), *Methods in Enzymology,* vol. 172. New York: Academic Press, 1989. The more recent volume updates volumes 31 and 32, and deals with methods of membrane separation, analysis and physical characterization.

Hawthorne, J. N., and G. B. Ansell (eds.), *New Comprehensive Biochemistry,* vol. 4: *Phospholipids.* Amsterdam: Elsevier, 1982. Detailed collection of review articles on the structures, functions, and biosynthesis of the phospholipids.

Helenius, A., and K. Simons, Solubilization of membranes by detergents. *Biochim. Biophys. Acta* 415:29, 1975. Lengthy review of the properties of detergents and their use for membrane solubilization.

Henderson, R., J. M. Baldwin, T. A. Ceska, F. Zemlin, E. Beckmann, and K. H. Downing, Model for the structure of bacteriorhodopsin based on high resolution electron cryo-microscopy. *J. Mol. Biol.,* 213:899, 1990. The structure of bacteriorhodopsin is determined to near-atomic resolution in this article.

Jain, M. K., *Introduction to Biological Membranes,* 2d ed. New York: Wiley, 1988. Lucid introduction to the organization and functions of biological membranes.

Jay, D., and L. Cantley, Structural aspects of the red cell anion exchange protein. *Ann. Rev. Biochem.* 55:511, 1986. Review of techniques that have been used to determine the intramembrane structure of the band 3 (anion-channel) protein of erythrocytes.

Jennings, M. L., Topography of membrane proteins. *Ann. Rev. Biochem.* 58:999, 1989. Review of membrane protein structure and methods of structure prediction and determination.

Razin, S., and S. Rottem (eds.), *Current Topics in Membranes and Transport,* vol. 17: *Membrane Lipids of Prokaryotes.* New York: Academic Press, 1982. A comprehensive summary of lipid structure and function in bacteria. Other recent volumes in this continuing series contain up-to-date review articles on topics of current interest in membrane structure and function.

Rees, D. C., H. Komiya, T. O. Yeates, J. P. Allen, and G. Feher, The bacterial photosynthetic reaction center as a model for membrane proteins. *Ann. Rev. Biochem.* 58:607, 1989. Discusses theoretical and practical implications of the RC structure on the possible structures of other membrane proteins.

Singer, S. J., The molecular organization of membranes. *Ann. Rev. Biochem.* 43:805, 1974. Excellent review of the earlier literature on membrane structure through 1973.

Vance, D. E., and J. E. Vance (eds.), *Biochemistry of Lipids, Lipoproteins and Membranes.* Amsterdam: Elsevier, 1991. Compilation of articles on lipid and membrane structure, metabolism, and function.

Problems

1. "Individuals whose diet is devoid of plant tissue may develop essential fatty acid deficiency." Defend or refute this statement.
2. Compare the relative efficiency of extraction of free fatty acid, fatty acid methylester, and triacylglycerol into organic solvent. Which component(s) have a pH dependency of extraction? Why?
3. Both triacylglycerols and phospholipids have fatty acid ester components, but only one group can be considered amphipathic. Indicate which is amphipathic and explain why.
4. The term phosphatidylcholine (PC) defines a class of phospholipids. What portion of the phosphatidylcholine molecule is common to all members of the class? What portion of the molecule is variable among the PCs?
5. Use Haworth formulas to represent the structure of the glycosphingolipid referred to as Trihexosylceramide (D-Gal-α-1,4-D-Gal-β-1,4-D-Glc-β-1,1-ceramide). What type of linkage bonds the carbohydrate to the ceramide?
6. If phosphatidylcholine were dispersed in an aqueous buffer, what types of molecular associations would yield the most stable suspension?
7. Triton X-100 (M_r = 625, CMC = 0.24 mM) and sodium deoxycholate (M_r = 414, CMC = 4 mM) are each used to solubilize proteins from membranes before application of further isolation techniques.
   (a) Which of these detergents would be more easily removed by dialysis from an 0.1% (wt/vol) aqueous solution?
   (b) In which of these detergents would ion-exchange chromatography be more likely successful for protein purification?
   (c) In which of the detergents would gel exclusion chromatography be likely to approximate the true molecular weight of an integral membrane protein in 0.1% aqueous solution of the detergent? Why?
8. Quinones that are structurally similar to vitamins K_1 and K_2 (see fig. 11.24) are associated with the membranes of mitochondria and chloroplasts. How does the structure represented by vitamin K_2 explain the molecule's association with a hydrophobic membrane?
9. The relative orientation of polar and nonpolar amino acid side chains in integral membrane proteins is "inside out" relative to that of the amino acid side chains of water-soluble globular proteins. Explain.
10. What physical properties are conferred on biological membranes by phospholipids? How could the charge characteristics of the phospholipids affect binding of peripheral proteins to the membrane? What role might divalent metal ions play in the interaction of peripheral membrane proteins with phospholipids?
11. Predict the effects of the following on the phase transition temperature (T_m) and/or phospholipid mobility in pure dipalmitoylphosphatidylcholine vesicles (T_m = 41° C).
    (a) Raising the temperature from 30° to 50° C.
    (b) Introducing dipalmitoylphosphatidylcholine into the vesicle.
    (c) Introducing dimyristoylphosphatidylcholine into the vesicles.
    (d) Incorporating integral membrane proteins into the vesicles.
12. Differentiate between peripheral and integral membrane proteins with respect to location, orientation, and interactions that bind the protein to the membrane. What are some strategies used to differentiate peripheral and integral proteins by means of detergents or chelating agents?
13. Frequently, integral membrane proteins are glycosylated with complex carbohydrate arrays. Explain how glycosylation further enhances the asymmetric orientation of integral proteins.
14. In examining a cell extract from a culture of Gram negative bacteria, you identified a protein suspected to be periplasmic. Upon performing the osmotic shock experiment described in the text, you find a small amount (app. 10%) of the total enzymatic activity in the extracellular fraction. What control or other experiments might you do to convince yourself that the protein is periplasmic? If you had only data from the experiment described, what other possibilities could explain your results?
15. You are characterizing a protein in a membrane fraction that was dissolved in octylglucoside. You have estimated the molecular weight to be approximately 60,000. However, upon exhaustive treatment to remove most of the detergent, the protein elutes from a 100,000-M_r cutoff gel exclusion column in the void (excluded) volume. What can you conclude from these data?
16. Integral transmembrane proteins often contain helical segments of the appropriate length to span the membrane. These helices are composed of hydrophobic amino acid residues. In transmembranous proteins with multiple segments that span the membrane, you may find some hydrophilic residue side chains. Why are hydrophilic side chains not favored in single-span membrane proteins? How may the hydrophilic side chains be accommodated in multiple-span proteins?

Catalysis

3

PART

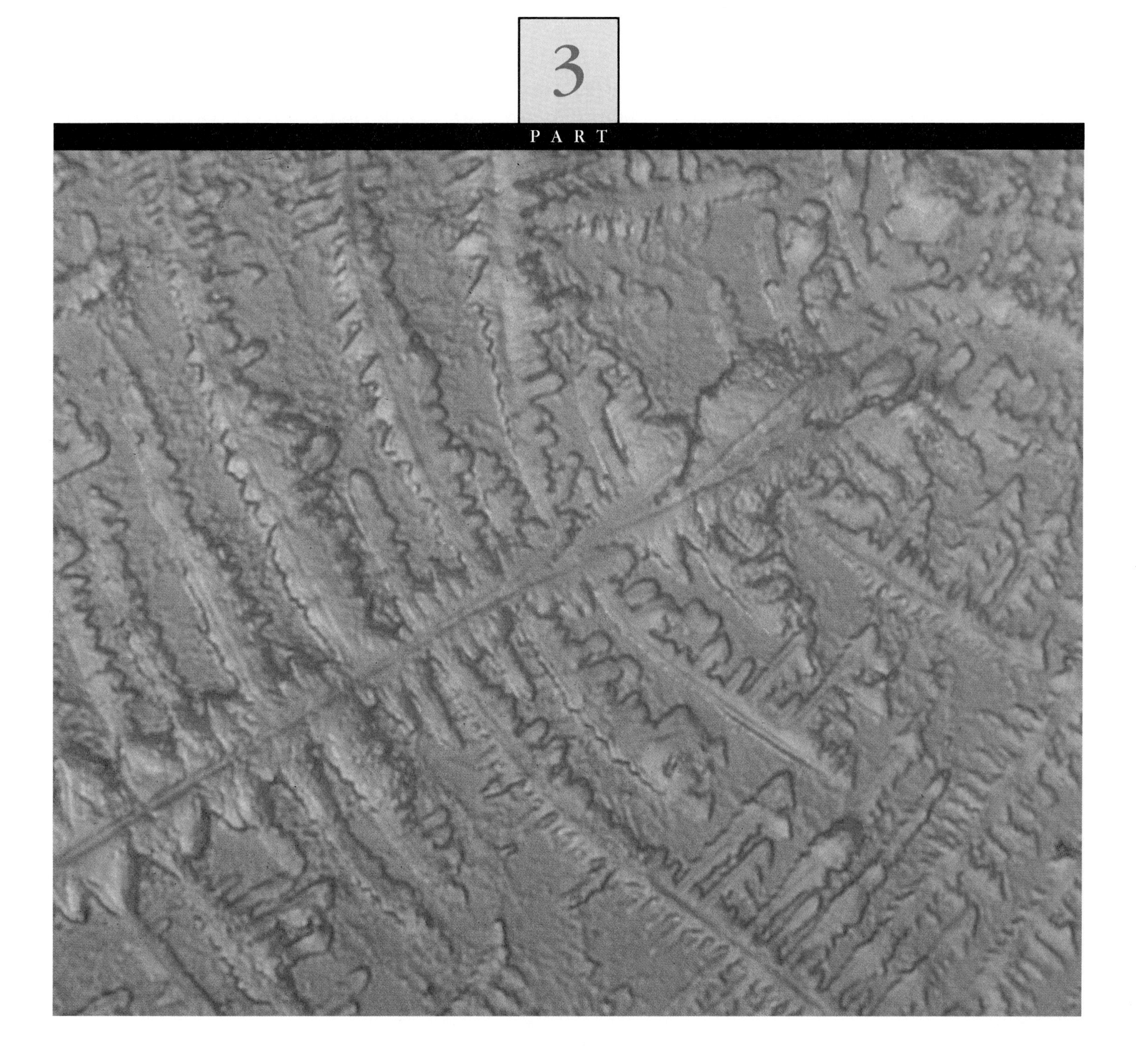

Light micrograph (160×) of human insulin crystals. The insulin was synthesized in bacteria using plasmids prepared by standard recombinant DNA technology methods. (© Visuals Unlimited.)

Most biochemical reactions are catalyzed by protein enzymes that function under very mild conditions so as not to disrupt the delicate fabric of the cell. The surface of the enzyme bears specific sites for binding the substrates and reactive groups that actually execute the reactions (chapters 8 and 9). Inhibitors also bind to the enzyme surface and either influence the binding of substrate or the activity of the enzyme. Enzyme kinetics is studied for two reasons: (1) it is a practical concern to determine the activity of the enzyme under different conditions; (2) frequently the analysis of enzyme kinetics gives information about the mechanism of enzyme action.

Most enzymes spontaneously process substrates when present, as long as inhibitory factors do not prevent this from happening. A few enzymes, known as regulatory enzymes, do not react spontaneously with their substrate(s) unless other metabolic signals give them the go-ahead. These metabolic signals usually consist of small molecules that bind to the enzyme and influence the structure of the active site (chapter 10). Regulatory enzymes function at strategic locations in metabolic pathways, usually at the first step in a linear pathway and at the first steps after a branchpoint in a forked pathway.

Frequently enzymes act in concert with small-molecule coenyzmes or cofactors, which supplement the reactive groups on the enzyme (chapter 11). Coenzymes or cofactors are distinguished from substrates by the fact that they function as catalysts. They are distinguishable from inhibitors or activators in that they participate directly in the reaction.

Enzyme Kinetics

Enzymes are biochemical catalysts. By a "catalyst" we mean a substance that accelerates a chemical reaction without itself undergoing any net change. The catalytic activities of some enzymes are extraordinary. For example, carbonic anhydrase, an enzyme found in red blood cells, catalyzes the reaction

$$CO_2 + H_2O \rightarrow H_2CO_3 \quad (1)$$

In the presence of the enzyme, this reaction occurs about 10^7 times more rapidly than it does in the absence of the enzyme. One molecule of carbonic anhydrase can hydrate about 10^6 molecules of CO_2 a second. Catalase, an enzyme that occurs in many different tissues, is even faster. It catalyzes the reaction

$$2\ H_2O_2 \rightarrow 2\ H_2O + O_2 \quad (2)$$

Each molecule of catalase can decompose more than 10^7 molecules of H_2O_2 a second! Catalase and carbonic anhydrase are among the fastest enzymes known. At the other extreme, lysozyme, which catalyzes the hydrolysis of glycosidic bonds in bacterial cell walls, succeeds in splitting a bond only about once every two seconds.

In this and the next two chapters, we will explore how enzymes work and how cells are able to regulate their activities. We will focus first on the kinetics of enzymatic reactions. By examining how the rates of enzymatic reactions depend on pH and temperature and on the concentrations of the reactants, products and inhibitors, we can learn a great deal about how enzymes operate.

The Discovery of Enzymes

The existence of catalysts in biological materials was recognized as early as 1835 by Jons Berzelius, who discovered several elements, introduced the way of writing chemical symbols, and coined the term "catalysis." Berzelius found that potatoes contained something that catalyzed the breakdown of starch, and he suggested that all natural products are formed under the influence of such catalysts. But the chemical nature of biological catalysts was unknown, and it remained a mystery for many years. In the period from 1850 to 1860, Louis Pasteur demonstrated that fermentation, the anaerobic breakdown of sugar to CO_2 and ethanol, occurred in the presence of living cells, and did not occur in a flask that was capped after any cells that it contained had been killed by heat. Then, in 1897, Eduard Buchner discovered quite by accident that fermentation was catalyzed by a clear juice that he had prepared by grinding yeast with sand and filtering out the unbroken cells. Looking for a way to preserve the juice, Buchner had added sugar. His reasoning was that cooks used sugar to preserve jam. It probably was a disappointment to him that the sugar was broken down rapidly and the mixture frothed with CO_2. In spite of his disappointment, Buchner's discovery made it possible to explore metabolic processes such as fermentation in a greatly simplified system, without having to deal with the complexities of cell growth and multiplication, and without the barriers imposed by cell walls or membranes. Arthur Harden and William Young soon showed that yeast extracts contained two different types of molecules, both of which were necessary in order for fermentation to occur. Some were small, dialyzable, heat-stable molecules such as inorganic phosphate; others were much larger, nondialyzable molecules, the enzymes, which were destroyed easily by heat.

Although early investigators surmised that enzymes might be proteins, the point remained in dispute until 1927, when James Sumner succeeded in purifying and crystallizing the enzyme urease from beans. In the 1930s, John Northrup isolated and characterized a series of digestive enzymes, generalizing Sumner's conclusion that enzymes are proteins. Since then, thousands of different enzymes have been purified, and the structures of many of them have been solved to atomic resolution; almost all of these molecules have proved to be proteins. Surprisingly, however, recent work has shown that some RNA molecules have enzymatic activity.

One of the most striking features of enzymes is their specificity. Each enzyme catalyzes only one type of reaction, showing a high selectivity for both reactants and products. The proteolytic enzyme trypsin, for example, catalyzes the cleavage of polypeptide chains next to a lysine or arginine residue (chapter 1). Chymotrypsin catalyzes the same type of reaction, but is specific for bonds following a phenylalanine, tyrosine, or tryptophan. Enzymes also are very specific with regard to the stereochemistry of a reaction. For example, fumarase, which adds water to the double bond of fumaric acid, always generates the L isomer of malic acid.

Enzyme Terminology

The reactants that enter into an enzymatic reaction are termed substrates. Many enzymes require additional small molecules called cofactors for their activity. Cofactors can be simple inorganic ions such as Mg^{2+}, or complex organic molecules known as coenzymes. In many cases the cofactor binds tightly to a specific site on the enzyme. An enzyme lacking an essential cofactor is called an apoenzyme, and the intact enzyme with the bound cofactor is called a holoenzyme.

Enzymes often are known by common names obtained by adding the suffix *-ase* to the name of the substrate or to the reaction that they catalyze. Thus glucose oxidase is an enzyme that catalyzes the oxidation of glucose; glucose-6-phosphatase catalyzes the hydrolysis of phosphate from glucose-6-phosphate; and urease catalyzes the hydrolysis of urea. Common names also are used for some groups of enzymes. For example, an enzyme that transfers a phosphate group from ATP to another molecule is usually called a "kinase," instead of the more formal "phosphotransferase."

Table 8.1
Classification of Two Representative Enzymes

Number	Systematic Name	Common Name	Reaction
1	Oxidoreductases (enzymes catalyzing oxidation-reduction reactions)		
1.1	Acting on CH—OH group of donors		
1.1.1	With NAD^+ or $NADP^+$ as acceptor		
1.1.1.1	Alcohol:NAD oxidoreductase	Alcohol dehydrogenase	Alcohol + $NAD^+ \leftrightharpoons$ aldehyde or ketone + NADH + H^+
1.1.3	With O_2 as acceptor		
1.1.3.4	β-D-glucose:oxygen oxidoreductase	Glucose oxidase	β-D-glucose + $O_2 \leftrightharpoons$ D-glucono-*d*-lactone + H_2O_2

A systematic scheme for classifying enzymes has been adopted by the International Union of Biochemistry. In this scheme, each enzyme is designated by four numbers separated by periods: the main class, the subclass, the sub-subclass, and the serial number of the enzyme in its sub-subclass. There are six main classes: (1) oxidoreductases, (2) transferases, (3) hydrolases, (4) lyases, (5) isomerases, and (6) ligases or synthases. Oxidoreductases catalyze oxidation-reduction reactions. Transferases catalyze the transfer of a functional group from one molecule to another. Hydrolases catalyze bond cleavage by the introduction of water. Lyases catalyze the removal of a group to form a double bond, or the addition of a group to a double bond. Isomerases catalyze intramolecular rearrangements. And ligases catalyze reactions that join two molecules. Table 8.1 illustrates the divisions of these main classes into subdivisions with two examples.

Basic Aspects of Chemical Kinetics

Before we dig into enzyme kinetics, we need to discuss some of the basic principles that apply to the kinetics of both enzymatic and nonenzymatic reactions. Let's first consider a simple nonenzymatic reaction in which a single reactant (A) is transformed into a product (P):

$$A \rightarrow P \quad (3)$$

To measure the rate, or velocity, of the reaction (v), we could plot the concentration of A as a function of time (fig. 8.1). The rate at any particular time t is

$$v = -\frac{d[A]}{dt} \quad (4)$$

where $d[A]/dt$ is the slope of the plot at that time. The minus sign is needed because v is defined by convention to be a positive number, whereas $d[A]/dt$ is always negative (A is disappearing with time). However, we might equally well choose to measure the increase in the concentration of the product as a function of time, in which case we could express the rate as $v = d[P]/dt$.

For a simple reaction of this type, the rate at any given time usually is found to be proportional to the remaining concentration of the reactant:

$$v = k[A] \quad (5)$$

The proportionality constant, k, is the rate constant. The rate constant is independent of the concentration of the reactant, but it can depend on other parameters such as temperature or pH, and as we will see it may be altered by a catalyst. It has dimensions of reciprocal seconds (s^{-1}). Combining equations (4) and (5), we have

$$\frac{d[A]}{dt} = -k[A] \quad (6)$$

This equation has the mathematical solution

$$[A] = [A]_0 e^{-kt} \quad \text{or} \quad \ln[A] = \ln[A]_0 - kt \quad (7)$$

where $[A]_0$ is the initial concentration of A and e is the base of the natural logarithm. Equation (7) says that [A] will decay exponentially from $[A]_0$ toward zero (see fig. 8.1). When $t = 1/k$, the reaction will have gone about 63% of the way to completion.

A reaction of this type is said to follow first-order kinetics, because the rate is proportional to the concentration of a single species raised to the first power (equation 5). An example is the decay of a radioactive isotope such as ^{14}C. The rate of decay at any time (the number of radioactive disintegrations per second) is simply proportional to the amount of ^{14}C that is present. The rate constant for this extremely slow nuclear reaction is 8×10^{-12} s^{-1}. Another example is the initial electron-transfer reaction that occurs when photosynthetic organisms are excited with light. In this case there actually are two reactants, the electron donor and the acceptor, but they are held close together on a protein so that they react as a unit; the excited complex simply decays spontaneously to a more stable state. This occurs with a rate constant of 3×10^{11} s^{-1}, which makes it one of the fastest reactions known. We will discuss this reaction in more detail in chapter 16.

A slightly more complicated situation arises if the reaction is reversible:

$$A \underset{k_r}{\overset{k_f}{\rightleftharpoons}} P \quad (8)$$

Figure 8.1

The kinetics of a first-order reaction in which a single reactant (A) is converted irreversibly to a product (P). The concentrations of A and P are plotted as functions of time. The rate (v) at any given time can be obtained from the slope of either curve: $v = -\frac{d[A]}{dt} = \frac{d[P]}{dt}$. The time-dependence of [A] is described by an exponential function, $[A] = [A]_0 e^{-kt}$, where $[A]_0$ is the initial concentration and k is the rate constant. When t is equal to $1/k$, [A] has decreased from $[A]_0$ to $[A]_0 \cdot e^{-1}$ (about $0.37[A]_0$).

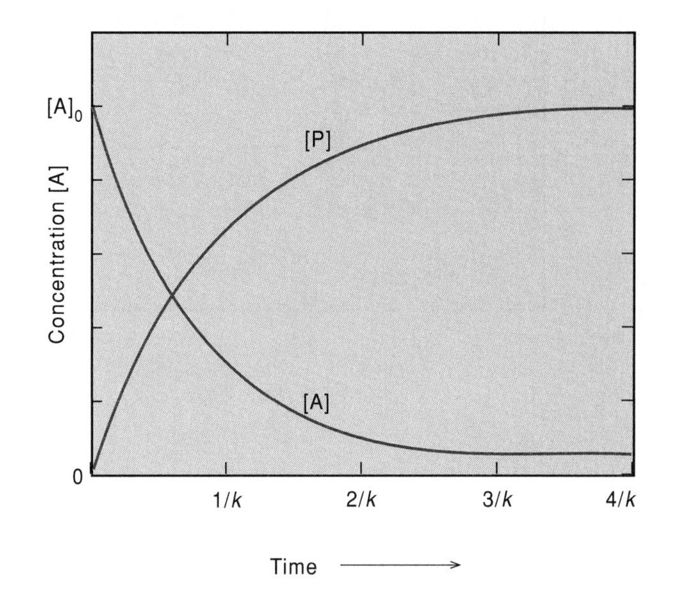

Here k_f is the rate constant for the forward reaction and k_r is that for the reverse. In a reversible reaction, the rate equation becomes

$$\frac{d[A]}{dt} = -k_f[A] + k_r[P] \quad (9)$$

The term $-k_f[A]$ is the same as in equation (6) and the term $k_r[P]$ describes the formation of A from P. In this case, [A] and [P] will proceed from their initial values ($[A]_0$ and $[P]_0$) to their final, equilibrium values, which generally will not be zero. At equilibrium, $d[A]/dt$ must go to zero, which requires that

$$k_f[A]_{eq} = k_r[P]_{eq} \quad (10)$$

where $[A]_{eq}$ and $[P]_{eq}$ are the equilibrium concentrations. Rearranging this expression gives

$$\frac{[P]_{eq}}{[A]_{eq}} = \frac{k_f}{k_r} = K_{eq} \quad (11)$$

where K_{eq} is the equilibrium constant. The solution to equation (9) is that [A] and [P] approach $[A]_{eq}$ and $[P]_{eq}$ exponentially with an overall rate constant that is the sum of k_f and k_r (fig. 8.2).

Now consider a reaction involving two reactants, A and B:

$$A + B \rightarrow P \quad (12)$$

Figure 8.2

The kinetics of a reversible first-order reaction. The solution to the rate equation for such a reaction is $([A] - [A]_{eq}) = ([A]_0 - [A]_{eq})\, e^{-k't}$, where $k' = k_f + k_r$ and $[A]_{eq} = [A]_0/(1 + k_f/k_r)$. The curve drawn here is obtained when $k_f = k_r$.

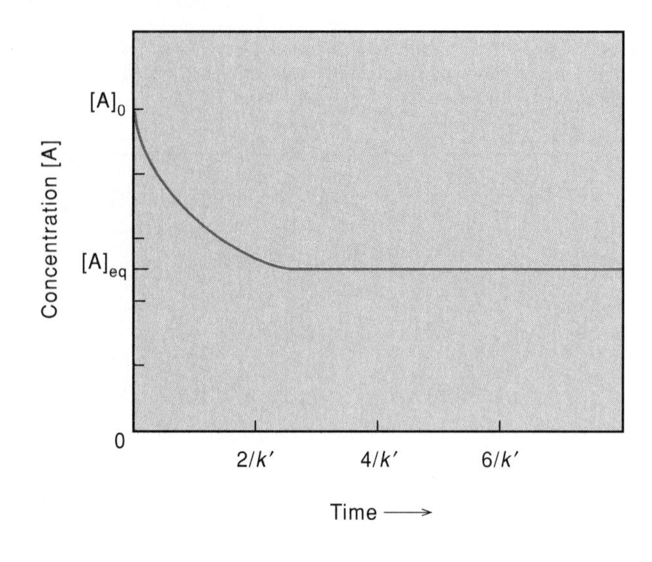

Figure 8.3

Curve 1: Kinetics of a second-order reaction between two reactants starting at the same concentration ($[A]_0$). The curve is a plot of the expression $[A] = 1/(kt + 1/[A]_0)$, which is the solution to the rate equation $d[A]/dt = -k[A]^2$. Curve 2: Kinetics of a second-order reaction when one reactant (B) is present in 100-fold excess. The rate is given by $v = k_{app}[A]$ with $k_{app} = k[B]$. The value of k was decreased by a factor of 100 relative to that used for curve 1. Curve 2 has essentially the same form as the curve for [A] in figure 8.1.

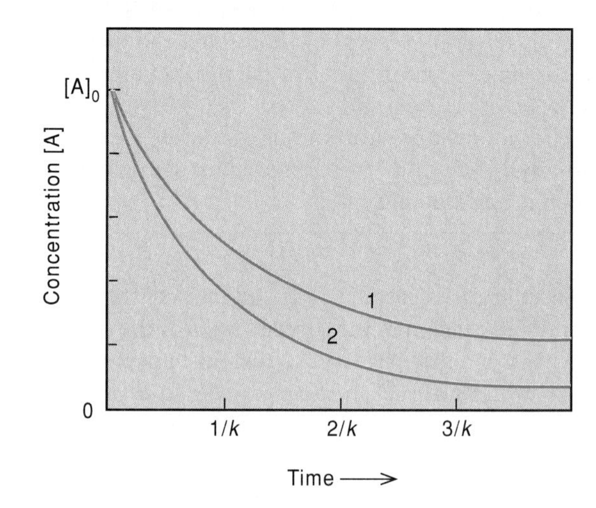

The rate of such a bimolecular reaction usually depends on the concentrations of both reactants:

$$\frac{d[A]}{dt} = -k[A][B] \tag{13}$$

The reaction is said to follow second-order kinetics, because its rate is proportional to a product of two concentrations. The kinetics are first order in either [A] or [B] alone, but second order overall. The rate constant for a second-order reaction has dimensions of $\text{M}^{-1}\,\text{s}^{-1}$. If the concentrations of A and B are the same at the start of the reaction, they will remain the same as they both decrease toward zero. The rate then would be proportional to $[A]^2$ (see curve 1 in figure 8.3). On the other hand, if B is present in a large excess, then its concentration will not change very much while [A] decreases. In this case, the rate will be approximately proportional to [A] (curve 2 in figure 8.3).

Molecular Interpretations of Rate Constants: A Critical Amount of Energy Is Needed for the Reactants to Reach the Transition State

Part of the rationale behind equation (13) is simply that reactants A and B have to collide in order to react. Collisions occur as a result of random diffusion of the reactants in the solution. The number of collisions that occur per second is proportional to the product of the two concentrations, and the second-order rate constant k includes the proportionality factor for this relationship. But not every collision will result in a reaction. The rate constant also must include a factor that gives the fraction of the collisions that are effective. If every collision does result in a reaction, so that this second factor is 1, the rate constant for the reaction of two small molecules in aqueous solution will typically be about $10^{11}\ \text{M}^{-1}\,\text{s}^{-1}$. Because of their large masses, proteins diffuse relatively slowly, so the frequency at which a protein and a small molecule will collide is lower than the collision frequency for two small molecules. The maximum possible value of the rate constant for a reaction between a protein and a small molecule thus is typically on the order of 10^8 to 10^9 $\text{M}^{-1}\,\text{s}^{-1}$.

What determines the fraction of the collisions that result in a reaction? A partial answer is that the colliding species must have a certain critical energy in order to surmount a barrier that separates the reactants from the products. This principle is illustrated schematically in figure 8.4. The surface in figure 8.4*a* represents the energy of a system in which a proton can be bound to either of two molecules. Suppose that the proton is initially on molecule A, and we are interested in how rapidly it moves to molecule B. As the proton moves from one place to the other, its electrostatic interactions with molecule A become less favorable, while its interactions with B improve. The energy of the system goes through a maximum when the proton is at an intermediate position. At this point, the system is said to be in the transition state. The probability that a collision will lead

Figure 8.4

Schematic energy diagrams for a reaction in which a proton moves from one molecule to another. In the perspective drawing (*a*), coordinates in the plane at the bottom represent the location of the proton. The proton moves in three-dimensional space, but only its positions in two of these dimensions can be indicated in the drawing. The energy of the system for any particular set of coordinates is represented by the distance of the cuplike surface above the plane. The two minima in the energy surface indicate the positions of the proton when it is bound optimally to one molecule or the other. The best route along the surface from one of these minima to the other goes through a pass or saddle point. Drawing (*b*) shows a plot of the energy as a function of distance along the optimal route over this pass. The activation energy of the reaction (ΔE_a) is the difference between the energies at the pass and at the starting point.

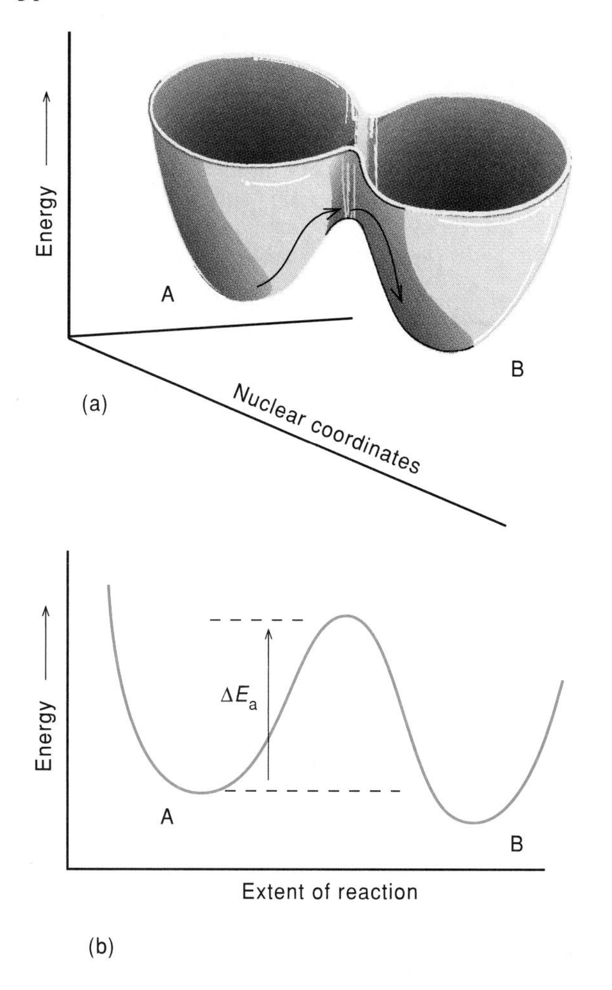

Figure 8.5

The temperature dependence of a reaction rate, shown in the form of an Arrhenius plot. The natural logarithm of the rate constant is plotted as a function of $1/T$, where T is the absolute temperature. (Temperature decreases from left to right in the graph.)

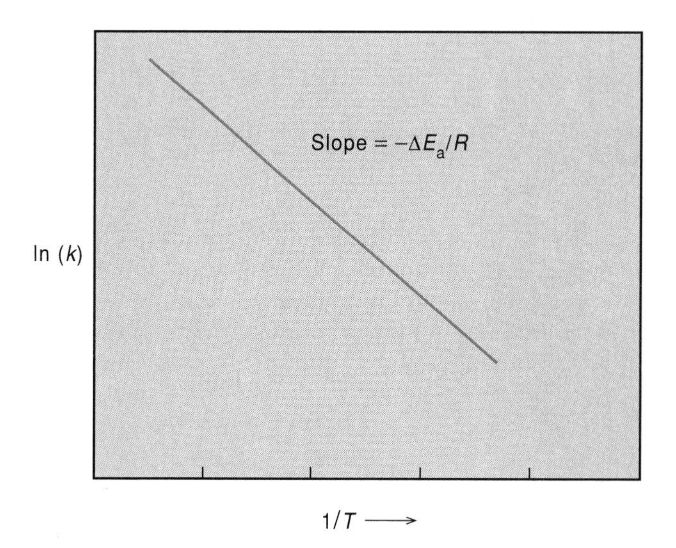

to a reaction depends, in part, on the probability that the molecules collide with enough energy to reach this state. The difference between the lowest possible energy of the reactants and the energy of the transition state is termed the activation energy.

The probability that the collision complex of the reactants will have enough energy to reach the transition state can be calculated from an expression that was first applied to chemical kinetics by Svante Arrhenius. Assume that in order to reach the transition state the energy of the collision complex must exceed the minimum possible energy by an amount ΔE_a. The fraction of the complexes whose energy meets this requirement is proportional to $e^{-\Delta E_a/RT}$, where R is the gas constant (2 cal mole^{-1} deg^{-1}) and T is the absolute temperature. We therefore can write the rate constant as

$$k = k_c e^{-\Delta E_a/RT} \quad \text{or} \quad \ln k = \ln k_c - \Delta E_a/RT \tag{14}$$

where k_c is a proportionality constant. Arrhenius pointed out that equation (14) provides a simple explanation for why many reactions speed up dramatically with increasing temperature. A plot of ln k versus $1/T$, which is termed an Arrhenius plot, frequently is found to be linear and to have a negative slope, as shown in figure 8.5. According to equation (14), the slope of such a plot is $-\Delta E_a/R$.

In figure 8.4, note that the nuclear geometry of the molecules in the transition state is intermediate between the optimal geometries of the reactants and the products. To reach the transition state from either the reactant side or the product side requires a nuclear distortion of the molecules, and this distortion must be in the right direction to connect the reactants with the products. In fact, we can view the transition state as a state in which the geometries of the reactants and the products have become the same.

The requirement for a certain nuclear geometry in the transition state is an additional factor that is superimposed on the requirement for the activation energy, ΔE_a. These two requirements can be combined by replacing ΔE_a in equation (14) by an activation free energy, $\Delta G^‡$:

$$k = k_0 e^{-\Delta G^‡/RT} \tag{15}$$

In this expression, k_0 is an intrinsic rate constant for the conversion of the transition state into the products. The factor $e^{-\Delta G^‡/RT}$ can be interpreted as an effective equilibrium constant

Activation Free Energies, Enthalpies, and Entropies

In the transition-state theory, the concentration of the reactant in the transition state (A^*) is related to the total concentration of the reactant by an effective equilibrium constant, $K^\ddagger$:

$$[A^*] = K^\ddagger[A] \tag{B1}$$

If the rate of the reaction is given by $k[A]$, and we equate this to $k_0[A^*]$, the observed rate constant k must be equal to $k_0K^\ddagger$. $K^\ddagger$ can be related to the standard free energy change ($\Delta G^\ddagger$) associated with the formation of A^* by using equation (13) of chapter 2:

$$\Delta G^\ddagger = -RT \ln K^\ddagger$$

or

$$K^\ddagger = e^{-\Delta G^\ddagger/RT} \tag{B2}$$

Combining equation (B2) with equation (16) of this chapter gives:

$$k = k_0 e^{-\Delta G^\ddagger/RT} = k_0 e^{-(\Delta H^\ddagger - T\Delta S^\ddagger)/RT} \tag{B3}$$

$$= k_0[e^{-\Delta H^\ddagger/RT}][e^{\Delta S^\ddagger/R}] \tag{B4}$$

or

$$\ln k = [\ln k_0 + \Delta S^\ddagger/R] - \Delta H^\ddagger/RT \tag{B5}$$

Equation (B5) indicates that the slope of an Arrhenius plot ($-\Delta E_a/R$) is actually $-\Delta H^\ddagger/R$. The activation enthalpy ($\Delta H^\ddagger$) usually is similar to the difference in energy between the reactants and the transition state, but it also takes into account any changes in volume. The activation entropy ($\Delta S^\ddagger$) contributes to the intercept of the Arrhenius plot on the ordinate.

---

for the formation of the transition state from the reactants (see box 8A). From the definition of free energy (equation 9 of chapter 2), we also can write

$$\Delta G^\ddagger = \Delta H^\ddagger - T\Delta S^\ddagger \tag{16}$$

where $\Delta H^\ddagger$ and $\Delta S^\ddagger$ are the changes in enthalpy and entropy that are needed in order to reach the transition state. Thus $\Delta G^\ddagger$ includes the change in entropy associated with arranging the nuclei of the molecules in the particular way that is required in the transition state. If reaching this state requires going to a more restricted nuclear geometry, as is generally the case, $\Delta S^\ddagger$ will be negative and therefore will decrease the rate constant.

Similar considerations apply to a unimolecular chemical reaction of the type expressed by equation (3). Although there are no collisions between separate reactants in that case, the requirements that the reactant have sufficient energy and undergo appropriate nuclear distortions are just the same. The magnitude of the rate constant depends on how frequently these conditions are met in the course of the random thermal fluctuations of the molecule and the solvent.

Although many reactions can be sped up by increasing the temperature, this is not a very useful option for living organisms. Many organisms have little or no control over the ambient temperature. In addition, most species can survive only within a rather narrow range of temperatures, partly because proteins and other macromolecules become denatured at higher temperatures. Finally, changing the temperature is not a very selective way to control reaction rates, since the great majority of chemical reactions have positive activation energies; all of these reactions will speed up when the temperature is raised.

From equation (15), it appears that there are two other possible ways to increase the rate constant for a reaction, in addition to changing the temperature: increase k_0 or decrease $\Delta G^\ddagger$. However, changing k_0 is not a very realistic option because there usually is no way that it can be done to any significant extent. $\Delta G^\ddagger$ is another matter. Because rate constants depend *exponentially* on $-\Delta G^\ddagger/RT$, relatively small changes in $\Delta G^\ddagger$ can change reaction rates by many orders of magnitude. At physiological temperature, it takes a decrease of only 1.36 kcal/mole to speed up a reaction by a factor of 10, and a decrease by 8.16 kcal/mole will increase the rate by a factor of 10^6. The formation of a single H bond can release anywhere from 4 to 10 kcal/mole. In addition, because $\Delta G^\ddagger$ depends critically on the detailed structures of the reactants and products and the detailed nature of the reaction, it clearly affords an opportunity to control reaction rates with a great deal of specificity. An enzyme, then, must work by decreasing the activation free energy for the specific reaction that it catalyzes (fig. 8.6). This could occur if the structure of the active site is, in some way, complementary to the structure of the transition state. If, for example, the formation of the transition state requires the accumulation of negative electrical charge on a particular atom of the substrate, the free energy of the transition state could be lowered if a positive charge was placed nearby on the enzyme.

From figure 8.6, note that an enzyme does not alter the free energies of the substrates or the products of the reaction. Thus an enzyme cannot change the overall equilibrium constant of a reaction. An enzyme, or any catalyst, for that matter, affects only the speed with which a reaction approaches equilibrium.

Figure 8.6

An enzyme speeds up a reaction by decreasing $\Delta G^{\ddagger}$. The enzyme does not change the free energy of the substrate (S) or product (P); it lowers the free energy of the transition state. The two vertical arrows indicate the activation free energies ($\Delta G^{\ddagger}$) of the catalyzed and uncatalyzed reactions. This figure neglects the enzyme-substrate and enzyme-product complexes that are intermediates in the reaction.

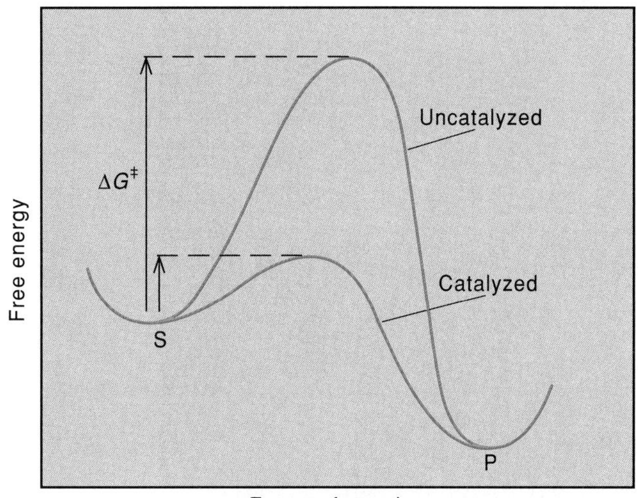

Figure 8.7

The rates of many enzyme-catalyzed reactions have a hyperbolic dependence on the substrate concentration. V_{max} is the maximum rate, and K_m is the substrate concentration at which the rate is half-maximal.

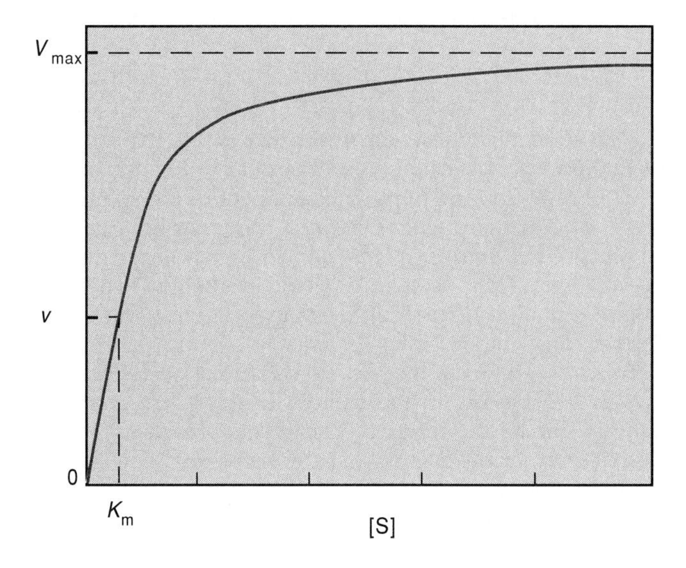

Kinetics of Enzyme-Catalyzed Reactions

The rate equations that we discussed in the preceding section imply that the velocity of an uncatalyzed reaction would increase indefinitely with an increase in the concentration of the reactants. With enzyme-catalyzed reactions, something very different is observed. The rate usually increases linearly with the substrate concentration at low concentrations, but then levels off at high concentrations (fig. 8.7). This behavior is described as a hyperbolic dependence on concentration, because the curve has the form of a hyperbola. The explanation is straightforward. In order for an enzyme to affect $\Delta G^{\ddagger}$, the substrate must bind to a special site on the protein, the active site (fig. 8.8). At very low concentrations of substrate, the active sites of most of the enzyme molecules in the solution will be unoccupied. Under these conditions, increasing the substrate concentration will bring more enzyme molecules into play, and the reaction will speed up. At high concentrations, on the other hand, most of the enzyme molecules will have their active sites occupied, and the observed rate will depend only on the rate at which the bound reactants are converted into products. Further increases in the substrate concentration then will have little effect.

Techniques for measuring rates of enzyme-catalyzed reactions are described in box 8B.

The Michaelis-Menten Equation Provides a Means to Analyze Enzyme-Catalyzed Reactions in Terms of Rate Constants

We can analyze the kinetics of an enzymatic reaction in more detail as follows. Consider a reaction involving a single substrate (S), one product (P), and an enzyme (E). Suppose that the substrate binds to the enzyme with rate constant k_1 to give an enzyme-substrate complex (ES). Suppose further that ES is converted to P with rate constant k_2, but that it also can dissociate to release S with rate constant k_{-1}:

$$E + S \underset{k_{-1}}{\overset{k_1}{\rightleftharpoons}} ES \overset{k_2}{\rightarrow} E + P \quad (17)$$

We can ignore any reversibility of the step in which ES is converted to E and P, provided that we measure the rate quickly enough after we mix E and S so that the concentration of P is still very small.

Let's express the velocity of the reaction in terms of the rate of formation of P:

$$v = \frac{d[P]}{dt} = k_2[ES] \quad (18)$$

The problem is to relate [ES], the concentration of the enzyme-substrate complex, to the concentration of the substrate, [S]. We can do this by writing a rate equation that describes the formation and breakdown of ES:

$$\frac{d[ES]}{dt} = k_1[E][S] - k_{-1}[ES] - k_2[ES] \quad (19)$$

Figure 8.8

Thermolysin, an enzyme that hydrolyzes peptide bonds, with a structural analog of a substrate bound at the active site. The enzyme is shown in orange; the substrate analog (phosphoramidon), in yellow. In the active site, the substrate interacts with a Zn^{2+} ion (white) and amino acid residues that participate in the catalytic mechanism. The side chains of two key residues, Glu 143 and His 231, are shown in turquoise. (Based on the crystal structure described by D. E. Tronrud, A. F. Monzingo, and B. W. Matthews.) The mechanism of action of thermolysin is discussed in chapter 9.

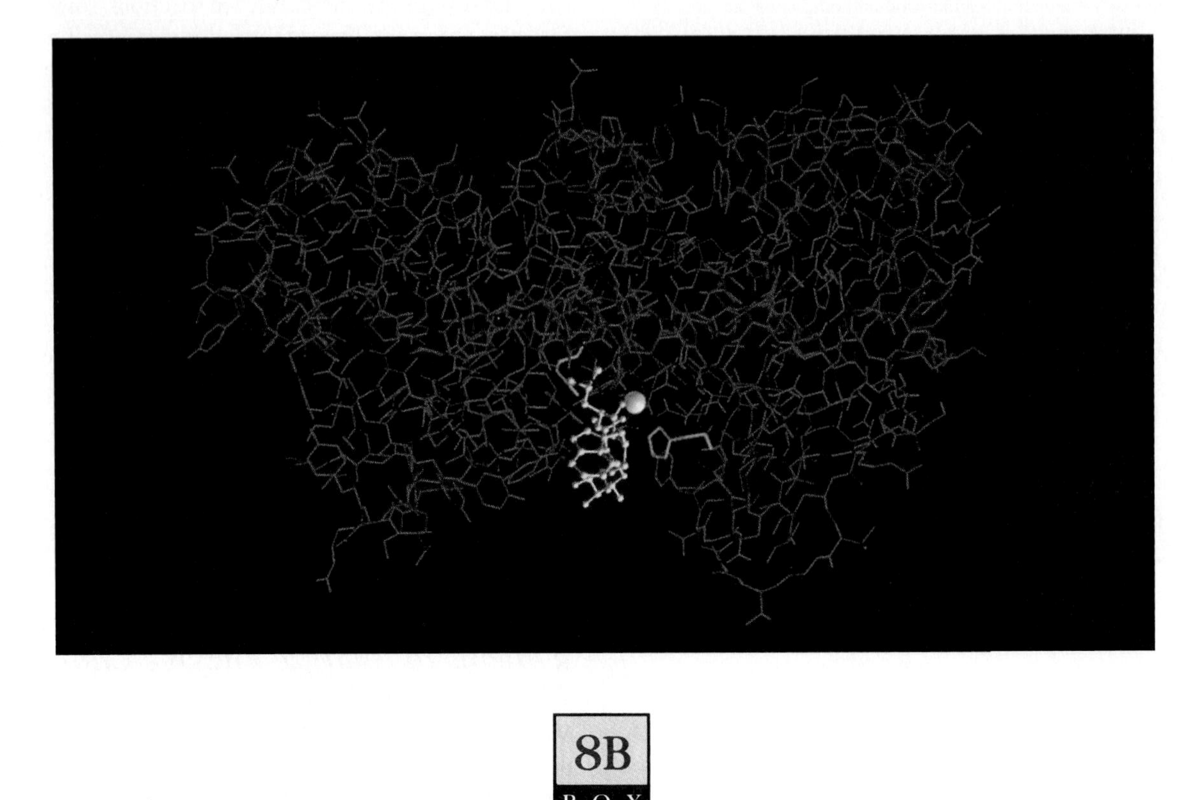

8B
BOX

Techniques for Measuring Enzymatic Reaction Rates

In a typical measurement of enzyme kinetics, we prepare a solution containing all components of the reaction mixture except one. The missing component could be either the enzyme or a substrate. We then add this material, allow the reaction to proceed for a fixed period of time (usually about a minute), and measure the change in the concentration of a reactant or product.

Spectroscopic analytical techniques lend themselves well to kinetic studies, because the measurements can be made essentially instantaneously. In many cases, it is not necessary to stop the reaction whose concentration is to be determined. For example, NADH, has an optical absorption band at 340 nm and a fluorescence emission band at 450 nm. Neither of these properties is shared by the oxidized form of the molecule, NAD^+. The kinetics of any reaction in which NADH is oxidized to NAD^+ can be measured from the change of absorbance at 340 nm or fluorescence at 450 nm.

Measurements of reactions that occur in less than a few seconds require special techniques, because it ordinarily takes several seconds to add the limiting component to the solution and mix the solution thoroughly. One way to circumvent this problem is to prepare two solutions, one containing the enzyme and the other containing the substrate, and to place them in two separate syringes. A pneumatic device then is used to inject the contents of both syringes rapidly into a common chamber.

To study a process that occurs on a time scale faster than 0.01 s, it is necessary to find some way other than mechanical mixing to initiate the reaction. The best approach usually is to create one of the reactants abruptly by exposing the solution to a brief flash of light.

In most studies of enzyme-catalyzed reactions the concentration of the enzyme is very low compared with the concentration of the substrate. Under these conditions, the conversion of some or even all of the enzyme into ES will cause only a relatively minute decrease in [S]. When the enzyme is first added to the solution of the substrate, there will be a brief period while [ES] increases and [E] decreases, but the system will soon reach a steady state in which [ES] is relatively constant (fig. 8.9). In

Figure 8.9

Concentrations of free enzyme (E), substrate (S), enzyme-substrate complex (ES), and product (P) over the time course of a reaction. The shaded portion of the top graph is shown in expanded form in the bottom graph. After a brief initial period (usually less than a few seconds) the concentration of ES remains approximately constant for an extended period. The steady-state approximation is applicable during this second period. Most measurements of enzyme kinetics are made in the steady state. The measurement still must be made quickly enough so that the substrate concentration is present at approximately its initial concentration, [P] is close to zero, and the rate of the reaction ($-d[S]/dt$ or $d[P]/dt$) is more or less constant. At later times, it is necessary to consider the reverse reaction, P → S.

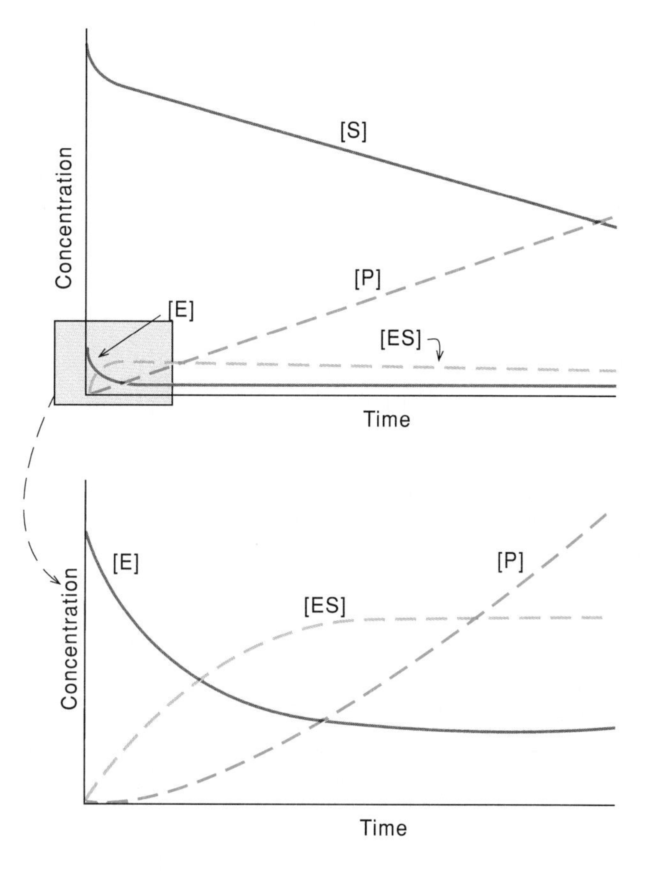

the steady state, the rates of formation and breakdown of ES have become essentially equal, and $d[\mathrm{ES}]/dt \approx 0$. From equation (19), we then have

$$k_1[\mathrm{E}][\mathrm{S}] = k_{-1}[\mathrm{ES}] + k_2[\mathrm{ES}] \quad (20)$$

or

$$[\mathrm{ES}] = [\mathrm{E}][\mathrm{S}]k_1/(k_{-1} + k_2) \quad (21)$$

In equations (19)–(21), note that [E] refers to the concentration of free enzyme, not the total concentration. But if we represent the total concentration of enzyme by $[\mathrm{E_T}]$, we can write

$$[\mathrm{E}] = [\mathrm{E_T}] - [\mathrm{ES}] \quad (22)$$

Combining equations (21) and (22) gives

$$[\mathrm{ES}] = \frac{\{[\mathrm{E_T}] - [\mathrm{ES}]\}[\mathrm{S}]k_1}{k_{-1} + k_2} \quad (23)$$

This expression can be simplified by defining a single constant, K_m, made up from the three rate constants:

$$K_m = \frac{k_{-1} + k_2}{k_1} \quad (24)$$

With this substitution, equation (23) becomes

$$[\mathrm{ES}] = \frac{\{[\mathrm{E_T}] - [\mathrm{ES}]\}[\mathrm{S}]}{K_m} \quad (25)$$

which can be rearranged to

$$[\mathrm{ES}] = \frac{[\mathrm{E_T}][\mathrm{S}]}{[\mathrm{S}] + K_m} \quad (26)$$

Combining equations (18) and (26) gives

$$v = \frac{k_2[\mathrm{E_T}][\mathrm{S}]}{[\mathrm{S}] + K_m} \quad (27)$$

The product $k_2[\mathrm{E_T}]$ in the numerator of equation (27) is the *maximum* rate of the reaction at the particular concentration of the enzyme designated by $[\mathrm{E_T}]$. Suppose that [S] is infinitely large ($>> K_m$). The denominator of equation (27) then becomes approximately equal to [S], and v approaches $k_2[\mathrm{E_T}]$. If we define another constant $V_{max} = k_2[\mathrm{E_T}]$, equation (27) can be written as

$$v = \frac{V_{max}[\mathrm{S}]}{[\mathrm{S}] + K_m} \quad (28)$$

Equation (28) accounts for the hyperbolic relationship between v and [S]. It is known as the Michaelis-Menten equation because it was first derived, in a slightly simpler form, by Leonor Michaelis and Maude Menten in 1913. The constant K_m is called the Michaelis constant. To see the meaning of K_m, suppose that $[\mathrm{S}] = K_m$. The denominator in equation (28) then is equal to 2[S], and $v = V_{max}/2$. Thus, K_m is the substrate concentration at which the velocity is half-maximal (see fig. 8.7). Note that K_m has the same dimensions as a concentration (M), because k_{-1} and k_2, the two rate constants in the numerator of equation (24), are first-order rate constants and have units of $\mathrm{s^{-1}}$, whereas k_1 is a second-order rate constant with units of $\mathrm{M^{-1}\,s^{-1}}$.

If [S] is much smaller than K_m, the denominator of equation (28) becomes approximately equal to K_m, and v is approximately equal to $V_{max}[\mathrm{S}]/K_m$. So the Michaelis-Menten equation also accounts for the linear dependence of v on [S] at very low substrate concentrations (see fig. 8.7).

A common way of plotting data that fit the Michaelis-Menten equation is to plot $1/v$ versus $1/[\mathrm{S}]$. Such a plot is known as a Lineweaver-Burk plot or a double-reciprocal plot. Taking the reciprocals of both sides of equation (28) gives

$$\frac{1}{v} = \frac{1}{V_{max}} + \frac{K_m}{V_{max}}\frac{1}{[\mathrm{S}]} \quad (29)$$

This expression indicates that a plot of $1/v$ versus $1/[\mathrm{S}]$ will fit a straight line with a slope of K_m/V_{max} (fig. 8.10). The intercept of the line on the ordinate occurs at $1/v = 1/V_{max}$, and the intercept on the abscissa occurs at $1/[S] = -1/K_m$. V_{max} and K_m thus can be determined simply from the graph. However, most experimenters now use a computer program to fit data directly to equation (28) because this approach provides a better analysis of the experimental uncertainties.

The Michaelis Constant Is a Function of Three or More Rate Constants

We have seen that K_m is the substrate concentration at which an enzyme-catalyzed reaction occurs at half its maximal rate. K_m thus provides an indication of the substrate concentration range at which the enzyme is most effective at increasing the rate of the reaction. K_m values for a number of different enzyme-substrate pairs are given in table 8.2. The values typically range between 10^{-6} and 10^{-1} M. With enzymes that act on several different substrates, K_m can vary substantially from substrate to substrate. With chymotrypsin, for example, the K_m for glycyltyrosinamide is about 50 times that for *N*-benzoyltyrosinamide (see table 8.2).

Does K_m give us any information on how tightly a particular substrate binds to the active site of an enzyme? Not necessarily. In equation (24), we defined K_m as $(k_{-1} + k_2)/k_1$. Note that the numerator in this expression includes both k_{-1}, the rate constant for dissociation of the enzyme-substrate complex, and k_2, the rate constant for the conversion of ES into products. If k_{-1} is much greater than k_2, K_m will be approximately equal to k_{-1}/k_1. The ratio k_{-1}/k_1 is the dissociation constant of ES, which is the reciprocal of the equilibrium constant for the formation of ES. (See equation 11.) Thus, in some cases, K_m is approximately equal to the dissociation constant of the enzyme-substrate complex. Under these circumstances, a smaller K_m means a smaller dissociation constant, which implies tighter binding to the enzyme. But whether or not this is a valid approximation depends on the enzyme and the substrate. In general, we can say only that the dissociation constant must be smaller than K_m.

The relationship between K_m and the dissociation constant of the enzyme-substrate complex becomes even more tenuous if the reaction mechanism is more complex than that shown in equation (17). Suppose, for example, that there are two different, interconvertible complexes on the enzyme, one involving the substrate and the other involving the products:

$$E + S \underset{k_{-1}}{\overset{k_1}{\rightleftharpoons}} ES \underset{k_{-2}}{\overset{k_2}{\rightleftharpoons}} EP \xrightarrow{k_3} E + P \quad (30)$$

It turns out that the rate of such a reaction still follows the Michaelis-Menten equation (equation 28), and still gives a linear Lineweaver-Burk plot (equation 29), but the observed K_m and V_{max} are made up of more complicated algebraic combinations of the individual rate constants. The equation for K_m is

$$K_m = \frac{k_{-1}k_3 + k_{-1}k_{-2} + k_2k_3}{k_1(k_2 + k_{-2} + k_3)} \quad (31)$$

Similarly, V_{max} for this mechanism is given by

$$V_{max} = k_2k_3[E_T]/(k_2 + k_{-2} + k_3) \quad (32)$$

If k_3 is much greater than k_2 and k_{-2}, these expressions collapse to the expressions that were derived for the simpler mechanism. The rate-determining step of the reaction then would be the conversion of ES to EP.

Figure 8.10

The Michaelis-Menten equation accounts for the hyperbolic dependence of velocity on substrate concentration. A plot of the reciprocal of the rate ($1/v$) as a function of the reciprocal of the substrate concentration ($1/[S]$) fits a straight line. Extrapolating the line to its intercept on the ordinate (infinite substrate concentration) gives $1/V_{max}$. Extrapolating to the intercept on the abscissa gives $-1/K_m$.

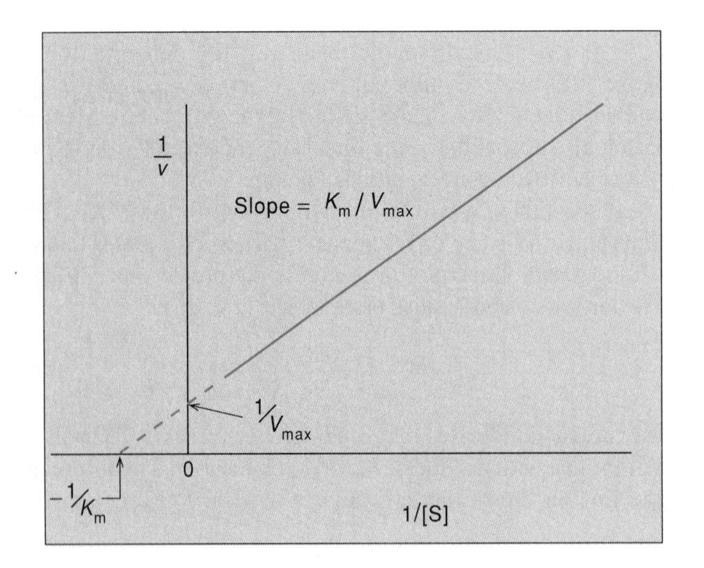

Table 8.2
The Michaelis Constants for Some Enzymes

Enzyme and Substrate	K_m (M)
Catalase	
H_2O_2	1.1
Hexokinase	
Glucose	1.5×10^{-4}
Fructose	1.5×10^{-3}
Chymotrypsin	
N-Benzoyltyrosinamide	2.5×10^{-3}
N-Formyltyrosinamide	1.2×10^{-2}
N-Acetyltyrosinamide	3.2×10^{-2}
Glycyltyrosinamide	1.2×10^{-1}
Aspartate aminotransferase	
Aspartate	9.0×10^{-4}
α-Ketoglutarate	1.0×10^{-4}
Fumarase	
Fumarate	5.0×10^{-6}
Malate	2.5×10^{-5}

If additional intermediate complexes are added to the mechanism, the Michaelis-Menten equation continues to hold, but the relationships of K_m and V_{max} to the microscopic rate constants become more and more complex. One lesson here is that kinetic measurements can never prove that a particular reaction mechanism is correct, because many different mechanisms could result in the same observed kinetics. Kinetic measurements can, however, often be used to rule out a possible mechanism, and thus to distinguish between several alternative mechanisms.

The Specificity Constant Is Usually the Best Index of Enzyme Effectiveness

The turnover number of an enzyme, k_{cat}, is the maximum number of moles of substrate that are converted to product each second, per mole of enzyme (or per mole of active sites if the enzyme has more than one active site). Because the maximum rate is obtained at high substrate concentrations, when all the active sites are occupied with substrate, k_{cat} is a measure of how rapidly an enzyme can operate once the active site is filled. It is given simply by $k_{cat} = V_{max}/[ET]$. As was the case with K_m, the relationship of k_{cat} to microscopic rate constants such as k_2 and k_3 depends on the details of the reaction mechanism. Some representative turnover numbers are given in table 8.3.

Under physiological conditions, enzymes usually do not operate at saturating substrate concentrations. More typically, the ratio of the substrate concentration to K_m is in the range of 0.01 to 1.0. We have seen that the rate of an enzyme-catalyzed reaction at a low substrate concentration is given by $v = V_{max}[S]/K_m$. Under these conditions, the number of moles of substrate converted to product per second per mole of enzyme is $(V_{max}[S]/K_m)/[E_T]$ which is the same as $(k_{cat}/K_m)[S]$. The ratio k_{cat}/K_m is therefore a measure of how rapidly an enzyme can work at low [S]. This ratio is referred to as the specificity constant. Values of k_{cat}/K_m for some particularly active enzymes are given in table 8.4.

The specificity constant, k_{cat}/K_m, is useful for comparing the relative abilities of different compounds to serve as a substrate for the same enzyme. If the concentrations of two substrates are the same, and are small relative to the values of K_m, the ratio of the rates with the two substrates is equal to the ratio of the specificity constants.

Another use of the specificity constant is to compare the rate of an enzyme-catalyzed reaction with the rate at which the random diffusion of the enzyme and substrate brings the two molecules into collision. We mentioned earlier that if every collision between a protein and a small molecule resulted in a reaction, the maximum possible value of the second-order rate constant would typically be on the order of 10^8 to 10^9 $M^{-1}\ s^{-1}$. Some of the values of k_{cat}/K_m given in table 8.4 are in this range. These enzymes have achieved an astonishing state of perfection. The reactions they catalyze proceed at nearly the maximum possible speed, given a fixed, low concentration of substrate and given the restriction that the enzyme and substrate have to find each other by diffusion. The only ways to go much faster would

Table 8.3
Values of k_{cat} for Some Enzymes

Enzyme	k_{cat} (s^{-1})
Catalase	40,000,000
Carbonic anhydrase	1,000,000
Acetylcholinesterase	14,000
Penicillinase	2,000
Lactate dehydrogenase	1,000
Chymotrypsin	100
DNA polymerase I	15
Lysozyme	0.5

Table 8.4
Enzymes for Which k_{cat}/K_m Is Close to the Diffusion-controlled Association Rate

Enzyme	Substrate	k_{cat} (s^{-1})	K_m (M)	k_{cat}/K_m ($M^{-1}\ s^{-1}$)
Acetylcholinesterase	Acetylcholine	1.4×10^4	9×10^{-5}	1.6×10^8
Carbonic	CO_2	1×10^6	0.012	8.3×10^7
anhydrase	HCO_3^-	4×10^5	0.026	1.5×10^7
Catalase	H_2O_2	4×10^7	1.1	4×10^7
Crotonase	Crotonyl-CoA	5.7×10^3	2×10^{-5}	2.8×10^8
Fumarase	Fumarate	800	5×10^{-6}	1.6×10^8
	Malate	900	2.5×10^{-5}	3.6×10^7
Triosephosphate isomerase	Glyceraldehyde 3-phosphate	4.3×10^3	4.7×10^{-4}	2.4×10^8
β-Lactamase	Benzylpenicillin	2.0×10^3	2×10^{-5}	1×10^8

Source: From *Enzyme Structure and Mechanism*, 3rd Edition by Alan Fersht. Copyright © 1985 by W. H. Freeman & Company. Reprinted with permission.

be to have the substrate generated right on the enzyme or in its immediate vicinity, so that little diffusional motion is necessary, or to decrease the size of the enzyme, so that it can diffuse more rapidly. The first of these possibilities arises when two or more enzymes are combined in a multienzyme complex. The product of one enzymatic reaction then can be released close to the active site of the next enzyme. It appears to be more difficult to decrease the sizes of enzymes because a certain amount of tertiary structure is necessary in order to create the proper geometry of the active site. In addition, the large sizes of many enzymes are dictated by the need for secondary binding sites for other molecules that act to regulate enzymatic activity.

Kinetics of Enzymatic Reactions Involving Two Substrates

Enzymes that catalyze reactions with two or more substrates work in a variety of ways. In some cases, the intermolecular reaction occurs when all of the substrates are bound in a common enzyme-substrate complex; in others, the substrates bind and react one at a time. A frequent application of kinetic measurements is to distinguish between such alternatives.

Consider a reaction in which two substrates, S_1 and S_2, are converted to products P_1 and P_2. One possible way for the reaction to occur is for S_1 to bind to the enzyme first, forming the binary complex ES_1. This could be followed by the binding of S_2 to give the ternary complex ES_1S_2, which then undergoes conversion to the products:

$$E \xrightleftharpoons{S_1} ES_1 \xrightleftharpoons{S_2} ES_1S_2 \xrightarrow{P_1 + P_2} E \quad (33)$$

This alternative is referred to as an ordered mechanism. Another possibility is that the two substrates can bind to the enzyme in either order:

$$E \xrightleftharpoons{S_1} ES_1 \xrightleftharpoons{S_2} ES_1S_2; \quad E \xrightleftharpoons{S_2} ES_2 \xrightleftharpoons{S_1} ES_1S_2; \quad ES_1S_2 \xrightarrow{P_1 + P_2} E \quad (34)$$

This alternative is a random mechanism. The ordered mechanism can be viewed as the limiting case of a random mechanism in which the upper path is much more favorable than the lower. Hexokinase (chapter 13) and citrate synthase (chapter 14) exhibit random, or nearly random, mechanisms. Lactate dehydrogenase (chapter 10) and aspartate carbamoyltransferase (chapter 10) use an ordered pathway, or at least display a marked preference for one route over the other.

In either the random or the ordered mechanism, substrates S_1 and S_2 both have to bind to the enzyme before either of the products is released. A different type of mechanism would be for one substrate, say S_1, to bind to the enzyme and be converted to P_1, leaving the enzyme in an altered form, E′. S_2 then could bind to E′ and be converted to P_2, returning the enzyme to its original form:

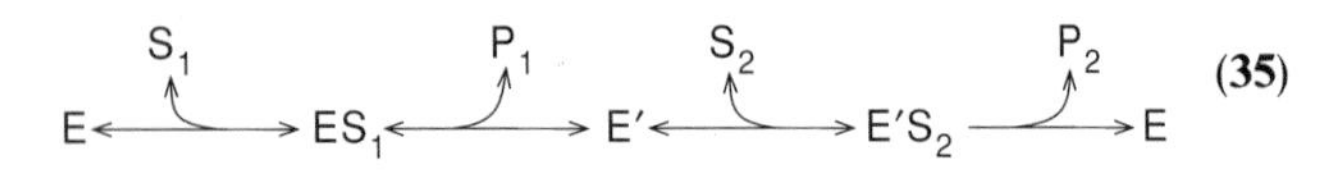

(35)

This is called a Ping-Pong mechanism to emphasize the bouncing of the enzyme between two states, E and E′. Ping-Pong pathways are commonly observed with enzymes that contain tightly bound coenzymes, such as pyridoxal phosphate or flavin groups (see chapter 11 for the structures of these coenzymes). The interconversion of the enzyme between the two forms usually involves a modification of the coenzyme. In the case of pyridoxal enzymes, the coenzyme switches between an aldehyde (pyridoxal phosphate) and an amine (pyridoxamine phosphate). An example is glutamate transaminase (chapters 11 and 19). In some enzymes with bound flavin coenzymes, the flavin alternates between oxidized and reduced states.

Kinetic equations for these different mechanisms can be worked out in the same way as those for reactions involving only one substrate. Techniques for doing so are described in the references given at the end of the chapter; here we will simply give some of the results. For the Ping-Pong mechanism (equation 35), the double-reciprocal form of the final kinetic expression is

$$\frac{1}{v} = \frac{1}{V_{max}}\left(1 + \frac{K_{m2}}{[S_2]}\right) + \frac{K_{m1}}{V_{max}}\frac{1}{[S_1]} \quad (36)$$

where K_{m1} is the Michaelis constant for S_1, and K_{m2} is that for S_2. This expression is similar to that for a single-substrate reaction (equation 29), except that the first term on the right is multiplied by the factor $1 + (K_{m2}/[S_2])$. If we measure the rate of the reaction as a function of $[S_1]$, keeping $[S_2]$ constant, a plot of $1/v$ versus $1/[S_1]$ will be linear, but the intercept on the ordinate (the apparent V_{max}) will depend on $[S_2]$. Increasing $[S_2]$ will increase the apparent V_{max} (fig. 8.11). From a series of such plots, measured at different values of $[S_2]$, we could find the true V_{max}, in addition to K_{m1} and K_{m2}. As with a reaction with one substrate but involving several steps (equations 30 to 32), the two values of K_m and V_{max} for a two-substrate reaction are made up of combinations of the microscopic rate constants.

The ordered and random mechanisms (equations 33 and 34) both give a kinetic expression of the form

$$\frac{1}{v} = \frac{1}{V_{max}}\left(1 + \frac{K_{m2}}{[S_2]} + \frac{K_{m1}}{[S_1]} + \frac{K_{m2}}{[S_2]}\frac{K_{d1}}{[S_1]}\right) \quad (37)$$

where K_{d1} is the dissociation constant for ES_1. A plot of $1/v$ versus $1/[S_1]$ at constant $[S_2]$ is still linear, but now both the slope and the intercept depend on $[S_2]$ (fig. 8.12). Again, all of the macroscopic constants can be obtained from a series of such plots measured at different S_2 concentrations. However, because of the symmetry of equation (37), additional measurements have to be made in order to determine whether the mechanism is ordered, and if so whether S_1 or S_2 binds to the enzyme first. In some cases, the substrate that binds first can be shown to bind to the enzyme to give a stable enzyme-substrate complex in the absence of the other substrate. Lactate dehydrogenase, for example, will bind NAD^+ in the absence of lactate, but will not bind lactate in the absence of NAD^+.

All of the kinetic models that we have discussed lead to hyperbolic dependences of the rate on the concentrations of the substrates, and to linear double-reciprocal plots. There are, however, many enzymes that behave in more complex ways. In

Figure 8.11

Double-reciprocal plots ($1/v$ versus $1/[S_1]$) for the Ping-Pong mechanism. Measurements made at different values of $[S_2]$ give a set of parallel straight lines. K_{m1}, K_{m2} and V_{max} can be obtained by replotting the intercepts as a function of $1/[S_2]$.

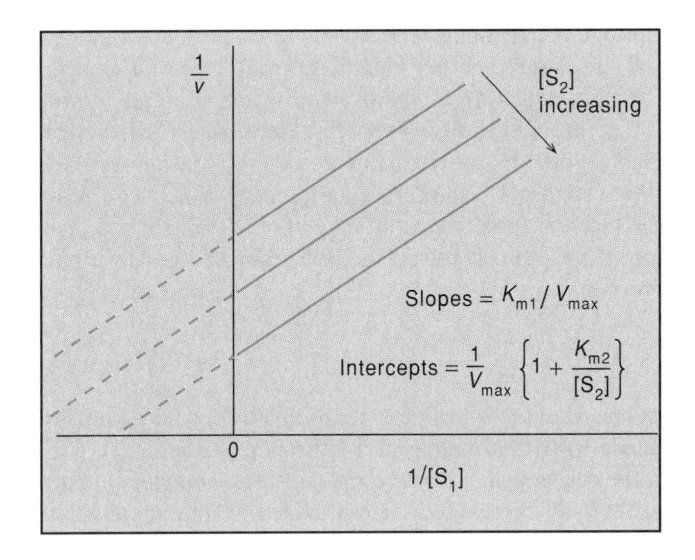

Figure 8.12

Double-reciprocal plots for an ordered or random mechanism. Measurements made at different fixed values of $[S_2]$ give a set of lines that intersect to the left of the ordinate. The two values of K_m, as well as V_{max} and K_{d1}, can be obtained by replotting the slopes and intercepts of these lines as functions of $1/[S_2]$. Ordered and random mechanisms can be distinguished by making such measurements for the reverse reaction ($P_1 + P_2 \rightarrow S_1 + S_2$) in addition to the forward reaction.

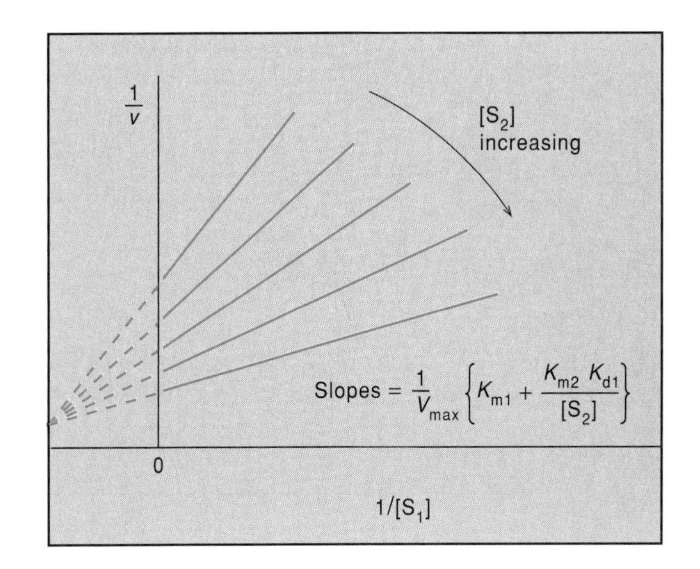

enzymes that have multiple subunits, the binding of substrate to one subunit may increase or decrease the activity of another subunit, just as the binding of O_2 to one of the subunits of hemoglobin increases the affinity of the other subunits for O_2 (see chapter 5). Such effects are seen in an important group of enzymes called allosteric enzymes, which are discussed in chapter 10.

Effects of Temperature, pH, and Isotopic Substitutions

With most enzymes, the turnover number increases with temperature until a temperature is reached where the enzyme is no longer stable (fig. 8.13). Above this point, there is a precipitous drop in activity that usually is irreversible. At lower temperatures, where the enzyme is stable, the apparent activation energy of the reaction (ΔE_a or $\Delta H^{\ddagger}$) can be obtained from an Arrhenius plot of ln k_{cat} versus $1/T$. The activation energy varies considerably from case to case, but with many enzymes a 10° rise in temperature increases k_{cat} by about a factor of 2, which translates into an apparent activation energy on the order of 12 kcal/mole. If k_{cat} is equal simply to k_2, as it is in the simple mechanism described by equation (17), $\Delta H^{\ddagger}$ can be related to the formation of the transition state between ES and the products. More generally, as we saw in equation (32), k_{cat} is made up of a combination of microscopic rate constants, and it is difficult to relate the activation energy to any particular step. If the mechanism involves a sequence of several steps, the kinetics are usually most sensitive to the slowest, or rate-determining, step.

Figure 8.13

This graph shows how the activity of a typical enzyme depends on temperature. The turnover number (k_{cat}) increases with temperature until a point is reached where the enzyme is no longer stable. The apparent activation energy of the rate-determining step could be obtained by replotting the data for low temperatures in an Arrhenius plot. The temperature at which k_{cat} is greatest should not be interpreted as the "optimum temperature" for the enzyme, because the position of the maximum depends partly on how quickly the experimenter is able to assay the enzyme's activity. (Denaturation of the enzyme occurs continuously during the measurement and increases in rate with increasing temperature.)

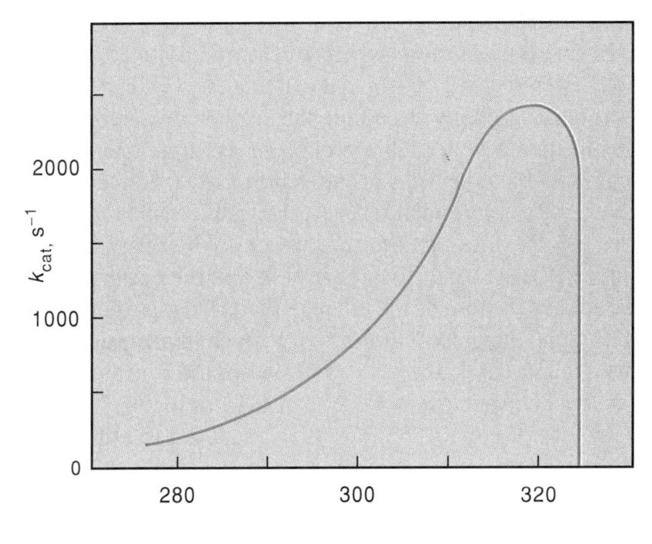

Figure 8.14

Enzyme activity (k_{cat}/K_m) as a function of pH for three different enzymes. The optimum pH usually is a characteristic of the enzyme and not the particular substrate. Often the pH sensitivity is an indication of an ionizable group at the active site, but it can also reflect changes in the tertiary structure of the enzyme.

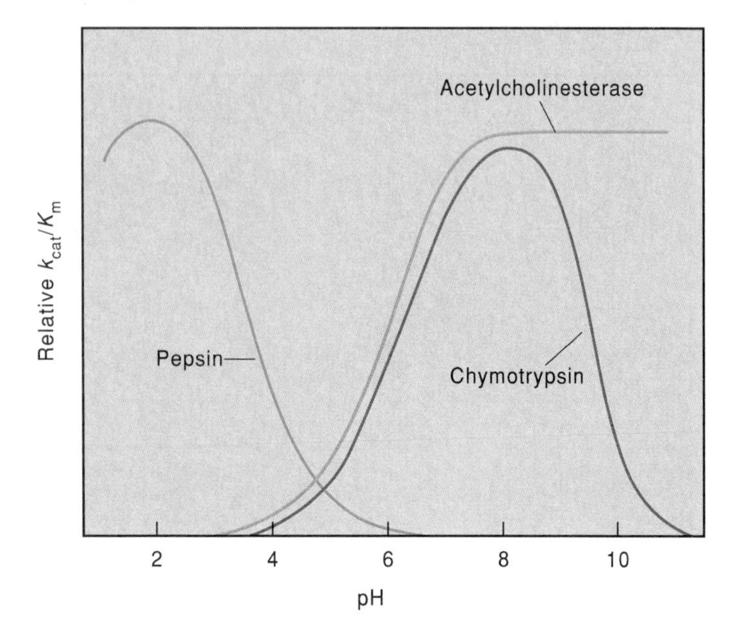

One way to explore the nature of the rate-determining step in an enzymatic reaction is to investigate the effect of replacing a particular H atom of the substrate by deuterium (^2H, or D). If a bond to this atom is broken in the critical step, replacing the H atom by the heavier D usually will result in a decrease in k_{cat} by a factor of between 3 and 10. Similar experiments can be done with isotopes of O or N, although the effects usually are smaller than those obtained with D.

Enzymes, like other proteins, are stable over only a limited range of pH. Outside this range, changes in the charges on ionizable amino acid residues result in modifications of the tertiary structure of the protein and eventually lead to denaturation. But within the range where an enzyme is stable, both k_{cat} and K_m often depend on pH. The effects of pH can reflect the pK_a of ionizing groups on either the enzyme or a substrate. A substrate that has an amine group, for example, may bind to the enzyme best when this group is protonated. In many cases, however, the pH dependence reflects ionizable residues that form part of the active site on the enzyme, or are essential for maintaining the structure of the active site, and the optimum pH is a characteristic more of the enzyme than of the particular substrate. Thus the maximum activity of chymotrypsin always occurs around pH 8, the activity of pepsin peaks around pH 2, and acetylcholinesterase works best at pH 7 or higher (fig. 8.14). The activity of papain, on the other hand, is essentially independent of pH between 4 and 8.

The sensitivity of acetylcholinesterase to pH (see fig. 8.14) probably can be attributed to the imidazole group of a critical histidyl residue, which must be unprotonated in order for the enzyme to operate. The histidine probably removes a proton from the bound substrate in the rate-determining step. Chymotrypsin's bell-shaped activity curve reflects the pK_a values of *two* critical groups, one of which must be protonated and the other unprotonated. As we will see in chapter 9, an unprotonated histidyl residue is essential in the catalytic step, and the free amino group of the N-terminal isoleucine apparently must be protonated. Although the isoleucine itself is not part of the active site, deprotonating its amino group causes a change in the tertiary structure of the protein, which evidently disrupts the structure of the site. The decrease in k_{cat}/K_m for chymotrypsin at low pH is due to a decrease in k_{cat}, whereas the decrease at high pH results from an increase in K_m. The activity of the enzyme fumerase exhibits a similar bell-shaped dependence on pH, but in this case both limbs of the curve reflect changes in k_{cat}.

Enzyme Inhibition

Most enzymes are sensitive to inhibition by specific agents that interfere with the binding of a substrate at the active site or with the conversion of the enzyme-substrate complex into products. Study of these effects can provide information about how an enzyme operates. In many cases, an inhibitor is found to resemble the substrate structurally, and to bind reversibly at the same site on the enzyme. This effect is called competitive inhibition because the inhibitor and the substrate compete for binding (fig. 8.15*a*); the inhibitor is prevented from binding if the site is already occupied by the substrate. This type of inhibition is exemplified by the effect of malonic acid on the enzyme succinate dehydrogenase. Malonate, whose structure resembles that of the substrate, succinate, competes with it for binding at the active site (fig. 8.16).

Inhibitors of a different sort can bind at separate sites where they do not compete directly with the substrate. Instead, they act by interfering with the reaction of the enzyme-substrate complex. An inhibitor that binds to an enzyme whether or not the active site is occupied by the substrate is termed a noncompetitive inhibitor (see fig. 8.15*b*). A third possibility is that the inhibitor binds only after formation of the enzyme-substrate complex (see fig. 8.15*c*). This effect is called uncompetitive inhibition. Uncompetitive inhibition is most common in reactions involving more than one substrate.

Kinetic measurements are useful for distinguishing between different types of inhibition as well as providing quantitative information on the effectiveness of various inhibitors. This information is essential for an understanding of how cells regulate their enzymatic activities. Comparisons of the effects of a series of inhibitors also can help in mapping the structure of an enzyme's active site, and such studies are a key step in the rational design of therapeutic drugs. Finally, competitive inhibitors are useful in x-ray crystallographic studies for pinpointing the active site in a crystal structure, and thus revealing how the surrounding amino acid residues interact with a bound molecule. Crystallographic studies usually cannot be carried out with the substrate itself because the enzyme-substrate complex is converted too rapidly into products.

Figure 8.15

Types of enzyme inhibition. (*a*) A competitive inhibitor competes with the substrate for binding at the same site on the enzyme. (*b*) A noncompetitive inhibitor binds to a different site, but blocks the conversion of the substrate to products. (*c*) An uncompetitive inhibitor binds only to the enzyme-substrate complex. (E = enzyme; S = substrate.)

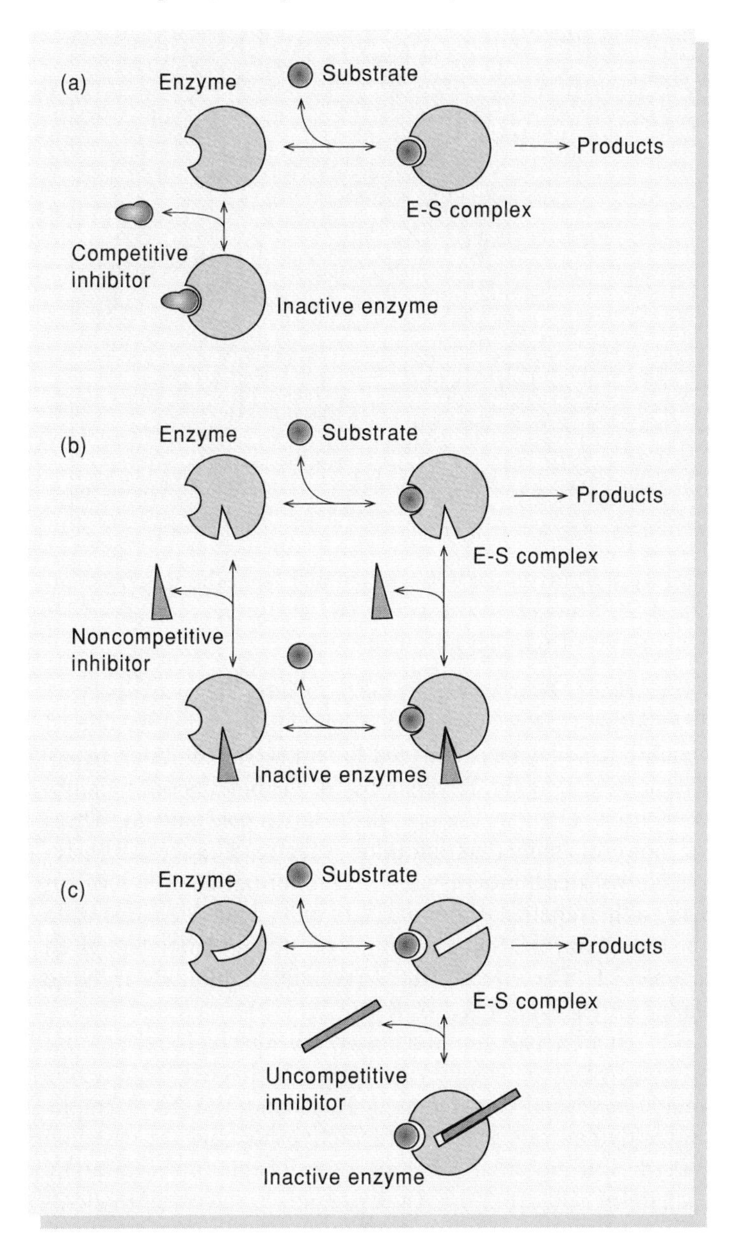

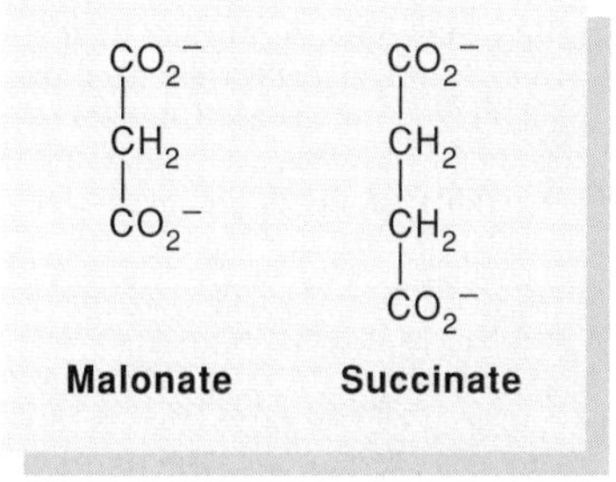

Competitive Inhibition An expression describing enzyme kinetics in the presence of a competitive inhibitor can be derived straightforwardly. Consider the simple, one-substrate reaction that we treated earlier:

$$\mathrm{E} + \mathrm{S} \underset{k_{-1}}{\overset{k_1}{\rightleftharpoons}} \mathrm{ES} \overset{k_2}{\rightarrow} \mathrm{E} + \mathrm{P} \tag{17'}$$

Figure 8.16

Malonate is a competitive inhibitor of succinate dehydrogenase. It is structurally similar to the substrate, succinate.

Malonate: CO_2^-–CH_2–CO_2^-

Succinate: CO_2^-–CH_2–CH_2–CO_2^-

We now have the additional feature that the enzyme can react reversibly with the inhibitor (I) to give an inactive complex (EI):

$$\mathrm{E} + \mathrm{I} \underset{k_{-1}}{\overset{k_1}{\rightleftharpoons}} \mathrm{EI} \tag{38}$$

The derivation proceeds just as in equations (18)–(29) above, except that in place of equation (22) we have

$$[\mathrm{E}] = [\mathrm{E_T}] - [\mathrm{ES}] - [\mathrm{EI}] \tag{39}$$

The inhibitor simply decreases the amount of free enzyme that is available to react with S. The concentration of the enzyme-inhibitor complex depends on the concentration of the free inhibitor and on a dissociation constant, K_I:

$$\frac{[\mathrm{E}][\mathrm{I}]}{[\mathrm{EI}]} = \frac{k_{-1}}{k_1} = K_I \tag{40}$$

As a result, in place of equations (28) and (29) we end up with

$$v = \frac{V_{max}[\mathrm{S}]}{[\mathrm{S}] + K_m(1 + [\mathrm{I}]/K_I)} \tag{41}$$

and

$$\frac{1}{v} = \frac{1}{V_{max}} + \frac{K_m}{V_{max}}\frac{1}{[\mathrm{S}]}\left(1 + \frac{[\mathrm{I}]}{K_I}\right) \tag{42}$$

According to equation (42), a plot of $1/v$ versus $1/[\mathrm{S}]$ will be linear and will pass through the same intercept on the ordinate as the plot obtained in the absence of the inhibitor ($1/V_{max}$). This is equivalent to saying that the effect of the inhibitor disappears at high substrate concentration, which is just what we might have expected. The slope of the double-reciprocal plot, however, depends on the product $(K_m/V_{max})(1 + [\mathrm{I}]/K_I)$, instead of simply K_m/V_{max} (fig. 8.17). By measuring the slope as a function of [I] we can determine K_I.

Competitive inhibitors can be designed to take advantage of the fact that an enzyme stabilizes the transition state of a reaction more than it does the initial enzyme-substrate complex. An inhibitor that is structurally similar to the transition state often will bind to the enzyme particularly tightly. Such an inhibitor is termed a transition-state analog. For example, many enzymes that hydrolyze phosphate diesters are very sensitive to inhibition by derivatives of vanadium in which the vanadium atom is surrounded by five oxygens. These inhibitors appear to mimic a transition state in which phosphorus has a similar pentacovalent geometry.

Figure 8.17

Competitive inhibition. A series of double-reciprocal plots ($1/v$ versus $1/[S]$) measured at different concentrations of the inhibitor (I) all intersect at the same point ($1/V_{max}$) on the ordinate. The slopes of the plots, and the intercepts on the abscissa, are simple, linear functions of $[I]/K_I$, where K_I is the dissociation constant of the inhibitor-enzyme complex.

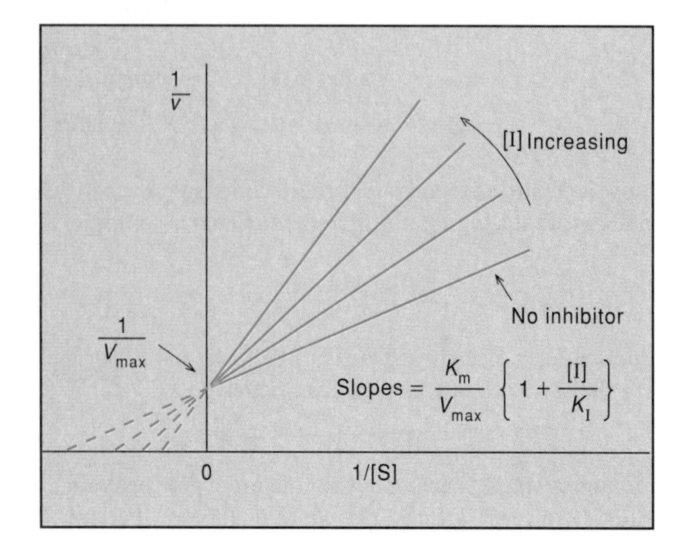

Figure 8.18

Noncompetitive inhibition. The double-reciprocal plots pass through different points on the ordinate, but intersect at the same point ($-1/K_m$) on the abscissa. The slopes and the intercepts on the ordinate are linear functions of $[I]/K_I$.

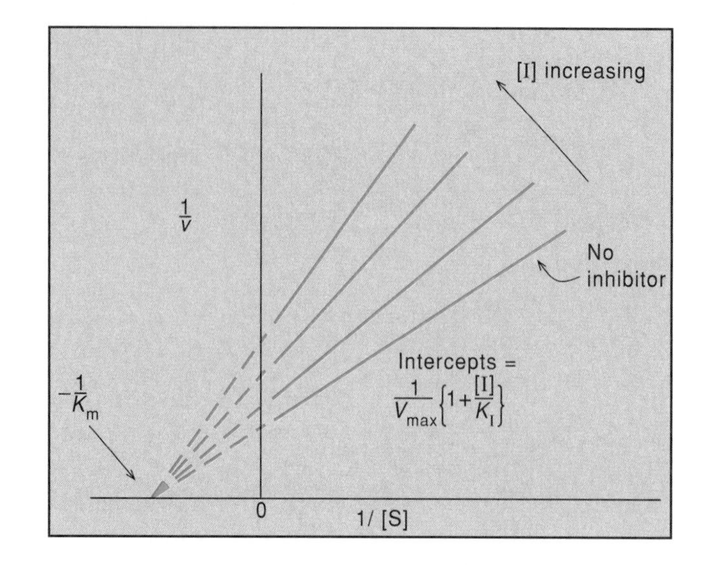

Noncompetitive Inhibition Noncompetitive inhibition can be treated in the same manner except that the inhibitor now can react with the enzyme even if S is already bound, so that in addition to equation (38) we have:

$$ES + I \underset{k_{-IS}}{\overset{k_{IS}}{\rightleftharpoons}} ESI \tag{43}$$

In the simplest situation, the binding of S has no effect at all on the binding of I, so that the dissociation constant of ESI is the same as that of ES (K_i). The final double-reciprocal expression for this case is

$$\frac{1}{v} = \left(\frac{1}{V_{max}} + \frac{K_m}{V_{max}}\frac{1}{[S]}\right)\left(1 + \frac{[I]}{K_I}\right) \tag{44}$$

This expression indicates that a noncompetitive inhibitor decreases the maximum velocity, but does not affect K_m. The inhibitor removes a certain fraction of the enzyme from operation, no matter what the concentration of the substrate is. Plots of $1/v$ versus $1/[S]$ in the presence of different concentrations of the inhibitor intersect at the same point on the abscissa ($-1/K_m$), but pass through the ordinate at different points (fig. 8.18). Again, K_I can be found by measuring the intercepts on the ordinate as a function of [I]. If the dissociation constant for S from ESI differs from that of ES, the double-reciprocal plots will intersect above or below the abscissa.

Uncompetitive Inhibition An uncompetitive inhibitor leads to a double-reciprocal kinetic expression of the form

$$\frac{1}{v} = \frac{1}{V_{max}}\left(1 + \frac{[I]}{K_I}\right) + \frac{K_m}{V_{max}}\frac{1}{[S]} \tag{45}$$

Here K_I pertains to the dissociation of I from the ternary complex, ESI. Plots of $1/v$ versus $1/[S]$ at different values of [I] give a series of parallel lines (fig. 8.19).

Irreversible Inhibitors and Affinity Labels The various types of inhibition that we have been discussing are all reversible. If the inhibited enzyme is dialyzed to remove the inhibitor, its activity increases again. Reversibility of the binding of the inhibitor is implicit in our use of a dissociation constant, K_I. There are, however, numerous inhibitors that react essentially irreversibly with enzymes, usually by the formation of a covalent bond to the functional group of an amino acid side chain or to a bound coenzyme. Some examples of such inhibitors are given in table 8.5. The effect of an irreversible inhibitor can be to change either V_{max} or K_m, or both.

Irreversible inhibitors often provide clues to the nature of the active site on an enzyme. Enzymes that are inhibited by organic mercurial compounds or by iodoacetate, for example, frequently have a cysteine in the active site, and the cysteinyl sulfhydryl group often plays an essential role in the catalytic mechanism (fig. 8.20). An example is glyceraldehyde-3-phosphate dehydrogenase, in which the catalytic mechanism begins with a reaction of the cysteine with the aldehyde substrate to form a thiohemiacetal. The mechanism of action of this enzyme will be discussed in more detail in chapter 13. Diisopropylfluorophosphate reacts irreversibly with a critical serine residue in many proteolytic enzymes, including trypsin and chymotrypsin (see fig. 8.20). The reaction of the serine group destroys the catalytic activity. The mechanism of action of trypsin and chymotrypsin is discussed in the following chapter.

In the case of glyceraldehyde-3-phosphate dehydrogenase, further studies have confirmed that the active site does

Figure 8.19

Uncompetitive inhibition. The double-reciprocal plots are parallel.

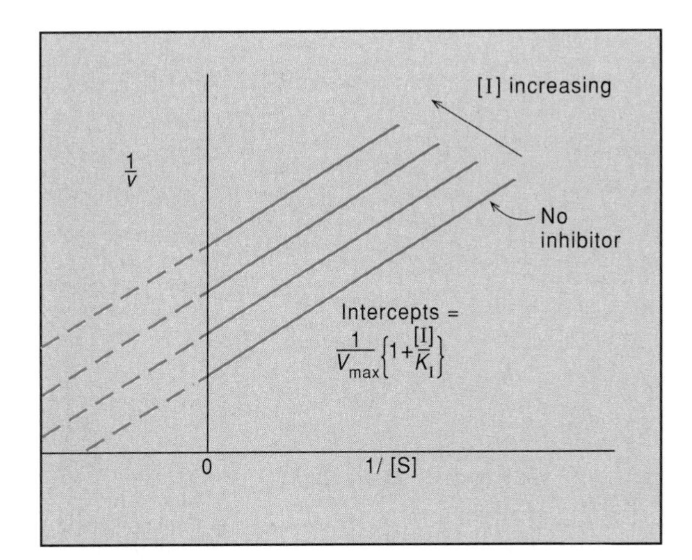

Figure 8.20

Iodoacetamide is an irreversible inhibitor of many enzymes that contain a cysteine residue in the active site. Diisopropylfluorophosphate is an irreversible inhibitor of trypsin, chymotrypsin, and several related enzymes. It reacts with a serine residue at the active site.

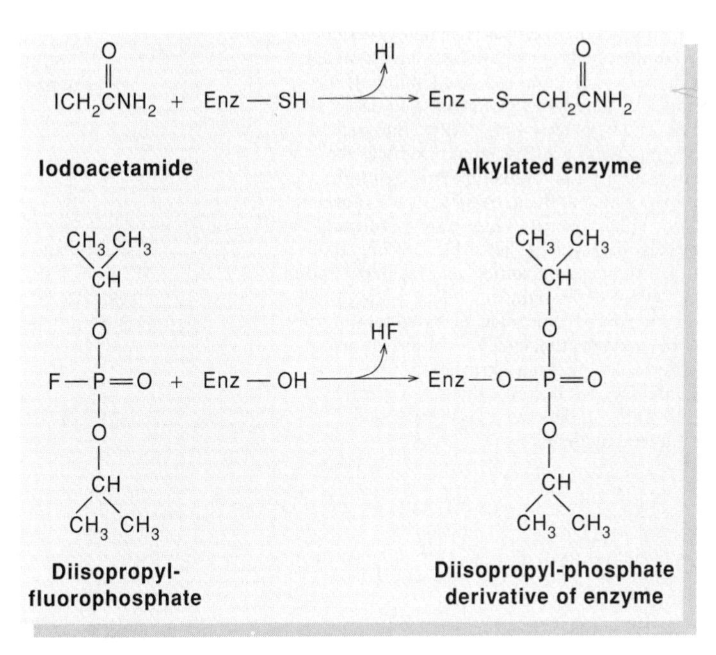

Table 8.5
Some Inhibitors of Enzymes that Form Covalent Linkages with Functional Groups on the Enzyme

Inhibitor	Enzyme Group that Combines with Inhibitor
Cyanide	Fe, Cu, Zn, other transition metals
p-Mercuribenzoate	Sulfhydryl
Diisopropylfluorophosphate	Serine hydroxyl
Iodoacetate	Sulfhydryl, imidazole, carboxyl, thioether

contain an essential cysteine residue, and the same is true of the essential serine in trypsin and chymotrypsin. But in exploring a new enzyme it is important to keep in mind that the chemical modification of an amino acid side chain generally causes some perturbation of the secondary or tertiary structure of a protein. A reaction involving an amino acid residue well outside the active site thus could have a long-range disruptive effect that alters the structure of the active site sufficiently to inhibit the enzyme. This possibility is less of a concern with a reversible, competitive inhibitor because in that case the competition with the substrate supports the conclusion that the inhibitor binds directly in the active site.

An irreversible inhibitor often can be designed for the active site of a particular enzyme by incorporating a reactive group in a molecule that resembles a substrate. For example, 3-bromoacetol phosphate is a structural analog of dihydroxyacetone phosphate, which is a substrate for triosephosphate isomerase (fig. 8.21*a*). The inhibitor binds to the active site of the enzyme, and then reacts irreversibly with the carboxyl group of a nearby glutamic acid residue. Binding to the active site greatly increases the selectivity of the inhibitor for reaction with this particular residue, in preference to glutamyl residues elsewhere in the protein. Labeling the inhibitor with a radioisotope such as ^{14}C or ^{3}H facilitates the identification of the derivatized amino acid residue after the protein has been split into smaller peptides, making it possible to locate the reactive residue in the amino acid sequence. Such a reagent is called an affinity label. Photoaffinity labels are a particularly useful group of reagents of this type, in which the covalent attachment to the protein can be triggered by light after the reagent has bound to the enzyme (see fig. 8.21*b*). Another related technique is to use a reagent that is not intrinsically reactive, but becomes reactive after it has been modified chemically by the enzyme itself (see fig. 8.21*c*). Such a reagent is termed a mechanism-based inhibitor or suicide substrate, to emphasize that the enzyme brings about its own inhibition.

Metal-ion Chelators Enzymes that require metal ions as cofactors often are inhibited by chelators that bind to the metal. Examples of such metalloenzymes are lactate dehydrogenase from muscle and aldolase from yeast, both of which contain Zn^{2+}. Chelators inhibit aldolase by removing the required metal. The inhibition is not reversed simply by dialysis, but it can be reversed by adding Zn^{2+} to the depleted enzyme. In the case of lactate dehydrogenase the enzyme holds the Zn^{2+} more tightly, and the metal-chelator complex remains attached to the inhibited enzyme.

Figure 8.21

Affinity labels and suicide inhibitors. (*a*) 3-bromoacetol phosphate is a structural analog of dihydroxyacetone phosphate. It binds to the active site of triosephosphate isomerase, and then reacts to form a covalent bond with the carboxyl group of a nearby glutamyl residue. Bromoketone groups have been incorporated into many molecules to make similar affinity labels for other enzymes. (*b*) A photoaffinity label can be made by attaching a diazoacetyl group to a molecule (R) that resembles the substrate for a particular enzyme. After the reagent binds to the active site, it is exposed to light. This causes it to break down, forming a carbene derivative. The carbene reacts rapidly with any of several amino acid residues to form a covalent bond to the enzyme. (*c*) Vinylglycine can be used as a mechanism-based inhibitor for some enzymes that catalyze modifications of amino acids. It is not intrinsically a reactive compound, but is converted into a reactive allyl-imine in the course of the reaction catalyzed by the enzyme.

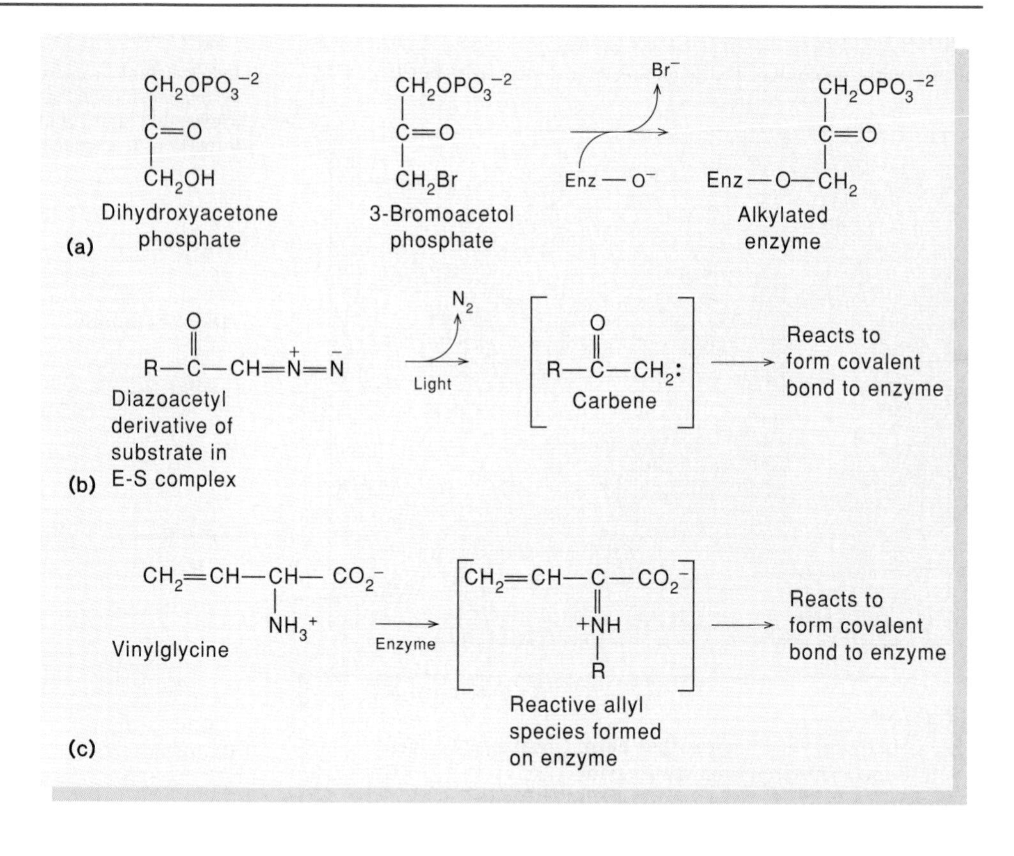

Summary

Enzymes are biological catalysts. Most enzymes are proteins and are highly specific in the reactions they catalyze. Kinetic analysis is one of the most broadly used tools for characterizing enzymatic reactions. In this chapter we have focused on the following points.

1. The rates of both enzymatic and nonenzymatic reactions are functions of the frequency of collisions between the reacting species and the fraction of the collisions that produce products. The former is determined by the temperature and the concentrations of the reactants; the latter depends on the temperature and the activation free energy, $\Delta G^{\ddagger}$. $\Delta G^{\ddagger}$ can be interpreted as the free energy that is needed to convert the reactants to a transition state. A catalyst increases the reaction rate by lowering $\Delta G^{\ddagger}$.
2. Enzymes have localized catalytic sites. The substrate (S) binds at the active site to form an enzyme-substrate complex (ES). Subsequent steps transform ES into an enzyme-product complex, and dissociation of the product regenerates the free enzyme. The overall speed of the reaction is proportional to the concentration of ES. Shortly after the enzyme and substrate are mixed, the concentration of ES becomes approximately constant and remains so for a significant period of time known as the steady state.
3. Enzymatic reaction rates usually are studied by leaving one essential ingredient out of the reaction and then adding this component and measuring the formation of product or the disappearance of a reactant with time. Special techniques are necessary to measure very fast reactions. It is common to measure the rate as a function of the concentrations of substrates and products, and to examine its dependence on pH and temperature.
4. The rate (v) of an enzymatic reaction in the steady state usually has a hyperbolic dependence on the concentration of the substrate. It is proportional to [S] at low concentrations but approaches a maximum (V_{max}) at high concentrations, when the enzyme is fully charged with substrate. The Michaelis constant, K_m, is the substrate concentration at which the rate is half-maximal. K_m and V_{max} often can be obtained from a plot of $1/v$ versus 1/[S]. If the dissociation of ES occurs rapidly relative to the conversion of the complex into products, K_m is approximately equal to the dissociation constant for the complex. Under other conditions, K_m is a function of all the rate constants involved in the formation of ES and its conversion to products. In general, K_m is larger than the dissociation constant.
5. The turnover number, k_{cat}, is the maximum number of moles of substrate converted to product per unit time per mole of enzyme, and is equal to V_{max} divided by the enzyme concentration. The specificity constant, k_{cat}/K_m, is a measure of how rapidly an enzyme can work at low substrate concentrations. This is usually the best index of the effectiveness of an enzyme with different substrates.
6. Enzymes that catalyze reactions of two or more substrates work in a variety of ways that can be distinguished by kinetic analysis. Some enzymes bind the substrates in a fixed order; others bind their substrates in random order. Sometimes the binding of one substrate gives a partial reaction before the second substrate binds.

7. Many enzymes are sensitive to inhibition by specific agents that interfere with the binding of substrate at the active site or with conversion of the enzyme-substrate complex into products. Study of these effects can provide information about how an enzyme operates.
8. Reversible inhibitors are classified as competitive, noncompetitive, or uncompetitive. A competitive inhibitor competes with substrate for binding to the enzyme. Consequently, a sufficiently high concentration of substrate can eliminate the effect of a competitive inhibitor. A noncompetitive inhibitor binds to the enzyme at a separate site and interferes with the reaction regardless of whether or not the active site is occupied by substrate. An uncompetitive inhibitor binds to the enzyme-substrate complex, but not to the free enzyme. The different forms of reversible inhibition are distinguishable by measuring the rate as a function of the concentrations of the substrate and the inhibitor.
9. Irreversible inhibitors often provide information on the active site by forming covalent complexes that can be characterized.

Selected Readings

Advances in Enzymology. New York: Academic Press. An ongoing, annually published volume containing monographs on selected topics.

Boyer, P. D. (ed.), *The Enzymes.* New York: Academic Press. A continuing series with more than 16 volumes of monographs on selected enzymes. See particularly the chapter entitled "Steady State Kinetics" by W. W. Cleland in vol. 2.

Fersht, A., *Enzyme Structure and Mechanism,* 2d ed. New York: Freeman, 1985.

Frost, A. A., and R. G. Pearson, *Kinetics and Mechanism* (2d ed.). New York: Wiley, 1961. An excellent introduction to general chemical kinetics.

Purich, D. L., *Contemporary Enzyme Kinetics and Mechanism.* New York: Academic Press, 1983. A good source of detailed information on how to analyze kinetic data, and on effects of temperature, pH, and inhibitors. The chapters are selected from several volumes of *Methods in Enzymology* (New York: Academic Press), an ongoing series of monographs.

Segal, I. H., *Enzyme Kinetics.* New York: Wiley, 1975.

Problems

1. Explain what is meant by the order of a reaction, using the reaction below as an example. What is the reaction order for each reactant? (Consider the forward and reverse reaction.) What is it for the overall reaction?

$$A + B \rightleftharpoons 2C$$

2. In a first order reaction a substrate is converted to product so that 87% of the substrate is converted in 7 min. Calculate the first-order rate constant. In what time would 50% of the substrate be converted to product?
3. K_m is frequently equated with K_S, the [ES] dissociation constant. However, there is usually a disparity between those values. Why? Under what conditions are K_m and K_S equivalent?
4. Differentiate between the enzyme-substrate complex and the transition-state intermediate in an enzymatic reaction.
5. An enzyme was assayed with a substrate concentraton of twice the K_m value. The progress curve of the enzyme (product produced per minute) is shown here. Give two possible reasons why the progress curve becomes nonlinear.

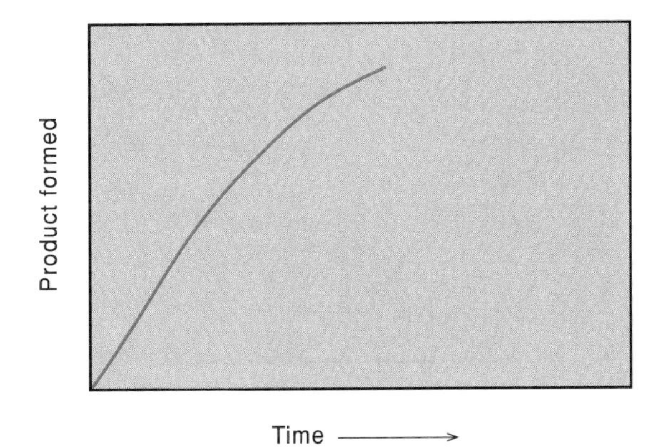

6. What is the steady-state approximation and under what conditions is it valid?
7. Assume that an enzyme-catalyzed reaction follows Michaelis-Menten kinetics with a K_m of 1 μM. The initial velocity is 0.1 μM/min at 10 mM substrate. Calculate the initial velocity at 1 mM, 10 μM, and 1 μM substrate. If the substrate concentration were increased to 20 mM, would the initial velocity double? Why or why not?
8. If the K_m for an enzyme is 1.0×10^{-5} M and the K_I of a competitive inhibitor of the enzyme is 1.0×10^{-6} M, what concentration of inhibitor is necessary to lower the reaction rate by a factor of 10 when the substrate concentration is 1.0×10^{-3} M? 1.0×10^{-5} M? 1.0×10^{-6} M?
9. Assume that an enzyme-catalyzed reaction follows the scheme shown:

$$E + S \underset{k_2}{\overset{k_1}{\rightleftharpoons}} ES \underset{k_4}{\overset{k_3}{\rightleftharpoons}} E + P$$

where $k_1 = 10^9$ M^{-1} s^{-1}, $k_2 = 10^5$ s^{-1}, $k_3 = 10^2$ s^{-1}, $k_4 = 10^7$ M^{-1} s^{-1}, and $[E_T]$ is 0.1 nM. Determine the value of each of the following.
   (a) K_m
   (b) V_{max}
   (c) Turnover number
   (d) Initial velocity when $[S]_0$ is 20 μM.

10. A colleague has measured the enzymatic activity as a function of reaction temperature and obtained the data shown in this graph. He insists on labeling point A as the "temperature optimum" for the enzyme. Try, tactfully, to point out the fallacy of that interpretation.

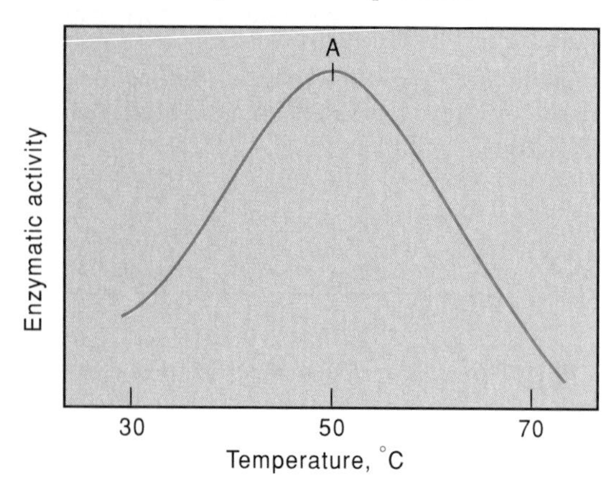

11. You have isolated an enzyme that catalyzes a bimolecular reaction:

$$A + B \rightleftharpoons P + Q$$

The initial velocity data yielded intersecting double-reciprocal plots with [A] varied at fixed [B] or [B] varied at fixed [A]. Which kinetic pattern—sequential (ordered or random), or Ping-Pong—might you rule out?

12. You have isolated a tetrameric NAD^+-dependent dehydrogenase. You incubate this enzyme with iodoacetamide in the absence or presence of NADH (at ten times the K_m concentration) and you periodically remove aliquots of the enzyme for activity measurements and amino acid composition analysis. The results of the analyses are shown.
    (a) What can you conclude about the reactivities of the cysteinyl and histidyl residues of the protein?
    (b) Which residue could you implicate in the catalytic active site? On what do you base the choice? Are the data conclusive concerning the assignment of a residue to the active site? Why or why not?
    (c) After 1 h you dilute the enzyme incubated with iodoacetamide but no NADH. Would you expect the enzyme activity to be restored? Explain.

13. The following initial velocity data were obtained for an enzyme.

[S] (nM)	Velocity (M s^{-1}) × 10^7
0.10	0.96
0.125	1.12
0.167	1.35
0.250	1.66
0.50	2.22
1.0	2.63

Each assay at the indicated substrate concentration was initiated by adding enzyme to a final concentration of 0.01 nM. Derive K_m, V_{max}, k_{cat}, and the specificity constant.

14. You have measured the initial velocity of an enzyme in the absence of inhibitor and with inhibitor A or inhibitor B. In each case, the inhibitor is present at 10 μM. The data are shown in the table.

[S] (mM)	Velocity (M s^{-1}) × 10^7 Uninhibited	Velocity (M s^{-1}) × 10^7 Inhibitor A	Velocity (M s^{-1}) × 10^7 Inhibitor B
0.333	1.65	1.05	0.794
0.40	1.86	1.21	0.893
0.50	2.13	1.43	1.02
0.666	2.49	1.74	1.19
1.0	2.99	2.22	1.43
2.0	3.72	3.08	1.79

    (a) Determine K_m and V_{max} of the enzyme.
    (b) Determine the type of inhibition imposed by inhibitor A and calculate $K_{I(S)}$.
    (c) Determine the type of inhibition imposed by inhibitor B and calculate $K_{I(S)}$.

15. Irreversible inactivation of an enzyme by a compound may be confused with noncompetitive inhibition. Why? How could you distinguish between a reversible noncompetitive inhibitor and an irreversible inactivator? Enzyme supply is not limiting.

	(No NADH Present)			(NADH Present)		
Time (min)	Activity (U/mg)	His (Residues/mole)	Cys (Residues/mole)	Activity (U/mg)	His (Residues/mole)	Cys (Residues/mole)
0	1,000	20	12	1,000	20	12
15	560	18.2	11.4	975	20	11.4
30	320	17.3	10.8	950	20	10.8
45	180	16.7	10.4	925	19.8	10.4
60	100	16.4	10.0	900	19.6	10.0

9

CHAPTER

Mechanisms of Enzyme Catalysis

In chapter 8 we saw that enzymes can increase the rates of reactions by many orders of magnitude. We noted that enzymes work under mild conditions of temperature, pH, and pressure, and that they are highly specific in the types of reactions they catalyze and in the particular substrates they accept. In this chapter we will explore the mechanisms of several enzyme-catalyzed reactions in greater detail. Our goal is to relate the activity of each of these enzymes to the structure of the active site, where the functional groups of amino acid side chains, the polypeptide backbone, or bound cofactors must interact with the substrates in such a way as to favor the formation of the transition state. We will be exploring enzyme catalytic mechanisms in many subsequent chapters as well, but usually in less detail than here.

Five Themes that Recur in Discussing Enzymatic Reactions

Several broad themes recur frequently in discussing enzymatic reaction mechanisms. Among the most important of these are (1) the proximity effect, (2) electrostatic effects, (3) general-acid and general-base catalysis, (4) nucleophilic or electrophilic catalysis by enzymatic functional groups, and (5) structural flexibility. For all known enzymes at least one of these themes is relevant, and in most cases more than one. We will start by discussing the five themes in general terms, and then see how they apply to some representative enzymes.

The Proximity Effect: Enzymes Bring Reacting Species Close Together

The idea of the proximity effect is that an enzyme can accelerate a reaction between two species simply by holding the two reactants close together in an appropriate orientation. It has long been known that intramolecular reactions between groups that are tied together in a single molecule are usually much faster than the corresponding intermolecular reactions between two independent molecules. The cyclization of succinic acid to form succinyl anhydride (equation 1), for example, is much more rapid than the formation of acetic anhydride from two molecules of acetic acid (equation 2):

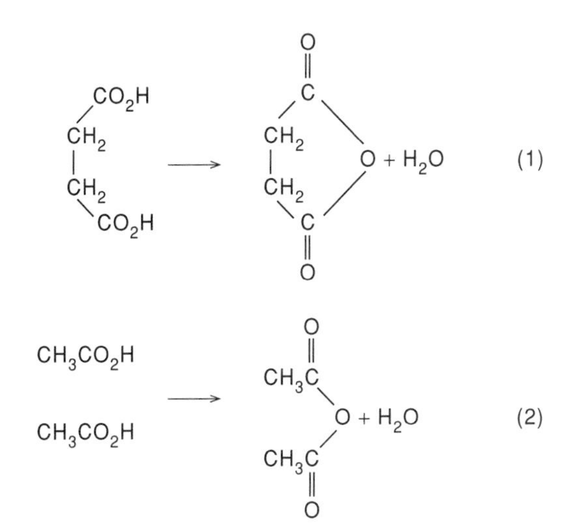

It is not possible to compare the rate constants for these two reactions directly, because they are expressed in different units. The intramolecular reaction (equation 1) is kinetically first order, while the intermolecular reaction (equation 2) is second order. But suppose that one of the two reactants in the intermolecular reaction is present in great excess over the other reactant, so that the process is effectively first order in the concentration of the second, limiting reactant (see fig. 8.3). Then we can ask what the molarity of the more abundant species would have to be, in order to make the effective first-order rate constant of the intermolecular reaction the same as the measured first-order rate constant for the intramolecular reaction. For the reactions of succinic and acetic acids (equations 1 and 2), the answer is 3×10^5 M. This is far above any concentration that could be obtained, even if the intermolecular reaction were carried out in pure acetic acid. (The concentration of CH_3COOH in glacial acetic acid is only 17.5 M.) In the related reaction

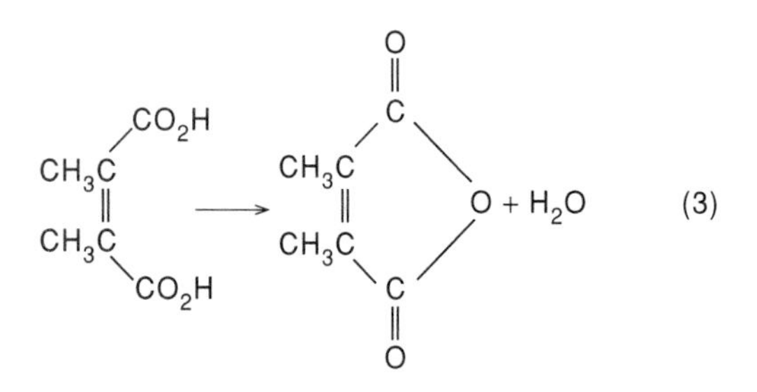

the result is even more dramatic. For the corresponding intermolecular reaction to match this intramolecular process the concentration of the fixed reactant would have to be 2×10^{12} M!

These examples from organic chemistry show that tying two reactants together in a single molecule can have an enormous effect on the rate of a reaction. This effect is, for the most part, due simply to differences between the entropy changes that accompany the inter- and intramolecular reactions. The formation of the product involves a much larger loss of translational and rotational entropy in the intermolecular molecular reaction than it does in the corresponding intramolecular reaction. A negative change in entropy increases both the overall free energy change in the reaction ($\Delta G = \Delta H - T\,\Delta S$), and the activation free energy ($\Delta G^{\ddagger} = \Delta H^{\ddagger} - T\,\Delta S^{\ddagger}$) for the formation of the transition state (see equation 9 in chapter 2 and equation 16 in chapter 8). In the intramolecular reaction much of this entropy decrease has already occurred during the preparation of the reactant.

Enzymes that catalyze intermolecular reactions take advantage of the proximity effect by binding the reactants close together in the active site, so that the reactive groups are oriented appropriately for the reaction. Once the substrates are fixed in this way, the subsequent reaction behaves kinetically like an intramolecular process. The entropy decrease associated with the formation of the transition state has been moved to an earlier step, the binding of the substrates to form the enzyme-substrate complex. This step often is driven by an enthalpy decrease associated with electrostatic interactions between polar or charged groups of the substrates and the enzyme. There are, however, exceptions to this generalization, particularly in reactions involving hydrophobic substrates. As we discussed in chapter 2, the removal of a hydrophobic molecule from aqueous solution is favored by an entropy increase, and the binding of hydrophobic substrates to enzymes can be driven in this way.

General-Base and General-Acid Catalysis Provide Ways of Avoiding the Need for Extremely High or Low pH

Chemical bonds are formed by electrons, and the rearrangement or breakage of bonds requires the migration of electrons. In broad terms, reactive chemical groups can be said to function either as electrophiles or as nucleophiles. Electrophiles are electron-deficient substances that react with electron-rich substances; nucleophiles are electron-rich substances that react with electron-deficient substances. The task of a catalyst often is to make a potentially reactive group more reactive by increasing its intrinsic electrophilic or nucleophilic character. In many cases the simplest way to do this is to add or remove a proton. As an example, consider the hydrolysis of an ester (fig. 9.1). Because the electronegativity of the oxygen atom in the C=O group is greater than that of the carbon, the oxygen has a fractional negative charge, δ^-, and the carbon has a fractional positive charge, δ^+. Hydrolysis of an ester in neutral aqueous solution can occur if the oxygen atom of H_2O, acting as a nucleophile, attacks the positively charged carbon. The initial product is an intermediate in which the carbon atom has four substituents in a tetrahedral arrangement. The reaction is completed by the rapid breakdown of the tetrahedral intermediate to release the alcohol.

Water is intrinsically a comparatively weak nucleophile, and its reaction with esters in the absence of a catalyst is very slow. The hydrolysis of esters occurs much more rapidly at high pH, when the negatively charged hydroxide ion replaces water as the reactive nucleophile (see fig. 9.1*a*). But the nucleophilic character of water itself also can be increased by interaction with a basic group other than OH^- (see fig. 9.1*b*). The base offers a pair of electrons to one of the protons of the water, and thus increases the electron density on the oxygen.

The term general base is used to describe any substance that is capable of binding a proton in aqueous solution. Enzymes use a variety of functional groups to fill this role. There are two factors that make free hydroxide ions unsuitable for enzymatic catalysis, and that dictate the choice of a general base. First, the low concentration of OH^- limits its availability at physiological pH. In contrast, proteins contain numerous functional groups that can serve as general bases at moderate pH, or even under mildly acidic conditions. The only requirement is that the base start out mainly in its unprotonated form, which will be the case as long as the ambient pH is above the pK_a of the conjugate acid. This condition can easily be met by selecting a basic group from among the ionizable or polar amino acid side chains, from an amino-terminal $-NH_2$ group or a carboxyl-terminal carboxylate ion, or from the oxygen or nitrogen atom

Figure 9.1

Several ways that the hydrolysis of an ester can occur. A colored, curved arrow represents the movement of an electron pair from an electron donor to an acceptor. (*a*) Catalysis by free hydroxide ion. (*b*) General-base catalysis. (*c*) General-acid catalysis.

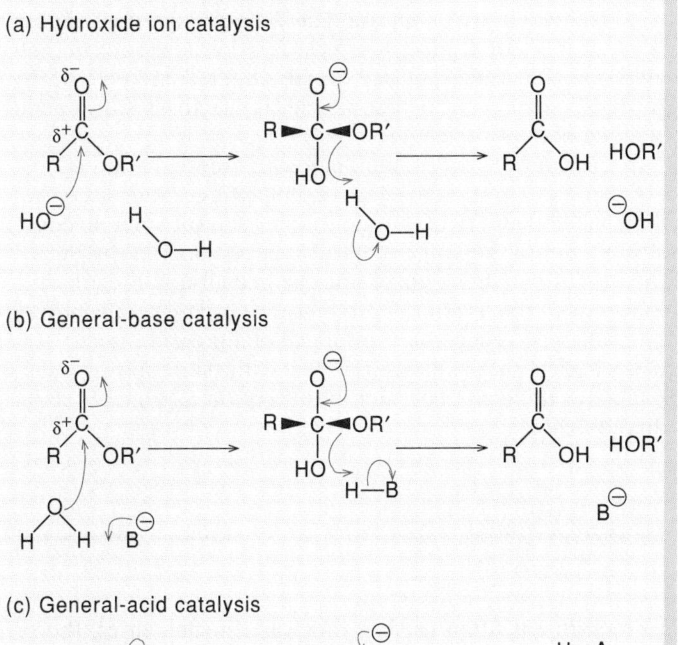

of a peptide bond (see table 3.3). The pK_a of any of these groups can vary over a considerable range, depending on the local environment in the enzyme. The second advantage of using a general base instead of OH^- is that a basic group that is provided by the protein can be positioned precisely with respect to the substrate in the active site, allowing the proximity effect to come into play. Free hydroxide ions tend to be much more mobile. In exceptional cases in which a hydroxide ion acts as a nucleophile in an enzymatic reaction, it usually is tightly bound to a metal ion.

The hydrolysis of an ester also can be catalyzed by an acid (see fig. 9.1*c*). The acid donates a proton to the oxygen of the ester's C=O group, increasing the positive charge on the carbon and increasing the susceptibility of the ester to attack by a nucleophile. Again, the term general acid is used to refer to any substance that is capable of releasing a proton, and enzymes almost always use such proton donors in preference to free protons or hydronium ions, presumably because a general acid can operate at moderate pH and is easy to fix in position. In this case, the requirement is that the pH be below the pK_a.

An important point to note in figure 9.1 is that the same general acid or base that catalyzes the formation of the tetrahedral intermediate also can participate in the decomposition of the intermediate. When a general acid (HA) donates a proton to the ester oxygen, it becomes the conjugate base (A^-), which can retrieve the proton as the intermediate breaks down. When a general base (B^-) removes a proton from water, it becomes the conjugate acid (BH), which can provide a proton to the alcohol. Note also that general-acid and general-base catalysis are not mutually exclusive: They could both occur in a concerted manner in the same step of a reaction.

Electrostatic Interactions Can Promote the Formation of the Transition State

The frequent use of general acids and general bases in enzymatic reaction mechanisms illustrates the underlying principle that enzymes act by stabilizing the distribution of electrical charge in transition states. In the enzymatic hydrolysis of an ester, the key transition state probably is structurally similar to the tetrahedral intermediates shown in figure 9.1. To form such an intermediate, electrons must move from the attacking nucleophile, through the carbon atom of the C=O group, to the oxygen of the C=O. There is thus a net movement of negative charge from the nucleophile to the substrate. In the absence of a general acid or base, a charge approaching +1 would appear on the nucleophile and a charge approaching −1 would appear on the C=O oxygen. A general base can stabilize this new distribution of charge by offering electrons to the nucleophile, so that some of the positive charge moves to the base. By providing a proton to the C=O oxygen, a general acid can delocalize the negative charge at this end of the system. But there are other ways that an enzyme could achieve a similar stabilization. Suppose that the active site included a positively charged amino acid side chain, such as that of lysine or arginine, located near the oxygen atom of the C=O group. A fixed positive charge in this region would favor the formation of the tetrahedral intermediate, even if there were no transfer of a proton from the charged species to the oxygen. A fixed negative charge in the region of the nucleophile would have a similar effect. The interactions of such fixed charges are termed electrostatic effects. The magnitude of electrostatic effects in proteins is discussed in box 9A.

Electrostatic interactions can be significant even between groups whose net formal charge is zero. This is because charge distributions within molecular groups are not uniform, but rather varies from atom to atom. We alluded to this point earlier in discussing the partial charges on the oxygen and carbon atoms of an ester (see fig. 9.1). Similar considerations apply to other functional groups, including even methylene and methyl groups: The electron distributions around the nuclei leave each atom with a small net positive or negative charge, even though the overall sum of these charges is zero. In an alcoholic $-CH_2OH$ group, for example, the oxygen atom has a negative charge of approximately −0.4 atomic charge units, and the hydrogen has a charge of about +0.4.

The Magnitude of Electrostatic Effects in Proteins

The energy of the electrostatic interaction between two charges Q_1 and Q_2 separated by a distance r Å is (in kcal/mole)

$$V = 332\frac{Q_1Q_2}{r} \quad \text{(B1)}$$

From this expression, it is clear that electrostatic effects can be appreciable even at relatively large distances; the interaction energy of a set of opposite charges 10 Å apart would be −33.2 kcal/mole. However, equation (B1) refers to charges in a vacuum. In a polar solvent such as water, electrostatic interactions are weaker because they are screened by the dielectric effect of the solvent. To take this into account, equation (B1) is often replaced by an expression of the form

$$V = 332\frac{Q_1Q_2}{\epsilon r} \quad \text{(B2)}$$

where ϵ is the dielectric constant. The dielectric constant of pure water at 25°C is 78. Dielectric effects arise because solvent molecules near a charged species become oriented and electrically polarized, so that each charged species is effectively surrounded by a cloud of opposite charges.

In the interior of a protein, molecular reorientation is relatively restricted, so that the effective dielectric constant is considerably smaller than that of the surrounding solvent water. Effective values of ϵ are in the range of 2 to 10, depending on the details of the structure in the region of the charged groups. Electrostatic interactions in the interior of proteins thus can be comparatively strong. Charged groups on the surface of a protein interact less strongly because of the dielectric effect of the surrounding water; the effective dielectric constant in this region is probably about 40 in most cases.

As a reacting substrate is transformed into a transition state, the changing charges on its atoms interact with the charges on all of the other atoms in the surrounding protein, and also with the charges on any nearby water molecules. The energy difference between the initial state and the transition state thus depends critically on the details of the protein structure. We will see illustrations of this principle in the serine proteases and the other enzymes that we discuss later in the chapter. Modern computational techniques, when taken with the wealth of structural information that has become available from x-ray crystallography and other biophysical studies, have made it possible to calculate the contributions that various components of an enzyme's active site make to the activation free energy ΔG^{+}, and to predict quantitatively how ΔG^{+} might be altered by modifications of the protein. These predictions can be tested experimentally by modifying the gene that encodes the protein, a technique termed "site-directed mutagenesis" (see chapter 27). This combination of biophysical, computational, and molecular biological techniques has opened exciting new frontiers for exploring the detailed mechanisms of enzymic catalysis.

Enzymatic Functional Groups Provide Nucleophilic and Electrophilic Catalysts

Another strategy for catalyzing the hydrolysis of an ester or an amide is to replace water by a stronger nucleophilic group that is part of the enzyme's active site. The $HOCH_2$— group of a serine residue is often used in this way. In such cases, the reaction of the serine with the substrate splits the overall reaction into a two-step process. Instead of immediately yielding the free carboxylic acid, the breakdown of the initial tetrahedral intermediate yields an intermediate ester that is covalently attached to the enzyme:

$$R-\overset{O}{\overset{\|}{C}}-NHR' + HOCH_2-\text{Enzyme} \rightarrow R-\overset{O}{\overset{\|}{C}}-OCH_2-\text{Enzyme} + R'NH_2 \quad \textbf{(4)}$$

The acyl-enzyme ester intermediate must be hydrolyzed by a second reaction, in which water becomes the nucleophile:

$$R-\overset{O}{\overset{\|}{C}}-OCH_2-\text{Enzyme} + H_2O \rightarrow R-\overset{O}{\overset{\|}{C}}-OH + HOCH_2-\text{Enzyme} \quad \textbf{(5)}$$

The proteolytic enzymes trypsin, chymotrypsin, and elastase, discussed in a later section, all work in this way. The two-step pathway requires that the intermediate be more susceptible to nucleophilic attack by water than the original ester or amide. This is likely if the original substrate is an amide, because amides are generally less reactive than esters.

Nucleophilic groups on enzymes participate in a variety of other types of reactions in addition to hydrolytic reactions. An example is acetoacetic acid decarboxylase, which catalyzes the reaction

$$CH_3-\overset{\overset{\large O}{||}}{C}-CH_2-CO_2H \rightarrow CH_3-\overset{\overset{\large O}{||}}{C}-CH_3 + CO_2 \quad (6)$$

The reaction proceeds by the formation of a Schiff base intermediate, in which the substrate is covalently attached to the ε-amino group of a lysine residue at the enzyme's active site:

$$CH_3-\overset{\overset{\large O}{||}}{C}-CH_2-CO_2H + \text{Enzyme}-NH_2 \rightarrow$$

$$CH_3-\overset{\overset{\large N-\text{Enzyme}}{||}}{C}-CH_2CO_2H + H_2O \quad (7)$$

This intermediate is formed by a nucleophilic attack of the amino group on the carbonyl carbon, followed by the splitting out of water. Protonation of the nitrogen atom of the Schiff base introduces a positive charge that pulls electrons from the nearby carbon–carbon bond, causing decarboxylation (fig. 9.2). This is an extreme example of an electrostatic effect: The enzyme introduces a charged group, not just nearby in the active site, but into the substrate itself! Aldolase and transaldolase, two enzymes that catalyze steps in the breakdown of carbohydrates, use lysine residues in a similar manner.

A basic feature of both of the mechanisms outlined in equations (4)–(7), and of other instances of nucleophilic catalysis by enzymes, is the formation of an intermediate state in which the substrate is covalently attached to a nucleophilic group on the enzyme. In addition to the $-CH_2OH$ group of serine and the ε-amino group of lysine, the $-CH_2SH$ of cysteine is often used as a nucleophile. The carboxylate of aspartate or glutamate participates in reactions involving the hydrolysis of ATP, and the imidazole group of histidine can play a similar role. Some enzymes take advantage of bound coenzymes such as thiamine, biotin, pyridoxamine, or tetrahydrofolate to obtain additional nucleophilic reagents (see chapter 11).

There also are numerous enzymes that use bound metal ions to form complexes with substrates. In these enzymes, the metal ion generally serves as an *electrophilic,* rather than a nucleophilic, functional group. Carbonic anhydrase, for example, contains a Zn^{2+} ion that binds one of the substrates, hydroxide ion, as a ligand. The bound OH^- reacts with the other substrate, CO_2. In alcohol dehydrogenase, and in the proteolytic enzymes thermolysin and carboxypeptidase A, a Zn^{2+} ion in the active site forms a complex with the carbonyl oxygen atom of the aldehyde or peptide substrate. The withdrawal of electrons by the Zn^{2+} increases the partial positive charge on the carbonyl carbon atom, and thus promotes the reaction of the carbon with a nucleophile. We will discuss such enzymes in more detail in a later section.

Figure 9.2

In acetoacetic acid decarboxylase, the positive charge of a protonated Schiff base intermediate pulls electrons from a nearby carbon–carbon bond, thereby releasing CO_2.

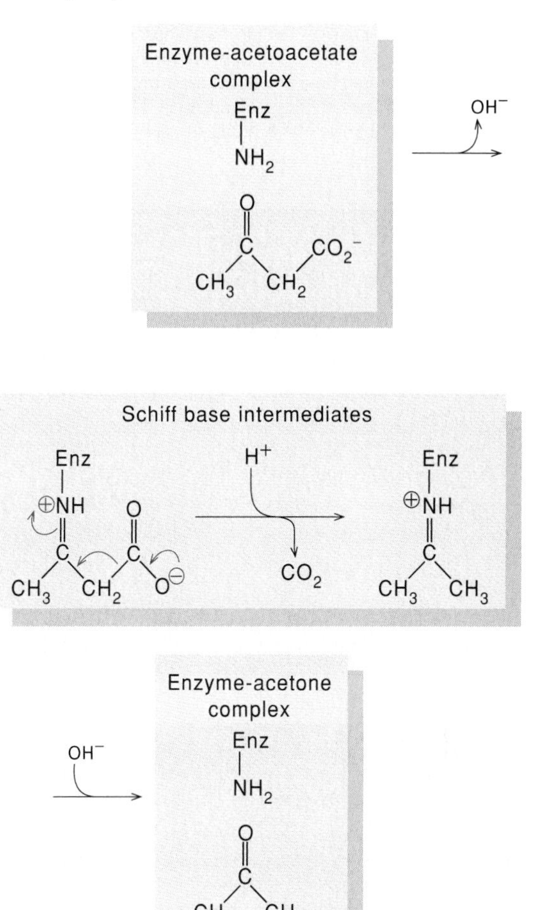

Structural Flexibility Can Increase the Specificity of Enzymes

Although precise positioning of the reactants is a fundamental aspect of enzyme catalysis, some enzymes undergo major structural rearrangements when they bind substrates or inhibitors. An example is hexokinase, which catalyzes the transfer of a phosphate group from ATP to glucose:

$$\text{ATP} + \text{glucose} \rightarrow \text{ADP} + \text{glucose-6-phosphate} \quad (8)$$

When hexokinase binds glucose, it undergoes a structural reorganization that brings together the elements of the active site (fig. 9.3). The enzyme literally closes like a set of jaws around the substrate! Such a structural change is often referred to as an induced fit.

Figure 9.3

Models of the crystallographic structure of hexokinase in the "open" (*a*) and "closed" (*b*) conformations. The enzyme (shown in blue) adopts the open conformation in the absence of substrates, but switches to the closed conformation when it binds glucose (red). Hexokinase also has been crystallized with a bound analog of ATP. In the absence of glucose, the enzyme with the bound ATP analog remains in the open conformation. The structural change caused by glucose would result in the formation of additional contacts between the enzyme and ATP. This could explain why the binding of glucose enhances the binding of ATP. (Courtesy of Dr. Thomas A. Steitz.)

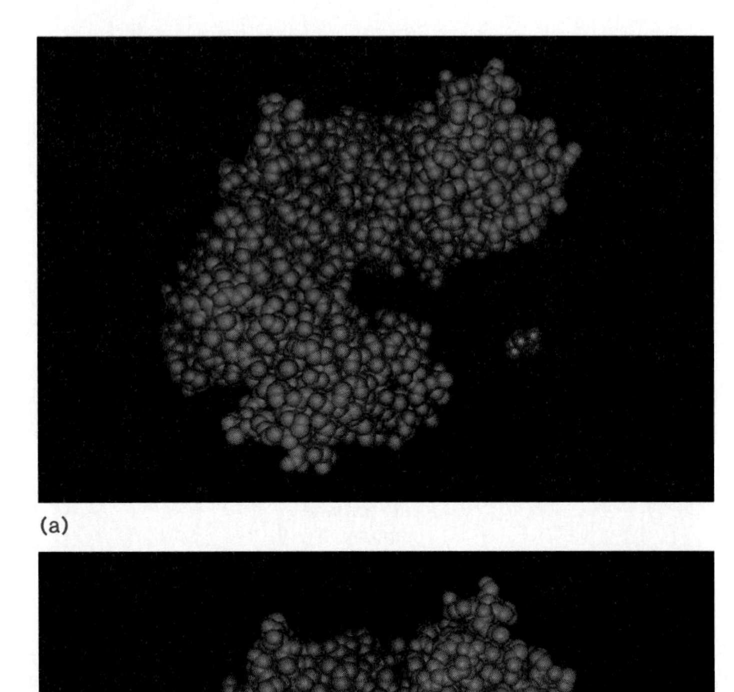

(a)

(b)

Carboxypeptidase A, which we will discuss in more detail later, is another enzyme that undergoes a major structural change when it binds its substrate. In this case, the rearrangement of the protein effectively pulls the hydrophobic part of the substrate out of the aqueous solution by surrounding it with nonpolar portions of the protein. Enfolding a substrate in this way can be beneficial in several ways. First, it can serve to maximize the favorable entropy change associated with removing a hydrophobic molecule from water. Second, it should allow the enzyme to control and intensify the electrostatic effects that promote the formation of the transition state. The substrate is forced to respond to the directed electrostatic fields from the enzyme's functional groups, instead of the disordered fields from the solvent.

Structural changes also can help to explain the high specificity of some enzymatic reactions. In hexokinase, for example, the structural change induced by glucose promotes the binding of the other substrate, ATP (see fig. 9.3). ATP does not bind to the enzyme properly unless glucose is already present in the catalytic site. If ATP were to bind in the absence of glucose, the enzyme might have a tendency to catalyze the transfer of phosphate from ATP to water, resulting in a wasteful loss of ATP:

$$ATP + H_2O \rightarrow ADP + P_i \quad \textbf{(9)}$$

Hexokinase does not catalyze this side-reaction; it waits for glucose to bind first.

As another example, the enzyme serine hydroxymethylase (see chapter 18) catalyzes the removal of formaldehyde from serine, forming glycine. In a second step, the enzyme transfers the formaldehyde to a bound coenzyme, tetrahydrofolic acid (THF). If the removal of formaldehyde from serine proceeded even in the absence of THF, the formaldehyde might be set free in the cytosol, where it could enter into other, undesirable reactions. (Free formaldehyde is highly reactive and is toxic to cells.) To prevent this potential disaster, serine hydroxymethylase does not catalyze the first reaction until after THF is bound. The binding of the coenzyme causes the enzyme to fold so that the first step can occur, even though the THF is not yet involved directly in the chemistry.

The structural changes that occur in hexokinase, carboxypeptidase A, and serine hydroxymethylase bring home the point that enzyme crystal structures give static snapshots of molecules that, in many cases, actually are highly flexible. In solution, the structure of an enzyme undergoes fluctuations that vary widely in amplitude and frequency from place to place in the protein. Vibrations and rotations involving only a few atoms occur on time scales of 10^{-13} to 10^{-11} s. Somewhat larger motions, such as the flipping of the aromatic ring of a tyrosine or tryptophan, typically occur on scales of 10^{-9} to 10^{-8} s. Major reorganizations may take 10^{-6} to 10^{-3} s. All of these types of motions can be important in catalysis.

Detailed Mechanisms of Enzyme Catalysis

In the foregoing sections, we have discussed five themes that are related to enzyme reaction mechanisms. We will now examine several representative enzymes in finer detail. We will focus on enzymes for which crystal structures have been obtained, because the most decisive advances in our understanding of enzyme reaction mechanisms have come by inspecting such structures. Crystals of many enzymes have been shown to be enzymatically active, and it appears that in most cases the three-dimensional structures of crystalline enzymes are close to the structures of the proteins in solution. It is important to keep in mind, however, that crystal structures provide pictures of enzymes in relatively stable states. To fill in the intermediates and transition states between these resting states requires a variety of other techniques, including studies of related nonenzymatic reaction mechanisms.

There are over 1,500 known enzymes, each with its own unique structure, specificity, and catalytic mechanism. However, the situation is less complicated than this number might suggest, because many enzymes can be grouped in families that share certain basic features. In some cases the enzymes that make up a family appear to have diverged from a common evolutionary ancestor. Family members that arose in this way are apt to retain similar secondary and tertiary structures, and they typically have the same amino acid residue at between 20 and 50% of the corresponding positions in their primary sequences. In other cases enzymes with diverse ancestral origins appear to have converged on structural features that are well suited for catalyzing particular types of reactions. Such enzymes resemble each other in their active sites, but may have little in common elsewhere in their structures.

Serine Proteases Are a Diverse Group of Enzymes that Use a Serine Residue for Nucleophilic Catalysis

The serine proteases are a large family of proteolytic enzymes that use the reaction mechanism for nucleophilic catalysis outlined in equations (4) and (5), with a serine residue as the reactive nucleophile. The best known members of the family are three closely related digestive enzymes, trypsin, chymotrypsin, and elastase. These enzymes are synthesized in the mammalian pancreas as inactive precursors termed zymogens. They are secreted into the small intestine, where they are activated by proteolytic cleavage in a manner that will be described in chapter 10. Many of the enzymes that participate in blood coagulation also are serine proteases; these enzymes circulate in the blood as inactive zymogens and are activated by proteolytic cleavage when blood vessels are damaged (see chapter 10). The serine protease family also includes many enzymes from bacteria and other nonmammalian organisms.

In the digestive system, trypsin, chymotrypsin, and elastase work as a team. They are all endopeptidases, which means that they cleave protein chains at internal peptide bonds, but each preferentially hydrolyzes bonds adjacent to a particular type of amino acid residue (fig. 9.4). Trypsin cuts just past the carbonyl groups of basic residues (lysine or arginine); chymotrypsin cuts next to aromatic residues (phenylalanine, tyrosine, or tryptophan); elastase is less discriminating, but prefers small, hydrophobic residues such as alanine.

About half of the amino acid residues of trypsin are identical to the corresponding residues in chymotrypsin, and about a quarter of the residues are conserved in all three of the pancreatic endopeptidases (fig. 9.5). The structural similarities of trypsin, chymotrypsin, and elastase are even more evident in the crystal structures. As is shown in figure 9.6, the folding of the polypeptide chain is essentially the same in all three enzymes, with the only substantial variations occurring in the external loops. These enzymes are classical illustrations of diverging evolution from a common ancestor. The structures of some of the bacterial serine proteases also are homologous to those of the mammalian enzymes. On the other hand, subtilisin,

Figure 9.4

Trypsin, chymotrypsin, and elastase, three members of the serine protease family, catalyze the hydrolysis of proteins at internal peptide bonds adjacent to different types of amino acids. Trypsin prefers lysine or arginine residues; chymotrypsin, aromatic side chains; and elastase, small, nonpolar residues. Carboxypeptidases A and B, which are not serine proteases, cut the peptide bond at the carboxyl-terminal end of the chain. Carboxypeptidase A preferentially removes aromatic residues; carboxypeptidase B, basic residues.

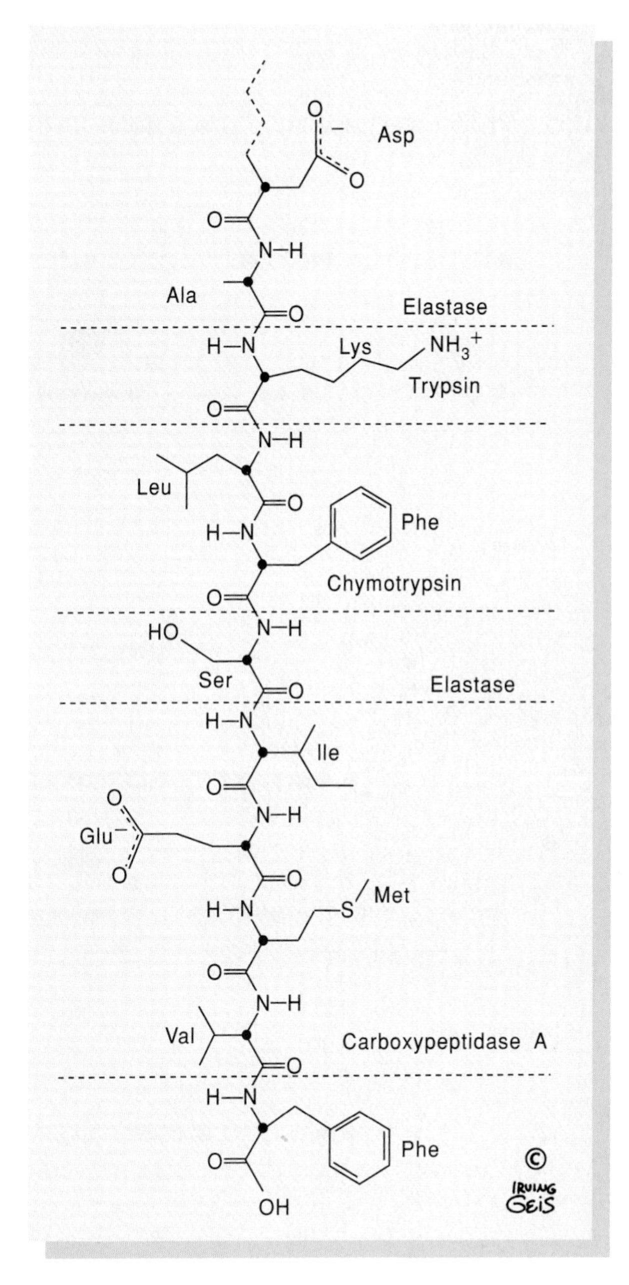

Figure 9.5

Schematic diagrams of the amino acid sequences of chymotrypsin, trypsin, and elastase. Each circle represents one amino acid. Amino acid residues that are identical in all three proteins are in solid color. The three proteins are of different lengths, but have been aligned to maximize the correspondence of the amino acid sequences. All of the sequences are numbered according to the sequence in chymotrypsin. Long connections between nonadjacent residues represent disulfide bonds. Locations of the catalytically important histidine, aspartate and serine residues are marked. The links that are cleaved to transform the inactive zymogens to the active enzymes are indicated by parenthesis marks. After chymotrypsinogen is cut between residues 15 and 16 by trypsin, and is thus transformed into an active protease, it proceeds to digest itself at the additional sites that are indicated; these secondary cuts have only minor effects on the enzyme's catalytic activity.

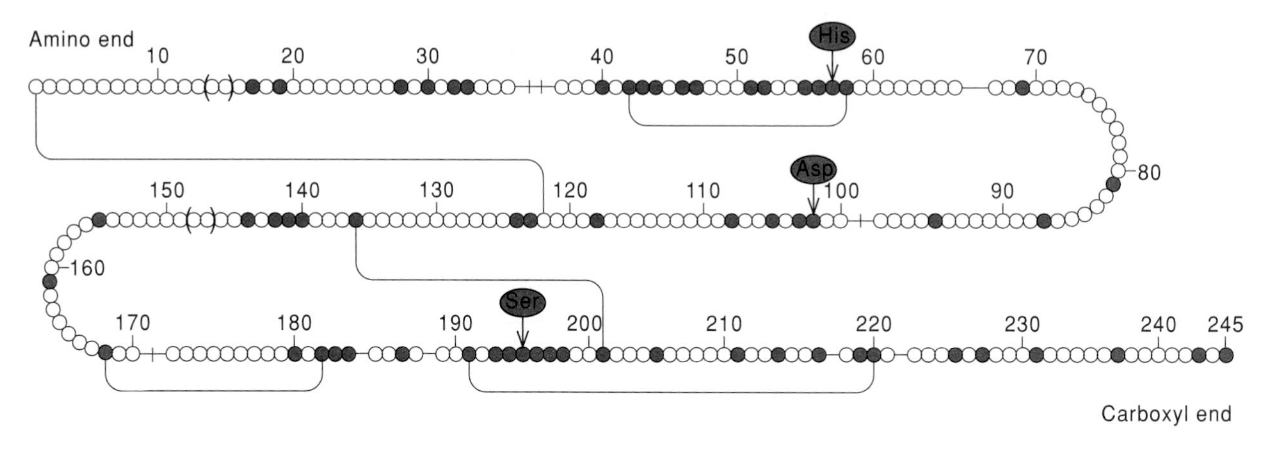

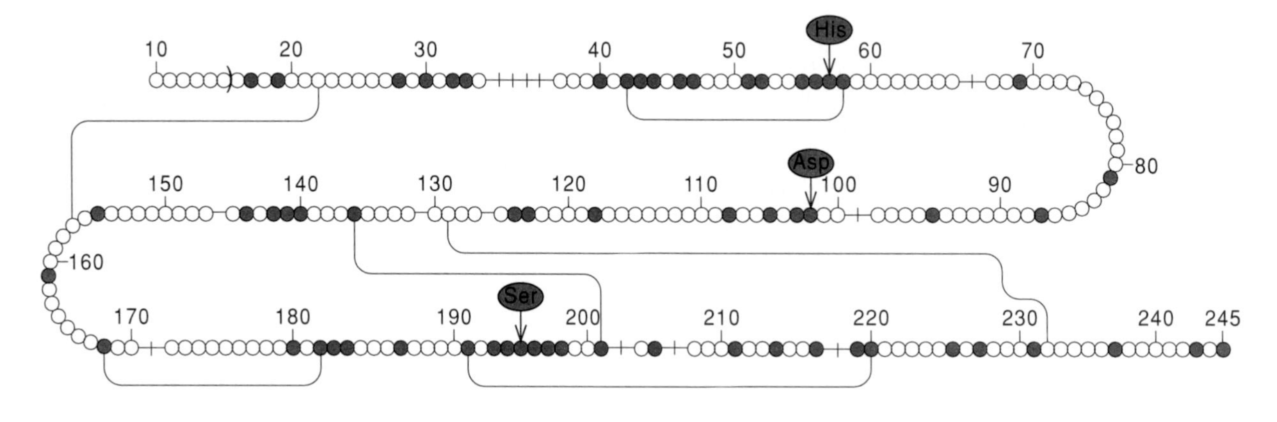

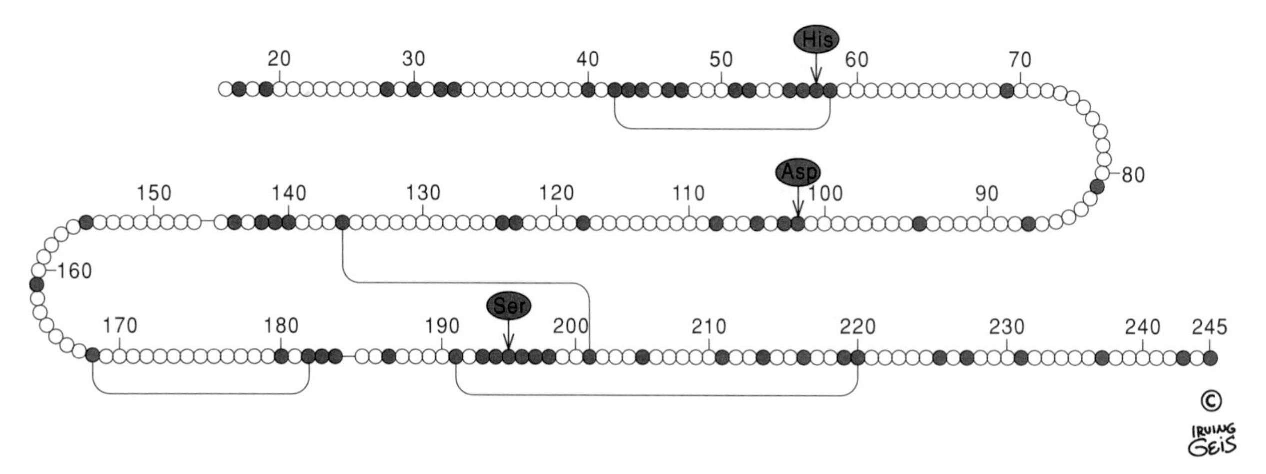

Figure 9.6

Crystal structures of (*a*) trypsin, (*b*) chymotrypsin, and (*c*) elastase. The yellow ribbons superimposed on the structures show the similar folding patterns of the α-carbon chains in the three structures. Three residues that play important roles in the catalytic mechanism (Asp 101, His 57, and Ser 195) are shown in red; other residues are in blue. The structure for trypsin includes a bound inhibitor, benzamidine, in yellow. The positively charged nitrogen atom of benzamidine fits into the bottom of the pocket that determines the specificity of trypsin for positively charged amino acid residues (lysine and arginine). (Based on the crystal structures described by W. Bode, P. Schwager, and J. Walter for trypsin, H. Tsukada and D. Blow for chymotrypsin, and L. C. Sieker and D. L. Hughes for elastase.)

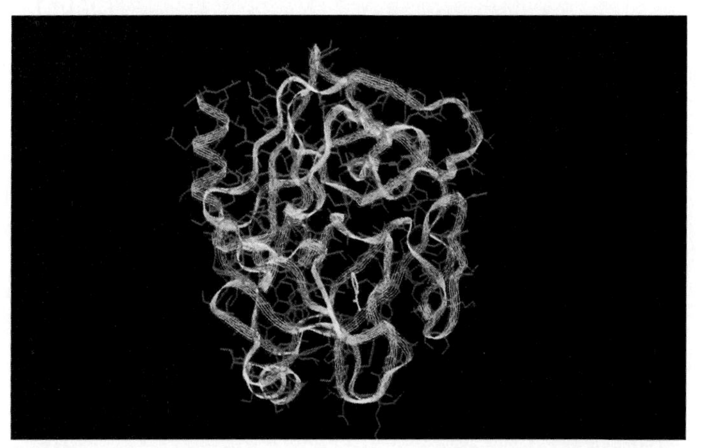

(a)

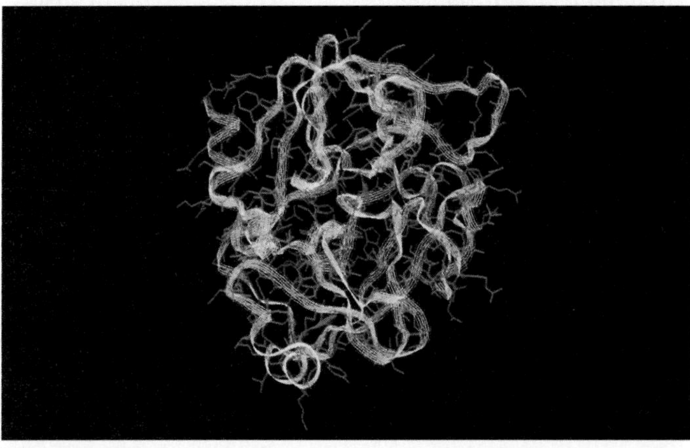

(b)

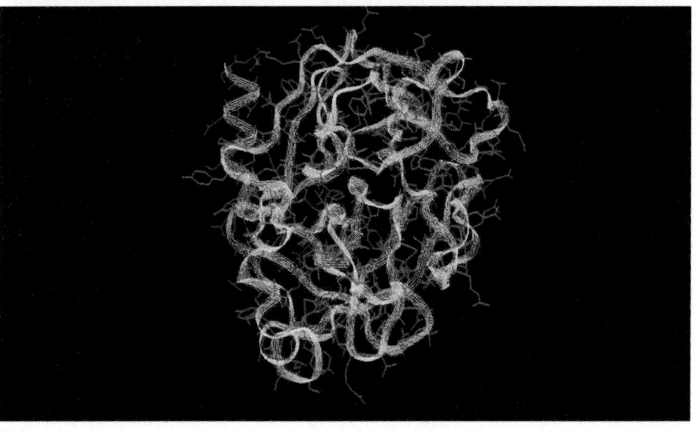

(c)

Figure 9.7

(*a*) A space-filling model of the crystal structure of trypsin, seen from the same perspective as in figure 9.6*a*. Hydrogen atoms are not shown. The coloring of the amino acid residues and the inhibitor (benzamidine) is as in figure 9.6. (*b*) A close-up view of the active site of trypsin with bound benzamidine. The amide NH hydrogens of Ser 195 and Gly 193, the OH hydrogen of Ser 195, and the imidazole NH hydrogen of His 57 are shown in green; other atoms are colored as in figure 9.6. In the oxyanion intermediate, the negatively charged oxygen atom of the substrate interacts with the amide hydrogens of Ser195 and Gly193. Benzamidine does not have an analogous oxygen atom.

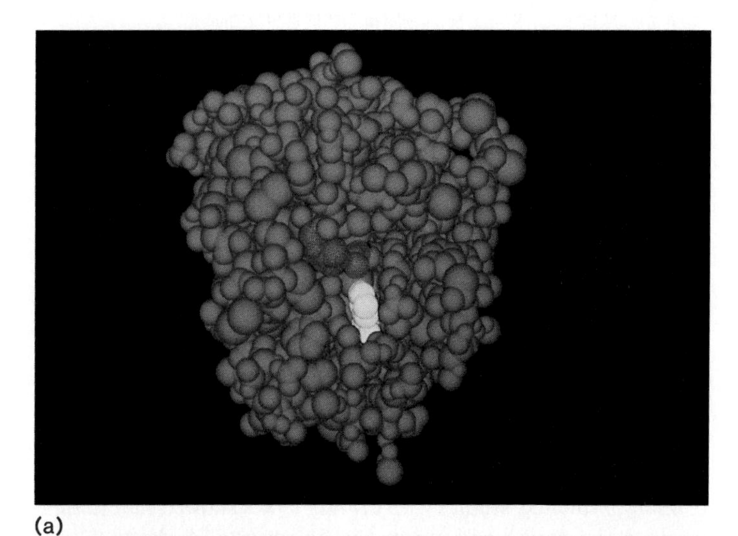

(a)

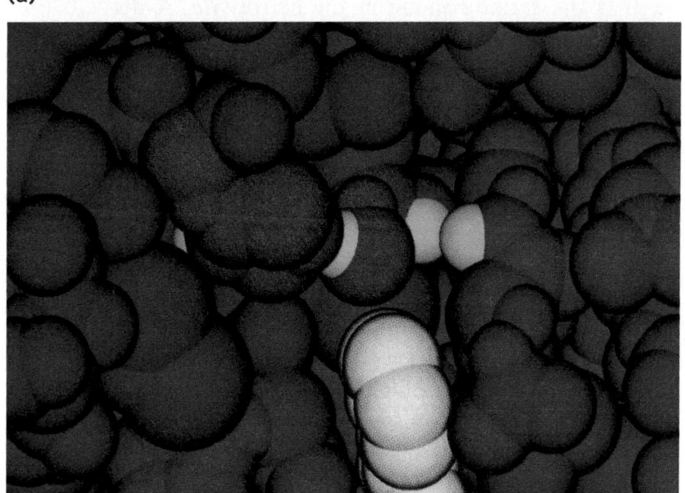

(b)

a serine protease obtained from *Bacillus subtilis,* has an amino acid sequence that seems totally unrelated to the mammalian sequences. Its three-dimensional structure also is very different from those of the mammalian enzymes (see fig. 5.23). Subtilisin therefore is likely to have joined the serine protease family by convergent evolution. Remarkably, there is a small set of critical amino acid residues that come together in the folded structure to form the essential elements of the active site in all of these proteases. In the chymotrypsin numbering system, these are His 57, Asp 102 and Ser 195. The locations of these residues in the three-dimensional structures of trypsin, chymotrypsin, and elastase can be seen in figures 9.6 and 9.7.

That Ser 195 played an important role in the catalytic mechanism was known from early studies on the enzyme inhibitor diisopropylfluorophosphate (see chapter 8). This inhibitor reacts irreversibly with chymotrypsin or trypsin to form an inactive derivative in which the diisopropylphosphate group is covalently attached to the serine residue (see fig. 8.20). The derivative, a phosphate ester of the serine, is similar in structure to the acyl-enzyme

$$(\mathrm{R} - \overset{\overset{\displaystyle \mathrm{O}}{\|}}{\mathrm{C}} - \mathrm{OCH_2} - \mathrm{Enzyme})$$

intermediate in equations (4) and (5). A variety of other inhibitors have been found to react in a parallel manner with this particular serine residue. As a rule, these inhibitors do not react with other serines in the enzyme, or with serines in enzymes that are not part of the serine protease family. Exceptions to the rule are a number of enzymes that catalyze the hydrolysis of esters; these enzymes have a similar, reactive serine residue, and their catalytic mechanism appears to be very similar to that of the serine proteases. The reactivity of Ser 195 thus is not a property of serine residues in general, but depends on the special surroundings of this residue in the protein. We will see shortly that it is the juxtaposition of Ser 195 with His 57 and Asp 102 that makes this serine especially reactive.

Although His 57 is far removed from Ser 195 in the primary sequence, studies with affinity labels showed that it must be near the serine residue in the active site. A derivative of phenylalanine containing a reactive chloromethyl ketone group was found to inhibit chymotrypsin irreversibly by reacting with the histidine. The inhibition can be prevented by the presence of other aromatic molecules that bind competitively at the active site.

The initial evidence for the formation of an acyl-enzyme ester intermediate came from studies of the kinetics with which chymotrypsin hydrolyzed various analogs of its normal polypeptide substrates. The enzyme turned out to hydrolyze esters, as well as peptides and simpler amides. Of particular interest was the reaction with the ester *p*-nitrophenyl acetate. This substrate is well suited for kinetic studies because one of the products of its hydrolysis, *p*-nitrophenol, has a characteristic yellow color in aqueous solution, whereas *p*-nitrophenyl acetate itself is colorless. The change in the absorption spectrum makes it easy to follow the progress of the reaction spectrophotometrically. When rapid mixing techniques were used to add the substrate to the enzyme, it was found that an initial burst of *p*-nitrophenol was released within the first few seconds, before the reaction settled down to a constant rate (fig. 9.8). The amount of *p*-nitrophenol that appeared in the burst was approximately equal to the amount of enzyme present in the solution. These observations suggested that the overall enzymatic reaction occurs in two distinct steps, as shown in figure 9.9. In the first step, *p*-nitrophenol is released and the acetyl group is transferred to the enzyme, forming an acyl-enzyme intermediate. In the second step, the intermediate is hydrolyzed, and acetate is released. Diisopropylfluorophosphate prevents the

Figure 9.8

p-Nitrophenol formation as a function of time during the hydrolysis of *p*-nitrophenyl acetate by chymotrypsin. A rapid initial burst of *p*-nitrophenol is followed by a slower, steady-state reaction. The amount of *p*-nitrophenol released in the burst is approximately equal to the amount of enzyme present.

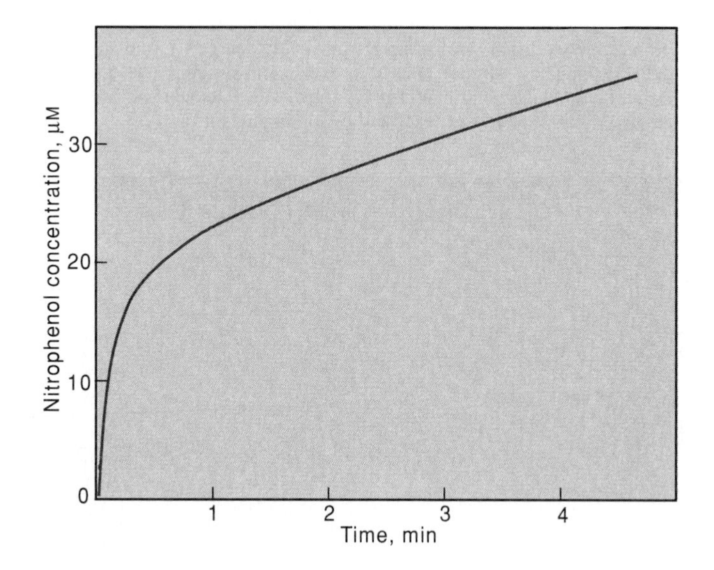

initial burst of *p*-nitrophenol, as well as the subsequent steady-state reaction, a fact suggesting that the enzymatic group that forms the ester intermediate is Ser 195.

When the crystal structures of trypsin, chymotrypsin, and elastase with bound substrate analogs were solved, the substrate analogs were indeed found to be located close to Ser 195 (see figs. 9.6 and 9.7). Histidine 57, the second of the three residues mentioned above, is located nearby, in an orientation suggesting that the OH group of Ser 195 forms a hydrogen bond to the imidazole side chain of the histidine. Aspartic acid 102 sits on the opposite edge of the imidazole ring, where its negatively charged carboxylate group could interact with the proton on the other nitrogen of the ring. The side chains of the aspartate and histidine residues thus appear to be oriented so as to facilitate removal of the proton from the serine's OH group:

$$-\overset{\overset{\displaystyle \mathrm{O}}{\|}}{\mathrm{C}}-\mathrm{O^-} \cdots \mathrm{HN} \diagdown_{\mathrm{C}} \diagup \mathrm{N} \cdots \mathrm{HOCH_2} - \quad (\text{imidazole ring: } \mathrm{C{=}C}) \tag{10}$$

Because withdrawing the proton would increase the nucleophilic character of the oxygen, this arrangement explains why Ser 195 is exceptionally reactive. The same arrangement of aspartate, histidine, and serine residues has been found in all of the serine proteases that have been examined. It is often referred to as a "charge relay system." The dependence of the enzyme kinetics on pH agrees with this picture. Enzymatic activity appears to depend on the presence of a basic group with a pK_a of about 6.8, which is in the range consistent with a histidine side chain. The k_{cat} decreases abruptly if this group is

Figure 9.9

Steps in the hydrolysis of *p*-nitrophenyl acetate by chymotrypsin. In the hydrolysis of this and most other esters, the breakdown of the acyl-enzyme intermediate is the rate-determining step. In the hydrolysis of peptides and amides, the rate-determining step usually is the formation of the acyl-enzyme intermediate. This makes the transient formation of the intermediate more difficult to study, because the intermediate breaks down as rapidly as it forms.

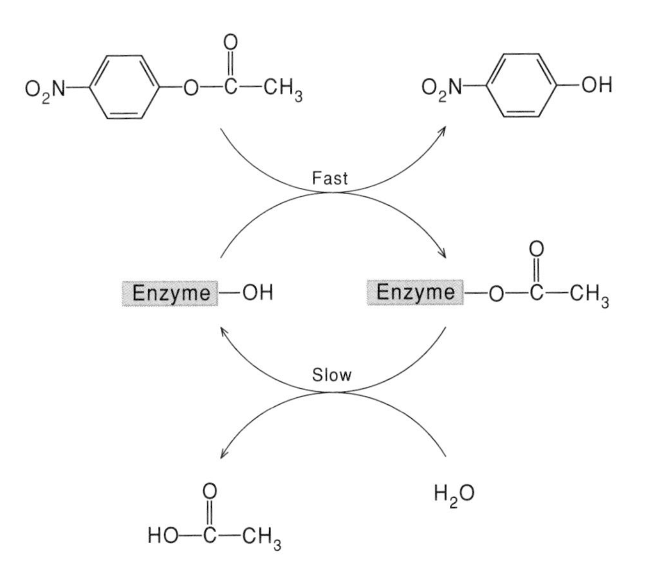

Figure 9.10

The turnover number (k_{cat}) and the Michaelis constant (K_m) as a function of pH for the hydrolysis of *N*-acetyl-L-tryptophanamide by chymotrypsin at 25°C. The decrease in k_{cat} as the pH is lowered between 8 and 6 probably reflects the protonation of His 57. The increase in K_m above pH 9 probably reflects the deprotonation of Ile 16, which results in the rotation of Gly 193 out of the substrate-binding site (see fig. 10.1).

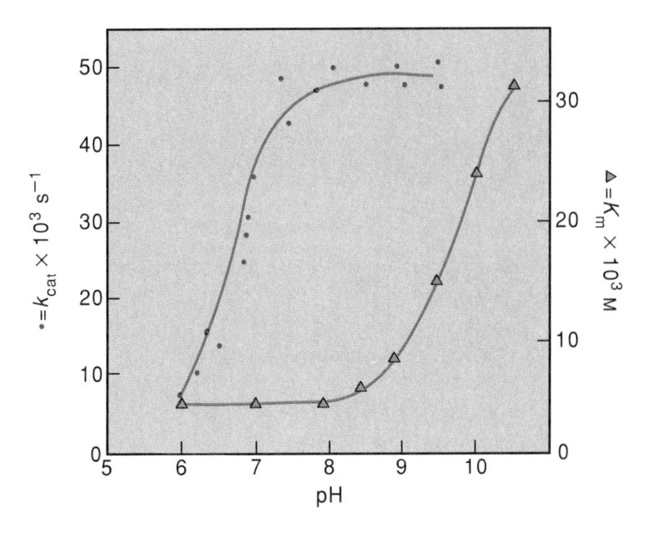

protonated (fig. 9.10). This is because protonating His 57 prevents the histidine from forming a hydrogen bond to Ser 195, and thus greatly decreases the nucleophilic reactivity of the serine.

The crystal structures also provided a simple explanation for the different substrate specificities of trypsin, chymotrypsin, and elastase. In both trypsin and chymotrypsin, the side chain of the substrate fits snugly into a pocket (see figs. 9.6*a*, 9.6*b*, and 9.7). At the far end of the pocket in trypsin is the carboxylate group of an aspartic acid residue. The negative charge of the carboxylate would favor the binding of the positively charged side chain of lysine or arginine. In chymotrypsin the aspartic acid residue is replaced by serine, creating a less polar environment that suits the side chain of tyrosine, phenylalanine, or tryptophan. In elastase (see fig. 9.6*c*), the binding pocket is obstructed by the bulky side chains of a valine and a threonine residue, so that it can accommodate only small substrates such as alanine.

Another important feature of the substrate binding site in serine proteases is that the carbonyl oxygen atom of the scissile peptide bond is hydrogen-bonded to one or more NH groups. In trypsin, chymotrypsin, and elastase, these are the amide NH groups of Ser 195 and of Gly 193. The interaction with the two protons would favor an increase in the negative charge on the oxygen, facilitating the formation of a tetrahedral intermediate state, as we discussed earlier in connection with general-acid catalysis and electrostatic effects. Figure 9.7*b* shows a space-filling model of this part of the active site in trypsin.

The overall reaction mechanism of chymotrypsin is sketched in figure 9.11. Part (*a*) shows the enzyme-substrate complex, with an aromatic side chain of the substrate seated in the binding pocket and the carbonyl oxygen atom hydrogen-bonded to the amide NH hydrogens of Gly 193 and Ser 195. (The crystal structure suggests that several additional hydrogen bonds contribute to fixing the substrate in an optimal orientation on the enzyme; these have been left out of the figure for clarity.) Aspartate 102 and His 57 are aligned as described above, with Ser 195 forming a hydrogen bond with the histidine. Part (*b*) shows a tetrahedral intermediate state. Serine 195 has released its proton to His 57 and launched a nucleophilic attack on the carbonyl carbon of the substrate. The histidine residue thus acts as a general-base catalyst in the formation of the intermediate. The movement of negative charge to the carbonyl oxygen creates what is often called an "oxyanion," which is stabilized largely by electrostatic interactions with the amide protons of Ser 195 and Gly 193. The tetrahedral oxyanion intermediate is not stable enough to be isolated, and the evidence for its existence is inferential. In part (*c*), this fleeting intermediate has decomposed to form the more stable acyl-enzyme intermediate, in which the serine is linked as an ester to the carboxylic part of the substrate and the amine product has been released. Histidine 57 probably facilitates this decomposition by acting as a general acid.

To complete the reaction, the breakdown of the acyl-enzyme probably occurs as shown in parts (*d*), (*e*), and (*f*) of figure 9.11. The steps here are essentially a reversal of the steps through parts (*a*), (*b*), and (*c*), except that water replaces the amino part of the substrate. The reaction probably proceeds by way of a tetrahedral intermediate (*e*) that is similar to one shown in (*b*) except for this substitution.

Figure 9.11

The probable mechanism of action of chymotrypsin. The six panels show the initial enzyme-substrate complex (*a*), the first tetrahedral (oxyanion) intermediate (*b*), the acyl-enzyme (ester) intermediate with the amine product departing (*c*), the same acyl-enzyme intermediate with water entering (*d*), the second tetrahedral (oxyanion) intermediate (*e*), and the final enzyme-product complex (*f*). In the transition states between these intermediates, there probably is a more even distribution of negative charge between the different oxygen atoms attached to the substrate's central carbon atom.

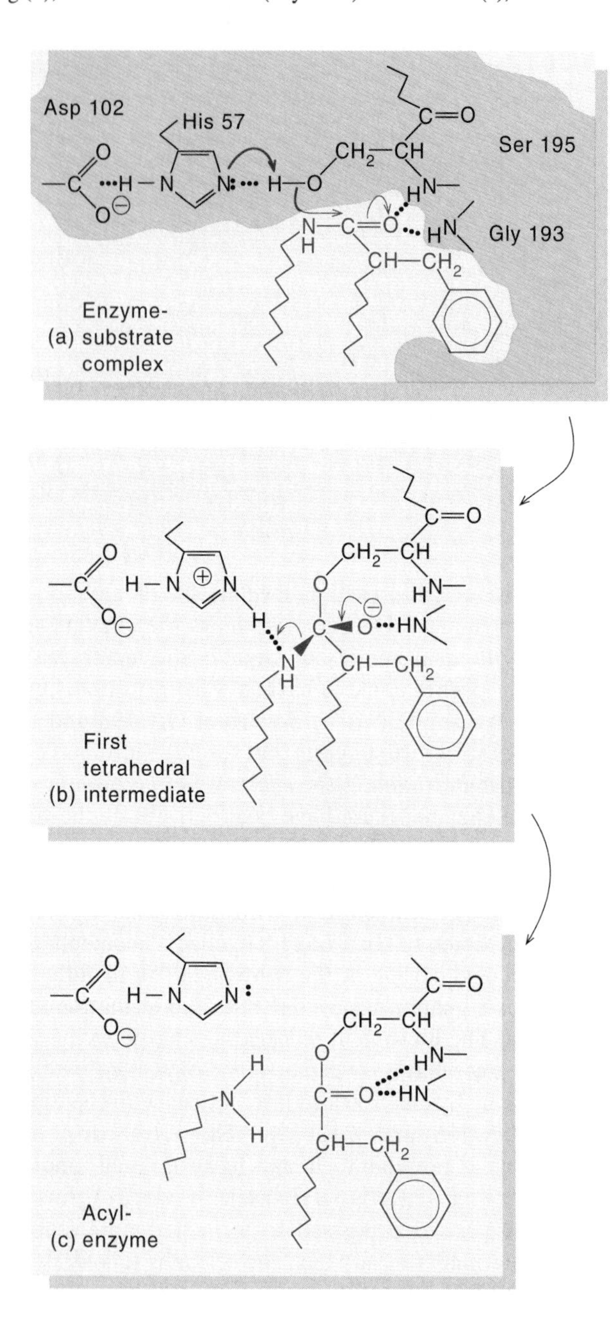

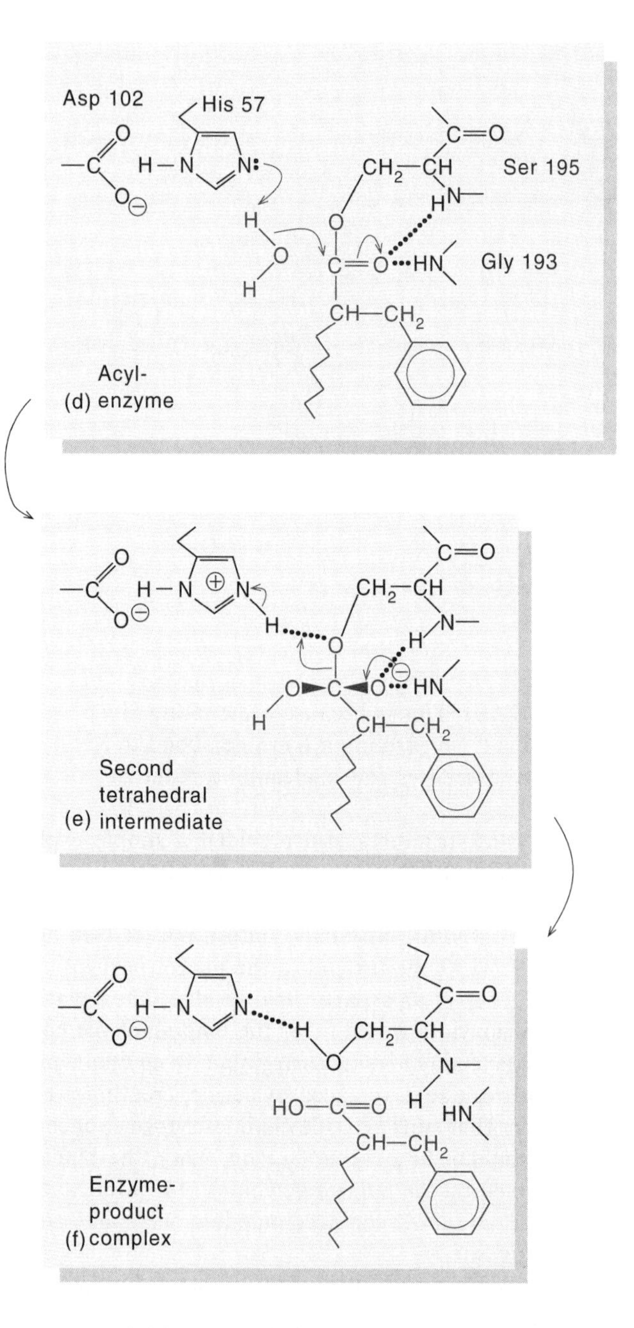

In figure 9.11, note that Asp 102 does not actually remove a proton from His 57 in the formation of either of the tetrahedral intermediates. Recent calculations indicate that the effect of the aspartate's negatively charged carboxyl group probably is best viewed simply as an electrostatic effect that favors the movement of positive charge from Ser 195 to His 57. The stabilization of the oxyanion intermediate by the amide protons of Gly 193 and Ser 195 can be described in a similar manner, as a favorable electrostatic interaction of the negatively charged oxygen with nearby atoms that tend to have positive charges, rather than as the actual transfer of a proton to the oxygen. The carboxyl group of Asp 102 also would help to align the imidazole ring of His 57 in the proper orientation for removing the proton from Ser 195.

One way to test a scheme such as that shown in figure 9.11 is to investigate the effects of modifying the amino acid residues that play important roles. We have already mentioned the inhibitory effect of the reaction of Ser 195 with diisopropylfluorophosphate. To examine the importance of His 57, this residue also was modified by methylation, which would disrupt the interaction with Asp 102. This treatment decreased the activity of chymotrypsin by a factor of more than 10^3. The importance of Asp 102 was tested recently by site-directed mutagenesis of trypsin and subtilisin. Replacing the aspartate by asparagine reduced k_{cat} by a factor of about 10^4.

Zinc Provides an Electrophilic Center in Some Proteases

In the preceding discussion, we described a group of closely related proteases secreted by the pancreas—trypsin, chymotrypsin, and elastase. The pancreas also secretes two proteases that hydrolyze oligopeptides one residue at a time from the C-terminal end, carboxypeptidases A and B. Carboxypeptidase A prefers aromatic residues (see fig. 9.4) and carboxypeptidase B has a preference for basic residues, providing a complementarity similar to that of chymotrypsin and trypsin. The two carboxypeptidases also resemble chymotrypsin and trypsin in being secreted as zymogens that are processed to form the active enzymes in the intestine. The carboxypeptidases are considerably more effective catalysts than the serine proteases. Values of k_{cat} for carboxypeptidase A with its best peptide substrates are on the order of 100 times greater than the k_{cat} values for trypsin with *its* best peptide substrates.

Carboxypeptidases A and B are members of a family of proteolytic enzymes that contain bound zinc. The zinc proteases also include enzymes that digest collagen (collagenases), and a number of bacterial enzymes. The most thoroughly studied of the bacterial zinc proteases is thermolysin, an enzyme obtained from *Bacillus thermoproteolyticus*. Thermolysin hydrolyzes internal peptide bonds adjacent to hydrophobic residues; collagenases cut internal bonds preferably adjacent to glycine residues. As in the case of serine proteases, the zinc protease family appears to have resulted from a combination of divergent and convergent evolution. Carboxypeptidases A and B have very similar structures indicative of a common ancestry; thermolysin has a very different structure overall, but resembles the carboxypeptidases in the amino acid residues that bind the zinc and in a few critical residues that probably participate in the enzymatic mechanism.

In the crystal structure of thermolysin, the Zn^{2+} ion is bound to the imidazole rings of two histidine residues and the carboxylate of a glutamic acid (fig. 9.12*a*). The fourth ligand of the Zn^{2+} is a molecule of water. Another glutamic acid residue (Glu 143) and a third histidine (His 231) are located nearby (see fig. 8.8). Thermolysin also has been crystallized with several different bound inhibitors, one of which, phosphoramidon, contains an amide bond between phosphoric acid and an amino acid (see figs. 8.8 and 9.12*b*). Because the oxygen and nitrogen substituents of the phosphorus atom are arranged in a tetra-

Figure 9.12

Functional groups in the active site of thermolysin. (*a*) The Zn^{2+} ion is bound to two histidines and a glutamic acid, and has a molecule of water as its fourth ligand. (*b*) Phosphoramidon, an inhibitor that probably mimics the tetrahedral intermediate formed during hydrolysis of a peptide, binds to the Zn^{2+} by displacing the molecule of water. Residue Glu 143 forms a hydrogen bond with the hydroxyl group of the bound inhibitor, and His 231 is located close to the NH group.

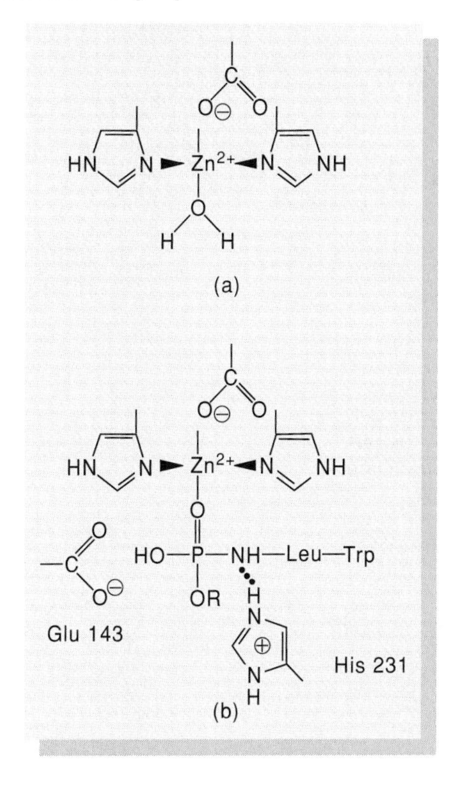

hedron, phosphoramidon seems likely to mimic the tetrahedral intermediate that would be formed in the course of the hydrolysis of a peptide (see box 9B). In the crystal, one of the oxygen atoms of the phosphate is linked directly to the Zn^{2+} ion, displacing the molecule of water. Glutamic acid 143 appears to form a hydrogen bond to the hydroxyl oxygen, and one of the nitrogen atoms of His 231 forms a hydrogen bond with the inhibitor's NH group (see fig. 9.12*b*).

The structure of the enzyme-phosphoramidon complex suggests that the enzymatic hydrolysis of a peptide by thermolysin proceeds as shown in figure 9.13. In this scheme, binding of the peptide carbonyl oxygen atom to the Zn^{2+} withdraws electrons from the carbonyl group, increasing its susceptibility to nucleophilic attack by water. Glutamic acid 143 acts as a general base to remove a proton from the water. Histidine 231 then could act as a general acid, providing a proton to the amine group as the tetrahedral intermediate decomposes.

Several variations on this mechanism have been proposed. One suggestion is that the water molecule that attacks the peptide carbonyl is the one that starts out as a ligand of the Zn^{2+}, and that the Zn^{2+} goes through a stage in which it has five ligands—both the peptide oxygen and the water, in addition

Figure 9.13

The reaction mechanism of thermolysin probably involves polarization of the peptide carbonyl group by interaction with the Zn^{2+} ion and general-base catalysis by Glu 143 to form a tetrahedral intermediate. His 231 probably acts as a general acid to promote the decomposition of the intermediate.

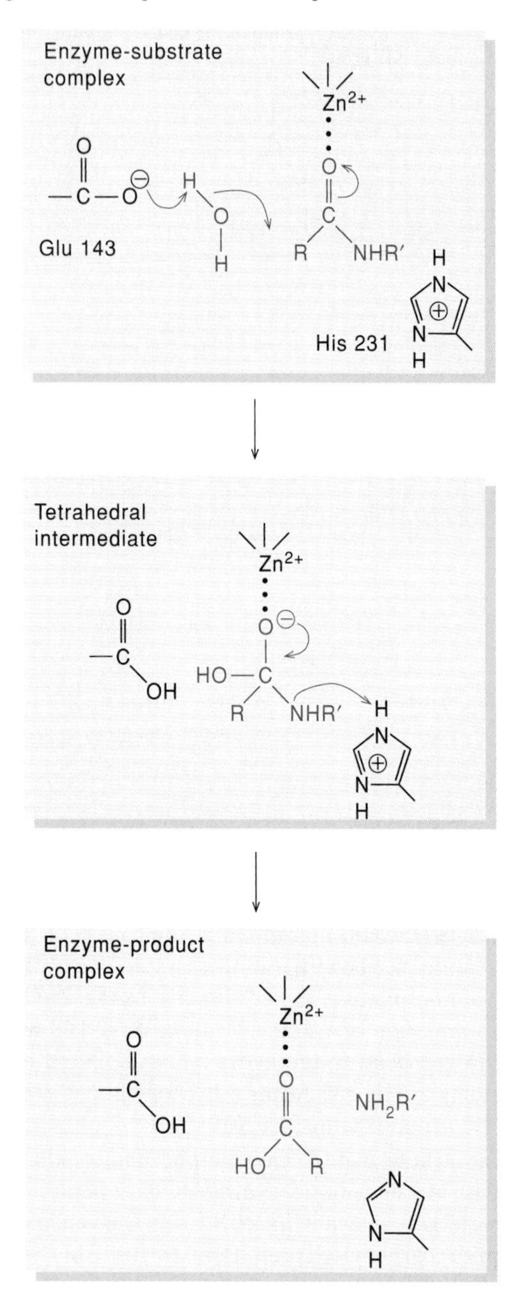

to the original two histidines and glutamic acid. It also has been suggested that the general acid that facilitates breakdown of the tetrahedral intermediate is Glu 143, rather than His 231. At present, the experimental data do not distinguish clearly between these alternatives. However, the central feature of all the proposed mechanisms is the electrostatic effect of the Zn^{2+} on the peptide C=O group.

The Zn^{2+} atom of carboxypeptidase A or B resembles that of thermolysin in having two histidines (His 69 and His 196), a glutamic acid (Glu 72), and a molecule of water as its ligands. Again a second glutamic acid residue (Glu 270) is found nearby, in a very similar position with respect to the metal ion. Carboxypeptidase A also has been crystallized with a bound inhibitor, glycyl-L-tyrosine. This dipeptide is hydrolyzed only very slowly, possibly because it interacts with the enzyme in a somewhat more complicated manner than the oligopeptides that are the normal substrates. The free N-terminal amino group of the dipeptide is linked indirectly to the carboxyl group of Glu 270 by way of a hydrogen-bonded molecule of water, as shown in figure 9.14. This could prevent the glutamic acid from participating as a general base in the hydrolytic reaction. In a normal oligopeptide substrate, the N-terminal amino group would be too far away to interact with the glutamic acid in this way.

As in thermolysin, the carbonyl oxygen of the peptide binds directly to the Zn^{2+} atom of carboxypeptidase A, displacing the water (see fig. 9.14). The active site also includes an arginine residue (Arg 145), which interacts closely with the negatively charged, C-terminal carboxyl group of glycyl-L-tyrosine and presumably would do the same with a better substrate. This explains the specificity of the carboxypeptidases for pruning off carboxyl-terminal amino acid residues. Carboxypeptidase A differs from thermolysin in that there is no histidine imidazole close enough to the bound dipeptide to serve as a general acid in the hydrolysis. The phenolic OH group of Tyr 248 *is* located nearby (see fig. 9.14), but we will see shortly that the role of the tyrosine residue appears to be rather different.

A remarkable finding in the carboxypeptidase A crystal structures, which was not seen with the serine proteases, is that the binding of the substrate analog (glycyl-L-tyrosine) causes the protein to undergo a major structural rearrangement (fig. 9.15). Arginine 145 and Glu 270 both move by about 2 Å to come into position near the substrate, and Tyr 248 swings down by about 12 Å so that its phenolic group is close to the NH group of the scissile bond. At least four molecules of H_2O are expelled in the process, so that the substrate finds itself surrounded largely by hydrophobic groups of the enzyme instead of by water. The binding of the substrate to the enzyme thus is favored by the entropy increase associated with removing the substrate's aromatic side chain from water, in addition to the electrostatic interactions between the terminal carboxyl group and Arg 145.

The most plausible mechanism for the operation of the carboxypeptidases is essentially the same as the scheme we just discussed for thermolysin: The Zn^{2+} ion pulls electrons from the oxygen atom of the peptide bond, increasing the positive charge on the carbonyl carbon, and Glu 270 acts as a general base, increasing the nucleophilic character of a molecule of water by removing a proton. Tyrosine 248 then might serve as a general acid to facilitate the decomposition of the tetrahedral intermediate state. Several experimental observations support this scheme. First, the pH dependence of the kinetics indicates that

Transition-State Analogs

Important clues to an enzyme mechanism often can be obtained by studying the inhibitory effects of compounds that resemble possible intermediates or transition states. Since the enzyme must stabilize the transition state of the reaction, a compound that mimics this state might be expected to bind to the enzyme particularly tightly. Such transition-state analogs frequently do prove to be strong competitive inhibitors. Thus the inhibitory effect of phosphoramidon on thermolysin supports the view that the enzymatic reaction proceeds by way of a state or intermediate in which a tetrahedral carbon atom is surrounded by a nitrogen and several oxygen atoms. A closely related example is the compound

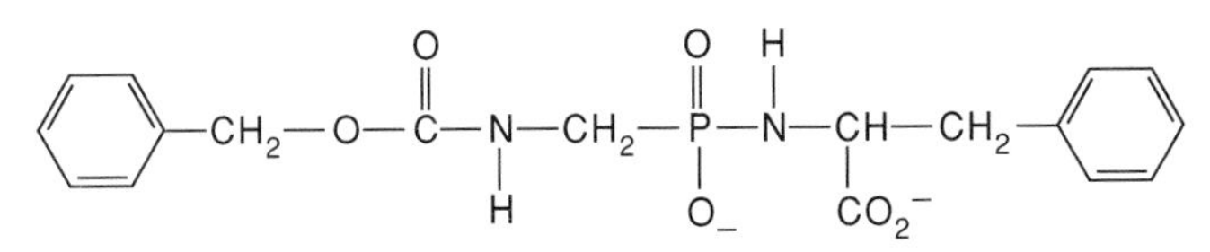

which is a potent competitive inhibitor of carboxypeptidase A. Here again, the tetrahedral phosphorus atom resembles the tetrahedral carbon in the intermediate that probably forms in the course of the enzymatic reaction. Similarly, we will see in this chapter that uridine vanadate, which contains a vanadium atom surrounded by five oxygen atoms, is a competitive inhibitor of ribonuclease. This behavior is consistent with the formation of an intermediate in which a phosphorus atom of the substrate takes on a pentavalent geometry (see fig. 9.22).

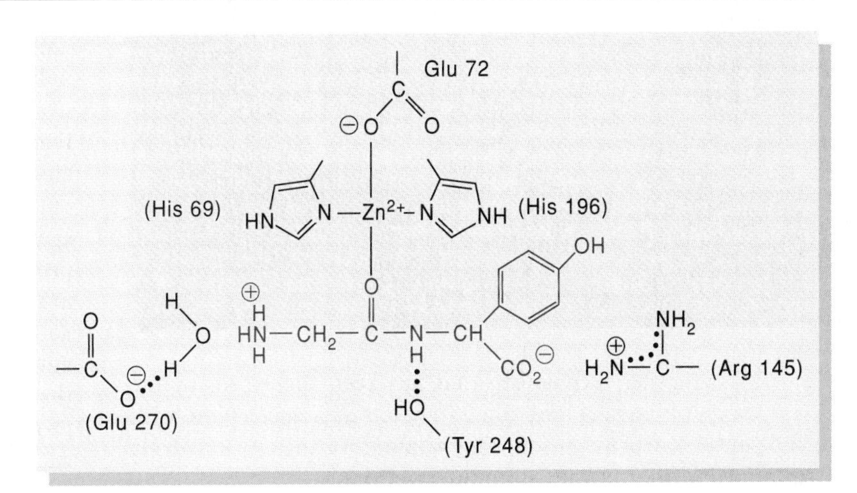

Figure 9.14

The functional groups surrounding the Zn^{2+} ion in carboxypeptidase A are similar to those in thermolysin. Glycyl-L-tyrosine, shown here in color, is a competitive inhibitor of carboxypeptidase A. The tyrosine side chain of glycyl-L-tyrosine fits into a hydrophobic pocket on the enzyme. The peptide oxygen binds directly to the Zn^{2+} ion, the carboxyl group forms a salt bridge with Arg 145, and the amide NH forms a hydrogen bond with Tyr 248. The salt bridge between the amino group and Glu 270 probably explains why glycyl-L-tyrosine is not a good substrate, because it would prevent the glutamic acid residue from acting as a general base.

both basic and acidic functional groups are involved in the reaction. The rate of the reaction decreases if a group with an apparent pK_a of about 6.2 is protonated, or if a group with an apparent pK_a of about 9.0 is deprotonated. The former group probably can be identified with Glu 270, and the latter could be Tyr 248. Second, chemical modifications of either Glu 270 or Tyr 248 destroy or modify the enzymatic activity. Acetylation of the tyrosine, for example, makes the enzyme essentially inactive. But when Tyr 248 was changed to phenylalanine by site-directed mutagenesis, it was found to decrease k_{cat} by only a factor of about 2, indicating that the tyrosine phenolic group is not essential for enzymatic activity. The substitution of phenylalanine for Tyr 248 did increase the K_m, suggesting that the tyrosine contributes primarily to the binding of the substrate.

Figure 9.15

Crystal structures of carboxypeptidase A with no substrate bound (*a*) and with bound glycyl-L-tyrosine (*b*). The glycyl-L-tyrosine or, in (*a*), the molecule of water that binds to the zinc in place of the substrate is shown in yellow. The side chains of Glu 270 and Tyr 248 are in orange, and the histidine ligands of the zinc are in red. Note how the large change in the position of the tyrosine, together with smaller changes in other residues, folds the enzyme around the substrate. The figures are based on the crystal structures described by W.N. Lipscomb.

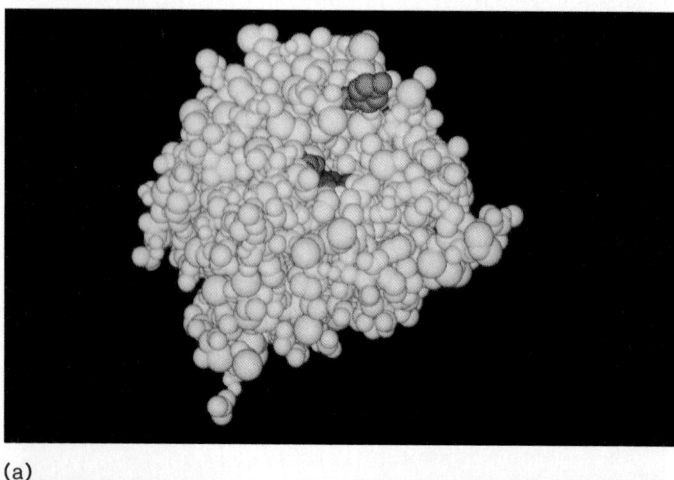

(a)

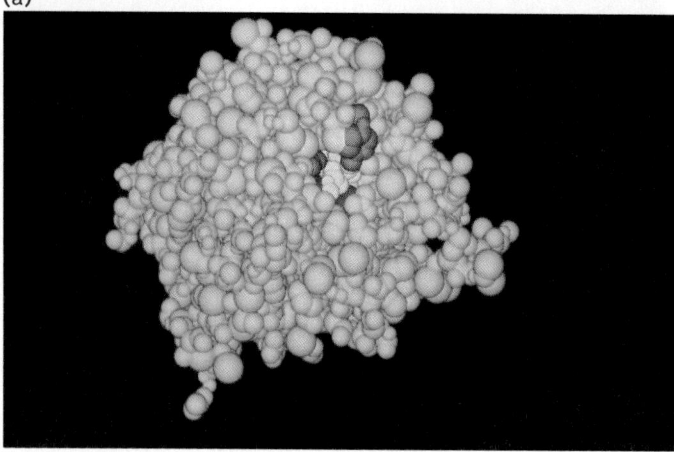

(b)

Another mechanism that has been suggested for the carboxypeptidases is that Glu 270 reacts as a nucleophile with the peptide carbonyl group. This would form an acyl-anhydride attached to the enzyme as an intermediate:

$$\text{R—C(=O)—NHR}' + \text{HO—C(=O)—Enzyme} \rightarrow \text{R—C(=O)—O—C(=O)—Enzyme} + \text{R}'\text{NH}_2 \quad (11)$$

$$\text{R—C(=O)—O—C(=O)—Enzyme} + \text{H}_2\text{O} \rightarrow \text{R—C(=O)—OH} + \text{HO—C(=O)—Enzyme} \quad (12)$$

There is, however, little evidence in favor of this scheme. For example, reagents that might be expected to react with an anhydride intermediate do not alter the course of the enzymatic reaction. Also, if the second step of the mechanism (equation 12) is reversible, the enzyme should catalyze an exchange of ^{18}O back and forth between H_2O and the carboxylate group of the product (see box 9C); attempts to detect such an exchange have given negative results.

Ribonuclease A: An Example of Concerted Acid–Base Catalysis

Ribonucleases are a widely distributed family of enzymes that hydrolyze RNA by cutting the P—O ester bond attached to a ribose 5′ carbon. The reaction occurs in two steps, with a 2′,3′-phosphate cyclic diester intermediate (fig. 9.16). The intermediate can be identified relatively easily, because its breakdown is much slower than its formation. Ribonucleases do not hydrolyze DNA, which lacks the 2′-hydroxyl group needed for the formation of the cyclic intermediate. The best-studied of the ribonucleases, the pancreatic enzyme ribonuclease A, is specific for a pyrimidine base (uracil or cytosine) on the 3′ side of the phosphate bond that is cleaved.

When the amino acid sequence of bovine ribonuclease A was determined in 1960 by Sanford Moore and William Stein, it was the first enzyme and only the second protein to be sequenced. Ribonuclease A also was one of the first enzymes whose three-dimensional structure was elucidated by x-ray diffraction, and it was the first to be synthesized completely from its amino acids. The synthetic protein proved to be enzymatically indistinguishable from the native enzyme.

Ribonuclease A (RNase A) is a relatively small protein, with a molecular weight of 13,680 and a single polypeptide chain of 124 amino acid residues. An early discovery that turned out to be extremely useful was that the protein could be cleaved specifically between residues 20 and 21 by the bacterial serine protease subtilisin. The resulting two polypeptides could be separated and purified. They were found to be enzymatically inactive individually, but to regain the complete activity of the native enzyme when they were recombined. This work showed that there are strong, noncovalent interactions that can hold protein chains together even when one of the peptide links is cut. It also made it possible to modify specific amino acid residues of the two polypeptide chains independently, and to explore how each residue contributed to the reassembly of the protein and the recovery of enzymatic activity.

RNase A is completely inhibited if either of two different histidine residues (His 12 or His 119) is modified by carboxymethylation with iodoacetate, a fact suggesting that both of these histidines play important roles in the active site (fig. 9.17). In support of this conclusion, the reaction of iodoacetate with His 12 or His 119 is inhibited by cytidine-3′-phosphate or other small molecules that bind at the active site. Lysine 41 has been implicated similarly in the active site by the observation that the reaction of fluorodinitrobenzene with the ε-amino group of this residue destroys enzymatic activity (see fig. 9.17). A

9C BOX

Isotopic Exchange Reactions

One way to detect an intermediate state in an enzymatic reaction is to look for a transfer of an isotopic label between the solvent and a reactant or product. Suppose that the reaction scheme of equations (11) and (12) is correct for carboxypeptidase A, and consider a reversal of these reactions. If the enzyme is incubated with both of the products ($R'NH_2$ and R — COOH), a small amount of

$$R - \overset{\overset{O}{\|}}{C} - NHR'$$

will be formed, depending on the equilibrium constant for the overall reaction. If the enzyme is incubated with R — COOH alone,

$$R - \overset{\overset{O}{\|}}{C} - NHR'$$

cannot be formed but the reverse reaction should still proceed as far as the anhydride intermediate,

$$R - \overset{\overset{O}{\|}}{C} - O - \overset{\overset{O}{\|}}{C} - \text{Enzyme}$$

If the incubation is continued, the intermediate will be formed and broken down again many times. The formation of the anhydride requires the splitting out of H_2O, so whenever the intermediate breaks down, the elements of H_2O must be reincorporated into R — COOH and the carboxylate group on the enzyme. Thus if the enzyme is incubated with R — COOH in water that is labeled with ^{18}O, the isotope will be incorporated into the carboxylate group. This prediction can be tested by isolating the R — COOH and examining it in a mass spectrometer. For carboxypeptidase A, no such isotopic exchange was detected. This result favors the view that Glu 270 in carboxypeptidase does not react with the substrate to form an anhydride intermediate, and more likely functions as a general base.

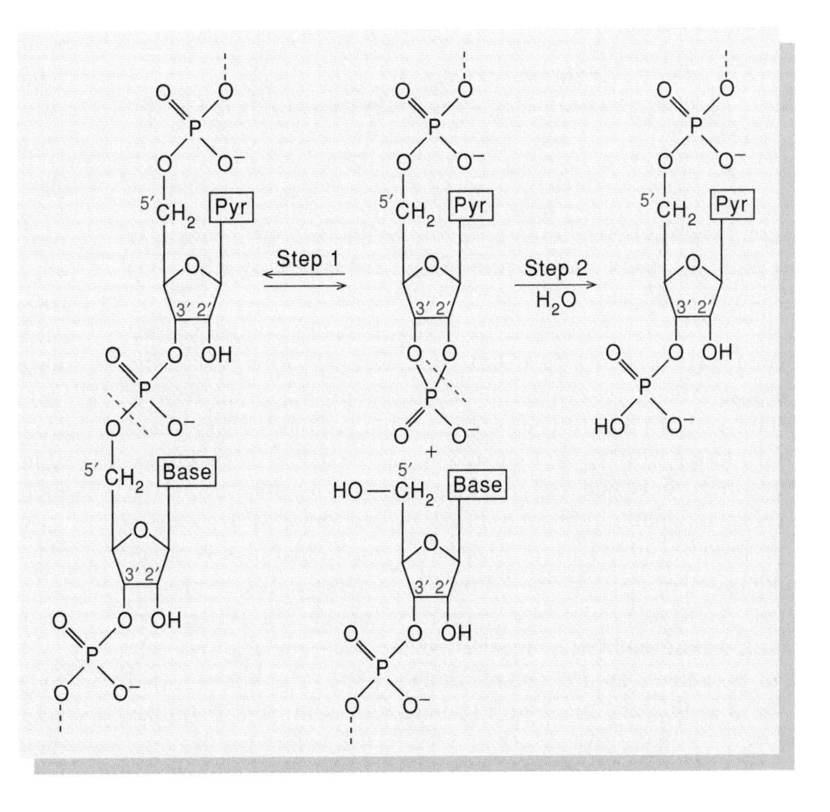

Figure 9.16

A portion of an RNA chain, indicating points of cleavage by pancreatic ribonuclease. "Pyr" refers to a pyrimidine; "Base" can be either a purine or a pyrimidine. The 2′, 3′ and 5′ carbon atoms are labeled. The enzymatic reaction proceeds in two steps, with a cyclic 2′, 3′-phosphate diester as an intermediate.

Figure 9.17

When RNase A is treated with iodoacetate ($ICH_2CO_2^-$), the two major products obtained are carboxymethylated derivatives of His 12 and His 119. Both of these enzymes are severely inhibited. This result suggests that both His 12 and His 119 are important in the active site. The enzyme also is completely inhibited by the reaction of Lys 41 with fluorodinitrobenzene.

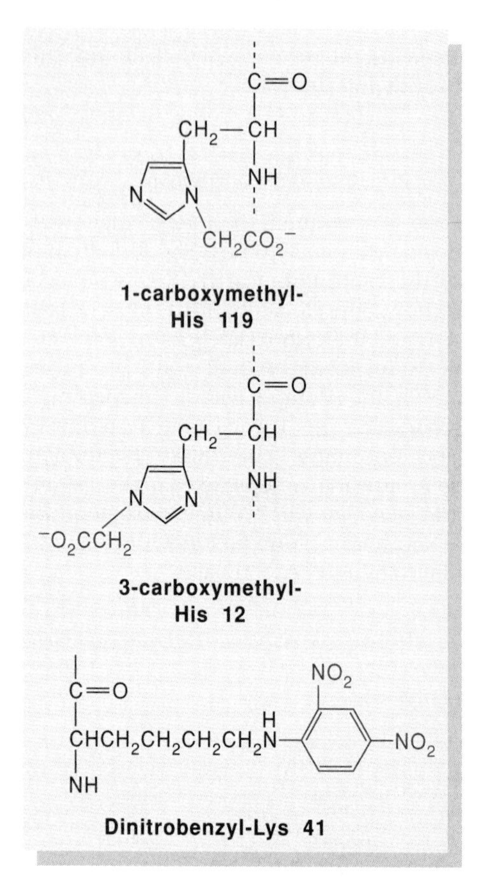

Figure 9.18

The dependence of k_{cat} of RNase A on pH. The bell-shaped curve suggests that one histidine residue must be in the protonated state and another must be unprotonated. Similar pH dependences are found for the hydrolysis of either RNA or pyrimidine nucleoside-2′, 3′-phosphate cyclic diesters, a fact indicating that the two steps of the overall reaction both require concerted general-acid and general-base catalysis.

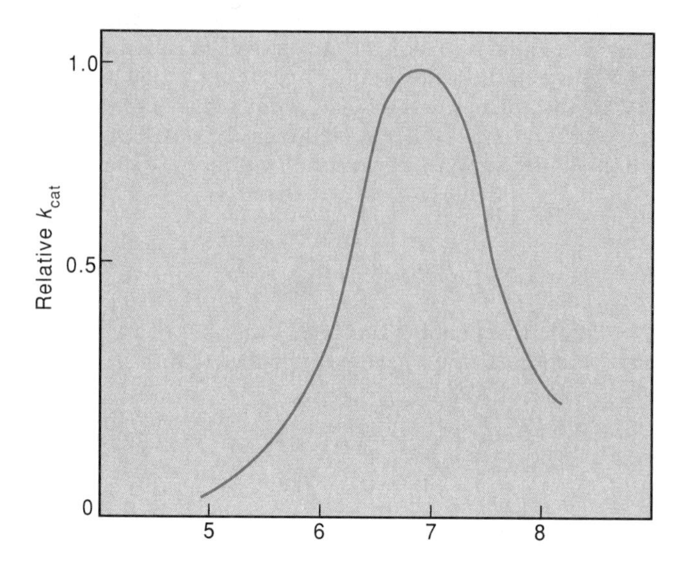

second lysine (Lys 7) also appears to be important for substrate binding, although the enzyme retains some activity when this residue is modified.

The enzymatic activity of RNase A shows a bell-shaped dependence on pH, with an optimum near pH 7 (fig. 9.18). The operation of the enzyme appears to require that a dissociable group with a pK_a of about 6.3 be in the protonated form, and that a group with a pK_a of about 8 be unprotonated. These groups have been identified as His 12 and His 119 by measuring the nuclear magnetic resonance (NMR) spectrum of the enzyme as a function of pH. The spectral peaks due to histidine residues undergo characteristic shifts as the imidazole ring is protonated. Analysis of the NMR spectra, though not straightforward, was aided by the fact that bovine RNase A contains only four histidine residues, one of which (His 12) can be separated from the others by cutting the protein with subtilisin. Histidines 12 and 119 have pK_a values of 5.8 and 6.2 in the free enzyme, but both pK_a values shift to a more alkaline range when the enzyme binds its substrate.

The pH dependence of the kinetics, taken with the information that histidines 12 and 119 are essential for enzymatic activity, suggests that the two histidines participate in the reaction by concerted general-base and general-acid catalysis. This supposition is supported by the crystal structure of the enzyme. Figure 9.19 shows a model of the crystal structure of RNase S (the active enzyme obtained by cutting ribonuclease with subtilisin and recombining the two polypeptides) with a bound substrate analog, "UpCH₂A." The analog is a competitive inhibitor of the enzyme. It resembles the dinucleotide UpA but cannot be hydrolyzed because it has a methylene carbon in place of the oxygen between the P and the 5′-CH_2 of the ribose of adenosine (fig. 9.20). Similar crystal structures have been obtained with a variety of other substrate analogs. In all of the structures, His 12 and His 119 are found on opposite sides of the substrate's phosphate group, with His 12 positioned close to the 2′ OH group. Lysine residues 7, 14, and 66 also are located nearby, where their positively charged amino groups would have favorable electrostatic interactions with the negatively charged phosphate. Near His 119 is the carboxyl group of an acidic residue, Asp 121. The electrostatic effect of Asp 121 thus would favor the transfer of a proton from a molecule of water to His 119, much as Asp 102 in chymotrypsin induces a proton to move from Ser 195 to His 57.

The pyrimidine ring on the 3′ side of the substrate fits snugly into a groove on RNase A. Valine 43 is located on one side of the pyrimidine ring, and Phe 120 on the other. The specificity of the enzyme for pyrimidine nucleotides appears to result largely from hydrogen bonding to Thr 45. The threonine side chain can form a pair of complementary hydrogen bonds with either uracil or cytosine (fig. 9.21). Crystallographic studies have

Figure 9.19

A model of the binding of the substrate analog $UpCH_2A$ (blue) to RNase S. (The structure of $UpCH_2A$ is shown in figure 9.20.) RNase S is an active enzyme, although it differs from the native enzyme because of a peptide bond cleavage between residues 20 and 21 (light dashed line). The two histidines and the lysine that are crucial to the active site (His 12, His 119, and Lys 41) are indicated in gray. The dotted lines indicate some of the hydrogen bonds that maintain the protein structure.

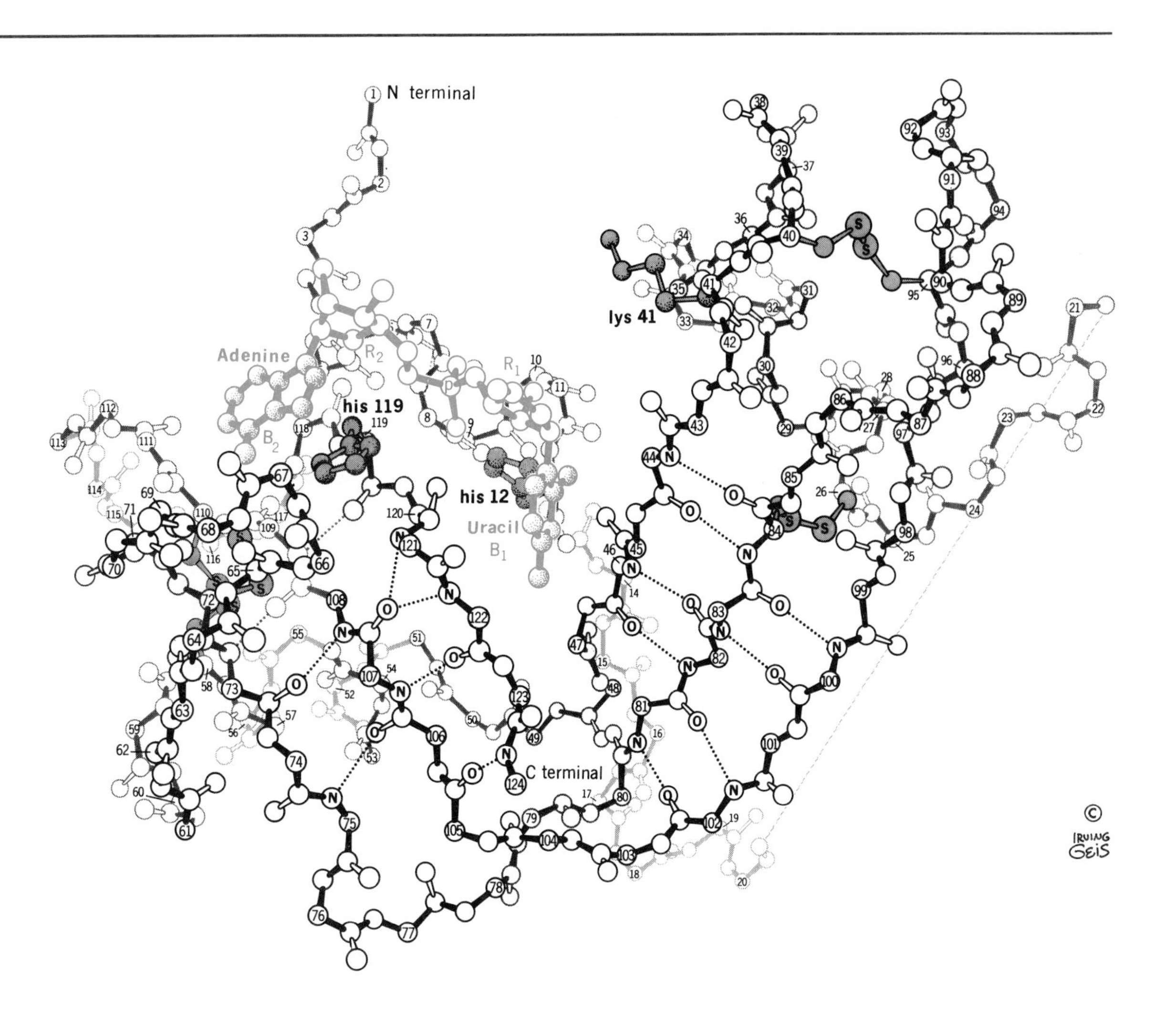

Figure 9.20

Structure of $UpCH_2A$, a competitive inhibitor of RNase A. $UpCH_2A$ differs from the dinucleotide substrate UpA in having a methylene group instead of oxygen between the phosphorus and C-5′ of the adenosine.

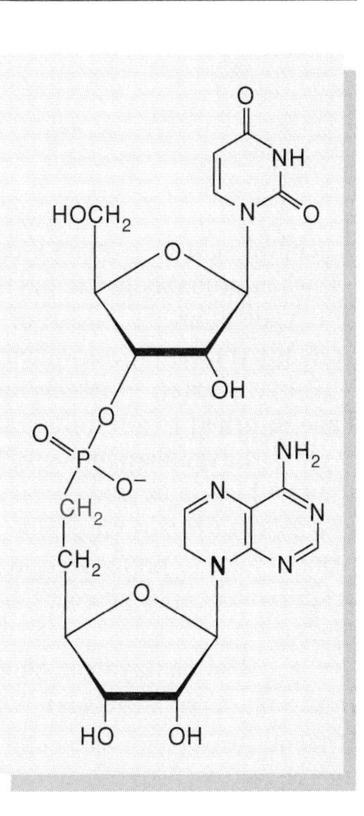

Figure 9.21

The side chain and amide NH group of Thr 45 in RNase A can form a pair of complementary hydrogen bonds with either uracil or cytosine. These probably account for the specificity of RNase A for pyrimidine nucleotides.

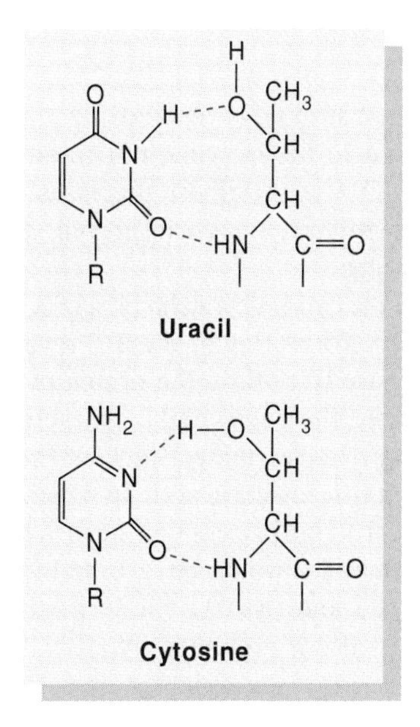

Figure 9.22

The probable catalytic mechanism of RNase A. In the formation of the 2′, 3′-cyclic ester, His 12 acts as a general base, and His 119 acts as a general acid. The increase in negative charge on the phosphate's oxygen atoms in the transition state is favored by electrostatic interactions with Lys 7 and Lys 41. In the breakdown of the cyclic ester, His 119 acts as a general base and His 12 acts as a general acid. The reaction proceeds through two short-lived intermediates in which the phosphorus atom is pentavalent, in addition to the more stable cyclic ester intermediate. Py = pyrimidine.

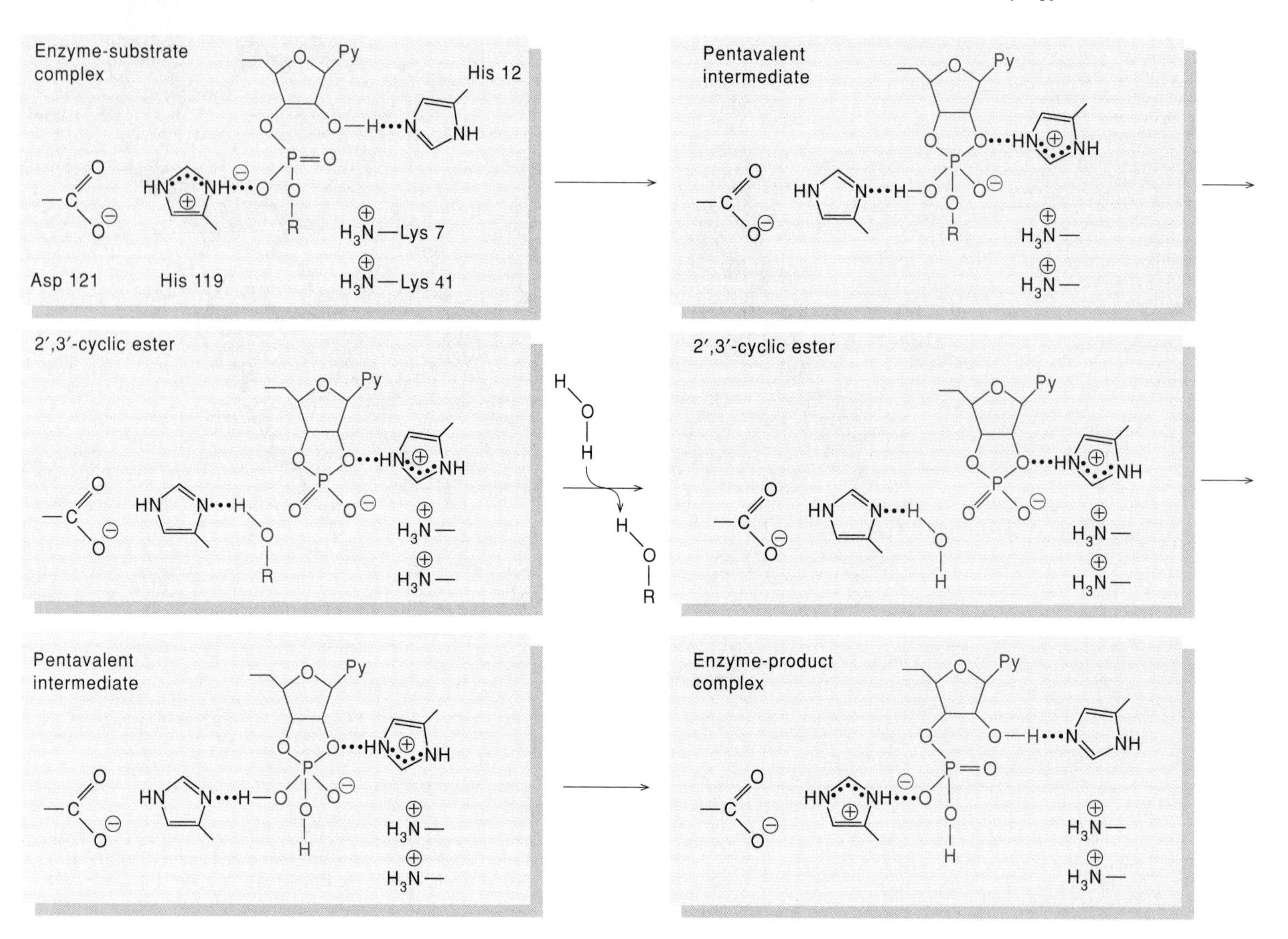

shown that if the pyrimidine is replaced by a purine, the compound can still bind to the enzyme, but the distance between His 12 and the 2′ OH of the ribose increases by about 1.5 Å. This increase evidently is enough to prevent the catalytic reaction from occurring.

The probable catalytic mechanism of RNase A is shown in figure 9.22. In the formation of the 2′,3′-cyclic ester, His 12 acts as a general base to remove the proton of the 2′ hydroxyl group. This increases the nucleophilic character of the oxygen atom, precipitating a nucleophilic attack on the phosphate and creating a transient intermediate state in which the phosphorus atom is pentavalent. The increase in negative charge on the phosphate's oxygen atoms would be favored by electrostatic interactions with Lys 7 and Lys 41. The formation of the pentavalent intermediate also is promoted by His 119, which acts as a general acid to protonate one of the phosphate oxygens. His 119 then probably participates in the transfer of a proton to the 5′ oxygen atom. Expulsion of the protonated 5′ OH group converts the transient pentavalent intermediate into the more stable 2′,3′-cyclic ester, in which the substituents of the phosphorus have returned to a tetrahedral geometry.

In the breakdown of the 2′,3′-cyclic ester, the roles of the two histidines are reversed. Histidine 119 now acts as general base to remove a proton from H_2O, and His 12 acts as a general acid to protonate the substrate's 2′ oxygen. Again, the reaction probably proceeds by way of a pentavalent intermediate.

The pentavalent phosphoryl intermediates that figure in the reactions of RNase A probably have the geometry of a trigonal bipyramid, as shown in the upper part of figure 9.23.

Figure 9.23

Two routes leading to phosphoryl group transfer. Both routes pass through a pentavalent intermediate, in which three of the oxygen atoms lie in a plane with the phosphorus atom. In the "in-line" mechanism, the entering and leaving groups ($^-OR'$ and RO^-) are on opposite sides of this plane. In the "adjacent" mechanism the entering and leaving groups are in neighboring positions. The in-line mechanism results in stereochemical inversion; the adjacent mechanism gives retention. Three of the oxygen atoms have been labeled with numbers to indicate their stereochemical relationships in the figure. Experimentally, the stereochemical outcome of the RNase reaction can be tested by using a substrate that is labeled stereospecifically with a combination of ^{17}O and ^{18}O. Substrates in which one of the oxygens is replaced by sulfur also can be used. (The adjacent mechanism requires one additional step not shown here: before it can depart, the leaving group has to move to an axial orientation perpendicular to the plane. This rearrangement is called *pseudorotation*. Several of the substituents move in a concerted manner in the pseudorotation step, so that their stereochemical relationships are maintained.)

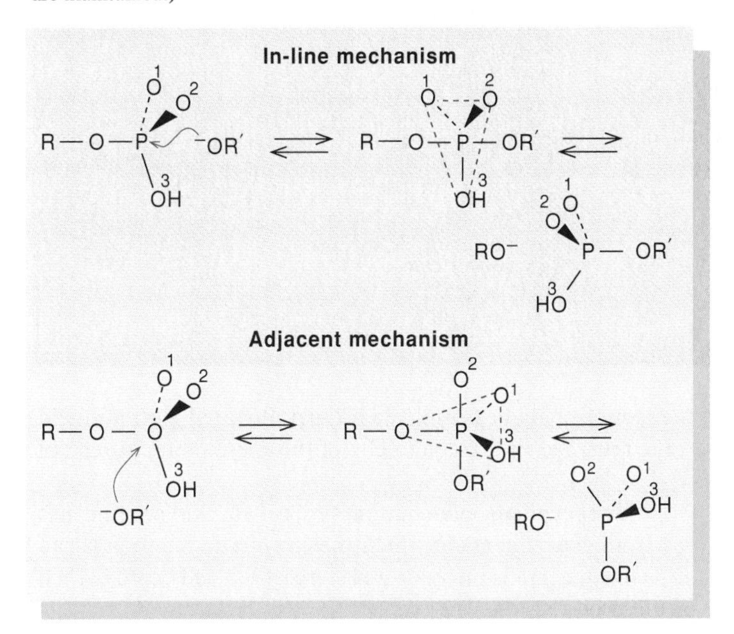

Figure 9.24

The crystal structure of the complex between RNase A and uridine vanadate. Uridine vanadate, which contains a vanadium atom surrounded by five oxygens, probably mimics the pentavalent transition state formed by a phosphoryl substrate. It binds to RNase A about 10^4 times more tightly than the corresponding uridine 2′, 3′-cyclic phosphate ester. The V atom of uridine vanadate is in purple, and the other atoms of the inhibitor in yellow. A smoothed representation of the α-carbon chain of the protein is shown in blue, with the side chains of histidines 12 and 119 indicated in red. (Based on the crystal structure described by A. Wlodawer.)

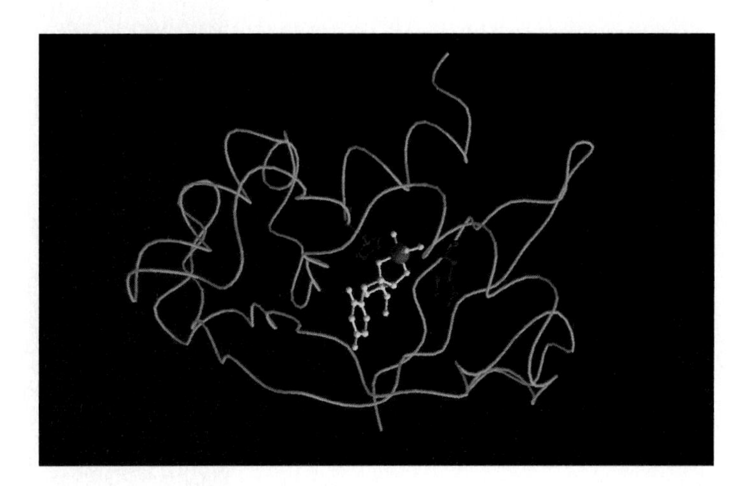

Three of the phosphorus atom's substituents lie in a plane. The entering and leaving oxygen atoms are on either side of the plane, forming a straight line with the phosphorus. One observation that supports the formation of such pentavalent intermediates is that RNase A is inhibited strongly by uridine vanadate, in which the vanadium atom is surrounded by five oxygens in a similar geometry. Figure 9.24 shows the structure of the complex of the enzyme with the inhibitor. The disposition of histidines 12 and 119 on either side of the vanadate group in the crystal structure is consistent with the linear arrangements of the entering and leaving oxygen atoms on either side of the phosphorus atom in figures 9.22 and 9.23. Uridine vanadate can be viewed as a transition-state analog (box 9B).

The "in-line" reaction mechanism also can be distinguished experimentally from the alternative "adjacent" mechanism (see fig. 9.23) by studying substrates in which one of the oxygen atoms is replaced by sulfur and another is labeled stereospecifically with ^{18}O. An in-line mechanism results in stereochemical inversion of the configuration around the phosphorus, whereas an adjacent arrangement of the entering and departing groups results in retention of the configuration (see fig. 9.23). The reactions catalyzed by RNase A and a number of other ribonucleases have been shown to proceed with inversion. Many of the other enzymes that transfer phosphoryl groups from one substrate to another exhibit retention.

Triosephosphate Isomerase Has Approached Evolutionary Perfection

Triosephosphate isomerase catalyzes the interconversion of dihydroxyacetone phosphate and glyceraldehyde-3-phosphate:

$$\begin{array}{ccc} CH_2OH & & H-C=O \\ | & & | \\ C=O & \rightleftarrows & H-C-OH \\ | & & | \\ CH_2OPO_3^{2-} & & CH_2OPO_3^{2-} \\ \text{Dihydroxyacetone} & & \text{Glyceraldehyde-3-} \\ \text{phosphate} & & \text{phosphate} \end{array} \quad (13)$$

The reaction requires moving a proton from carbon 1 of dihydroxyacetone phosphate to carbon 2, moving another proton from the oxygen atom at C-1 to the oxygen at C-2, and moving a pair of electrons in the same direction. The interconversion of the two triosephosphates is an essential step in the catabolism of carbohydrates. Triosephosphate isomerase is among the enzymes that have approached evolutionary perfection in the sense that the specificity constant k_{cat}/K_m is greater than $10^8\ M^{-1}\ s^{-1}$, which is near the limit set by the rates of diffusion of the substrate and the protein (see table 8.4). The enzyme achieves this high value for the specificity constant by having both a relatively high k_{cat} and a relatively low K_m.

Figure 9.25

Crystal structure of triosephosphate isomerase from chicken muscle. This figure shows one of the two identical subunits of the enzyme. The side chains of Glu 165, His 95, and Lys 13 are in purple, red, and yellow, respectively. (Based on the crystal structure described by D. W. Banner, A. C. Bloomer, G. A. Petsko, D. C. Phillips, and I. A. Wilson.)

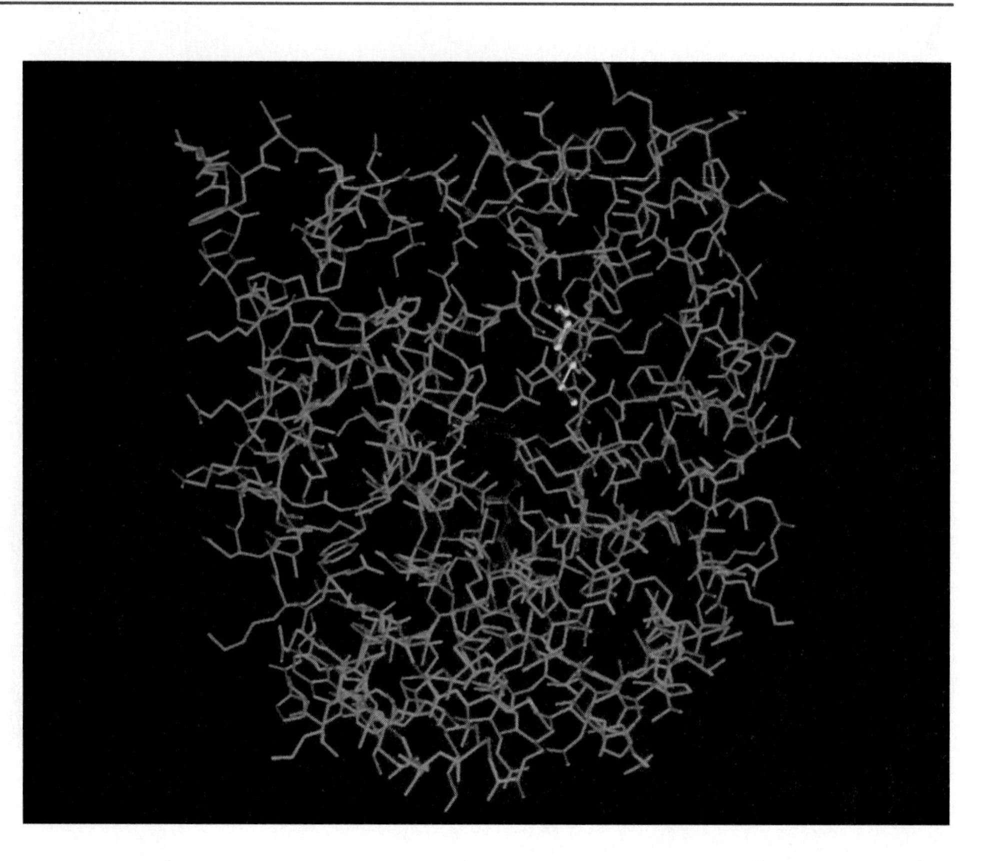

Crystal structures have been obtained for triosephosphate isomerase purified both from yeast and from chicken muscle. Although the proteins from the two organisms differ at about half of the positions in their amino acid sequences, their three-dimensional structures are nearly identical, and the same sets of amino acids have been implicated in their catalytic mechanisms. The overall structure is a β barrel, with eight parallel β sheets linked by eight α helices (fig. 4.24 and fig. 9.25). In both enzymes, a glutamic acid residue (Glu 165) reacts with affinity labels that probably bind at the active site, such as chloroacetone phosphate. (Chloroacetone phosphate is the same as dihydroxyacetone phosphate except that a chlorine atom replaces the hydroxyl group on C-1.) Crystal structures of the complexes of the enzyme with bound dihydroxyacetone phosphate or a competitive inhibitor show that the carboxyl group of Glu 165 is located close to C-1 of the substrate. When Glu 165 was changed to aspartate by site-directed mutagenesis, the rate of the isomerization reaction catalyzed by the enzyme was decreased approximately 1,000-fold. This is a remarkable effect, considering that the substitution of Asp for Glu probably changes the position of the carboxyl group by less than 1 Å.

The rate of the isomerase reaction decreases if a basic group is protonated by bringing the pH below 6.5. It seems clear that this is the carboxyl group of Glu 165, because the reaction of Glu 165 with affinity labels drops out in parallel with the enzymatic activity. This suggests that, in the enzymatic reaction, Glu 165 acts as a general base to remove the proton from C-1 of the dihydroxyacetone phosphate, converting the substrate into an ene-diolate intermediate as shown in figure 9.26. The reaction could be completed if the protonated glutamic acid residue returned the proton to C-2 of the ene-diolate, instead of to C-1.

Independent evidence in support of this scheme has come from experiments in which the enzymatic reaction is run in D_2O. Deuterium is incorporated stereospecifically onto carbon 2 of the product:

$$\begin{array}{c} CH_2OH \\ | \\ C=O \\ | \\ CH_2OPO_3^{-2} \end{array} \xrightarrow{D_2O} \begin{array}{c} H-C=O \\ | \\ D-C-OH \\ | \\ CH_2OPO_3^{-2} \end{array} \qquad (14)$$

In the absence of the enzyme, the carbon-bound hydrogen atom on glyceraldehyde-3-phosphate does not exchange with deuterium atoms of the solvent at any significant rate. (We are not concerned here with an exchange of the proton that is bound to the alcohol oxygen in the reactant or product. This proton does exchange rapidly with the solvent in either the presence or the absence of an enzyme, but the deuteron that is incorporated here can be removed immediately by putting the product back in H_2O.) The incorporation of the deuterium isotope can be explained as follows: After a proton is transferred from C-1 of the substrate to the carboxyl group of Glu 165, it has an opportunity to escape into the solution and to be replaced by a deuteron. A proton attached to a carboxylic acid oxygen usually exchanges very rapidly with the solvent. The deuteron on Glu 165 then can be transferred to C-2 of the product.

Figure 9.26

The probable reaction mechanism of triosephosphate isomerase. The γ-carboxylate group of Glu 165 acts as a general base to remove a proton from C-1 of the substrate, dihydroxyacetone phosphate (DHAP). This generates a planar ene-diolate intermediate that has two tautomeric forms. After an interconversion of ene-diolate tautomers, the protonated glutamic acid residue returns a proton to C-2. The electrostatic effects of His 95 and Lys 12 facilitate the formation of the negatively charged intermediate. A possible variation on the mechanism would be for His 95 or Lys 12 to act as a general acid. If this happened, the intermediate would be a neutral ene-diol. However, the pH-dependence of the kinetics, and the hydrogen-bonding pattern of His 95 in the crystal structure are more consistent with the scheme shown here.

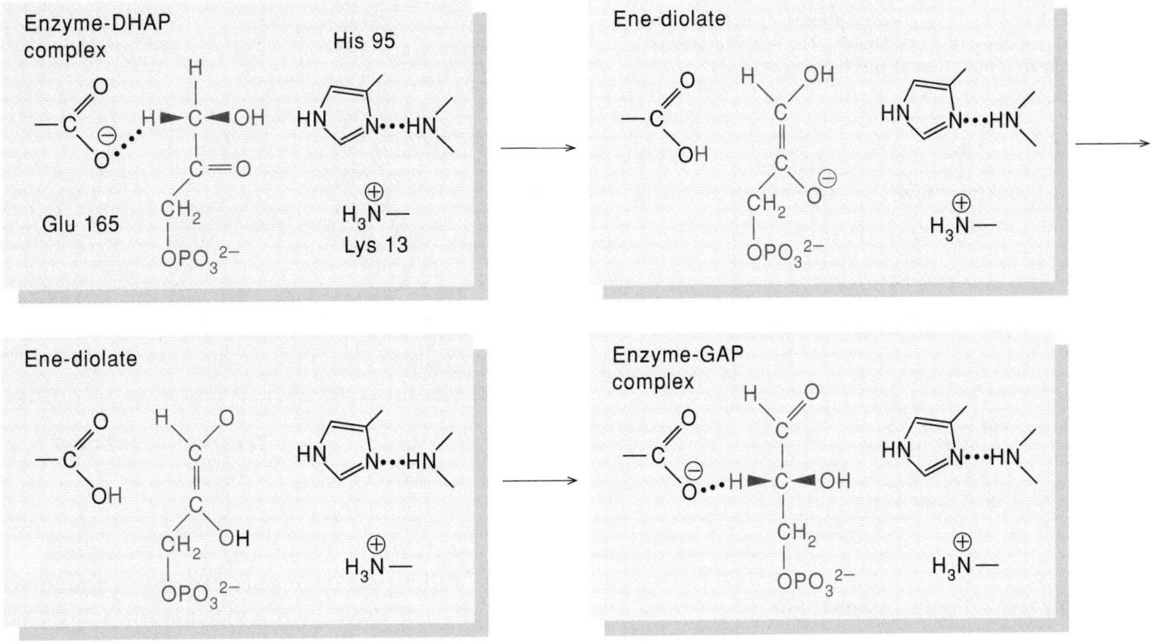

Additional studies of the kinetics of such exchange processes indicate that the proton probably moves from Glu 165 directly back to the ene-diolate in one step, rather than hopping first to another acid–base group and proceeding from there to the ene-diolate. This is consistent with the crystal structure, which shows the carboxylate group of Glu 165 in a good position to interact with a proton on either C-1 or C-2 of the substrate.

We have not yet considered the itinerary of the proton that must move from one of the oxygen atoms to the other. In the crystal structure there are two other nearby acid–base groups that could participate in the travels of this proton; these are His 95 and Lys 13 (see fig. 9.25). (The latter number is for the chicken enzyme; in yeast, the homologous lysine is residue 12.) His 95 appears to play an important role in the enzymatic reaction, because replacing this residue with glutamine by site-directed mutagenesis decreases k_{cat} by a factor of about 400. Considering the position of His 95 in the crystal structure, a possible scheme for the overall reaction would be for this residue to act together with Glu 165 to provide concerted acid–base catalysis in both steps of the reaction. In the formation of the intermediate state, Glu 165 would remove a proton from C-1 of dihydroxyacetone phosphate, and His 95 would add a proton to the carbonyl oxygen. This means that the intermediate would be a neutral ene-diol, rather than the ene-diolate ion shown in figure 9.26. In the breakdown of the ene-diol, Glu 165 would return a proton to C-2, and His 95 would extract a proton from the oxygen atom on C-1. Histidine 95 is in a better position than Lys 13 to do this because the ϵ-amino group of the lysine, though close to the oxygen atom on C-2, seems too far away from the C-1 oxygen.

There are several difficulties with the idea that His 95 works in this way. First, if His 95 has to donate a proton to the substrate, the rate of the reaction should decrease when the histidine is deprotonated by raising the pH. In fact, k_{cat} is essentially independent of pH from 6.5 up to almost 10. A pK_a in the region of 10 would seem more consistent with a lysine residue than with a histidine. In addition, as is indicated in figure 9.26, His 95 is hydrogen-bonded to an amide NH group, which means that it is unlikely to be protonated in the pH range where the enzyme is active.

These observations suggest that the roles of His 95 and Lys 12 may not be to act as general acids, but rather to stabilize the negative charge on an ene-diolate intermediate by electrostatic effects. Several additional observations are consistent with this view. The carbonyl bond of dihydroxyacetone phosphate appears to be strongly polarized when the substrate binds to the enzyme, because its infrared absorption band shifts to a markedly longer wavelength and its reactivity with sodium borohydride is greatly increased. The most likely mechanism for triosephosphate isomerase thus appears to involve a combination of the electrostatic effects of His 95 and Lys 12 with the general-base catalysis by Glu 165, as shown in figure 9.26.

Figure 9.27

The substrates of triosephosphate isomerase can be labeled specifically with deuterium. A kinetic isotope effect of about 3 is obtained with [1(R)-^{2}H]-dihydroxyacetone phosphate (*a*). This indicates that the formation of the ene-diolate intermediate is the rate-limiting step in the conversion of dihydroxyacetone phosphate to glyceraldehyde-3-phosphate. Little or no isotope effect is obtained when the reaction is run in the opposite direction, starting with [2-^{2}H]-glyceraldehyde-3-phosphate (*b*). The breakdown of the ene-diolate must be rate-limiting in this direction. Note that although the two hydrogen atoms on C-1 of dihydroxyacetone phosphate are chemically identical, they are stereochemically distinguishable. The enzyme always removes the one that has been replaced by deuterium in (*a*).

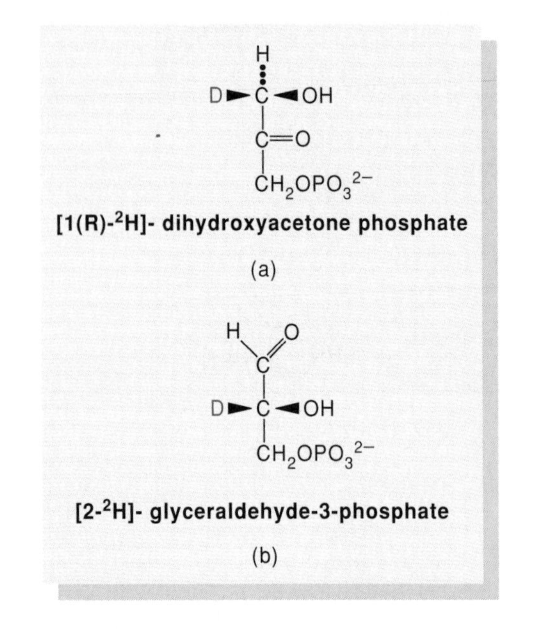

Figure 9.28

The calculated free energy profile of the reaction catalyzed by triosephosphate isomerase. (Enz = enzyme; DHAP = dihydroxyacetone phosphate; GAP = glyceraldehyde-3-phosphate.) The free energy changes associated with binding of DHAP and GAP to the enzyme are calculated on the assumption that DHAP and GAP are present at concentrations of 40 μM.

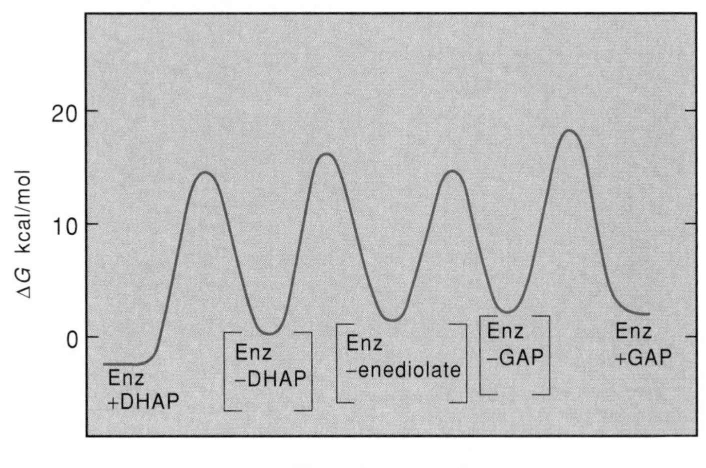

Assuming that the scheme shown in figure 9.26 is correct, how could we determine whether the formation or the breakdown of the ene-diolate intermediate is the rate-limiting step in the overall reaction? The best approach is to study the kinetics using substrates that are specifically labeled with deuterium. With [1(R)-^{2}H]-dihydroxyacetone phosphate (fig. 9.27*a*), k_{cat} is about three times smaller than the k_{cat} obtained with the ^{1}H analog. An isotope effect of this magnitude is consistent with the view that the bond holding the H or deuterium atom to C-1 must be broken in the rate-limiting step. This means that the formation of the ene-diolate is rate-limiting. Similar measurements with [2-^{2}H]-glyceraldehyde-3-phosphate (see fig. 9.27*b*) show that there is little or no isotope effect on k_{cat} for the reverse direction, from glyceraldehyde-3-phosphate to dihydroxyacetone phosphate. In this direction, the formation of the ene-diolate intermediate must be faster than its breakdown.

By measuring the kinetics of the isotopic exchange reactions catalyzed by the enzyme, along with the overall kinetics of the isomerization reaction in both directions, Jeremy Knowles and his colleagues have obtained rate constants for the individual steps of the catalytic mechanism. Figure 9.28 shows the complete pathway in the form of a free energy profile. In this diagram, the free energy changes accompanying the binding of the reactant or product to the enzyme are calculated on the assumption that dihydroxyacetone phosphate and glyceraldehyde-3-phosphate are both present at a concentration of 40μM, which is approximately their concentration in muscle cells. At this low concentration of substrate, when the overall rate of the reaction is described by the specificity constant k_{cat}/K_m, the rate-limiting step appears to be the formation of the enzyme-substrate complex. This step occurs with a rate constant of about 10^8 M^{-1} s^{-1} and evidently is set by the rate of diffusion of the enzyme and substrate. It can be slowed down by increasing the viscosity of the solution. Because the activation free energies for the additional steps in the interconversion of the enzyme-substrate and enzyme-product complexes are low enough so that these steps do not limit the overall rate of the reaction, there is no evolutionary pressure to increase the rate constants for these steps any further. Furthermore, evolutionary changes that led to tighter binding of the substrate could actually be harmful, because lowering the free energy of the enzyme-substrate complex could result in an increase in the activation free energy for one or more of the steps in the conversion of this complex into the product. Given the fixed, low concentration of its substrate, triosephosphate isomerase appears to have reached perfection!

Figure 9.29

Cell-wall polysaccharide, the substrate of lysozyme. (*a*) Conventional drawing of the hexose rings. (*b*) Drawing showing the conformations of the rings. In (*b*), alternating hexose units are flipped over by 180° relative to their neighbors; this is the preferred conformation for polysaccharides with ß(1,4) linkages (see chapter 6). The bond that is cleaved by lysozyme is indicated.

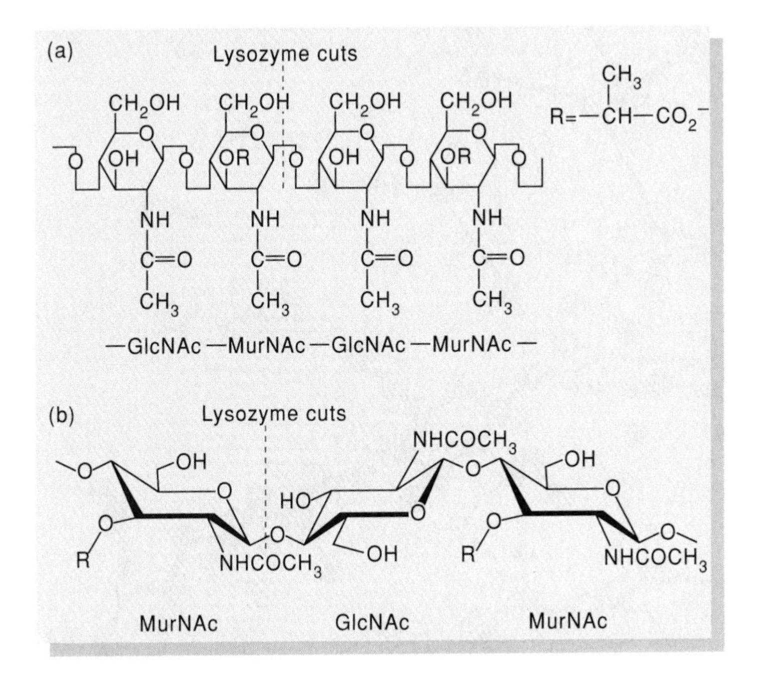

Table 9.1
Rates of Reaction and Cleavage Patterns Shown by Different Substrates of Lysozyme

Compound	Turnover Number k_{cat} (s^{-1})	Cleavage Pattern
$(GlcNAc)_3$	8.3×10^{-6}	$X_1 \downarrow X_2 \downarrow X_3$
$(GlcNAc)_4$	6.6×10^{-5}	$X_1 \downarrow X_2 \downarrow X_3 \downarrow X_4$
$(GlcNAc)_5$	0.033	$X_1 \downarrow X_2 \downarrow X_3 \downarrow X_4 \downarrow X_5$
$(GlcNAc)_6$	0.25	$X_1 — X_2 — X_3 — X_4 \downarrow X_5 — X_6$
(GlcNAc-MurNAc)$_3$	0.50	$X_1 — X_2 — X_3 — X_4 \downarrow X_5 — X_6$

Lysozyme Hydrolyzes Complex Polysaccharides Containing Five or More Residues

Lysozyme catalyzes the hydrolysis of polysaccharide chains that form structural elements of bacterial cell walls. It was discovered first in the whites of chicken eggs, and subsequently was found to occur widely in biological tissues and secretions such as tears. Related glycosidases have been obtained from a variety of organisms including fungi, bacteria, and the bacteriophage T4. The enzyme from hen eggwhite, a small, readily purified protein with a molecular weight of 14,500, was the first enzyme to have its crystal structure solved by x-ray diffraction. Although essentially nothing was known at the time about how lysozyme worked, the structure quickly suggested a likely mechanism.

Lysozyme's principal substrate, the bacterial cell-wall polysaccharide, is a polymer of alternating *N*-acetylglucosamine (GlcNAc) and *N*-acetylmuramic acid (MurNAc) residues connected by $\beta(1,4)$ glycosidic linkages (fig. 9.29). In the cell walls, the MurNAc residues are cross-linked by short polypeptides to form a two-dimensional network termed a peptidoglycan (see chapter 6). Lysozyme cuts the polysaccharide chain at C-1 of a MurNAc residue. It also will hydrolyze some shorter oligosaccharides such as (GlcNAc-MurNAc)$_3$ or $(GlcNAc)_6$, but it does not accept $(MurNAc)_6$ or other homopolymers of MurNAc. Thus the active site appears to be specific for a GlcNAc residue next to the bond that is cleaved. Very short oligomers of GlcNAc, such as $(GlcNAc)_3$, bind to the enzyme, but are poor substrates (table 9.1).

The crystal structure of eggwhite lysozyme was solved both with and without bound $(GlcNAc)_3$. Although longer oligosaccharides were hydrolyzed too rapidly to afford crystal structures, it was possible to build a model for the enzyme with bound $(GlcNAc)_6$ by starting with the structure of the complex with $(GlcNAc)_3$. Figure 9.30 shows a drawing of the model. The oligosaccharide sits in a shallow crevice that contains recognizable binding sites for six hexose units. These are labeled A through F in the figure. All of the sites appear to be tailored to bind the acetamide

$$CH_3\overset{O}{\overset{\|}{C}}NH—$$

side chains, but sites A, C, and E would not be spacious enough to accommodate the lactyl

$$HO—\overset{O}{\overset{\|}{C}}—\overset{CH_3}{\overset{|}{CH}}—O—$$

group of MurNAc. The restrictions are particularly severe in site C because of the bulky side chain of Ile 98. This suggests that the normal substrate of repeating (GlcNAc-MurNAc) units would bind with GlcNAc units occupying sites A, C, and E and with MurNAc units in sites B, D, and F, as shown schematically in figure 9.31. When taken with the observation that lysozyme hydrolyzes $(GlcNAc)_6$ but not $(GlcNAc)_3$, and the fact that (GlcNAc-MurNAc)$_3$ is hydrolyzed at C-1 of a MurNAc residue (see fig. 9.29 and table 9.1), the model indicates that the glycosidic linkage that is cleaved must fall between sites D and E.

Figure 9.30

A model of the complex between lysozyme and a substrate $(GlcNAc)_6$. The crevice that forms the active site runs horizontally across the molecule. The substrate is shown in darker color. Hexose rings A, B and C are in the positions occupied by the corresponding rings of the competitive inhibitor $(GlcNAc)_3$, as seen in the crystal structure of the enzyme-inhibitor complex. The positions of rings D, E and F were inferred by model building. The side chains of some of the amino acid residues that appear to interact with the substrate are indicated without color. The binding sites at positions A, C and E appear to be too cramped to accommodate MurNAc residues. This establishes the way that the (MurNAc-GlcNAc) units of the cell-wall polysaccharide probably bind to the enzyme, and indicates that the locus of cleavage in the active site is between positions D and E.

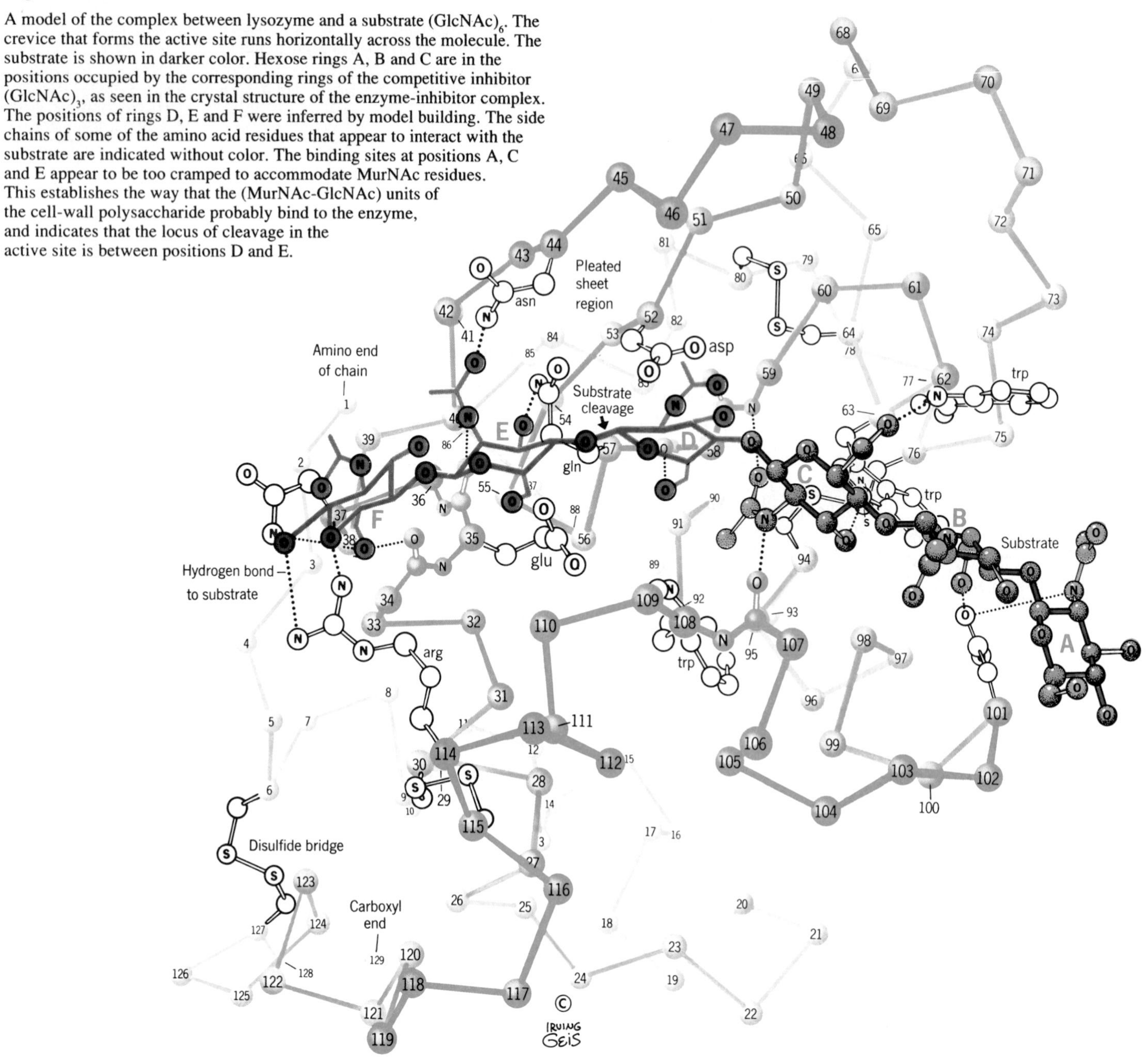

There was another reason for focusing attention on the region of the enzyme between binding sites D and E. The amino acid side chains here included two potentially reactive groups, the carboxyl groups of Asp 52 and Glu 35. When the model for the $(GlcNAc)_6$ substrate was positioned so that sites A–E were all occupied, the glycosidic bond that would be cleaved was located in between the two carboxyl groups (see fig. 9.30). Homologous acidic amino acid residues subsequently were found in the active site of bacteriophage T4 lysozyme, and mutations that perturb the activity of the bacteriophage enzyme were found to cluster in this region. In eggwhite lysozyme, chemical conversion of the carboxyl group of Asp 52 to $-CH_2OH$ destroys enzymatic activity. Studies with affinity labels also have implicated Asp or Glu residues in other bacterial and fungal glucosidases.

Lysozyme works best under mildly acidic conditions; its pH optimum is about 5. The pH dependence of the kinetics indicates that the reaction depends on an acidic group with a pK_a of about 6 and on a basic group with a pK_a of about 4.5. The latter group was identified as Asp 52 by studying how specific chemical modification of this residue affects the pH titration curve of the protein. The group with a pK_a of about 6 is probably Glu 35.

Figure 9.31

A schematic diagram showing the specificity of the hen eggwhite lysozyme for its cell-wall polysaccharide substrate. Six subsites (A–F) on the enzyme bind the hexose units. Alternate sites (A, C, and E) interact with the acetamide side chains (a). These sites are unable to accommodate MurNAc residues with their lactyl side chains (Lac). The glycosidic linkage that is cleaved is between sites D and E.

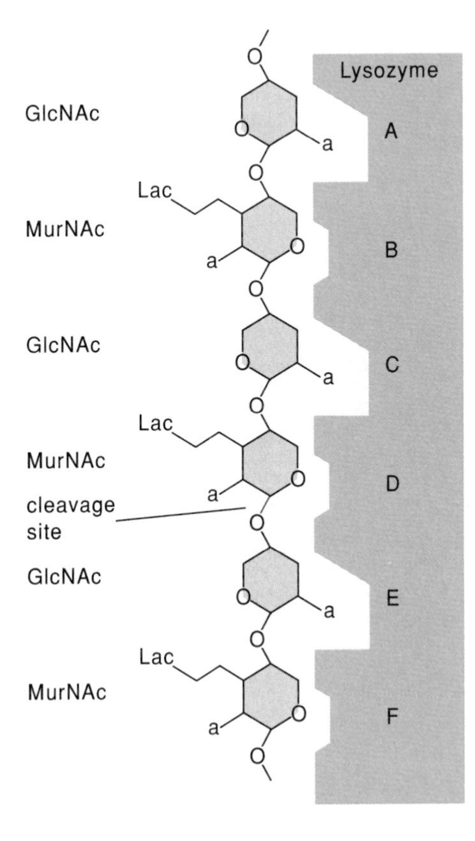

Figure 9.32

A direct nucleophilic (S_N2) attack by a hydroxyl ion on the ß-glycosidic bond in the cell-wall polysaccharide would leave the MurNAc product with an α-hydroxyl group. The observed product has a ß-hydroxyl group. This suggests that the reaction involves an intermediate step in which a derivative of the MurNAc residue remains associated with the enzyme while the alcohol product is replaced by water, as shown in figure 9.33.

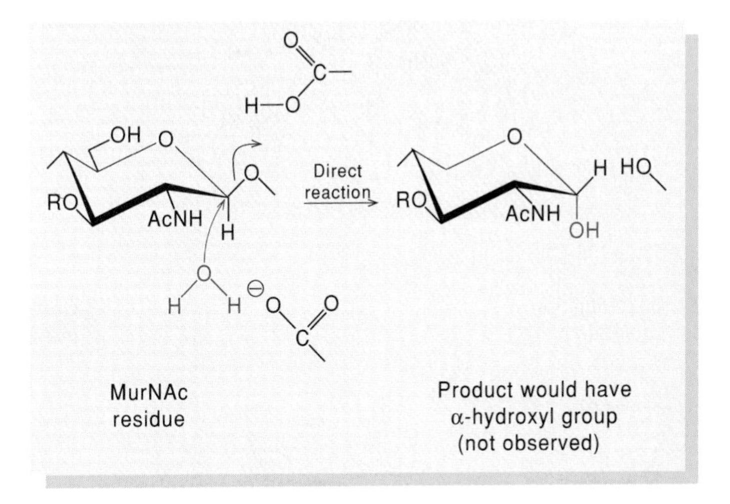

Given the information that the enzymatic reaction requires the carboxyl group of Glu 35 to be protonated and that of Asp 52 to be unprotonated, the first reaction mechanism that comes to mind might be a direct nucleophilic attack by H_2O, aided by concerted general-acid and general-base catalysis (fig. 9.32). In this scheme, Asp 52 could remove a proton from the water molecule, and Glu 35 could provide a proton to the departing alcoholic group. But this mechanism is at odds with the stereochemistry of the reaction. If H_2O or HO^- attacked C-1 of the MurNAc residue from one side of the tetrahedral carbon atom, and the alcoholic group of GlcNAc departed from the other side, the result would be an inversion of configuration around the C-1 carbon. The β-glycosidic bond would be replaced by a hydroxyl group in the α configuration. Contrary to this expectation, the product is found to have the hydroxyl group in the β configuration. This indicates that the reaction probably proceeds in two distinct steps. The first step evidently removes the alcohol from the β side of the MurNAc residue, but leaves an intermediate derivative of the MurNAc residue associated with the enzyme. Water or HO^- then must enter the active site on the same side of the MurNAc residue, replacing the alcohol.

Figure 9.33 shows a mechanism that meets these requirements. In this scheme, the intermediate is a carboxonium ion, in which a positive charge is distributed between C-1 of the MurNAc and the attached oxygen atom. The formation of this species would involve general-acid catalysis by Glu 35, supported by a favorable electrostatic interaction between the carboxonium ion and the negatively charged carboxylate of Asp 52. In the breakdown of the carboxonium intermediate, Glu 35 could act as a general base to remove a proton from water.

Another plausible mechanism would be for Asp 52 to launch a direct nucleophilic attack on the glycosidic bond. This could generate an enzyme-bound, carboxylic ester of the MurNAc as an intermediate. As yet, the experimental information is not sufficient to distinguish decisively between this mechanism and the one shown in figure 9.33, although measurements of kinetic isotope effects favor the latter.

The original formulation of the carboxonium-intermediate mechanism (see fig. 9.33) included an additional feature that was suggested by the crystallographic structure. In the model of $(GlcNAc)_6$ bound to the enzyme, it seemed difficult to squeeze even a GlcNAc residue into site D without some distortion of the hexose ring. Experimental measurements of the binding energies for a series of substrates also indicated that, whereas the binding of a GlcNAc residue to any of sites A, B, C, E, or F was energetically favorable, the binding to site D was very weak, or even energetically unfavorable. These observations suggested that the tight binding of hexose units to sites A, B, C, E, and F forced the hexose unit at site D to sit uncomfortably in a distorted geometry. The geometric distortion appeared to push the hexose ring from its normal boat configuration into a chair configuration. In the chair configuration, carbons 1 and 2 and the internal oxygen atom of the pyranose ring all lie

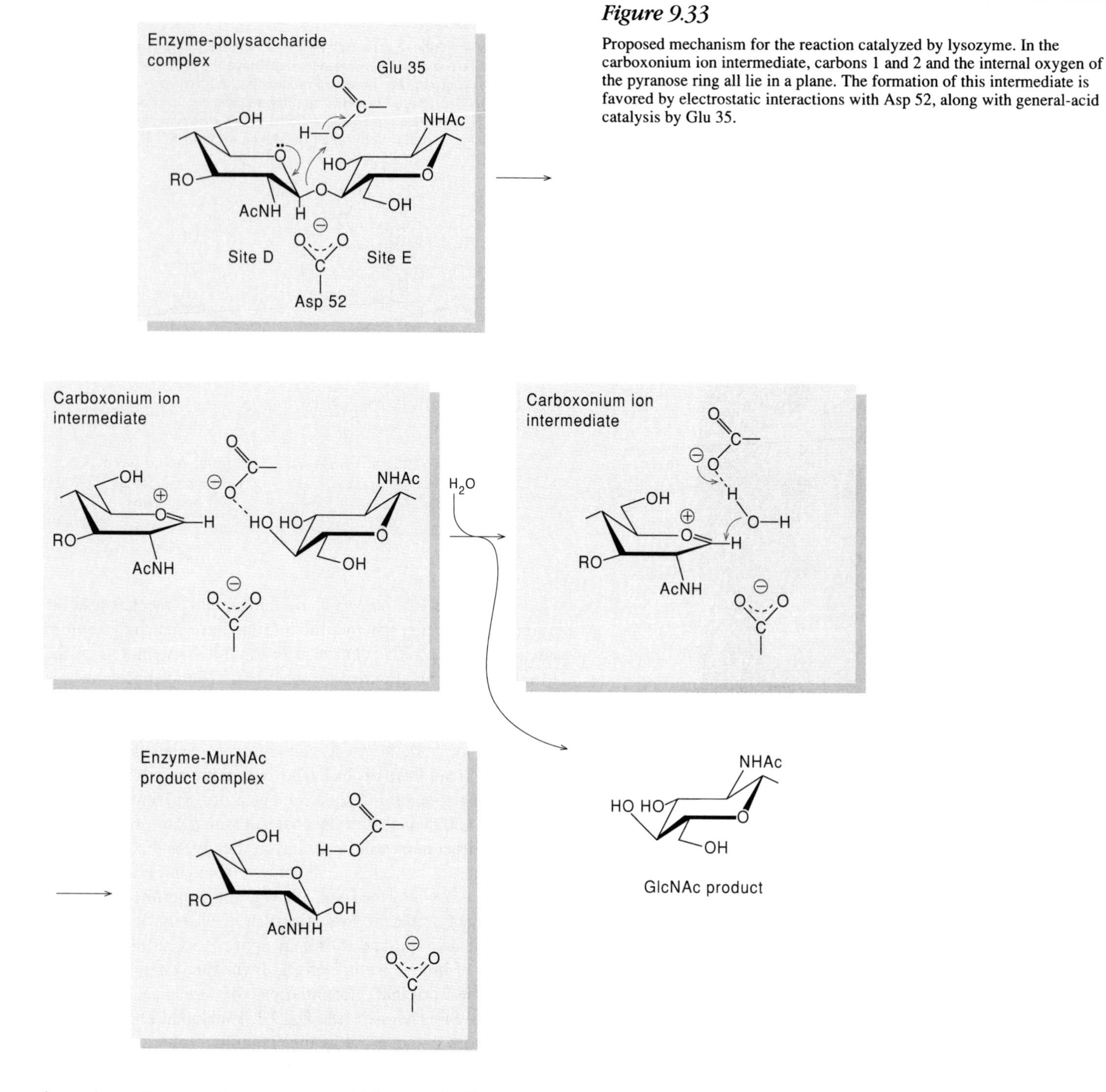

Figure 9.33

Proposed mechanism for the reaction catalyzed by lysozyme. In the carboxonium ion intermediate, carbons 1 and 2 and the internal oxygen of the pyranose ring all lie in a plane. The formation of this intermediate is favored by electrostatic interactions with Asp 52, along with general-acid catalysis by Glu 35.

in a plane. Because these atoms would have a similar, planar configuration in the carboxonium intermediate (see fig. 9.33), it was suggested that geometric "strain" in the bound substrate contributed significantly to pushing the substrate in the direction of the transition state. The enzyme thus would decrease the activation free energy for the reaction (ΔG^{+}) partly by raising the energy of the bound substrate.

Although the notion that an enzyme can exert a geometric strain on a bound substrate is an appealing one, it is not supported by computer modeling that takes into account the flexibility of the enzyme. Calculations of the energies of the enzyme-substrate complex in various conformations indicate that the hexose unit that binds to side D on lysozyme probably remains in the boat conformation and is not significantly strained.

Figure 9.34

Structures of NAD^+ and NADH.

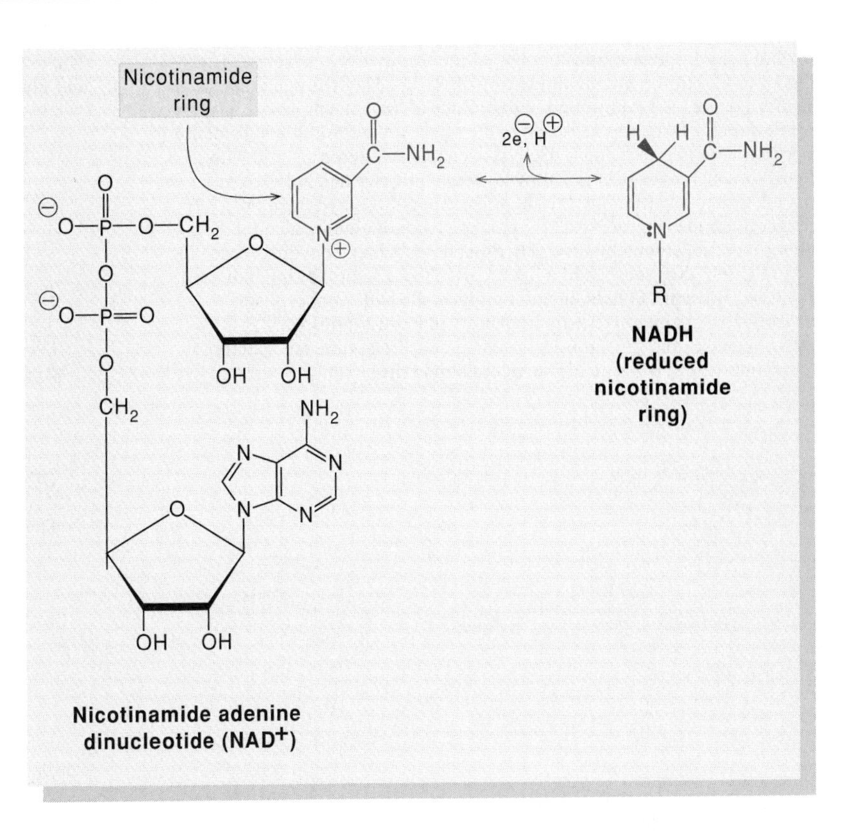

The experimental finding that the binding of a hexose unit to this site is energetically unfavorable can be attributed to the displacement of two water molecules that are bound to the carboxylate of Asp 52 in the free enzyme. Thus the largest contribution to lowering $\Delta G^{\ddagger}$ probably comes from the electrostatic effect of Asp 52, and not from geometric strain in the substrate.

Alcohol Dehydrogenase and Lactate Dehydrogenase: Enzymes that Catalyze the Transfer of Electrons

Numerous enzymes catalyze the transfer of electrons from one substrate to another. In many of these oxidation-reduction reactions, one or more protons also are removed from the substrate that becomes oxidized and are added to the substrate that is reduced. The oxidation of an alcohol to an aldehyde, for example, requires the removal of two protons along with two electrons:

$$RCH_2OH \rightleftharpoons R-\overset{\displaystyle H}{\overset{|}{C}}=O + 2\,e^- + 2\,H^+ \qquad (15)$$

The most common electron acceptor in the biological oxidation of alcohols is nicotinamide adenine dinucleotide (NAD^+), a coenzyme whose structure is shown in figure 9.34. When NAD^+ is reduced to nicotinamide adenine dinucleotide hydrogen (NADH), it picks up one proton and two electrons:

$$NAD^+ + 2\,e^- + H^+ \rightleftharpoons NADH \qquad (16)$$

The oxidation of an alcohol by NAD^+ thus results in the transfer of two electrons and the net release of one proton:

$$RCH_2OH + NAD^+ \rightleftharpoons R-\overset{\displaystyle H}{\overset{|}{C}}=O + NADH + H^+ \qquad (17)$$

Because the reaction written in equation (17) is formally equivalent to the removal of two hydrogen atoms from the alcohol, enzymes that catalyze such oxidation-reduction reactions generally are called dehydrogenases. The term "oxidase," which also might be a reasonable name for an oxidation-reduction enzyme, is reserved for reactions in which the oxidant is molecular oxygen (O_2).

The atomic or subatomic particles that a dehydrogenase actually moves from one substrate to another are usually not neutral hydrogen atoms ($H^{\bullet}$), as the name "dehydrogenase" suggests, nor are they necessarily free electrons and protons, as equations (15) and (16) suggest. In the reduction of NAD^+ a proton is released to the solution, but the species that is transferred from the reducing substrate to the nicotinamide ring of the coenzyme is probably a hydride ion (H^-), which is a proton with two electrons. This conclusion is based largely on model studies of nonenzymatic reactions of molecules resembling the coenzymes. In addition, it has been shown that in the enzymatic reactions, the hydride ion is transferred stereospecifically to a particular side of the nicotinamide ring, without equilibrating with protons of the solvent (see figs. 9.35 and 11.7). The strategy of all NAD^+-dependent dehydrogenases appears to be to orient

Figure 9.35

Alcohol dehydrogenase transfers one of the two hydrogens from C-1 of ethanol to C-4 of the nicotinamide ring of NAD^+. The hydrogen that is transferred does not equilibrate with protons of the solvent, and it probably moves directly from one substrate to the other in the form of a hydride ion (a proton with two electrons). Dehydrogenases are stereospecific with regard to which hydrogen is removed from a primary alcohol such as ethanol, and also with regard to the orientation of the hydrogen on the nicotinamide ring of NADH. Some enzymes, including alcohol dehydrogenase, soluble malate dehydrogenase, lactate dehydrogenase, and isocitrate dehydrogenase, add a hydrogen to the side of the ring indicated here (the "A" side). Others, including glyceraldehyde-3-phosphate dehydrogenase, glutamate dehydrogenase, glucose-6-phosphate dehydrogenase, and glycerol-3-phosphate dehydrogenase, are specific for the opposite ("B") side of the nicotinamide ring.

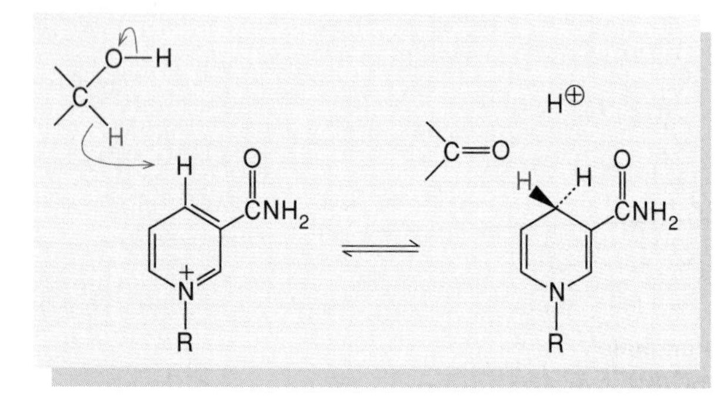

Figure 9.36

The reactions catalyzed by malate dehydrogenase, lactate dehydrogenase, and glyceraldehyde-3-phosphate dehydrogenase.

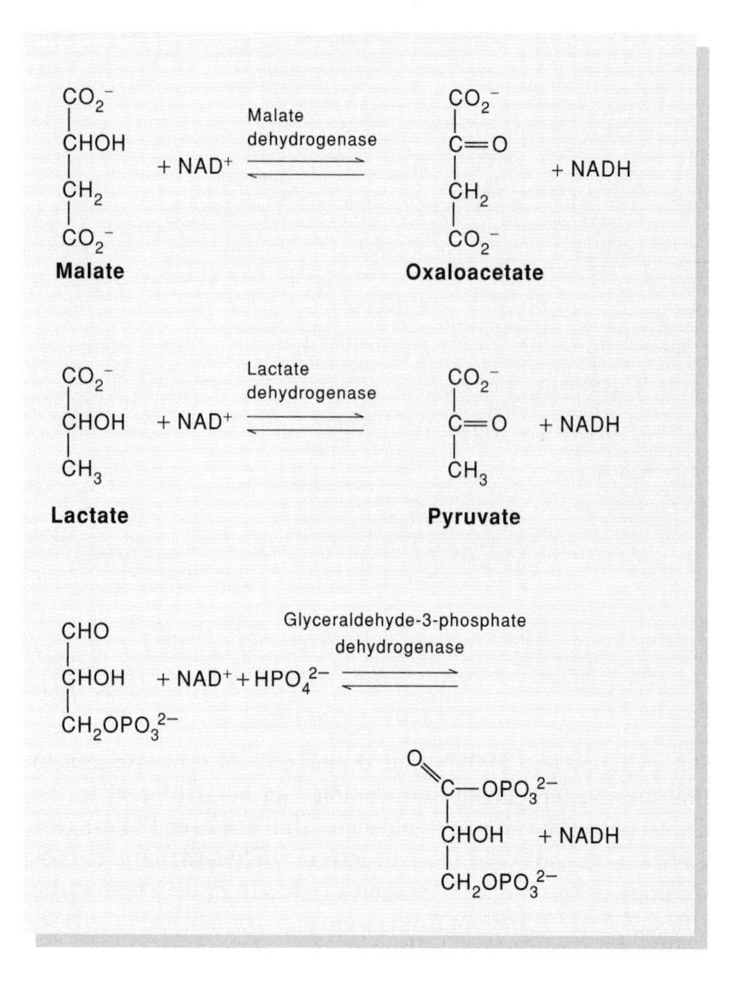

the coenzyme and substrate on the enzyme surface so that C-4 of the nicotinamide ring is close to the reactive hydrogen of the substrate, and to provide a functional group that can facilitate a redistribution of electrical charge in the substrate's carbon–oxygen bond.

The crystal structures of several dehydrogenases have been determined in the presence and absence of bound coenzymes and competitive inhibitors. The best-known structures are those of alcohol dehydrogenase, which catalyzes the interconversion of ethanol and acetaldehyde (equation 17 with R $= CH_3$), glyceraldehyde-3-phosphate dehydrogenase, lactate dehydrogenase, and soluble malate dehydrogenase. The last three enzymes catalyze the oxidation of glyceraldehyde-3-phosphate to glyceric acid 1,3-bisphosphate, lactic acid to pyruvic acid, and malic acid to oxaloacetic acid (fig. 9.36). All four enzymes play important roles in carbohydrate metabolism. (The term "soluble" attached to malate dehydrogenase is used to distinguish a cytosolic enzyme from the enzyme that catalyzes the same reaction in mitochondria.)

Lactate dehydrogenase and the soluble malate dehydrogenase probably arose from a common ancestor, because their overall structures are very similar. There is more variability in the structures of alcohol dehydrogenase and glyceraldehyde-3-phosphate dehydrogenase. In all four enzymes, however, the region of the protein that binds NAD^+ contains similar structural elements. The nucleotide-binding domain is made up of six strands of parallel β sheet connected by α-helical stretches running antiparallel to the β strands (fig. 9.37). NAD^+ binds to the enzyme in an extended conformation near the ends of the β strands, although the details of its contacts with enzyme vary among the different dehydrogenases. The negatively charged pyrophosphate group of the coenzyme usually is sandwiched between a positively charged residue such as arginine and the end of an α helix (see fig. 9.37). The helix is oriented so that its internal hydrogen bonds create a net excess of positive charge close to the pyrophosphate.

As we discussed in chapter 8, there are several possible pathways for an enzymatic reaction involving two substrates: (1) a random pathway, in which either substrate can bind to the enzyme first; (2) an ordered pathway, in which one substrate must bind before the other; and (3) a reactive-intermediate pathway, in which the first substrate binds and reacts with the enzyme to form one of the products, leaving the enzyme in a modified form that then reacts with the second substrate. Kinetic studies have shown that the reaction mechanism of lactate dehydrogenase follows an ordered sequence. In the oxidation of lactate, NAD^+ binds to the enzyme first, followed by lactate. The transfer of the hydride ion then occurs rapidly in

Figure 9.37

Crystal structures of the nucleotide-binding regions of (*a*) lactate dehydrogenase and (*b*) alcohol dehydrogenase. Bound NAD^+ is shown in yellow in (*b*), and a covalently linked adduct of NAD^+ and the substrate lactate ("S-lac-NAD") in (*a*). The arginine residue that interacts with the pyrophosphate of the NAD^+ is in red; other amino acid residues are in brown. The yellow ribbons show the folding of the α-carbon chain. The figures are based on crystal structures described by U. M. Grau and M. S. Rossmann for lactate dehydrogenase, and H. Eklund, J.-P. Sanama, L. Wallen, C.-I. Branden, A. Akeson, and T. A. Jones for alcohol dehydrogenase.

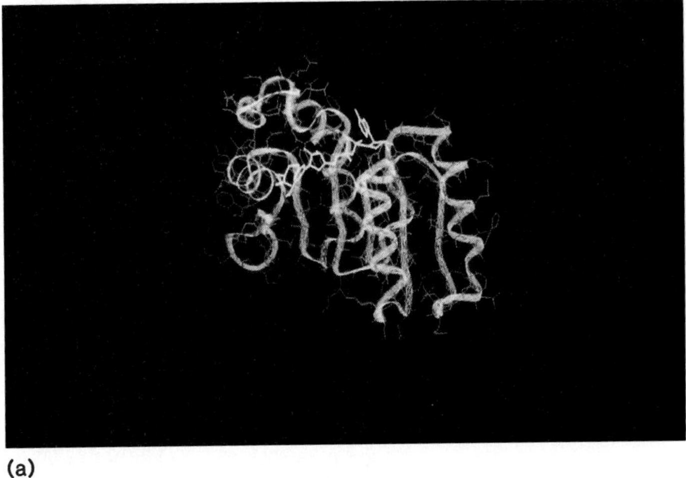

(a)

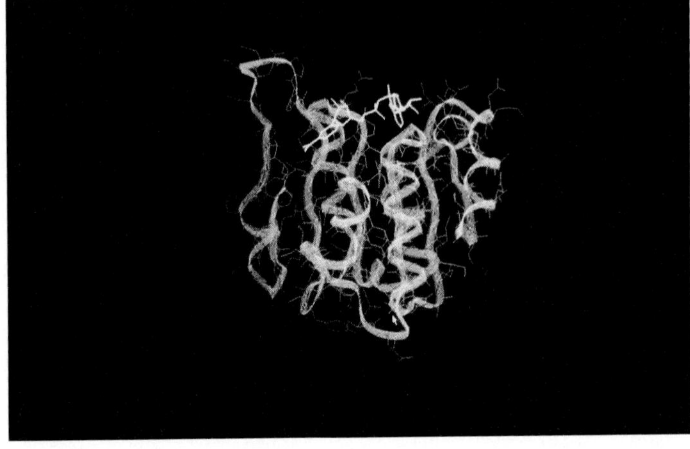

(b)

Figure 9.38

The catalytic mechanism of lactate dehydrogenase. His 195 probably acts as a general base in the conversion of lactate to pyruvate, and as a general acid in the conversion of pyruvate to lactate. Arg 171 probably serves to orient the substrate.

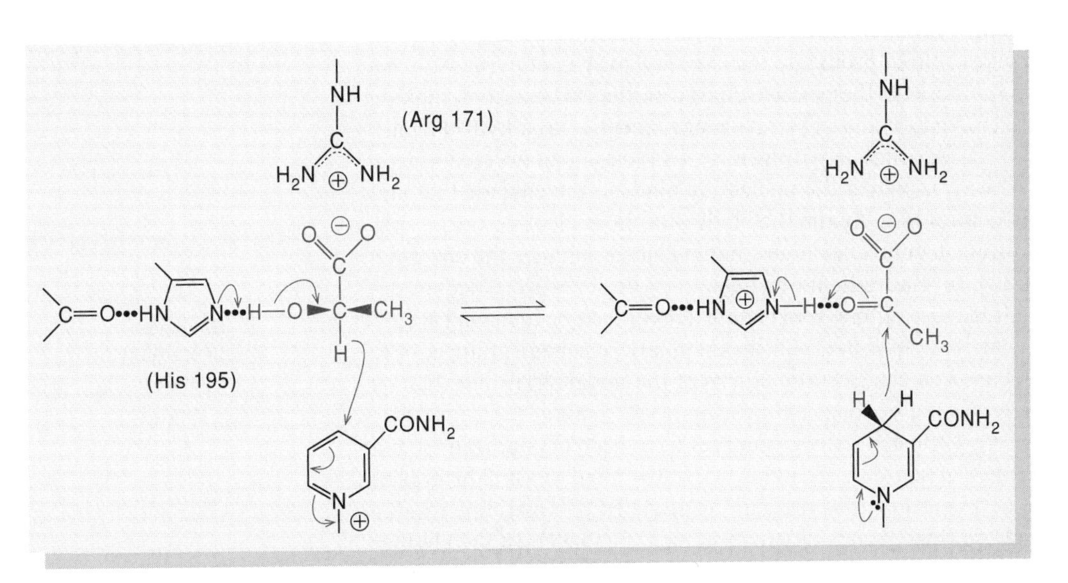

either direction, giving an equilibrium mixture of the two ternary complexes, enzyme-NAD^+-lactate and enzyme-NADH-pyruvate. Finally, pyruvate dissociates from the enzyme, followed by NADH. The rate-limiting step in the overall steady-state reaction is the dissociation of NADH. The obligatory sequences of the binding and dissociation steps can be explained by structural changes that result from the binding of NAD^+ or NADH. The binding of either form of the nucleotide causes a major structural rearrangement that brings together components of the binding site for lactate or pyruvate. The structural changes include a particularly large movement of the loop of the protein between two of the β strands. Arginine 101, the residue that interacts with the pyrophosphate of NAD^+ or NADH, is in the middle of this loop (see fig. 9.37*a*). In the free enzyme the loop extends out into the solvent; in the ternary complexes it is pulled in around the coenzyme and substrate.

The substrate-binding site of lactate dehydrogenase contains an arginine residue (Arg 171) that serves to orient lactate or pyruvate by forming an ion pair with the carboxylate group, and a histidine (His 195) that forms a hydrogen bond to the alcohol or keto group. In the oxidation of lactate to pyruvate, His 195 probably acts as a general base to remove a proton from the substrate's hydroxyl group, as shown in figure 9.38.

Figure 9.39

Functional groups at the active site of alcohol dehydrogenase from horse liver. (*a*) In the absence of substrate, the ligands of the Zn^{2+} ion are two cysteines, a histidine, and a molecule of water that is hydrogen-bonded to Ser 48. (*b*) The substrate ethanol probably binds to the Zn^{2+} as the alcoholate anion, displacing the molecule of water and forming a hydrogen bond to Ser 48.

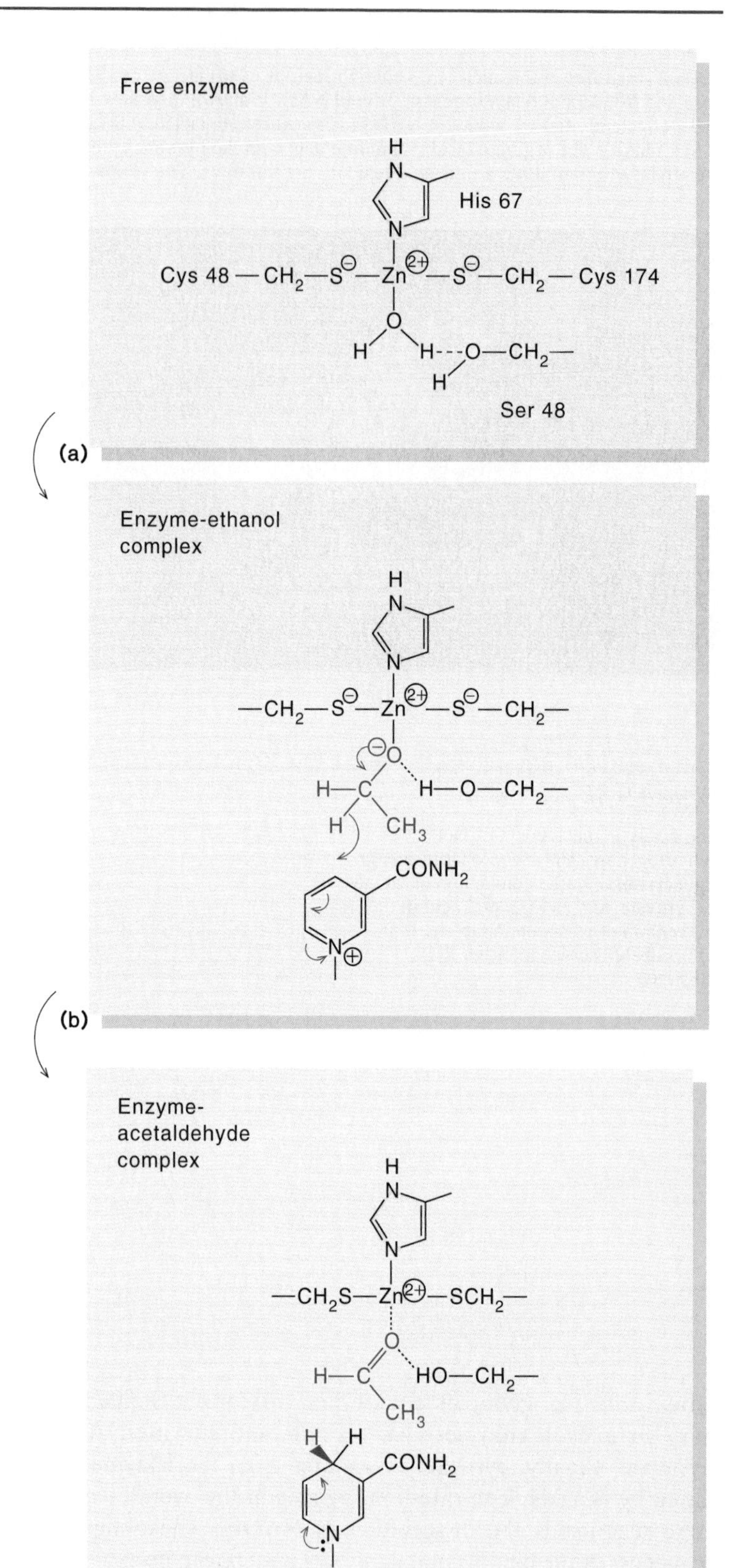

This would increase the electron density on the adjacent carbon atom, easing the release of the hydride ion. In the reverse reaction, the imidazolium group of the histidine would act as a general acid to protonate the carbonyl oxygen of pyruvate. This would increase the positive charge on the carbon atom, making the enzyme-bound substrate more receptive to a hydride ion.

Alcohol dehydrogenase also follows an ordered kinetic pathway, in which the nucleotide binds to the enzyme first and dissociates last. Again, the binding of the nucleotide causes substantial structural changes that evidently enhance the enzyme's ability to bind ethanol or acetaldehyde. However, alcohol dehydrogenase differs from lactate dehydrogenase in having a Zn^{2+} ion at its active site, and its enzymatic mechanism appears to be based on electrostatic effects of the metal ion rather than on general-acid or -base catalysis.

Alcohol dehydrogenases have been crystallized from a variety of sources, including horse liver and yeast. The crystal structures show that the Zn^{2+} is attached to a histidine residue and the thiol groups of two cysteines (fig. 9.39). In the absence of the substrate, the fourth ligand of the Zn^{2+} is a molecule of water that forms a hydrogen bond to a nearby serine residue. When the substrate adds to the enzyme, the alcoholic or carbonyl oxygen atom appears to coordinate directly to the Zn^{2+}, displacing the molecule of water. Studies of the pH dependence of binding of a variety of alcohols indicate that the alcohol probably binds in the form of the alcoholate anion, as shown in figure 9.39. The negative charge on the oxygen atom makes the alcoholate a better donor of a hydride ion for the reduction of NAD^+, compared with the undissociated alcohol. For the reaction in the opposite direction, the electrostatic effect of the Zn^{2+} ion would facilitate the reduction of acetaldehyde.

Summary

In this chapter we have looked at mechanisms of enzyme catalysis, first from a general standpoint, and then in some detail. The following points are the most important.

1. Like ordinary chemical catalysts, enzymes associate directly with the reacting species and interact with them in a manner that lowers the free energy of the transition state. Enzyme-substrate interactions usually are highly specific, and can depend on critically positioned amino acid side chains, the atoms of the polypeptide backbone, or bound cofactors such as metal ions.
2. All known enzymatic reaction mechanisms depend on one or more of the following themes: proximity effects (enzymes hold the reactants close together in an appropriate orientation); electrostatic effects (charged, polar, or polarizable groups of the enzyme are positioned to favor the redistribution of electrical charges that occur as the substrate evolves into the transition state); general-acid or general-base catalysis (acidic or basic groups of the enzyme donate or remove protons, and often do first one and then the other); nucleophilic or electrophilic catalysis (nucleophilic or electrophilic functional groups of the enzyme or a cofactor react with complementary groups of the substrate to form covalently-linked intermediates); and structural flexibility (changes in the protein structure can increase the specificity of enzymatic reactions by insuring that substrates bind or react in an obligatory order and by sequestering bound substrates in pockets that are protected from the solvent).
3. Trypsin, chymotrypsin, and elastase are very similar in structure, but have substrate-binding pockets that are tailored for different types of amino acid side chains. The active site of each enzyme contains three critical residues: serine, histidine, and aspartate. The side chains of these residues are oriented so that the serine hydroxyl group becomes a strong nucleophilic reagent for attacking the substrate's peptide carbonyl group adjacent to the preferred amino acid. This reaction displaces the amine member of the peptide bond, and generates an acyl-ester intermediate, in which the carboxyl member is linked covalently to the serine group of the enzyme. Formation and hydrolysis of the acyl-ester intermediate probably are promoted both by electrostatic interactions that stabilize tetrahedral transition states and by general-acid and general-base catalysis.
4. Thermolysin and carboxypeptidase A contain a bound Zn^{2+} ion, which interacts with the carbonyl oxygen of the peptide bond that is to be cleaved. The electrostatic effect of the Zn^{2+} probably facilitates formation of a tetrahedral transition state in the direct nucleophilic attack of H_2O or OH^- at the carbonyl carbon. Acidic and basic amino acid side chains again may serve as general-acid and general-base catalysts in the reaction. Carboxypeptidase A undergoes a substantial change in structure when it binds its substrate.
5. Ribonuclease A hydrolyzes RNA adjacent to pyrimidine bases. The reaction proceeds through a 2′,3′-phosphate cyclic diester intermediate. The formation and breakdown of the cyclic diester appear to be promoted by concerted general-base and general-acid catalysis by two critical histidine residues, and by electrostatic interactions with two lysines. These reactions proceed through pentavalent phosphoryl intermediates. The geometry of the oxygens surrounding the phosphorus atom in these intermediates resembles the geometry of vanadate compounds that act as inhibitors of the enzyme.
6. Triosephosphate isomerase interconverts dihydroxyacetone phosphate and glyceraldehyde-3-phosphate. A glutamic acid residue probably acts as a general base to remove a proton from the substrate, forming an ene-diolate intermediate. Nearby histidine and lysine side chains would stabilize this intermediate electrostatically. Triosephosphate isomerase appears to have reached evolutionary perfection, in the sense that it catalyzes its reaction at the maximum possible rate, given the concentration of the substrate in the cell. The rate-limiting step is the collision of the enzyme and substrate by diffusion through the solution.
7. Lysozyme hydrolyzes complex polysaccharides containing five or more hexose residues. Its substrate-binding site includes a crevice that can accommodate six such residues. The enzyme probably first releases the truncated polysaccharide from one side of the bond that is cleaved, leaving the polysaccharide on the other side bound noncovalently to the enzyme in the form of a positively charged carboxonium intermediate. The formation of this intermediate is promoted by general-acid catalysis by a glutamic acid residue, and by the electrostatic effect of a nearby aspartate.
8. Lactate dehydrogenase and alcohol dehydrogenase catalyze the reversible transfer of a hydride ion (H^-) to NAD^+ from, respectively, lactate and ethanol. Both enzymes bind their substrates in an obligatory order. NAD^+ binds first, causing a structural change that sets up the binding site for the alcohol. The alcohol substrate then binds with the hydrogen that will be transferred close to C-4 of the nicotinamide ring. In lactate dehydrogenase, a histidine residue probably acts as a general base to remove a proton from the —OH group. This would facilitate the release of the negatively charged hydride ion. In alcohol dehydrogenase, a Zn^{2+} ion attached to the enzyme probably encourages the ethanol to bind in the form of the alcoholate anion, which again is a good donor of a hydride ion.

Selected Readings

Albery, W. J., and J. R. Knowles, Free-energy profile of the reaction catalyzed by triosephosphate isomerase. *Biochem.* 15:5588, 5627, 1976.

Blackburn, P., and S. Moore, Pancreatic ribonuclease. *The Enzymes* 15:317, 1982.

Blow, D., Structure and mechanism of chymotrypsin. *Acc. Chem. Res.* 9:145, 1976.

Branden, C. -I., H. Jornvall, H. Eklund, and B. Furugren, Alcohol dehydrogenases. *The Enzymes* 11:104, 1975.

Breslow, R., How do imidazole groups catalyze the cleavage of RNA in enzyme models and enzymes? Evidence from "negative catalysis." *Acc. Chem. Res.* 24:317, 1991.

Eklund, H., B. V. Plapp, J. -P. Samama, and C. -I. Branden, Binding of substrate in a ternary complex of horse liver alcohol dehydrogenase. *J. Biol. Chem.* 257:14349, 1982.

Fersht, A., *Enzyme Structure and Mechanism,* 2d ed. New York: Freeman, 1985.

Fersht, A. R., D. M. Blow, and J. Fastrez, Leaving group specificity in chymotrypsin-catalyzed hydrolysis of peptides: A stereochemical interpretation. *Biochem.* 12:2035, 1973.

Findlay, D., D. G. Herries, A. P. Mathias, B. R. Rabin, and C. A. Ross, The active site and mechanism of pancreatic ribonuclease. *Nature* 190:781, 1961.

Holbrook, J. J., A. Liljas, S. J. Steindel, and M. G. Rossmann, Lactate dehydrogenase. *The Enzymes* 11:191, 1975.

Holmes, M. A., D. E. Tronrud, and B. W. Matthews, Structural analysis of the inhibition of thermolysin by an active-site-directed irreversible inhibitor. *Biochem.* 22:236, 1983.

Imoto, T., L. N. Johnson, A. C. T. North, D. C. Phillips, and J. A. Rupley, Vertebrate lysozymes. *The Enzymes* 7:665, 1972.

Kelly, J. A., A. R. Sielecki, B. D. Sykes, M. N. G. James, and D. C. Phillips, X-ray crystallography of the binding of the bacterial cell wall trisaccharide NAM-NAG-NAM to lysozyme. *Nature* 282:875, 1979.

Kraut, J., How do enzymes work? *Science* 242:533, 1988.

Lolis, E., T. Alber, R. C. Davenport, D. Rose, F. C. Hartman, and G. A. Petsko, Structure of yeast triosephosphate isomerase at 1.9-Å resolution. *Biochem.* 29:6609, 1990.

Markley, J. L., Correlation proton magnetic resonance studies at 250 MHz of bovine pancreatic ribonuclease. I. Reinvestigation of histidine peak assignments. *Biochem.* 14:3546, 1975.

Page, M. I. (ed.), *Enzyme Mechanisms.* London: Royal Society of Chemistry, 1987. See particularly the chapters by M. I. Page (Theories of Enzyme Catalysis), A. L. Fink (Acyl Group Transfer—The Serine Proteinases), D. S. Auld (Acyl Group Transfer—Metalloproteinases), P. M. Cullis (Acyl Group Transfer—Phosphoryl Transfer), M. L. Sinnott (Glycosyl Group Transfer), and J. P. Richard (Isomerization Mechanisms through Hydrogen and Carbon Transfer).

Reeke, G. N., J. A. Hartsuck, M. L. Ludwig, F. A. Quiocho, T. A. Steitz, and W. N. Lipscomb, The structure of carboxypeptidase A: Some results at 2.0-Å resolution, and the complex with glycyl tyrosine at 2.8-Å resolution. *Proc. Natl. Acad. Sci. USA* 58:2220, 1967.

Rose, I. A., Mechanism of the aldose-ketose isomerase reactions. *Adv. Enzymol.* 43:491, 1975.

Shoham, M., and T. Steitz, Crystallographic studies and model building of ATP at the active site of hexokinase. *J. Mol. Biol.* 140:1, 1980.

Warshel, A., G. Naray-Szabo, F. Sussman, and J. -K. Hwang, How do serine proteases really work? *Biochem.* 28:3629, 1989.

Problems

1. (a) In what ways are the mechanistic features of chymotrypsin, trypsin, and elastase similar?
   (b) If the mechanisms of these enzymes are similar, what features of the enzyme active site dictate substrate specificy?
2. (a) If you monitor the lactate dehydrogenase reaction by the formation of NADH (increase in absorbance at 340 nm), should increasing pH make it easier to measure the dehydrogenation of lactate to pyruvate?
   (b) Would a chemical trapping agent for pyruvate serve the same purpose at lower pH values? Why?
   (c) Dehydrogenase activity can be measured by reoxidizing the NADH and reducing a tetrazolium dye. The reduced dye is intensely colored. What are the advantages of measuring the LDH reaction by means of the tetrazolium dye system?
3. For many enzymes, V_{max} is dependent on pH. At what pH would you expect V_{max} of RNase to be optimal? Why?
4. If a lysine were substituted for the aspartate in the trypsin side chain binding crevice, would you expect the enzyme to be functional? If it were functional, what effect would you predict the substitution to have on substrate specificity?
5. Carboxypeptidase A preferentially cleaves C-terminal aromatic residues from proteins. When the aromatic substrate side chain is bound, water is expelled from the active site. How does the release of water stabilize binding of substrate in the active site?
6. You have isolated a metalloenzyme that preferentially cleaves basic amino acids from the carboxyl terminal of proteins. Would you expect the enzyme to retain an arginine in the active site as does carboxypeptidase A? Why or why not? What other residues would you predict to be in the substrate binding site for the new enzyme? How would these residues dictate cleavage specificity?
7. RNase can be completely denatured by boiling or by treatment with chaotropic agents (e.g., urea), yet can refold to its fully active form upon cooling or removal of the denaturant. By contrast, when enzymes of the trypsin family and carboxypeptidase A are denatured, they do not regain full activity upon renaturation. What aspects of trypsin and carboxypeptidase A structure preclude their renaturation to the fully active form?

8. (a) The amino acids in the active site of the protease papain are shown. Predict a feasible reaction mechanism for papain.

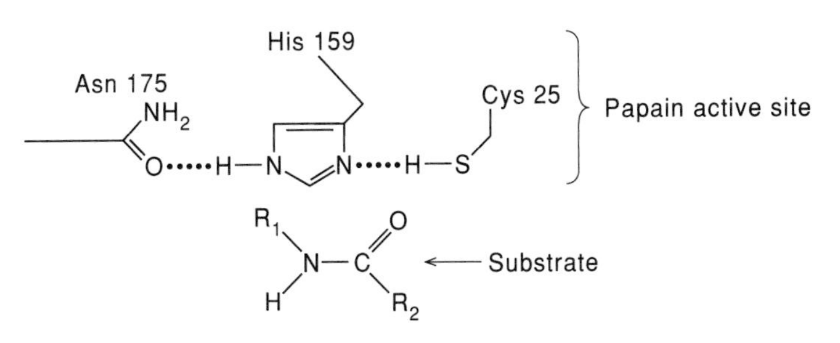

(b) *N*-ethylmaleimide (NEM) reacts rapidly with cysteine thiolate anion via a Michaelis addition. What is the product of the reaction between NEM and cysteine? Would you expect the rate of R–SH reaction with *N*-ethylmaleimide to be more rapid at pH 5 or pH 7.5? Why?

(c) Would you expect cysteine 25 (see part a) to be more reactive with *N*-ethylmaleimide than any of the other cysteine residues in the protein? Explain.

9. Why do structural analogs of the transition-state intermediate of an enzyme inhibit the enzyme competitively and with low K_I values?

10. Transition-state analogs of a specific chemical reaction have been used to elicit antibodies with catalytic activity. These catalytic antibodies have great promise as experimental tools as well as having commercial value. Why is it reasonable to assume that the binding site for the transition-state analog on the antibody would mimic the enzyme active site? What difficulties might be encountered if a catalytic antibody were sought for a reaction requiring a cofactor (coenzyme)?

11. Superoxide dismutases catalyze the reaction

$$O_2^- + O_2^- + 2H^+ \rightarrow O_2 + H_2O_2$$

The catalytic mechanism of the superoxide dismutases involves active site transition metals (Cu, Fe, Mn) that undergo valence changes in the catalytic cycle. Write equations that represent the catalytic cycle of each of these transition metals. (Remember, in each case you must finish with the catalyst in the same state as when you began.) (See Fridovich, I. 1986. *Adv. Enzymol.* 58:61–97.)

12. The superoxide dismutase isolated from most sources has an isoelectric point around 5.
    (a) What problem does the acidic pI of superoxide dismutase present to the catalytic disproportionation of superoxide anion at pH 7? (The pK_a of $HO_2\cdot$ is 4.8.)
    (b) How might strategically placed basic residues assist in catalysis?
    (c) Would the acidic pI of the enzyme or presence of basic residues at the active site impede release of product from the active site? Why?

13. Using site-directed mutagenesis techniques, you have isolated a series of recombinant enzymes in which specific lysine residues were replaced with aspartate residues. The enzymatic assay revealed the following.

Enzyme Form	Activity (U/mg)
Native enzyme	1,000
Recombinant Lys 21 → Asp 21	970
Recombinant Lys 86 → Asp 86	100
Recombinant Lys 101 → Asp 101	970

(a) What might you infer about the role(s) of Lys 21, 86, and 101 in the catalytic mechanism of the native enzyme?
(b) Speculate on the location of Lys 21 and Lys 101. Would you expect these residues to be conserved in an evolutionary sense?
(c) Would you expect Lys 86 to be evolutionarily conserved? Why or why not?

14. You have isolated a microorganism capable of metabolizing the deoxysugar shown below. You find that the compound is phosphorylated by a specific kinase on the C-1 OH group and then the bond between C-3–C-4 is cleaved, yielding dihydroxyacetone phosphate and acetaldehyde. You also find that the cleavage is catalyzed by a zinc-containing enzyme.

CH_2OH
|
C=O
|
CHOH
|
CHOH
|
CH_3

(a) The mechanism of cleavage of the sugar includes generation of a carbanion on C-3. Explain how zinc might stabilize the carbanion. (Consider the role of zinc in the carboxypeptidase A mechanism.)
(b) Would you expect the nonphosphorylated deoxysugar to be a substrate? Why or why not?

15. During catalysis, covalent chemical bonds are broken and formed. In that sense, almost all enzymes perform covalent catalysis. However, the terms "covalent" and "noncovalent" have particular meanings in an enzyme mechanism. Define the difference between those terms as they apply to catalysis.

Regulation of Enzyme Activities

Perhaps the characteristic that most distinguishes living things from inanimate objects is their ability to control their own metabolic activities. Living cells achieve this control by regulating both the concentrations and the activities of specific enzymes. Because the great majority of enzymes are proteins, enzyme concentrations depend on a balance between the rates of protein synthesis and degradation. The synthesis of specific proteins is regulated at several steps, including the synthesis of mRNA and the translation of the RNA message into protein. Rates of protein degradation are controlled by varying the concentrations or activities of other, degradative enzymes, and by modifying particular proteins in a way that tags them for destruction. These controls on protein synthesis and degradation will be discussed in chapters 28, 29, and 30.

Because protein synthesis and turnover are relatively slow processes, adjustments of their rates are used mainly to respond to long-term changes of conditions. The levels of several enzymes that participate in fatty acid biosynthesis, for example, increase markedly within a few days after mammals are switched to a diet that is rich in carbohydrates but low in fats. These enzymes also increase for periods of weeks or months during lactation. But living organisms often need to respond quickly to more sudden changes in conditions, and they do so by regulating enzyme *activities*. This will be the focus of the present chapter.

In principle, the activities of many enzymes could be altered by changes in pH. Cells do take advantage of this possibility in a few cases. Lysozyme, for example, is most active in the pH region of 5 that is characteristic of some extracellular secretions, and is much less active in the intracellular pH region near 7. The activity of lysozyme thus remains low until the enzyme is secreted. But this is not a very practical solution to the problem of regulating the activity of an enzyme that must remain in the cell, because in most cells the intracellular pH must be held within rather narrow limits. There are two strategies that are much more widely applicable. The first is to modify the covalent structure of the enzyme in such a way as to alter either K_m or k_{cat}. The second strategy is to use an inhibitor or activator molecule, an *effector,* that binds reversibly to the enzyme and, again, alters either the K_m or k_{cat}. Such an effector may bind either at the active site itself, or at some more distant site on the enzyme. In the latter case, it is termed an allosteric effector, from the Greek *allos* ("other") and *stereos* ("solid," or "space"), and enzymes that are regulated by such effectors are called allosteric enzymes. Although allosteric effectors usually are small molecules such as ATP, some proteins are inhibited or activated when they bind to another protein.

In the following sections, we will first consider some of the general features of covalent and allosteric control mechanisms, and then turn to a more detailed discussion of how several important enzymes are regulated.

Partial Proteolysis: An Irreversible Covalent Modification

In chapter 9, we mentioned that the pancreas secretes trypsin, chymotrypsin, elastase, and the carboxypeptidases as inactive zymogens, which are activated extracellularly when other proteases cleave them at specific peptide bonds. Trypsin is activated when the intestinal enzyme enterokinase cuts off an N-terminal hexapeptide. Trypsin in turn activates chymotrypsin by cutting it at the N-terminal end between Arg 15 and Ile 16. Delaying the activation in this way prevents the digestive enzymes from destroying the pancreatic cells in which they are synthesized.

The crystal structures of chymotrypsin and its zymogen precursor chymotrypsinogen have provided an explanation for the increase in catalytic activity that results from trimming off the N-terminal end of the zymogen. The sluggishness of chymotrypsinogen can be attributed primarily to a much higher K_m for peptide substrates in the zymogen than in the active enzyme. Although the charge relay system of Asp 102, His 57, and Ser 195 has a similar structure in chymotrypsinogen and chymotrypsin, the substrate-binding pocket is not properly formed in the zymogen (fig. 10.1*a*). The NH group of Gly 193 is not in position to form a hydrogen bond with the carbonyl oxygen of the substrate. The importance of this bond for the activity of the enzyme was discussed in chapter 9 (see figs. 9.7 and 9.11). The major constraint preventing the completion of the binding pocket in the zymogen appears to be a hydrogen bond between the negatively charged carboxylate group of the neighboring residue, Asp 194, and His 40. Rotation of Gly 193 into the correct orientation occurs when the zymogen is cleaved between Arg 15 and Ile 16 and the new N-terminal —NH_3^+ group of Ile 16 forms a salt bridge with Asp 194 (see fig. 10.1*b*). A similar salt bridge between the N-terminal group and Asp 194 occurs in trypsin.

The enzymes that participate in blood clotting also are activated by partial proteolysis, and again this serves to keep them in check until they are needed. The blood-coagulation system involves a cascade of at least seven serine proteases, each of which activates the subsequent enzyme in the series (fig. 10.2). Because each molecule of enzyme that is activated can, in turn, activate many molecules of the next enzyme, initiation of the process by factors that are exposed in damaged tissue leads explosively to the conversion of prothrombin to thrombin, the final serine protease in the series. Thrombin then cuts another protein, fibrin, into peptides that stick together to form a clot.

Figure 10.1

A schematic drawing of the structural changes that occur when chymotrypsinogen is converted to chymotrypsin. In chymotrypsinogen, the carboxylate group of Asp 194 forms a salt bridge to His 40; in chymotrypsin, the bridge goes to the new N-terminal—NH_3^+ group of Ile 16. This change evidently allows Gly 193 to swing around so that its amide NH comes closer to the NH of Ser 195. The two amide groups form essential hydrogen bonds to the substrate in the enzyme-substrate complex (see figs. 9.7 and 9.11).

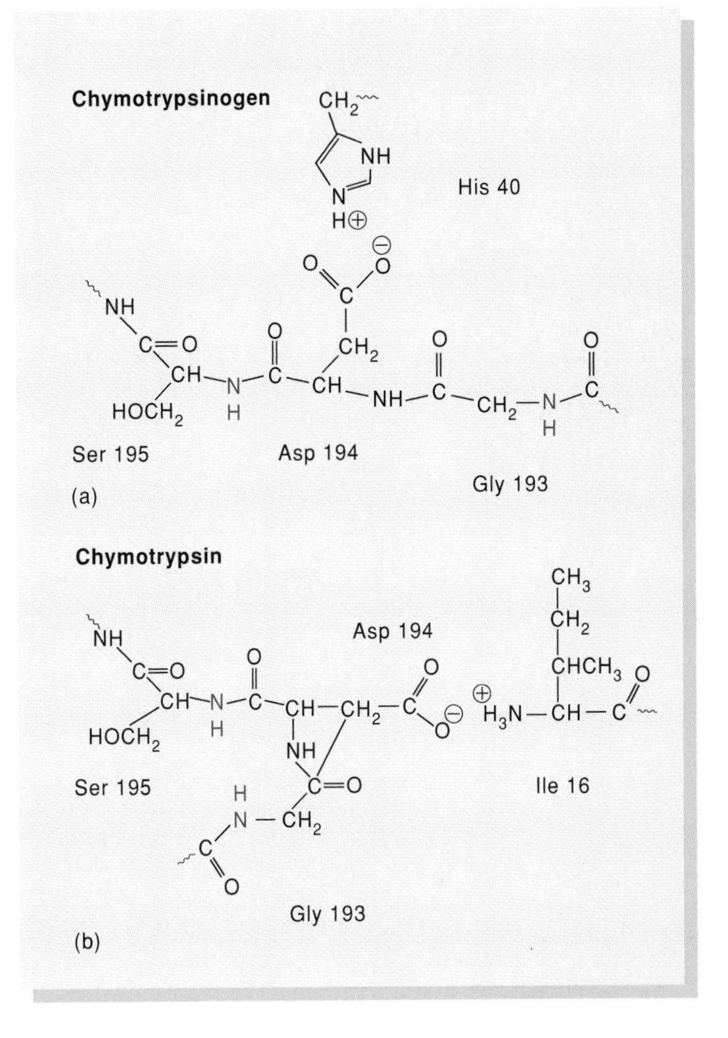

Table 10.1 lists some other enzymes that are activated by partial proteolysis. In general, the peptide bond that is cleaved is located in a loop connecting two different domains of the protein, and cutting this bond probably relieves a constraint that interferes with the formation of the active site, much as it does in chymotrypsin or trypsin.

Figure 10.2

The blood-coagulation cascade. Each of the curved red arrows represents a proteolytic reaction, in which a protein is cleaved at one or more specific sites. With the exception of fibrinogen, the substrate in each reaction is an inactive zymogen; except for fibrin, each product is an active protease that proceeds to cleave another member in the series. Many of the steps also depend on interactions of the proteins with Ca^{2+} ions and phospholipids. The cascade starts when factor XII and prekallikrein come into contact with materials that are released or exposed in injured tissue. (The exact nature of these materials is still not fully clear.) When thrombin cleaves fibrinogen at several points, the trimmed protein (fibrin) polymerizes to form a clot.

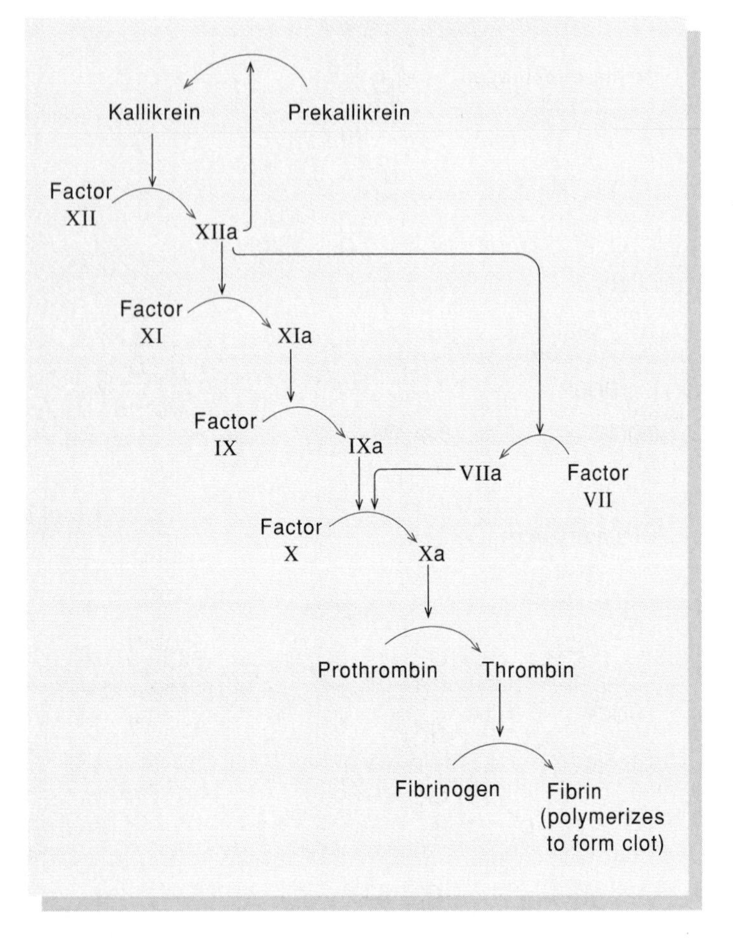

The activation of an enzyme by partial proteolysis is an irreversible process. Once the enzyme is cut, it remains active until it is degraded or inhibited by some other means. This is fine for the digestive enzymes, but it clearly raises a problem in blood coagulation: There must be a mechanism to prevent the clot from spreading away from the site of the injury and taking over the entire blood supply. Several mechanisms work to this end. The activated enzymes are diluted by the flow of the blood, degraded in the liver, and inhibited by other blood proteins that bind tightly to the active enzymes and occlude their active sites. Blood coagulation thus depends on a delicate balance between a rapid, irreversible mechanism of enzyme activation, and rapid, irreversible mechanisms for disposing of the active enzymes. We will return later to the role that specific inhibitory proteins play in maintaining this balance.

Table 10.1
Some Enzymes that Are Regulated by Partial Proteolysis

Digestive Enzymes
Trypsin, chymotrypsin, carboxypeptidase A and B, elastase, pepsin, phospholipase
Blood-Coagulation Enzymes
Factors VII, IX, X, XI, and XIII, kallikrein, thrombin
Enzymes Involved in Dissolving Blood Clots
Plasminogen, plasminogen activator
Enzymes Involved in Programmed Development
Chitin synthetase, cocoonase, collagenase

Phosphorylation, Adenylylation, and Disulfide Reduction: Reversible Covalent Modifications

A type of covalent modification that is used more frequently than partial proteolysis is phosphorylation of the side chains of serine, threonine, or tyrosine residues. Phosphorylation differs from partial proteolysis in being reversible. The introduction and removal of the phosphate group are catalyzed by separate enzymes (phosphorylation by a protein kinase, and dephosphorylation by a phosphatase), which are themselves generally under metabolic regulation (fig. 10.3).

In eukaryotic organisms, phosphorylation is used to control the activities of literally hundreds of enzymes, a few of which are listed in table 10.2. These enzymes generally are phosphorylated, or dephosphorylated, in response to *extracellular* signals such as hormones or growth factors. The adrenal hormone epinephrine, for example, is transmitted through the blood to muscle and adipose tissue when there is an immediate need for muscular exertion. On reaching the target tissue, epinephrine initiates a chain of events that lead to the activation of a protein kinase. The kinase (cAMP-dependent protein kinase) then catalyzes the phosphorylation of one or more specific enzymes, depending on the tissue. Some enzymes are activated when they are phosphorylated, and inactivated when the phosphate is removed; others are inactivated by phosphorylation. In adipose tissue, phosphorylation switches on triacylglycerol lipase, an enzyme that breaks down esters of fatty acids (triacylglycerols). In muscle, phosphorylation activates glycogen phosphorylase, an enzyme that breaks down glycogen, and it stops the synthesis of glycogen by switching off glycogen synthase.

Phosphorylation also can increase or decrease an enzyme's sensitivity to particular allosteric effectors. Phosphorylation of glycogen synthase greatly increases its sensitivity to inhibition by glucose-6-phosphate or P_i. Phosphorylation of glycogen phosphorylase makes this enzyme much less dependent

Figure 10.3

Phosphorylations of serine side chains in enzymes are catalyzed by kinases, and dephosphorylation by phosphatases. Threonine and tyrosine side chains undergo similar reactions, but these are less common than the phosphorylation of serine.

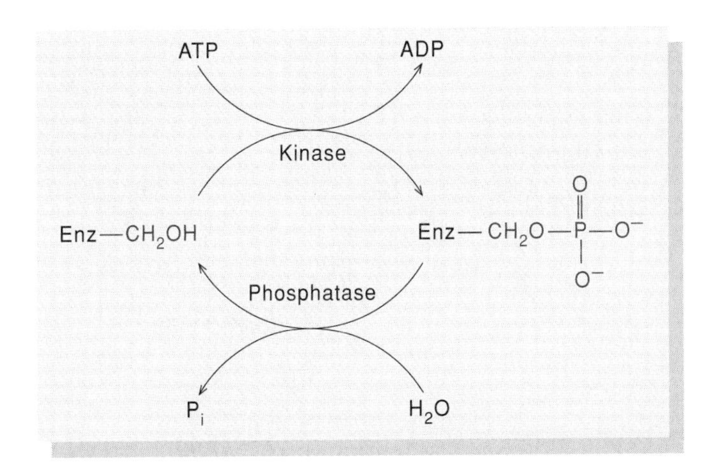

Table 10.2 Some Enzymes that Are Regulated by Phosphorylation
Enzymes of Carbohydrate Metabolism
Glycogen phosphorylase
Phosphorylase kinase
Glycogen synthase
Phosphofructokinase-2
Pyruvate kinase
Pyruvate dehydrogenase
Enzymes of Lipid Metabolism
Hydroxymethylglutaryl-CoA reductase
Acetyl-CoA carboxylase
Triacylglycerol lipase
Enzymes of Amino Acid Metabolism
Branched-chain ketoacid dehydrogenase
Phenylalanine hydroxylase
Tyrosine hydroxylase

on the allosteric activator AMP. Thus a covalent modification triggered by an extracellular signal can, in some cases, override the influence of intracellular allosteric regulators; the response is relatively insensitive to the exact concentrations of the local agents. In other cases, variations in the concentrations of intracellular allosteric effectors can cause the response to the covalent modification to vary, depending on the metabolic state of affairs in the cell.

As yet, glycogen phosphorylase is the only enzyme for which crystal structures of both the phosphorylated and the nonphosphorylated forms are known. Because glycogen phosphorylase also is regulated by allosteric effectors, we will postpone a discussion of this complex enzyme until after we have considered some simpler examples of allosteric regulation. The manner in which hormones lead to an increase in the activity of cAMP-dependent protein kinase by increasing the concentration of its allosteric effector 3′,5′-cyclicAMP (cAMP) is discussed in chapters 13 and 24.

The covalent addition of an adenylyl (adenosine monophosphate) group to a tyrosine residue is another form of reversible, covalent modification (fig. 10.4). In *E. coli,* adenylylation is used to regulate glutamine synthase, an enzyme that plays a major role in nitrogen metabolism (see chapter 18). The tyrosine residue that accepts the adenylyl group on glutamine synthase is located close to the active site. The addition of the bulky and negatively charged adenylyl group inhibits the enzyme, perhaps simply by occluding the active site.

Other groups that can be attached covalently to enzymes include fatty acids, isoprenoid alcohols such as farnesol, and carbohydrates. Although such modifications appear to be common, our understanding of how cells use them to regulate enzymatic activities is still fragmentary.

A reversible covalent modification used extensively in plants is the reduction of cystine disulfide bridges to sulfhydryls. Many of the enzymes that participate in photosynthetic carbohydrate synthesis are activated when a disulfide bond is reduced to produce two cysteines (table 10.3). Some of the enzymes involved in carbohydrate breakdown are inactivated by the same mechanism. The reductant for this reaction is a small protein called thioredoxin, which contains two nearby cysteine residues (fig. 10.5). In plants, reduced thioredoxin becomes available as a result of electron-transfer reactions that are driven by sunlight, and it serves as a signal to switch carbohydrate metabolism from carbohydrate breakdown to synthesis. In one of the enzymes that are regulated (phosphoribulokinase), the cysteines that are freed when the disulfide bridge is reduced are spaced 39 amino acid residues apart, and one of them probably forms part of the catalytic active site. In two other cases (NADP-malate dehydrogenase, and fructose-1,6-bisphosphatase), the cysteines are spaced only four or five residues apart, but both appear to be located at some distance from the catalytic site. Exactly how the structural changes that result from the reduction are transmitted to the active sites in these enzymes is not yet known.

Covalent modifications of specific enzymes allow a cell to regulate its metabolic activities more rapidly, and in much more intricate ways, than would be possible by changing the absolute concentrations of the same enzymes. They still do not provide a mechanism for a truly instantaneous response to a change in conditions, however, because each such modification requires the action of another enzyme, which must itself be subject to regulation. There also is a lag in responding to the removal or inactivation of the enzyme that causes the modification, because reversing the modification requires still another enzyme.

Figure 10.4

The transfer of the adenylyl group from ATP to a tyrosine residue is used to regulate some enzymes. The other product of the adenylylation reaction (*top*) is inorganic pyrophosphate. Hydrolysis of the adenylyl-tyrosine ester bond releases AMP (*bottom*).

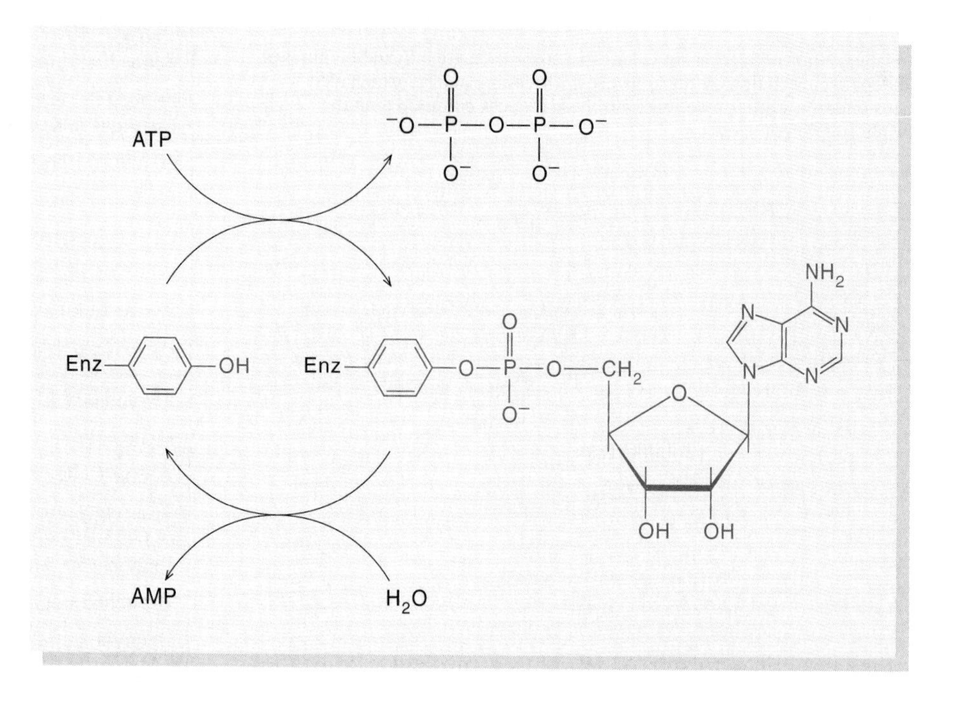

Table 10.3
Some Enzymes that Are Regulated by Disulfide-Reduction in Plants

Activated by Reduction
Fructose-1,6-bisphosphatase
Sedoheptulose-1,7-bisphosphatase
Glyceraldehyde-3-phosphate dehydrogenase
NADP-malate dehydrogenase
Phosphoribulokinase
Thylakoid ATP-synthase
Inhibited by Reduction
Phosphofructokinase

Figure 10.5

Reduction of the disulfide bond of cystine is used to activate enzymes of photosynthetic carbohydrate biosynthesis in plants. The reductant is a small protein called thioredoxin. Thioredoxin also serves as a reductant for the biosynthesis of deoxynucleotides in animals and microorganisms as well as in plants.

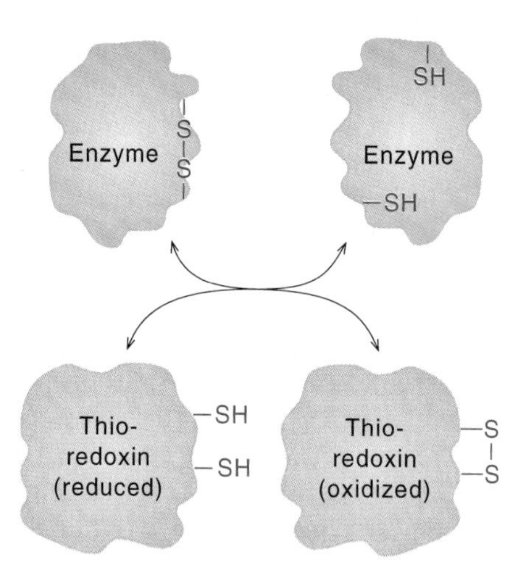

Allosteric Regulation

Whereas eukaryotic organisms generally use phosphorylation to handle responses to extracellular signals, both prokaryotic and eukaryotic organisms commonly use allosteric regulation in responding to changes in conditions within a cell. A typical circumstance that might demand such a response would be a surplus or deficit of ATP, a particular amino acid, or some other metabolic intermediate that the cell is equipped to synthesize or consume. Allosteric regulation enables a cell to adjust a specific enzymatic activity almost instantaneously in response to a change in the concentration of a particular metabolic intermediate because, unlike covalent modification, it does not require the action of an intermediate enzyme. If the intermediate acts as an allosteric effector, the activity of the enzyme can increase (or decrease) as soon as the concentration of the effector rises, and can decrease (or increase) again as soon as the concentration falls.

Regulation of enzymes by allosteric effectors is considerably more common than regulation by compounds that bind directly at the active site. There probably are two main reasons for this. First, whereas an agent that binds at the active site will usually act as an inhibitor, a compound that binds at an allosteric site can act as either an inhibitor or an activator, depending on the structure of the enzyme. Second, a substance that binds to an allosteric site does not need to have any structural relationship to the substrate. Consider, for example, the metabolic pathway of histidine biosynthesis in plants and bacteria. The pathway requires nine enzymes that work one after another. If histidine is already present in abundant supply, it is advantageous for a cell to cut off the entire pathway at the first step to avoid wasting energy or accumulating the products of the intermediate steps. The first step in the pathway is the reaction between phosphoribosyl pyrophosphate and ATP, neither of which even vaguely resembles histidine, and yet the enzyme that catalyzes this step is strongly inhibited by histidine. Binding of histidine to an allosteric site on the enzyme causes a structural change that is transmitted to the active site.

The control of the pathway leading to histidine illustrates a recurrent theme in enzyme regulation: Enzymes that are regulated often occupy key positions in metabolic pathways. Typical control points are the first step of a pathway, or the first step of a branch leading to an alternate product. Many key enzymes are regulated by ATP, ADP, AMP, or P_i. The relative concentrations of these materials provide a cell with an index as to whether energy is abundant or in short supply. Because ATP, ADP, AMP, and P_i often are chemically unrelated to the substrate of the enzyme that must be regulated, they generally bind to an allosteric site rather than to the active site.

The Kinetics of Allosteric Enzymes Typically Exhibit a Sigmoidal Dependence on Substrate Concentration

In chapter 5 we saw that the binding of glycerate-2,3-bisphosphate to hemoglobin causes a decrease in the affinity of the protein for O_2, and that the binding of O_2 to any one of the four subunits increases the affinity of the other subunits for O_2. These effects can be related to cooperative changes in the tertiary and quaternary structure of the protein. The binding of O_2 alters the interactions between the different subunits in such a way that the entire protein tends to flip into a state with increased O_2 affinity; binding of glycerate-2,3-bisphosphate favors a transition in the opposite direction. When Jacques Monod, Jeffreys Wyman, and Jean-Pierre Changeux first advanced the idea of allosteric enzymes in 1963, they suggested that these enzymes might often contain multiple subunits, and that the changes in catalytic activity resulting from the binding of allosteric effectors might reflect alterations in quaternary structure. This suggestion has turned out to be remarkably accurate. Although there is no reason why an enzyme consisting of a single subunit cannot be sensitive to allosteric effectors, most of the enzymes that are regulated in this way do have multiple subunits, and in many cases the changes in enzymatic activity can be related to interactions among the subunits.

One indication that allosteric effectors often involve cooperative interactions within multisubunit proteins is that many allosteric enzymes do not obey the classical Michaelis-Menten kinetic equation. A plot of the rate of reaction as a function of substrate concentration is not hyperbolic, as described by the Michaelis-Menten equation, but rather sigmoidal, resembling the curve for the cooperative binding of O_2 to hemoglobin (see fig. 5.3). Further, allosteric effectors often cause the kinetics to change from one of these forms to the other, much as glycerate-2,3-bisphosphate, bicarbonate, and protons affect the degree of cooperativity in the binding of O_2 to hemoglobin. Figure 10.6 shows an illustration of these effects for phosphofructokinase, an enzyme that catalyzes the formation of fructose-1,6-bisphosphate by transferring a phosphate group from ATP to fructose-6-phosphate:

$$\text{Fructose-6-phosphate} + \text{ATP} \xrightarrow{\text{phosphofructokinase}} \text{fructose-1,6-bisphosphate} + \text{ADP} \quad (1)$$

In the presence of 1.5-mM ATP, the kinetics have a sigmoidal dependence on the concentration of the substrate fructose-6-phosphate; at very low concentrations of ATP, or in the presence of the allosteric effector AMP, the kinetics become hyperbolic.

Figure 10.6

Kinetics of the reaction catalyzed by phosphofructokinase. In the presence of 1.5 mM ATP, the rate has a sigmoidal dependence on the concentration of the substrate fructose-6-phosphate. Although ATP also is a substrate for the reaction, the sigmoidal kinetics seen under these conditions are associated with the binding of ATP to an inhibitory allosteric site. The kinetics become hyperbolic if a low concentration of AMP is added.

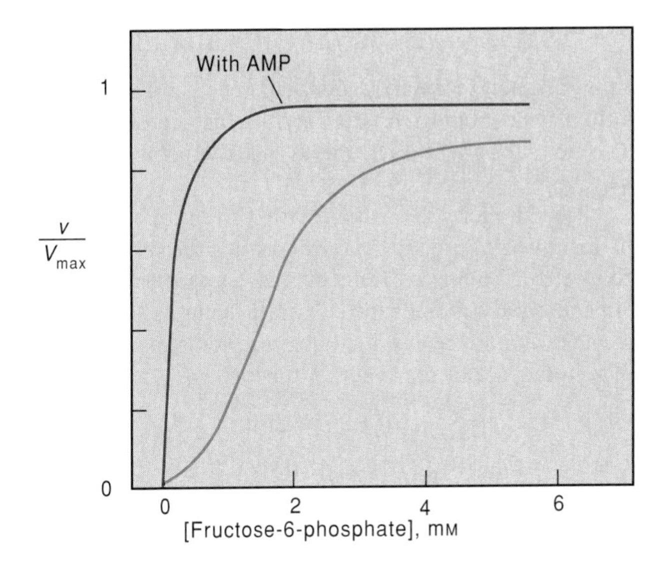

Phosphofructokinase has four identical subunits. To explore how sigmoidal kinetics can arise in such an enzyme, consider an enzyme that has just two such subunits, each with its own catalytic active site. The binding of a substrate to the enzyme can be schematized as follows:

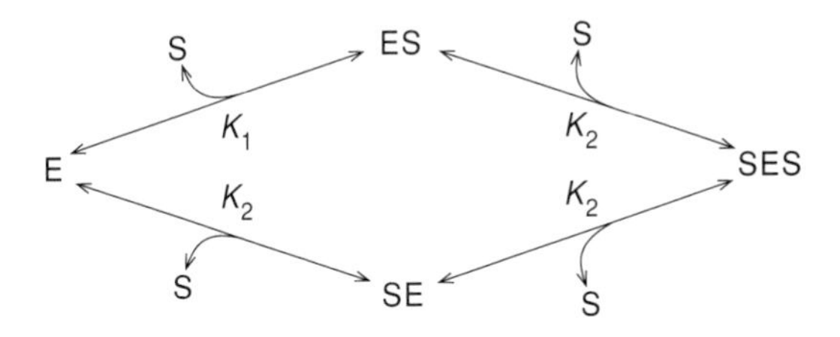

Here ES and SE represent the binary complexes of the enzyme (E) with substrate (S) on the two different subunits, and SES represents the ternary complex in which both binding sites are occupied. Using the dissociation constants K_1 and K_2, we can relate the concentrations of any of these complexes to the concentrations of the free enzyme and substrate:

$$[ES] = [SE] = [E][S]/K_1 \quad (2)$$

$$[SES] = [ES][E]/K_2 = [SE][S]/K_2 = [[E][S]/K_1][[S]/K_2] \quad (3)$$

K_1 and K_2 are not necessarily identical because the binding of substrate to one subunit could affect the dissociation constant for the other subunit. In fact, this is just the point we want to explore.

For simplicity, let's assume that the rate constant k_{cat} for the formation of products at either site is the same, whether only that site or both sites are occupied. Suppose also that the binding and dissociation of the substrate occur rapidly relative to the conversion of the enzyme-substrate complexes into products. The total rate of the reaction then will be

$$v = k_{cat}\{[ES] + [SE] + 2[SES]\}$$
$$= 2k_{cat}\{[E][S]/K_1 + [[E][S]/K_1][[S]/K_2]\} \quad (4)$$

The factor of 2 reflects the fact that the formation of products occurs independently (and with the same rate constant) at the two sites. Because the total enzyme concentration $[E_T]$ is [E] + [ES] + [SE] + [SES], the maximum rate of the reaction is

$$V_{max} = 2k_{cat}[E_T] = 2k_{cat}\{[E] + [E][S]/K_1 + [E][S]/K_1 + [[E][S]/K_1][[S]/K_2]\} \quad (5)$$

Combining equations (4) and (5) gives

$$\frac{v}{V_{max}} = \frac{\frac{[S]}{K_1}\left[1 + \frac{[S]}{K_2}\right]}{\left[1 + \frac{[S]}{K_1}\right] + \frac{[S]}{K_1}\left[1 + \frac{[S]}{K_2}\right]} \quad (6)$$

If the binding of S to one subunit does *not* affect binding to the other, then $K_2 = K_1$, and equation (6) reduces to

$$\frac{v}{V_{max}} = \frac{[S]}{K_d + [S]} \quad (7)$$

where $K_d = K_1 = K_2$. This is simply the Michaelis-Menten equation for the limiting case in which substrate release is much faster than the conversion of the enzyme-substrate complex into products. (See equations 24 and 28 in chapter 8.) A plot of v versus [S] according to equation (7) is hyperbolic. Such a plot is shown again as the solid curve in figure 10.7. The dashed curve in figure 10.7 is a similar plot of equation (6) for the case $K_2 = K_1/25$. To facilitate comparison with the solid curve, K_1 is taken to be $5K_d$ and K_2 to be $K_d/5$, where K_d has the same value for the two curves. This means that the overall free energy change for the formation of the ternary complex SES from E + 2S is the same. (The overall equilibrium constant for this process is $(1/K_1)(1/K_2)$, or $1/K_d^2$. Remember that the standard free energy change is $-2.3RT$ times the $\log_{10}$ of the equilibrium constant, where T is the temperature and $R = 2$ cal/mole/degree.) Note that the dashed curve is decidedly sigmoidal in shape. It starts out rising more slowly than the solid curve, rises steeply in the region of $[S] \approx K_d$, where $v/V_{max} \approx 0.5$, and then continues rising more slowly.

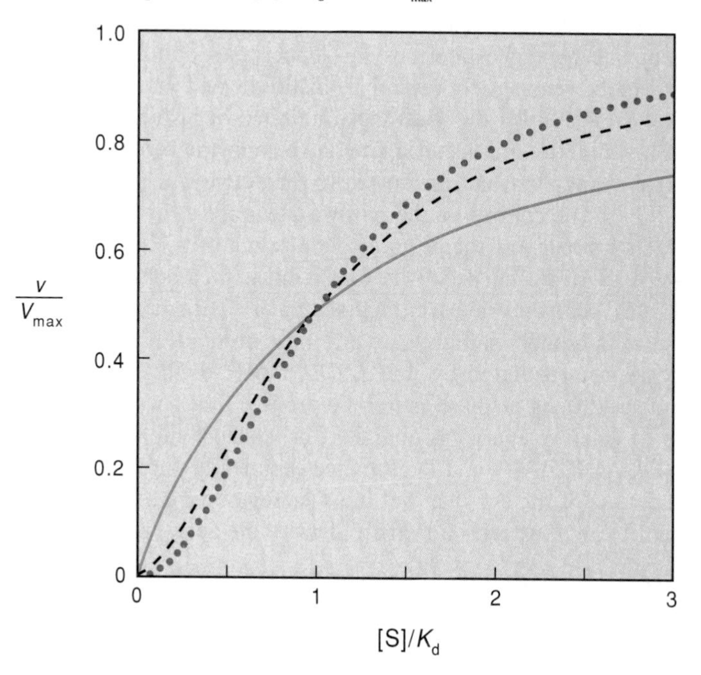

Figure 10.7

Kinetics of an enzyme with two identical subunits. The curves were calculated with equation (6). (Substrate dissociation is assumed to be rapid relative to the catalytic step.) The abscissa is the ratio of the substrate concentration [S] to K_d, where K_d is the geometric mean of K_1 and K_2 ($K_d = \sqrt{K_1 K_2}$). K_d is taken to be the same for all three curves. For the solid curve, the dissociation constants K_1 and K_2 for the first and second molecule of substrate were assumed to be identical ($K_1 = K_2 = K_d$); equation (6) then reduces to equation (7). For the dashed curve, $K_2 = K_1/25$ ($K_1 = 5K_d$ and $K_2 = K_d/5$). For the dotted curve, $K_2 = 10^{-4}K_1$ ($K_1 = 100K_d$ and $K_2 = K_d/100$); equation (6) then reduces to equation (8). The three curves intersect at the point where $[S] = K_d$ and $v = V_{max}/2$.

The ratio of 25 between the dissociation constants K_1 and K_2 used in this illustration means that the standard free energy change for binding the second molecule of substrate to the enzyme is only 1.9 kcal/mole more favorable than that for binding the first molecule. This is the order of magnitude of the free energy change associated with forming a single hydrogen bond in the interior of a protein. Evidently, a sigmoidal kinetic curve similar to the dashed curve in figure 10.7 might be obtained if binding the substrate caused a relatively minor change in the conformation of the protein. In chapter 9, we noted that the binding of a substrate or coenzyme to hexokinase, carboxypeptidase A, or lactate dehydrogenase results in pronounced conformational changes that can be seen in the crystal structures. But in order to yield sigmoidal kinetics, it is essential that events occurring at two different binding sites for the substrate be coupled: Binding at one site must cause a decrease in the

Figure 10.8

Symmetry model for allosteric transitions of a dimeric enzyme. The model assumes that the enzyme can exist in either of two different conformations (T and R), which have different dissociation constants for the substrate (K_T and K_R). Structural transitions of the two subunits are assumed to be tightly coupled, so that both subunits must be in the same state. L is the equilibrium constant (T)/(R) in the absence of substrate. If the substrate binds much more tightly to R than to T ($K_R << K_T$), the binding of a molecule of substrate to either subunit will pull the equilibrium between T and R in the direction of R. Because both subunits must go to the R state, the binding of the second molecule of substrate will be promoted.

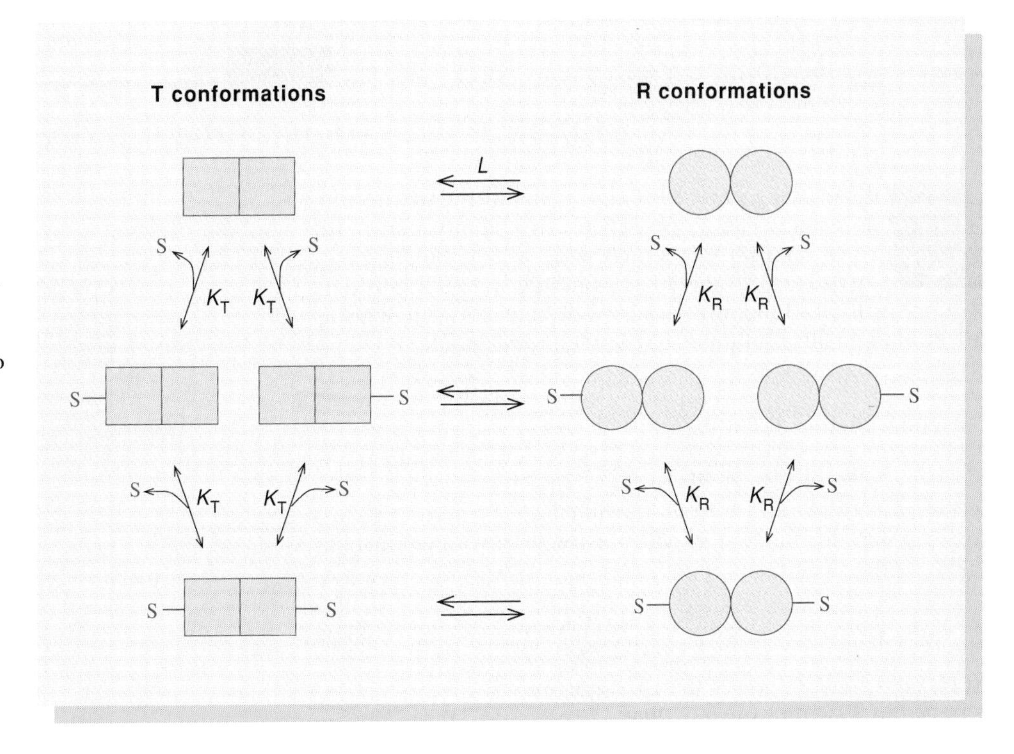

dissociation constant at the other site. If there is no such coupling, the kinetics will follow equation (7) and will be hyperbolic even if the enzyme does have multiple subunits. This situation is observed with many enzymes, including several that we discussed in chapter 9. Thus, although lactate dehydrogenase exists as a tetramer of four identical subunits, and triosephosphate isomerase exists as a dimer, the kinetics in both cases are hyperbolic.

The dotted curve in figure 10.7 shows a plot of equation (6) with $K_2 = 10^{-4} K_1$, which is equivalent to a difference of 5.5 kcal/mole between the $\Delta G°$ values for binding the first and second molecules of substrate. When the ratio of the dissociation constants is this large, the kinetics can be described equally well by the simpler expression

$$\frac{v}{V_{max}} = \frac{[S]^2}{K_1K_2 + [S]^2} \quad (8)$$

This is the limiting form of equation (6) for the situation $K_2 << [S] << K_1$. It is essentially the same as the equation that we used in chapter 5 to describe the cooperative binding of O_2 to hemoglobin, except that the exponent for [S] was larger than 2 in that case because the binding of O_2 to hemoglobin involves cooperative interactions among four subunits. The general equation of this form, with an arbitrary exponent n, is called the Hill equation, and the exponent is referred to as the Hill coefficient (see box 10A).

Although we have considered an enzyme with just two subunits, it is not uncommon for enzymes that are regulated allosterically to have four or even more subunits. The active form of acetyl-CoA carboxylase, for example, consists of linear strings of ten or more identical subunits. In general, the larger the number of subunits that interact in a manner such that the binding of substrate to one subunit promotes binding to others, the more steeply the enzyme's kinetics will depend on [S] in the region where $v/V_{max} \approx 0.5$. This point is developed in more detail in box 10A. We will return to sigmoidal enzyme kinetics in chapter 12, where we consider their implications for the regulation of metabolic pathways. At the moment, our concern is to explore what types of structural changes can couple the binding of substrates on different subunits of a protein and how this cooperativity can be modified by allosteric effectors.

The Symmetry Model Provides a Useful Framework for Relating Conformational Transitions to Allosteric Activation or Inhibition

Several theoretical models have been developed for relating changes in dissociation constants to changes in the tertiary and quaternary structures of oligomeric proteins. Two such models were discussed in chapter 5 in connection with the oxygenation of hemoglobin (see fig. 5.15), and one of them is shown again in a slightly different form in figure 10.8. The basic idea is that the protein's subunits can exist in either of two distinct conformations, R and T. The substrate is assumed to bind more tightly to the R form than to the T form, which means that binding of the substrate favors the transition from the T conformation to R. The conformational transitions of the individual subunits are assumed to be tightly linked, so that if one subunit flips from T

10A BOX

Graphical Evaluation of the Hill Coefficient

Cooperative interactions in proteins with multiple subunits often can be described conveniently by the Hill equation. Consider a protein, E, with n identical subunits, each of which has a binding site for a substrate, S. Suppose that the binding of the first molecule of S strongly favors the binding to all n subunits, giving the overall reaction

$$\mathrm{E} + n\,\mathrm{S} \rightleftharpoons \mathrm{ES}_n \tag{B1}$$

The concentration of ES_n then will be simply

$$[\mathrm{ES}_n] = [\mathrm{E}][\mathrm{S}]^n/K_\mathrm{h} \tag{B2}$$

where K_h is the product of the individual dissociation constants for all n steps leading to ES_n. The fraction of the protein that has taken up the substrate is

$$y \approx \frac{[\mathrm{ES}_n]}{[\mathrm{E}] + [\mathrm{ES}_n]} = \frac{[\mathrm{S}]^n}{K_\mathrm{h} + [\mathrm{S}]^n} \tag{B3}$$

This is the general form of the Hill equation.

Figure 1 shows plots of equation (B3) for $n = 1$, 2, and 4. The curve for $n = 1$ has the same hyperbolic shape as the solid curve in figure 10.7, and the curve for $n = 2$ has essentially the same sigmoidal shape as the dotted curve in that figure. Increasing the value of n to 4 makes the curve rise still more steeply in the region where $[\mathrm{S}] \approx K_\mathrm{h}$.

The Hill equation can be rearranged to

$$\frac{y}{1-y} \approx \frac{[\mathrm{S}]^n}{K_\mathrm{h}} \tag{B4}$$

By taking the logarithm of both sides of this expression we find

$$\log\frac{y}{1-y} \approx n \log[\mathrm{S}] - \log K_\mathrm{h} \tag{B5}$$

A plot of $\log[y/(1-y)]$ versus $\log[\mathrm{S}]$ thus should give a straight line with a slope of n. A set of such plots for $n = 1$, 2, and 4 is shown in figure 2. For an enzymatic reaction in which binding and dissociation of the substrate are rapid relative to the catalytic step, the quantity y is equivalent to v/V_{max}.

Equation (B3), like equation (8), is an approximation and not an exact expression; the denominator neglects the concentrations of the intermediate species that have some, but not all, of the binding sites occupied. However, plots like that shown in illustration 2 often are used to find an effective value of the Hill coefficient as a phenomenological measure of the cooperativity in the binding. The binding of O_2 to hemoglobin, for example, can be described reasonably well by taking n to be approximately 2.8. This value is obtained by plotting $\log[y/(1-y)]$ versus log [S] and taking the slope at the point where $y = 0.5$. In the case of phosphofructokinase, which has four subunits, the dependence of the rate on the concentration of fructose-6-phosphate at 1 mM ATP is described well by the Hill equation with $n \approx 3.8$.

Figure 1

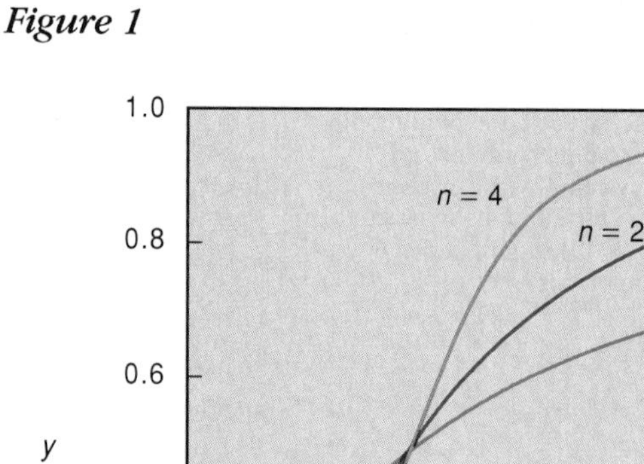

Figure 2

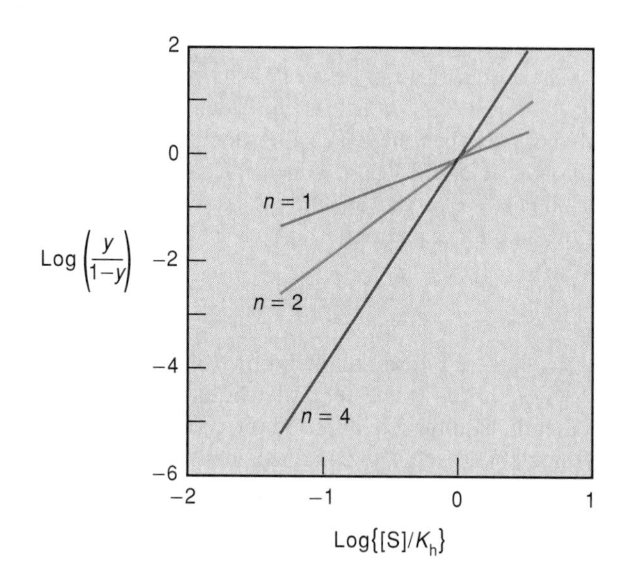

to R the others must do the same. The binding of the first molecule of substrate thus promotes the binding of the second. Because the concerted transition of all of the subunits from T to R or back preserves the overall symmetry of the protein, this model is called the symmetry model. The symmetry model can be elaborated to include allosteric activators by assuming that these compounds also bind preferentially to the R state and thus stabilize the conformation that is more effective at binding the substrate. Allosteric inhibitors would act by stabilizing the T state.

Given the symmetry model, the fraction of the binding sites that will be occupied at any given substrate concentration can be described mathematically with an expression that includes the substrate dissociation constants K_R and K_T for the two conformations and the equilibrium constant between the T and R conformations in the absence of substrate, $L = (T)/(R)$. Thus the symmetry model attempts to explain the difference between the two dissociation constants K_1 and K_2 in equation (6) by introducing a third independent parameter.

Considering that equation (6) would fit the experimental data for a dimeric enzyme satisfactorily with only two parameters, what do we gain by using a more complicated equation? The usefulness of the symmetry model is that it can easily be generalized to enzymes that have a larger number of subunits. Although the general expression for the extent of binding is more complex than equation (6), it still contains only four independent parameters: K_R, K_T, L, and the total number of subunits (n). But this simplicity comes with a cost: We have to assume that the entire oligomeric protein is restricted to just two different conformational states. Would it not be more realistic to assume that the protein also can adopt a variety of intermediate states? The answer depends in part on the particular protein, and in part on our goals in using any type of model. Models that allow additional conformational states have been developed, but they of course require even more independent parameters. The sequential model shown in figure 5.15*b* is one of these more elaborate models. Although the sequential model is more general than the symmetry model, and thus is probably more realistic, the available experimental data in most cases do not justify a distinction between the two.

The symmetry model is useful even if it does oversimplify the situation, because it provides a conceptual framework for discussing the relationships between conformational transitions and the effects of allosteric activators and inhibitors. In the following sections we will consider three oligomeric enzymes that are under metabolic control, and we will see that substrates and allosteric effectors do tend to stabilize each of these enzymes in one or the other of two distinctly different conformations.

Phosphofructokinase: Allosteric Control of Glycolysis Is Consistent with the Symmetry Model

Phosphofructokinase catalyzes the transfer of a phosphate group from ATP to the —OH group on carbon 1 of fructose-6-phosphate (equation 1). This is the major site of regulation of glycolysis, the metabolic pathway by which glucose breaks down to pyruvate (see chapter 13). As we saw in figure 10.6, the kinetics of phosphofructokinase are strongly cooperative with respect to the substrate fructose-6-phosphate. The kinetics are noncooperative with respect to the other substrate, ATP, at low concentrations, but at concentrations in the range of 0.5 to 1 mM or higher, ATP acts as an inhibitor. The inhibitory effect results from binding to an allosteric site that is distinct from the substrate-binding site for ATP. Phosphofructokinase also is inhibited by phosphoenolpyruvate and by citrate, a key intermediate in two other metabolic pathways that embark from pyruvate. On the other hand, it is *stimulated* by ADP, AMP, GDP, cAMP, fructose-2,6-bisphosphate, and a variety of other compounds, depending to some extent on the organism from which the enzyme is purified. Most if not all of the activators bind to the same allosteric site as ATP, and may work simply by preventing the inhibitory effect of ATP. The effects of ATP, ADP, and AMP are such that phosphofructokinase is restrained when the cell's needs for energy have been satisfied, as reflected in a high ratio of [ATP] to [ADP] and [AMP], and is unleashed when additional energy is needed. The effects of cAMP and fructose-2,6-bisphosphate relate to the hormonal control of carbohydrate metabolism in higher organisms, and will be discussed further in chapters 13 and 24. Our focus here will be on how fructose-6-phosphate and some of the major allosteric effectors alter the structure of the enzyme.

Phosphofructokinase was one of the first enzymes to which Monod and his colleagues applied the symmetry model of allosteric transitions. It contains four identical subunits, each of which has both an active site and an allosteric site. The very high cooperativity of the kinetics suggests that the enzyme can adopt two different conformations (T and R) that have similar affinities for ATP but differ markedly in their affinity for fructose-6-phosphate. The binding for fructose-6-phosphate is calculated to be about 2,000 times tighter in the R conformation than in T. When fructose-6-phosphate binds to any one of the subunits, it appears to cause all four subunits to flip from the T conformation to R, just as the symmetry model specifies. The allosteric effectors ADP, GDP, and phosphoenolpyruvate do not alter the maximum rate of the reaction, but change the dependence of the rate on the fructose-6-phosphate concentration in a manner suggesting that they change the equilibrium constant (L) between the T and R conformations.

Philip Evans and his co-workers have determined the crystal structures of phosphofructokinase from two species of bacteria, *E. coli* and *Bacillus stearothermophilus*. By crystallizing the enzyme in the presence and absence of the substrate and several allosteric effectors, they obtained detailed views of both the T and R conformations. This work led to an explanation of why phosphofructokinase appears to be constrained largely to all-or-nothing transitions between these two states, rather than adopting a series of intermediate conformations.

Figure 10.9 shows the crystal structure of one of the four identical subunits of phosphofructokinase from *B. stearothermophilus*. In the complete enzyme, the subunits are disposed symmetrically about three mutually perpendicular axes.

Figure 10.9

Structure of phosphofructokinase from *Bacillus stearothermophilus*. The figure shows the α-carbon chain of one of the four identical subunits of the enzyme. It is based on the crystal structure described by P. R. Evans and P. J. Hudson. The enzyme was crystallized in the R conformation in the presence of the substrate fructose-6-phosphate (yellow) and the allosteric activator ADP (blue). A molecule of ADP at the catalytic site is shown in red. Mg^{2+} ions bound to the ADP molecules are in white.

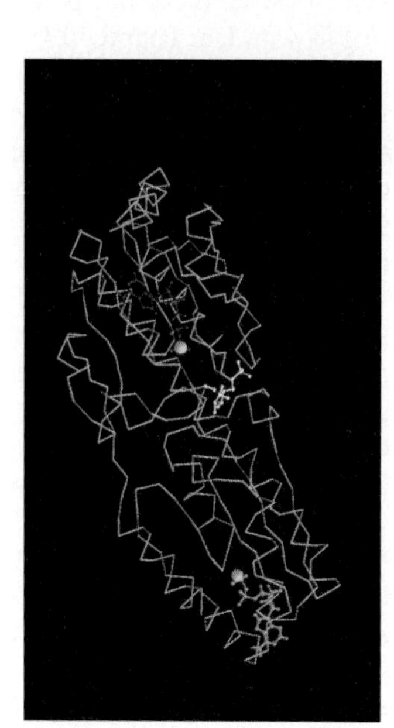

Each of these axes is a twofold symmetry axis, which means that rotating the entire structure by one-half of a full circle around the axis results in an identical structure. This effect is shown diagrammatically in figure 10.10. Because ADP is a product of the enzymatic reaction as well as an allosteric activator, it binds at both the catalytic and allosteric sites. The locations of both sites can be seen in figure 10.9 and are indicated in figure 10.10. In each subunit, the catalytic site for fructose-6-phosphate is at the interface of the subunit with one of its neighbors, and the allosteric site is at the interface with a different neighbor.

In the transition between the T and R conformations, the four subunits rotate by about 7° with respect to each other (see fig. 10.10). This rotation is associated with coupled rearrangements of the structures at the interfaces between adjacent subunits. Figure 10.11 shows how these rearrangements affect the binding site for fructose-6-phosphate. The most significant structural change in this region is an inversion of the orientation of the side chains of Glu 161 and Arg 162. In the R structure (fig. 10.11*b*), Arg 162 forms a hydrogen bond to the phosphate

Figure 10.10

Outlines of phosphofructokinase in the T (solid lines) and R (dashed lines) conformations. The enzyme contains four identical subunits (A, B, C, and D). The locations of the catalytic and allosteric sites are indicated in the two subunits closest to the viewer (A and D). The binding sites for fructose-6-phosphate (F6P) are at the interface of these subunits; the allosteric sites are at the interfaces of A with B, and of D with C. Two of the three perpendicular symmetry axes are labeled p and q. A 180° rotation about axis q interchanges the positions of subunits A and C, and also interchanges B and D. A similar rotation about p interchanges A with D, and B with C. (Source: T. Schirmer and P. R. Evans, "Structural basis of the allosteric behaviour of phosphofructokinase" in *Nature*, 343:140, 1990. Copyright © 1990 Macmillan Magazines Ltd., London, England.)

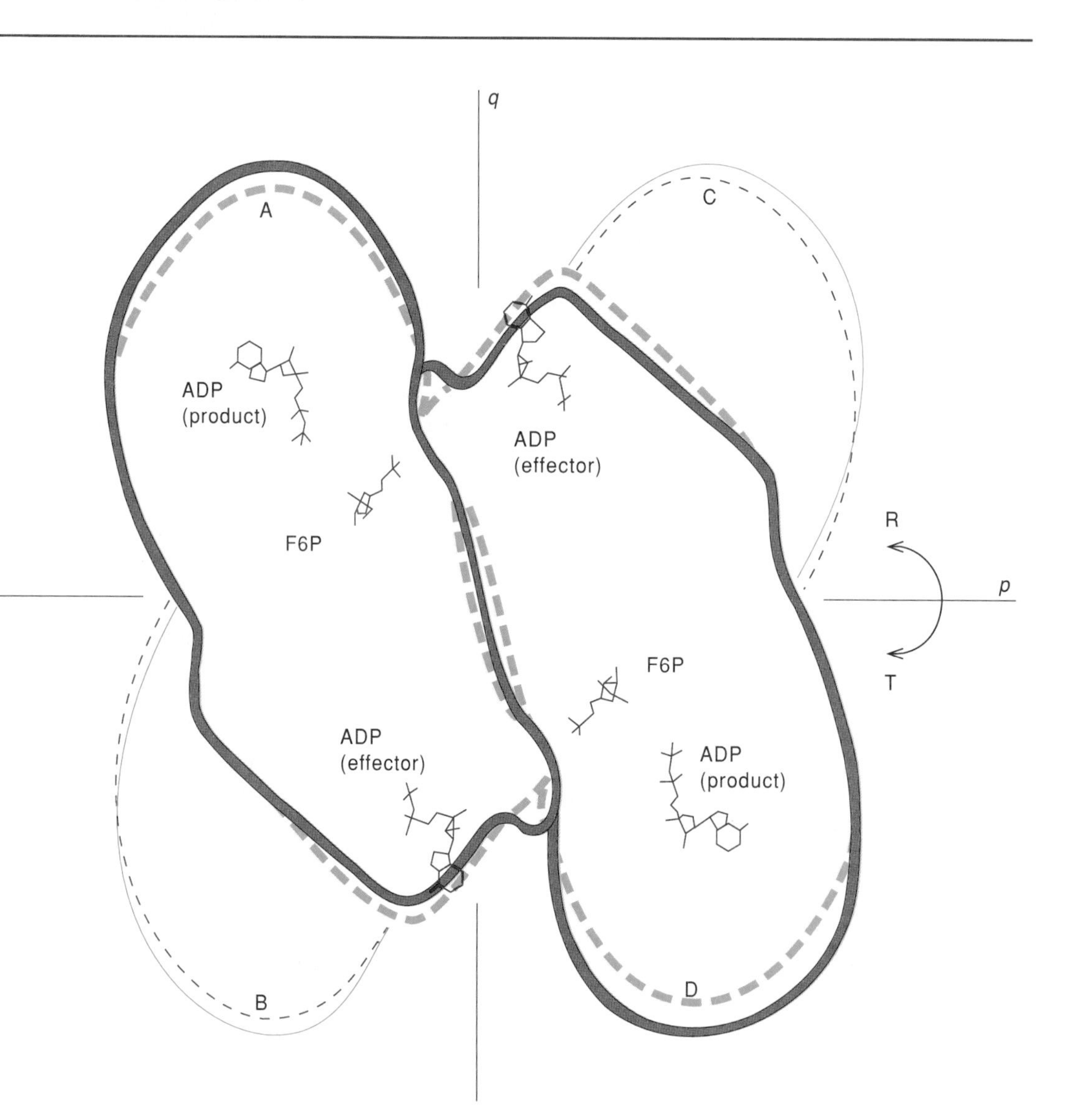

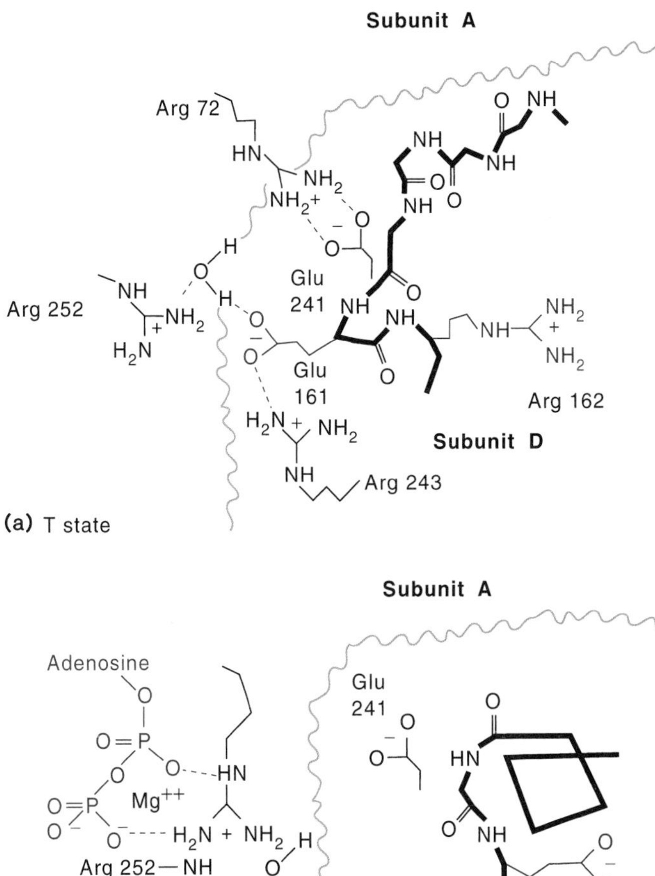

Figure 10.11

Interface between subunits A and D of phosphofructokinase near the catalytic site in (*a*) the T and (*b*) the R structures. Crystals of the enzyme in the R state were obtained in the presence of fructose-6-phosphate and ADP (see fig. 10.9); crystals in the T state were obtained in the presence of a nonphysiological allosteric inhibitor, 2-phosphoglycolate. The wavy line represents part of the boundary between subunits A and D. The heavy line indicates the polypeptide backbone. The side chains of Glu 161 and Arg 162 are shown in color. Note the inversion of the positions of these side chains in the two structures. (Source: T. Schirmer and P. R. Evans, "Structural basis of the allosteric behaviour of phosphofructokinase" in *Nature* 343:140, 1990. Copyright © 1990 Macmillan Magazines Ltd., London, England.)

group of fructose-6-phosphate, while Glu 161 points in the opposite direction. In the T structure (fig. 10.11*a*), Arg 162 points away from the binding site, while Glu 161 inserts a negative charge into the site, where it forms a hydrogen bond with Arg 243. The change in the orientations of the negatively charged Glu 161 and the positively charged Arg 162 probably accounts for most of the difference between the dissociation constants for fructose-6-phosphate in the two states. Note that although the molecule of fructose-6-phosphate that binds in the R state is located on subunit A in figure 10.11*b*, Glu 161 and Arg 162 are residues of subunit D. Arginines 252 and 243 also contribute to the binding site from opposite sides of the boundary. Structural changes that occur on one of the subunits thus are intricately linked to changes on the other. The substrate binding site for ATP, on the other hand, is made up of residues from only one subunit (subunit A in figure 10.11). This may explain why the binding of fructose-6-phosphate to the enzyme is strongly cooperative, whereas the binding of ATP as a substrate is not cooperative.

In the T structure, Glu 161 and Arg 162 are located at the end of a stretch of polypeptide that winds up into a helical turn in the transition to the R structure (see fig. 10.11). This coiling is linked to a significant structural change that occurs in an adjacent region of the interface between the A and D subunits. The interface here includes a pair of antiparallel β strands, each of which is hydrogen-bonded to a parallel strand in its own subunit, as shown in figure 10.12. In the T structure (fig. 10.12*a*), the two antiparallel β strands from the different subunits are hydrogen-bonded together directly across the interface. In the R structure (fig. 10.12*b*), the strands have moved apart, and the region between them is filled by a row of six hydrogen-bonded water molecules.

The insertion of water at the interface between subunits A and D is an essential component of the rotation of the subunits with respect to each other, and it would appear to be an all-or-none effect. Intermediate conformations in which only some of the water molecules are present would have a less extensive network of hydrogen bonds, and thus probably would be less stable than either the R or the T conformation. The same might be said of the winding of the helical turn between residues 155 and 161; intermediates in which the helical turn is partially unwound probably would be destabilized by steric crowding, or would force the structure to expand in a way that would leave empty spaces in other regions. These considerations, taken with the close coupling between the individual subunits, seem to explain why all of the subunits undergo concerted transitions from the R to the T state or back, without giving appreciable concentrations of intermediate states.

Although the crystal structures explain why the binding of fructose-6-phosphate is strongly cooperative, they leave unclear why different allosteric effectors tend to stabilize the protein in different conformational states. The allosteric binding site, like the site for fructose-6-phosphate, lies at the interface of different subunits, but the binding of either an activator or

Figure 10.12

Hydrogen bonds of the peptide backbone and the side chain of threonine 245 at the interface between subunits A and D of phosphofructokinase, in (*a*) the T and (*b*) the R structures. Note the additional molecules of water between the two subunits in the R structure. The portions of the polypeptide chains shown here are almost contiguous with those shown in figure 10.11. (Source: T. Schirmer and P. R. Evans, "Structural basis of the allosteric behaviour of phosphofructokinase" in *Nature* 343:140, 1990. Copyright © 1990 Macmillan Magazines Ltd., London, England.)

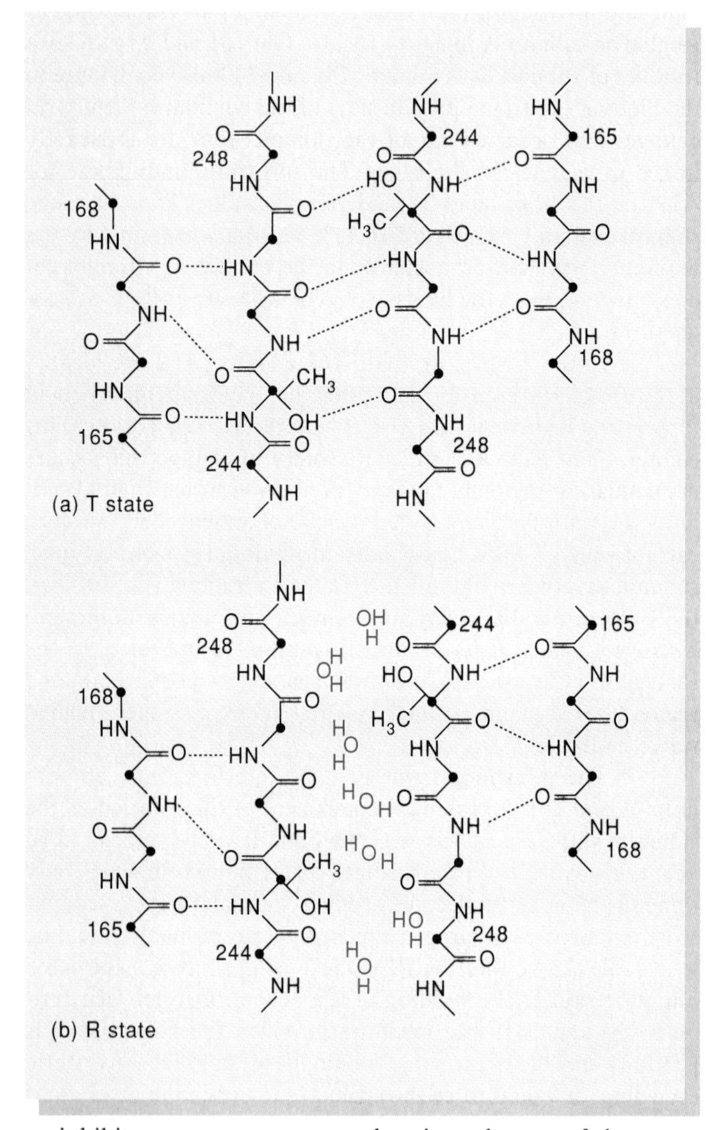

an inhibitor appears to cause only minor changes of the structure in this region of the protein. The side chain of a glutamic acid residue does rotate by about 140° to take up different positions, depending on whether the allosteric effector is an activator or an inhibitor. Whether this rearrangement is sufficient to account for a large shift in the equilibrium between the R and T conformations is unclear. However, the crystal structures that have been elucidated so far include only a single example of an allosteric inhibitor, and this was not ATP but rather a nonphysiological inhibitor, 2-phosphoglycolate. Further studies with ATP or other inhibitors should provide a broader basis for identifying structural changes that are essential to the allosteric inhibition.

Aspartate Carbamoyltransferase: Allosteric Control of Pyrimidine Biosynthesis

Aspartate carbamoyltransferase, or aspartate transcarbamylase, as it is frequently called, catalyzes the transfer of a carbamoyl group

$$H_2N{-}\overset{\displaystyle O}{\overset{\|}{C}}{-}$$

from carbamoyl phosphate to the amino group of aspartic acid to form carbamoyl aspartate (fig. 10.13). This step commits aspartate to the biosynthetic pathway for pyrimidines. Aspartate carbamoyltransferase is regulated by feedback inhibition by cytidine triphosphate (CTP) and uridine triphosphate (UTP), the end products of the pathway, and it is stimulated by ATP. The opposing effects of CTP, UTP, and ATP serve to keep the biosynthesis of pyrimidines in balance with that of purines. This balance is important because cells need purines and pyrimidines in approximately equal amounts for the synthesis of nucleic acids.

The kinetics and physical properties of aspartate carbamoyltransferase from *E. coli* have been studied in considerable detail by Howard Schachman and his colleagues. The rate of the reaction has a sigmoidal dependence on the concentration of aspartate, as shown in figure 10.14. CTP shifts the kinetic curve to the right. The enzyme thus is inhibited strongly by CTP at low concentrations of aspartate but not at high concentrations. ATP reverses the effect of CTP, or in the absence of CTP eliminates the positive cooperativity altogether, making the kinetics hyperbolic instead of sigmoidal. UTP (not shown in the figure) acts similarly to CTP.

Direct evidence for the allosteric nature of these effects came from the finding that the regulatory behavior disappeared when the enzyme was treated with an organic mercurial compound such as *p*-hydroxymercuribenzoate (see fig. 10.14). After this treatment, the binding of aspartate no longer showed positive cooperativity, and ATP and CTP had no effect on its activity. Exposure to mercurials was found to cause the enzyme to dissociate into two types of fragments, one of which retained the enzymatic activity but was no longer affected by CTP or ATP. The other fragment bound CTP and ATP, but had no enzymatic activity. In the native enzyme, each molecule is a complex of six catalytic (*c*) subunits, and six regulatory (*r*) subunits (fig. 10.15). The fragments that result from treatment with mercurials consist of trimers of the *c* subunit and dimers of *r*. When the intact c_6r_6 complex was reconstituted from the c_3 and r_2 fragments, the enzyme regained its sigmoidal kinetics and its sensitivity to CTP and ATP. These observations provided a conclusive demonstration that the regulatory agents bind at an allosteric site on the enzyme, and not at the active site. The two sites are on totally different subunits!

The binding of substrate analogs causes changes in several physical and chemical properties of aspartate carbamoyltransferase. An analog that has been particularly useful for studying these effects is *N*-phosphonacetyl-L-aspartate (PALA). PALA is structurally similar to a covalently linked adduct of

Figure 10.13

The reaction catalyzed by aspartate carbamoyltransferase, and the feedback inhibition of this enzyme in *E. coli* by the end product of the pathway, CTP. The series of small arrows represents additional reaction steps in the pathway from carbamoyl aspartate to CTP. These steps are discussed in chapter 20. The upward arrow with the negative sign indicates the feedback inhibition.

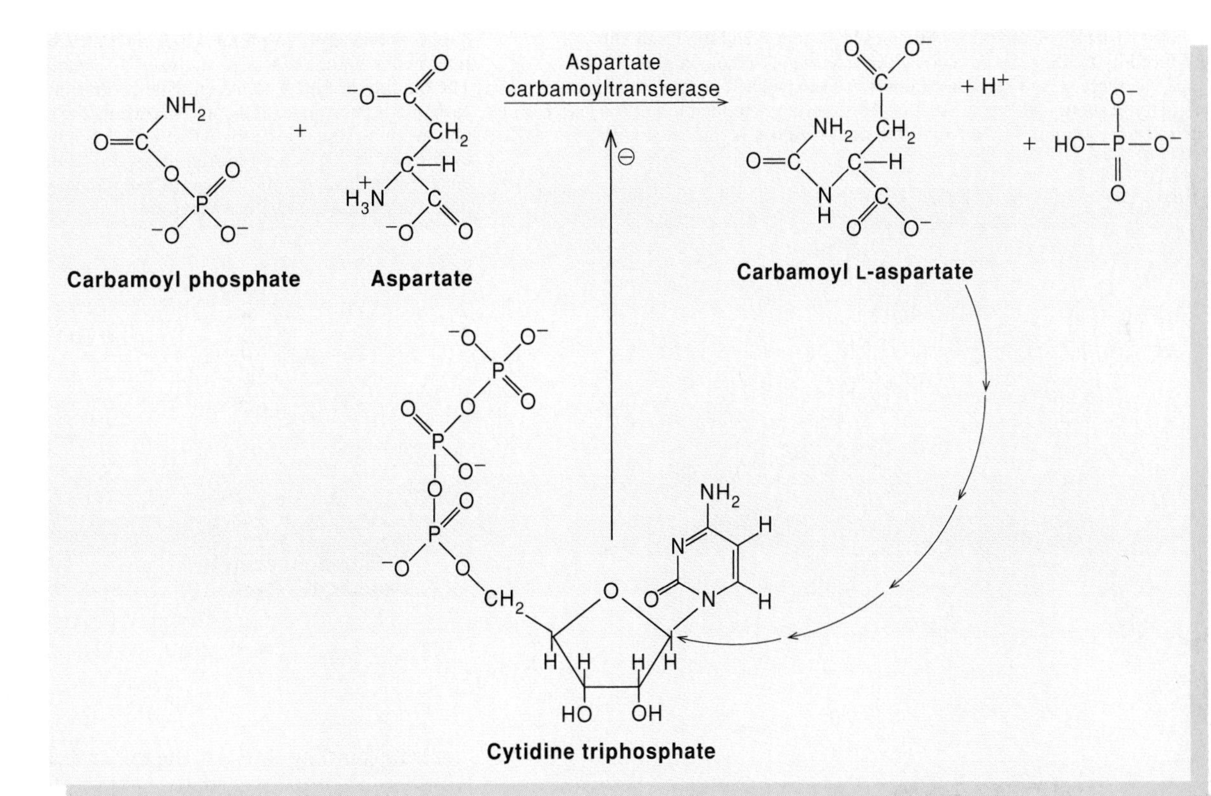

Figure 10.14

Effects of CTP, ATP, and mercurials on the rate of the reaction catalyzed by aspartate carbamoyltransferase. In the absence of CTP and ATP, the sigmoidal kinetics show positive cooperativity with respect to aspartate. CTP augments the positive cooperativity; ATP reverses the effect of CTP. An organic mercurial, or ATP in the absence of CTP, eliminates the cooperativity, converting the curve from sigmoidal to hyperbolic.

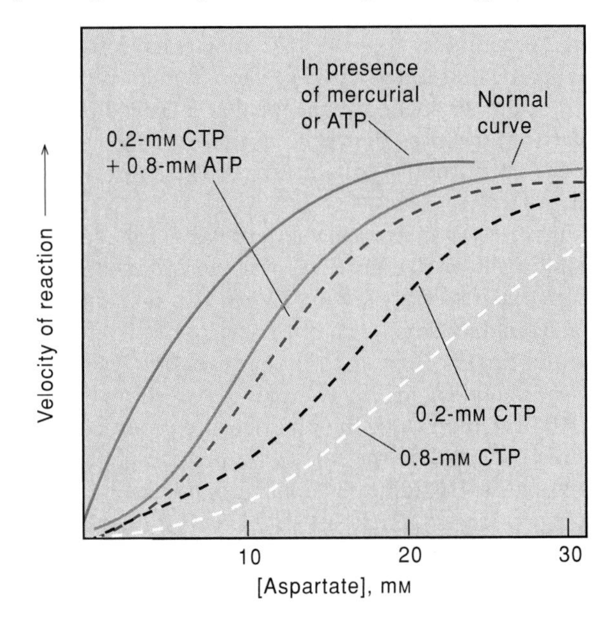

carbamoyl phosphate and aspartate, and thus resembles an intermediate that is likely to be formed in the course of the enzymatic reaction (fig. 10.16). It binds to the enzyme in a highly cooperative manner, with a Hill coefficient of 2.0, and its binding is promoted by ATP and opposed by CTP. (ATP decreases the Hill coefficient to 1.4, and CTP raises it to 2.3.) The binding of PALA causes a decrease in the sedimentation and diffusion coefficients of the enzyme, indicating that the protein expands or changes shape. It also causes a decrease in the chemical reactivity of a cysteine residue in each *c* subunit, and an increase in the reactivity of several cysteines in each *r* subunit. CTP opposes these effects.

The changes in sedimentation coefficient and chemical reactivity caused by PALA can be interpreted by the model that the enzyme exists in two distinct conformations (T and R), and that the binding of PALA to only one or two of the *c* subunits in the c_6r_6 complex causes the entire enzyme to flip from the T to the R state. The dissociation constant for PALA is higher in the T state. The equilibrium constant $L = (T)/(R)$ has been calculated to be 250 in the absence of substrates and allosteric effectors, 70 in the presence of ATP alone, and 1,250 in the presence of CTP alone. Thus ATP shifts the equilibrium toward the conformational state that favors the binding of the substrate, and CTP shifts the equilibrium in the direction of weaker binding.

Figure 10.15

Subunit structure of aspartate carbamoyltransferase and the fragments produced by treating the enzyme with mercurials. In the complete enzyme (top), the three sets of regulator dimers are sandwiched between two trimers of catalytic subunits (see fig. 10.17). The approximate location of the active site in each *c* subunit of the trimer facing the viewer is indicated with a *c*.

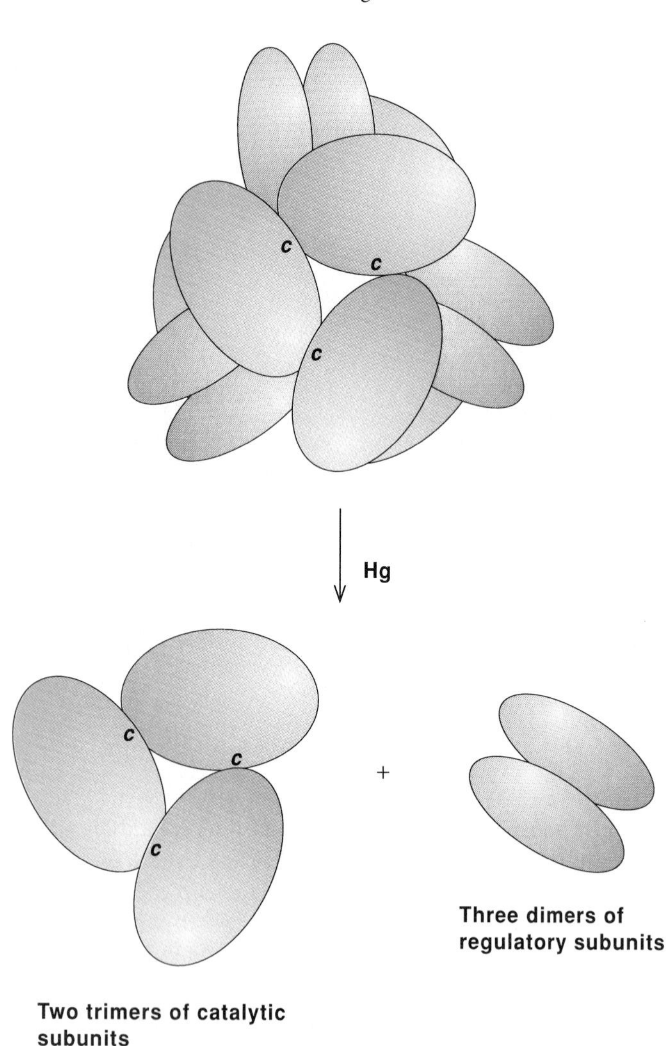

Figure 10.16

N-phosphonacetyl-L-aspartate (PALA) is structurally similar to a likely intermediate in the reaction catalyzed by aspartate carbamoyltransferase. The binding of PALA to the enzyme is prevented competitively by carbamoyl phosphate, and PALA prevents the binding of aspartate. These observations support the view that PALA binds at the catalytic site. The binding of PALA is not blocked by aspartate alone, probably because the enzyme mechanism follows an ordered pathway in which carbamoyl phosphate must bind before aspartate.

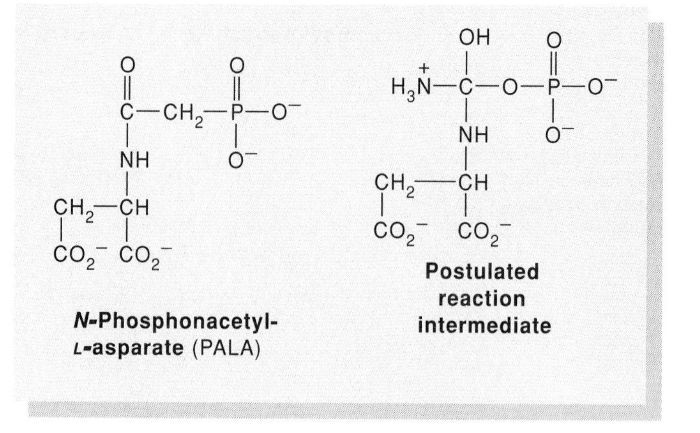

The crystallization of aspartate carbamoyltransferase with and without bound PALA or CTP, and the elucidation of the crystal structures by William Lipscomb and his colleagues, led to detailed pictures of the structural changes that accompany the transition of the enzyme between the R and T states. Figures 10.17 and 10.18 show the crystal structures from two different perspectives. In both the R and the T conformations, the six *c* subunits are arranged in two equilateral trimers, one of which is inverted and stacked on top of the other (see figs. 10.15 and 10.17). The three *c* units within each trimer are related to each other by a threefold axis of symmetry. (Rotating the structure by one-third of a circle about this axis results in an identical structure.) The substrate-binding site on each *c* subunit is located in a pocket between two domains of the polypeptide. At the end of one of these domains the *c* subunit interacts with another *c* subunit in the same trimer, close to *its* substrate-binding site. In the other domain, it interacts more extensively with a *c* subunit in the other c_3 trimer. The six *r* subunits are arranged in three sets of dimers that form another equilateral triangle about the same threefold axis of symmetry. Like the *c* subunits, each *r* subunit is folded into two domains: a peripheral domain where it interacts with its companion *r* subunit in the dimer, and a smaller domain that interacts with two adjacent *c* subunits. In the latter region, the *r* subunit binds an atom of Zn^{2+} that evidently plays a purely structural role in the enzyme. The binding site for the allosteric effectors CTP and ATP is located in the peripheral domain of the *r* subunit, at a considerable distance from the active sites.

In the transition from the T to the R conformation, the two *c* trimers rotate slightly with respect to each other about the threefold symmetry axis, so that they come into a more eclipsed alignment (see fig. 10.17). The *r* dimers rotate with respect to each other about a perpendicular axis, and appear to act as a lever that moves the two *c* trimers apart by about 12 Å and opens up a cavity at the center of the entire structure (see fig. 10.18).

Figure 10.19 shows some of the details of the substrate-binding site in the R structure. As was mentioned above, the binding site consists of a pocket between two domains of a *c* subunit, and residues from each of these domains interact with the substrate. Some of the key residues are Arg 105 and His 134 from one domain, and Arg 167 and Arg 229 from the other. Arginine 105 interacts with the phosphonate group of PALA, and presumably would do the same with the phosphate of carbamoyl phosphate. Histidine 134 appears to provide a hydrogen bond to the peptide oxygen atom of PALA. If it does the same to the corresponding oxygen of carbamoyl phosphate, it could serve as a general acid in the catalytic mechanism. Arginines 167 and 229 interact with the two carboxylate groups of PALA, and presumably would interact similarly with aspartate.

Figure 10.17

Structures of aspartate carbamoyltransferase in the T conformation (*a*), and the R conformation (*b*), viewed along the threefold symmetry axis. The enzyme contains two c_3 clusters and three r_2 clusters. The α-carbon chains of one of the c_3 groups are shown in aqua; those of the other c_3 group are in blue. One of the *r* subunits in each of the r_2 groups is shown in orange; the other, in red. The enzyme was crystallized in the T form in the absence of substrate or allosteric effectors; the R structure was obtained with bound PALA. The PALA molecules are seen in yellow in (*b*). Zinc ions bound to the *r* subunits are shown in white. (Based on the crystal structures described by W. N. Lipscomb and his colleagues.)

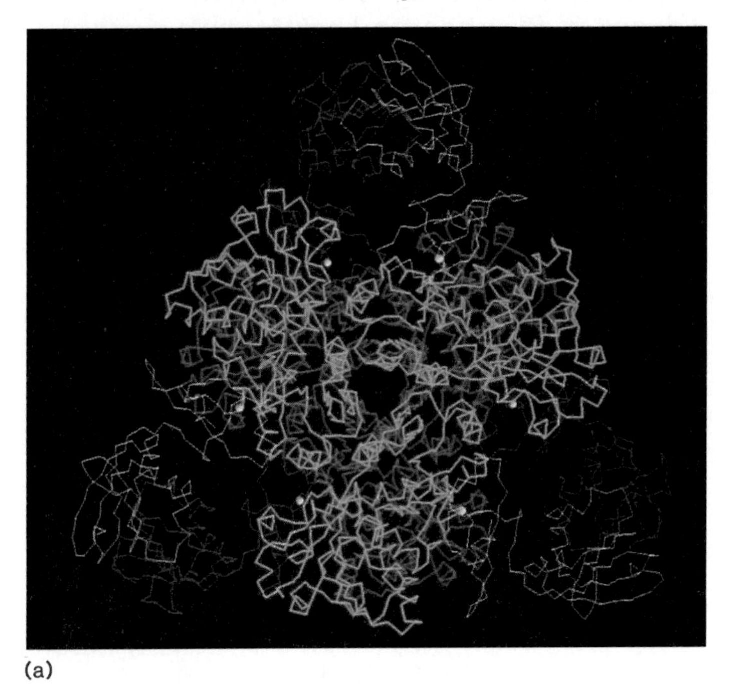

(a)

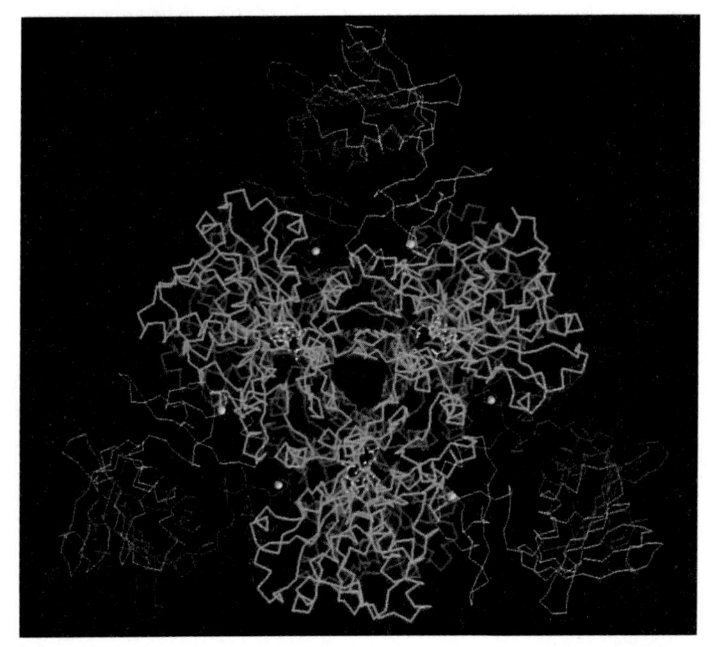

(b)

Figure 10.18

Structures of aspartate carbamoyltransferase in the T conformation (*a*), and the R conformation (*b*), viewed along an axis perpendicular to the threefold symmetry axis. The structures and the color coding are the same as in figure 10.17. Note the expansion of the cavity between the upper and lower c_3 groups in the R structure.

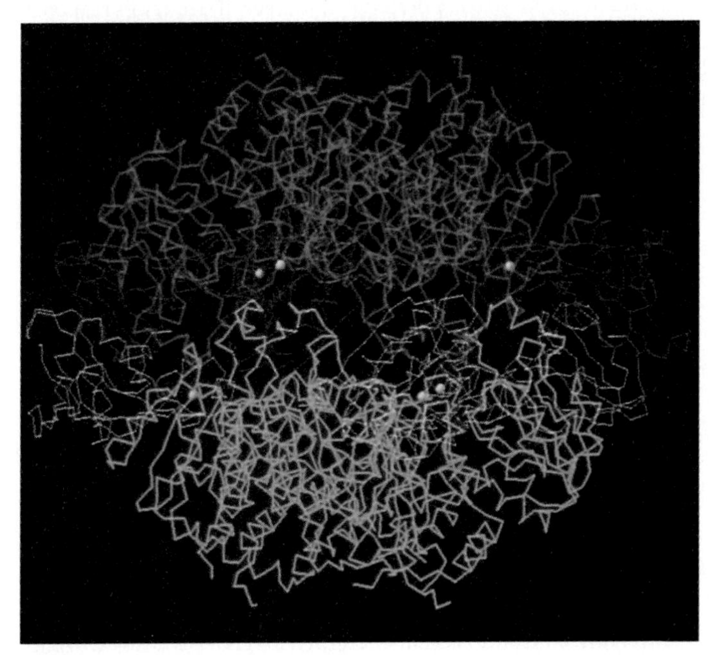

(a)

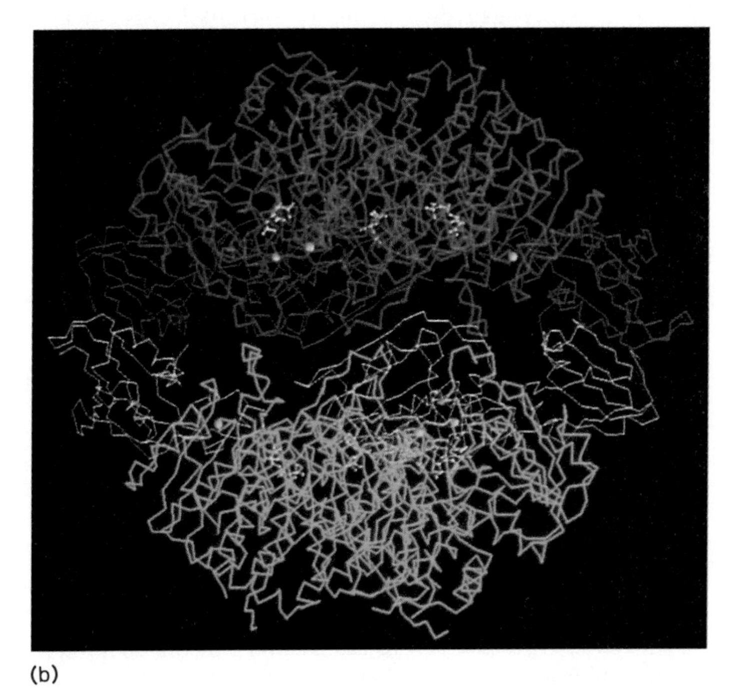

(b)

Figure 10.19

The binding of PALA to the R conformation of aspartate carbamoyltransferase. Arg 105 and His 134 are provided by one domain of a *c* subunit, and Arg 167 and Arg 229 by the other domain. Ser 80 and Lys 84 are part of a loop of protein from a different *c* subunit. The bound PALA is shown in color.

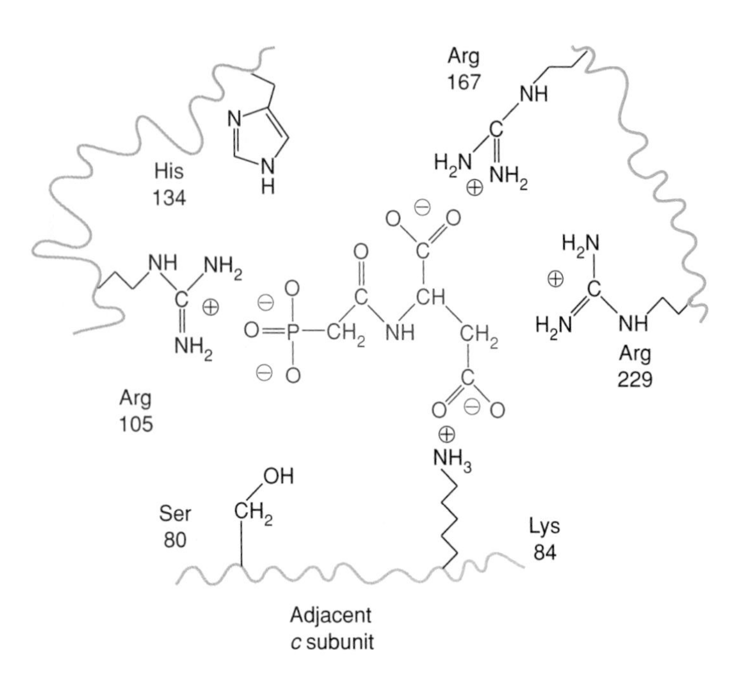

In addition to these residues, the active site also includes two residues from a different *c* subunit, Ser 80 and Lys 84 (see fig. 10.19). Serine 80 appears to be hydrogen-bonded to the phosphonate group of PALA, and Lys 84 interacts with one of the carboxylate groups. As with phosphofructokinase, the location of the active site at the interface between two subunits provides a clue to how the binding of substrate to one subunit can affect the binding at another.

The structural transition to the T state disrupts the active site in two major ways. First, the domain of the *c* subunit that includes Arg 105 and His 134, which must interact with carbamoyl phosphate, is pulled away from the domain that must interact with aspartate, because some of the residues in both domains are tied up in alternative sets of hydrogen bonds. Arginine 105 is hydrogen-bonded to Glu 50 in the same domain, instead of to the substrate, and His 134 interacts with a residue in the other *c* trimer. In addition, the loop of the *c* subunit that contains Ser 80 and Lys 84 is pulled out of the active site by a set of hydrogen bonds involving still another *c* subunit. A baroque network of interrelationships thus links the catalytic sites of all the different *c* subunits in the complex.

The transition between the R and T states also involves large changes in the conformation of the *r* subunits. These changes include both the peripheral domain where CTP or ATP binds and the domain that interfaces with the *c* subunits (see figs. 10.17 and 10.18). However, it still is not clear how the binding of CTP to the peripheral domain tips the conformational equilibrium constant L in favor of T, whereas ATP, which binds to the same site as CTP, favors the formation of R. The crystal structures of the enzyme with bound ATP or CTP are very similar, both at the catalytic site and at the allosteric site.

Glycogen Phosphorylase: Control of Glycogen Breakdown both by Allosteric Effectors and by Phosphorylation

Glycogen phosphorylase catalyzes the removal of a terminal glucose residue from glycogen. The glycosidic bond is cleaved by a reaction with inorganic phosphate ion (a "phosphorolysis") instead of simply by hydrolysis, so that the product is glucose-1-phosphate instead of free glucose:

$$\text{Glycogen with } n \text{ glucose units} + P_i \longrightarrow \text{glycogen with } n - 1 \text{ glucose units} + \text{glucose-1-phosphate} \quad (9)$$

This is the first step in the metabolic breakdown of glycogen to pyruvate.

In the early 1940s, Carl and Gerty Cori discovered that phosphorylase exists in two forms, *a* and *b*, which differ greatly in their catalytic activities. As shown in figure 10.20, phosphorylase *b* has virtually no activity in the absence of AMP. It is activated by AMP with an apparent dissociation constant of about 40 μM, but the activation is inhibited competitively by ATP. At the concentrations of AMP and ATP that prevail in resting muscle tissue, phosphorylase *b* is essentially inactive. Phosphorylase *a*, on the other hand, has about 80% of its maximal activity in the absence of AMP, and becomes fully active at very low concentrations of AMP. It also is relatively insensitive to inhibition by ATP (see fig. 10.20).

The Coris found that the interconversion of phosphorylase *a* and *b* is catalyzed by another enzyme, and subsequent work by Earl Sutherland showed that this process is under hormonal control. In muscle, the conversion of phosphorylase *b* to *a* occurs in response to epinephrine; in liver, it occurs in response to glucagon, a hormone elaborated by the pancreas. However, the structural basis for the difference between the two forms of the enzyme remained unknown until the late 1950s, when Edwin Krebs and Edmond Fischer showed that phosphorylase *a* and *b* differ by a single covalent modification: Phosphorylase *a* has a phosphate on Ser 14. Krebs and Fischer also showed that the kinase that catalyzes the addition of the phosphate group to the serine residue (phosphorylase kinase) is itself regulated by a phosphorylation catalyzed by another kinase (the cAMP-dependent protein kinase).

The main effect of AMP on either phosphorylase *b* or phosphorylase *a* is to decrease the K_m for P_i. The K_m for glucose-1-phosphate also is increased for the reaction in the reverse direction. These changes can be interpreted much as we have interpreted the actions of allosteric effectors on phosphofructokinase and aspartate carbamoyltransferase, on the model that the enzyme can exist in either of two different conformational states (R and T) with different affinities for the substrate. However, phosphorylase presents the additional complexity that the equilibrium constant (L) between the two conformational states can be altered by a covalent modification of the enzyme. In the

Figure 10.20

The rate of the reaction catalyzed by glycogen phosphorylase, as a function of the concentration of its main allosteric activator, AMP. The curves shown in color were obtained in the presence of ATP. Phosphorylase *b* (lower two curves) is almost completely inactive in the absence of AMP. Its activity is half-maximal at an AMP concentration of about 40 μM. ATP greatly increases the concentration of AMP required for activity. Phosphorylase *a* (upper two curves) has about 80% of its maximal activity in the absence of AMP, and reaches full activity at very low AMP concentrations; it also is relatively insensitive to inhibition by ATP. (Source: N. B. Madsen in *The Enzymes,* 3d ed., vol. XVII, Edited by P. D. Boyer and E. G. Krebs. Copyright © 1986 Academic Press, New York, N.Y.)

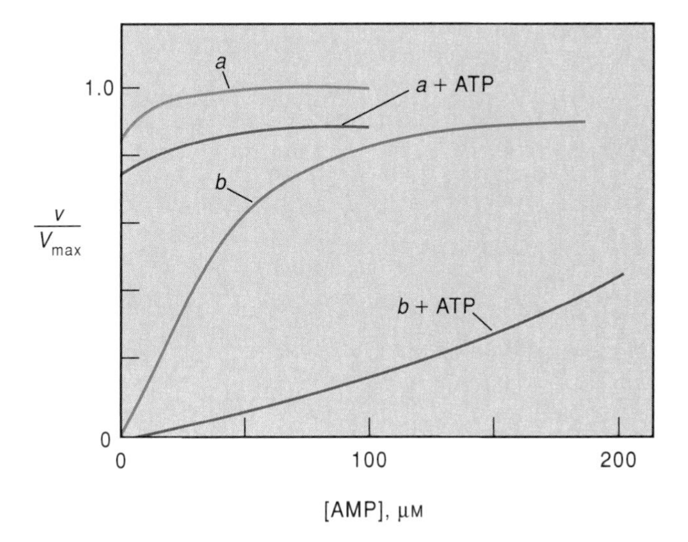

absence of substrates, (T)/(R) is estimated to be greater than 3,000 in phosphorylase *b,* but to decrease to about 10 in phosphorylase *a.* The phosphorylation thus appears to reduce the free energy difference between the two conformations by at least 3.5 kcal/mole.

In addition to activation by AMP and inhibition by ATP, both forms of phosphorylase are sensitive to inhibition by glucose or glucose-6-phosphate. Glucose inhibits by binding at the catalytic site; glucose-6-phosphate binds predominantly at the same allosteric site as AMP and ATP. There also is a separate inhibitory allosteric site that binds adenine, adenosine, or (much more weakly) AMP.

Louise Johnson and her co-workers have determined structures for both the T and the R forms of muscle phosphorylase *b,* and also of the R form of phosphorylase *a.* In parallel with this work, Robert Fletterick and his co-workers determined the crystal structure for the T form of muscle phosphorylase *a.* The crystal structures provide an incisive look at the structural changes that accompany the transition from the T to the R form and the conversion of nonphosphorylated form of the enzyme to the phosphorylated.

In keeping with the complexity of its allosteric and covalent regulation, phosphorylase is a large, complex enzyme. It consists of a dimer of two identical subunits, each with a molecular weight of about 97,400 (fig. 10.21). The catalytic site is buried near the center of each subunit, at the end of a tunnel about 15 Å in length. The tunnel opens to a concave surface

Figure 10.21

(*a*) Ribbon diagram of the crystal structure of phosphorylase *a* in the R state. The view is along the twofold rotational symmetry axis of the dimer, with the allosteric sites and the phosphoserine (Ser-P) on each subunit facing forward. Regions where the positions of the C_{α} carbons differ by more than 1 Å between the R and T states are shown in orange for subunit 1 (*bottom*) and in pink for subunit 2 (*top*). More fixed regions of the polypeptide chain are in green for subunit 1, and in blue for subunit 2. The N-terminal residues (10–23) and the C-terminal residues (837–842) are in white (residues 1–9 are disordered, and cannot be seen in the crystal structure). The bound pyridoxal-phosphate (PLP) indicates the location of the catalytic site. The allosteric effector site is occupied by AMP. The first two helices at the N-terminal end (labeled α_1 and α_2 in subunit 1) are connected by a loop (Cap) that forms one of the interfaces between the two subunits. (*b*) Ribbon diagram of phosphorylase *b* in the T state. The orientation of the dimer is the same as in (*a*). Regions where the C_{α} positions differ by more than 1Å between the R and T states are represented in red for subunit 1 (*bottom*) and in yellow for subunit 2 (*top*); more fixed regions are in cyan and purple. The N-terminal residues and the C-terminal residues are in white. PLP is at the catalytic site and AMP at the allosteric effector site, as in (*a*). Maltopentaose is bound at the glycogen-storage site. There also is a molecule of glucose-1-phosphate bound at the catalytic site, and (not clearly visible) a second molecule of AMP at the nucleoside inhibitor site. (From D. Barford, S.-H. Hu, and L. N. Johnson, "Structural mechanism for glycogen phosphorylase control by phosphorylation and AMP," *J. Mol. Biol.* 218:233, 1991. © 1991 Academic Press LTD., London England.)

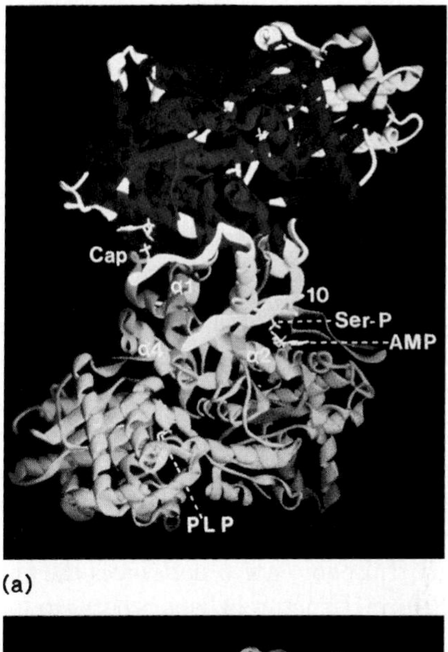

(a)

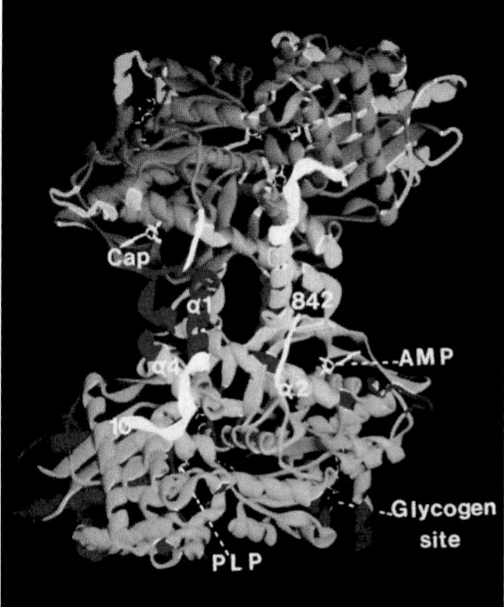

(b)

Figure 10.22

Pyridoxal phosphate, a prosthetic group in glycogen phosphorylase, is covalently attached to a lysine side chain of the enzyme. The phosphate group of the pyridoxal phosphate probably acts as a general acid to transfer a proton to inorganic phosphate in the enzymatic mechanism. (Other parts of the pyridoxal molecule can be modified chemically with little effect on the enzymatic activity, but modifying the phosphate group destroys the activity.) Other biochemical reactions of pyridoxal phosphate are discussed in chapter 11.

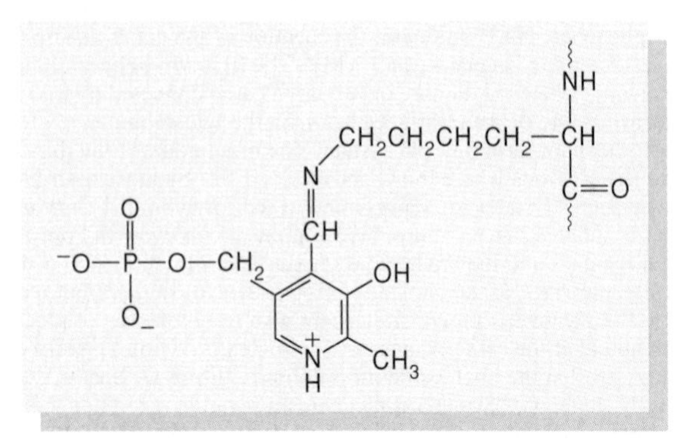

whose curvature is similar to the curvature of a glycogen particle. A binding site for glycogen can be recognized by the location of a small oligosaccharide (maltoheptaose or maltopentaose) in the crystal structure. Because the glycogen-attachment site is about 30 Å from the catalytic site, the enzyme evidently clings to one branch of the glycogen particle while it chews on another branch. Near the catalytic site there is a covalently bound molecule of pyridoxal phosphate (fig. 10.22), which appears to participate in the catalytic mechanism by acting as a general acid.

The binding site for the allosteric effectors AMP, ATP, and glucose-6-phosphate is about 30 Å from the catalytic site, at one of the interfaces between the two subunits (see fig. 10.21). Serine 14, the locus of the covalent modification that converts phosphorylase *b* to *a*, is in the same region of the molecule, about 15 Å from the allosteric site.

When phosphorylase *b* undergoes the transition from the T to the R form, structural changes occur in the N-terminal region of each subunit (see fig. 10.21). There also are significant changes in a loop between residues 282 to 286, which connects two α-helical chains (fig. 10.23). In the T form, this loop is well ordered, and is located so that it obstructs the substrate-binding site for phosphate. In the R form, the loop is pulled out of the phosphate site, and is disordered. Some of the details of the structural changes in the active site are shown in figure 10.24. In the R form, the substrate phosphate ion is bound to Arg 569, the amide NH of Gly 135, Lys 574 (not shown in the figure), and the pyridoxal phosphate. In the T form, the side chain of Asp 283 sits in this region, and Arg 569 is pulled away by hydrogen bonding to the amide NH of Pro 281. The replacement of the positively charged arginine side chain by the negatively charged aspartate explains why the affinity for phosphate is much lower in the T form than in the R form.

Figure 10.23

(*a*) In the T state of phosphorylase *b*, a loop of the polypeptide chain between residues 282 and 286 obstructs the active site. Asp 283, near the middle of this loop, inserts its negatively charged side chain into the binding site for phosphate. The locations of the aspartate and the pyridoxal phosphate prosthetic group (PLP) are indicated in red. (*b*) In the R state, the subunits rotate with respect to each other. The loop from residues 282 to 286 is disordered, and Asp 283 leaves the phosphate-binding site. (Source: D. Barford and L. N. Johnson, "The allosteric transition of glycogen phosphorylase" in *Nature*, 340:609, 1989. Copyright © 1989 Macmillan Magazines Ltd., London, England.)

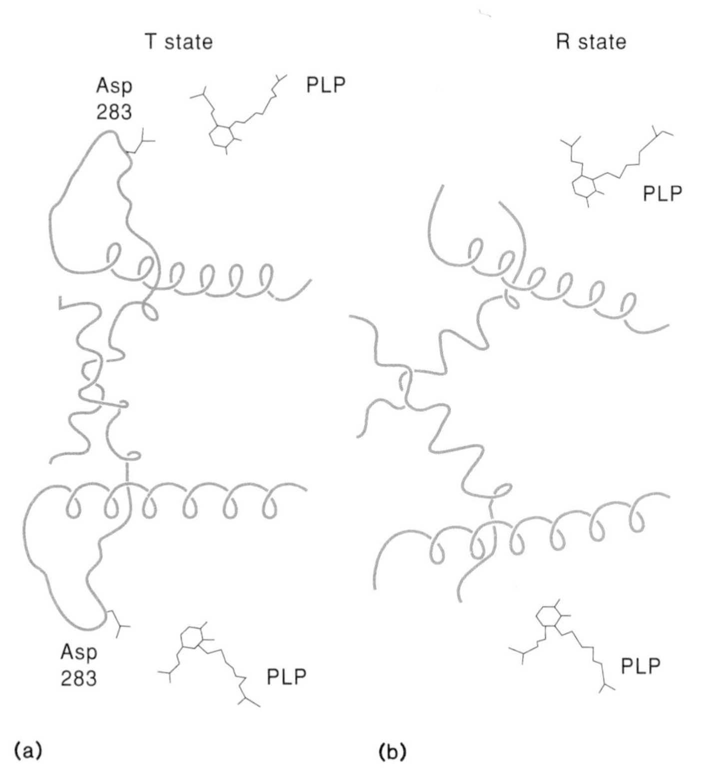

The transformation from phosphorylase *b* to phosphorylase *a* includes structural changes that tighten the interactions between the two subunits of the enzyme. Figures 10.21 and 10.25 show some of these changes. In phosphorylase *a*, the phosphate group attached to each Ser 14 is hydrogen-bonded to Arg 69 of its own subunit, but also to Arg 43 of the other subunit. There also are hydrogen bonds between Arg 10 and Leu 115 of different subunits, and between Glu 72 and Asp 42. All of these intersubunit hydrogen bonds are missing in phosphorylase *b*. Instead, there is a bond between Arg 43 and leucine in the same subunit, and one intersubunit hydrogen bond between His 36 and Asp 838. The N-terminal portion of the chain is ejected from this region of the structure in phosphorylase *b*, and is replaced by residues from the C-terminal end, including Asp 838. At the nearby allosteric effector site, the pulling together of the two subunits in phosphorylase *a* enhances the binding of AMP, but disfavors the binding of the inhibitor glucose-6-phosphate.

Figure 10.24

Residues in the region of the substrate-binding site for phosphate in phosphorylase *b* in the T state (*a*) and the R state (*b*). The side chain of Asp 283 leaves the binding site in the R structure, and the side chain of Arg 569 becomes available to interact with the phosphate. (Source: D. Barford and L. N. Johnson, "The allosteric transition of glycogen phosphorylase" in *Nature,* 340:609, 1989. Copyright © 1989 Macmillan Magazines Ltd., London, England.)

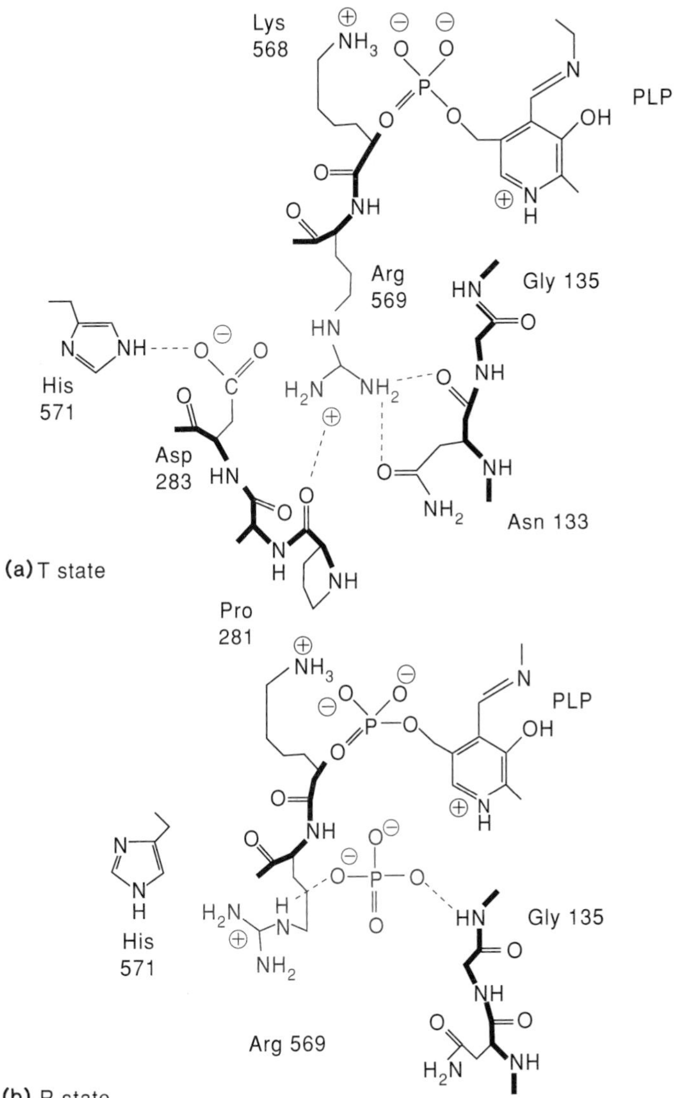

Figure 10.25

Structural changes that accompany the conversion of phosphorylase *b* to phosphorylase *a*. This figure shows the interface between the two subunits in the region of Ser 14, the residue that gains a phosphate group in phosphorylase *a*. Portions of the polypeptide backbone and the side chains of some residues from one of the subunits are drawn in color; the other subunit is drawn in black. A larger number of hydrogen bonds link the two subunits in phosphorylase *a* (*top*) than in phosphorylase *b* (*bottom*). (Source: S. R. Sprang et al., "Structural changes in glycogen phosphorylase induced by phosphorylation" in *Nature,* 336:215, 1988. Copyright © 1989 Macmillan Magazines Ltd., London, England.)

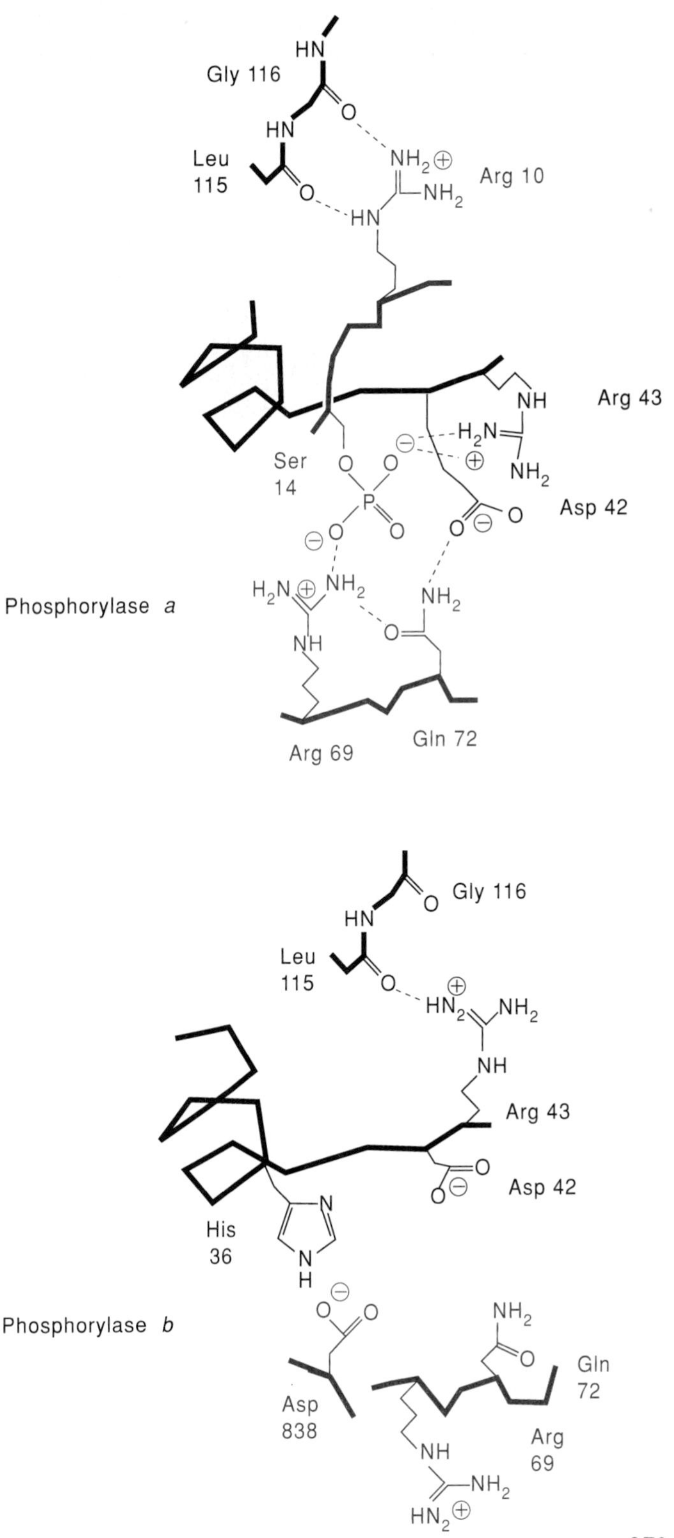

Calmodulin Mediates Regulation of Some Enzymes by Ca^{2+}

Phosphorylase kinase, the enzyme that converts phosphorylase *b* to phosphorylase *a,* also is controlled by both covalent modification and allosteric effectors. When muscle cells are stimulated by epinephrine, or liver cells by glucagon, the cAMP-dependent protein kinase catalyzes the phosphorylation of phosphorylase kinase at multiple serine residues. (The modification of more than one serine actually is typical of enzymes

Figure 10.26

The structure of calmodulin, a small protein that binds Ca^{2+} and regulates the activities of many enzymes in response to changes in the intracellular Ca^{2+} concentration. The bound Ca^{2+} ions are shown in yellow. (Based on the crystal structure described by Y. S. Babu, C. E. Bugg, and W. J. Cook.)

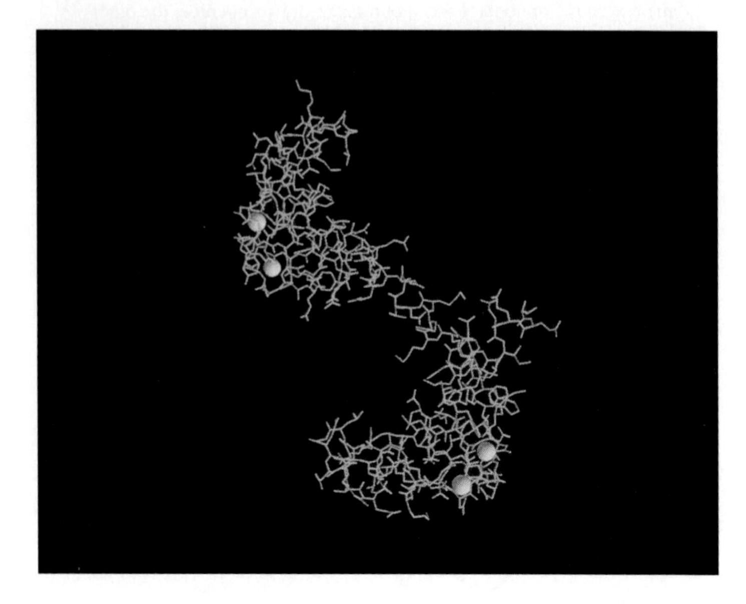

that are controlled by phosphorylation; phosphorylase is unusual in that only a single serine residue is modified.) Muscle phosphorylase kinase contains four nonidentical subunits (α, β, γ, and δ), which form an $(\alpha\beta\gamma\delta)_4$ tetramer with a total molecular weight of about 1.2×10^6. The catalytic sites are on the γ subunits, whereas the serines that become phosphorylated are on the α and β subunits. The δ subunit has turned out to be of particular interest because it is a polypeptide called calmodulin that also is associated with many other enzymes. Calmodulin appears to modify the activities of these enzymes in response to changes in the intracellular concentration of Ca^{2+} ions.

Calmodulin occurs in virtually all eukaryotic cells. As shown in figure 10.26, it is a relatively small protein (M_r = 17,000) that has two globular domains connected by an extended α helix. Each of the globular regions contains two subdomains that form binding sites for Ca^{2+} ions. Ca^{2+} binds to calmodulin with a dissociation constant of about 10^{-6} M, causing the globular domains to undergo conformational changes that evidently can be transmitted to the enzyme with which the calmodulin is associated.

The role of calmodulin in muscle phosphorylase kinase appears to be to enable the activation of the enzyme whenever the Ca^{2+} concentration rises above a level of about 1 μM. Ca^{2+} is released from intracellular stores when muscles are stimulated neurally, and it plays a critical role in initiating muscular contraction. The binding of Ca^{2+} to the calmodulin associated with phosphorylase kinase serves to switch on glycogen breakdown in synchrony with contraction, with the ultimate effect of providing energy to support the contraction. Calmodulin-Ca^{2+} complexes also activate an enzyme in the cell membrane to pump Ca^{2+} out of the cell.

Table 10.4
Some Enzymes that Are Regulated by Ca^{2+}-Calmodulin

Enzyme
Muscle phosphorylase kinase
Myosin light-chain kinase
Ca^{2+}-ATPase (transmembrane Ca^{2+} pump)
Ca^{2+}/calmodulin-dependent protein kinase

Some of the other enzymes that are regulated by calmodulin and Ca^{2+} are listed in table 10.4. The regulatory systems that use calmodulin resemble the systems that use the cAMP-dependent protein kinase in that they generally respond to signals that come from outside the cell. However, calmodulin also is essential in yeast, which are unicellular organisms. The roles that it plays in these organisms are not fully understood.

Inhibitory Proteins that Block the Actions of Proteolytic Enzymes

In discussing the activation of digestive enzymes and blood-coagulation enzymes by partial proteolysis, we noted the importance of keeping these enzymes in an inactive state unless they are needed. Tissues, blood, and other biological fluids of both vertebrate and invertebrate organisms contain a large array of specific inhibitory proteins that participate in this task. Similar inhibitory proteins have been purified from bacteria, especially from various species of *Streptomyces*. They also are abundant in plants, where they probably play a role in combatting insect pests.

Protease inhibitors typically bind very tightly to the active site of a particular protease. Bovine pancreatic trypsin inhibitor, for example, binds to trypsin with a dissociation constant of about 10^{-13} M. Crystal structures have been obtained for several such enzyme-inhibitor complexes, including the complexes of trypsin with soybean trypsin inhibitor and bovine pancreatic trypsin inhibitor, and the complex of subtilisin with *Streptomyces* subtilisin inhibitor. Figure 10.27 shows the crystal structure of the complex of bovine trypsin with the bovine pancreatic trypsin inhibitor. Inspection of the crystal structure in the region of the active site reveals that the inhibitor presents itself to the enzyme as a substrate. The carbonyl carbon atom of Lys 15 of the inhibitor is close to the OH oxygen of trypsin's reactive Ser 195, and the carbonyl oxygen atom of the lysine points toward the NH hydrogen atoms of Ser 195 and Gly 193. The carbonyl carbon thus seems poised to react with the serine oxygen, but for reasons that are not entirely clear, the reaction occurs only at a very low rate. The inhibitor behaves like a substrate that reacts with an extremely low K_m, but also with a very low k_{cat}. This also applies to the other serine-protease inhibitors that have been studied. A possible explanation for the low value of k_{cat} is that the enzyme-inhibitor complex is too rigid to complete the structural distortion that is needed for hydrolysis of the peptide bond.

Specific inhibitory proteins also have been found for enzymes other than proteases, including α-amylases, deoxyribonucleases, protein kinases, and phospholipases. However, these inhibitors are greatly outnumbered by the protein inhibitors that act on proteases, and perhaps this is not surprising. Because proteins are the natural substrates of proteases, it may take only minor modifications to turn them into inhibitors.

Figure 10.27

The complex of trypsin with bovine pancreatic trypsin inhibitor. The enzyme is in purple; the inhibitor, in red. (Based on the crystal structure described by R. Huber and J. Deisenhofer.)

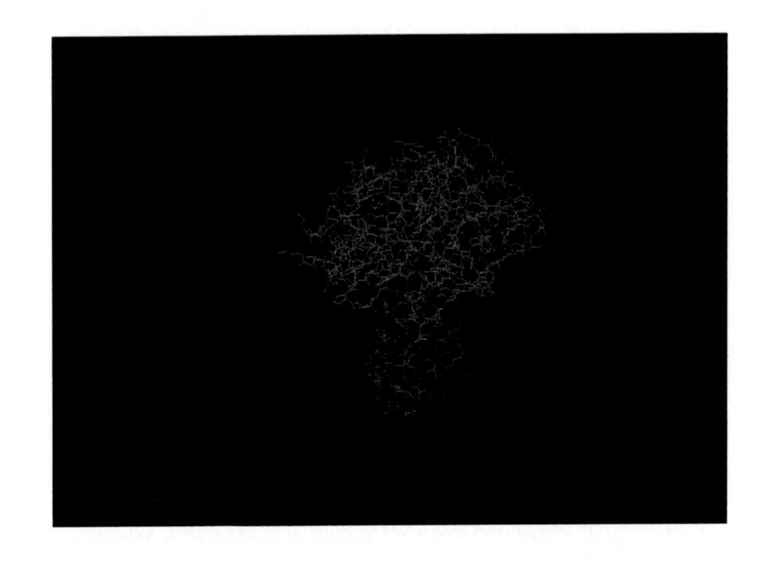

Summary

In this chapter we have been concerned with the regulation of enzyme activities. The chief points in our discussion are the following.

1. Cells regulate their metabolic activities by modulating the rates of synthesis and degradation of enzymes and by adjusting the activities of specific enzymes. Enzyme activities vary in response to changes in pH, temperature, and the concentrations of substrates or products, but also can be controlled by covalent modifications of the protein structure or by interactions with activators or inhibitors.
2. Partial proteolysis, an irreversible process, is used to activate proteases and other digestive enzymes after their secretion, and to switch on cascades of enzymes that cause blood coagulation or dissolve blood clots. Common types of reversible covalent modification include phosphorylation of serine or tyrosine residues, adenylylation, and disulfide reduction.
3. Allosteric effectors are inhibitors or activators that bind to enzymes at sites distinct from the active sites. Allosteric regulation allows cells to adjust enzyme activities rapidly and reversibly in response to changes in the concentrations of substances that are structurally unrelated to the enzyme substrates or products.
4. The initial steps in a biosynthetic pathway commonly are inhibited by the end products of the pathway, and numerous enzymes are regulated by ATP, ADP, or AMP.
5. The activities of allosteric enzymes typically show a sigmoidal dependence on substrate concentration, rather than the hyperbolic dependence predicted by the Michaelis-Menten equation. These enzymes usually have multiple subunits, and the sigmoidal kinetics can be ascribed to cooperative interactions of the subunits. Binding of the substrate to one subunit changes the dissociation constant for substrate on another subunit. The extent of the cooperativity can be described phenomenologically by the Hill equation or by equations based on the symmetry model or the more general sequential model.
6. The symmetry model postulates that the enzyme can exist in either of two conformations, T and R. It is assumed that the substrate binds more tightly to the R conformation than to T, that the binding of the substrate or an allosteric effector can change the equilibrium between these conformations, and that cooperative interactions among the subunits make all of the subunits switch from one conformation to the other in a concerted manner. Although the symmetry model generally oversimplifies the situation, it provides a useful conceptual framework that is consistent with the behavior of many allosteric enzymes.
7. Phosphofructokinase, the key regulatory enzyme of glycolysis, has four identical subunits. It exhibits sigmoidal kinetics with respect to fructose-6-phosphate and is inhibited by ATP and stimulated by ADP and numerous other metabolites. The crystal structures show that the four subunits rotate with respect to each other in the transition between the R and T conformations. Accompanying this rotation, there is a rearrangement of the binding site for fructose-6-phosphate, which is located at an interface between different subunits. At another interface, antiparallel β strands of two subunits are hydrogen-bonded together in the T structure but are separated by a row of water molecules in the R structure. This structural feature explains the cooperative nature of the conformational transition in all of the subunits: Intermediate structures probably would be much less stable than either the R or the T structure.

8. Aspartate carbamoyltransferase, the first enzyme in the biosynthesis of pyrimidines from aspartate, is inhibited allosterically by CTP, an end product of the pathway, and is stimulated by ATP. It has six identical catalytic subunits and six regulatory subunits. The different types of subunits can be separated by treating the enzyme with mercurials. Binding of a substrate analog (PALA) to one of the catalytic subunits causes the entire c_6r_6 complex to flip from T to R; binding of CTP to a regulatory subunit favors the T conformation. Again, the crystal structures show that the substrate-binding sites are located at interfaces between subunits, and that the T $\rightarrow$ R transition results in a rotation of the subunits and brings together components of the substrate-binding site.
9. Glycogen phosphorylase breaks down glycogen to glucose-1-phosphate. It exists in two forms that differ by a covalent modification. Phosphorylase *a* is phosphorylated on a serine residue and is the more active form under typical cellular conditions. Phosphorylase *b*, which lacks the phosphate, is less active under cellular conditions because of strong allosteric inhibition by ATP, although it is stimulated allosterically by AMP. The enzyme consists of two large, identical subunits. The binding site for allosteric effectors is at the interface; the substrate-binding site is about 30 Å away. In the T $\rightarrow$ R transition there is a repositioning of arginine and aspartate residues at the binding site for the substrate phosphate. Conversion of phosphorylase *b* to phosphorylase *a* tightens the interactions between the subunits and favors the transition to the R form.
10. Some proteins are activated or inhibited by interactions with other proteins. Calmodulin, a Ca^{2+}-binding protein, activates numerous proteins in response to changes in intracellular Ca^{2+} concentrations. Proteases generally are held in check by specific inhibitor proteins.

Selected Readings

Babu, Y. S., J. S. Sack, T. J. Greenough, C. E. Bugg, A. R. Means, and W. J. Cook, Three-dimensional structure of calmodulin. *Nature* 315:37, 1985.

Barford, D., S. -H. Hu, and L. N. Johnson, Structural mechanism for glycogen phosphorylase control by phosphorylation and AMP. *J. Mol. Biol.* 218:233, 1991.

Barford, D., and L. N. Johnson, The allosteric transition of glycogen phosphorylase. *Nature* 340:609, 1989.

Cséke, C., and B. B. Buchanan, Regulation of the formation and utilization of photosynthate in leaves. *Biochem. Biophys. Acta.* 853:43, 1986.

Furie, B., and B. C. Furie, The molecular basis of blood coagulation. *Cell* 53:505, 1988.

Laskowski, M., Jr., and I. Kato, Protein inhibitors of proteinases. *Ann. Rev. Biochem.* 49:593, 1980.

Lipscomb, W. N., Structure and function of allosteric enzymes. *Chemtracts-Biochem. Mol. Biol.* 2:1, 1991.

Perutz, M. F., Mechanisms of cooperativity and allosteric regulation in proteins. *Quart. Revs. Biophys.* 22:139–51, 1989.

Schachman, H. R., Can a simple model account for the allosteric transition of aspartate transcarbamoylase? *J. Biol. Chem.* 263:18583, 1988.

Schirmer, T., and P. R. Evans, Structural basis of the allosteric behaviour of phosphofructokinase. *Nature* 343:140, 1990.

Sprang, S. R., K. R. Acharya, E. J. Goldsmith, D. I. Stuart, K. Varvill, R. J. Fletterick, N. B. Madsen, and L. N. Johnson, Structural changes in glycogen phosphorylase induced by phosphorylation. *Nature* 336:215, 1988.

Stevens, R. C., J. E. Gouaux, and W. N. Lipscomb, Structural consequences of effector binding to the T state of aspartate carbamoyltransferase: Crystal structures of the unligated and ATP- and CTP-complexed enzymes at 2.6 Å resolution. *Biochem.* 29:7691, 1990.

Problems

1. Show by writing the reactions how the combination of a protein kinase and the corresponding phosphoprotein phosphatase, if unregulated, theoretically forms a futile cycle.
2. The cAMP-dependent protein kinases phosphorylate specific Ser (Thr) residues on target proteins. Given the availability of serine and threonine residues on the surface of globular proteins, how might a protein kinase select the "correct" residues to phosphorylate?
3. Assume that the flow diagram shown represents an amino acid biosynthetic pathway where G, J, and H are amino acids and A is a common precursor. Products G, H, and J are required by the cell. Enzymes catalyzing the steps are numbered. Suggest a plausible scheme for the regulation of specific enzymes by their products.

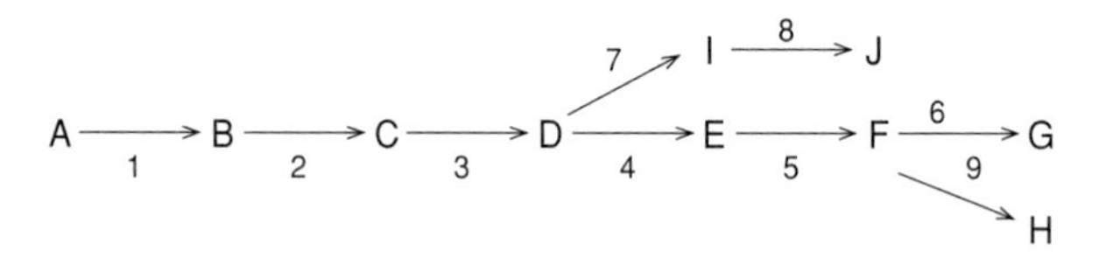

4. Aspartate carbamoyltransferase is an allosteric enzyme in which the active sites and the allosteric effector binding sites are on different subunits. Explain how it might be possible for an allosteric enzyme to have both kinds of sites on the same subunit.

5. ATP is both a substrate and an inhibitor of the enzyme phosphofructokinase (PFK). Although the substrate fructose-6-phosphate binds cooperatively to the active site, ATP does not bind cooperatively. Explain how ATP may be both a substrate and an inhibitor of PFK.
6. Calculate the substrate concentration [S] in terms of K_m for a hyperbolically responding enzyme when the velocity is 10% V_{max} or 90% V_{max}. What is the ratio of substrate concentrations that affect the ninefold velocity change ($S_{0.9}/S_{0.1}$)? How would the ($S_{0.9}/S_{0.1}$) ratio differ for an allosterically responding enzyme?
7. The substrate concentration yielding half-maximal velocity is equal to K_m for hyperbolically responding enzymes. Is this relationship true for allosterically or sigmoidally responding enzymes? How do positive or negative allosteric effectors change the substrate concentration required for half-maximal velocity?
8. Examine the relationship of aspartate carbamoyltransferase (ACTase) activity to aspartate concentration shown in the figure. Estimate the ($S_{0.9}/S_{0.1}$) ratio for the reaction under the following conditions: (a) normal curve, (b) plus 0.2-mM CTP, (c) plus 0.8-mM ATP. Do these ratios differ significantly? Explain.

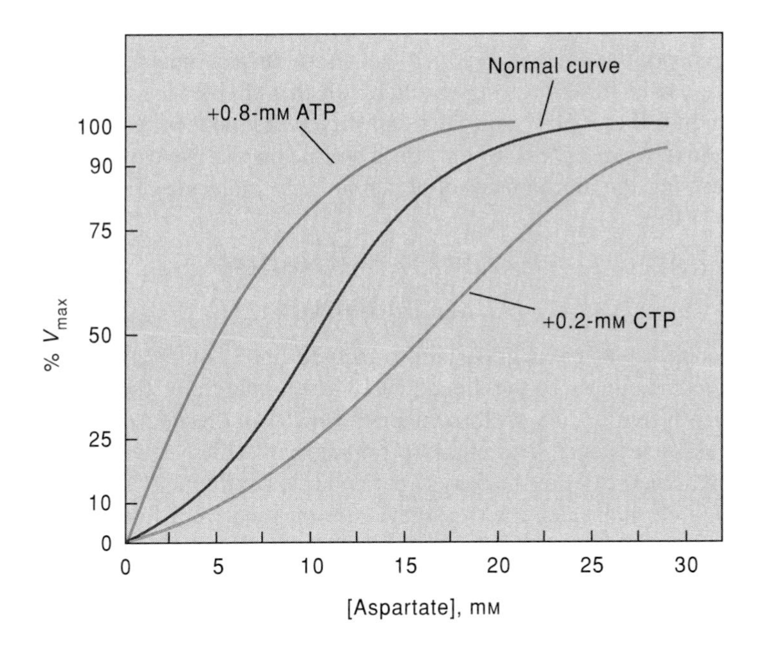

9. If you separated hemoglobin into dimers of ($\alpha + \beta$) subunits, would you expect the dimers to bind more or less O_2 at low O_2 tensions? Explain. What effect would 2,3-bisphosphoglycerate have on oxygen binding? (Review O_2 binding to hemoglobin in chapter 5.)
10. Treatment of ACTase with mercurials causes loss of allosteric regulation by ATP and CTP and eliminates positive cooperativity with aspartate. Can you suggest a strategy to "lock" an allosterically regulated enzyme into the active form without dissociating the subunits?
11. NAD^+ binding to the dimeric liver alcohol dehydrogenase (LADH) causes rearrangement of the active site residues to foster binding of the substrate, yet LADH shows no cooperativity in binding either NAD^+ or substrate. What element of the conformational change is lacking in the LADH that is present in allosteric enzymes?
12. Light-dependent activation of key enzymes in photosynthetic CO_2 fixation involves activation by thioredoxin-mediated reduction of critical disulfides on the enzymes. Write a reaction linking the reductant ferredoxin (a single-electron donor) to the reduction of protein disulfides using thioredoxin as an intermediate.
13. Assuming that the thioredoxin-dependent activation of proteins resulting from disulfide reduction is reversible, what possibilities exist for the "deactivation" of the enzymes? What are the metabolic and energetic ramifications to the cell in utilizing irreversibly activated or inactivated enzymes?
14. In some instances, protein kinases are activated or inhibited by low-molecular-weight modifiers. Explain how a metabolite may more effectively regulate an enzyme by modifying a protein kinase rather than directly inhibiting the target enzyme.
15. Cite some advantages in having calmodulin tightly associated with the Ca^{2+}-regulated enzyme rather than in an uncomplexed form in the cell.

11

CHAPTER

Vitamins, Coenzymes, and Metal Cofactors

he types of chemical reactions that can be catalyzed by proteins alone are limited by the chemical properties of the functional groups found in the side chains of nine amino acids: the imidazole ring of histidine; the carboxyl groups of glutamate and aspartate; the hydroxy groups of serine, threonine, and tyrosine; the amino group of lysine; the guanidinium group of arginine; and the sulfhydryl group of cysteine. These groups can act as general acids and bases in catalyzing proton transfers and as nucleophilic catalysts in group transfer reactions.

Many metabolic reactions involve chemical changes that could not be brought about by the structures of the amino acid side chain functional groups in enzymes acting by themselves. In catalyzing these reactions, enzymes act in cooperation with other smaller organic molecules or metallic cations, which possess special chemical reactivities or structural properties that are useful for catalyzing reactions. In this chapter we will introduce these small molecules and survey the range of reactions that they catalyze. Our emphasis will be on the principles underlying the mechanisms of action of these molecules.

Vitamins—Essential Nutrients Required in Small Amounts

People knew it was important to include small amounts of certain substances in the diet as early as the middle of the eighteenth century, long before the biochemistry of these substances was understood. Vitamins are organic molecules essential in small quantities for healthy nutrition in rats or humans. The list of such molecules grew as they were purified from foodstuffs and shown to cure various disorders in animals maintained on deficient diets. The name "vitamine" was given in 1911 to the first vitamin to be isolated, thiamine. When it became clear that a number of essential organic micronutrients were not amines, the *-e* was dropped.

Vitamins are divided into water-soluble and lipid-soluble groups. In addition to vitamins there are vitaminlike nutrients that are required in small amounts by the organism and frequently function in similar capacities in the organism. These compounds are not classified as vitamins because rats and humans have a limited capacity to synthesize them, provided that the diet contains the essential precursors. Table 11.1 lists both vitamins and vitaminlike nutrients.

Most Coenzymes Are Modified Forms of Vitamins

In this chapter we will be concerned, not primarily with vitamins *per se,* but with coenzymes. Most coenzymes are modified forms of vitamins. The modifications take place in the organism

Table 11.1
Vitamins and Vitaminlike Nutrients

Vitamin	Function
Water-Soluble Vitamins	
Thiamine (B_1)	Precursor of the coenzyme thiamine pyrophosphate. Deficiency can cause beriberi.
Riboflavin (B_2)	Precursor of the coenzymes flavin mononucleotide and flavin adenine dinucleotide. Deficiency leads to growth retardation.
Pyridoxine (B_6)	Precursor of the coenzyme pyridoxal phosphate. Deficiency causes dermatitis in rats.
Nicotinic acid (niacin)	Precursor of the coenzymes nicotinamide adenine dinucleotide and nicotinamide adenine dinucleotide phosphate. Deficiency leads to pellagra.
Pantothenic acid	Precursor of coenzyme A (CoA). Deficiency leads to dermatitis in chickens.
Biotin	Precursor of the coenzyme biocytin. Deficiency leads to dermatitis in humans.
Folic acid	Precursor of the coenzyme tetrahydrofolic acid. Deficiency causes anemias.
Vitamin B_{12}	Precursor of the coenzyme deoxyadenosyl cobalamin. Deficiency leads to pernicious anemia.
Vitamin C	Cosubstrate in the hydroxylation of proline in collagen. Deficiency leads to scurvy.
Lipid-Soluble Vitamins	
Vitamin A	Vision, growth, and reproduction (see chapter 36).
Vitamin D	Regulation of calcium and phosphate metabolism (see chapter 24).
Vitamin E	Antisterility factor in rats.
Vitamin K	Important for blood coagulation.
Vitaminlike Nutrients	
Inositol	Mediator of hormone action (see chapters 22 and 24).
Choline	Important for integrity of cell membranes and lipid transport (see chapter 22).
Carnitine	Essential for transfer of fatty acids to mitochondria (see chapter 17).
α-Lipoic acid	Coenzyme in the oxidative decarboxylation of keto acids (this chapter).
p-Aminobenzoate (PABA)	Component of folic acid (this chapter).
Coenzyme Q (ubiquinones)	Important for electron transport in mitochondria (see chapter 15).

after ingestion of the vitamins. Coenzymes act in concert with enzymes to catalyze biochemical reactions. Tightly bound coenzymes are sometimes referred to as prosthetic groups. A coenzyme usually functions as a major component of the active site on the enzyme, which means that understanding the mechanism of coenzyme action usually requires a complete understanding of the catalytic process.

Water-Soluble Vitamins and Their Coenzymes

As we have noted, some vitamins are soluble in water, while others are soluble in lipids. In the following sections we will survey the range of biochemical reactions in which water-soluble coenzymes participate.

Thiamine Pyrophosphate (TPP) Is Involved in C — C and C—X Bond Cleavage

The structure of thiamine pyrophosphate is given in figure 11.1. The vitamin, thiamine or vitamin B_1, lacks the pyrophosphoryl group. Thiamine pyrophosphate (TPP) is the essential coenzyme involved in the actions of enzymes that catalyze cleavages

Figure 11.1

(*a*) Structure of thiamine pyrophosphate and (*b, c*) the bonds it cleaves or forms. Reactive part of the coenzyme and the bonds subject to cleavage in (*b*) and (*c*) are indicated in color.

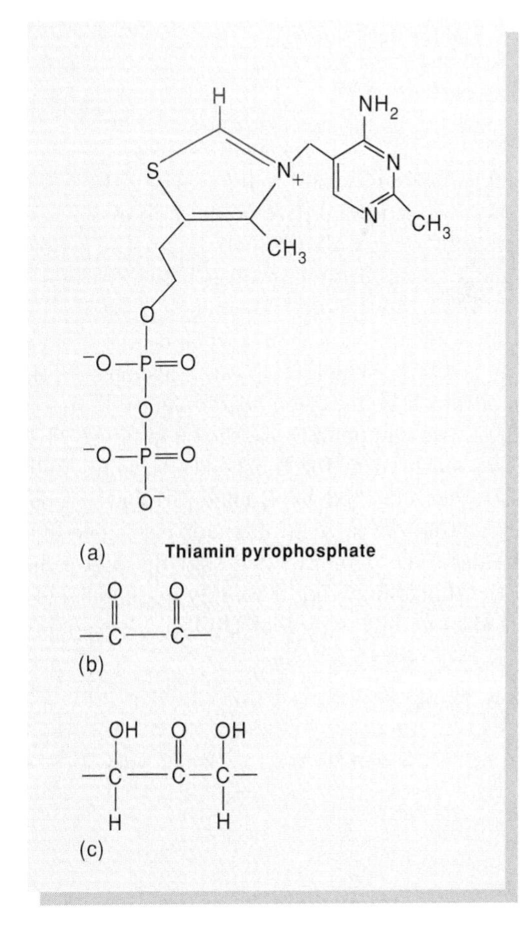

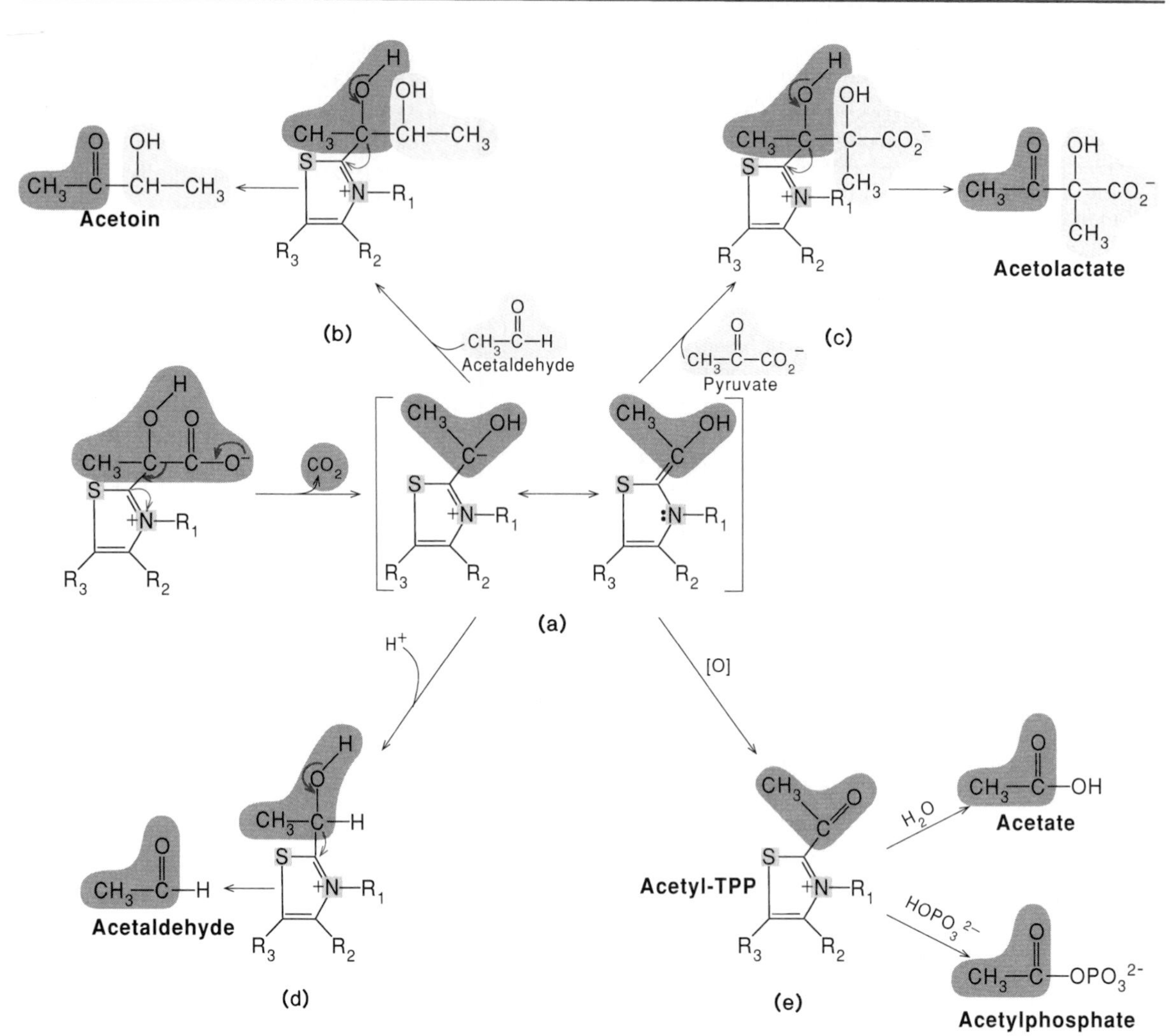

Figure 11.2

Mechanism of thiamine pyrophosphate action. Intermediate (*a*) is represented as a resonance-stabilized species. It arises from the decarboxylation of the pyruvate-thiamine pyrophosphate addition compound shown at left of (*a*) and in equation (2). It can react as a carbanion with acetaldehyde, pyruvate, or H^+ to form (*b*), (*c*), or (*d*), depending on the specificity of the enzyme. It can also be oxidized to acetyl-thiamine pyrophosphate (*e*) by other enzymes, such as pyruvate oxidase. The intermediates (*b*) through (*e*) are further transformed to the products shown by the actions of specific enzymes.

of the bonds indicated in color in figure 11.1. The bond scission in figure 11.1*b* is representative of those in many α-keto acid decarboxylations, nearly all of which require the action of TPP. The phosphoketolase reaction involves both cleavages shown in figure 11.1*c*, while the transketolase reaction (see fig. 13.22) involves the cleavage of the carbon–carbon bond but not the elimination of —OH. Acetolactate and acetoin arise by the formation of the carbon–carbon bond in figure 11.1*c* (the structures of these two compounds are shown in figure 11.2).

The mechanism for the bond cleavages indicated in figure 11.1*b* was clarified by Ronald Breslow. In one of the earliest applications of nuclear magnetic resonance to biochemical mechanisms, he demonstrated that the proton bonded to C—2 in the thiazolium ring is readily exchangeable with the protons of H_2O and deuterons of D_2O in a base-catalyzed reaction:

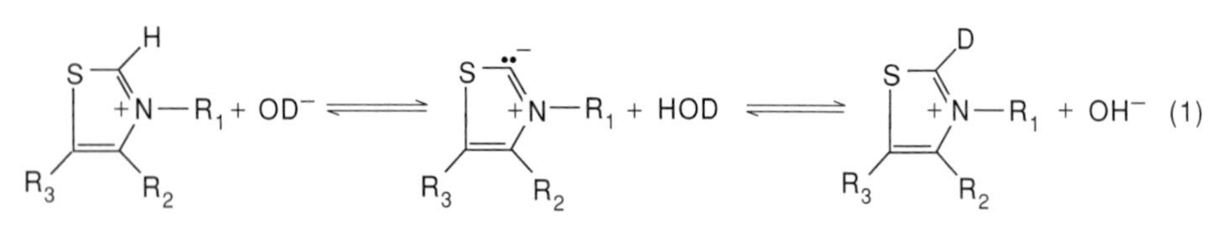

The active intermediate shown in equation (1) undergoes nucleophilic addition to the bond of polar carbonyl groups in substrates to produce intermediates such as

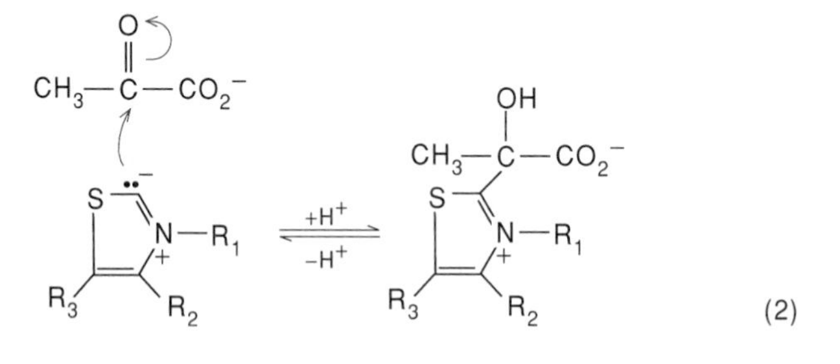

Figure 11.3

Structures of vitamin B_6 derivatives and the bonds cleaved or formed by the action of pyridoxal phosphate (*a*). The reactive part of the coenzyme is shown in color in (*a*). The bonds shown in color in (*d*) are the types of bonds in substrates that are subject to cleavage.

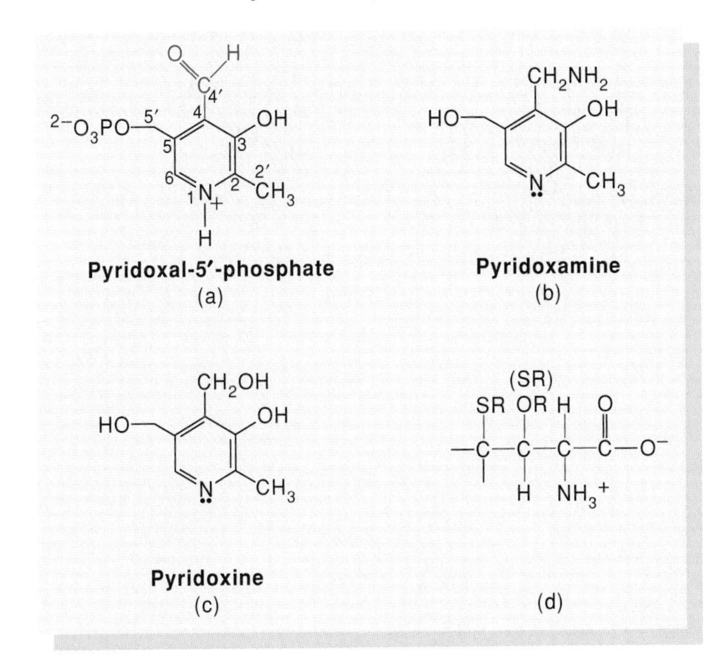

Intermediates of this type have the necessary chemical reactivity for cleaving the bonds indicated in figure 11.1*b* and *c*. The decarboxylated product of the pyruvate adduct shown in equation (2) is resonance-stabilized by the thiazolium ring (fig. 11.2*a*). This intermediate may be protonated to α-hydroxyethyl thiamine pyrophosphate (fig. 11.2*d*); alternatively, it may react with other electrophiles, such as the carbonyl groups of acetaldehyde or pyruvate, to form the species in figure 11.2*b* and *c;* or it may be oxidized to acetyl-thiamine pyrophosphate (fig. 11.2*e*). The fate of the intermediate depends on the reaction specificity of the enzyme with which the coenzyme is associated.

Pyridoxal-5′-Phosphate Is Required for a Variety of Reactions with α-Amino Acids

Pyridoxal-5′-phosphate is the coenzyme form of vitamin B_6, and has the structure shown in figure 11.3. The name vitamin B_6 is applied to any of a group of related compounds lacking the phosphoryl group, including pyridoxal, pyridoxamine, and pyridoxine.

Pyridoxal-5′-phosphate participates in many reactions with α-amino acids, including transaminations, α-decarboxylations, racemizations, α,β eliminations, β,γ eliminations, aldolizations, and the β decarboxylation of aspartic acid. The following equations illustrate several reactions in which pyridoxal-5′-phosphate acts as a coenzyme.

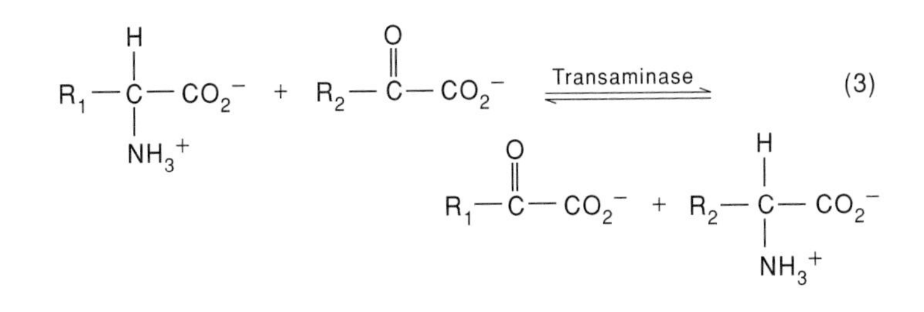

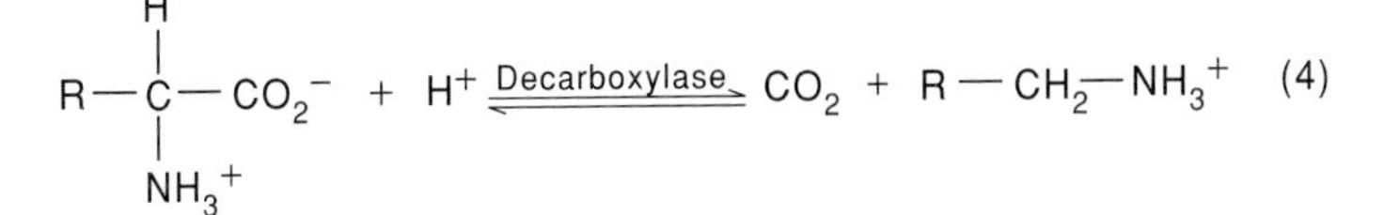

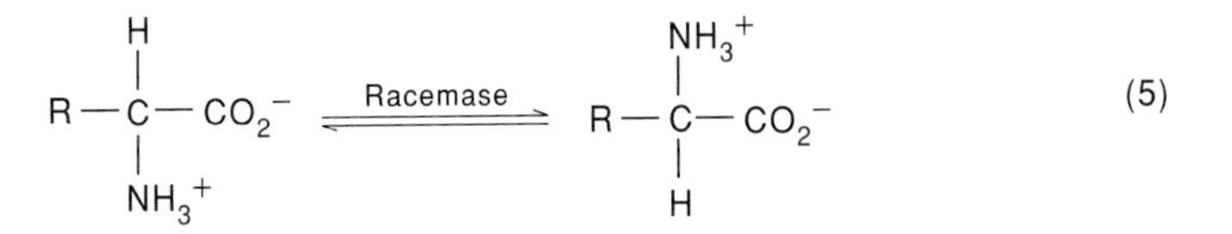

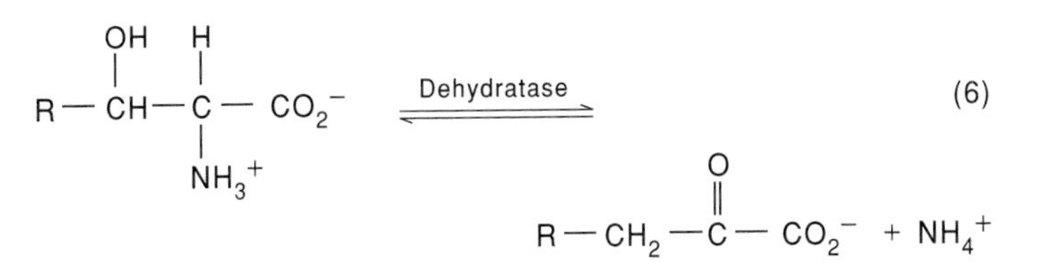

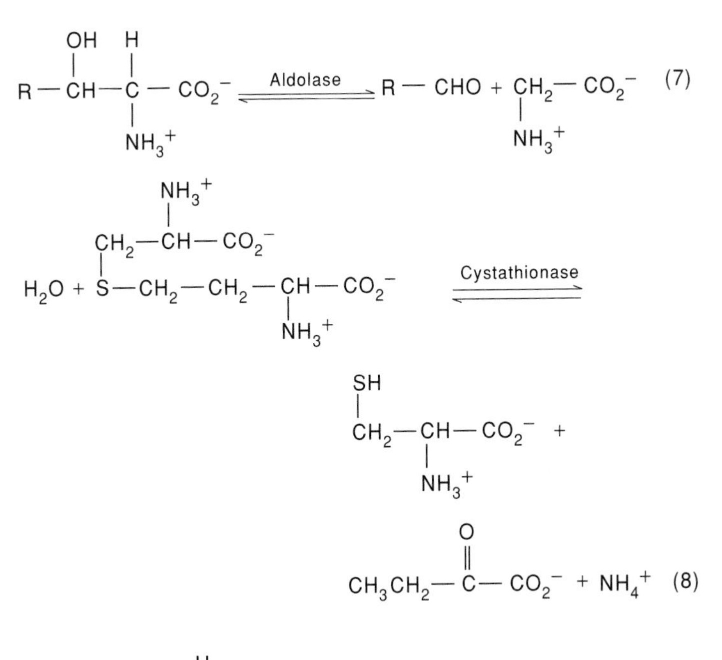

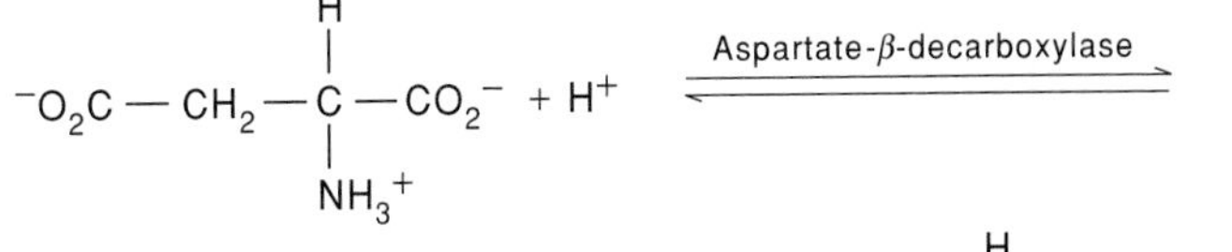

Figure 11.4

Structures of catalytic intermediates in pyridoxal-phosphate–dependent reactions. The initial aldimine intermediate resulting from Schiff's base formation between the coenzyme and the α-amino group of an amino acid (*a*). This aldimine is converted to the resonance-stabilized intermediate (*b*) by loss of a proton at the alpha carbon. Further enzyme-catalyzed proton transfers to intermediates (*c*) and (*d*) may occur, depending upon the specificity of a given enzyme. The enzymes use their general acids and bases to catalyze these proton transfers.

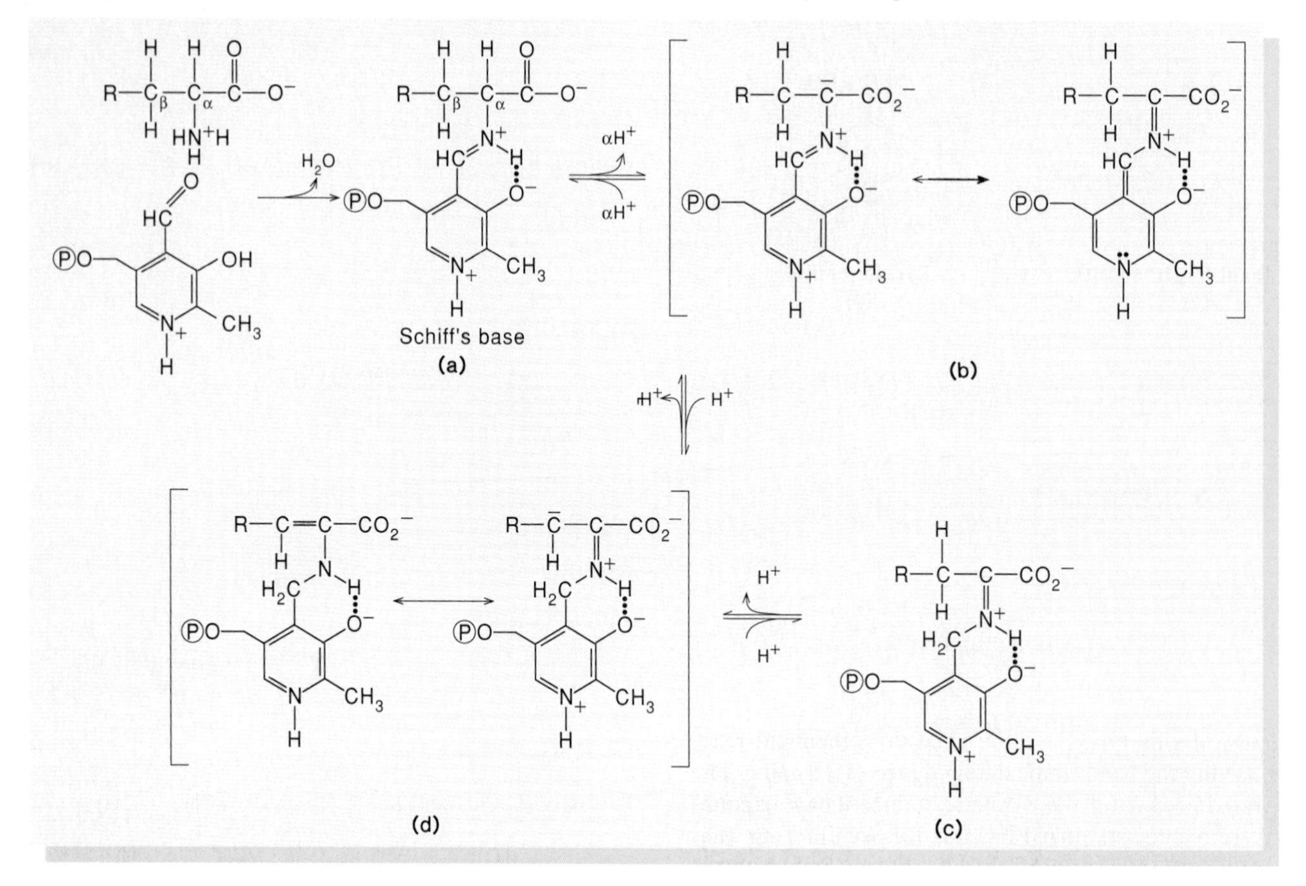

These equations involve bond cleavages of the type shown in color in figure 11.3*d*. Pyridoxal-5′-phosphate promotes these heterolytic bond cleavages by stabilizing the resulting electron pairs at the α- or β-carbon atoms of α-amino acids. To do this, the aldehyde group of the coenzyme first reacts with the α-amino group of an amino acid to produce an aldimine (fig. 11.4*a*) or Schiff's base, which is internally stabilized by H bonding. Loss of the α hydrogen as H^+ produces a resonance-stabilized species (fig. 11.4*b*) in which the electron pair is delocalized into the pyridinium system. This active intermediate may undergo further reactions at the carbon to form products determined by the reaction specificity of the enzyme. If, for example, the enzyme is a racemase, the species resulting from the loss of the proton from the α carbon may accept a proton from the opposite side to produce, ultimately, the enantiomer of the amino acid.

When the substrate is substituted at the β carbon with a potential leaving group, such as —OH, —SH, —OPO_3^{3-} (see fig. 11.3*d*), the corresponding α-carbanion intermediate (see fig. 11.4*b*) can eliminate the group. This is an essential step in α,β eliminations, such as the dehydratase reaction (equation 6; also see fig. 19.11). Upon hydrolysis, the elimination intermediate produces pyridoxal-5′-phosphate and the substrate-derived enamine, which spontaneously hydrolyzes to ammonia and an α-keto acid.

The full series of intermediates in a transamination is shown in figure 11.5*a*. After protonation at the aldimine carbon of pyridoxal-5′-phosphate (step 3), hydrolysis (step 4) forms an α-keto acid and pyridoxamine-5′-phosphate. The reverse of this sequence with a second α-keto acid (steps 5 through 8) completes the transamination reaction.

An intermediate analogous to that in figure 11.4*b*, but generated from glycine and so lacking the β and γ carbons, can react as a carbanion with an aldehyde to produce a β-hydroxy-α-amino acid. These reactions are catalyzed by aldolases such as threonine aldolase or serine hydroxymethyl transferase (see fig. 19.11).

β-Decarboxylases (fig. 11.5*b*) generate intermediates analogous to that in figure 11.4*b* by catalyzing the elimination of CO_2 instead of H^+ from the intermediate in figure 11.4*a* (step 4, fig. 11.5*b*). Protonation of the α-carbanionic intermediates by protons from H_2O, followed by hydrolysis of the resulting imines, produces the amines corresponding to the replacement of the carboxylate group in the substrate by a proton (steps 5 through 8, fig. 11.5*b*).

The stability of the resonance hybrid (see fig. 11.4*b*) accounts for the catalytic action of pyridoxal-5′-phosphate in the reactions shown in equations (3) through (7).

Returning to the intermediate in figure 11.4*c*, we see that elimination of a β proton produces a β-carbanion (fig. 11.4*d*) that is stabilized by resonance with the neighboring protonated imine. This carbanion can eliminate a good leaving group from the carbon, exemplified by the γ-cystathionase-catalyzed elimination of cysteine from cystathionine in equation (8). The β decarboxylation of aspartate (equation 9) proceeds by elimination of a β-carbanionic intermediate like that in figure 11.4*d* from the ketimine, analogous to the intermediate produced by loss of the α proton from the aldimine of aspartate with pyridoxal-5′-phosphate.

The fundamental biochemical function of pyridoxal-5′-phosphate is the formation of aldimines with α-amino acids that stabilize the development of carbanionic character at the α and β carbons of α-amino acids in intermediates such as those in figures 11.4*b* and *c*. Enzymes acting alone cannot stabilize these carbanions and so cannot, by themselves, catalyze reactions requiring their formation as intermediates.

Nicotinamide Coenzymes Are Used in Reactions Involving Hydride Transfers

Nicotinamide adenine dinucleotide (NAD^+) is one of the two coenzymatic forms of nicotinamide (fig. 11.6). The other is nicotinamide adenine dinucleotide phosphate ($NADP^+$), which differs from NAD^+ by the presence of a phosphate group at C-2′ of the adenosyl moiety.

The nicotinamide coenzymes are biological carriers of reducing equivalents, i.e., electrons. The most common function of NAD^+ is to accept two electrons and a proton (H^- equivalent) from a substrate undergoing metabolic oxidation to produce NADH, the reduced form of the coenzyme. This then diffuses or is transported to the terminal-electron transfer sites of the cell and reoxidized by terminal-electron acceptors, O_2 in aerobic organisms, with the concomitant formation of ATP (chapter 15). Equations (10), (11), and (12) are typical reactions in which NAD^+ acts as such an acceptor.

$$NAD^+ + CH_3CH_2OH \underset{}{\overset{\text{Alcohol dehydrogenase}}{\rightleftharpoons}} CH_3{-}\overset{\overset{\displaystyle O}{\|}}{C}H + NADH + H^+ \qquad (10)$$

$$^-O_2C(CH_2)_2\overset{\overset{\displaystyle NH_3^+}{|}}{C}HCO_2^- + NAD^+ + H_2O \overset{\text{Glutamate dehydrogenase}}{\rightleftharpoons} {}^-O_2C(CH_2)_2\overset{\overset{\displaystyle O}{\|}}{C}CO_2^- + NADH + NH_4^+ + H^+ \qquad (11)$$

$$HPO_4^{2-} + {}^{2-}O_3POCH_2\overset{\overset{\displaystyle OH}{|}}{C}H{-}\overset{\overset{\displaystyle O}{\|}}{C}H + NAD^+ \overset{\text{Glyceraldehyde-3P dehydrogenase}}{\rightleftharpoons} {}^{2-}O_3POCH_2\overset{\overset{\displaystyle OH}{|}}{C}H\overset{\overset{\displaystyle O}{\|}}{C}OPO_3^{2-} + NADH + H^+ \qquad (12)$$

The chemical mechanisms by which NAD^+ is reduced to NADH in equations (10) through (12) are probably similar, as represented in generalized forms in equation (13).

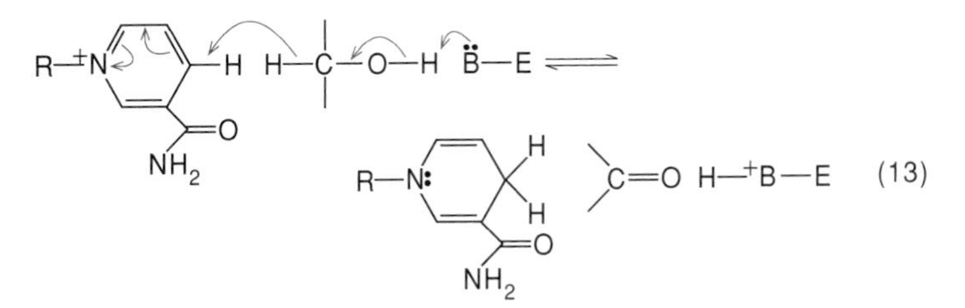

According to this formulation, the immediate oxidation product in equation (11), where $-NH_2$ replaces $-OH$ in equation (13), would be the imine of α-ketoglutarate, which would quickly undergo hydrolysis to α-ketoglutarate and ammonia in aqueous solution. The oxidation of an aldehyde group catalyzed by glyceraldehyde-3-phosphate dehydrogenase [equation (12)] also can be understood on the basis of this formulation once it is realized that there is an essential —SH group at the active site that is transiently acylated during the course of the reaction. The —SH group reacts with the aldehyde group of glyceraldehyde-3-phosphate according to equation (14), forming a thiohemiacetal which becomes oxidized. The resulting acyl-enzyme then reacts with phosphate to produce glycerate-1,3-bisphosphate.

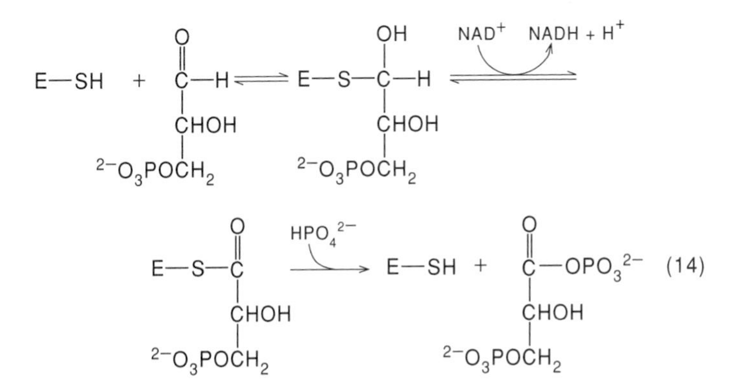

Figure 11.5

Mechanisms of action of pyridoxal phosphate: (*a*) in glutamate-oxaloacetate transaminase, and (*b*) in aspartate ß-decarboxylase.

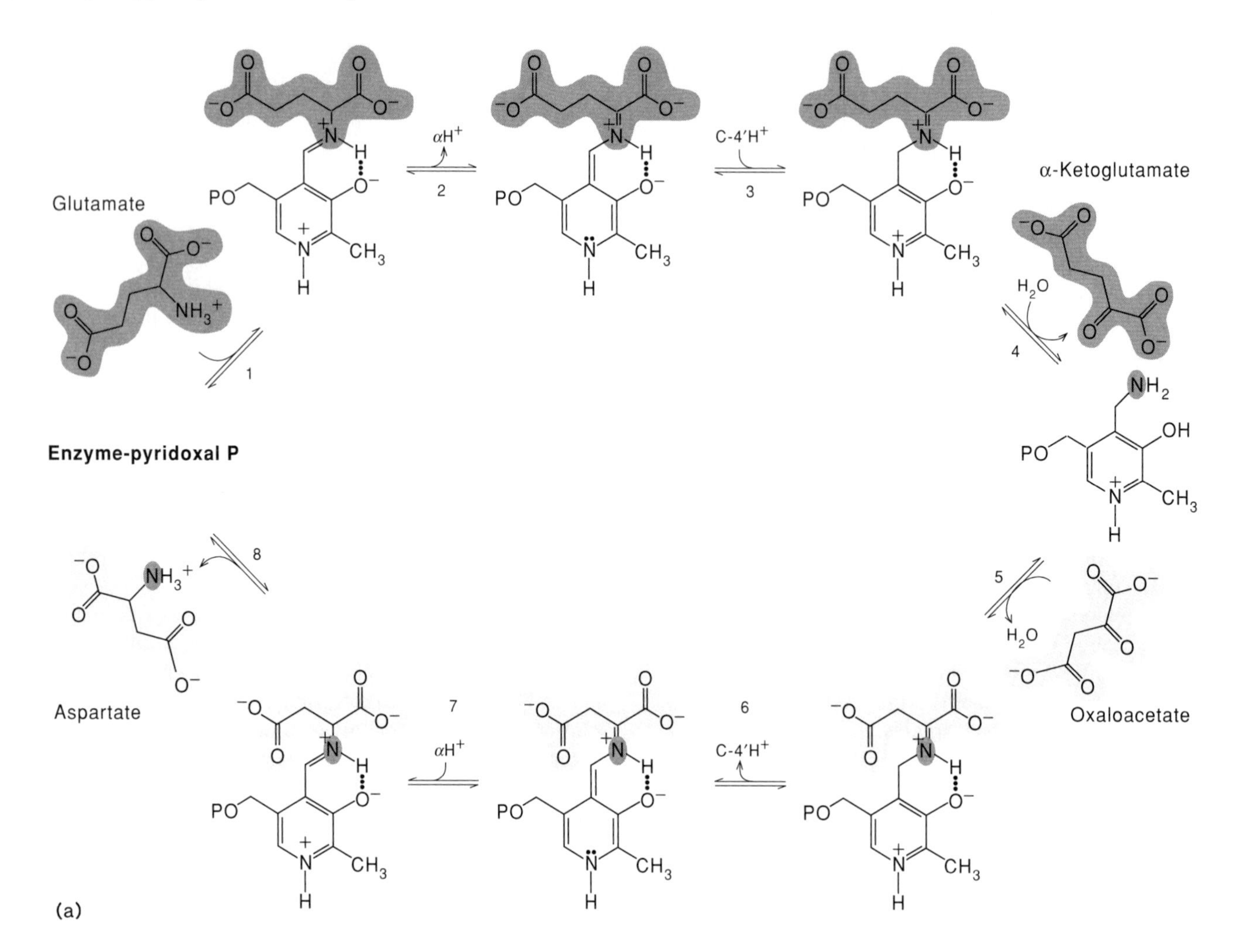

Figure 11.6

Structures of nicotinamide and nicotinamide coenzymes. The reactive sites of the coenzymes are shown in color.

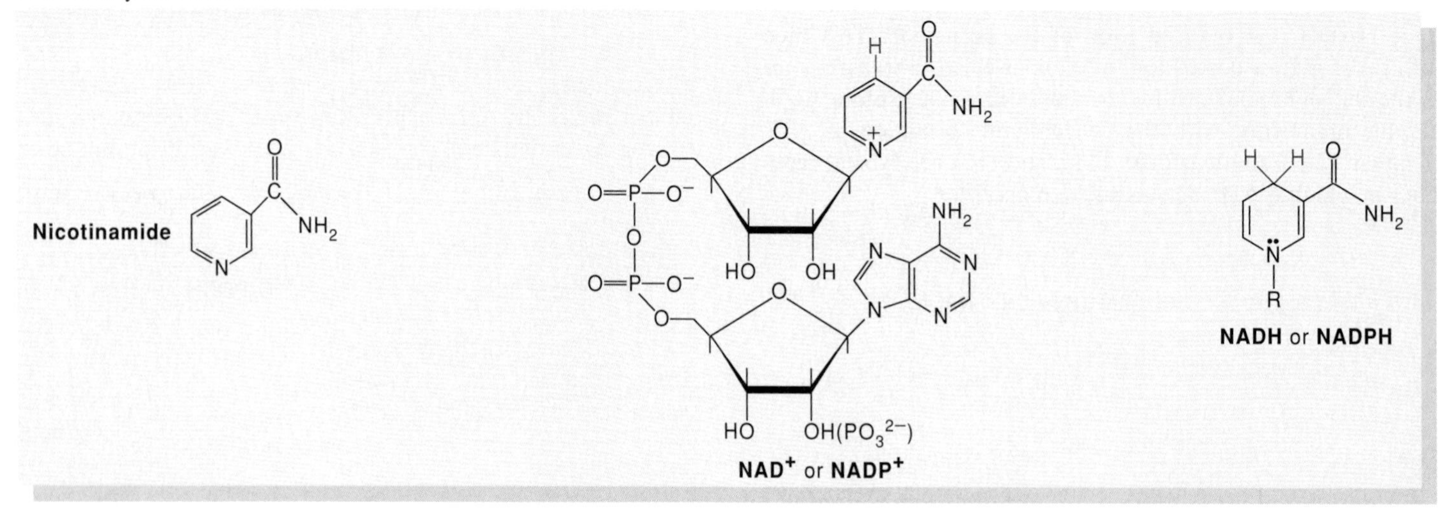

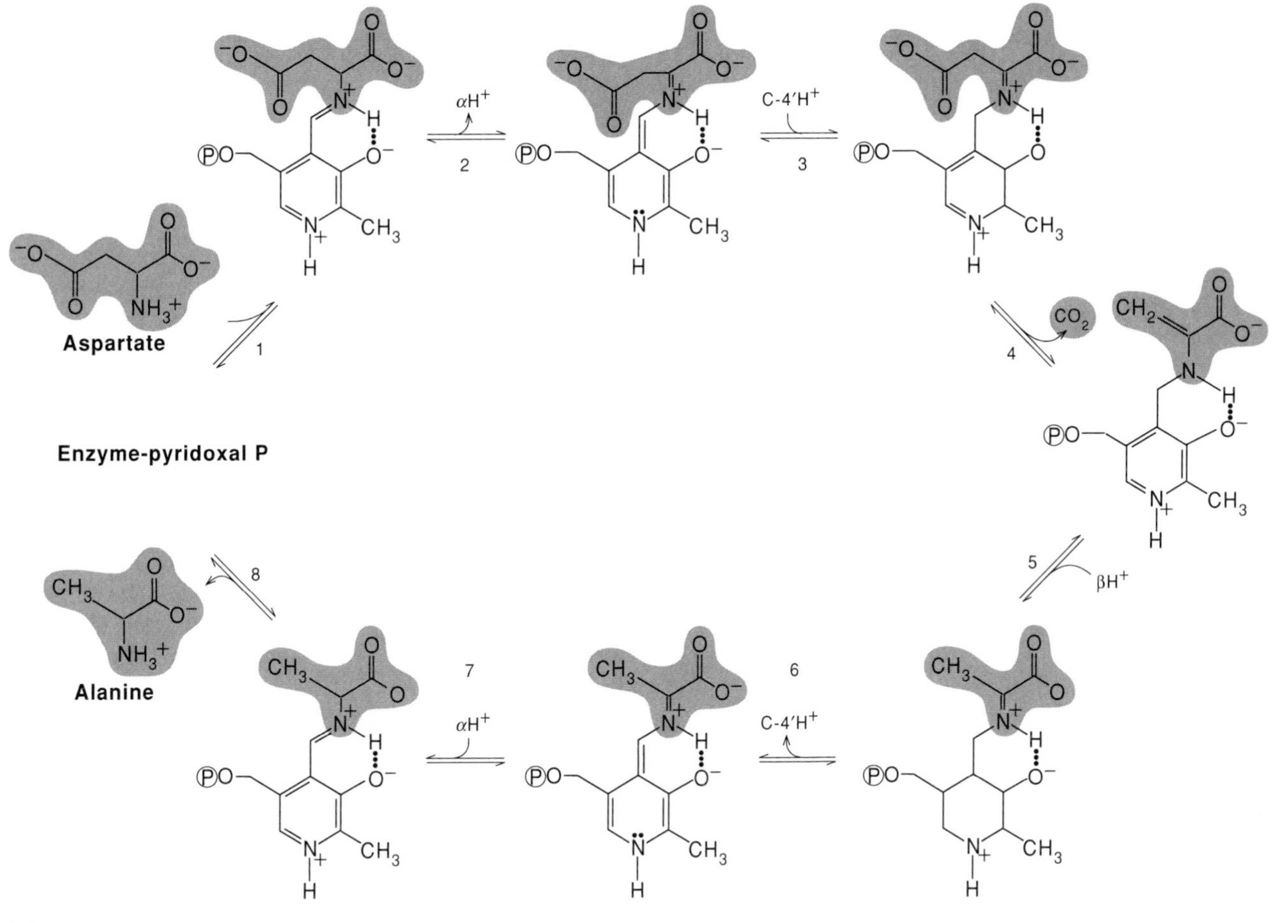

Equation (13) implies that the hydrogen atom and two electrons are transferred in a concerted process, i.e., as a hydride equivalent, with the quaternary nitrogen in the pyridinium ring serving as an electron sink. The hydrogen atom is certainly transferred directly; this reaction is discussed in the following section.

In a series of classic experiments, Westheimer, Vennesland, and co-workers demonstrated that hydrogen transfer between NAD^+ and substrates such as those in equations (10) through (12) are direct hydrogen transfers and occur with stereospecificity. The experiments with alcohol dehydrogenase were the first to establish these points, and they were the earliest to define the remarkable stereospecificity of enzymatic action at prochiral centers involving chemically equivalent hydrogen atoms. (For a discussion of prochiral centers, see box 11A.) Once the stereospecificity of hydrogen transfer to NAD^+ and the absolute configurations of the molecules were established, it became possible to formulate these processes as set forth in equations (15) through (17) for alcohol dehydrogenase.

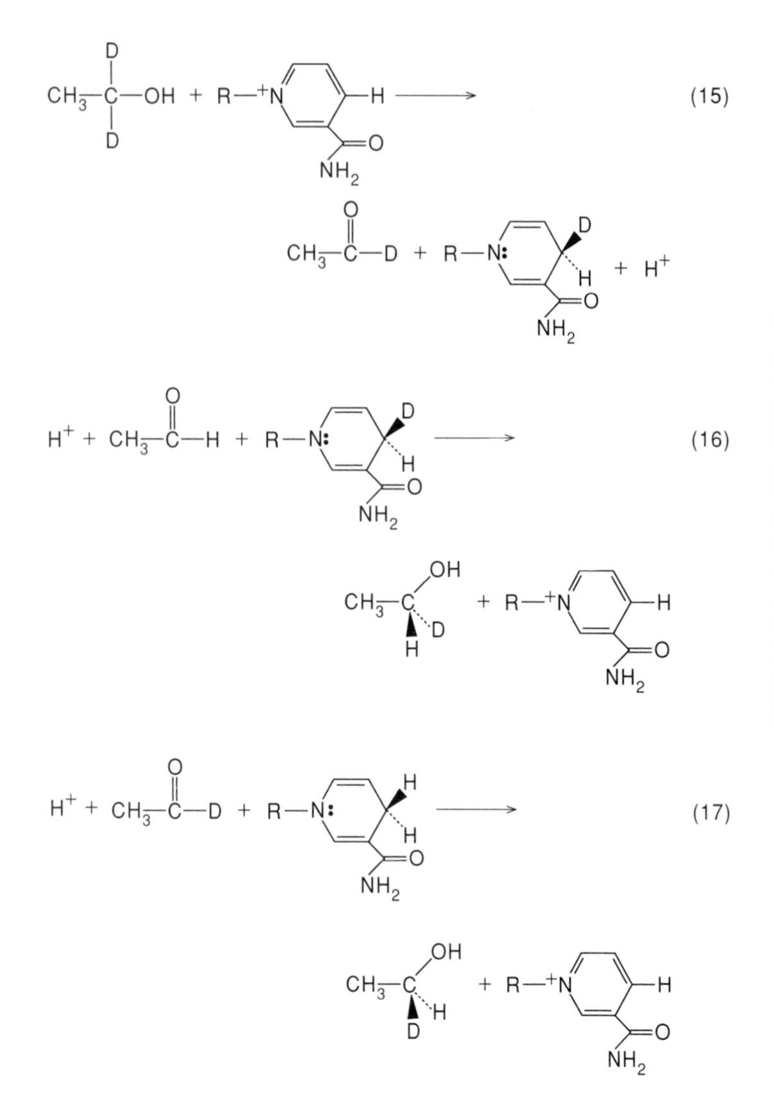

Figure 11.7

Stereospecificity of hydrogen transfer in nicotinamide coenzymes. This example shows that a highly stereoselective complex with substrates can result in a stereospecific reaction even when the substrate has a plane of symmetry. The arrows represent hypothetical enzyme-binding interactions. Because of the way in which the coenzyme and the substrate are bound and oriented toward one another, one hydrogen is predisposed for transfer, the other is not. The subscripted stereochemical symbols R and S are explained in box 11A.

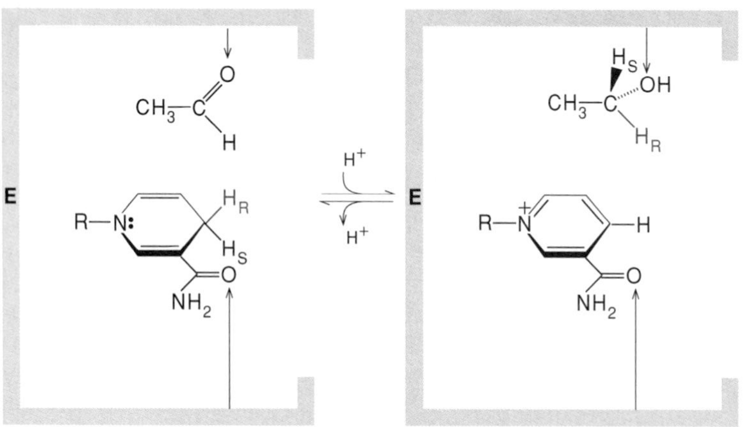

The tracing of deuterium through these transformations established quite clearly that enzymes are stereospecific in abstracting chemically equivalent hydrogens from prochiral centers and in transferring hydrogens specifically to one face of planar molecules, even in molecules as small as acetaldehyde. This specificity is thought to be a natural consequence of the fact that enzymes are asymmetrical molecules that form highly stereoselective complexes with their substrates, even some that have planes of symmetry or are themselves planar molecules. The situation is illustrated schematically in figure 11.7, which shows how specific binding interactions can lead to stereospecific hydrogen transfer between acetaldehyde and NADH.

In NADH, the two hydrogens bonded to nicotinamide C-4 are chemically equivalent, so that from a purely chemical standpoint either could be transferred to an aldehyde or ketone. It is their topographic inequivalence that leads to stereospecificity in enzymatic reactions: however, the enzymes do not all exhibit the same stereospecificity for catalyzing hydrogen transfer from this center or in forming this center from NAD^+.

Those enzymes such as alcohol dehydrogenase which catalyze transfer of the pro-A hydrogen in NADH (see fig. 11.7) are known as *R*-side-specific enzymes, and those transferring the other hydrogen, the pro-S hydrogen, are known as *S*-side-specific enzymes.

In addition to acting as a cellular electron carrier, NAD^+ also acts as a true coenzyme with certain enzymes. Enzymes are sometimes confronted with the problem of catalyzing such reactions as epimerizations, aldolizations, and eliminations on substrates lacking the intrinsic chemical reactivities required for these reactions to occur at significant rates. Sometimes such reactivities can be introduced into the substrate by oxidizing an appropriate alcohol group to a carbonyl group, and the enzyme is then found to contain NAD^+ as a tightly bound coenzyme. NAD^+ functions coenzymatically by transiently oxidizing the key alcohol group to the carbonyl level, producing an oxidatively activated intermediate whose further transformation is catalyzed by the enzyme. In the last step, the carbonyl group is reduced back to the hydroxyl group by the transiently formed NADH. A reaction of this type is illustrated in figure 11.8 for the enzyme UDP-galactose-4-epimerase, which contains tightly bound NAD^+.

Flavins Are Used in Reactions Where One or Two Electron Transfers Are Involved

Flavin adenine dinucleotide (FAD) (fig. 11.9) and flavin mononucleotide (FMN) are the coenzymatically active forms of vitamin B_2, riboflavin. Riboflavin is the N^{10}-ribityl isoalloxazine portion of FAD, which is enzymatically converted into its coenzymatic forms first by phosphorylation of the ribityl C-5′ hydroxy group to FMN and then by adenylylation to FAD. FMN and FAD appear to be functionally equivalent coenzymes, and the one that is involved with a given enzyme appears to be a matter of enzymatic binding specificity.

Figure 11.8

Mechanism of NAD^+ action in UDPgalactose-4-epimerase. There is no net oxidation or reduction. Only the intermediate is oxidized.

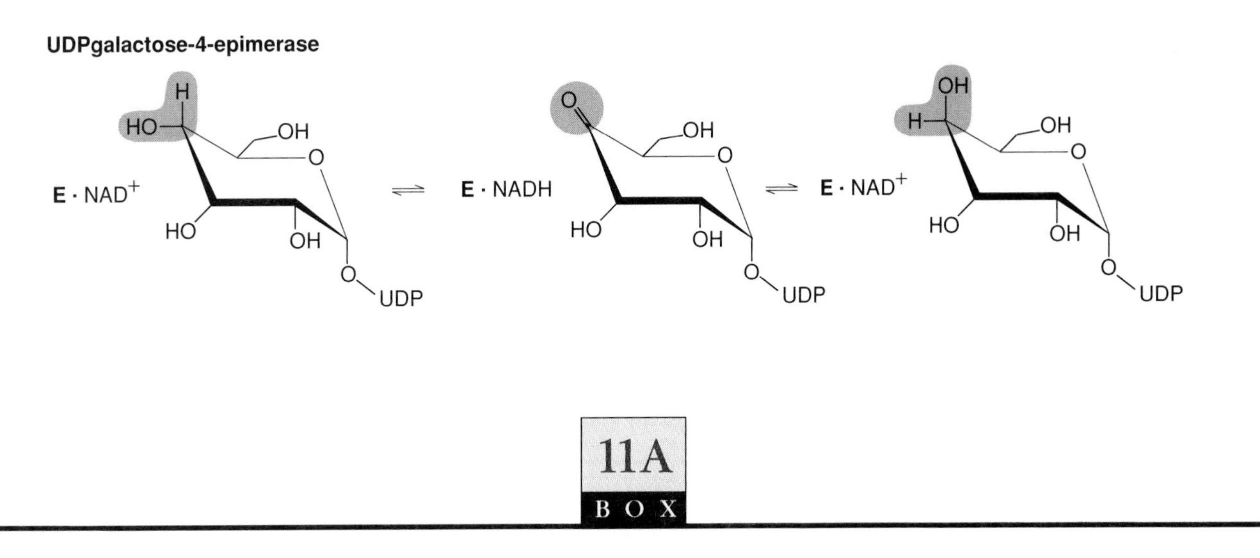

11A
BOX

The R, S System for Naming Stereoisomer Configurations

"Chiral" and "prochiral" are derived from the Greek word χειρ, meaning "hand," and as we have seen, they refer to "handedness" in stereoisomeric or potentially stereoisomeric centers. In earlier chapters we encountered the use of the prefixes L and D for distinguishing between stereoisomers with left-handed and right-handed twist, respectively. Another set of stereochemical symbols frequently used for designating two stereoisomers is the pair R and S. These symbols are assigned as follows.

For a tetrahedral carbon (or other tetrahedral atom), the four different substituents are assigned relative priorities by applying rules that generally accord higher priority to groups having the larger summation of atomic weights. (You will find these rules set out in the article by G. Popják that is listed in the Selected Readings at the end of this chapter.) Once the priorities are assigned, the atom is viewed from the side opposite the lowest priority group, and the symbol R, for "rectus," is assigned if the remaining groups appear in clockwise order from highest to lowest priority. The symbol S, for "sinister," is assigned if they appear in counterclockwise order.

For atoms whose substituent groups are $a < b < c < d$ in order of increasing priority, the following structures are those of the R and S isomers:

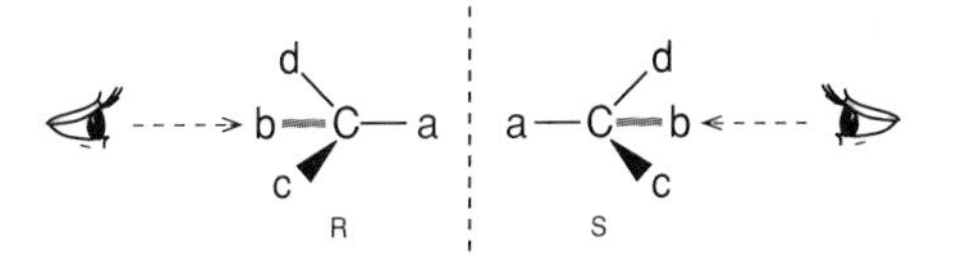

Atoms in which two of the groups are identical are said to be prochiral, since the elevation of one of the identical groups to a higher priority would lead to a chiral center. Carbon-1 in ethanol and nicotinamide C-4 in NADH are prochiral centers, since they each have two hydrogens and two other substituents. In figure 11.7, the two hydrogens are subscripted R or S for the configurational designations that would result from according higher priority to one or the other. One means of granting such priority is by the use of heavy isotopes of hydrogen. Thus deuterium has a higher priority than protium, so that the configuration of deuterioethanol in equation (16) is R, while in equation (17) it is S.

The catalytically functional portion of the coenzymes is the isoalloxazine ring, specifically N-5 and C-4a (see fig. 11.9*b*), which are thought to be the immediate locus of catalytic function, although the entire chromophoric system extending over N-5, C-4a, C-10, N-1, and C-2 should be regarded as an indivisible catalytic entity, as are the nicotinamide, pyridinium, and thioazolium rings of NAD^+, pyridoxal phosphate, and thiamine pyrophosphate.

Flavin-containing enzymes are known as flavoproteins and, when purified, normally contain their full complements of FAD or FMN. The bright yellow color of flavoproteins is due to the isoalloxazine chromophore in its oxidized form. In a few flavoproteins, the coenzyme is known to be covalently bonded to the protein by means of a sulfhydryl or imidazole group at the C-8 methyl group and in at least one case at C-6. In most flavoproteins, the coenzymes are tightly but noncovalently bound, and many can be resolved into apoenzymes that can be reconstituted to holoenzymes by readdition of FAD or FMN.

Figure 11.9

Structures of the vitamin riboflavin (*a*) and the derived flavin coenzymes (*b*). Like NAD^+ and $NADP^+$, the coenzyme pair FMN and FAD are functionally equivalent coenzymes, and the coenzyme involved with a given enzyme appears to be a matter of enzymatic binding specificity. The catalytically functional portion of the coenzymes is shown in color.

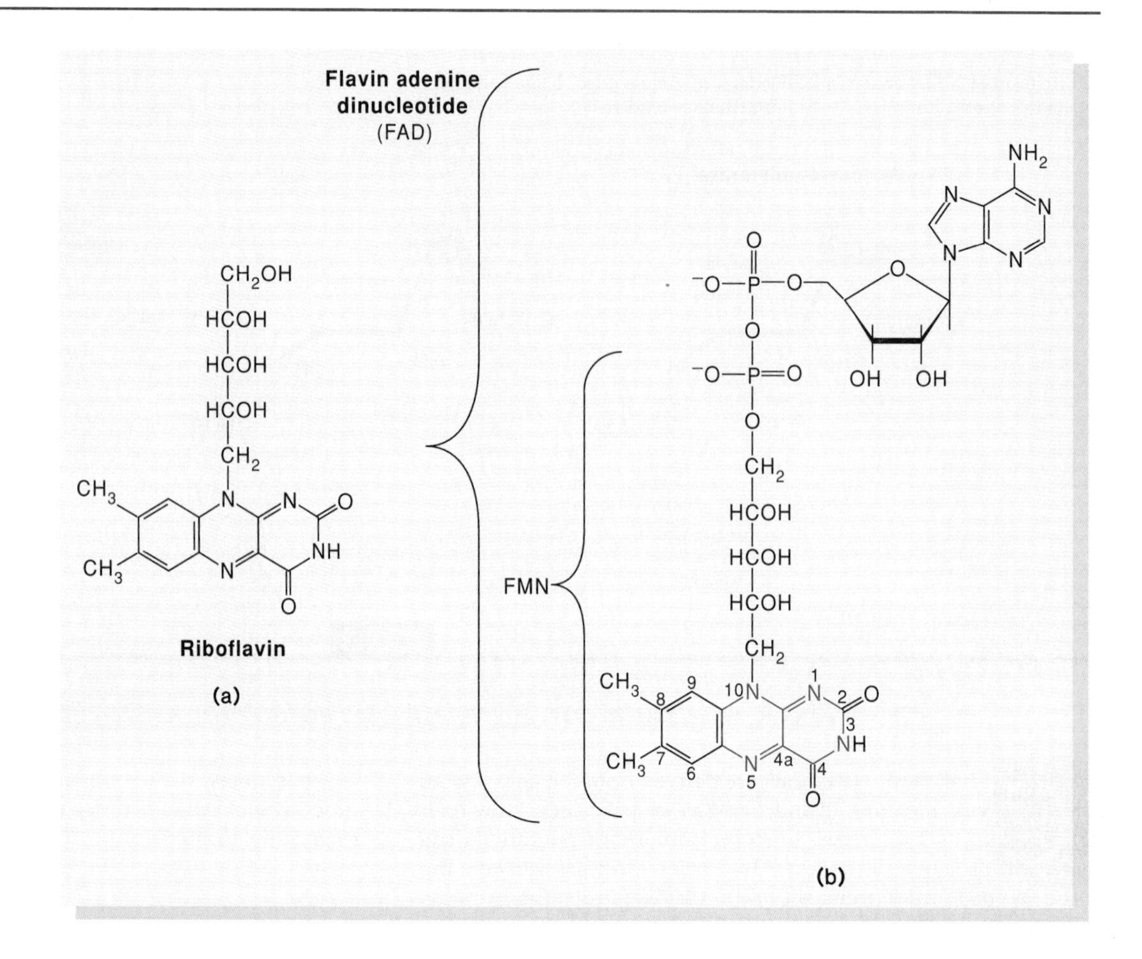

Flavin coenzymes exist in three spectrally distinguishable oxidation states that account in part for their catalytic functions; the yellow oxidized form, the red or blue one-electron-reduced form, and the colorless two-electron-reduced form. Their structures are depicted in figure 11.10. These and other less well-defined forms often have been detected spectrally as intermediates in flavoprotein catalysis.

Flavins are very versatile redox coenzymes. Flavoproteins are dehydrogenases, oxidases, and oxygenases that catalyze a variety of reactions on an equal variety of substrate types. Since these classes of enzymes do not consist exclusively of flavoproteins, it is difficult to define catalytic specificity for flavins. Biological electron acceptors and donors in flavin-mediated reactions can be two-electron acceptors, such as NAD^+ or $NADP^+$, or a variety of one-electron acceptor systems, such as cytochromes (Fe^{2+}/Fe^{3+}) and quinones, and molecular oxygen is an electron acceptor for flavoprotein oxidases as well as the source of oxygen for oxygenases. The only obviously common aspect of flavin-dependent reactions is that all are redox reactions.

Typical reactions catalyzed by flavoproteins are listed in table 11.2, which groups flavoproteins into those that do not utilize molecular oxygen as a substrate and those that do. You can best appreciate the significance of this difference when you realize that $FADH_2$, a likely intermediate in many flavoprotein reactions, spontaneously reacts with O_2 to produce H_2O_2.In the case of the dehydrogenases, therefore, either $FADH_2$ is not an intermediate or it is somehow prevented from reacting with O_2. Among the dehydrogenases are several that utilize the two-electron acceptor substrates NAD^+ or $NADP^+$, and it is reasonable to suppose that the two-electron reduction of NAD^+ by an intermediate $E \cdot FADH_2$ might be involved. Also listed in table 11.2 are other dehydrogenases for which the electron acceptors from $E \cdot FADH_2$ are not given. These enzymes are membrane-bound and transfer electrons directly to membrane-bound acceptors, mainly one-electron acceptors such as quinones and cytochromes (Fe^{2+}/Fe^{3+}). The stability of the flavin semiquinone, FAD · and FMN · in figure 11.10, gives flavins the capability to interact with one-electron acceptors in electron-transport systems.

The other classes of flavoproteins in table 11.2 interact with molecular oxygen either as the electron-acceptor substrates in redox reactions catalyzed by oxidases or as the substrate sources of oxygen atoms for oxygenases. Molecular oxygen also serves as an electron acceptor and source of oxygen for metalloflavoproteins and dioxygenases, which are not listed in table

Figure 11.10

Oxidation states of flavin coenzymes. The flavin coenzymes exist in three spectrally distinguishable oxidation states that account in part for their catalytic functions. They are the yellow oxidized form, the red or blue one-electron-reduced form, and the colorless two-electron-reduced form.

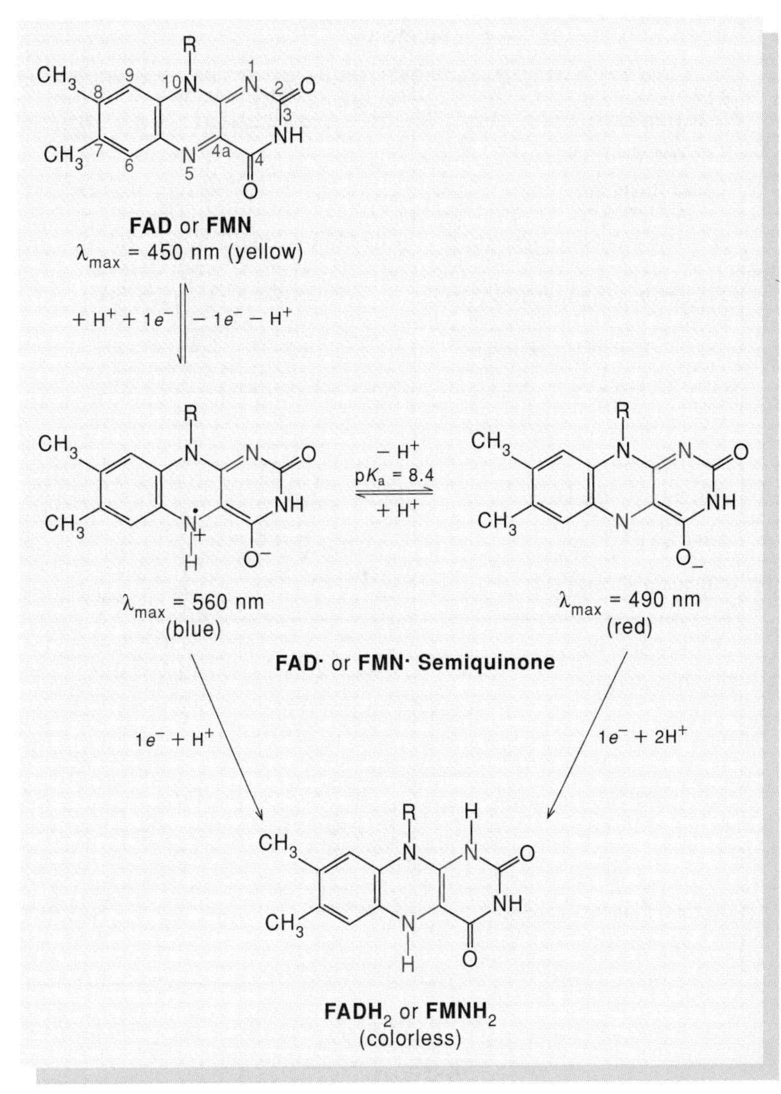

11.2. These enzymes catalyze more complex reactions, involving catalytic redox components such as metal ions and metal-sulfur clusters in addition to flavin coenzymes.

The mechanisms of action of flavin coenzymes are currently under active investigation. A recurrent theme appears to be the probable involvement of $FADH_2$ or $FMNH_2$ as transient intermediates in a variety of flavoprotein reactions. Figure 11.11 illustrates a reasonable catalytic pathway for the first enzyme listed in table 11.2; this reaction shows the likely involvement of E · $FADH_2$ in each case. The mechanisms by which E · FAD is reduced to E · $FADH_2$ by NADPH in the forward direction and by glutathione in the reverse direction are undoubtedly different. The mechanism shown is one recently proposed.

The biochemical importance of flavin coenzymes appears to be their versatility in mediating a variety of redox processes, including electron transfer and the activation of molecular oxygen for oxygenation reactions. The detailed mechanisms of oxygen activation are not well understood. An especially important manifestation of their redox versatility is their ability to serve as the switch point from the two-electron processes, which predominate in cytosolic carbon metabolism, to the one-electron transfer processes, which predominate in membrane-associated terminal electron-transfer pathways. In mammalian cells, for example, the end products of the aerobic metabolism of glucose are CO_2 and NADH (see chapter 14). The terminal electron-transfer pathway is a membrane-bound system of cytochromes, nonheme iron proteins, and copper-heme proteins—all one-electron acceptors that transfer electrons ultimately to O_2 to produce H_2O and NAD^+ with the concomitant production of ATP from ADP and P_i. The interaction of NADH with this pathway is mediated by NADH dehydrogenase, a flavoprotein that couples the two-electron oxidation of NADH with the one-electron reductive processes of the membrane.

Table 11.2
Reactions Catalyzed by Flavoproteins

Flavoprotein	Reaction
Dehydrogenases	
Glutathione reductase	$H^+ + GSSG + NADPH \rightleftharpoons 2\ GSH + NADP^+$
Acyl-CoA dehydrogenases	$RCH_2CH_2COSCoA + NAD^+ \rightleftharpoons RCH{=}CHCOSCoA + NADH + H^+$
Succinate dehydrogenase	$^-O_2CCH_2CH_2CO_2^- + E\cdot FAD \rightleftharpoons {}^-O_2CCH{=}CHCO_2^- + E\cdot FADH_2$
D-Lactate dehydrogenase	$CH_3{-}CHOH{-}CO_2^- + E\cdot FAD \rightleftharpoons CH_3{-}CO{-}CO_2^- + E\cdot FADH_2$
Oxidases	
Amino acid oxidases	$R{-}CH(NH_3^+){-}CO_2^- + O_2 + H_2O \longrightarrow R{-}CO{-}CO_2^- + H_2O_2 + NH_4^+$
Glucose oxidase	D-Glucose $+ O_2 \longrightarrow$ D-gluconolactone $+ H_2O_2$
Monoamine oxidase	$R{-}CH_2\overset{+}{N}H_3 + O_2 + H_2O \longrightarrow R{-}CHO + H_2O_2 + \overset{+}{N}H_4$
Monooxygenases	
Lactate oxidase	$CH_3{-}CHOH{-}CO_2^- + O_2 \longrightarrow CH_3{-}CO_2^- + CO_2 + H_2O$
Salicylate hydroxylase	$2H^+ +$ (salicylate: OH, CO_2^- on benzene ring) $+ O_2 + NADH \longrightarrow$ (catechol: OH, OH on benzene ring) $+ CO_2 + NAD^+ + H_2O$
Ketone monooxygenase	$H^+ +$ (cyclopentanone) $+ O_2 + NADPH \longrightarrow$ (lactone) $+ NADP^+ + H_2O$

Figure 11.11

Mechanism of the flavin-dependent glutathione reductase reaction. The first steps, not shown, involve the reduction of FAD to $FADH_2$ by NADPH and the binding of glutathione (glutathione is a sulfhydryl compound, see figure 19.32). The mechanism by which oxidized glutathione is reduced by the E•$FADH_2$ is shown.

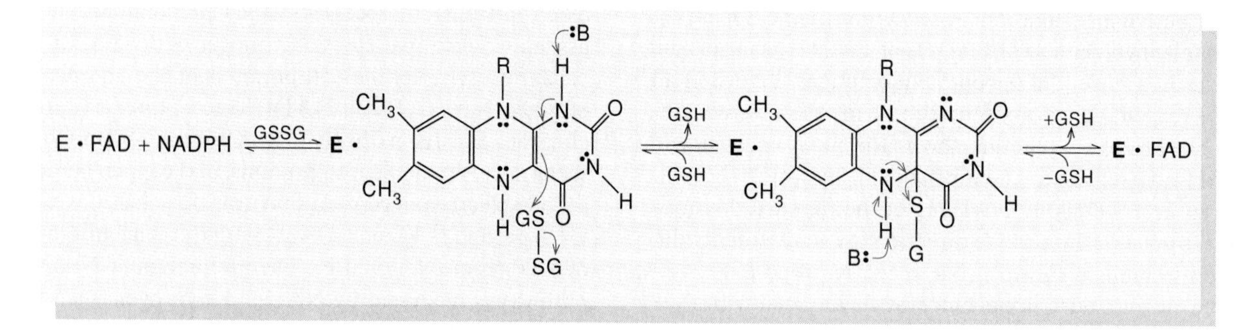

Figure 11.12

Structures of the vitamin pantothenic acid and coenzyme A. The terminal —SH is the reactive group in coenzyme A (CoASH).

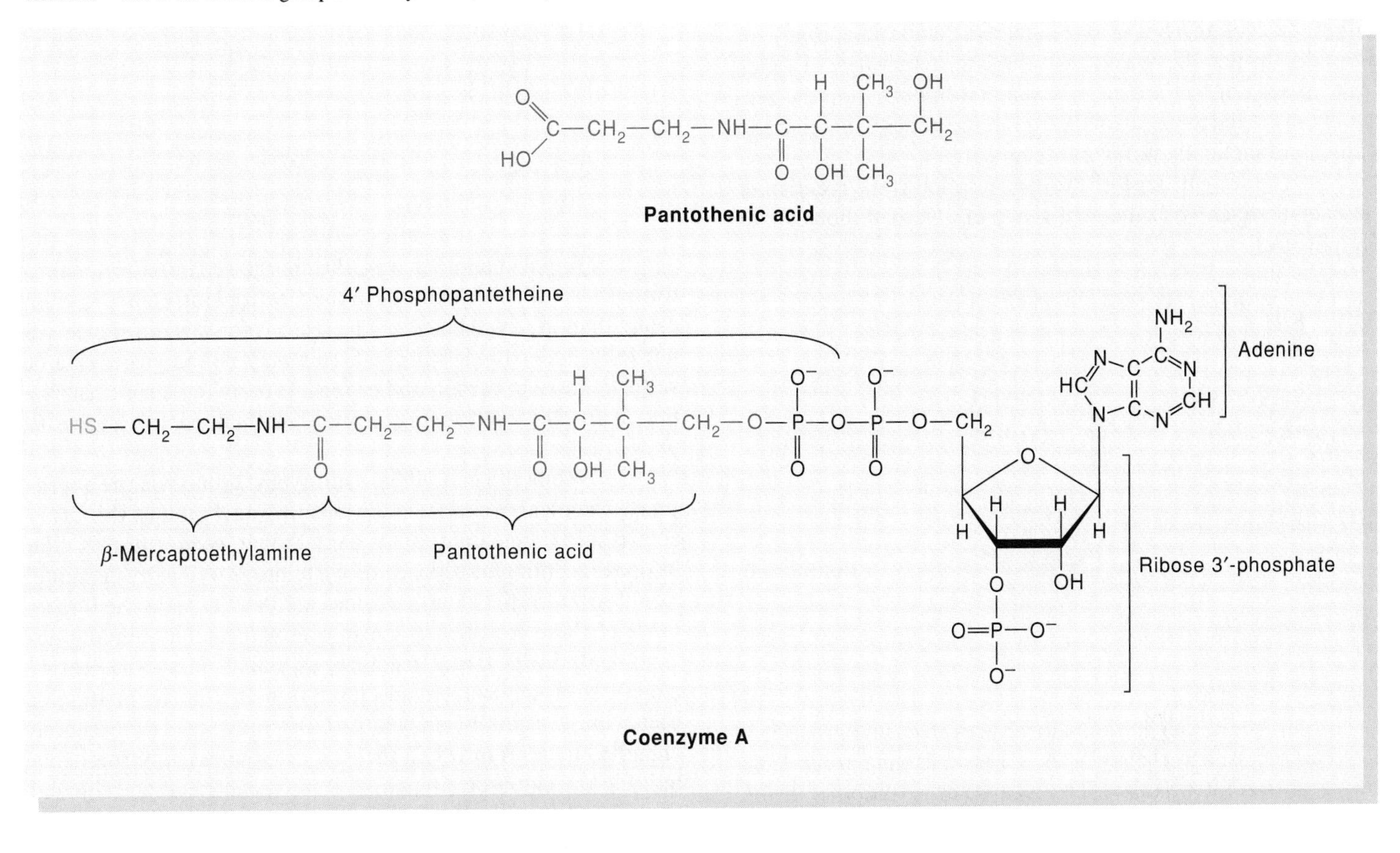

Reactions Requiring Acyl Activation Frequently Use Phosphopantetheine Coenzymes

4′-Phosphopantetheine coenzymes are the biochemically active forms of the vitamin pantothenic acid. In figure 11.12, 4′-phosphopantetheine is shown as covalently linked to an adenylyl group in coenzyme A; or it can also be linked to a protein such as a serine hydroxyl group in acyl carrier protein (ACP). It is also found bonded to proteins that catalyze the activation and polymerization of amino acids to polypeptide antibiotics. Coenzyme A was discovered, purified, and structurally characterized by Fritz Lipmann and colleagues in work for which Lipmann was awarded the Nobel Prize in 1953.

The sulfhydryl group of the β-mercaptoethylamine (or cysteamine) moiety of phosphopantetheine coenzymes is the functional group that is directly involved in the enzymatic reactions for which they serve as coenzymes. From the standpoint of the chemical mechanism of catalysis, it is the essential functional group, although it is now recognized that phosphopantetheine coenzymes have other functions as well. Many reactions in metabolism involve acyl-group transfer or enolization of carboxylic acids that exist as unactivated carboxylate anions at physiological pH. The predominant means by which these acids are activated for acyl transfer and enolization is esterification with the sulfhydryl group of pantetheine coenzymes.

The mechanistic importance of activation is exemplified by the condensation of two molecules of acetyl-coenzyme A to acetoacetyl-coenzyme A catalyzed by β-ketothiolase:

$$CH_3-\overset{O}{\overset{\|}{C}}-SCoA + CH_3-\overset{O}{\overset{\|}{C}}-SCoA \rightleftharpoons CH_3-\overset{O}{\overset{\|}{C}}-CH_2-\overset{O}{\overset{\|}{C}}-SCoA + CoASH \quad (18)$$

The two important steps of the reaction depend on both acetyl groups being activated, one for enolization and the other for acyl-group transfer. In the first step, one of the molecules must be enolized by the intervention of a base to remove an α proton, forming an enolate:

$$B\!: \; H-CH_2-\overset{O}{\overset{\|}{C}}-SCoA \rightleftharpoons [\overset{\delta^+}{B} \cdots H \cdots \overset{\delta^-}{CH_2} \cdots \overset{O}{\overset{\vdots}{C}}-SCoA] \rightleftharpoons \overset{+}{B}-H + CH_2 \cdots \overset{O^-}{\overset{\vdots}{C}}-SCoA \quad (19)$$

The enolate is stabilized by delocalization of its negative charge between the α carbon and the acyl oxygen atom, making it thermodynamically accessible as an intermediate. Moreover, this developing charge is also stabilized in the transition state preceding the enolate, so it is also kinetically accessible; that is, it is rapidly formed. If, by contrast, the same enolization reaction were carried out by the acetate anion, it would result in the generation of a second negative charge in the enolate, an energetically and kinetically unfavorable process.

The second stage of the condensation is the reaction of the enolate anion with the acyl group of a second molecule of acetyl-CoA:

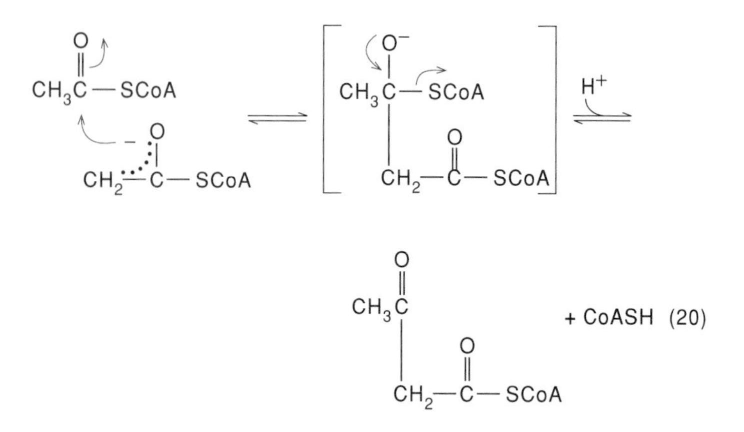

Nucleophilic addition to the neutral activated acyl group is a favored process, and coenzyme A is a good leaving group from the tetrahedral intermediate. The occurrence of this process with the acetate anion, i.e., acetate reacting with an enolate anion, again provides a sharp contrast with the process of equation (20), for it would entail the nucleophilic addition of an anion to an anionic center, generating a dianionic transition state, an unfavorable process from both thermodynamic and kinetic standpoints. Moreover, the resulting intermediate would not have a very good leaving group other than the enolate anion itself, so the transition-state energy for acetoacetate formation would be high. Finally, the K_{eq} for the condensation of 2 moles of acetate to 1 mole of acetoacetate is not favorable in aqueous media, whereas the condensation of 2 moles of acetyl-CoA to produce acetoacetyl-CoA and coenzyme A is thermodynamically spontaneous. The maintenance of metabolic carboxylic acids involved in enolization and acyl-group transfer reactions as coenzyme A esters provides the ideal lift over the kinetic and thermodynamic barriers to these reactions.

The foregoing discussion, in emphasizing the purely electrostatic energy barriers, does not address the question of whether there is an activation advantage in thiol esters relative to oxygen esters. Why thiol esters in preference to oxygen esters? Thiol esters are more readily enolized than oxygen esters. They are more "ketonelike" because of their electronic structures, in which the degree of resonance-electron delocalization from the sulfur atom to the acyl group resulting from overlapping of the occupied p orbitals of sulfur with the acyl-π bond is less than that of oxygen esters.

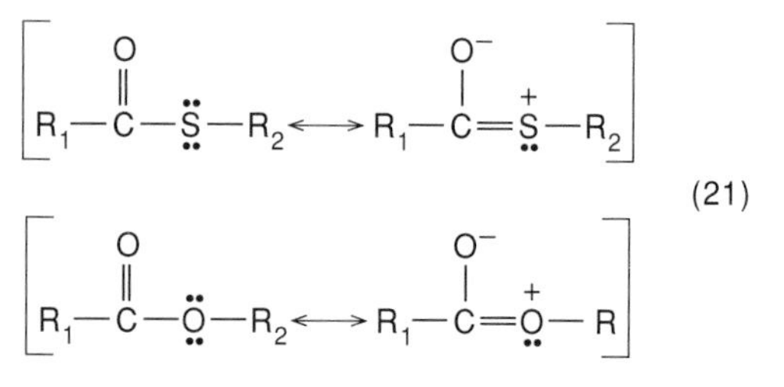

To put it another way, the charge-separated resonance form is a smaller contributor to the electronic structure in thiol esters than in oxygen esters. The reasons for this difference are not fully understood, but one factor may be the larger size of sulfur relative to carbon and oxygen, leading to a poorer energy match for the overlapping orbitals in thiol esters relative to oxygen esters.

While the pantetheine sulfhydryl group has the appropriate chemical properties for activating acyl groups, this characteristic is not unique to pantetheine coenzymes in the biosphere. Both glutathione and cysteine, as well as cysteamine, would serve, so the chemistry does not itself explain the importance of these coenzymes. Coenzyme A has many binding determinants in its large structure, especially in the nucleotide moiety, so it may serve a specificity function in the binding of coenzyme A esters by enzymes. It also may serve as a binding "handle" in cases in which the acyl group must have some mobility in the catalytic site, i.e., if it must enolize at one site and then diffuse a short distance to undergo an addition reaction to a ketonic group of a second substrate.

One system in which pantetheine almost certainly performs such a carrier role is the fatty acid synthase from *E. coli,* in which 4′-phosphopantetheine is a component of the acyl carrier protein (see chapter 17).

α-Lipoic Acid Is the Coenzyme of Choice for Reactions Requiring Acyl Group Transfers Linked to Oxidation-Reduction

α-Lipoic acid is the internal disulfide of 6,8-dithiooctanoic acid, whose structural formula is given in figure 11.13. It is the coupler of electron and group transfers catalyzed by α-keto acid dehydrogenase multienzyme complexes. The pyruvate and α-ketoglutarate dehydrogenase complexes are centrally involved in the metabolism of carbohydrates by the glycolytic pathway (chapter 13) and the tricarboxylic acid cycle (chapter 14). They catalyze two of the three decarboxylation steps in the complete oxidation of glucose, and they produce NADH and activated acyl compounds from the oxidation of the resulting ketoacids:

$$R-\overset{O}{\overset{\|}{C}}-CO_2^- + NAD^+ + CoASH \longrightarrow CO_2 + NADH + R-\overset{O}{\overset{\|}{C}}-SCoA \qquad (22)$$

Figure 11.13

Structure of the α-lipoyl enzyme, showing the reactive disulfide in color. The lipoic acid is commonly covalently bonded through amide linkage with the ε amino group of a lysine residue as shown.

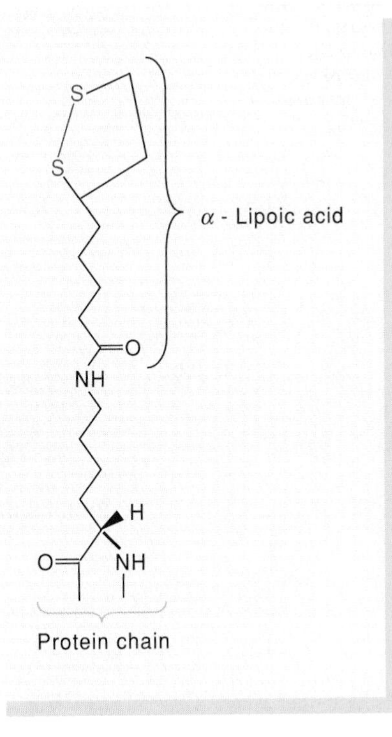

Figure 11.14

Interactions of α-lipoyl groups in the pyruvate dehydrogenase complex. The cubic structure represents the 24 subunits of dihydrolipoyl transacetylase, which constitutes the core of the complex. Two of the 48 lipoyl groups in the core are shown interacting with one of the 24 pyruvate decarboxylases (E_1•TPP) and one of the 12 dihydrolipoyl dehydrogenase (E_3•FAD) subunits. Note the interaction of the lipoyl groups in relaying electrons over the long distance between TPP and FAD. The lipoic acid must interact at active sites over distances too long to be covered by a single fully extended coenzyme. This problem is solved by use of a shuttle system involving two α-lipoyl groups for each transfer.

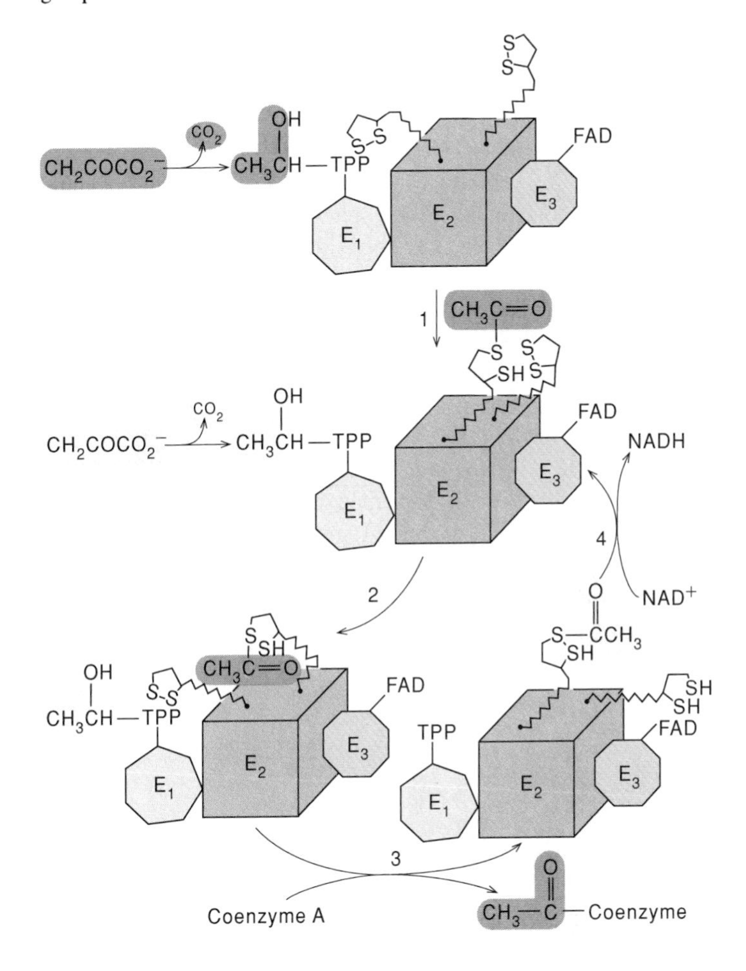

The chemical aspect of the coenzymatic action of α-lipoic acid is to mediate the transfer of electrons and activated acyl groups resulting from the decarboxylation and oxidation of α-keto acids within the complexes. In this process, lipoic acid is itself transiently reduced to dihydrolipoic acid, and this reduced form is the acceptor of the activated acyl groups. Its dual role of electron and acyl-group acceptor enables lipoic acid to couple the two processes.

The interactions of α-lipoic acid in the *E. coli* pyruvate dehydrogenase complex exemplify its coenzymatic functions. The complex consists of three proteins: a pyruvate decarboxylase, which is thiamine-pyrophosphate-dependent and designated $E_1 \cdot$ TPP; dihydrolipoyl transacetylase, designated E_2-lipoyl-S_2, which contains α-lipoic acid covalently bonded through amide linkage with the ε amino group of a lysine residue (see fig. 11.14); and dihydrolipoyl dehydrogenase, a flavoprotein designated $E_3 \cdot$ FAD. A single particle of the complex consists of at least 24 chains of each of the first two enzymes and 12 of the flavoprotein.

Lipoic acid must interact at active sites on all three enzymes. α-Lipoic acid bonded to E_2 is, as shown in figure 11.13, bonded through an amide linkage to a lysyl-ε-NH_2 group that places the reactive disulfide at the end of a flexible chain of atoms with rotational freedom about as many as ten single bonds. When fully extended, this chain is 1.4 nm (14 Å) long, giving α-lipoic acid the potential capacity to sweep out a space having a spherical diameter of 2.8 nm. This distance turns out to be inadequate to account fully for the transport of electrons in the complex because the average distance between TPP on E_1 and FAD on E_3 has been estimated at between 4.5 and 6.0 nm by fluorescence energy-transfer measurements. The problem of long-distance interactions is overcome in the complex by the fact that each E_2 subunit contains two lipoyl groups that interact with each other and with α-lipoyl groups on other subunits. This interaction facilitates the transport of electrons and acetyl groups through a network of α-lipoyl groups encompassing the entire core of E_2. Moreover, the lipoyl-bearing domains of E_2 are conformationally mobile; this mobility is thought to extend the "reach" of the lipoyl groups. The relay process, illustrated schematically in figure 11.14, shows how two or more *S*-acetyldihydrolipoyl groups can interact to transport electrons over the large distances separating the sites for TPP on pyruvate dehydrogenase and FAD on dihydrolipoyl dehydrogenase.

The coenzymatic capabilities of α-lipoyl groups result from a fusion of its chemical and physical properties, the ability to act simultaneously as both electron and acyl-group acceptor, the ability to span long distances to interact with sites separated by up to 2.8 nm, and the ability to act cooperatively with other α-lipoyl groups by disulfide interchange to relay electrons and acyl groups through distances that exceed its reach. This reaction is discussed in greater detail in chapter 14.

Biotin Mediates Carboxylations

The biotin structure shown in figure 11.15 is an imidazolone ring *cis*-fused to a tetrahydrothiophene ring substituted at position 2 by valeric acid. In carboxylase enzymes, biotin is covalently bonded to the proteins by an amide linkage between its carboxyl group and a lysyl-ϵ-NH_2 group in the polypeptide chain. This arrangement places the imidazolone ring at the end of a long flexible chain of atoms extending a maximum of about 1.4 nm from the α carbon of lysine.

Biotin is the essential coenzyme for carboxylation reactions involving bicarbonate as the carboxylating agent. Several reactions have been described in which ATP-dependent carboxylation occurs at carbon atoms activated for enolization by ketonic or activated acyl groups. One reaction is known where a nitrogen atom of urea is carboxylated.

A general formulation of the ATP-dependent carboxylation of an α carbon by ^{18}O-enriched bicarbonate is

$$RCH_2{-}\overset{O}{\overset{\|}{C}}{-}SCoA + ATP + HC^{18}O_3^- \xrightarrow{\text{Biotinyl carboxylase}} H^+ + R{-}\underset{C^{18}O_2^-}{\underset{|}{C}H}{-}\overset{O}{\overset{\|}{C}}{-}SCoA + ADP + H^{18}OPO_3^{2-} \quad (23)$$

The appearance of ^{18}O in inorganic phosphate verifies that the function of ATP in the reaction is essentially the "dehydration" of bicarbonate.

The ATP-dependent carboxylation of biotin by bicarbonate is believed to control the transient formation of carbonic-phosphoric anhydride, or "carboxyphosphate," as an active carboxylation intermediate:

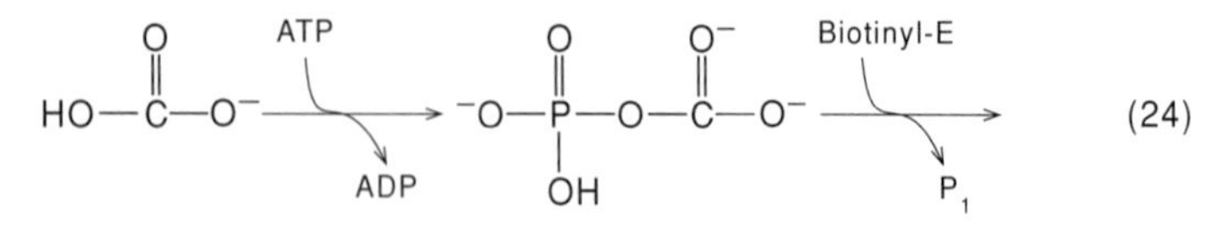

N^1-carboxybiotinyl-E

The coenzymatic function of biotin appears to be to mediate the carboxylation of substrates by accepting the ATP-activated carboxyl group and transferring it to the carboxyl acceptor substrate. There is good reason to believe that the enzymatic sites of ATP-dependent carboxylation of biotin are physically separated from the sites at which N^1-carboxybiotin transfers the carboxyl group to acceptor substrates, i.e., the transcarboxylase sites. In fact, in the case of the acetyl-CoA carboxylase from *E. coli* (see chapter 17), these two sites reside on two different subunits, while the biotinyl group is bonded to a third, a small subunit designated biotin carboxyl carrier protein. Transcarboxylase is also a multisubunit protein, one subunit being a small biotinyl protein. The carboxylation of acetyl-CoA by carboxybiotin is discussed in chapter 17.

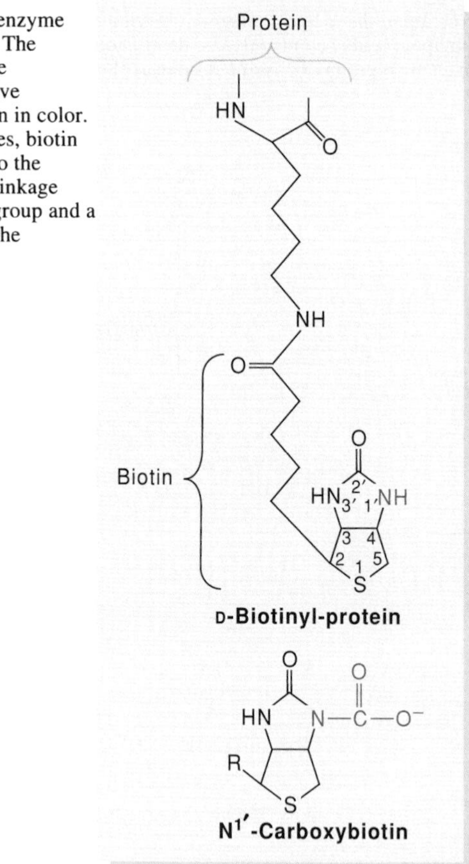

Figure 11.15

Structures of biotinyl enzyme and N^1-carboxybiotin. The reactive portions of the coenzyme and the active intermediate are shown in color. In carboxylase enzymes, biotin is covalently bonded to the proteins by an amide linkage between its carboxyl group and a lysyl-ε-NH_2 group in the polypeptide chain.

Biotin appears to have just the right chemical and structural properties to mediate carboxylation. It readily accepts activated carboxyl groups at N^1 and maintains them in an acceptably stable yet reactive form for transfer to acceptor substrates. Since biotin is bonded to a lysyl group, the N^1-carboxyl group is at the end of a 1.6-nm chain with bond rotational freedom about nine single bonds, giving it the capability to transport activated carboxyl groups through space from the carboxyl activation sites to the carboxylation sites.

Folate Coenzymes Are Used in Reactions for One-Carbon Transfers

Tetrahydrofolate and its derivatives N^5,N^{10}-methylenetetrahydrofolate, N^5,N^{10}-methenyltetrahydrofolate, N^{10}-formyltetrahydrofolate, and N^5-methyltetrahydrofolate are the biologically active forms of folic acid, a four-electron-oxidized form of

Figure 11.16

Structures and enzymatic interconversions of folate coenzymes. The reactive centers of the coenzymes are shown in red. The most active forms of the coenzyme contain oligo- or polyglutamyl groups.

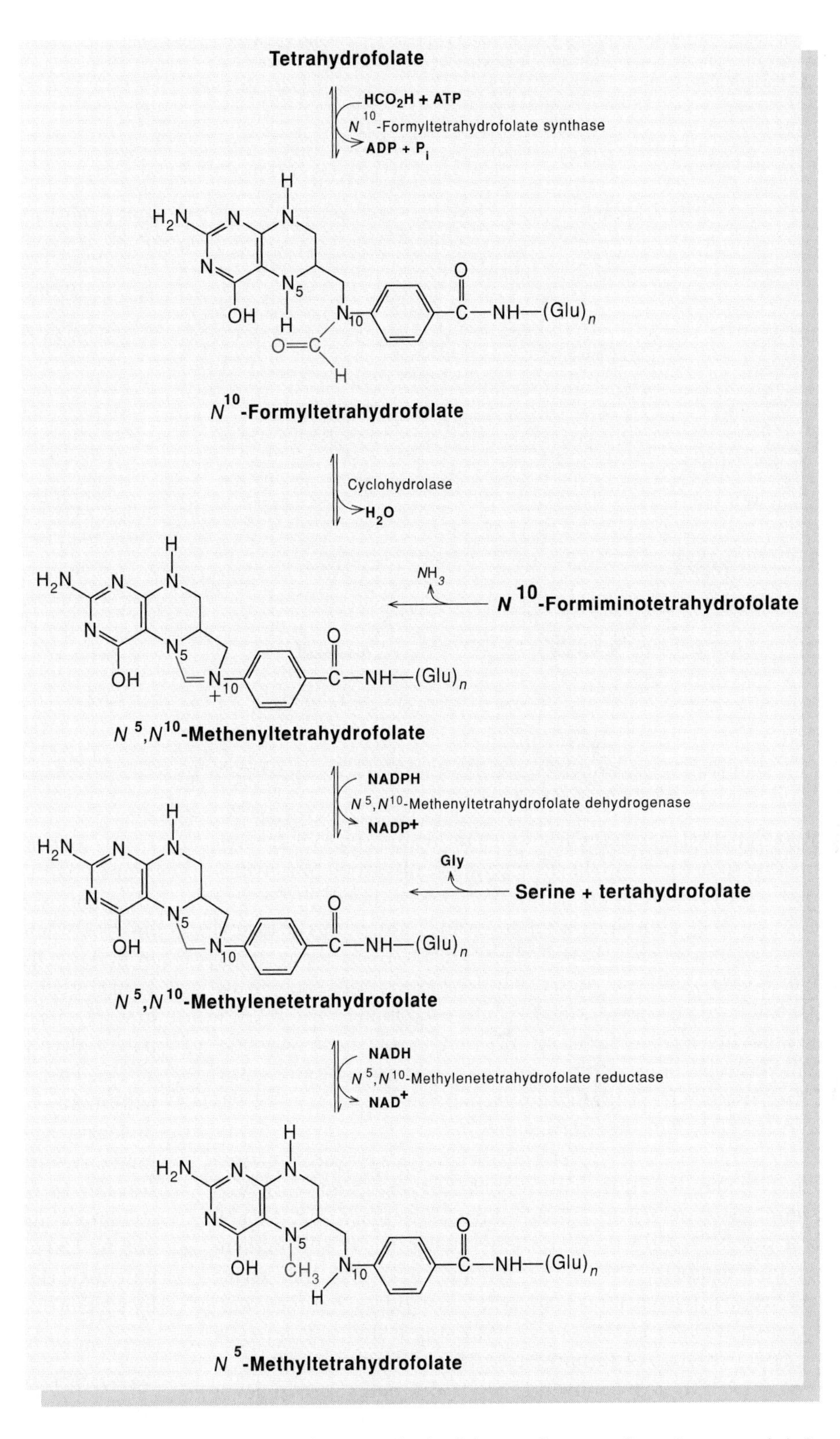

tetrahydrofolate. The structural formulas are given in figure 11.16, which also shows how they arise from tetrahydrofolate. The structures are shown glutamylated on the carboxyl group of the *p*-amino benzoyl group; the most active forms contain oligo- or polyglutamyl groups, linked through the γ-carboxyl groups.

The tetrahydrofolates do not function as tightly enzyme-bound coenzymes. Rather they function as cosubstrates for a variety of enzymes associated with one-carbon metabolism. N^{10}-Formyltetrahydrofolate is a formyl-group donor substrate for several transformylases. The methenyl derivative is an intermediate between N^{10}-formyltetrahydrofolate and

Figure 11.17

Involvement of folate coenzymes in one-carbon metabolism. Shown in red are the one-carbon units of the end products that originate with the reactive one-carbon units of the folate coenzymes.

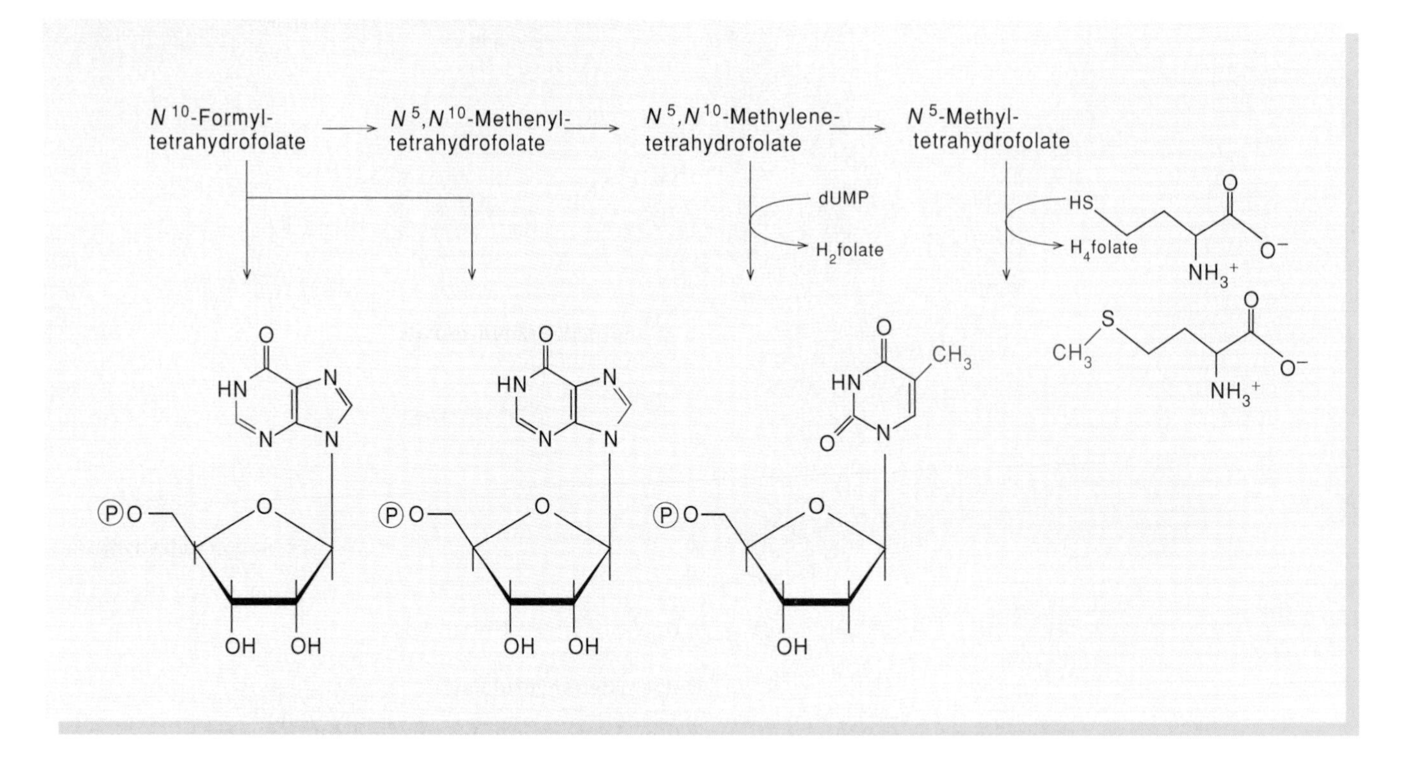

N^5,N^{10}-methylenetetrahydrofolate. In living cells, N^{10}-formyltetrahydrofolate is produced enzymatically from tetrahydrofolate and formate in an ATP-linked process in which formate is activated by phosphorylation to formyl phosphate: the formyl group of formyl phosphate is then transferred to N^{10} of tetrahydrofolate. N^{10}-Formyltetrahydrofolate is a formyl donor substrate for some enzymes and is interconvertible with N^5,N^{10}-methenyltetrahydrofolate by the action of cyclohydrolase. N^{10}-Formyltetrahydrofolate and N^5,N^{10}-methenyltetrahydrofolate also can be synthesized in nonenzymatic reactions of tetrahydrofolate with free formic acid.

N^5,N^{10}-Methylenetetrahydrofolate is a hydroxymethyl group donor substrate for several enzymes and a methyl group donor substrate for thymidylate synthase (fig. 11.17). It arises in living cells from the reduction of N^5,N^{10}-methenyltetrahydrofolate by NADPH and also by the serine hydroxymethyltransferase-catalyzed reaction of serine with tetrahydrofolate. It also can be synthesized nonenzymatically by direct reaction of tetrahydrofolate with formaldehyde.

N^5-Methyltetrahydrofolate is the methyl-group donor substrate for methionine synthase, which catalyzes the transfer of the five-methyl group to the sulfhydryl group of homocysteine. This and selected reactions of the other folate derivatives are outlined in figure 11.17, which emphasizes the important role tetrahydrofolate plays in nucleic acid biosynthesis by serving as the immediate source of one-carbon units in purine and pyrimidine biosynthesis.

Formaldehyde is a toxic substance that reacts spontaneously with amino groups of proteins and nucleic acids, hydroxymethylating them and forming methylene-bridge cross-links between them. Free formaldehyde therefore wreaks havoc in living cells and could not serve as a useful hydroxymethylating agent. In the form of N^5,N^{10}-methylenetetrahydrofolate, however, its chemical reactivity is attenuated but retained in a potentially available form where needed for specific enzymatic action. Formate, however, is quite unreactive under physiological conditions and must be activated to serve as an efficient formylating agent. As N^{10}-formyltetrahydrofolate and N^5,N^{10}-methenyltetrahydrofolate it is in a reactive state suitable for transfer to appropriate substrates. The fundamental biochemical importance of tetrahydrofolate is to maintain formaldehyde and formate in chemically poised states, not so reactive as to pose toxic threats to the cell but available for essential processes by specific enzymatic action.

Vitamin B_{12} Coenzymes Are Associated with Rearrangements on Adjacent Carbon Atoms

The principal coenzymatic form of vitamin B_{12} is 5′-deoxyadenosylcobalamin, whose structural formula is given in figure 11.18. The structure includes a cobalt–carbon bond between the 5′ carbon of the 5′-deoxyadenosyl moiety and the cobalt (III) ion of cobalamin. [Note: The metal oxidation state may be denoted either as Co^{3+} or as Co(III). The former stresses

Figure 11.18

Structure of 5′-deoxyadenosylcobalamin coenzyme (vitamin B_{12}). The reactive groups are shown in color.

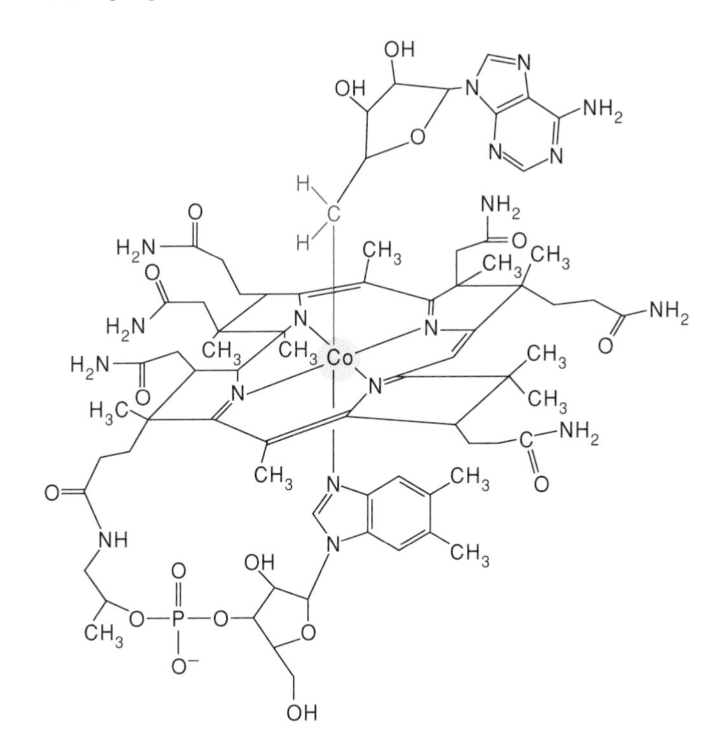

the free-ion character, while the latter stresses the bound character.] Vitamin B_{12} itself is cyanocobalamin, in which the cyano group is bonded to cobalt in place of the 5′-deoxyadenosyl moiety. Other forms of the vitamin have water (aquocobalamin) or the hydroxyl group (hydroxycobalamin) bonded to cobalt.

The vitamin was discovered in liver as the anti–pernicious anemia factor in 1926, but discovery of its complete structure had to await its purification, chemical characterization, and crystallization, which required more than twenty years. Even then the determination of such a complex structure proved to be an elusive goal by conventional approaches of that day and had to await the elegant x-ray crystallographic study of Lenhert and Hodgkin in 1961, for which Dorothy Hodgkin was awarded the Nobel Prize in 1964.

Most 5′-deoxyadenosylcobalamin-dependent enzymatic reactions are rearrangements that follow the pattern of equation (25), in which a hydrogen atom and another group (designated X) bonded to an adjacent carbon atom exchange positions, with the group X migrating from C_α to C_β:

$$a-\overset{\displaystyle b}{\underset{\displaystyle X}{C_\alpha}}-\overset{\displaystyle c}{\underset{\displaystyle H}{C_\beta}}-d \rightleftharpoons a-\overset{\displaystyle b}{\underset{\displaystyle H}{C_\alpha}}-\overset{\displaystyle c}{\underset{\displaystyle X}{C_\beta}}-d \quad (25)$$

Three specific examples of rearrangement reactions are given in equations (26) through (28). It is interesting and significant that the migrating groups —OH, —COSCoA, and —CH$(NH_3^+)CO_2^-$ have little in common and that the hydrogen atoms migrating in the opposite direction are often chemically unreactive.

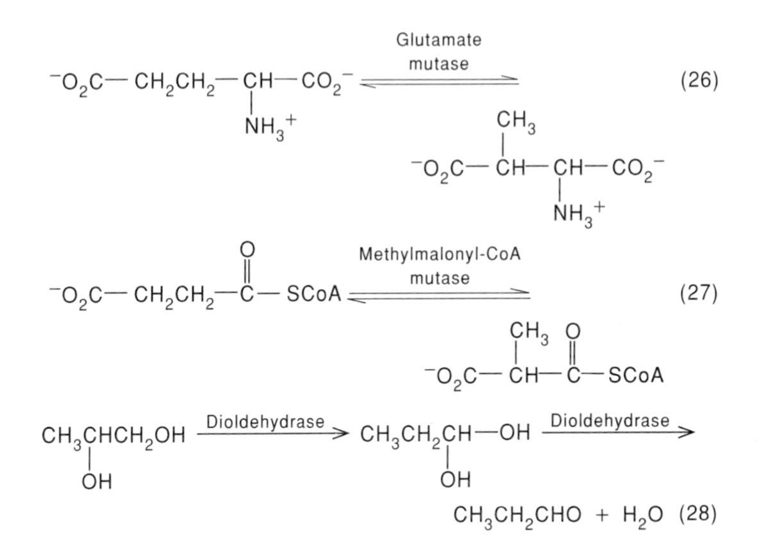

Indeed, hydrogen migrations in all the B_{12} coenzyme-dependent rearrangements proceed without exchange with the protons of water; i.e., isotopic hydrogen in substrates is conserved in the products. This fact plus spectroscopic evidence implicating Co(II) and organic radicals as catalytic intermediates in the reaction have led to the proposal of the mechanism illustrated in figure 11.19.

The reaction begins by homolytic cleavage of the Co—C bond (fig. 11.19), generating Co(II) and 5′-deoxyadenosyl free radical (step 1). The radical abstracts a hydrogen atom from the substrate, the migrating hydrogen in equation (25), generating 5′-deoxyadenosine and a substrate-derived free radical as intermediates (step 2). The substrate-radical undergoes rearrangement to a product-derived free radical, which abstracts a hydrogen atom to form the final product and regenerate the coenzyme (steps 3–6). Much evidence supports the involvement of the intermediates shown in figure 11.19; however, quantitative aspects of available data suggest the possible existence of additional species and a more complex hydrogen transfer process.

The most fundamental property of 5′-deoxyadenosylcobalamin leading to its unique action as a coenzyme is the weakness of the Co—C bond. This bond has a low dissociation energy, less than 30 kcal/mole, strong enough to be essentially stable in free solution but weak enough to be broken as a result of strain induced by multiple binding interactions between the enzymic binding sites and the adenosyl and cobalamin portions of the coenzyme. The radicals resulting from cleavage of this bond and abstraction of hydrogen from substrates undergo the rearrangements characteristic of B_{12}-dependent reactions.

Figure 11.19

Hypothetical partial mechanism of vitamin B_{12}-dependent rearrangements. The designations Co(III) and Co(II) refer to species that are spectrally and magnetically similar to Co^{3+} and Co^{2+}, respectively. Co(III) is diamagnetic and red, while Co(II) is paramagnetic (unpaired electron) and yellow. The metal does not undergo a change in electrostatic charge when the cobalt–carbon bond breaks homolytically (i.e., without charge separation), since one electron remains with the metal and the other with 5′-deoxyadenosine. One or more unknown intermediates, symbolized by the brackets, may be involved in the rearrangement.

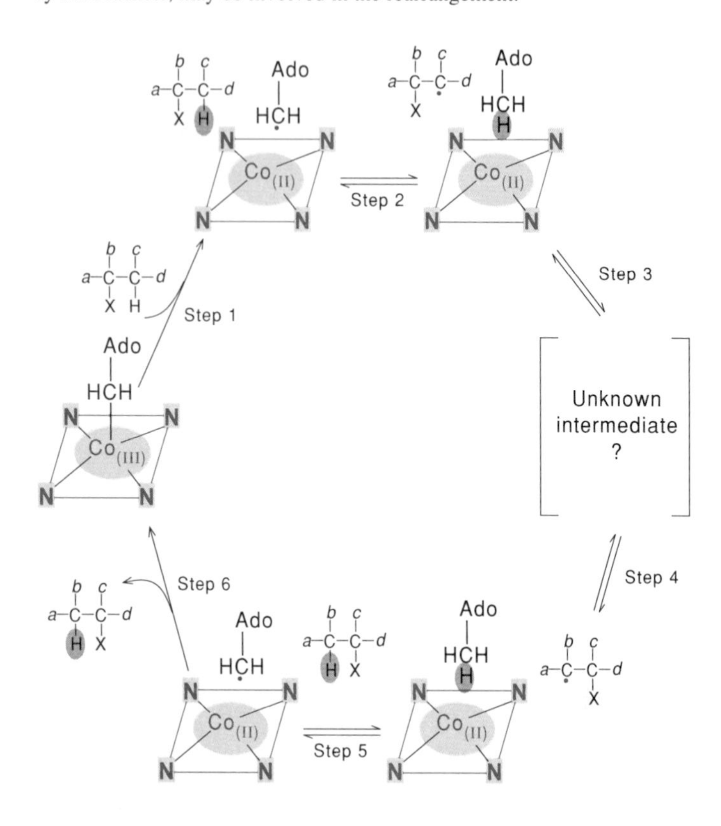

Figure 11.20

Structure of protoporphyrin IX. This coenzyme acts in conjunction with a number of different enzymes involved in oxidation and reduction reactions.

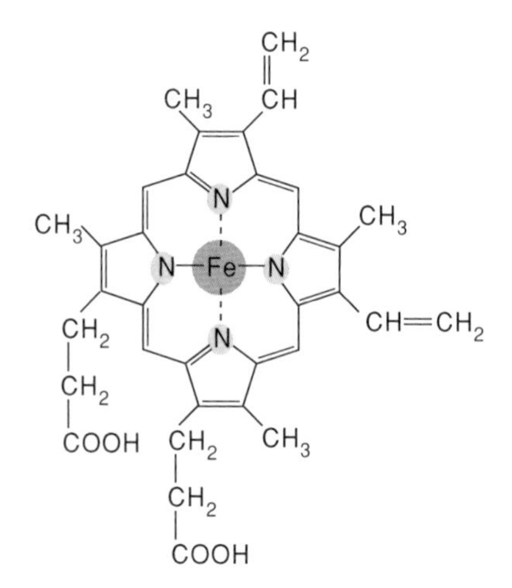

Figure 11.21

Structures of iron-sulfur clusters. Many redox enzymes contain iron-sulfur clusters that mediate one-electron transfer reactions.

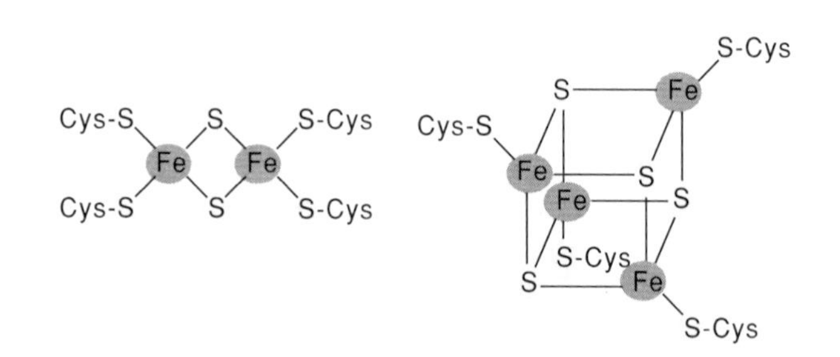

Iron-Containing Coenzymes Are Frequently Involved in Redox Reactions

Iron as a cofactor in catalysis is receiving increasing attention. The metal exists in two oxidation states; Fe^{2+} and Fe^{3+}. Iron complexes are nearly all octahedral, and practically all are paramagnetic (as a result of unpaired electrons in the 3*d* orbital). The most common form of iron in biological systems is heme. Heme groups (Fe^{2+}) and hematin (Fe^{3+}) most frequently involve a complex with protoporphyrin IX (fig. 11.20). They are the coenzymes (prosthetic groups) for a number of redox enzymes, including catalase, which catalyzes dismutation of hydrogen peroxide (equation (29)), and peroxidases, which catalyze the reduction of alkyl hydroperoxides by such reducing agents as phenols, hydroquinones, and dihydroascorbate (represented as AH_2 in equation (30)).

$$2\,H_2O_2 \rightleftharpoons 2\,H_2O + O_2 \quad (29)$$

$$R{-}O{-}O{-}H + AH_2 \longrightarrow A + R{-}O{-}H + H_2O \quad (30)$$

Heme proteins exhibit characteristic visible absorption spectra as a result of protoporphyrin IX; their spectra differ depending on the identities of the lower axial ligand donated by the protein and the oxidation state of the iron as well as the identities of the upper axial ligands donated by the substrates. Spectral data show clearly that the heme coenzymes participate directly in catalysis; however, the mechanisms of action of hemes are not as well understood as those of other coenzymes.

Many redox enzymes contain iron-sulfur clusters that mediate one-electron transfer reactions. The clusters consist of two or four irons and an equal number of inorganic sulfide ions clustered together with the iron, which is also liganded to cysteinyl-sulfhydryl groups of the protein (fig. 11.21). The enzyme nitrogenase, which catalyzes the reduction of N_2 to 2 NH_3 contains such clusters in which some of the iron has been replaced by molybdenum (see chapter 18). Electron-transferring proteins involved in one-electron transfer processes often contain iron-sulfur clusters. These proteins include the mitochondrial membrane enzymes NADH dehydrogenase and succinate dehydrogenase (chapter 14), which are flavoproteins, and the small-molecular-weight proteins ferredoxin, rubredoxin, adrenodoxin, and putidaredoxin (chapters 15, 16, and 23).

Heme coenzymes, iron-sulfur clusters, flavin coenzymes, and nicotinamide coenzymes cooperate in multienzyme systems to catalyze the chemically remarkable hydroxylations of hydrocarbons such as steroids (chapter 23). In these hydroxylation systems, the heme proteins constitute a family of proteins known as cytochrome P450, named for the wavelength corresponding to the most intense absorption band of the carbon monoxide-liganded heme, an inhibited form. The reactions catalyzed by these systems are represented in generalized form by equation (31), which also shows the fate of the two oxygens from $^{18}O_2$.

$$H^+ + NADPH + {}^{18}O_2 + R\!-\!CH_2\!-\!R \longrightarrow R\!-\!CH({}^{18}OH)\!-\!R + NADP^+ + H_2{}^{18}O \qquad (31)$$

One oxygen of O_2 is incorporated into the hydroxyl group of the product, whereas the other is incorporated into water. The enzymes usually include a cytochrome P450, an iron-sulfur cluster-containing protein such as adrenodoxin or putidaredoxin, and a flavoprotein reductase.

In the mechanism of oxygenation by these enzymes, the flavoproteins and iron-sulfur proteins supply reducing equivalents in one-electron units from NADPH to cytochrome P450, and reduced cytochrome P450 reacts directly with O_2. These reactions generate a Fe(III)-peroxide complex, shown in figure 11.22, which undergoes a further dehydration to an oxygenating species of cytochrome P450. The oxygenating species is thought to be an oxo-complex of Fe(IV), in which the porphyrin ring has been oxidized to a radical-cation by loss of an electron. The positive charge and unpaired electron in figure 11.22 are stabilized by delocalization through the conjugated π-electron system of the porphyrin ring (see fig. 11.20).

The most widely accepted mechanism for cytochrome P450-catalyzed oxygenation of substrates is illustrated for a hydrocarbon substrate in figure 11.23. The oxygenating species, the oxo-Fe(IV) radical-cation, abstracts a hydrogen atom from the hydrocarbon to form a hydrocarbon radical and a hydroxy-Fe complex. Hydrogen abstraction proceeds by the pairing of one electron from the oxo-Fe(IV) species and one electron from the carbon–hydrogen bond of the hydrocarbon. In the second step of oxygenation the hydrocarbon radical abstracts the hydroxy group from the hydroxy-Fe(III) complex by a pairing of the unpaired radical electron with one electron of the hydroxyl group, leaving cytochrome P450 in the +3 oxidation state, where it began in figure 11.22. The mechanism shown in figure 11.22 is known as the "rebound" mechanism of oxygenation. The second step of the mechanism is the rebound step, and the rate constant for this step in a microsomal cytochrome P450 has been estimated to be greater than 10^{10} s^{-1}. The magnitude of this rate constant explains why the radical and hydroxy-Fe intermediates have not been detected by presently available spectroscopic methods.

Figure 11.22

Steps in the formation of the oxygenating species of cytochrome P450. The oxygenating species of cytochrome P450 is generated by the transfer of two electrons in discrete steps from NADPH via the flavoprotein reductase and iron-sulfur clusters to the iron porphyrin complex, together with the reaction with oxygen. The Fe(III)-peroxide then undergoes a further expulsion of water to form the oxygenating species shown in brackets. The oxygenating species is not directly observed, since it reacts quickly, but it is thought to contain Fe(IV) and a delocalized radical-cation in the porphyrin ring.

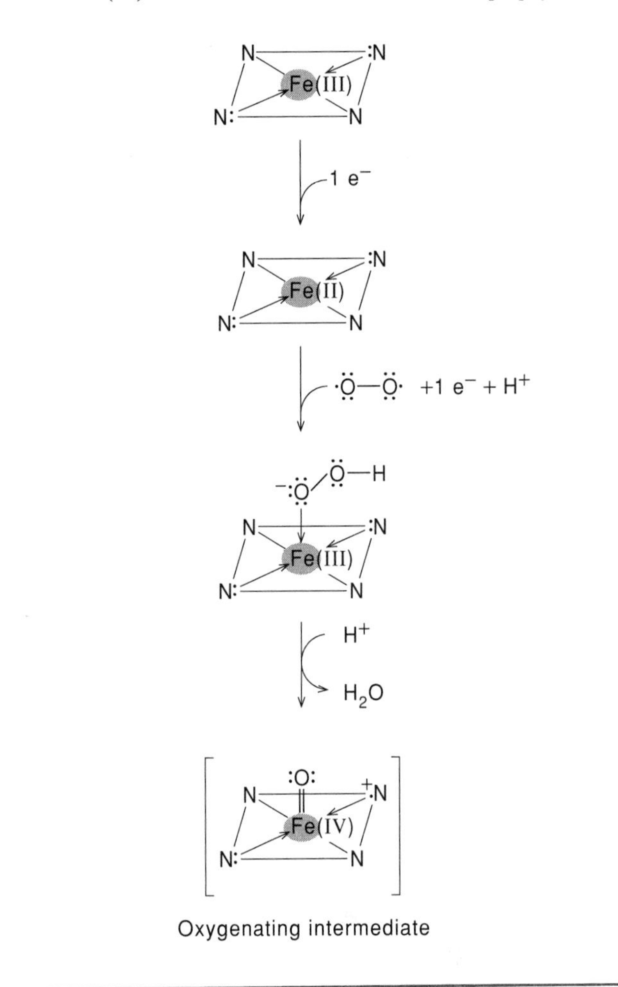

Figure 11.23

The rebound mechanism for oxygenation of hydrocarbons by cytochrome P450. The oxygenating species of cytochrome P450 in figure 11.22 is a highly reactive paramagnetic species that can abstract an unactivated and unreactive hydrogen from a hydrocarbon in the first step to form a hydrocarbon radical. In the rebound step the hydrocarbon radical abstracts a hydroxyl radical from the hydroxy-porphyrin intermediate.

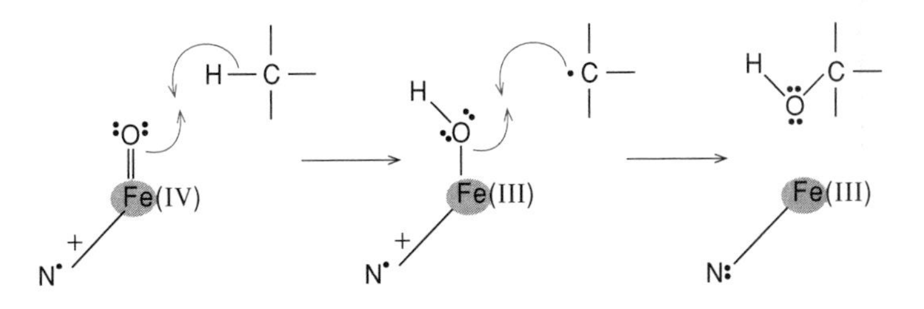

Table 11.3
Complexing Properties of Transition Metals of Biologic Importance

Metal and Valence State	Electron Configuration	Coordination Number	Stereo Chemistry	Examples
Manganese (Mn)				
Mn(II)	$3d^5$	4	Tetrahedral	Pyruvate decarboxylase (see chapters 11, 14, and 15)
		4	Square planar	
Mn(III)	$3d^4$	6	Octahedral	
Iron (Fe)	$3d^6$	4	Tetrahedral	
Fe(II)		5	Trigonal bipyramidal	
		6	Octahedral	Hemoglobin (chapter 5)
Fe(III)	$3d^5$	4	Tetrahedral	Iron sulfur coenzymes and cytochromes (see chapters 11 and 15)
		6	Octahedral	
Cobalt (Co)				
Co(I)	$3d^8$	4 5	Tetrahedral Trigonal bipyramidal	
Co(II)	$3d^7$	4 5 6	Tetrahedral; square planar Trigonal bipyramidal Octahedral	
Co(III)	$3d^6$	6	Octahedral	Vitamin B_{12} (this chapter)
Copper (Cu)				
Cu(I)	$3d^{10}$	2 4	Linear [L—M—L] Tetrahedral	
Cu(II)	$3d^9$	4 6	Tetrahedral; square pyramidal	Cytochrome (see chapter 15)
Zinc(Zn)				
Zn(II)	$3d^{10}$	4 5 6	Tetrahedral Trigonal bipyramidal Octahedral	Carboxypeptidase (chapter 9) Nucleic acid polymerase (chapter 26) Regulatory proteins (chapters 30 and 31)
Molybdenum (Mo)				
Mo(III)	$4d^3$	6 8	Octahedral Dodecahedral	
Mo(V)	$4d^1$	5 6 8	Trigonal bipyramidal Octahedral Dodecahedral	Nitrogenase (chapter 18)
Mo(VI)	$4d^0$	4 6	Tetrahedral Octahedral	

Metal Cofactors

In addition to cobalt and iron (discussed previously), other metals frequently function as cofactors in enzyme-catalyzed reactions. Like coenzymes, their usefulness is due to the fact that they offer something not available in amino acid side chains. The most important features of metals in this regard are their high concentrations of positive charge, their directed valences for interacting with two or more ligands, and their ability to exist in two or more valence states.

The alkali metal and alkaline earth metal ions are spherically symmetrical with respect to charge, so they do not usually show directed valences. There are notable exceptions, such as the case of Mg^{2+} in chlorophyll (see chapter 16), which adopts an approximately square planar distribution of orbitals. Alkali metals, Na^+ and K^+, with their single positive charges and lack of *d*-electronic orbitals for sharing, almost never make tight complexes with proteins. On rare occasions the alkaline earth divalent cations, Mg^{2+} and Ca^{2+}, can make strong complexes. The alkali metal and alkali earth cations are asymmetrically distributed in the organism, with K^+ and Mg^{2+} being concentrated in the cytosol and Na^+ and Ca^{2+} being concentrated in the organelles and the blood.

Most of the remaining metals found in biological systems are from the first transition series, in which the 3*d* orbitals are only partially filled. These metals all prefer structures with multiple coordination numbers, and, except for Zn^{2+}, which has a filled 3*d* orbital, they can exist in more than one oxidation state, a feature making them potentially useful in oxidation-reduction reactions. The stability of zinc's electronic state combined with its preference for a coordination number of 4 to 6 makes zinc singularly useful in enzyme reactions, where it acts as a Lewis acid, and in structural situations where it chelates nitrogen (usually in histidine side chains) and/or sulfur-containing amino acid side chains (usually cysteines) to rigidify the structural domain of a protein.

The involvement of transition metals in biochemical reactions is discussed in later chapters as indicated in table 11.3, which also indicates the coordination numbers and stereochemistries of the most significant oxidation states of these metals.

Lipid-Soluble Vitamins

We have seen that most water-soluble vitamins are converted by single or multiple steps to coenzymes. Our understanding of the lipid-soluble vitamins (see table 11.1), and of how they are utilized by the organism, is much less extensive. In this section we will briefly discuss the structure and functions of some lipid-soluble vitamins.

Vitamin D_3 (cholecalciferol) can be made in the skin from 7-dehydrocholesterol in the presence of ultraviolet light (see fig. 24.13). Vitamin D_3 is formed by the cleavage of ring B of 7-dehydrocholesterol. Vitamin D_3 made in skin or absorbed from the small intestine is transported to the liver and hydroxylated at C-25 by a microsomal mixed-function oxidase. 25-Hydroxyvitamin D_3 appears to be biologically inactive until it is hydroxylated at C-1 by a mixed-function oxidase in kidney mitochondria. The 1,25-dihydroxyvitamin D_3 is delivered to target tissues for the regulation of calcium and phosphate metabolism. The structure and mode of action of 1,25-dihydroxyvitamin D_3 is analogous to that of the steroid hormones (see chapter 24).

Figure 11.24

Structures of vitamins K_1 and K_2. K_1 is found in plants, K_2 in animals and bacteria.

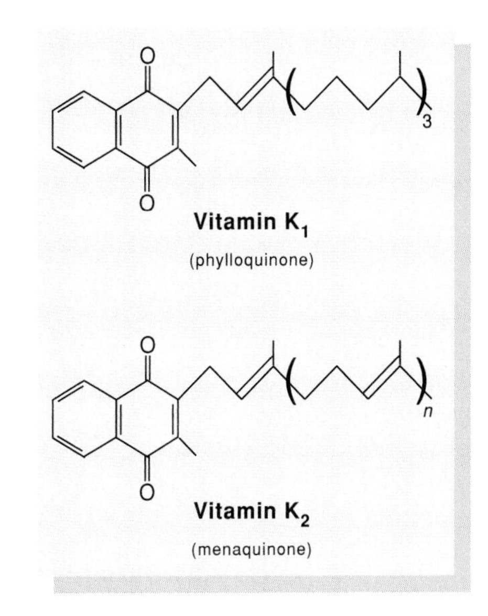

Vitamin K was discovered by Henrik Dam in Denmark in the 1920s as a fat-soluble factor important in blood coagulation (K is for "koagulation"). The structures of vitamins K_1 and K_2 (fig. 11.24) were elucidated by Edward Doisy. Vitamin K_1 is found in plants, vitamin K_2 in animals and bacteria. How this vitamin functions in blood coagulation eluded scientists until 1974, when vitamin K was shown to be needed for the formation of γ-carboxyglutamic acid (fig. 11.25) in certain proteins. γ-Carboxyglutamic acid specifically binds calcium, which is important for blood coagulation. Such modified glutamic acid residues appear to be important in many other processes involving calcium transport and calcium-regulated metabolic sequences.

Vitamin E (α-tocopherol) (fig. 11.26) was recognized in 1926 as an organic-soluble compound that prevented sterility in rats. The function of this vitamin still has not been clearly established. A favorite theory is that it is an antioxidant that prevents peroxidation of polyunsaturated fatty acids. Tocopherol certainly prevents peroxidation *in vitro,* and it can be replaced by other antioxidants. However, other antioxidants will not relieve all the symptoms of vitamin E deficiency.

Vitamin A (*trans*-retinol) is called an isoprenoid alcohol because it consists, in part, of units of a single five-carbon compound called isoprene:

Figure 11.25

Vitamin-K-dependent carboxylation of a glutamic acid residue in a protein. This reaction is essential to blood clotting.

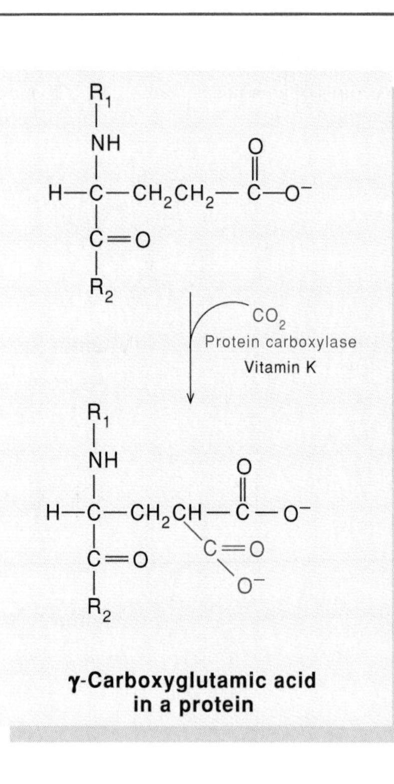

Figure 11.26

Structure of vitamin E (α-tocopherol).

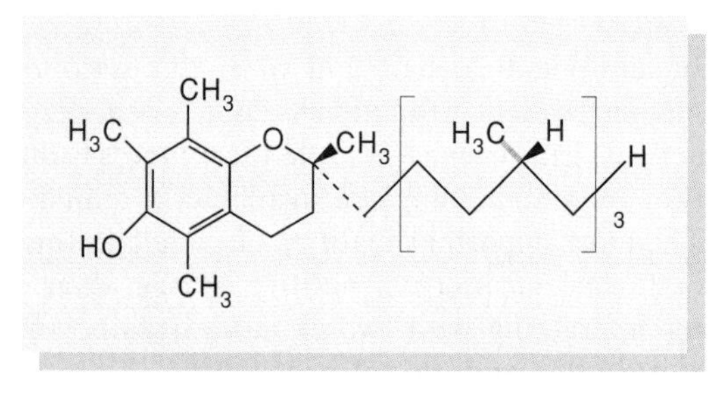

Isoprene is also a precursor of steroids and terpenes. This relationship will become clear when we examine the biosynthesis of these compounds (see chapter 23). Vitamin A is either biosynthesized from β-carotene (see fig. 36.4) or absorbed in the diet. Vitamin A is stored in the liver predominantly as an ester of palmitic acid. For many decades, it has been known to be important for vision and for animal growth and reproduction. The form of vitamin A active in the visual process is 11-*cis*-retinal, which combines with the protein opsin to form rhodopsin. Rhodopsin is the primary light-gathering pigment in the vertebrate retina (see chapter 36).

Summary

Coenzymes are molecules that act in cooperation with enzymes to catalyze biochemical processes that require functions that enzymes are otherwise chemically or physically ill-equipped to carry out. Our discussion in this chapter has focused on the following points.

1. Although a few coenzymes, such as hemes, lipoic acid, and iron-sulfur clusters, are biosynthesized in the body, most coenzymes are derivatives of the water-soluble vitamins.
2. Each coenzyme plays a unique chemical or physical role in the enzymatic processes of living cells, and the chemical structure and reactivity of each are suited to its biochemical function.
3. Thiamine pyrophosphate promotes the decarboxylation of α-keto acids and the cleavage of α-hydroxyl ketones by reacting with the ketone groups of substrates to produce intermediates in which electron pairs resulting from carbon–carbon bond cleavages can be stabilized by the thiazolium ring of the coenzyme.
4. Pyridoxal-5′-phosphate promotes decarboxylations, racemizations, transaminations, aldol cleavages, and α,β and β,γ eliminations in amino acid substrates in which electron pairs resulting from cleavages of covalent bonds are stabilized by the pyridine ring of the coenzyme.
5. Nicotinamide coenzymes act as intracellular electron carriers in transporting reducing equivalents from metabolic intermediates to the terminal electron-transport systems in cellular and mitochondrial membranes. Nicotinamide coenzymes also act as cocatalysts with enzymes by transiently oxidizing substrates to form reactive intermediates.
6. Flavin coenzymes are oxidation-reduction coenzymes th act as cocatalysts with enzymes in a large variety of biochemical redox reactions, most of which involve O_2.
7. Phosphopantetheine coenzymes facilitate the enolizatio and acyl-group transfer reactions of acyl groups to whicl they are bonded by thioester linkages.
8. Lipoic acid in α-keto acid dehydrogenase complexes mediates electron transfer and acyl-group transfer amon; the active sites in the complex by undergoing transient reduction and acylation.
9. Biotin mediates the carboxylation of substrates by undergoing transient carboxylation to N^1-carboxybiotin.
10. Phosphopantetheine, lipoic acid, and biotin, by virtue o their flexible chainlike structures, facilitate the physical translocation of chemically reactive species among catal sites or subunits.
11. Tetrahydrofolates are cosubstrates for a variety of enzym associated with one-carbon metabolism. Tetrahydrofolat maintain formaldehyde and formate in chemically poise states available for essential processes by specific enzym action.
12. Heme coenzymes play essential roles in a variety of enzymatic reactions involving reductions of peroxides. Iron-sulfur clusters, composed of Fe and S in equal numbers together with cysteinyl side chains of proteins, mediate electron-transfer processes in a variety of enzymatic reactions, including the reduction of N_2 to 2 NH_3. Nicotinamide, flavin, and heme coenzymes act cooperatively with iron-sulfur clusters in multienzyme systems to catalyze hydroxylations of hydrocarbons.

13. 5′-Deoxyadenosylcobalamin (vitamin B_{12} coenzyme) is involved as the essential cofactor in intramolecular rearrangements in which an unreactive hydrogen exchanges positions with a group bonded to an adjacent carbon. These are radical rearrangements in which the coenzyme initiates the formation of substrate radicals by virtue of the weakness of the cobalt–carbon bond in the coenzyme. Homolytic scission of this bond generates a 5′-deoxyadenosyl-5′-radical that initiates radical formation by abstraction of a hydrogen atom.
14. Metals often perform the role of enzyme cofactors. Examples include iron in porphyrins and iron-sulfur compounds. The most important features of metals as cofactors are their high concentrations of positive charge, their directed valences for interacting with two or more ligands, and their ability to exist in two or more valence states.
15. A number of vitamins are lipid-soluble, including vitamins D, K, E, and A. In general, less is known about the mechanisms of action of lipid-soluble vitamins and their derivatives.

Selected Readings

Bruice, T. C., and S. J. Benkovic, *Bioorganic Chemistry,* vols. 1 and 2. Menlo Park, Calif.: Benjamin, 1966. A detailed discussion of the mechanisms of bioorganic reactions, including those involving coenzymes.

DiMarco, A. A., T. A. Bobik, and R. S. Wolfe, Unusual coenzymes of methanogenesis. *Ann. Rev. Biochem.* 59:355 (1990). A review of the recently characterized coenzymes required for the biosynthesis of methane in methanogenic bacteria.

Dolphin, D. (ed.), *B_{12}*, vols. 1 and 2. New York: Wiley-Interscience, 1982. Chemistry and mechanism of action of Vitamin B_{12}.

Dolphin, D., R. Poulson, and O. Avamovic (eds.), *Pyridoxal Phosphate, Part A & Part B.* New York: Wiley-Interscience, 1987. Chemistry and mechanism of action of pyridoxal phosphate.

Dolphin, D., R. Poulson, and O. Avamovic (eds.), *Pyridine Nucleotide Coenzymes, Part A & Part B.* New York: Wiley-Interscience, 1987. Chemical, biochemical, and medical aspects of pyridine nucleotide coenzymes.

Frausto de Silva, J. J. K., and R. J. P. Williams, *The Biological Chemistry of the Elements.* Oxford: Clarendon Press, 1991.

Frey, P. A., The importance of organic radicals in enzymatic cleavage of unactivated C—H bonds. *Chem. Rev.* 90:1343, 1990. A brief review of coenzymes required to cleave unreactive C—H bonds.

Jencks, W. P., *Catalysis in Chemistry and Enzymology.* New York: McGraw-Hill, 1969. A detailed analysis of mechanisms of enzymatic and nonenzymatic reactions, including those involving coenzymes.

Knowles, J. R., The mechanism of biotin-dependent enzymes. *Ann. Rev. Biochem.* 58:195, 1989. Review of the chemical mechanism of biotin-dependent carboxylation reactions.

Ortiz de Montellano, P. R. (ed.), *Cytochrome P-450: Structure, Mechanism and Biochemistry.* New York: Plenum, 1986. A treatise on the structure and function of cytochrome P450 monooxygenases.

Phipps, D. A., *Metals and Metabolism.* Oxford Chemistry Series. Oxford: Clarendon Press, 1976. Examines the importance of metal ions in metabolic processes.

Popják, G., Stereospecificity of enzymic reactions. In P. D. Boyer (ed.), *The Enzymes,* vol. 2. New York: Academic Press, 1970. P. 115.

Walsh, C. T., *Enzymatic Reaction Mechanisms.* San Francisco: Freeman, 1977. Provides an up-to-date discussion of the mechanisms of enzymatic reactions. An indepth treatment of coenzymes.

Problems

1. What structural features of biotin and lipoic acid allow these cofactors to be covalently bound to a specific protein in a multienzyme complex yet participate in reactions at active sites on other enzymes of the complex?
2. The following reactions are catalyzed by pyridoxal-5′-phosphate-dependent enzymes. Write a reaction mechanism for each, showing how pyridoxal-5′-phosphate is involved in catalysis.

(a) $CH_3-C(=O)-COO^- + R-CH(NH_3^+)-COO^- \rightarrow CH_3-CH(NH_3^+)-COO^- + R-C(=O)-COO^-$

(b) $H_2N-(CH_2)_4-CH(NH_2)-CO_2^- \rightarrow CO_2 + H_2N-(CH_2)_5-NH_2$

(c) $^-O_2C-CH_2-CH(NH_2)-CO_2^- \rightarrow CO_2 + CH_3-CH(NH_2)-CO_2^-$

3. $NADP^+$ differs from NAD^+ only by phosphorylation of the C-2′ OH group on the adenosyl moiety. The redox potentials differ only by about 5 mV. Why do you suppose it is necessary for the cell to employ two such similar redox cofactors?
4. Thiamine-pyrophosphate-dependent enzymes catalyze the reactions shown below. Write a chemical mechanism that shows the catalytic role of the coenzyme.

(a) $CH_3-C(=O)-CO_2H \rightarrow CO_2 + CH_3-C(=O)-H$

(b) $Ⓟ OCH_2-(CHOH)_2-CH(OH)-C(=O)-CH_2OH + HOPO_3^{2-} \rightarrow Ⓟ O-CH_2-(CHOH)_2-CHO + CH_3-C(=O)-OPO_3^{2-} + H_2O$

5. Malate synthase catalyzes the condensation of acetyl-CoA with glyoxalate to form L-malate.
   (a) Write a chemical reaction illustrating the reaction catalyzed by malate synthase.
   (b) Explain the chemical basis for using acetyl-CoA rather than acetate in the condensation reaction.
   (c) The product released from the enzyme is L-malate rather than malyl-CoA. Explain the role of thioester hydrolysis in the condensation reaction.
6. Write the mechanism showing how NAD(P)$^+$ is involved in the following reactions.
   (a) Malate dehydrogenase:

$$\mathrm{HOOC{-}CH_2{-}CHOH{-}COOH} + \mathrm{NAD^+} \rightarrow \mathrm{HOOC{-}CH_2{-}C(=O){-}COOH} + \mathrm{NADH} + \mathrm{H^+}$$

   (b) Malate dehydrogenase (decarboxylating; malic enzyme)

$$\mathrm{HOOC{-}CH_2{-}CHOH{-}COOH} + \mathrm{NADP^+} \rightarrow \mathrm{CH_3{-}C(=O){-}COOH} + \mathrm{NADPH} + \mathrm{CO_2} + \mathrm{H^+}$$

7. What chemical features allow flavins (FAD, FMN) to mediate electron transfer from NAD(P)H to cytochromes or iron-sulfur proteins?
8. Write the mechanisms that show the involvement of biotin in the following reactions.

(a) $\mathrm{CH_3{-}C(=O){-}SCoA} + \mathrm{HCO_3^-} + \mathrm{ATP} \rightarrow \mathrm{HO_2C{-}CH_2{-}C(=O){-}SCoA} + \mathrm{ADP} + \mathrm{HOPO_3^{2-}}$

(b) $\mathrm{CH_3{-}CH_2{-}C(=O){-}SCoA} + \mathrm{HO_2C{-}CH_2{-}C(=O){-}CO_2H} \rightleftarrows \mathrm{HO_2C{-}CH(CH_3){-}C(=O){-}SCoA} + \mathrm{CH_3{-}C(=O){-}CO_2H}$

9. (a) What metabolic advantage is gained by having flavin cofactors covalently or tightly bound to the enzyme?
   (b) Would covalently bound NAD$^+$ (NADP$^+$) be a metabolic advantage or disadvantage?
10. Given the amino acid of the general structure

$$\mathrm{CH_3{-}CH(X){-}CH(NH_3^+){-}COO^-}$$

we could use pyridoxal-5′-phosphate to eliminate X, decarboxylate the amino acid, or oxidize the α carbon to a carbonyl with formation of pyridoxamine-5′-phosphate. The metabolic diversity afforded by PLP, unchanneled, could wreak havoc in the cell. What other components are required to channel the PLP-dependent reaction along specific reaction pathways?

11. Amino acid oxidases catalyze the flavin-dependent reaction

$$\mathrm{R{-}CH(NH_3^+){-}COO^-} + \mathrm{O_2} + 2\,\mathrm{H^+} \rightarrow \mathrm{R{-}C(=O){-}COO^-} + \mathrm{H_2O_2} + \mathrm{NH_4^+}$$

What advantages are gained by the cell in using the PLP-dependent transamination reaction rather than the FAD-dependent deamination reaction to convert α-amino acids to the corresponding α-keto acids?
12. How is flavin adenine dinucleotide involved in the following reaction? Show possible mechanisms.

$$\alpha\text{-D-glucose} + \mathrm{O_2} \rightarrow \text{D-gluconolactone} + \mathrm{H_2O_2}$$

13. For each of the following enzymatic reactions, identify the coenzyme involved.

(a) $\mathrm{R{-}CH(NH_3^+){-}COOH} + \mathrm{H_2O} + \mathrm{O_2} \rightarrow \mathrm{R{-}C(=O){-}COOH} + \mathrm{NH_4^+} + \mathrm{H_2O_2}$

(b) $\mathrm{HO{-}CH_2{-}CH(NH_3^+){-}CO_2H} \rightarrow \mathrm{CH_3{-}C(=O){-}CO_2H} + \mathrm{NH_4^+}$

(c) $\mathrm{CH_3{-}CH_2{-}C(=O){-}SCoA} + \mathrm{HCO_3^-} + \mathrm{ATP} \rightarrow \mathrm{HO_2C{-}CH(CH_3){-}C(=O){-}SCoA} + \mathrm{ADP} + \mathrm{P_i}$

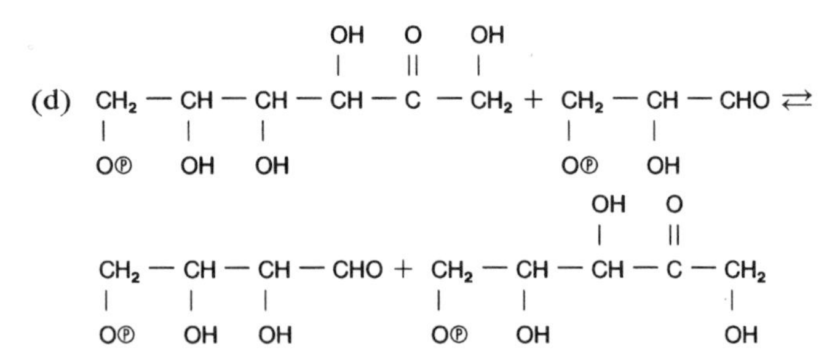

14. Lipid-soluble polycyclic aromatic hydrocarbons are in part excreted from the body conjugated to one of several hydrophilic substituents, principally glucuronic acid. What role does the liver microsomal cytochrome P450 (polysubstrate monooxygenase) system play in converting a polycyclic aromatic hydrocarbon to forms that could be conjugated to glucuronic acid?
15. Some bacterial toxins use NAD$^+$ as a true substrate rather than as a coenzyme. The toxins catalyze the transfer of ADP-ribose to an acceptor protein. Examine the structure of NAD$^+$ and indicate which portion of the molecule is transferred to the protein. What is the other product of the reaction?

A Student's Guide to Methods of Biochemical Analysis

iochemistry is an experimental science, in which a great variety of techniques are used. We provide examples of the use of various methods throughout this text, usually in connection with the accounts of specific experiments. This approach has the advantage of allowing you to appreciate immediately the relevance of a particular technique. It does mean, however, that the examples are scattered throughout the text. Our aim here is to supply a guide to the locations of those examples within the framework of a unifying outline, so that you can see at a glance which methods of analysis are most useful for certain purposes and so that you can easily find the places in the text where those methods are described. In the outline, you are referred to the figures or boxes that contain such descriptions; you will sometimes find additional information in the adjoining text.

Our descriptions of techniques are brief. There are a number of excellent printed sources where you can find more complete descriptions, as well as comprehensive surveys of available techniques. For example, Academic Press has published more than two hundred books in their series *Methods in Enzymology,* and Oxford Press has published more than eighty books in their series *Biochemistry: A Practical Approach.* In addition, you will find descriptions of the procedures for manipulating DNA in the remarkable three-volume set *Molecular Cloning: A Laboratory Manual,* by J. Sambrook, E. F. Fritsch, and T. Maniatis, 2d ed. (Cold Spring Harbor, N.Y.: Cold Spring Harbor Laboratory, 1989).

I. Separation of major cellular components
Comparison of differential centrifugation with isopycnic centrifigution (fig. 7.20).
Separation of periplasmic proteins and inner and outer membranes of a Gram negative bacterium (fig. 7.23).

II. Subfractioning of major cellular components into their constituent molecules
Methods of protein purification are described in the latter part of chapter 5, which covers differential precipitation with $(NH_4)_2SO_4$ (box 5B), column fraction procedures (box 5C), gel electrophoresis (box 5D), isoelectric focusing (box 5D), and differential centrifugation (box 5E). The following, related material appears in other chapters.
Analysis of protein mixtures by polyacrylamide gel electrophoresis (fig. 7.31).
Analysis of ribosomal proteins by two-dimensional gel electrophoresis (box 28A).
Method for lipid fractionation (box 7A).
Detergent solubilization of biological membranes (fig. 7.29).
Separation of N-linked carbohydrates by lectin-affinity chromatography (fig. 6.26).

III. Structural characterization of individual cellular components

A. **Amino acids and protein primary structure**
Determination of p*K* for amino acids (fig. 3.2, 3.3, and 3.4).
pH titration curve for a protein (fig. 3.6).
Column method for analysis of amino acid mixtures (fig. 3.14).
Determination of amino acid composition of a protein (figs. 3.15 and 3.16).
Dansyl chloride method for N-terminal amino acid determination (box 3B).
Determination of the amino acid sequence of the polypeptide chain(s) of a protein (figs. 3.17 through 3.22).

Separation of amino acid derivatives by thin-layer chromatography (fig. 3.23).
Measurement of ultraviolet absorption in solution (box 3A).
Determination of essential amino acids in rats and humans (table 18.2).

B. **Properties of intact proteins**
Most methods of determining protein conformation are described in chapter 4. These include predicting protein tertiary structure from the primary structure (fig. 4.37), X-ray diffraction methods for determining the structure of fibrous proteins (box 4A) and of globular proteins (box 4B), different ways of visualizing molecular structures of proteins (box 4C), and radiation techniques for examining protein structure (box 4D). The following, related material appears in other chapters.
Measuring the binding of inducer to repressor by equilibrium dialysis (box 30B).
Domain-swap experiment demonstrating bifunctional nature of a DNA-binding protein (fig. 31.5).
Electron microscopic examination of striated muscle indicating the structural changes that accompany muscular contraction (fig. 5.17).
Determination of the size of protein molecules in solution by sedimentation and diffusion (box 4E).

C. **Lipids and membranes**
Isolation and analysis of phospholipids (box 7A).
Hydrophobic interaction chromatography (fig. 7.30).
Hydropathy plot to determine segments of a polypeptide chain most likely to be membrane-bound (figs. 7.37 and 32.10).
Differential scanning calorimetry for estimating stability of a membrane structure (fig. 7.39).
Determination of the rate of diffusion of a molecule across a semipermeable membrane (figs. 32.2, 32.3 and 32.4).

D. **Carbohydrates**
Use of glycosidases for structural analysis of oligosaccharides (fig. 6.27).
Polarimetry (box 6A).
Nuclear magnetic resonance spectroscopy for structural analysis of oligosaccharides (box 6B).

E. **Nucleic acids**
Use of ultraviolet absorbance to study DNA denaturation and renaturation (fig. 25.22).
Use of reassociation kinetics to study sequence complexity of DNA (figs. 25.25 and 25.26).
Use of enzyme probes to investigate bacterial chromosome structure (fig. 25.28).
Assessment of supercoiling by ultracentrifugation (fig. 25.19) or electrophoresis (fig. 25.21).
X-ray diffraction analysis of DNA (box 25A).
Techniques using nucleic acid renaturation (box 25D).
Equilibrium density-gradient centrifugation for analysis of DNAs of different densities (box 25B).
Chromosome analysis by gel electrophoresis (box 25C).
Southern blot analysis to characterize DNA sequences in spliced immunoglobulin genes (fig. 33.4; also see fig. 27.27).
Hybridization followed by CsCl density-gradient sedimentation for structural analysis of phage transcripts (fig. 28.2).
Determination of DNA protein-binding sites by chemical protection (fig. 28.15).
Determination of intron size and location by selective annealing and electron microscopy (fig. 28.25).
Footprinting technique for determining location of site of protein binding to DNA (box 28B).
Techniques for manipulating DNA (see listing under part VI of this outline).

IV. **Characterization of individual reactions *in vitro***
Chapter 8 is concerned with kinetic methods for analyzing enzyme-catalyzed reactions. These include determination of the Michaelis constant, the turnover number, and the specificity constant, special kinetic considerations for reactions involving two substrates, and analysis of the effects of temperature, pH, and various types of inhibitors. Chapters 9 and 10 are concerned with methods for analyzing mechanisms of enzyme-catalyzed reactions. Chapter 10 deals exclusively with the unique properties of regulatory enzymes. The following, related material appears in other chapters.
Calculation of the equilibrium constant, K_{eq}, when the numbers of reactants and products are not equal (box 13A).
Calculation of the equilibrium constant for a reaction from the thermodynamic data (chapter 2).
Apparatus for measuring the difference between the standard redox potentials, E^0, of two redox couples (fig. 15.10).
Titration curves for redox couples (fig. 15.11).
Assay for the activity of fatty acid synthase (box 17A).
In vitro studies to determine properties of the recA enzyme (fig. 26.34).
Instrument for recording action potentials (figs. 35.2 and 35.3).
Isotope labeling technique for determining stereospecificity of an enzyme-catalyzed reaction (fig. 11.7 and associated text).

V. ***In vitro* or *in vivo* characterization of reactions in pathways**
Methods for pathway analysis are reviewed at the end of chapter 12. The following, related material appears in other chapters.
Difference spectra method for examining the redox states of respiratory carriers in intact mitochondria (fig. 15.13).

Inhibitors used to analyze electron transport in mitochondria (fig. 15.14).
Determination of the P/O ratio in respiring mitochondria (fig. 15.21).
Use of uncouplers to study the mechanism of ATP synthesis in respiring mitochondria (fig. 15.23).
Use of enzyme inhibitors to study ATP synthesis in mitochondria (fig. 15.24).
In vitro methods for studying the mechanism of ATP synthesis in respiring mitochondria (figs. 15.26, 15.27, 15.29, 15.32, and 15.34).
Analysis of O_2 production by illumination of chloroplasts (figs. 16.16, 16.20, and 16.23).
Spectral methods for analysis of photosystems I and II (fig. 16.21).
Manipulation of pH to study ATP synthesis of chloroplasts (fig. 16.26).
Spectral methods for the analysis of phytochrome behavior (fig. 16.23).
Use of equilibrium density-gradient centrifugation to determine overall mode of DNA replication (figs. 26.3 and 26.6).
Use of autoradiography to determine mode of DNA replication in eukaryotic chromosomes (figs. 26.4 and 26.6).
Use of pulse-chase labeling to follow precursor-product relationship in DNA synthesis (fig. 26.9).
Use of synthetic polynucleotides for determining genetic code assignments (fig. 29.7).
In vitro assay for translation release factors (fig. 29.20).
Device for recording action potentials (figs. 35.2 and 35.3).
Use of polarized light to determine orientation of optical pigments in rod cells (fig. 36.14).
Tissue transplantation technique used to establish the mechanism for immunologic tolerance (fig. 33.14).
Pathway analysis by genetic complementation (box 18A).
Use of isotopes as tracers to delineate the valine and isoleucine pathways (box 18C).
Use of isotopes and mutants to demonstrate that indole is not a true intermediate in the tryptophan pathway (box 18D).
Precursors of purine ring determined by isotope labeling (fig. 20.12).
Use of antibiotics to determine pathway for cell-wall biosynthesis in bacteria (figs. 21.16, 21.17, and 21.19).
Use of inhibitors to determine stages in N-linked carbohydrate synthesis (table 21.3).
Use of mutants to determine stages in N-linked and O-linked carbohydrate synthesis (table 21.4).
Assay for transit of glycoproteins between Golgi membranes (box 21D).
Use of substrates and inhibitors for demonstrating the circular nature of the Krebs cycle (fig. 14.3).
Demonstration of the prochiral nature of citrate in the Krebs cycle (fig. 14.9).

VI. Genetic methods

Chapter 27 is concerned with techniques of DNA manipulation. These include the Maxam-Gilbert and Sanger methods for sequencing DNAs and RNAs, the synthesis of oligonucleotides, the amplification of sequences by the polymerase chain reaction, the use of restriction enzymes to cut DNA into well-defined fragments, procedures for cloning DNA into *E. coli,* yeast, mammals, and plants, site-directed mutagenesis, and the use of recombinant DNA techniques for mapping the globin gene family and the cystic fibrosis gene. The following, related material appears in other chapters.
DNA transformation (fig. 25.2).
Demonstration that nucleic acids carry the genetic information in viruses (figs. 25.3 and 25.4).
Pathway analysis by genetic complementation (box 18A).
Use of mutants to determine stages in N-linked and O-linked carbohydrate synthesis (table 21.4).
Procedure for constructing a congenic strain (fig. 33.16).
Tissue transplantation technique to establish the mechanism for immunologic tolerance (fig. 33.14).
Genetic analysis of *lac* operon expression by mutant studies on haploid and diploid cells (table 30.1).
Mutant studies on expression of the tryptophan operon of *E. coli* (table 30.2).
Nuclear transplantation assay for totipotent nucleus (fig. 31.12).
Genetic analysis of the interaction between regulatory proteins involved in early *Drosophila* development (fig. 31.31).
Mutant analysis of genes responsible for characteristics of specific body segments (fig. 31.33).
Genetic analysis of genes regulating galactose metabolism in yeast (table 31.1).
Determination of the location of developmental gene expression in the *Drosophila* embryo by *in situ* labeling with specific probes followed by autoradiography (box 31A).
Analysis of key events in early *Drosophila* development through detection of specific transcripts at various times and locations (fig. 31.29).
Use of mutants to determine the significance of DNA polymerase I of *E. coli* (chapter 26).
Determination of the crucial amino acid residues in the hemoglobin molecule by observations on the invariant residues found in all hemoglobins (fig. 5.13) and the amino acid residues that lead to pathologic conditions (fig. 5.14).

Common Abbreviations in Biochemistry

A	adenine
ACh	acetylcholine
ACTH	adrenocorticotropic hormone
ADP	adenosine-5′-diphosphate
AIDS	acquired immune deficiency syndrome
Ala (A)	alanine
AMP	adenosine-5′-monophosphate
Asn (N)	asparagine
Asp (D)	aspartic acid
ATP	adenosine-5′-triphosphate
BChl	bacteriochlorophyll
bp	base pair
BPG	D-2,3-bisphosphoglycerate
C	cytosine
Cys (C)	cysteine
CAP	catabolite gene activator protein
CDP	cytidine-5′-diphosphate
CMP	cytidine-5′-monophosphate
CTP	cytidine-5′-triphosphate
CoA or CoASH	coenzyme A
CoQ	coenzyme Q (ubiquinone)
cAMP	adenosine 3′, 5′-cyclic monophosphate
cGMP	guanosine 3′, 5′-cyclic monophosphate
cyt	cytochrome
d	2′-deoxy-
DHAP	dihydroxyacetone phosphate
DHF	dihydrofolate
DHFR	dihydrofolate reductase
DMS	dimethyl sulfate
DNA	deoxyribonucleic acid
cDNA	complementary DNA
DNase	deoxyribonuclease
DNP	2,4-dinitrophenol
ER	endoplasmic reticulum
FAD	flavin adenine dinucleotide (oxidized form)
$FADH_2$	flavin adenine dinucleotide (reduced form)
FBP	fructose-1,6-bisphosphate
fMet	*N*-formylmethionine
FMN	flavin mononucleotide (oxidized form)
$FMNH_2$	flavin mononucleotide (reduced form)
F1P	fructose-1-phosphate
F6P	fructose-6-phosphate
G	guanine
G protein	guanine-nucleotide binding protein
GDP	guanosine-5′-diphosphate
Gly (G)	glycine
GMP	guanosine-5′-monophosphate
Gln (Q)	glutamine
Glu (E)	glutamic acid
GSH	glutathione
GSSG	glutathionine disulfide
GTP	guanosine-5′-triphosphate
Hb	hemoglobin
HDL	high-density lipoprotein
HETPP	hydroxyethylthiamine pyrophosphate
HGPRT	hypoxanthine-guanosine phosphoribosyl transferase
His (H)	histidine
HIV	human immunodeficiency virus
HMG-CoA	β-hydroxy-β-methylglutaryl-CoA
HPLC	high-performance liquid chromatography
IDL	intermediate-density lipoprotein
IF	initiation factor
IgG	immunoglobulin G
Ile (I)	isoleucine
IMP	inosine-5′-monophosphate
IP_1	inositol-1-phosphate
IP_3	inositol-1,4,5-triphosphate
IPTG	isopropylthiogalactoside
K_m	Michaelis constant
kb	kilobase pair
kDa	kilodaltons

LDL	low-density lipoprotein
Leu (L)	leucine
Lys (K)	lysine
Man	mannose
MHC	major histocompatability complex
Met (M)	methionine
NAD^+	nicotinamide-adenine dinucleotide (oxidized form)
NADH	nicotinamide-adenine dinucleotide (reduced form)
$NADP^+$	nicotinamide-adenine dinucleotide phosphate (oxidized form)
NADPH	nicotinamide-adenine dinucleotide phosphate (reduced form)
NDP	nucleoside-5′-diphosphate
NAM	*N*-acetylmuramic acid
NMR	nuclear magnetic resonance
NTP	nucleoside-5′-triphosphate
Phe (F)	phenylalanine
P_i	inorganic orthophosphate
PEP	phosphoenolpyruvate
PFK	phosphofructokinase
PG	prostaglandin
2PG	2-phosphoglycerate
3PG	3-phosphoglycerate
PIP_2	phosphatidylinositol-4,5-bisphosphate
PK	pyruvate kinase
PLP	pyridoxal-5-phosphate
PP_i	inorganic pyrophosphate
Pro (P)	proline
PRPP	phosphoribosylpyrophosphate
PS	photosystem
Q	ubiquinone or plastoquinone
QH_2	ubiquinol or plastoquinol
RER	rough endoplasmic reticulum
RF	release factor or replicative form
RFLP	restriction-fragment length polymorphism
RNA	ribonucleic acid
hnRNA	heterogeneous nuclear RNA
mRNA	messenger RNA
rRNA	ribosomal RNA
snRNA	small nuclear RNA
tRNA	transfer RNA
snRNP	small ribonucleoprotein
RNase	ribonuclease
Ru1,5P	ribulose-1,5-bisphosphate
Ru5P	ribulose-5-phosphate
R5P	ribose-5′-phosphate
RSV	Rous sarcoma virus
s	Svedberg constant
SAM	*S*-adenosylmethionine
SDS	sodium dodecyl sulfate
Ser (S)	serine
S7P	sedoheptulose-7-phosphate
SRP	signal recognition particle
T	thymine
THF	tetrahydrofolate
Thr (T)	threonine
TLC	thin-layer chromatography
TMV	tobacco mosaic virus
TPP	thiamine pyrophosphate
Trp (W)	tryptophan
TTP	thymidine-5′-triphosphate
Tyr (Y)	tyrosine
U	uracil
UDP	uridine-5′-diphosphate
UDPG	UDP-glucose
UMP	uridine-5′-monophosphate
UQ	ubiquinone
Val (V)	valine
VLDL	very-low-density lipoprotein
XMP	xanthosine-5′-monophosphate
Xu5P	xylulose-5′-phosphate

Appendix A
Some Major Discoveries in Biochemistry

n this appendix we list, in chronological order, some of the most important discoveries made in biochemistry in the past two centuries. It is impossible, for reasons of space, to give credit to every worker who has made a significant contribution, but it is possible to identify certain events as milestones, and thus to show how progress in this field has accelerated with the passage of time.

1770–1774
Priestly showed that oxygen is produced by plants and consumed by animals.

1773
Rouelle isolated urea from urine.

1828
Wohler synthesized the first organic compound, urea, from inorganic components.

1838
Schleiden and Schwann proposed that all living things are composed of cells.

1854–1864
Pasteur proved that fermentation is caused by microorganisms.

1864
Hoppe-Seyler crystallized hemoglobin.

1866
Mendel demonstrated the segregation and independent assortment of alleles in pea plants.

1893
Ostwald showed that enzymes are catalysts.

1905
Knoop deduced the β oxidation mechanism for fatty acid degradation.

1907
Fletcher and Hopkins showed that lactic acid is formed quantitatively from glucose during anaerobic muscle contraction.

1910
Morgan discovered sex-limited inheritance in *Drosophila.*

1912
Warburg postulated a respiratory enzyme for the activation of oxygen.

1913
Michaelis and Menten developed a kinetic theory of enzyme action.

1922
McCollum showed that lack of vitamin D causes rickets.

1926
Sumner crystallized the first enzyme, urease.

1926
Jansen and Donath isolated vitamin B_1 (thiamine) from rice polishings.

1926–1930
Svedberg invented the ultracentrifuge and determined the molecular weights of macromolecules.

1928
Levene showed that nucleotides are the building blocks of nucleic acids.

1928
Szent-Gyorgyi isolated ascorbic acid (Vitamin C).

1928–1933
Warburg deduced the iron-prophyrin presence in the respiratory enzyme.

1931
Engelhardt discovered that phosphorylation is coupled to respiration.

1932
Warburg and Christian discovered the "yellow enzyme," a flavoprotein.

1933
Krebs and Henseleit discovered the urea cycle.

1933
Embden and Meyerhof demonstrated the intermediates in the glycolytic pathway.

1935
Schoenheimer and Rittenberg first used isotopes as tracers in the study of intermediary metabolism.

1935
Stanley first crystallized a virus, tobacco mosaic virus.

1937
Krebs discovered the citric acid cycle.

1937
Warburg showed how ATP formation is coupled to the dehydrogenation of glyceraldehyde-3-phosphate.

1938
Hill found that cell-free suspensions of chloroplasts yield oxygen when illuminated in the presence of an electron acceptor.

1939
C. Cori and G. Cori demonstrated the reversible action of glycogen phosphorylase.

1939
Lipmann postulated the central role of ATP in the energy-transfer cycle.

1939–1946
Szent-Gyorgyi discovered actin and actin-myosin complex.

1940
Beadle and Tatum deduced the one gene–one enzyme relationship.

1942
Bloch and Rittenberg discovered that acetate is the precursor of cholesterol. Subsequently Woodward and Bloch determined the complete pathway for cholesterol biosynthesis.

1943
Chance applied spectrophotometric methods to the study of enzyme–substrate interactions.

1943
Martin and Synge developed partition chromatography.

1944
Avery, MacLeod, and McCarty demonstrated that bacterial transformation is caused by DNA.

1947–1950
Lipmann and Kaplan isolated and characterized coenzyme A.

1948
Leloir discovered the role of uridine nucleotides in carbohydrate metabolism.

1948
Hogeboom, Schneider, and Palade refined the differential centrifugation method for fractionation of cell parts.

1948
Kennedy and Lehninger discovered that the tricarboxylic acid cycle, fatty acid oxidation, and oxidative phosphorylation all take place in mitochondria.

1950–1953
Chargaff discovered the base equivalences in DNA.

1951
Pauling and Corey proposed the α-helix structure for α-keratins.

1951
Lynen postulated the role of coenzyme A in fatty acid oxidation.

1952
Palade, Porter, and Sjostrand perfected thin sectioning and fixation methods for electron microscopy of intracellular structures.

1952–1954
Zamecnik and his colleagues developed the first cell-free systems for the study of protein synthesis.

1953
Sanger and Thompson determined the complete amino acid sequence of insulin.

1953
Horecker, Dickens, and Racker elucidated the 6-phosphogluconate pathway of glucose catabolism.

1953
Watson and Crick and Wilkins determined the double-helix structure of DNA.

1955
Ochoa and Grunberg-Manago discovered polynucleotide phosphorylase.

1956
Kornberg discovered the first DNA polymerase.

1956
Umbarger reported that the end product isoleucine inhibits the first enzyme in its biosynthesis from threonine.

1956
Ingram showed that normal and sickle-cell hemoglobin differ in a single amino acid residue.

1956
Anfinsen and White concluded that the three-dimensional conformation of proteins is specified by their amino acid sequence.

1957
Hoagland, Zamecnik, and Stephenson isolated tRNA and determined its function.

1957
Sutherland discovered cyclic AMP.

1958
Weiss, Hurwitz, and Stevens discovered DNA-directed RNA polymerase.

1958
Meselson and Stahl demonstrated that DNA is replicated by a semiconservative mechanism.

1960
Kendrew reported the x-ray analysis of the structure of myoglobin.

1961
Jacob and Monod proposed the operon hypothesis.

1961
Jacob, Monod, and Changeux proposed a theory of the function and action of allosteric enzymes.

1961
Mitchell postulated the chemiosmotic hypothesis for the mechanism of oxidative phosphorylation.

1961
Nirenberg and Matthaei reported that polyuridylic acid codes for polyphenylalanine.

1961
Marmur and Doty discovered DNA renaturation.

1962
Racker isolated F_1 ATPase from mitochondria and reconstituted oxidative phosphorylation in submitochondrial vesicles.

1966
Maizel introduced the use of sodium dodecylsulfate (SDS) for high-resolution electrophoresis of protein mixtures.

1966
Gilbert and Muller-Hill isolated the lac repressor.

1968
Meselson and Yuan discovered the first DNA restriction enzyme. Shortly thereafter Smith and Wilcox discovered the first restriction enzyme that cuts DNA at a specific sequence.

1969
Zubay and Lederman developed the first cell-free system for studying the regulation of gene expression.

1973
Cohen, Chang, Boyer, and Helling reported the first DNA cloning experiments.

1975
Sanger and Barrell developed rapid DNA-sequencing methods.

1977
Starlinger discovered the first DNA insertion element.

1977
Splicing of RNA simultaneously discovered in Broker's and Sharp's laboratories.

1978
Shortles and Nathans did the first experiments in directed mutagenesis.

1978
Tonegawa demonstrated DNA splicing for an immunoglobulin gene.

1981
Cech discovered RNA self-splicing.

1981
Steitz determined the structure of CAP protein.

1981–1982
Palmiter and Brinster produced transgenic mice.

1983
Mullis amplified DNA by the polymerase chain reaction (PCR) method.

1984
Schwartz and Cantor developed pulsed field gel electrophoresis for the separation of very large DNA molecules.

Appendix B

Answers to Selected Problems

Chapter 2

1. Reactions within the cell occur under a much more restricted range of temperature, pressure, pH, and reactant concentration than is possible *in vitro.*
3. Intensive thermodynamic parameters are independent of the amount of material in the state (e.g., temperature, density), whereas extensive parameters depend on the amount of material (energy, mass).
5. Free energy changes are state functions independent of pathway. A reaction mechanism defines a specific reaction path. You can predict free energy changes on the basis of a reaction mechanism but not vice versa.
7. The reaction will proceed toward oxaloacetate formation in the cell if the concentration of products is kept low. Oxidation of NADH by the mitochondrial electron-transport system and utilization of oxaloacetate in the formation of citrate shifts the malate–oxaloacetate reaction toward oxaloacetate production.
9. **(a)** For reaction (P1), $\Delta G°' = -2.4$ kcal/mole; **(b)** $\Delta G°'$ of ATP hydrolysis is -7.9 kcal/mole.
11. A minimum of 2.7 moles of protons is required per mole of ADP phosphorylated.
13. **(a)** If the ratio of NADH/NAD^+ is 1:10, the ratio of lipoamide/dihydrolipoamide is 0.98. **(b)** If the NADH/NAD^+ ratio is 10:1, the ratio is 9.8×10^{-3}.
15. K'_{eq} is 1.5×10^{-2} with pyrophosphatase absent; 4.8×10^{3} with pyrophosphatase present.

Chapter 3

1. **(a)** $K_a = 1.11 \times 10^{-4}$; $pK_a = 3.95$; **(b)** pH = 3.47.
3. 1 mmole of NaOH
5. pH 2: Arg, His, Lys; pH 7: Arg, Asp, Glu, Lys; pH 12: Arg, Asp, Glu, Tyr, Cys
7. The two principal forms differ in degree of protonation of the R group (ϵ-amino group). Each will have a fully deprotonated carboxylate (net negative charge) and an unprotonated α-amino group (neutral). The equilibrium mixture will contain 0.028 mmoles protonated and 0.072 mmoles unprotonated ϵ-amino group.
9. Polyhistidine will bear little positive charge at pH 7.5 and even less at pH 10, hence the polymer will be insoluble. At pH 5.5, the fraction of imidazole groups protonated will be greater and the polymer more soluble.
11. At pH 6.1 the negatively charged glutamate will bind to the basic column, arginine will bind to the acidic column, and alanine will pass unretarded through both columns.
13. NH_2-Ala-Ala-Lys-Ala-Ala-Phe-Ala
15. NH_2-Ala-Ser-Lys-Phe-Gly-Lys-Tyr-Asp

Chapter 4

1. Consider the entropic effect of decreasing water organization by moving the hydrophobic residue side chains from an aqueous to a nonaqueous environment.
3. The α helix is a rodlike element that cannot easily change direction. Loops, β bends, and "random" structure break the helical structure and allow these directional changes.
5. An α helix broken at the Pro-Asn-Ala region with the hydrophobic residues on the exterior should insert into the membrane.
7. The metal at the active site must be aligned in reasonably precise geometry. Conservation of residues around the metal ligands preserves the geometry of metal binding.
9. The two proteins in question may share common epitopes and cross-react. Alternatively, the "pure" protein may have been contaminated with a trace of the other protein.
11. Dimer of 100,000-M_r units. Each 100,000-M_r unit is a dimer of a 25,000-M_r and a 75,000-M_r protein joined covalently by disulfide bonds.
13. Consider right-hand twist of extended α helices, right-hand crossovers, and right-hand twist of the β-sheet structure.
15. A_6, A_5B, A_4B_2, A_3B_3, A_2B_4, A_5B, B_6 held together by ionic or hydrophobic forces or by covalent (disulfide) bonds.

Chapter 5

1.

Step	Specific Activity (U/mg)	Yield (%)	Purification (n-fold)
Cell extract	0.039	100	1
$((NH_4)_2SO_4)$ fractionation	0.091	85	2.3
Heat treatment	0.12	73	3.1
DEAE chromatography	4.3	62	110
CM-cellulose	29	50	740
Bio-Gel A	32	41	820

The DEAE chromatography (36-fold purification by this step) yields greatest purification. The enzyme was 0.12% of the protein in the initial extract.

3. The buffer pH must be between above pH 6 but below pH 9 to assure that the protein is negatively charged and the diethylaminoethyl groups are positively charged.
5. The volume following the heat treatment step is too large to allow efficient chromatography by gel exclusion (a column containing approximately 60 liters of resin would be required).
7. Contaminant B can be separated from both contaminant A and the ribonuclease by gel exclusion chromatography. Contaminant A and the ribonuclease can be separated by ion-exchange chromatography.
9. The characteristic absorbance features at 280 nm and 288 nm are consistent with the presence of tryptophan as a component of the amino acid composition of protein A, whereas protein B has little if any tryptophan.
11. The molecular weight is approximately 39,000.
13. Increasing the net positive charge is accomplished either by (a) substitution of a neutral amino acid for an acidic amino acid or (b) substitution of a basic amino acid for a neutral amino acid.
15. The quaternary structure is a tetramer of approximately 16,000-M_r subunits associated through ionic or electrostatic interactions.
17. [Hb] = 4.8 mM and is approximately equal to the concentration of GBP.

Chapter 6

1. D-mannose is the 2-epimer of D-glucose; D-galactose is the 4-epimer of D-glucose.
3. Carbohydrates as branched structures with multiple linkage sites conceivably could have attained the catalytic function now provided by proteins. However, there may be functional groups provided by amino acids (arginine, isoleucine, for example) that would be difficult to attain with carbohydrates.
5. Pure α or β anomers of the sugars are thermodynamically less stable than the mixture and mutarotate via opening and reclosing the sugar ring.
7. Cellulose is a linear polymer of $\beta(1,4)$-linked Glc residues. Polymer strands H bond with each other to form stable, insoluble fibrils suited for structural roles. Starch and glycogen are polymers of glucose in $\alpha(1,4)$ linkages with $\alpha(1,6)$ branches. The secondary structure is a hydrated helical coil with multiple nonreducing termini. These termini in glycogen provide the "rabid" release of stored glucose upon metabolic demand.
9. 2,3,4,6-tetra-O-methyl-D-mannose; 2,4,6-tri-O-methyl-D-glucose; 2,3,4-tri-O-methyl-D-galactose
11. D-Man $\alpha(1,6)$ / D-Man $\beta(1,4)$ > D-Gal (1,4) D-Glc
13. Carbohydrates are linked to the amide N of Asn in the sequence Asn-X-Ser(Thr). Other Asn residues not in that configuration will be exceedingly poor candidates for glycosylation.

Chapter 7

1. Plants provide a source of linolenic and linoleic acids, some of which is present in meat and milk of herbivores.
3. Triacylglycerols are neutral lipids, readily soluble in organic but not aqueous media. Phospholipids are amphipathic because the polar phosphoglyceryl portion is soluble in aqueous media, while the fatty acids esterified to the glyceryl portion are hydrophobic.
5. Review Haworth representations of Gal and Glc. The linkage to ceramide is a β-glycosidic linkage.
7. Triton X-100 is present above the critical micellar concentration and will probably not pass through the dialysis membrane. It is neutral and could be used in the ion-exchange chromatography. Sodium deoxycholate is present below CMC, should dialyze easily, and could be used for the gel exclusion experiment. The ionic nature of NaDOC would limit its use in ion exchange.
9. The hydrophobic amino acid side chains on the exterior of the integral membrane protein interact with the hydrophobic lipid of the membrane exterior and are stable in the nonaqueous environment. These residues pack in the interior, hydrophobic environment of globular proteins.
11. (a) Temperature above phase transition, greater lipid mobility. (b) Unsaturated fatty acid will decrease T_m. (c) Short-chain fatty acids will decrease T_m. (d) Integral proteins will increase and broaden temperature range of the phase change.
13. Glycosyl residues are hydrophilic and will not readily pass through the hydrophobic interior of a membrane.
15. Removal of the detergent has caused aggregation of the detergent-solubilized protein.

Chapter 8

1. Reaction order is the power to which a reactant concentration is raised in defining the rate equation. The example is first order in A and B, second order overall, second order in C.
3. K_m is $(k_2 + k_3)/k_1$, whereas K_S is k_2/k_1. K_m is equal to the [ES] dissociation constant K_S when the value of k_3 is much smaller than the value of k_2.
5. Nonlinearity develops because substrate concentration is less than saturating and velocity is proportional to substrate concentration. Substrate depletion causes decreased velocity; product concentration increases and reverse reaction becomes significant.
7. When [S] is 10^{-3} M, $v_0 = V_{max}$ (10^{-7} M min^{-1}).
   When [S] is 10^{-5} M, $v_0 = 9.0 \times 10^{-8}$ M min^{-1} (90% V_{max}).
   When [S] is 10^{-6} M, $v_0 = 5.0 \times 10^{-8}$ M min^{-1} (50% V_{max}).
   When [S] is 2×10^{-2} M, $v_0 = V_{max}$, thus no increase in rate.

9. (a) $K_m = 10^{-4}$ M; (b) $V_{max} = 10^{-8}$ M s^{-1}; (c) $k_{cat} = 100\ s^{-1}$; (d) $v_0 = 1.7 \times 10^{-9}$ M s^{-1}
11. Intersecting initial velocity plots are consistent with sequential (ordered or random) but not Ping-Pong mechanism.
13. $V_{max} = 3.3 \times 10^{-7}$ M s^{-1}; $K_m = 2.4 \times 10^{-4}$ M; $k_{cat} = 3.3 \times 10^4\ s^{-1}$; specificity constant $= 1.4 \times 10^8\ M^{-1}\ s^{-1}$
15. Irreversible inactivating reagents decrease the amount of active enzyme without altering the kinetic constants of the remaining active enzyme. Lineweaver-Burk analysis of the data reveals an apparent slope and *y*-intercept effect that may be confused with noncompetitive inhibition. Enzyme reversibly inhibited is reactivated upon dilution, while irreversibly inactivated enzyme is not.

Chapter 9

1. Each enzyme uses the "catalytic triad" (Asp-His-Ser) at the active site and an enzyme-bound intermediate is formed. The binding site for the side chain of the residue contributing the carboxyl to the bond that is cleaved differs in each case.
3. RNase activity depends on acid–base reaction catalyzed by two histidyl residues at the active site. Removal of the C′-2 OH proton is best accomplished by an unprotonated imidazole (pH > p*K*) and the addition of a proton to the C-5′ ($-O^-$) is accomplished by a protonated imidazole (pH < p*K*). The best compromise is pH at p*K* of imidazole (pH 6), where each of the histidines exists with the maximal fraction of each imidazole in the correct form. correct form.
5. Release of water during the conformational change initiated by binding the substrate provides a hydrophobic environment to stabilize the nonpolar side chain of the substrate.
7. Renaturation of denatured protein is dictated by the primary structure of the protein. The trypsin family of enzymes and carboxypeptidase A are synthesized as proenzymes that are proteolytically activated. The proteolyzed, active enzymes have primary structures different from the gene product and are not active upon renaturation. In addition, zinc is a cofactor required for carboxypeptidase A activity.
9. The structure of the transition-state analog is complementary to the structure of the active site. The analog thus binds tightly to the active site.
11. For the iron-dependent form, as an example,

$$Fe^{3+} + O_2^- \rightarrow Fe^{2+} + O_2$$

$$Fe^{2+} + O_2^- + 2\ H^+ \rightarrow Fe^{3+} + H_2O_2$$

13. (a) Substitution of Asp for Lys 86 markedly decreases activity and several explanations are possible: The lysine is a critical residue either at or near the active site or is essential for maintaining a catalytically competent conformation of the enzyme. (b) Lysines 21 and 101 are probably outside the catalytic site and may not be evolutionarily conserved. Their replacement with aspartate yielded no great change in enzymatic activity. (c) Lysine 86 is essential for enzymatic activity and would be conserved.
15. An intermediate covalently bound to the active site is an essential part of the mechanism in covalent catalysis. In noncovalent catalysis, the intermediates are bound through ionic, hydrophobic, or hydrogen bonds.

Chapter 10

1. Consider the reactions

$$\text{Protein(Ser)} + \text{ATP} \rightarrow \text{Protein(Ser-P}_i\text{)} + \text{ADP}$$

$$\text{Protein(Ser-P}_i\text{)} + \text{HOH} \rightarrow \text{Protein(Ser)} + \text{P}_i$$

$$\text{Sum: ATP} + \text{HOH} \rightarrow \text{ADP} + \text{P}_i$$

3. If G, H, and J are essential end products, each would be likely to inhibit its own production without inhibiting production of the others. The committed step (enzyme 1) may be cumulatively inhibited by each of the products. G and H might cumulatively inhibit enzyme 4. J could inhibit enzymes 7 and 8.
5. ATP binds at the active site and is a substrate, and also binds at an allosteric site, inhibiting the enzyme.
7. The substrate concentration required for half-maximal activity ($S_{0.5}$) of an allosterically regulated enzyme will depend on the cumulative effects of allosteric activators and/or inhibitors also present. Hence $S_{0.5}$ may be decreased with allosteric activators and may be increased with allosteric inhibitors.
9. Cooperativity in the tetrameric hemoglobin depends on the interaction of the two α and two β subunits. Separation of the tetramer into two α-β dimers would destroy the cooperativity exhibited by the tetramer. There should be more O_2 bound at lower O_2 tensions. 2,3-Bisphosphoglycerate should not stabilize the deoxyform of the dimer. The O_2-binding curve for the dimer should resemble that of myoglobin.
11. Multisubunit complexes respond cooperatively by transmitting the conformational change at one subunit to an adjacent subunit with subsequent changes in substrate binding. There may be local conformational changes in a subunit, as with LADH, that do not alter substrate binding to adjacent subunits.
13. The sulfhydryl groups on the activated enzymes must be reoxidized to the disulfide form. Whether the disulfide is formed by an enzymatically catalyzed process or occurs spontaneously via low-molecular-weight oxidants is not known. Irreversibly modified enzymes must in principle be degraded to allow recycling of the amino acids. Although the cell can reclaim the amino acids, there is still a large metabolic expense in *de novo* synthesis of the protein.
15. Free calmodulin would respond to Ca^{2+} signal by forming a Ca^{2+}-calmodulin complex, which then must collide with and bind to a target protein. This scenario requires the binding of three to four Ca^{2+} atoms per molecule of calmodulin, then a bimolecular reaction of Ca^{2+}-calmodulin with the target protein. Calmodulin complexed to the target protein is a single entity that must collide with Ca^{2+} atoms. The latter reaction should be kinetically less complex and would occur with a higher probability.

Chapter 11

1. The carboxyl group on the acyl substituent of biotin and lipoic acid forms an amide bond with the ϵ-amino groups of lysine in the cofactor-containing subunit. The cofactors are bound to flexible extended structures that allow the chemically reactive portion of the cofactor to move between several catalytic centers present on different subunits.

3. The two pyridine nucleotides (NADH, NADPH), although functionally similar, are specific coenzymes for dehydrogenases. The dehydrogenase-dependent specificity provides a mechanism frequently used to differentiate anabolic and catabolic processes. Pathways may be regulated by either NADH/NAD^+ or NADPH/$NADP^+$ ratios.

5. (a)

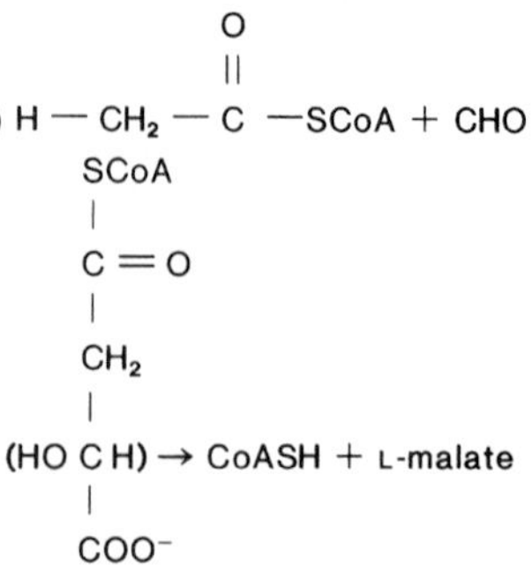

$$H-CH_2-\overset{\overset{\displaystyle O}{\|}}{C}-SCoA + CHO-COO^- \rightarrow \begin{array}{c} SCoA \\ | \\ C=O \\ | \\ CH_2 \\ | \\ (HO\ C\ H) \\ | \\ COO^- \end{array} \rightarrow CoASH + \text{L-malate}$$

(b) The thioester activates the carboxyl group of acetate and stabilizes the enolate formed upon enzyme-catalyzed abstraction of the α proton. Abstraction of an α proton from acetate, forming the dianion, is improbable. **(c)** Coenzyme A is a good leaving group and hydrolysis of the thioester is thermodynamically favorable. The products malate and CoASH are more stable when compared with the malylthioester.

7. Flavins are fully reduced upon accepting two electrons but may transfer electrons to one-electron acceptors (iron-sulfur proteins and hemoproteins). The free radical remaining after transfer of a single electron from FADH is resonance-stabilized by the isoalloxazine ring system.
9. (a) Reduced flavins in solution rapidly reduce O_2 to superoxide and H_2O_2, metabolites that are toxic to the cell. Enzyme-bound flavins are usually shielded from rapid oxidation by O_2. **(b)** Tightly bound NAD(P) is an advantage to enzymes catalyzing rapid $H:^-$ removal and readdition in a stereospecific fashion. Freely diffusing NADH is an advantage in transferring reducing equivalents among various enzyme-catalyzed reactions.
11. The products of the oxidation, NH_4^+ and H_2O_2, are toxic to most cells. In PLP-dependent transamination the amino group is retained in organic molecules of low toxicity.
13. (a) Flavin; **(b)** PLP; **(c)** biotin; **(d)** thiamine pyrophosphate.
15. NAD^+ is cleaved at the glycosidic bond between nicotinamide and the ribose ring. Adenosine diphosphate ribose is transferred to the acceptor and nicotinamide is released.

Chapter 12

1. Thermodynamics dictates the feasibility of a reaction and the direction in which a reaction proceeds to reach equilibrium, but does not specify the rate of the reaction or the reaction pathway. Kinetics describes that rate at which a reaction approaches equilibrium. The most thermodynamically favorable reaction is of no biological value if it is not adequately catalyzed.
3. Using different high-energy intermediates, or different portions of the same high energy intermediate, and different forms of reductant allows the cell ample sites to control catabolism and anabolism differentially.
5. The "committed step" in a reaction sequence steers the metabolite to a sequence of reactions whose intermediates have no other function in the cell. Control of the committed step prevents wasteful accumulation of these single-purpose intermediates and obviates the necessity of controlling each enzyme in a pathway.

Chapter 13

1. (a) Bacteria are adept at fermenting numerous sugars. **(b)** The NADH formed by glyceraldehyde-3-phosphate dehydrogenase is used to reduce pyruvate to lactate. There is no net oxidation, but the products are of lower free energy than substrates. Energy difference between products and substrates is available to drive the phosphorylation of ADP. **(c)** The culture could also be decarboxylating pyruvate to acetaldehyde and CO_2. The acetaldehyde is subsequently reduced to ethanol, using the NADH from the glyceraldehyde-3-phosphate dehydrogenase reaction. **(d)** Fluoride forms an inhibitor (fluorophosphate) of enolase. Inhibition of enolase would increase the amount of 2-phosphoglycerate, but the amount of PEP would decline as a result of pyruvate kinase activity.
3. (a) Hydrolysis rather than phosphorolysis of the thioester bound to glyceraldehyde-3-phosphate dehydrogenase releases 3-phosphoglycerate. **(b)** There would be no net yield of ATP if glycolysis began with glucose. **(c)** The obligate aerobe could oxidize the pyruvate, with the formation of sufficient ATP for growth.
5. Arsenate substitutes for phosphate at the glyceraldehyde-3-phosphate dehydrogenase reaction and forms a transient arsenocompound that spontaneously hydrolyzes, regenerating the arsenate. Arsenate acts catalytically rather than stoichiometrically in dispelling the energy of the thioester bond. Arsenate is ineffective at the pyruvate kinase step because the active phosphate was derived initially from ATP, not orthophosphate.
7. If the $\Delta G^{\circ\prime}$ for the hexokinase reaction is -5.0 kcal/mole, the K'_{eq} is 4,800. The glucose concentration in equilibrium with 1 mM glucose-6-phosphate must be at least 42 nM, a concentration easily attainable in the cell. However, the K_m of hexokinase for glucose is about 100 μM. The glucose concentration in actively metabolizing cells must be several orders of magnitude greater than 40 nM to have reasonable hexokinase activity.
9. (a) Alcohol dehydrogenase reduces acetaldehyde to ethanol to regenerate NAD^+ from the NADH produced in glycolysis. **(b)** Alcohol dehydrogenase oxidizes ethanol to acetaldehyde. Acetaldehyde is subsequently oxidized to acetyl CoA_2.
11. Consider enolase, aldolase, and phosphohexoisomerase.
13. (a) Review pyruvate carboxylase and PEP carboxykinase reactions. **(b)** $\Delta G^{\circ\prime}$ overall is app. -1.0 kcal/mole if the hydrolysis of PEP releases 14 kcal/mole and if hydrolysis of ATP (GTP) releases 7.5 kcal/mole. If the overall free

energy change for the coupled reactions is as calculated, the formation of PEP from pyruvate is metabolically feasible.

15. (a) PFK-2 will be active when the insulin/glucagon ratio is high (for example, when the animal is fed a carbohydrate-rich meal), or when adrenalin is released. FBPase-2 will be active when the insulin/glucagon ratio is low (blood glucose is low). (b) If fructose-2,6-bisphosphate were still present although not being actively synthesized, it would continue to stimulate PFK-1 and inhibit FBPase-1. (c) Activation of FBPase-2 causes the hydrolysis of fructose-2,6-bisphosphate, alleviates inhibition of FBPase-1, and prevents activation of PFK-1. (d) Insulin/glucagon ratio decreases. Thus FBPase-2, FBPase-1, pyruvate carboxylase, and PEPCK activities should increase. PFK-2 and PFK-1 activities will decline. (e) The administered insulin will restore the insulin/glucagon ratio. The cAMP level in the liver cell will decrease, as will the activities of FBPase-2, FBPase-1, and PEP carboxykinase.

Chapter 14

1. (a) False. Lipoamide transacetylase catalyzes reduction of the disulfide on lipoamide concomitantly with oxidation and transfer of the hydroxyethyl group from thiamine pyrophosphate. Dihydrolipoamide dehydrogenase catalyzes oxidation of dihydrolipoamide and reduction of NAD^+. (b) False. Hydrolysis of acetyl-CoA thioester should yield as much free energy as succinyl-CoA hydrolysis, a process coupled to ADP phosphorylation (succinate thiokinase). (c) False. The methyl group of acetyl-CoA could be derived from pyruvate, from β oxidation of long-chain fatty acids, or from amino acid metabolism. (d) False. If aconitase failed to discriminate between the ($-CH_2-COO^-$) groups, half the CO_2 would arise from oxaloacetate and half from the acetate carboxylate. The two CO_2 molecules released by oxidative decarboxylation of isocitrate and α-ketoglutarate are derived from the carboxyl groups of oxaloacetate with which the acetyl-CoA was condensed. (e) False. Malate can easily be dehydrated to fumarate by reversal of the fumarase reaction.
3. (a) Glyceraldehyde-3-phosphate, C-3; (b) acetyl-CoA, methyl of acetate; (c) citrate, C-2 (derived from methyl of acetyl-CoA); (d) α-ketoglutarate, C-4; (e) succinate, C-2, C-3; (f) malate, both C-2 and C-3.
5. (a) Hydroxypyruvate $\rightarrow$ B $\rightarrow$ D $\rightarrow$ C $\rightarrow$ A $\rightarrow$ pyruvate (b) Hydroxypyruvate + NADH + $H^+ \rightarrow$ pyruvate + NAD^+ + HOH (c) A = phosphoenolpyruvate; B = glycerate; C = 2-phosphoglycerate; D = 3-phosphoglycerate. A, C, and D are glycolytic intermediates. (d) ATP used to phosphorylate glycerate is regenerated from PEP.
7. Lipoic acid bound to the acyltransferase in the multienzyme complexes (a) provides a greater coenzyme concentration at the active site than may be seen if the cofactor must diffuse into and away from the complex; (b) decreases the diffusion path required for a product of one active site to move to the next catalytic center.
9. If the complexes were deficient in dihydrolipoamide dehydrogenase, dihydrolipoamide would be reoxidized to lipoamide more slowly than necessary to maintain activity of the complexes. ATP production would be lowered as a consequence of reduced TCA cycle activity.
11. In bacteria, acetate is used as a source of energy via the TCA cycle and oxidative phosphorylation and as a carbon source for synthesis of intermediates for cell growth and division. Competition between ICDH and isocitrate lyase for isocitrate is shifted to anabolism by phosphorylation of ICDH and shifted to catabolism upon dephosphorylating (reactivating) ICDH. The fraction of ICDH in the active form dictates the level of isocitrate available to isocitrate lyase. The isocitrate lyase is apparently not regulated at the metabolite level.
13. In plants, the TCA cycle and glyoxalate bypass are located in different subcellular organelles. The TCA cycle is located in mitochondria and the glyoxalate bypass is located in glyoxosomes. Communication between the compartments includes flow of succinate to the mitochondria for oxidation to malate.
15. In mammalian liver, acetyl-CoA provides reducing equivalents by its oxidation in the TCA cycle and subsequent oxidative phosphorylation to provide ATP. Acetyl-CoA is an obligatory activator of the pyruvate carboxylase, whose activity provides oxaloacetate for gluconeogenesis.

Chapter 15

1. Iron-sulfur protein: one electron, no H^+ transferred, no semiquinone; flavoprotein: isoalloxazine ring, two electrons, one H^+/electron, semiquinone formed; quinone: two electrons, one H^+/electron, semiquinone formed.
3. (a) $\Delta E^{\circ\prime} = +110$ mV, $\Delta G^{\circ\prime} = -2.5$ kcal/mole (b) $\Delta E^{\circ\prime} = +580$ mV, $\Delta G^{\circ\prime} = -53$ kcal/mole O_2 reduced. (c) $\Delta E^{\circ\prime} = +270$ mV, $\Delta G^{\circ\prime} = -12$ kcal/mole (d) $\Delta E^{\circ\prime} = -430$ mV, $\Delta G^{\circ\prime} = +20$ kcal/mole
5. At +300 mV, the cytochrome *c* couple will be 91% oxidized and 9% reduced.
7. (a) Data are consistent with diffusion of cytochrome *c* between complex III and the cytochrome oxidase. Electrons are probably donated and accepted from the same area on the cytochrome *c* structure. (b) The lysines surrounding the heme crevice on cytochrome *c* should interact electrostatically with anionic groups (Glu, Asp) spatially located on complex III and cytochrome oxidase to facilitate interaction with the cytochrome.
9. (a) P/O = 2.3; (b) P/O = 0.2; (c) P/O = 2.7; (d) P/O = 0.5.
11. (a) Tightly coupled mitochondria exhibit an obligatory codependence of electron transfer and phosphorylation of ADP. In "perfectly" coupled mitochondria, no electron transport to O_2 should occur in the absence of ADP and phosphate. (b) Respiratory control is a function of coupled mitochondria. Oxidation of NADH and succinate is blocked when ATP levels are high; glycolysis and TCA cycle are inhibited. When ATP demand is high, ADP levels will stimulate the oxidation of succinate and NADH. Low ATP levels and high NAD^+/NADH ratio will stimulate glycolysis and the TCA cycle. (c) Brown-fat mitochondria contain the uncoupler thermogenin and the electron-transport system is partially uncoupled. The oxidation of fat leads to nonshivering thermogenesis, a process important to animals aroused from hibernation.

13. (a) Membrane potential is app. 180 mV, negative on matrix side of membrane. **(b)** An oxidative phosphorylation uncoupler allows H^+ equilibration across the membrane, collapsing both membrane potential and H^+ gradient.
15. (a) Transport of H^+ is in the same orientation in these SMPs as in the mitochondria, but the cytoplasmic aspect of the inner mitochondrial membrane is facing the interior of the vesicle. **(b)** Mechanical or chemical uncoupling allows free equilibration of H^+ across the SMP membrane. Succinate is oxidized without vectorial accumulation of H^+. **(c)** Atractyloside inhibits ADP transport into mitochondria. The SMP have the F_o-F_1 complex exposed to bulk solvent and do not have a transport barrier to ADP. The ADP/ATP transporter is not required, so atractyloside would have no effect in this system. SMP oxidize extravesicular NADH but intact mitochondria do not.

Chapter 16

1. (a) 33 kcal/Ein; **(b)** excited state reduction potential is −1.0 volt.
3. If oxidation of compound A requires operation of both photosystem I and photosystem II, oxidation of compound A will diminish if chloroplasts are illuminated with red light (>680 nm) and will be inhibited by DCMU. If compound A is oxidized by photosystem I, DCMU will not inhibit the oxidation, nor will the action spectrum of compound A oxidation exhibit this phenomenon of "red drop."
5. Light energy absorbed by antenna chlorophyll molecules is transferred to a photocenter trap (P_{680} or P_{700}) where an electron is transferred initially to pheophytin, then to bound plastoquinone. If the potential is decreased to 0 mV, the quinone pool will be reduced, electron transfer from pheophytin will diminish, and the steady-state level of activated chlorophyll will increase. The alternative energy dissipation pathway, fluorescence, increases.
7. The quantum yield of O_2 evolution from plants as a function of light absorbed parallels the absorption spectrum until the wavelength impinged is greater than approximately 700 nm. Although the chloroplasts absorb wavelengths greater than 700 nm, O_2 evolution declines because the oxygenic photosystem II is no longer activated at these wavelengths.
9. Photophosphorylation is the process of ATP generation driven by the proton gradient generated during the light-driven electron transfer (Z scheme). Noncyclic photophosphorylation requires net electron transfer from a donor (HOH) to an acceptor ($NADP^+$) with vectorial H^+-translocation. Cyclic photophosphorylation is driven by a proton gradient generated by electron transfer from photosystem I through the quinone pool and the electron-transfer system, back to photosystem I.
11. CO_2 and O_2 each bind to the catalytic site of the ribulose bisphosphate carboxylase/oxygenase (RubisCO) and react with the ribulose bisphosphate. The enzyme uses either CO_2 or O_2 as (alternative) substrate, and O_2 competes with CO_2 for the bound ribulose bisphosphate.
13. CO_2 is released by oxidative decarboxylation of malate in the bundle sheath cells of C_4 plants, causing the CO_2/HCO_3 concentration to exceed the CO_2/HCO_3^- concentration in the mesophyll cell. The larger concentration of CO_2 increases the fraction of ribulose bisphosphate that is carboxylated rather than oxygenated. Net photorespiration is lower in C_4 plants as compared with C_3 plants.
15. (a) $NADP^+$ is reduced to NADPH by the noncyclic electron flow in the chloroplast to $NADP^+$-ferredoxin oxidoreductase. **(b)** 6 moles of ATP are used to phosphorylate 6 moles of glycerate-3-phosphate to glycerate-1,3-bisphosphate. An additional 3 moles of ATP are consumed during phosphorylation of 3 moles of ribulose-5-phosphate to 3 moles of ribulose-1,5-bisphosphate. **(c)** The ATP consumed by the Calvin cycle remains unaltered, but 2 additional moles of ATP are used to regenerate PEP from pyruvate catalyzed by the pyruvate, phosphate dikinase. **(d)** NADPH is used in the mesophyll cell to reduce oxaloacetate to malate, but the NADPH is regenerated upon oxidative decarboxylation of malate to CO_2 and pyruvate.

Chapter 17

1. (a) Oxidation of 1 mole of glucose yields 32 moles of ATP. 350 kcal of energy are stored as ATP or 1.9 kcal/gm. **(b)** Oxidation of 1 mole of palmitate yields 106 moles of ATP. 1200 kcal of energy are stored as ATP, or 4.7 kcal/g. **(c)** Lipids are more highly reduced than are carbohydrates and supply more reducing equivalents to the electron-transport system than do carbohydrates. **(d)** Lipids have approximately 2.5 times greater energy storage per gram than do carbohydrates and are stored as compact, hydrophobic globules. Storage of an equivalent energy as carbohydrate would require at least 2.5 times the mass, not considering the water of hydration that would accompany the carbohydrate.
3. Carnitine acyltransferases (CAT) catalyze the reversible formation of fatty acyl carnitine esters from the acyl-CoA thioesters. The carnitine esters, but not the CoA derivatives, are transported across the inner mitochondrial membrane.
5. Propionyl-CoA, a product of β oxidation of odd-chain-length fatty acids and of pristanic acid, is metabolized to L-methylmalonyl-CoA in two steps. The L-methylmalonyl-CoA is substrate for the vitamin-B_{12}-dependent mutase. The product, succinyl-CoA, is metabolized in the TCA. A deficiency of vitamin B_{12} could limit the mutase reaction.
7. (a) Ketone body formation in liver supplies an easily transported, water-soluble, energy-rich metabolite that can be used in lieu of glucose in many nonhepatic tissues. **(b)** β-hydroxybutyrate supplies an additional hydride (2 reducing equivalents) compared with acetoacetate. **(c)** Consider the reactions catalyzed by β-hydroxybutyrate dehydrogenase, 3-ketoacyl-CoA transferase, and thiolase.
9. The NADPH-dependent dienoyl-CoA reductase reduces the Δ^2-*trans* $\Delta^{4,7}$-*cis* decatrienoyl-CoA produced after removal of the first 4 acetyl-CoA units from linolenyl-CoA in β oxidation. The product, Δ^3-*trans,* Δ^7-*cis* decadienoyl-CoA, is isomerized to the Δ^2-*trans,* Δ^7-*cis* decadienoyl-CoA by the enoyl-CoA isomerase.
11. (a) The ^{14}C-labeled methyl group of acetyl-CoA will be C-16 of palmitate. **(b)** Only one deuterium atom from each labeled malonyl-CoA will remain in the reduced lipid chain. Carbons 2, 4, 6, 8, 10, 12, and 14 of palmitic acid will each have one deuterium label. **(c)** The ^{14}C label will be lost by decarboxylation and no label will remain in the palmitate.

13. Glucose-6-phosphate dehydrogenase, 6-phosphogluconate dehydrogenase, and the $NADP^+$-malic enzyme are sources of the 14 moles of NADPH required for biosynthesis of palmitate.
15. The thioesterase activity of the fatty acid synthase prefers the palmitoyl acyl carrier protein thioester as substrate.

Chapter 18

1. Glutamine, the product of glutamine synthase, is the source of nitrogen required in the synthesis of a number of diverse, structurally unrelated compounds synthesized by different pathways. Total inhibition of the glutamine synthase by a single product would in turn inhibit the synthesis of all compounds requiring glutamine.
3. In the absence of serine hydroxymethyltransferase, the mutant would be unable to synthesize glycine. Glycine supplementation would be required. Metabolism of glycine contributes to the one-carbon pool, as does methionine. Addition of methionine should decrease the demand for glycine contribution to the one-carbon pool and stimulate growth of the mutant.
5. Pyruvate, from glycolysis of glucose, is carboxylated to oxaloacetate or oxidized to acetyl-CoA. These metabolites enter the Krebs cycle, are metabolized to α-ketoglutarate and oxaloacetate, then transaminated to aspartate or glutamate. Asn, Gln, and Pro are synthesized from Asp or Glu. The cycle replenishes intermediates via the anaplerotic reactions (e.g., carboxylation of pyruvate to form oxaloacetate).
7. Serine carbon: α carboxyl: C-3 or C-4 of glucose; α carbon: C-2 or C-5 of glucose; β carbon: C-1 or C-6 of glucose.
9. Glutamate is phosphorylated and reduced to L-glutamyl-γ-semialdehyde. The aldehyde condenses with the α-amino group on the same molecule to form Δ^1-pyrroline-5-carboxylate. Reduction of the latter compound yields proline. Transamination implies that the α-keto acceptor and the α-amino donor are separate molecules.
11. Tryptophan synthase catalyzes the cleavage of indole glycerol phosphate, with retention of indole as an enzyme-bound intermediate and release of glyceraldehyde-3-phosphate. The enzyme-bound indole replaces the hydroxyl group of serine to form tryptophan.
13. Hydroxyproline is formed by a posttranslational modification of proline residues in the protein. The ^{14}C-labeled hydroxyproline is not incorporated directly into the collagen because there is no genetic codon to specify the incorporation of hydroxyproline.

Chapter 19

1. Pyridoxal phosphate deficiency would curtail the transamination of α-amino group amino acids to recipient α-ketoacids. Disposition of the amino group via the urea cycle and subsequent catabolism of the carbon skeleton would be diminished.
3. Glutaminase catalyzes the hydrolytic cleavage of the amide group from glutamine, leaving glutamate. Glutamate dehydrogenase catalyzes the NAD^+-dependent oxidative deamination of glutamine to ammonium ion, α-ketoglutarate and NADH.
5. The α-ketoacid pyruvate, a product of glycolysis, is transaminated to form alanine and as such transports the α-amino groups, from amino acids catabolized in muscle, to the liver. In the liver, transamination of alanine with α-ketoglutarate yields pyruvate, a gluconeogenic substrate, and glutamate, an amino donor in urea biosynthesis.
7. Glutamate is oxidatively deaminated to α-ketoglutarate, a TCA cycle intermediate subsequently oxidized to oxaloacetate. Aspartate is transaminated directly to oxaloacetate. The TCA cycle regenerates oxaloacetate, preventing the complete oxidation of oxaloacetate directly. Oxaloacetate is reduced to L-malate, transported to the cytoplasm, and oxidatively decarboxylated by the $NADP^+$-dependent malic enzyme to pyruvate and NADPH. Pyruvate is completely oxidized in the mitochondria.
9. The amidino group of arginine is required for creatine biosynthesis. Creatine phosphate, a phosphoramide, is a "high-energy" phosphate reservoir for the muscle.
11. Each of these amino acids is catabolized to succinyl-CoA, a TCA cycle intermediate. Succinate is oxidized to L-malate, shuttled to the cytoplasm, and oxidized to oxaloacetate. Oxaloacetate is a gluconeogenic substrate.
13. Phenylalanine and tyrosine are catabolized to 4-hydroxyphenylpyruvate, whose further metabolism is blocked by the dioxygenase deficiency. A diet containing only maintenance levels of phenylalanine and deficient in tyrosine is recommended to prevent accumulation of the 4-hydroxyphenylpyruvate.
15. (a) L-glutathione is synthesized in successive steps catalyzed by γ-glutamyl cysteine synthase and glutathione synthase. Glutathione synthesis is directed by the substrate specificity of these enzymes. (b) Decreased glutathione synthesis would increase the probability of oxidative damage to the cell.

Chapter 20

3. Humans have two carbamoyl syntheses, one in the mitochondria (CPase I) and one in the cytosol (CPase II). The mitochondrial enzyme is used in urea and arginine synthesis, while the cytosolic enzyme (part of the multienzyme complex) is dedicated to the synthesis of pyrimidines and is the first enzyme in that pathway. Bacteria have only one CPase and the regulation is at the first committed enzyme in pyrimidine biosynthesis, ATCase—a form of regulation that makes biochemical sense.
5. Large amounts of dATP would accumulate and inhibit ribonucleotide reductase. The lymphocytes could not make the large amounts of dNTPs needed for DNA synthesis and cell division.
7. Because Lesch-Nyhan patients lack the enzyme hypoxanthine-guanine phosphoribosyltransferase, they accumulate high levels of PRPP, which stimulates purine biosynthesis to high levels, leading to large amounts of uric acid being made. The brain may not have high levels of *de novo* purine biosynthesis and probably relies on the salvage pathway enzymes for its purine nucleotides.

9. Antifolates indirectly inhibit the synthesis of thymidylate and prevent DNA synthesis; therefore cells that are growing rapidly may be killed by methotrexate. Normal cells that are actively dividing (in the cell cycle) and replicating their DNA are also sensitive to the toxic effects of antifolates. This is why we see harmful side effects of anticancer drugs.
11. GTP requires the hydrolysis of eight high-energy bonds, while CTP requires the hydrolysis of five high-energy bonds.
13. Allopurinol inhibits xanthine oxidase, preventing the oxidation of 6-mercaptopurine so that it is not as rapidly degraded.

Chapter 21

1. Lactose synthase is a galactosyltransferase using UDP-galactose and glucose to make the disaccharide lactose. The enzyme is composed of two peptides, the galactosyltransferase and α-lactalbumin. The galactosyltransferase is found in all tissues in the body, but the α-lactalbumin (which has no catalytic activity) is found only in milk. α-Lactalbumin modifies substrate specificity of the galactosyltransferase to use glucose so that lactose can be produced.
4. Sugars have large numbers of hydroxyl groups that form glycosidic bonds with the anomeric carbons of other sugars and can generate a large number of structures from a limited number of monosaccharides (over 80 different glycosidic linkages are known).
7. In mutants that lack Dol-P-Man synthase, the glycosylphosphatidylinositol (GPI) anchor is not made and the glycoproteins that would be on the plasma membrane are secreted outside the cell (not attached).
10. Patients with I-cell disease do not phosphorylate the mannose residues on the glycoproteins that are lysosome-bound. These "lysosomal hydrolases" are therefore secreted.

Chapter 22

1. Phospholipids are essential to living organisms, including humans. Any defects would be lethal in the early stages of development and would never be observed.
3. The lung surfactant dipalmitoylphosphatidylcholine is synthesized by removing the fatty acid in the C-2 position and then reacylating with palmitoyl-CoA. This process is called remodeling.
5. Some eukaryotic membranes contain significant amounts of ether-linked glycerophospholipids (for example, 50% of all phospholipids in heart tissue are ether-linked in the C-1 position). These phospholipids are called plasmalogens when they have a hydrocarbon chain linked to the SN-1 carbon by a vinyl ether linkage and are called alkylacylglycerophospholipids when linked by an ether linkage. Archaebacteria have ether-linked lipids in their cell membranes.
7. This person could still make PGE_2 because linoleic acid (which is considered an essential fatty acid in animals) can be desaturated and elongated to form arachidonic acid.
9. Arachidonic acid is stored in membranes as phospholipids with C_{20} polyunsaturated fatty acids in the SN-2 position. Phospholipase A_2 releases arachidonic acid, which is then used to synthesize prostaglandins, which induce inflammation.

Chapter 23

1. The major regulation occurs with HMG-CoA reductase. This enzyme is regulated at three levels: gene expression, rate of degradation of HMG-CoA reductase, and phosphorylation/dephosphorylation of HMG-CoA reductase. High cholesterol levels reduce HMG-CoA reductase mRNA, which reduces synthesis and increases degradation, and the enzyme already present is phosphorylated to reduce its activity. LDL receptors will be reduced in number.
3. Any of the enzymes in the synthesis of acetoacetate (acetoacetyl-CoA thiolase, HMG-CoA synthase, and HMG-CoA lyase) might be defective. Different isozymes are used for ketone bodies and cholesterol biosynthesis; therefore normal cholesterol biosynthesis would not be a clue for the deficient enzyme.
5. The patient could be treated with lovastatin and a bile-acid-binding resin, which together could lower serum cholesterol levels about 50%. Lovastatin inhibits HMG-CoA reductase, lowering cholesterol biosynthesis. Bile-acid-binding resins increase the excretion of bile acids, which stimulates the liver to convert more cholesterol to bile acids.
9. HDL and other lipoproteins would have a reduced level of cholesterol esters, but serum levels of free cholesterol would be elevated.

Chapter 24

2. Nothing will happen! Normally you would expect a person to go into insulin shock as a result of a drop in blood glucose levels, but a starving person's brain is using ketone bodies for energy and the blood glucose is already low.
4. Progestin and estrogen inhibit secretion of FSH and LH by the pituitary and thus prevent follicular growth and ovulation.
6. The patient most probably has a primary defect.
9. Vitamin D can be considered both a hormone and a vitamin. Its mode of action is like that of other steroid hormones and it is synthesized in the body. It can be given in the diet, for example, in supplemented milk, and would then be called a vitamin.

Chapter 25

2. Investigators used two different strains of TMV that had identifiable proteins. They made different reconstituted viruses and found that the progeny viral proteins were always determined by the type of RNA in the reconstituted virus.
4. At the pH extremes, some of the bases have either positive charges (low pH) or negative charges (high pH), which are disruptive of duplex structure.
6. 7-M urea, 90% formamide, pure water, and the T4 gene-32-encoded protein lower the T_m of the DNA. Increased salt raises the T_m of the DNA duplex. Urea and formamide disrupt hydrophobic forces that hold the duplex together (it was earlier thought that these compounds disrupted hydrogen bonding, but water is a better hydrogen-bonding reagent). Higher salt shields the anionic phosphate groups from each other, limiting the electrostatic repulsion of the DNA strands. The T4 protein binds to single-strand regions and pulls the equilibrium to the denatured state.

8. These proteins would bind to the Z DNA and pull the equilibrium to the Z structure. Therefore, even if the B form predominated in pure DNA at equilibrium, the binding of protein to the Z DNA would cause this region of the DNA to be in the Z configuration.
10. A is a DNA-RNA hybrid duplex.
12. T4 DNA will renature 25 times faster than *E. coli* DNA. Unsheared T4 DNA will anneal about 20 times faster than sheared DNA.

Chapter 26

2. The location of the origin is close to gene *c*, which has the highest number of gene copies. The pattern observed agrees with a bidirectional mode of replication.
4. Both ϕX174 and *E. coli* DNA replication are inhibited by rifampicin, which inhibits RNA polymerase. ϕX174 is inhibited during the replication of the RF since it is dependent on the synthesis of a viral protein A, which generates the 3′-OH by cleavage of the DNA. The synthesis of this protein is dependent on the transcription of gene *A* by RNA polymerase. The initiation of replication in *E. coli* is dependent on RNA polymerase to synthesize the RNA primer at *oriC*.
6. It would take about 33 min to replicate the chromosome, since you have a bidirectional mode of replication with two replication forks. If you had multifork replication (one round of replication starting before the other finished), then a 20-min division time would be possible.
9. This experiment by Cairns and DeLucia showed that DNA Pol I was not the primary DNA polymerase responsible for replicating DNA in *E. coli* and suggested that another enzyme was involved.
12. Reverse transcriptase is the unique viral protein that converts the viral RNA into DNA. A specific tRNA serves as a primer (all template-dependent DNA polymerases require a primer, even if they have a RNA template).
14. Cytosine spontaneously deaminates to uracil, which would then lead to a point mutation if not removed. The thymine base allows the organism to identify uracil produced by spontaneous deamination, which would not be possible if you had A-U base pairs in DNA.

Chapter 27

2. The indirect methods of sequencing would not detect many of the posttranscriptional modifications found in RNA, such as 2′-O-methylation, various base methylations, etc.
6. Digest the circular viral DNA with *Eco*R1 and ligate the structure with DNA ligase. This digestion would generate a hybrid virus with only one 72-bp sequence.
9. Human serum albumin is synthesized in the liver; therefore the levels of mRNA for serum albumin would be high. The cDNA made from liver would have many copies of albumin cDNA, so that the job of cloning this sequence would be easier. All human tissues should contain the albumin gene in one copy per haploid amount of DNA, so any tissue would serve to establish a genomic library.
11. If a fragment of DNA of interest has linked to it the thymidine kinase gene (*tk*), then if a cell has taken up this DNA it will also contain the *tk* gene. If the cell line used for cloning contains a mutant (tk^-) that does not allow it to grow in a selection media (HAT, or medium containing hypoxanthine, amethopterin, and thymidine), then only cells taking up the DNA with the *tk* marker will grow.

Chapter 28

1. The base composition of the newly synthesized RNA would be: C = 20%, G = 25%, U = 15%, and A = 40%.
3. mRNA has a short half-life (a couple of minutes), while tRNA and rRNA are very stable (half-life measured in hours) and accumulate to make up most of the RNA in the cell (95%).
5. The D and T loops in tRNA interact with each other to form the tertiary structure, leaving only the anticodon with a single-stranded loop able to be cleaved by RNase.
11. The 5′ terminus is capped (G^mpppX—), generating a guanosine nucleoside with 2′ and 3′ -OH groups and a 5′-5′ pyrophosphate linkage. The other end of the mRNA has poly(A) added, so the 3′ end is adenosine. Therefore most of the mRNA in the cell has a guanosine and an adenosine 2′ and 3′ -OH and no typical triphosphate ending (pppN—).
15. If an intron or part of an intron containing a stop signal for translation was not removed, this mRNA would be longer but would yield a shorter polypeptide. Alternative splicing would remove the intron or use an alternative splice site, generating a shorter mRNA and a longer polypeptide.

Chapter 29

2. The large size of tRNAs (larger than a three-nucleotide anticodon) results from the requirements of specific charging of the tRNA by its aminoacyl tRNA synthase. The enzyme must recognize its tRNA and attach the correct amino acid (a correspondence sometimes called the second genetic code). Also, the structure of the ribosome requires some unique spacing between the codon–anticodon interaction and the site of peptide bond formation.
4. If tRNAs were not charged correctly, the genetic code would have a high error rate. The correct charging of the tRNA with the correct amino acid is just as important as the codon–anticodon interaction that defines the genetic code.
12. PEST sequences are regions rich in proline, serine, and acetic amino acids that are common to short-lived eukaryotic proteins. Regulatory proteins generally have short half-lives, while structural proteins are more stable, as is required by their function in the cell.
16. The incorporation of the valine analog prevents the formation of the active tetramer structure of hemoglobin. This unassembled globin would be a preferred target for reaction with ubiquitin, which would lead to rapid degradation.

Chapter 30

2. (a) Both the wild type and z^- would express a low level of acetylase with no treatment. (b) When lactose is added, only the wild type would produce a lot of the acetylase. (c) IPTG would induce a high level of acetylase in both strains.
6. IPTG does not have to be metabolized by galactosidase to form an active inducer. It is stable and builds up to high levels in the cell.

8. At the enzyme level, the end product of the histidine pathway (histidine) inhibits the first enzyme in the pathway by simple linear feedback inhibition. Transcription is regulated by attenuation; no feedback repression is observed with this operon.
12. Bacterial genes are regulated in a reversible manner, depending on metabolic needs, while viral genes are turned on only once, in a predesignated sequence.
14. The expression of the *crp* gene is negatively autoregulated by cAMP-CAP. A divergent promoter is activated by cAMP-CAP, and the RNA produced binds to the 5′ end of the *crp* mRNA, producing a rho-independent terminator. The binding of this antisense RNA stops the transcription of the *crp* gene.

Chapter 31

2. In eukaryotes, transcription and processing of the mRNA occurs in the nucleus while translation occurs in the cytosol. Attenuation control is unlikely because you can't form a transcriptional-translational complex.
6. Hypomethylation of the 5′ flanking sequences is correlated with the expression of some genes in eukaryotes. 5-Azacytidine can be incorporated into DNA and generate hypomethylated regions in the DNA, leading to gene activation.
9. Histones bind to DNA by electrostatic interaction of basic amino acids with the negative charge on the phosphate in the deoxyribose phosphate backbone of the DNA duplex. While many regulatory proteins initially interact (nonspecific binding) with DNA by charge interactions, high-affinity binding (at specific binding sites) occurs by hydrophobic interactions, which are stabilized by high salt.
12. The gene for dihydrofolate reductase can be amplified to a high copy number, giving a large gene dose. The large amount of dihydrofolate reductase produced binds the methotrexate, leaving some active enzyme to meet the cell's need for tetrahydrofolate. A DNA probe for dihydrofolate reductase gene could be used to measure the amount of specific DNA amplified (the "dot blot" procedure with genomic DNA could be used to compare normal and resistant cells).
14. The enhancer sequence in the DNA can bind to tissue specific protein(s). The DNA can then loop to allow the enhancer proteins to come in close contact with the promoter of the gene being stimulated, even at great distances (5,000 bp).

Chapter 32

1. **(b)** $D = 5.7 \times 10^{-6}$ cm²/sec.
2. **(a)** $\Delta\Psi = +18$ mV; **(b)** from side 1 to side 2; **(c)** Side 1: 64.3 mM K^+, 50 mM Na^+, 144.3 mM Cl^-. Side 2: 85.7 mM K^+, 85.7 mM Cl^-. $\Delta\Psi = 7.5$ mV.
5. Sucrose uptake should be inhibited by a proton ionophore if uptake is by a proton symport. If a protein-binding system was operational, then membrane vesicles or cells subjected to osmotic shock would be defective in uptake. If a Na^+ symport was involved, then uptake would be dependent on extracellular Na^+. If a PTS was operational, then sucrose phosphorylation would be dependent on PEP and not ATP in a crude cell extract.
10. If Hg^{2+} inhibits the uptake of D-glucose, this suggests that the uptake is dependent on a protein permease (facilitated diffusion) that has —SH groups present.

Chapter 33

1. B cells mature in the bone marrow and then migrate to secondary lymphoid organs, while T cells mature in the thymus and then migrate to different areas of the body.
3. There is the delayed-type hypersensitivity response, in which T cells react with antigens and secrete lymphokines. A second type of T cell, known as a "killer cell," reacts with antigens bound to target cells and causes the death of the latter.
7. Irradiate the mouse from strain A with whole-body radiation to destroy the immune system and then transplant bone marrow and thymus from strain B. This new immune system is now tolerant to transplants from strain B mice.
10. Develop antibodies to the transition-state intermediate or a transition-state intermediate analog.
12. The thiol reagent would disrupt disulfide bonds and allow the peptides that form the antibody structure (four peptide chains) to dissociate.

Chapter 34

3. Expression of some genes is induced by hypomethylation of DNA on the 5′ side of the gene. Therefore some genes, not usually active in adult tissue, may be expressed. For example, some of the genes important in development can become oncogenes when expressed at the "wrong" time (oncofetal genes).
5. RNA viruses such as retroviruses can cause cancer by a reverse transcription mechanism, using reverse transcriptase to produce a DNA copy of their genome that can then integrate into the host-cell genome.
10. HIV can cause different cancers by destroying the immune system (T4 cells) and preventing immune surveillance.

Chapter 35

1. **(a)** $\Delta\Psi = -71$ mV; **(b)** $\Delta\Psi = -63$ mV.
3. Criteria are (i) excitation of proper postsynaptic cell, (ii) presence of the compound in presynaptic terminal, (iii) release upon stimulation, and (iv) receptors on postsynaptic membrane.
6. L-DOPA can pass the blood-brain barrier but dopamine cannot. Inhibitors of dopa carboxylase cannot pass the blood-brain barrier; therefore the DOPA is not converted into dopamine until it enters the brain.
8. The high rate of Na^+ flux can be accounted for only by a pore model, as opposed to the slower carrier-mediated mechanism.

Chapter 36

1. Compounds that absorb visible light have many alternating double bonds, i.e., conjugated double bonds.
4. Rhodopsin is converted to bathorhodopsin and then to rhodopsin.
5. The Schiff's base linkage between the retinal and opsin is protonated in metarhodopsin I but not in metarhodopsin II.
10. Cone cells contain 11-*cis*-retinal bound to proteins like opsin. There are three types of cones, which absorb blue, green, and yellow light, thereby allowing us to see color. Rod cells are connected to each bipolar cell and have multiple bipolar cells connected to each ganglion cell, thereby summing the signals from the rods. The cone cells are not connected in this manner, and therefore are not sensitive to dim light.

Glossary

A

A form. A duplex DNA structure with right-handed twisting in which the planes of the base pairs are tilted about 70° with respect to the helix axis.

Acetal. The product formed by the successive condensation of two alcohols with a single aldehyde. It contains two ether-linked oxygens attached to a central carbon atom.

Acetyl-CoA. Acetyl-coenzyme A, a high-energy ester of acetic acid that is important both in the tricarboxylic acid cycle and in fatty acid biosynthesis.

Actin. A protein found in combination with myosin in muscle and also found as filaments constituting an important part of the cytoskeleton in many eukaryotic cells.

Actinomycin D. An antibiotic that binds to DNA and inhibits RNA chain elongation.

Activated complex. The highest free energy state of a complex in going from reactants to products.

Active site. The region of an enzyme molecule that contains the substrate binding site and the catalytic site for converting the substrate(s) into product(s).

Active transport. The energy-dependent transport of a substance across a membrane.

Adenine. A purine base found in DNA or RNA.

Adenosine. A purine nucleoside found in DNA, RNA, and many cofactors.

Adenosine diphosphate (ADP). The nucleotide formed by adding a pyrophosphate group to the 5′-OH group of adenosine.

Adenosine triphosphate (ATP). The nucleotide formed by adding yet another phosphate group to the pyrophosphate group on ADP.

Adenylate cyclase. The enzyme that catalyzes the formation of cyclic 3′,5′-adenosine monophosphate (cAMP) from ATP.

Adipocyte. A specialized cell that functions as a storage depot for lipid.

Aerobe. An organism that utilizes oxygen for growth.

Affinity chromatography. A column chromatographic technique that employs attached functional groups that have a specific affinity for sites on particular proteins.

Alcohol. A molecule with a hydroxyl group attached to a carbon atom.

Aldehyde. A molecule containing a doubly bonded oxygen and a hydrogen attached to the same carbon atom.

Alleles. Alternative forms of a gene.

Allosteric enzyme. An enzyme whose active site can be altered by the binding of a small molecule at a nonoverlapping site.

Angstrom (Å). A unit of length equal to 10^{-8} cm.

Anomers. The sugar isomers that differ in configuration about the carbonyl carbon atom. This carbon atom is called the anomeric carbon atom of the sugar.

Antibiotic. A natural product that inhibits bacterial growth (is bacteriostatic) and sometimes results in bacterial death (is bacteriocidal).

Antibody. A specific protein that interacts with a foreign substance (antigen) in a specific way.

Anticodon. A sequence of three bases on the transfer RNA that pair with the bases in the corresponding codon on the messenger RNA.

Antigen. A foreign substance that triggers antibody formation and is bound by the corresponding antibody.

Antiparallel β-pleated sheet (β sheet). A hydrogen-bonded secondary structure formed between two or more extended polypeptide chains.

Apoactivator. A regulatory protein that stimulates transcription from one or more genes in the presence of a coactivator molecule.

Asexual reproduction. Growth and cell duplication that does not involve the union of nuclei from cells of opposite mating types.

Asymmetric carbon. A carbon that is covalently bonded to four different groups.

Attenuator. A provisional transcription stop signal.

Autoradiography. The technique of exposing film in the presence of disintegrating radioactive particles. Used to obtain information on the distribution of radioactivity in a gel or a thin cell section.

Autoregulation. The process in which a gene regulates its own expression.

Autotroph. An organism that can form its organic constituents from CO_2.

Auxin. A plant growth hormone usually concentrated in the apical bud.

Auxotroph. A mutant that cannot grow on the minimal medium on which a wild-type member of the same species can grow.

Avogadro's number. The number of molecules in a gram molecular weight of any compound (6.023×10^{23}).

B

B cell. One of the major types of cells in the immune system. B cells can differentiate to form memory cells or antibody-forming cells.

B form. The most common form of duplex DNA, containing a right-handed helix and about 10 (10.5 exactly) base pairs per turn of the helix axis.

β bend. A characteristic way of turning an extended polypeptide chain in a different direction, involving the minimum number of residues.

β oxidation. Oxidative degradation of fatty acids that occurs by the successive oxidation of the β-carbon atom.

β sheet. A sheetlike structure formed by the interaction between two or more extended polypeptide chains.

Base analog. A compound, usually a purine or a pyrimidine, that differs somewhat from a normal nucleic acid base.

Base stacking. The close packing of the planes of base pairs, commonly found in DNA and RNA structures.

Bidirectional replication. Replication in both directions away from the origin, as opposed to replication in one direction only (unidirectional replication).

Bilayer. A double layer of lipid molecules with the hydrophilic ends oriented outward, in contact with water, and the hydrophobic parts oriented inward toward each other.

Bile salts. Derivatives of cholesterol with detergent properties that aid in the solubilization of lipid molecules in the digestive tract.

Biochemical pathway. A series of enzyme-catalyzed reactions that results in the conversion of a precursor molecule into a product molecule.

Bioluminescence. The production of light by a biochemical system.

Blastoderm. The stage in embryogenesis when a unicellular layer at the surface surrounds the yolk mass.

Bond energy. The energy required to break a bond.

Branchpoint. An intermediate in a biochemical pathway that can follow more than one route in subsequent steps.

Buffer. A conjugate acid-base pair that is capable of resisting changes in pH when acid or base is added to the system. This tendency will be maximal when the conjugate forms are present in equal amounts.

C

cAMP. 3′,5′ cyclic adenosine monophosphate. The cAMP molecule plays a key role in metabolic regulation.

CAP. The catabolite gene activator protein, sometimes incorrectly referred to as the CRP protein. The latter term, in small letters (*crp*), should be used to refer to the gene but not to the protein.

Capping. Covalent modification involving the addition of a modified guanidine group in a 5′-5″ linkage. It occurs only in eukaryotes, primarily on mRNA molecules.

Carbohydrate. A polyhydroxy aldehyde or ketone.

Carboxylic acid. A molecule containing a carbon atom attached to a hydroxyl group and to an oxygen atom by a double bond.

Carcinogen. A chemical that can cause cancer.

Carotenoids. Lipid-soluble pigments that are made from isoprene units.

Catabolism. That part of metabolism that is concerned with degradation reactions.

Catabolite repression. The general repression of transcription of genes associated with catabolism that is seen in the presence of glucose.

Catalyst. A compound that lowers the activation energy of a reaction without itself being consumed.

Catalytic site. The site of the enzyme involved in the catalytic process.

Catenane. An interlocked pair of circular structures, such as covalently closed DNA molecules.

Catenation. The linking of molecules without any direct covalent bonding between them, as when two circular DNA molecules interlock like the links in a chain.

cDNA. Complementary DNA, made *in vitro* from the mRNA by the enzyme reverse transcriptase and deoxyribonucleotide triphosphates.

Cell commitment. That stage in a cell's life when it becomes committed to a certain line of development.

Cell cycle. All of those stages that a cell passes through from one cell generation to the next.

Cell line. An established clone originally derived from a whole organism through a long process of cultivation.

Cell lineage. The pedigree of cells resulting from binary fission.

Cell wall. A tough outer coating found in many plant, fungal, and bacterial cells that accounts for their ability to withstand mechanical stress or abrupt changes in osmotic pressure. Cell walls always contain a carbohydrate component and frequently also a peptide and a lipid component.

Chelate. A molecule that contains more than one binding site and frequently binds to another molecule through more than one binding site at the same time.

Chemiosmotic coupling. The coupling of ATP synthesis to an electrochemical potential gradient across a membrane.

Chimeric DNA. Recombinant DNA whose components originate from two or more different sources.

Chiral compound. A compound that can exist in two forms that are nonsuperimposable images of one another.

Chlorophyll. A green photosynthetic pigment that is made of a magnesium dihydroporphyrin complex.

Chloroplast. A chlorophyll-containing photosynthetic organelle, found in eukaryotic cells, that can harness light energy.

Chromatin. The nucleoprotein fibers of eukaryotic chromosomes.

Chromatography. A procedure for separating chemically similar molecules. Segregation is usually carried out on paper or in glass or metal columns with the help of different solvents. The paper or glass columns contain porous solids with functional groups that have limited affinities for the molecules being separated.

Chromosome. A threadlike structure, visible in the cell nucleus during metaphase, that carries the hereditary information.

Chromosome puff. A swollen region of a giant chromosome; the swelling reflects a high degree of transcription activity.

***Cis* dominance.** Property of a sequence or a gene that exerts a dominant effect on a gene to which it is linked.

Cistron. A genetic unit that encodes a single polypeptide chain.

Citric acid cycle. *See* tricarboxylic acid (TCA) cycle.

Clone. One of a group of genetically identical cells or organisms derived from a common ancestor.

Cloning vector. A self-replicating entity to which foreign DNA can be covalently attached for purposes of amplification in host cells.

Coactivator. A molecule that functions in conjunction with a protein apoactivator. For example, cAMP is a coactivator of the CAP protein.

Codon. In a messenger RNA molecule, a sequence of three bases that represents a particular amino acid.

Coenzyme. An organic molecule that associates with enzymes and affects their activity.

Cofactor. A small molecule required for enzyme activity. It could be organic in nature, like a coenzyme, or inorganic in nature, like a metallic cation.

Complementary base sequence. For a given sequence of nucleic acids, the nucleic acids that are related to them by the rules of base pairing.

Configuration. The spatial arrangement in which atoms are covalently linked in a molecule.

Conformation. The three-dimensional arrangement adopted by a molecule, usually a complex macromolecule. Molecules with the same configuration can have more than one conformation.

Consensus sequence. In nucleic acids, the "average" sequence that signals a certain type of action by a specific protein. The sequences actually observed usually vary around this average.

Constitutive enzymes. Enzymes synthesized in fixed amounts, regardless of growth conditions.

Cooperative binding. A situation in which the binding of one substituent to a macromolecule favors the binding of another. For example, DNA cooperatively binds histone molecules, and hemoglobin cooperatively binds oxygen molecules.

Coordinate induction. The simultaneous expression of two or more genes.

Cosmid. A DNA molecule with *cos* ends from λ bacteriophage that can be packaged *in vitro* into a virus for infection purposes.

Cot curve. A curve that indicates the rate of DNA-DNA annealing as a function of DNA concentration and time.

Cytidine. A pyrimidine nucleoside found in DNA and RNA.

Cytochromes. Heme-containing proteins that function as electron carriers in oxidative phosphorylation and photosynthesis.

Cytokinin. A plant hormone produced in root tissue.

Cytoplasm. The contents enclosed by the plasma membrane, excluding the nucleus.

Cytosine. A pyrimidine base found in DNA and RNA.

Cytoskeleton. The filamentous skeleton, formed in the cytoplasm, that is largely responsible for controlling cell shape.

Cytosol. The liquid portion of the cytoplasm, including the macromolecules but not including the larger structures, such as subcellular organelles or cytoskeleton.

D

D loop. An extended loop of single-stranded DNA displaced from a duplex structure by an oligonucleotide.

Dalton. A unit of mass equivalent to the mass of a hydrogen atom (1.66×10^{-24} g).

Dark reactions. Reactions that can occur in the dark, in a process that is usually associated with light, such as the dark reactions of photosynthesis.

***De novo* pathway.** A biochemical pathway that starts from elementary substrates and ends in the synthesis of a biochemical.

Deamination. The enzymatic removal of an amine group, as in the deamination of an amino acid to an α-keto acid.

Dehydrogenase. An enzyme that catalyzes the removal of a pair of electrons (and usually one or two protons) from a substrate molecule.

Denaturation. The disruption of the native folded structure of a nucleic acid or protein molecule; may be due to heat, chemical treatment, or change in pH.

Density-gradient centrifugation. The separation, by centrifugation, of molecules according to their density, in a gradient varying in solute concentration.

Dialysis. Removal of small molecules from a macromolecule preparation by allowing them to pass across a semipermeable membrane.

Diauxic growth. Biphasic growth on a mixture of two carbon sources in which one carbon source is used up before the other one is mobilized. For example, in the presence of glucose and lactose, *E. coli* will utilize the glucose before the lactose.

Difference spectra. Display comparing the absorption spectra of a molecule or an assembly of molecules in different states, for example, those of mitochondria under oxidizing or reducing conditions.

Differential centrifugation. Separation of molecules and/or organelles by sedimentation rate.

Differentiation. A change in the form and pattern of a cell and the genes it expresses as a result of growth and replication, usually during development of a multicellular organism. Also occurs in microorganisms (e.g., in sporulation).

Diploid cell. A cell that contains two chromosomes ($2N$) of each type.

Dipole. A separation of charge within a single molecule.

Directed mutagenesis. In a DNA sequence, an intentional alteration that can be genetically inherited.

Dissociation constant. An equilibrium constant for the dissociation of a molecule into two parts (e.g., dissociation of acetic acid into acetate anion and proton).

Disulfide bridge. A covalent linkage formed by oxidation between two SH groups either in the same polypeptide chain or in different polypeptide chains.

DNA. Deoxyribonucleic acid. A polydeoxyribonucleotide in which the sugar is deoxyribose; the main repository of genetic information in all cells and most viruses.

DNA cloning. The propagation of individual segments of DNA as clones.

DNA library. A mixture of clones, each containing a cloning vector and a segment of DNA from a source of interest.

DNA polymerase. An enzyme that catalyzes the formation of 3′-5′ phosphodiester bonds from deoxyribonucleotide triphosphates.

Domain. A segment of a folded protein structure showing conformational integrity. A domain could comprise the entire protein or just a fraction of the protein. Some proteins, such as antibodies, contain many structural domains.

Dominant. Describing an allele whose phenotype is expressed regardless of whether the organism is homozygous or heterozygous for that allele.

Double helix. A structure in which two helically twisted polynucleotide strands are held together by hydrogen bonding and base stacking.

Duplex. Synonymous with double helix.

Dyad symmetry. Property of a structure that can be rotated by 180° to produce the same structure.

E

Ecdysone. A hormone that stimulates the molting process in insects.

Edman degradation. A systematic method of sequencing proteins, proceeding by stepwise removal of single amino acids from the amino terminal of a polypeptide chain.

Eicosanoid. Any fatty acid with twenty carbons.

Electrophoresis. The movement of particles in an electrical field. A commonly used technique for analysis of mixtures of molecules in solution according to their electrophoretic mobilities.

Elongation factors. Protein factors uniquely required during the elongation phase of protein synthesis. Elongation factor G (EF-G) brings about the movement of the peptidyl-tRNA from the A site to the P site of the ribosome.

Eluate. The effluent from a chromatographic column.

Embryo. Plant or animal at an early stage of development.

Enantiomorphs. Isomers that are mirror images of one another.

Endergonic reaction. A reaction with a positive free energy change.

End-product (feedback) inhibition. The inhibition of the first enzyme in a pathway by the end product of that pathway.

Endocrine glands. Specialized tissues whose function is to synthesize and secrete hormones.

Endonuclease. An enzyme that breaks a phosphodiester linkage at some point within a polynucleotide chain.

Endopeptidase. An enzyme that breaks a polypeptide chain at an internal peptide linkage.

Endoplasmic reticulum. A system of double membranes in the cytoplasm that is involved in the synthesis of transported proteins. The rough endoplasmic reticulum has ribosomes associated with it. The smooth endoplasmic reticulum does not.

Energy charge. The fractional degree to which the AMP-ADP-ATP system is filled with high-energy phosphates (phosphoryl groups).

Enhancer. A DNA sequence that can stimulate transcription at an appreciable distance from the site where it is located. It acts in either orientation and either upstream or downstream from the promoter.

Entropy. The randomness of a system.

Enzyme. A protein that contains a catalytic site for a biochemical reaction.

Epimers. Two stereoisomers with more than one chiral center that differ in configuration at one of their chiral centers.

Equilibrium. In chemistry the point at which the concentrations of two compounds are such that the interconversion of one compound into the other compound does not result in any change in free energy.

Escherichia coli (E. coli). A Gram negative bacterium commonly found in the vertebrate intestine. It is the bacterium most frequently used in the study of biochemistry and genetics.

Established cell line. A group of cultured cells derived from a single origin and capable of stable growth for many generations.

Ether. A molecule containing two carbons linked by an oxygen atom.

Eukaryote. A cell or organism that has a membrane-bounded nucleus.

Excision repair. DNA repair in which a damaged region is replaced.

Excited state. An energy-rich state of an atom or a molecule, produced by the absorption of radiant energy.

Exergonic reaction. A chemical reaction that takes place with a negative change in free energy.

Exon. A segment within a gene that carries part of the coding information for a protein.

Exonuclease. An enzyme that breaks a phosphodiester linkage at one or the other end of a polynucleotide chain so as to release single or small nucleotide residues.

F

F factor. A large bacterial plasmid, known as the sex-factor plasmid because it permits mating between F^+ and F^- bacteria.

Facultative aerobe. An organism that can use molecular oxygen in its metabolism but that also can live anaerobically.

Fatty acid. A long-chain hydrocarbon containing a carboxyl group at one end. Saturated fatty acids have completely saturated hydrocarbon chains. Unsaturated fatty acids have one or more carbon–carbon double bonds in their hydrocarbon chains.

Feedback inhibition. *See* end-product inhibition.

Fermentation. The energy-generating breakdown of glucose or related molecules by a process that does not require molecular oxygen.

Fingerprinting. The characteristic two-dimensional paper chromatogram obtained from the partial hydrolysis of a protein or a nucleic acid.

Fluorescence. The emission of light by an excited molecule in the process of making the transition from the excited state to the ground state.

Frameshift mutations. Insertions or deletions of genetic material that lead to a shift in the translation of the reading frame. The mutation usually leads to nonfunctional proteins.

Free energy. That part of the energy of a system that is available to do useful work.

Furanose. A sugar that contains a five-membered ring as a result of intramolecular hemiacetal formation.

Futile cycle. *See* pseudocycle.

G

G_1 phase. That period of the cell cycle in which preparations are being made for chromosome duplication, which takes place in the S phase.

G_2 phase. That period of the cell cycle between S phase and mitosis (M phase).

Gametes. The ova and the sperm, haploid cells that unite during fertilization to generate a diploid zygote.

Gel exclusion chromatography. A technique that makes use of certain polymers that can form porous beads with varying pore sizes. In columns made from such beads, it is possible to separate molecules, which cannot penetrate beads of a given pore size, from small molecules that can.

Gene. A segment of the genome that codes for a functional product.

Gene amplification. The duplication of a particular gene within a chromosome two or more times.

Gene splicing. The cutting and rejoining of DNA sequences.

General recombination. Recombination that occurs between homologous chromosomes at homologous sites.

Generation time. The time it takes for a cell to double its mass under specified conditions.

Genetic map. The arrangement of genes or other identifiable sequences on a chromosome.

Genome. The total genetic content of a cell or a virus.

Genotype. The genetic characteristics of an organism (distinguished from its observable characteristics, or phenotype).

Globular protein. A folded protein that adopts an approximately globular shape.

Goldman equation. An equation expressing the quantitative relationship between the concentrations of charged species on either side of a membrane and the resting transmembrane potential.

Golgi apparatus. A complex series of double-membrane structures that interact with the endoplasmic reticulum and that serve as a transfer point for proteins destined for other organelles, the plasma membrane, or extracellular transport.

Gluconeogenesis. The production of sugars from nonsugar precursors such as lactate or amino acids. Applies more specifically to the production of free glucose by vertebrate livers.

Glycogen. A polymer of glucose residues in 1,4 linkage and 1,6 linkage at branchpoints.

Glycogenic. Describing amino acids whose metabolism may lead to gluconeogenesis.

Glycolipid. A lipid containing a carbohydrate group.

Glycolysis. The catabolic conversion of glucose to pyruvate with the production of ATP.

Glycoprotein. A protein linked to an oligosaccharide or a polysaccharide.

Glycosaminoglycans. Long, unbranched polysaccharide chains composed of repeating disaccharide subunits in which one of the two sugars is either *N*-acetylglucosamine or *N*-acetylgalactosamine.

Glycosidic bond. The bond between a sugar and an alcohol. Also the bond that links two sugars in disaccharides, oligosaccharides, and polysaccharides.

Glyoxylate cycle. A pathway that uses some of the enzymes of the TCA cycle and some enzymes whereby acetate can be converted into succinate and carbohydrates.

Glyoxysome. An organelle containing key enzymes of the glyoxylate cycle.

Gram molecular weight. For a given compound, the weight in grams that is numerically equal to its molecular weight.

Ground state. The lowest electronic energy state of an atom or a molecule.

Growth factor. A substance that must be present in the growth medium to permit cell proliferation.

Growth fork. The region on a DNA duplex molecule where synthesis is taking place. It resembles a fork in shape, since it consists of a region of duplex DNA connected to a region of unwound single strands.

Guanine. A purine base found in DNA or RNA.

Guanosine. A purine nucleoside found in DNA and RNA.

H

Hairpin loop. A single-stranded complementary region that folds back on itself and base-pairs into a double helix.

Half-life. The time required for the disappearance of one half of a substance.

Haploid cell. A cell containing only one chromosome of each type.

Heavy isotopes. Forms of atoms that contain greater numbers of neutrons (e.g., ^{15}N, ^{13}C).

Helix. A spiral structure with a repeating pattern.

Heme. An iron-porphyrin complex found in hemoglobin and cytochromes.

Hemiacetal. The product formed by the condensation of an aldehyde with an alcohol; it contains one oxygen linked to a central carbon in a hydroxyl fashion and one oxygen linked to the same central carbon by an ether linkage.

Henderson-Hasselbalch equation. An equation that relates the pK_a to the pH and the ratio of the proton acceptor (A^-) and the proton donor (HA) species of a conjugate acid-base pair.

Heterochromatin. Highly condensed regions of chromosomes that are not usually transcriptionally active.

Heteroduplex. An annealed duplex structure between two DNA strands that do not show perfect complementarity. Can arise by mutation, recombination, or the annealing of complementary single-stranded DNAs.

Heteropolymer. A polymer containing more than one type of monomeric unit.

Heterotroph. An organism that requires preformed organic compounds for growth.

Heterozygous. Describing an organism (a heterozygote) that carries two different alleles for a given gene.

Hexose. A sugar with a six-carbon backbone.

High-energy compound. A compound that undergoes hydrolysis with a high negative standard free energy change.

Histones. The family of basic proteins that is normally associated with DNA in most cells of eukaryotic organisms.

Holoenzyme. An intact enzyme containing all of its subunits with full enzymatic activity.

Homologous chromosomes. Chromosomes that carry the same pattern of genes, but not necessarily the same alleles.

Homopolymer. A polymer composed of only one type of monomeric building block.

Homozygous. Describing an organism (a homozygote) that carries two identical alleles for a given gene.

Hormone. A chemical substance made in one cell and secreted so as to influence the metabolic activity of a select group of cells located at other sites in the organism.

Hormone receptor. A protein that is located on the cell membrane or inside the responsive cell and that interacts specifically with the hormone.

Host cell. A cell used for growth and reproduction of a virus.

Hybrid (or chimeric) plasmid. A plasmid that contains DNA from two different organisms.

Hydrogen bond. A weak attractive force between one electronegative atom and a hydrogen atom that is covalently linked to a second electronegative atom.

Hydrolysis. The cleavage of a molecule by the addition of water.

Hydrophilic. Preferring to be in contact with water.

Hydrophobic. Preferring not to be in contact with water, as is the case with the hydrocarbon portion of a fatty acid or phospholipid chain.

Hydrophobic bonding. The association of nonpolar groups with each other in aqueous solution.

Hydroxyapatite. A calcium phosphate gel used, in the case of nucleic acids, to selectively absorb duplex DNA-RNA from a mixture of single-stranded and duplex nucleic acids.

I

Icosahedral symmetry. The symmetry displayed by a regular polyhedron that is composed of twenty equilateral triangular faces with twelve corners.

Imine. A molecule containing a nitrogen atom attached to a carbon atom by a double bond. The nitrogen is also covalently linked to a hydrogen.

Immunofluorescence. A cytological technique in which a specific fluorescent antibody is used to label an antigen. Frequently used to determine the location of an antigen in a tissue or a cell.

Immunoglobulin. A protein made in a B plasma cell and usually secreted; it interacts specifically with a foreign agent. Synonymous with antibody. It is composed of two heavy and two light chains linked by disulfide bonds. Immunoglobulins can be divided into five classes (IgG, IgM, IgA, IgD, and IgE) based on their heavy-chain component.

In vitro. Literally, "in glass," describing whatever happens in a test tube or other receptacle, as opposed to what happens in whole cells of the whole organism (*in vivo*).

Induced fit. A change in the shape of an enzyme that results from the binding of substrate.

Initiation factors. Those protein factors that are specifically required during the initiation phase of protein synthesis.

Intercalating agent. A chemical, usually containing aromatic rings, that can sandwich in between adjacent base pairs in a DNA duplex. The intercalation leads to an adjustment in the DNA secondary structure, as adjacent base pairs are usually close-packed.

Interferon. One of a family of proteins that are liberated by special host cells in the mammal in response to viral infection. The interferons attach to an infected cell, where they stimulate antiviral protein synthesis.

Intervening sequence. *See* intron.

Intron. A segment of the nascent transcript that is removed by splicing. Also refers to the corresponding region in the DNA. Synonymous with intervening sequence.

Inverted repeat. A chromosome segment that is identical to another segment on the same chromosome except that it is oriented in the opposite direction.

Ion-exchange resin. A polymeric resinous substance, usually in bead form, that contains fixed groups with positive or negative charge. A cation exchange resin has negatively charged groups and is therefore useful in exchanging the cationic groups in a test sample. The resin is usually used in the form of a column, as in other column chromatographic systems.

Isoelectric pH. The pH at which a protein has no net charge.

Isomerase. An enzyme that catalyzes an intramolecular rearrangement.

Isomerization. Rearrangement of atomic groups within the same molecule without any loss or gain of atoms.

Isozymes. Multiple forms of an enzyme that differ from one another in one or more of the properties.

K

K_m. *See* Michaelis constant.

Ketogenic. Describing amino acids that are metabolized to acetoacetate and acetate.

Ketone. A functional group of an organic compound in which a carbon atom is double-bonded to an oxygen. Neither of the other substituents attached to the carbon is a hydrogen. Otherwise the group would be called an aldehyde.

Ketone bodies. Refers to acetoacetate, acetone, and β-hydroxybutyrate made from acetyl-CoA in the liver and used for energy in nonhepatic tissue.

Ketosis. A condition in which the concentration of ketone bodies in the blood or urine is unusually high.

Kilobase. One thousand bases in a DNA molecule.

Kinase. An enzyme catalyzing phosphorylation of an acceptor molecule, usually with ATP serving as the phosphate (phosphoryl) donor.

Kinetochore. A structure that attaches laterally to the centromere of a chromosome; it is the site of chromosome tubule attachment.

Krebs cycle. *See* tricarboxylic acid (TCA) cycle.

L

Lampbrush chromosome. Giant diplotene chromosome found in the oocyte nucleus. The loops that are observed are the sites of extensive gene expression.

Law of mass action. The finding that the rate of a chemical reaction is a function of the product of the concentrations of the reacting species.

Leader region. The region of an mRNA between the 5′ end and the initiation codon for translation of the first polypeptide chain.

Lectins. Agglutinating proteins usually extracted from plants.

Ligase. An enzyme that catalyzes the joining of two molecules together. In DNA it joins 3′-OH to 5′ phosphates.

Linkers. Short oligonucleotides that can be ligated to larger DNA fragments, then cleaved to yield overlapping cohesive ends, suitable for ligation to other DNAs that contain comparable cohesive ends.

Linking number. The net number of times one polynucleotide chain crosses over another polynucleotide chain. By convention, right-handed crossovers are given a plus designation.

Lipid. A biological molecule that is soluble in organic solvents. Lipids include steroids, fatty acids, prostaglandins, terpenes, and waxes.

Lipid bilayer (*see* Bilayer). Model for the structure of the cell membrane based on the hydrophobic interaction between phospholipids.

Lipopolysaccharide. Usually refers to a unique glycolipid found in Gram negative bacteria.

Lyase. An enzyme that catalyzes the removal of a group to form a double bond, or the reverse reaction.

Lysogenic virus. A virus that can adopt an inactive (lysogenic) state, in which it maintains its genome within a cell instead of entering the lytic cycle. The circumstances that determine whether a lysogenic (temperate) virus will adopt an inactive state or an active lytic state are often subtle and depend on the physiological state of the infected cell.

Lysosome. An organelle that contains hydrolytic enzymes designed to break down proteins that are targeted to that organelle.

Lytic infection. A virus infection that leads to the lysis of the host cell, yielding progeny virus particles.

M

M phase. That period of the cell cycle when mitosis takes place.

Meiosis. Process in which diploid cells undergo division to form haploid sex cells.

Membrane transport. The facilitated transport of a molecule across a membrane.

Merodiploid. An organism that is diploid for some but not all of its genes.

Mesosome. An invagination of the bacterial cell membrane.

Messenger RNA (mRNA). The template RNA carrying the message for protein synthesis.

Metabolic turnover. A measure of the rate at which already existing molecules of the given species are replaced by newly synthesized molecules of the same type. Usually isotopic labeling is required to measure turnover.

Metabolism. The sum total of the enzyme-catalyzed reactions that occur in a living organism.

Metamorphosis. A change of form, especially the conversion of a larval form to an adult form.

Metaphase. That stage in mitosis or meiosis when all of the chromosomes are lined up on the equator (i.e., an imaginary line that bisects the cell).

Micelle. An aggregate of lipids in which the polar head groups face outward and the hydrophobic tails face inward; no solvent is trapped in the center.

Michaelis constant (K_m). The substrate concentration at which an enzyme-catalyzed reaction proceeds at one-half maximum velocity.

Michaelis-Menten equation (also known as the Henri-Michaelis-Menten equation). An equation relating the reaction velocity to the substrate concentration of an enzyme.

Microtubules. Thin tubules, made from globular proteins, that serve multiple purposes in eukaryotic cells.

Mismatch repair. The replacement of a base in a heteroduplex structure by one that forms a Watson-Crick base pair.

Missense mutation. A change in which a codon for one amino acid is replaced by a codon for another amino acid.

Mitochondrion. An organelle, found in eukaryotic cells, in which oxidative phosphorylation takes place. It contains its own genome and unique ribosomes to carry out protein synthesis of only a fraction of the proteins located in this organelle.

Mitosis. The process whereby replicated chromosomes segregate equally toward opposite poles prior to cell division.

Mobile genetic element. A segment of the genome that can move as a unit from one location on the genome to another, without any requirement for sequence homology.

Molecularity of a reaction. The number of molecules involved in a specific reaction step.

Monolayer. A single layer of oriented lipid molecules.

Mutagen. An agent that can bring about a heritable change (mutation) in an organism.

Mutagenesis. A process that leads to a change in the genetic material that is inherited in subsequent generations.

Mutant. An organism that carries an altered gene or change in its genome.

Mutarotation. The change in optical rotation of a sugar that is observed immediately after it is dissolved in aqueous solution, as the result of the slow approach of equilibrium of a pyranose or a furanose in its α and β forms.

Mutation. The genetically inheritable alteration of a gene or group of genes.

Myofibril. A unit of thick and thin filaments in a muscle fiber.

Myosin. The main protein of the thick filaments in a muscle myofibril. It is composed of two coiled subunits (M_r about 220,000) that can aggregate to form a thick filament, which is globular at each end.

N

Nascent RNA. The initial transcripts of RNA, before any modification or processing.

Negative control. Regulation of the activity by an inhibitory mechanism.

Nernst equation. An equation that relates the redox potential to the standard redox potential and the concentrations of the oxidized and reduced form of the couple.

Nitrogen cycle. The passage of nitrogen through various valence states, as the result of reactions carried out by a wide variety of different organisms.

Nitrogen fixation. Conversion of atmospheric nitrogen into a form that can be converted by biochemical reactions to an organic form. This reaction is carried out by a very limited number of microorganisms.

Nitrogenous base. An aromatic nitrogen-containing molecule with basic properties. Such bases include purines and pyrimidines.

Noncompetitive inhibitor. An inhibitor of enzyme activity whose effect is not reversed by increasing the concentration of substrate molecule.

Nonsense mutation. A change in the base sequence that converts a sense codon (one that specifies an amino acid) to one that specifies a stop (a nonsense codon). There are three nonsense codons.

Northern blotting. *See* Southern blotting.

Nuclease. An enzyme that cleaves phosphodiester bonds of nucleic acids.

Nucleic acids. Polymers of the ribonucleotides or deoxyribonucleotides.

Nucleohistone. A complex of DNA and histone.

Nucleolus. A spherical structure visible in the nucleus during interphase. The nucleolus is associated with a site on the chromosome that is involved in ribosomal RNA synthesis.

Nucleophilic group. An electron-rich group that tends to attack an electron-deficient nucleus.

Nucleosome. A complex of DNA and an octamer of histone proteins in which a small stretch of the duplex is wrapped around a molecular bead of histone.

Nucleotide. An organic molecule containing a purine or pyrimidine base, a five-carbon sugar (ribose or deoxyribose), and one or more phosphate groups.

Nucleus. In eukaryotic cells, the centrally located organelle that encloses most of the chromosomes. Minor amounts of chromosomal substance are found in some other organelles, most notably the mitochondria and the chloroplasts.

O

Okazaki fragment. A short segment of single-stranded DNA that is an intermediate in DNA synthesis. In bacteria, Okazaki fragments are 1,000–2,000 bases in length; in eukaryotes, 100–200 bases in length.

Oligonucleotide. A polynucleotide containing a small number of nucleotides. The linkages are the same as in a polynucleotide; the only distinguishing feature is the small size.

Oligosaccharide. A molecule containing a small number of sugar residues joined in a linear or a branched structure by glycosidic bonds.

Oncogene. A gene of cellular or viral origin that is responsible for rapid, unruly growth of animal cells.

Operon. A group of contiguous genes that are coordinately regulated by two *cis*-acting elements, a promoter and an operator. Found only in prokaryotic cells.

Optical activity. The property of a molecule that leads to rotation of the plane of polarization of plane-polarized light when the latter is transmitted through the substance. Chirality is a necessary and sufficient property for optical activity.

Organelle. A subcellular membrane-bounded body with a well-defined function.

Osmotic pressure. The pressure generated by the mass flow of water to that side of a membrane-bounded structure that contains the higher concentration of solute molecules. A stable osmotic pressure is seen in systems in which the membrane is not permeable to some of the solute molecules.

Oxidation. The loss of electrons from a compound.

Oxidative phosphorylation. The formation of ATP as the result of the transfer of electrons to oxygen.

Oxido-reductase. An enzyme that catalyzes oxidation-reduction reactions.

P

Palindrome. A sequence of bases that reads the same in both directions on opposite strands of the DNA duplex (e.g., GAATTC).

Pentose. A sugar with five carbon atoms.

Pentose phosphate pathway. The pathway involving the oxidation of glucose-6-phosphate to pentose phosphates and further reactions of pentose phosphates.

Peptide. An organic molecule in which a covalent amide bond is formed between the α-amino group of one amino acid and the α-carboxyl group of another amino acid, with the elimination of a water molecule.

Peptide mapping. Same as fingerprinting.

Peptidoglycan. The main component of the bacterial cell wall, consisting of a two-dimensional network of heteropolysaccharides running in one direction, cross-linked with polypeptides running in the perpendicular direction.

Periplasm. The space between the inner and outer membranes of a bacterium.

Permease. A protein that catalyzes the transport of a specific small molecule across a membrane.

Peroxisomes. Subcellular organelles that contain flavin-requiring oxidases and that regenerate oxidized flavin by reaction with oxygen.

Phenotype. The observable trait(s) that result from the genotype in cooperation with the environment.

Phenylketonuria. A human disease caused by a genetic deficiency in the enzyme that converts phenylalanine to tyrosine. The immediate cause of the disease is an excess of phenylalanine, which can be alleviated by a diet low in phenylalanine.

Pheromone. A hormonelike substance that acts as an attractant.

Phosphodiester. A molecule containing two alcohols esterified to a single molecule of phosphate. For example, the backbone of nucleic acids is connected by 5′-3′ phosphodiester linkages between the adjacent individual nucleotide residues.

Phosphogluconate pathway. Another name for the pentose phosphate pathway. This name derives from the fact that 6-phosphogluconate is an intermediate in the formation of pentoses from glucose.

Phospholipid. A lipid containing charged hydrophilic phosphate groups; a component of cell membranes.

Phosphorylation. The formation of a phosphate derivative of a biomolecule.

Photoreactivation. DNA repair in which the damaged region is repaired with the help of light and an enzyme. The lesion is repaired without excision from the DNA.

Photosynthesis. The biosynthesis that directly harnesses the chemical energy resulting from the absorption of light. Frequently used to refer to the formation of carbohydrates from CO_2 that occurs in the chloroplasts of plants or the plastids of photosynthetic microorganisms.

Pitch length (or pitch). The number of base pairs per turn of a duplex helix.

Plaque. A circular clearing on a lawn of bacterial or cultured cells, resulting from cell lysis and production of phage or animal virus progeny.

Plasma membrane. The membrane that surrounds the cytoplasm.

Plasmid. A circular DNA duplex that replicates autonomously in bacteria. Plasmids that integrate into the host genome are called episomes. Plasmids differ from viruses in that they never form infectious nucleoprotein particles.

Polar group. A hydrophilic (water-loving) group.

Polar mutation. A mutation in one gene that reduces the expression of a gene or genes distal to the promoter in the same operon.

Polarimeter. An instrument for determining the rotation of polarization of light as the light passes through a solution containing an optically active substance.

Polyamine. A hydrocarbon containing more than two amino groups.

Polycistronic messenger RNA. In prokaryotes, an RNA that contains two or more cistrons; note that only in prokaryotic mRNAs can more than one cistron be utilized by the translation system to generate individual proteins.

Polymerase. An enzyme that catalyzes the synthesis of a polymer from monomers.

Polynucleotide. A chain structure containing nucleotides linked together by phosphodiester (5′-3′) bonds. The polynucleotide chain has a directional sense with a 5′ and a 3′ end.

Polynucleotide phosphorylase. An enzyme that polymerizes ribonucleotide diphosphates. No template is required.

Polypeptide. A linear polymer of amino acids held together by peptide linkages. The polypeptide has a directional sense, with an amino and a carboxyl-terminal end.

Polyribosome (polysome). A complex of an mRNA and two or more ribosomes actively engaged in protein synthesis.

Polysaccharide. A linear or branched chain structure containing many sugar molecules linked by glycosidic bonds.

Porphyrin. A complex planar structure containing four substituted pyrroles covalently joined in a ring and frequently containing a central metal atom. For example, heme is a porphyrin with a central iron atom.

Positive control. A system that is turned on by the presence of a regulatory protein.

Posttranslational modification. The covalent bond changes that occur in a polypeptide chain after it leaves the ribosome and before it becomes a mature protein.

Primary structure. In a polymer, the sequence of monomers and the covalent bonds.

Primer. A structure that serves as a growing point for polymerization.

Primosome. A multiprotein complex that catalyzes synthesis of RNA primer at various points along the DNA template.

Prochiral molecule. A nonchiral molecule that may react with an enzyme so that two groups that have a mirror-image relationship to each other are treated differently.

Prokaryote. A unicellular organism that contains a single chromosome, no nucleus, no membrane-bound organelles, and has characteristic ribosomes and biochemistry.

Promoter. That region of the gene that signals RNA polymerase binding and the initiation of transcription.

Prophage. The silent phage genome. Some prophages integrate into the host genome; others replicate autonomously. The prophage state is maintained by a phage-encoded repressor.

Prophase. The stage in meiosis or mitosis when chromosomes condense and become visible as refractile bodies.

Proprotein. A protein that is made in an inactive form, so that it requires processing to become functional.

Prostaglandin. An oxygenated eicosanoid that has a hormonal function. Prostaglandins are unusual hormones in that they usually have effects only in that region of the organism where they are synthesized.

Prosthetic group. Synonymous with coenzyme except that a prosthetic group is usually more firmly attached to the enzyme it serves.

Protamines. Highly basic, arginine-rich proteins found complexed to DNA in the sperm of many invertebrates and fish.

Protein subunit. One of the components of a complex multicomponent protein.

Proteoglycan. A protein-linked heteropolysaccharide in which the heteropolysaccharide is usually the major component.

Protist. A relatively undifferentiated organism that can survive as a single cell.

Proton acceptor. A functional group capable of accepting a proton from a proton donor molecule.

Proton motive force (Δp). The thermodynamic driving force for proton translocation. Expressed quantitatively as $\Delta G_{H+}/F$ in units of volts.

Protooncogene. A cellular gene that can undergo modification to a cancer-causing gene (oncogene).

Pseudocycle. A sequence of reactions that can be arranged in a cycle but that usually do not function simultaneously in both directions. Also called a futile cycle, since the net result of simultaneous functioning in both directions would be the expenditure of energy without accomplishing any useful work.

Pulse-chase. An experiment in which a short labeling period is followed by the addition of an excess of the same, unlabeled compound to dilute out the labeled material.

Purine. A heterocyclic ring structure with varying functional groups. The purines adenine and guanine are found in both DNA and RNA.

Puromycin. An antibiotic that inhibits polypeptide synthesis by competing with aminoacyl-tRNA for the ribosomal binding site A.

Pyranose. A simple sugar containing the six-membered pyran ring.

Pyrimidine. A heterocyclic six-membered ring structure. Cytosine and uracil are the main pyrimidines found in RNA, and cytosine and thymine are the main pyrimidines found in DNA.

Pyrophosphate. A molecule formed by two phosphates in anhydride linkage.

Q

Quaternary structure. In a protein, the way in which the different folded subunits interact to form the multisubunit protein.

R

R group. The distinctive side chain of an amino acid.

R loop. A triple-stranded structure in which RNA displaces a DNA strand by DNA-RNA hybrid formation in a region of the DNA.

Rapid-start complex. The complex that RNA polymerase forms at the promoter site just before initiation.

Recombination. The transfer to offspring of genes not found together in either of the parents.

Redox couple. An electron donor and its corresponding oxidized form.

Redox potential (E). The relative tendency of a pair of molecules to release or accept an electron. The standard redox potential ($E°$) is the redox potential of a solution containing the oxidant and reductant of the couple at standard concentrations.

Regulatory enzyme. An enzyme in which the active site is subject to regulation by factors other than the enzyme substrate. The enzyme frequently contains a nonoverlapping site for binding the regulatory factor that affects the activity of the active site.

Regulatory gene. A gene whose principal product is a protein designed to regulate the synthesis of other genes.

Renaturation. The process of returning a denatured structure to its original native structure, as when two single strands of DNA are reunited to form a regular duplex, or an unfolded polypeptide chain is returned to its normal folded three-dimensional structure.

Repair synthesis. DNA synthesis following excision of damaged DNA.

Repetitive DNA. A DNA sequence that is present in many copies per genome.

Replica plating. A technique in which an impression of a culture is taken from a master plate and transferred to a fresh plate. The impression can be of bacterial clones or phage plaques.

Replication fork. The Y-shaped region of DNA at the site of DNA synthesis; also called a growth fork.

Replicon. A genetic element that behaves as an autonomous replicating unit. It can be a plasmid, phage, or bacterial chromosome.

Repressor. A regulatory protein that inhibits transcription from one or more genes. It can combine with an inducer (resulting in specific enzyme induction) or with an operator element (resulting in repression).

Resonance hybrid. A molecular structure that is a hybrid of two structures that differ in the locations of some of the electrons. For example, the benzene ring can be drawn in two ways, with double bonds in different positions. The actual structure of benzene is in between these two equivalent structures.

Restriction-modification system. A pair of enzymes found in most bacteria (but not eukaryotic cells). The restriction enzyme recognizes a certain sequence in duplex DNA and makes one cut in each unmodified DNA strand at or near the recognition sequence. The modification enzyme methylates (or modifies) the recognition sequence, thus protecting it from the action of the restriction enzyme.

Reverse transcriptase. An enzyme that synthesizes DNA from an RNA template, using deoxyribonucleotide triphosphates.

Rho factor. A protein involved in the termination of transcription of some messenger RNAs.

Ribose. The five-carbon sugar found in RNA.

Ribosomal RNA (rRNA). The RNA parts of the ribosome.

Ribosomes. Small cellular particles made up of ribosomal RNA and protein. They are the site, together with mRNA, of protein synthesis.

RNA (ribonucleic acid). A polynucleotide in which the sugar is ribose.

RNA polymerase. An enzyme that catalyzes the formation of RNA from ribonucleotide triphosphates, using DNA as a template.

RNA splicing. The excision of a segment of RNA, followed by a rejoining of the remaining fragments.

Rolling-circle replication. A mechanism for the replication of circular DNA. A nick in one strand allows the 3′ end to be extended, displacing the strand with the 5′ end, which is also replicated, to generate a double-stranded tail that can become larger than the unit size of the circular DNA.

S

S phase. The period during the cell cycle when the chromosome is replicated.

Salting in. The increase in solubility that is displayed by typical globular proteins upon the addition of small amounts of certain salts, such as ammonium sulfate.

Salting out. The decrease in protein solubility that occurs when salts such as ammonium sulfate are present at high concentrations.

Salvage pathway. A family of reactions that permits, for instance, nucleosides as well as purine and pyrimidine bases resulting from the partial breakdown of nucleic acids to be reutilized in nucleic acid synthesis.

Satellite DNA. A DNA fraction whose base composition differs from that of the main component of DNA, as revealed by the fact that it bands at a different density in a CsCl gradient. Usually repetitive DNA or organelle DNA.

Scissile. Capable of being cut smoothly or split easily.

Second messenger. A diffusible small molecule, such as cAMP, that is formed at the inner surface of the plasma membrane in response to a hormonal signal.

Secondary structure. In a protein or a nucleic acid, any repetitive folded pattern that results from the interaction of the corresponding polymeric chains.

Semiconservative replication. Duplication of DNA in which the daughter duplex carries one old strand and one new strand.

Sigma factor. A subunit of RNA polymerase that recognizes specific sites on DNA for initiation of RNA synthesis.

Single-copy DNA. A region of the genome whose sequence is present only once per haploid complement.

Somatic cell. Any cell of an organism that cannot contribute its genes to a subsequent generation.

SOS system. A set of DNA repair enzymes and regulatory proteins that regulate their synthesis so that maximum synthesis occurs when the DNA is damaged.

Southern blotting. A method for detecting a specific DNA restriction fragment, developed by Edward Southern. DNA from a gel electrophoresis pattern is blotted onto nitrocellulose paper; then the DNA is denatured and fixed on the paper. Subsequently the pattern of specific sequences in the Southern blot can be determined by hybridization to a suitable probe and autoradiography. A Northern blot is similar, except that RNA is blotted instead onto the nitrocellulose paper.

Splicing. *See* RNA splicing.

Sporulation. Formation from vegetative cells of metabolically inactive cells that can resist extreme environmental conditions.

Stacking energy. The energy of interaction that favors the face-to-face packing of purine and pyrimidine base pairs.

Steady state. In enzyme-kinetic analysis, the time interval when the rate of reaction is approximately constant with time. The term is also used to describe the state of a living cell where the concentrations of many molecules are approximately constant because of a balancing between their rate of synthesis and breakdown.

Stem cell. A cell from which other cells stem or arise by differentiation.

Stereoisomers. Isomers that are nonsuperimposable mirror images of each other.

Steroids. Compounds that are derivatives of a tetracyclic structure composed of a cyclopentane ring fused to a substituted phenanthrene nucleus.

Structural domain. An element of protein tertiary structure that recurs in many structures.

Structural gene. A gene encoding the amino acid sequence of a polypeptide chain.

Structural protein. A protein that serves a structural function.

Subunit. Individual polypeptide chains in a protein.

Supercoiled DNA. Supertwisted, covalently closed duplex DNA.

Suppressor gene. A gene that can reverse the phenotype of a mutation in another gene.

Suppressor mutation. A mutation that restores a function lost by an initial mutation and that is located at a site different from the initial mutation.

Svedberg unit (S). The unit used to express the sedimentation constant s: $1S = 10^{-13}$ s. The sedimentation constant s is proportional to the rate of sedimentation of a molecule in a given centrifugal field and is related to the size and shape of the molecule.

Synapse. The chemical connection for communication between two nerve cells or between a nerve cell and a target cell such as a muscle cell.

Synapsis. The pairing of homologous chromosomes, seen during the first meiotic prophase.

T

Tandem duplication. A duplication in which the repeated regions are immediately adjacent to one another.

TCA cycle. *See* tricarboxylic acid cycle.

Template. A polynucleotide chain that serves as a surface for the absorption of monomers of a growing polymer and thereby dictates the sequence of the monomers in the growing chain.

Termination factors. Proteins that are exclusively involved in the termination reactions of protein synthesis on the ribosome.

Terpenes. A diverse group of lipids made from isoprene precursors.

Tertiary structure. In a protein or nucleic acid, the final folded form of the polymer chain.

Tetramer. Structure resulting from the association of four subunits.

Thioester. An ester of a carboxylic acid with a thiol or mercaptan.

Thymidine. One of the four nucleosides found in DNA.

Thymine. A pyrimidine base found in DNA.

Topoisomerase. An enzyme that changes the extent of supercoiling of a DNA duplex.

Transamination. Enzymatic transfer of an amino group from an α-amino acid to an α-keto acid.

Transcription. RNA synthesis that occurs on a DNA template.

Transduction. Genetic exchange in bacteria that is mediated via phage.

Transfection. An artificial process of infecting cells with naked viral DNA.

Transfer RNA (tRNA). Any of a family of low-molecular-weight RNAs that transfer amino acids from the cytoplasm to the template for protein synthesis on the ribosome.

Transferase. An enzyme that catalyzes the transfer of a molecular group from one molecule to another.

Transformation. Genetic exchange in bacteria that is mediated via purified DNA. In somatic cell genetics the term is also used to indicate the conversion of a normal cell to one that grows like a cancer cell.

Transgenic. Describing an organism that contains transfected DNA in the germ line.

Transition state. The activated state in which a molecule is best suited to undergoing a chemical reaction.

Translation. The process of reading a messenger RNA sequence for the specified amino acid sequence it contains.

Transport protein. A protein whose primary function is to transport a substance from one part of the cell to another, from one cell to another, or from one tissue to another.

Tricarboxylic acid (TCA) cycle. The cyclical process whereby acetate is completely oxidized to CO_2 and water, and electrons are transferred to NAD^+ and FAD. The TCA cycle is localized to the mitochondria in eukaryotic cells and to the plasma membrane in prokaryotic cells. Also called the Krebs cycle.

Trypsin. A proteolytic enzyme that cleaves peptide chains next to the basic amino acids arginine and lysine.

Tryptic peptide mapping. The technique of generating a chromatographic profile characteristic of the fragments resulting from trypsin enzyme cleavage of the protein.

Tumorigenesis. The mechanism of tumor formation.

Turnover number. The maximum number of molecules of substrate that can be converted to product per active site per unit time.

U

Ultracentrifuge. A high-speed centrifuge that can attain speeds up to 60,000 rpm and centrifugal fields of 500,000 times gravity. Useful for characterizing and/or separating macromolecules.

Unidirectional replication. *See* bidirectional replication.

Unwinding proteins. Proteins that help to unwind double-stranded DNA during DNA replication.

Urea cycle. A metabolic pathway in the liver that leads to the synthesis of urea from amino groups and CO_2. The function of the pathway is to convert the ammonia resulting from catabolism to a nontoxic form, which is subsequently secreted.

UV irradiation. Electromagnetic radiation with a wavelength shorter than that of visible light (200–390 nm). Causes damage to DNA (mainly pyrimidine dimers).

V

van der Waals forces. Refers to two types of interactions, one attractive and one repulsive. The attractive forces are due to favorable interactions among the induced instantaneous dipole moments that arise from fluctuations in the electron charge densities of neighboring nonbonded atoms. Repulsive forces arise when noncovalently bonded atoms come too close together.

Viroids. Pathogenic agents, mostly of plants, that consist of short (usually circular) RNA molecules.

Virus. A nucleic-acid-protein complex that can infect and replicate inside a specific host cell to make more virus particles.

Vitamin. A trace organic substance required in the diet of some species. Many vitamins are precursors of coenzymes.

W

Watson-Crick base pairs. The type of hydrogen-bonded base pairs found in DNA, or comparable base pairs found in RNA. The base pairs are A-T, G-C, and A-U.

Wild-type gene. The form of a gene (allele) normally found in nature.

Wobble. A proposed explanation for base-pairing that is not of the Watson-Crick type and that often occurs between the 3′ base in the codon and the 5′ base in the anticodon.

X

X-ray crystallography. A technique for determining the structure of molecules from the x-ray diffraction patterns that are produced by crystalline arrays of the molecules.

Y

Ylid. A compound in which adjacent, covalently bonded atoms, both having an electronic octet, have opposite charges.

Z

Z form. A duplex DNA structure in which there is the usual type of hydrogen bonding between the base pairs but in which the helix formed by the two polynucleotide chains is left-handed rather than right-handed.

Zwitterion. A dipolar ion with spatially separated positive and negative charges. For example, most amino acids are zwitterions, having a positive charge on the α-amino group and a negative charge on the α-carboxyl group but no net charge on the overall molecule.

Zygote. A cell that results from the union of haploid male and female sex cells. Zygotes are diploid.

Zymogen. An inactive precursor of an enzyme. For example, trypsin exists in the inactive form trypsinogen before it is converted to its active form, trypsin.

Credits

Portions of this text have been adopted from Geoffrey Zubay, *Genetics.* Copyright © 1987 Benjamin/Cummings Publishing Company, Inc.

Line Art

Chapter 1

1.1: From Sylvia S. Mader, *Inquiry Into Life,* 6th ed. Copyright © 1991 Wm. C. Brown Communications, Inc., Dubuque, Iowa. All Rights Reserved. Reprinted by permission. **1.10, 1.12, 1.13, and 1.22:** Illustration copyright by Irving Geis. **1.15:** Illustration copyright by Irving Geis. Coordinates courtesy of D. C. Phillips, Oxford.

Chapter 3

3.1, 3.9, and 3.10: Illustration copyright by Irving Geis.

Chapter 4

4.1, 4.2, 4.4*a*, 4.5, 4.8, and 4.9*a*, *b*: Illustration copyright by Irving Geis. **4.11:** Line art courtesy of Karl A. Piez. **4.12–4.14, 4.18*c*, and 4.20:** Illustration copyright by Irving Geis. **4.21, 4.23, 4.24, and 4.28–4.34:** From Jane S. Richardson. **4.39*a* and 4.42:** From Lansing M. Prescott, et al., *Microbiology.* Copyright © 1990 Wm. C. Brown Communications, Inc. Dubuque, Iowa. All Rights Reserved. Reprinted by permission. **Box 4C: Figures 1–4:** Courtesy of Irving Geis.

Chapter 5

5.2, 5.7–5.11, 5.13 and 5.14: Illustration copyright by Irving Geis. **5.16:** From John W. Hole, Jr., *Human Anatomy and Physiology,* 5th ed. Copyright © 1990 Wm. C. Brown Communications, Inc., Dubuque, Iowa. All Rights Reserved. Reprinted by permission. **5.23 and 5.24:** From Jane S. Richardson. **5.25:** Illustration copyright by Irving Geis.

Chapter 6

6.16: Illustration copyright by Irving Geis. **6.21** (*top*): From B. Alberts, *Molecular Biology of the Cell.* Copyright © 1983 Garland Publishing Inc., New York, N.Y. Reprinted by permission.

Chapter 7

7.43*a*: Reprinted by permission from Dr. Vincent Marchesi. **7.43*b*:** From S. B. Shohet and S. E. Lux, *Hospital Practice 19* (Nov. 1984):90. Redrawn based on a drawing by Robert Margulies. Reprinted by permission.

Chapter 9

9.4, 9.5, 9.19, and 9.30: Illustration copyright by Irving Geis.

Chapter 15

15.27: Illustration copyright by Irving Geis.

Chapter 25

25.1 and 25.4: From Robert Tamarin, *Principles of Genetics,* 3d ed. Copyright © 1991 Wm. C. Brown Communications, Inc., Dubuque, Iowa. All Rights Reserved. Reprinted by permission. **25.9*a*, *b*:** Reprinted by permission from *Nature,* vol. 171, p. 737. Copyright © 1953 Macmillan Magazines Ltd. **25.13:** Illustration copyright by Irving Geis. **25.14:** From A. Wang et al., "Left-handed Double Helical DNA: Variations in the Backbone Conformation" in *Science,* 211. Copyright © 1981 by the American Association for the Advancement of Science. Reprinted by permission. **25.28:** From Geoffrey Zubay, *Genetics.* Copyright © 1987 Benjamin/Cummings Publishing Company Inc. Reprinted by permission of the author. **25.30*b*:** From Robert F. Weaver and Philip W. Hedrick, *Basic Genetics.* Copyright © 1991 Wm. C. Brown Communications, Inc., Dubuque, Iowa. All Rights Reserved. Reprinted by permission.

Chapter 26

26.12: From D. L. Ollis et al., "Structure of large fragment of *Escherichia coli* DNA polymerase I complexed with dTMP." Reprinted by permission from *Nature,* Vol. 313, pp. 762–766. Copyright © 1985 Macmillan Magazines Ltd. **26.13:** From C. M. Joyce and T. A. Steitz, "DNA polymerase I: From crystal structure to function via genetics" in *Trends in Biochemical Sciences,* 12:288, 1987. Copyright © 1987 Elsevier Trends Journals, Cambridge, England. Reprinted with permission. **26.27:** From J. Tooze, *The Molecular Biology of DNA Tumor Viruses.* Copyright © 1980 Cold Spring Harbor Press, New York. Reprinted by permission. **26.29:** From Robert F. Weaver and Philip W. Hedrick, *Basic Genetics.* Copyright © 1991 Wm. C. Brown Communications, Inc., Dubuque, Iowa. All Rights Reserved. Reprinted by permission. **26.33:** From Geoffrey Zubay, *Genetics.* Copyright © 1987 Benjamin/Cummings Publishing Co. Inc. Reprinted by permission of the author.

Chapter 27

27.11: From J. L. Marx, "Multiplying genes by leaps and bounds" in *Science,* 240:1408–1410, June 10, 1988. Copyright © 1988 by the American Association for the Advancement of Science. Reprinted by permission. **27.13:** From Robert F. Weaver and Philip W. Hedrick, *Basic Genetics.* Copyright © 1991 Wm. C. Brown Communications, Inc., Dubuque, Iowa. All Rights Reserved. Reprinted by permission. **27.34:** From J. R. Riordan et al., "Identification of the cystic fibrosis gene: Cloning and characterization of complementary DNA" in *Science,* 245:1066, Sept. 8, 1989. Copyright © 1989 by the American Association for the Advancement of Science. Reprinted by permission.

Chapter 28

28.2: From Geoffrey Zubay, *Genetics.* Copyright © 1987 Benjamin/Cummings Publishing Company Inc., Menlo Park, CA. Reprinted by permission of the author. **28.3*b*:** Illustration copyright by Irving Geis. **28.6:** From Robert F. Weaver and Philip W. Hedrick, *Genetics.* Copyright © 1989 Wm. C. Brown Communications, Inc., Dubuque, Iowa. All Rights Reserved. Reprinted by permission. **28.7:** Source: R. R. Gutell et al., "Comparative anatomy of 16S-like ribosomal RNA" in *Progress in Nucleic Acid Research and Molecular Biology,* 32:155–216, 1985. Copyright © 1985 Academic Press Inc., San Diego, CA. **Box 28A:** Courtesy of Bishwajit Nag and Robert R. Traut. **28.8 and 28.9:** From Robert F. Weaver and Philip W. Hedrick, *Genetics.* Copyright © 1989 Wm. C. Brown Communications, Inc., Dubuque, Iowa. All Rights Reserved. Reprinted by permission. **28.17*b*:** From Robert Tamarin, *Principles of Genetics,* 3d ed. Copyright © 1991 Wm. C. Brown Communications, Inc., Dubuque, Iowa. All Rights Reserved. Reprinted by permission.

Chapter 29

29.5*b*: From Quigley and Rich, *Science,* 185:436, 1974. Copyright © 1974 by the American Association for the Advancement of Science, Washington, DC. Reprinted by permission. **29.6:** From C. Bernabau and J. A. Lake, *Proceedings. National Academy of Sciences USA,* 79:3111–3115, 1982. Copyright © 1982 National Academy of Sciences, Washington, DC. Reprinted by permission of the author. **29.8:** From Robert F. Weaver and Philip W. Hedrick, *Genetics.* Copyright © 1989 Wm. C. Brown Communications, Inc., Dubuque, Iowa.

All Rights Reserved. Reprinted by permission. **29.11:** From L. H. Schulman and J. Abelson, "Recent excitement in understanding transfer RNA identity" in *Science,* 240:1591, June 17, 1988. Copyright © 1988 by the American Association for the Advancement of Science, Washington, DC. Reprinted by permission. **29.23:** From *Molecular Cell Biology,* 2nd edition. By James Darnell, Harvey Lodish, and David Baltimore. Copyright © 1990 by Scientific American Books. Reprinted with permission of W. H. Freeman and Company.

Chapter 31

31.16: From Geoffrey Zubay, *Genetics.* Copyright © 1987 Benjamin/Cummings Publishing Company Inc., Menlo Park, CA. Reprinted by permission of the author.

Chapter 32

32.16*b:* From E. Schneider and K. Altendorf, "Bacterial adenosine 5′-triphosphate synthase (F1FO): Purification and reconstitution of Fo complexes and biochemical and functional characterization of their subunits" in *Microbiological Reviews,* 51:477–497, 1987. Copyright © 1987 American Society for Microbiology, Washington, DC. Reprinted by permission. **32.31*c:*** From A. C. Steven et al., "Ultrastructure of a periodic protein layer in the outer membrane of *Escherichia coli."* Reproduced from the *Journal of Cell Biology,* 1977, vol. 72, p. 292 by copyright permission of the Rockefeller University Press. **32.32:** Source: M. S. Weiss et al., "The three-dimensional structure of porin from *Rhodobacter capsulatus* at 3 A resolution" in *FEBS Letters* 267:268–272, 1990. Copyright © 1990 Elsevier Science Publishers B.V., Amsterdam, Netherlands. **32.32:** Source: M. S. Weiss et al, "The three-dimensional structure of porin from *Rhodobacter capsulatus* at 3 A resolution" on *FEBS Letters* 267:268–272, 1990. © 1990 Elsevier Science Publishers B.V., Amsterdam, Netherlands.

Chapter 33

33.2, 33.6, 33.7, 33.10, and 33.16–33.19: From Geoffrey Zubay, *Genetics.* Copyright © 1987 Benjamin/Cummings Publishing Company Inc., Menlo Park, CA. Reprinted by permission of the author.

Chapter 34

34.1: From B. Alberts et al., *Molecular Biology of the Cell,* 2d ed. Copyright © 1989 Garland Publishing Inc., New York, NY. Reprinted by permission. **34.2:** © Scientific American, Inc., George V. Kelvin. **34.3:** From J. Yunis, "The chromosomal basis of human neoplasia" in *Science,* 221:227–236, Jan. 1, 1983. Copyright © 1983 by the American Association for the Advancement of Science. Reprinted by permission. **34.5:** From Robert F. Weaver and Philip W. Weaver, *Genetics.* Copyright © 1989 Wm. C. Brown Communications, Inc., Dubuque, Iowa. All Rights Reserved. Reprinted by permission. **34.6:** From Geoffrey Zubay, *Genetics.* Copyright © 1987 Benjamin/Cummings Publishing Company Inc., Menlo Park, CA. Reprinted by permission of the author. **34.7:** From L. Tong, et al., "Structure of the *ras* protein" in *Science,* 245:243, July 21, 1989. Copyright © 1989 by the American Association for the Advancement of Science. Reprinted by permission.

Chapter 35

35.10: From W. A. Catterall, "Structure and function of voltage-sensitive ion channels" in *Science* 242:50, Oct. 7, 1988. Copyright © 1988 American Association for the Advancement of Science. Reprinted by permission. **35.12:** Reprinted by permission from Suzanne Black. **35.22:** From N. Unwin et al., "Arrangement of the acetylcholine receptor subunits in the resting and desensitized states, determined by cryoelectron microscopy of crystallized *Torpedo* postsynaptic membranes." Reproduced from the *Journal of Cell Biology,* 1988, vol. 107, p. 1123 by copyright permission of the Rockefeller University Press. **35.23:** From M. Noda et al., "Structural homology of *Torpedo California* acetylcholine receptor subunits." Reprinted by permission from *Nature,* vol. 302, p. 528. Copyright © 1983 Macmillan Magazines Ltd.

Index

Page numbers followed by an "F" refer to illustrations; "T" indicates material in tables.

A

B

C

D

E

F

G

H

I

J

K

L

M

N

O

P

R

S

T

U

V

W

X

Y

Z